Past Climate Variability through Europe and Africa

Developments in Paleoenvironmental Research

VOLUME 6

Past Climate Variability through Europe and Africa

Edited by

Richard W. Battarbee

Environmental Change Research Centre, Department of Geography,
University College London, London, U.K.

Françoise Gasse

University of Aix-Marseille II,
Aix-en-Provence, France

and

Catherine E. Stickley

Environmental Change Research Centre, Department of Geography,
University College London, London, U.K.
and School of Earth, Ocean and Planetary Sciences,
Cardiff University, Cardiff, U.K.

Springer

A C.I.P. Catalogue record for this book is available from the Library of Congress.

ISBN 1-4020-2120-8 (HB)
ISBN 1-4020-2121-6 (e-book)

Published by Springer,
P.O. Box 17, 3300 AA Dordrecht, The Netherlands.

Sold and distributed in North, Central and South America
by Springer,
101 Philip Drive, Norwell, MA 02061, U.S.A.

In all other countries, sold and distributed
by Springer,
P.O. Box 322, 3300 AH Dordrecht, The Netherlands.

From North to South: Pechora River floodplain, northern Russia (Photo: Dan Hammarlund);
northern mixed forest (Photo: Dirk Verschuren); Lycian grave monuments, southwestern Turkey
(Photo: Dirk Verschuren); Tidé Berdégi Plateau, northern Chad (Photo: Dirk Verschuren); Kyambura
Reserve, western Rift Valley, Uganda (Photo: Dirk Verschuren); Kising'a-Rugaro Forest, Eastern Arc
Mountains, Tanzania (Photo: Robert Marchant and Cassian Mumbi)

Printed on acid-free paper

TABLE OF CONTENTS

DEVELOPMENTS IN PALEOENVIRONMENTAL RESEARCH
BOOK SERIES

http://www.springeronline.com/sgw/cda/frontpage/0,,4-40109-69-33113470-0,00.html
http://home.cc.umanitoba.ca/~mlast/paleolim/dper.html

Series Editors:

John P. Smol, Department of Biology, Queen's University Kingston, Ontario, Canada

William M. Last, Department of Geological Sciences, University of Manitoba, Winnipeg, Manitoba, Canada

---

Volume 1:   **Tracking Environmental Change Using Lake Sediments. Volume 1:**
             Basin Analysis, Coring, and Chronological Techniques
             Edited by W. M. Last and J. P. Smol
             Hardbound, ISBN 0-7923-6482-1, June 2001

A Continuation Order Plan is available for this series. A continuation order will bring delivery of each new volume immediately upon publication. Volumes are billed only upon actual shipment. For further information please contact the publisher.

LIST OF CONTRIBUTORS

Helge Arz,Universität Bremen, Geowissenschaft, 28334 Bremen, Germany.
e-mail: helge.arz@uni-bremen.de

Avner Ayalon, Geological Survey of Israel, 30 Malchei Israel St., Jerusalem, 95501, Israel.
e-mail: Ayalon@mail.gsi.gov.il

Keith Barber, Palaeoecology Laboratory, Department of Geography, University of Southampton, Southampton, SO17 1BJ, United Kingdom.
e-mail: Keith.Barber@soton.ac.uk

Philip A. Barker, Department of Geography, Lancaster University, Lancaster, LA1 4YB, United Kingdom.
e-mail: p.barker@lancs.ac.uk

Miryam Bar-Matthews, Geological Survey of Israel, 30 Malchei Israel St, Jerusalem, 95501, Israel
e-mail: matthews@mail.gsi.gov.il

Richard W. Battarbee, Environmental Change Research Centre, Department of Geography, University College London, 26 Bedford Way, London, WC1H 0AP, United Kingdom.
email: r.battarbee@geog.ucl.ac.uk

Raymonde Bonnefille, Centre Européen de Recherche et d'Enseignement, de Géosciences de l' Environnement (CEREGE), Europole de l'Arbois, B.P. 80, 13545 Aix-en-Provence cedex 04, France.
e-mail: bonnefille@cerege.fr

Pascale Braconnot, Laboratoire des Sciences du Climat et de l' Environnement, UMR CEA-CNRS 1572, CEA Saclay bat 709, 91191 Gif sur Yvette cedex, France.
e-mail: pasb@lsce.saclay.cea.fr

Simon Brewer, Centre Européen de Recherche et d'Enseignement, de Géosciences de l' Environnement (CEREGE), Europole de l'Arbois, B.P. 80, 13545 Aix-en-Provence cedex 04, France.
e-mail: brewer@cerege.fr

Keith R. Briffa, Climatic Research Unit, University of East Anglia, Norwich, NR4 7TJ, United Kingdom.
e-mail: k.briffa@uea.ac.uk

Erik T. Brown, Large Lakes Observatory, University of Minnesota Duluth, Duluth, MN 55812, USA.
e-mail: etbrown@d.umn.edu

Rachid Cheddadi, European Pollen Database, Place de la République, 13200 Arles, France.
e-mail: rachid.cheddadi@Ibhp.u-3mrs.fr

Russell Coope, Department of Geography, Royal Holloway, University of London, Egham, Surrey, TW20 0EX, United Kingdom.
e-mail: rcoope@rhul.ac.uk

Basil Davis, Department of Geography, University of Newcastle, Newcastle upon Tyne, NE1 7RU, United Kingdom.
e-mail: basil.davis@ncl.ac.uk

Djaira Djoret, Faculté des Sciences Exactes et Appliques, University of N'Djamena, BP 1027 N'Djamena, Republic of Chad.
e-mail: fsciences@sdntcd.undp.org

Abdelkader Dodo, Départemente de Géologie, Faculté des Sciences, Université Abdou Moumouni, BP 13316 Niamey, Niger.

Jean-Claude Duplessy, Laboratoire des Sciences du Climat et de l'Environnement, Laboratoire Mixte CNRS-CEA, Parc du CNRS 91 198, Gif-sur-Yvette cedex, France.
e-mail: jean-claude.duplessy@lsce.cnrs-gif.fr

Shane Doyle, Cambridge Group for the History of Population and Social Structure, University of Cambridge, Cambridge, United Kingdom.
e-mail: sdd20@cam.ac.uk

Lydie M. Dupont, Universität Bremen, Fachbereich Geowissenschaften, P.O. Box 30440, D-28334 Bremen, Germany.
e-mail: dupont@allgeo.uni-bremen.de

William M. Edmunds, British Geological Survey, Crowmarsh Gifford, Wallingford, Oxon, OX10 8BB, United Kingdom.
e-mail: wme@btopenworld.com

Hilaire Elenga, Faculté des Sciences, Université Marien Ngouabi, BP 69 Brazzaville, Congo.
e-mail: hilaire_elenga@yahoo.fr
and:
IRD, Centre de Pointe-Noire, BP 1286 Pointe-Noire, Congo.

Isabelle Farrera, Paléoenvironnements et Palynologie, Université Montpellier–2, 34095 Montpellier, France.
e-mail: farrera@isem.univ.mont2.fr

Michel Fontugne, Laboratoire des Sciences du Climat et de l'Environnement, Laboratoire Mixte CNRS-CEA, Parc du CNRS 91 198, Gif-sur-Yvette cedex, France.
e-mail: michel.fontugne@lsce.cnrs-gif.fr

Françoise Gasse, Centre Européen de Recherche et d'Enseignement, de Géosciences de l′ Environnement (CEREGE), Europole de l'Arbois, B.P. 80, F- 13454 Aix-en-Provence cedex 4, France.
e-mail: gasse@cerege.fr

Cheikh B. Gaye, International Atomic Energy Agency, Wagramerstasse 5, A1040 Vienna, Austria.
e-mail: c.b.gaye@iaea.org

Ibrahim B. Goni, University of Maiduguri, Maiduguri, Borno State, Nigeria.
e-mail: ibgoni@infoweb.abs.net

Sandy P. Harrison, Max-Planck-Institut für Biogeochemie, Postfach 100164, 07701 Jena, Germany.
e-mail: sandy.harrison@bgc-jena.mpg.de

Chris D. Hewitt, Hadley Centre for Climate Prediction and Research, Met Office, Bracknell, Berkshire, RG12 2SY, United Kingdom.
e-mail: chris.hewitt@metoffice.com

Philipp Hoelzmann, Max-Planck-Institut für Biogeochemie, Postfach 100164, 07701 Jena, Germany.
e-mail: philipp.hoelzmann@leco.de
Or:
LECO Instrumente GmbH, Marie Bernays-Ring 31, 41199 Moenchengladbach, Germany.

Thomas C. Johnson, Large Lakes Observatory, University of Minnesota Duluth, Duluth, MN 55812, USA.
e-mail: tcj@d.umn.edu

Sylvie Joussaume, Laboratoire des Sciences du Climat et de l' Environnement, Laboratoire Mixte CNRS-CEA, Parc du CNRS 91 198, Gif-sur-Yvette cedex, France.
e-mail: sylvie.joussaume@cnrs-dir.fr

Nejib Kallel, Faculté des Sciences de Sfax, B.P. 802, 3018 Sfax, Tunisia.
e-mail: nejib.kallel@fss.rnu.tn

Akio Kitoh, Climate Research Department, Meteorological Research Institute, Nagamine 1-1, Tsukuba, Ibaraki, 305-0052, Japan.
e-mail address: kitoh@mri.jma.go.jp

Nalan Koç, Norwegian Polar Institute, N-9296 Tromsø, Norway.
e-mail: Nalan.Koc@npolar.no

Atte Korhola, Department of Ecology and Systematics, Division of Hydrobiology, University of Helsinki, P.O. Box 17 (Arkadiankatu 7), FIN-0014, Finland.
e-mail: atte.korhola@helsinki.fi

John E. Kutzbach, Department of Atmospheric and Oceanic Sciences, University of Wisconsin – Madison, 1225 West Dayton Street, Madison, Wisconsin 53706, USA.
e-mail: jek@facstaff.wisc.edu

Laurent Labeyrie, Laboratoire des Sciences du Climat et de l'Environnement, Laboratoire Mixte CNRS-CEA, Parc du CNRS 91 198, Gif-sur-Yvette cedex, France.
e-mail: laurent.labeyrie@lsce.cnrs-gif.fr

Angela L. Lamb, School of Biological and Earth Sciences, Liverpool John Moores University, Liverpool, L3 3AF, United Kingdom.
e-mail: a.lamb@livjm.ac.uk

Henry F. Lamb, Institute of Geography and Earth Sciences, University of Wales, Aberystwyth, SY23 3DB, United Kingdom.
e-mail: hfl@aber.ac.uk

Julia A. Lee-Thorp, Department of Archaeology and Quaternary Research Centre, University of Cape Town, Private Bag Rondebosch 7701, South Africa.
e-mail: jlt@science.uct.ac.za

Dagnachew Legesse, Department of Geology and Geophysics, Addis Ababa University, P.O.Box 3434, Addis Ababa, Ethiopia.
e-mail: dagnu@geobs.aau.edu.et

Melanie J. Leng, NERC Isotope Geosciences Laboratory, Keyworth, Nottingham, NG12 5GG, United Kingdom.
e-mail: mjl@nigl.nerc.ac.uk

Dirk C. Leuschner, Institut für Geophysik und Geologie, Universität Leipzig, Talstraße 35, 04103 Leipzig, Germany.
e-mail: dcleu@mail.uni-mainz.de

Thomas Litt, Institute for Palaeontology, University of Bonn, Bonn, Germany.
e-mail: t.litt@uni-bonn.de

André F. Lotter, Laboratory of Palaeobotany and Palynology, University of Utrecht, Budapestlaan 4, 3584 CD Utrecht, The Netherlands.
e-mail: A.Lotter@bio.uu.nl

John J. Lowe, Department of Geography, Royal Holloway, University of London, Egham, Surrey, TW20 0EX, United Kingdom.
e-mail: J.Lowe@rhbnc.ac.uk

Juha P. Lunkka, Institute of Geosciences, P.O. Box 3000, 90014 University of Oulu, Finland.
e-mail: juha.pekka.lunkka@oulu.fi

Zhengyu Liu, Department of Atmospheric and Oceanic Sciences, University of Wisconsin – Madison, 1225 West Dayton Street, Madison, Wisconsin 53706, USA.
e-mail: zliu3@facstaff.wisc.edu

Jean Maley, Paléoenvironnements et Palynologie, Université Montpellier–2, 34095, Montpellier, France.
e-mail: jmaley@isem.univ-montp2.fr

Robert Marchant, Department of Botany, Trinity College, University of Dublin, Dublin 2, Ireland.
e-mail: marchantr@tcd.ie

Donatella Margri, Dipartimento di Biologia Vegetale, Università "La Sapienza", P.le Aldo Moro, 5, 00185 Roma, Italy.
e-mail: donatella.magri@uniroma1.it

Fabienne Marret, School of Ocean Sciences, University of Wales Bangor, Menai Bridge, Anglesey, LL59 5EY, United Kingdom.
e-mail: f.marret@bangor.ac.uk

James McManus, Large Lakes Observatory, University of Minnesota Duluth, Duluth, MN 55812, USA.
Or:
College of Ocean and Atmospheric Sciences, Oregon State University, Corvallis, OR 97331, USA.
e-mail: jmcmanus@coas.oregonstate.edu

Biancamaria Narcisi, ENEA - C.R. Casaccia, PO Box 2400, 00100 Roma AD, Italy.
e-mail: narcisi@casaccia.enea.it

Eric O. Odada, Department of Geology, University of Nairobi, PO Box 30197, Nairobi, Kenya.
e-mail: pass@uonbi.ac.ke

Daniel O. Olago, Department of Geology, University of Nairobi, PO Box 30197, Nairobi, Kenya.
e-mail: dolago@uonbi.ac.ke

Frank Oldfield, Department of Geography, University of Liverpool, Liverpool, L69 72T, United Kingdom.
e-mail: f.oldfield@onetel.com

Luc Ortlieb, UR PALEOTROPIQUE (Paléoenvironnements tropicaux et variabilité climatique), Institut de Recherche pour le Développement, Centre IRD-Ile de France, 32 avenue Henri-Varagnat, F-93143 Bondy cedex, France.
e-mail: Luc.Ortlieb@bondy.ird.fr

Bette L. Otto-Bliesner, Climate Change Research, National Center for Atmospheric Research , 1850 Table Mesa Drive, P.O. Box 3000, Boulder, Colorado 80307, USA.
e-mail: ottobli@ucar.edu

Tim C. Partridge, Climatology Research Group, University of the Witwatersrand, Private Bag 3, WITS 2050, Johannesburg, South Africa.
e-mail: tcp@iafrica.com

Martine Paterne, Laboratoire des Sciences du Climat et de l'Environnement, Laboratoire Mixte CNRS-CEA, Parc du CNRS 91 198, Gif-sur-Yvette cedex, France.
e-mail: martine.paterne@lsce.cnrs-gif.fr

Hans Renssen, Faculty of Earth and Life Sciences, Vrije Universiteit Amsterdam, De Boelelaan 1085, NL-1081 HV Amsterdam, The Netherlands.
e-mail: hans.renssen@falw.vu.nl

Neil Roberts, School of Geography, University of Plymouth, Plymouth, PL4 8AA, United Kingdom.
e-mail: cnroberts@plymouth.ac.uk

Peter Robertshaw, Department of Anthropology, California State University, San Bernardino, CA 92407-2397, USA.
e-mail: proberts@csusb.edu

Arlene Rosen, Institute of Archaeology, University College London, 31-34 Gordon Square, London, WC1H OPY, United Kingdom.
e-mail: a.rosen@ucl.ac.uk

Matti Saarnisto, Geological Survey of Finland, P.O. Box 96, FIN-02150 Espoo, Finland.
e-mail: matti.saarnisto@gsf.fi

Ulrich Salzmann, J.W. Goethe-Universität, Seminar für Vor- und Frühgeschichte, Archäologie und Archäobotanik Afrikas, Grüneburgplatz 1, 60323 Frankfurt /M., Germany.
Or:

Zentrum für Marine Tropenoekologie (ZMT) (Center for Tropical Marine Ecology), Fahrenheitstrasse 6, 28359 Bremen, Germany.
e-mail: ulrich.salzmann@zmt.uni-bremen.de

Ralph R. Schneider, Fachbereich Geowissenschaften, Universitat Bremen, Klagenfurter Strasse, 28359 Bremen, Germany.
e-mail: rschneid@uni-bremen.de

Louis Scott, Department of Botany and Genetics, University of the Free State, P.O. Box 339, Bloemfontein 9300, South Africa.
e-mail: Scottl.sci@mail.uovs.ac.za

James Scourse, School of Ocean Sciences, University of Wales Bangor, Menai Bridge, Anglesey, LL59 5EY, United Kingdom.
e-mail: j.scourse@bangor.ac.uk

Frank Sirocko, Institut für Geowissenschaften, Universität Mainz, Becherweg 21, 55099 Mainz, Germany.
e-mail: sirocko@mail.uni-mainz.de

Ian Snowball, Quaternary Geology, Geobiosphere Science Centre, Lund University, Solvegatan 12, SE-223 62 Lund, Sweden.
e-mail: ian.snowball@geol.lu.se

M. Adebisi Sowunmi, Department of Archaeology and Anthropology, University of Ibadan, Ibadan, Nigeria.
e-mail: sowunmi@skannet.com

Michael Staubwasser, Department of Earth Sciences, University of Oxford, Parks Road, 0X1 3PR, Oxford, United Kingdom.
e-mail: michael.staubwasser@earth.ox.ac.uk

Anthony C. Stevenson, Faculty of Law, Environment and Social Sciences, University of Newcastle, Newcastle upon Tyne, NE1 7RU, United Kingdom.
e-mail: tony.stevenson@ncl.ac.uk

F.Alayne Street-Perrott, Department of Geography, University of Wales Swansea, Swansea, SA2 8PP, United Kingdom.
e-mail: f.a.street-perrott@swansea.ac.uk

Jozef Syktus, CSIRO Division of Atmospheric Research, Private Bag No 1, Aspendale 3195, Victoria, Australia.
e-mail: jozef.syktus.dar.csiro.au

Michael R. Talbot, Geological Institute, University of Bergen, 5007 Bergen, Norway.
e-mail: michael.talbot@geol.uib.no

Pavel Tarasov, Department of Geography, Moscow State University, Vorobievy Gory, 119899 Moscow, Russia.
e-mail: paveltarasov@hotmail.com

David Taylor, Department of Geography, Trinity College, University of Dublin, Dublin 2, Ireland.
e-mail: taylord@tcd.ie

Simon F.B. Tett, Hadley Centre for Climate Prediction and Research, Met Office, Bracknell, Berkshire, RG12 2SY, United Kingdom.
e-mail: simon.tett@metoffice.com

Roy Thompson, School of Geosciences, The University of Edinburgh, Edinburgh, EH9 3JW, Scotland.
e-mail: roy@ed.ac.uk

Yves Travi, Laboratoire Hydrogeologie, Faculte des Sciences, 33 Rue Pasteur , 84000 Avignon, France.
e-mail: yves.travi@univ-avignon.fr

Mohammed Umer, Department of Geology and Geophysics, Addis Ababa University, P.O.Box 3434, Addis Ababa, Ethiopia.
e-mail: umermm@geol.aau.edu.et

Jef Vandenberghe, Faculty of Earth and Life Sciences, Vrije Universiteit, Amsterdam, The Netherlands.
e-mail: jef.vandenberghe@falw.vu.nl

Dirk Verschuren, Department of Biology, Ghent University, Ledeganckstraat 35, B-9000 Gent, Belgium.
e-mail: dirk.verschuren@UGent.be

Annie Vincens, Centre Européen de Recherche et d'Enseignement, de Géosciences de l´ Environnement (CEREGE), Europole de l'Arbois, B.P. 80, F-13454 Aix-en-Provence cedex 4, France.
e-mail: vincens@cerege.fr

Laurence Vidal, Centre Européen de Recherche et d'Enseignement, de Géosciences de l´ Environnement (CEREGE), Europôle de l'Arbois, B.P. 80, F- 13454 Aix-en-Provence cedex 4, France.
e-mail: vidal@cerege.fr

Hans Von Storch, GKSS-Forschungszentrum Geesthacht GmbH, Max-Planck-Straße, D-21502 Geesthacht, Germany.
e-mail: hans.von.storch@gkss.de

S.L. Weber, Royal Netherlands Meteorological Institute KNMI,
P.O. Box 201, NL-3730 AE De Bilt, The Netherlands.
e-mail: weber@knmi.nl

Bernd Zolitschka, GEOPOLAR, Institut für Geographie, Universität Bremen,
FB 8 Celsiusstraße FVG-M, 28359, Bremen, Germany.
e-mail: zoli@uni-bremen.de

Ludwig Zoller Geographisches Institut, University of Bonn (at present University of
Bayreuth), Germany.
e-mail: Ludwig.Zoeller@uni-bayreuth.de

Kamel Zouari, Ecole Nationale d'Ingenieurs, 3038 Sfax, Tunisia.
e-mail: kamel.zouari@rnu.tn

Gian-Maria Zuppi, Department of Environmental Science, University of Venice,
Ca' Foscari, Santa Marta, Dorsoduro 2137, 30123 Venice, Italy.
e-mail: zuppi@tin.it

1. INTRODUCTION

FRANÇOISE GASSE (gasse@cerege.fr)
Centre Européen de Recherche et d'Enseignement
de Géosciences de l' Environnement (CEREGE)
Europole de l'Arbois
B.P. 80, F-13454 Aix-en-Provence cedex 4
France

RICHARD W. BATTARBEE (r.battarbee@geog.ucl.ac.uk)
Environmental Change Research Centre
Department of Geography
University College London
26 Bedford Way
London, WC1H 0AP
UK

Introduction

This volume contains a series of papers that collectively summarise evidence for climate change in Africa and Europe over approximately the last 200,000 years. It is a product of PEP III, the pole-equator-pole transect through Europe and Africa, that has been defined by the IGBP-PAGES project for palaeoclimate study along with PEP II (Asia-Australasia) and PEP I, the Americas (Fig. 1). The PEP transects focus on two time intervals, time-stream 1, the Holocene (with an emphasis on the last 2000 years), and time-stream 2, the last two glacial cycles. The principal long-term objectives of PEP III are (i) to understand how and why climate has varied in the past along the transect; (ii) to assess how climate change and variability has affected natural ecosystems and human society in the past; and (iii) to provide a basis both for developing and testing climate models that are needed to forecast climate change in the future. The specific papers contained in this volume were presented at a conference on "Climate Variability through Europe and Africa" held in Aix-en-Provence in August, 2001. The volume presents an attempt to bring together in a coherent way our present understanding of past climate change along the PEP III transect, providing a basis for the future work needed to make progress towards these objectives.

This publication also complements other related syntheses that overview the scientific achievements of the PAGES community over the last decade, notably for the PAGES programme as a whole (Alverson et al. 2003), and for the PEP III sister transect, PEP I (Markgraf 2001) and PEP II (Dodson et al. 2004).

1

R. W. Battarbee et al. (eds) 2004. *Past Climate Variability through Europe and Africa.*
Springer, Dordrecht, The Netherlands.

Much of Europe lies under a temperate climate. Oceanic polar and sub-polar climates affect the extreme North of the transect only. Europe is strongly influenced by the North Atlantic weather patterns, especially the North Atlantic Oscillation (NAO), at the seasonal to inter-annual time-scales. However, there is no natural boundary between Europe and Asia. Consequently, a marked gradient from Atlantic to continental climate is observed across Europe, with an increase in thermal seasonal contrasts and aridity from West to East. In winter, the sub-tropical Azores anticyclone is centred around 30 °N while the sub-polar low-pressure cell (the Iceland Low) is deep and expanded southward over the warm North Atlantic. Southwesterlies and westerlies bring warmth and moisture to Western Europe, where mild, wet winters prevail from the Mediterranean to the Arctic. Local cyclogenesis bring heavy rains around the Mediterranean. Eastern Europe is influenced by the winter Siberian High responsible for drier and very cold winters. In summer, the Azores sub-tropical High shifts northward and becomes stronger. Westerlies and northwesterlies generate around the Azores anticyclone and flow over Northwestern Europe which remains cool. The Mediterranean region is dry and hot. Eastward, the Siberian High vanishes while a low pressure cell establishes over Central Asia. Summers in Eastern Europe are hot, and relatively wet due to the low atmospheric pressure and to the reactivation of perturbations along the westerly trajectory.

These general patterns are greatly modified by the topography. Three major mountain ranges (the Pyrenees, the Alps, the Caucasus) with peaks exceeding 4000 m.a.s.l. in elevation, act as climatic barriers for atmospheric circulation. Alpine climate dominates above 3000 m.a.s.l. in Southern Europe and at much lower elevation in the smaller ranges of higher latitudes. Another characteristic of the European portion of the PEP III transect is the presence of large inter- and intra-continental seas, the Baltic Sea, the Caspian Sea, the Black Sea and the Mediterranean Sea. These large waterbodies, which act as reservoirs for water recycling and/or as evaporative basins, greatly influence the regional climate.

Africa spans a large range of latitudes from 37 °N to 34.5 °S. The distribution of modern climates is more or less symmetric about the equator as a result of its relatively simple physiography. The northern and southern ends of the continent project into the belt of mid-latitude westerlies, and experience Mediterranean summer-dry climates, receiving most precipitation during winter from westerly cyclonic disturbances. These Mediterranean areas are both flanked by sub-tropical deserts, the Sahara north of the Equator, and the Namib coastal desert in Southwestern Africa, which are dominated by sub-tropical anticyclones throughout the year. These arid zones are separated by a wide belt of tropical climates. The tropical climate is governed by the seasonal migration of the Intertropical Convergence Zone (ITCZ; the meteorological equator) in response to changes in the location of maximum solar heating. This results in northern and southern belts of monsoonal climates with summer rains and winter drought, bracketing a humid equatorial zone characterised by a double rainfall maximum. The zonal temperature and moisture patterns are altered by highlands (the Atlas, Central Sahara and East Africa mountains, and the eastern escarpment of Southern Africa), which act as water towers for the surrounding lowlands. The zonal ("Walker") circulation along the equator, caused by East-West differences in surface and tropospheric temperatures, also greatly influences tropical climates. Rainfall fluctuations in many areas of the continent are statistically linked to the ENSO, but these may be more directly a response to sea-surface temperature (SST) fluctuations in the Indian and Atlantic Oceans which occur in the context of ENSO. The East-West climate asymmetry is linked

1. INTRODUCTION

FRANÇOISE GASSE (gasse@cerege.fr)
*Centre Européen de Recherche et d'Enseignement
de Géosciences de l'Environnement (CEREGE)
Europole de l'Arbois
B.P. 80, F-13454 Aix-en-Provence cedex 4
France*

RICHARD W. BATTARBEE (r.battarbee@geog.ucl.ac.uk)
*Environmental Change Research Centre
Department of Geography
University College London
26 Bedford Way
London, WC1H 0AP
UK*

Introduction

This volume contains a series of papers that collectively summarise evidence for climate change in Africa and Europe over approximately the last 200,000 years. It is a product of PEP III, the pole-equator-pole transect through Europe and Africa, that has been defined by the IGBP-PAGES project for palaeoclimate study along with PEP II (Asia-Australasia) and PEP I, the Americas (Fig. 1). The PEP transects focus on two time intervals, time-stream 1, the Holocene (with an emphasis on the last 2000 years), and time-stream 2, the last two glacial cycles. The principal long-term objectives of PEP III are (i) to understand how and why climate has varied in the past along the transect; (ii) to assess how climate change and variability has affected natural ecosystems and human society in the past; and (iii) to provide a basis both for developing and testing climate models that are needed to forecast climate change in the future. The specific papers contained in this volume were presented at a conference on "Climate Variability through Europe and Africa" held in Aix-en-Provence in August, 2001. The volume presents an attempt to bring together in a coherent way our present understanding of past climate change along the PEP III transect, providing a basis for the future work needed to make progress towards these objectives.

This publication also complements other related syntheses that overview the scientific achievements of the PAGES community over the last decade, notably for the PAGES programme as a whole (Alverson et al. 2003), and for the PEP III sister transect, PEP I (Markgraf 2001) and PEP II (Dodson et al. 2004).

1

R. W. Battarbee et al. (eds) 2004. *Past Climate Variability through Europe and Africa.*
Springer, Dordrecht, The Netherlands.

GASSE AND BATTARBEE

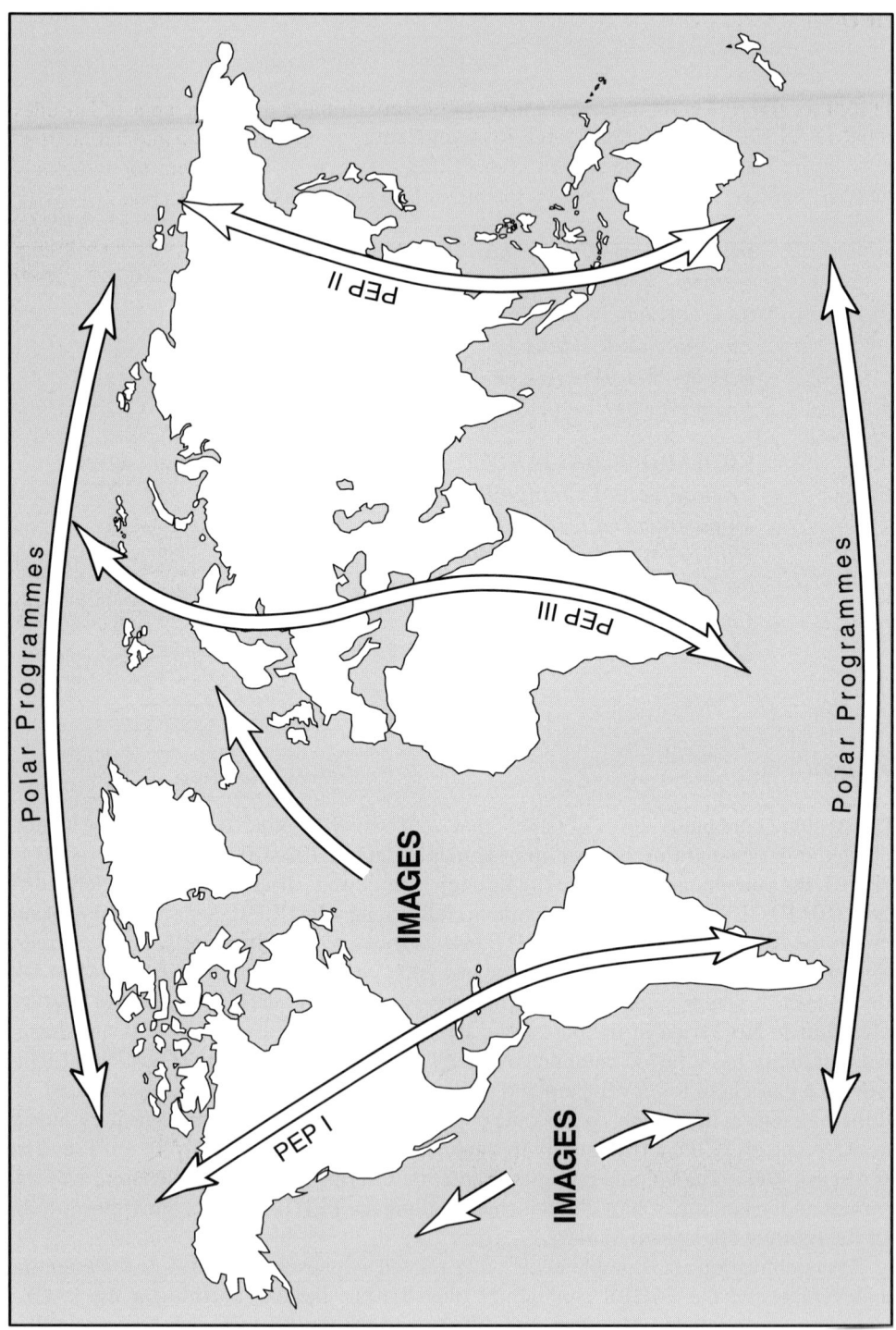

Figure 1. PEP III in the context of IGBP-PAGES.

The contemporary climate system

The PEP III transect exhibits a large variety of modern climates due to its range of latitudes (about 75 °N - 34.5 °S) creating a large zonal gradient of annual insolation and interactions between the major atmospheric circulation patterns and areas of different topographical complexity, land-sea distribution, and surrounding sea-surface conditions (Fig. 2).

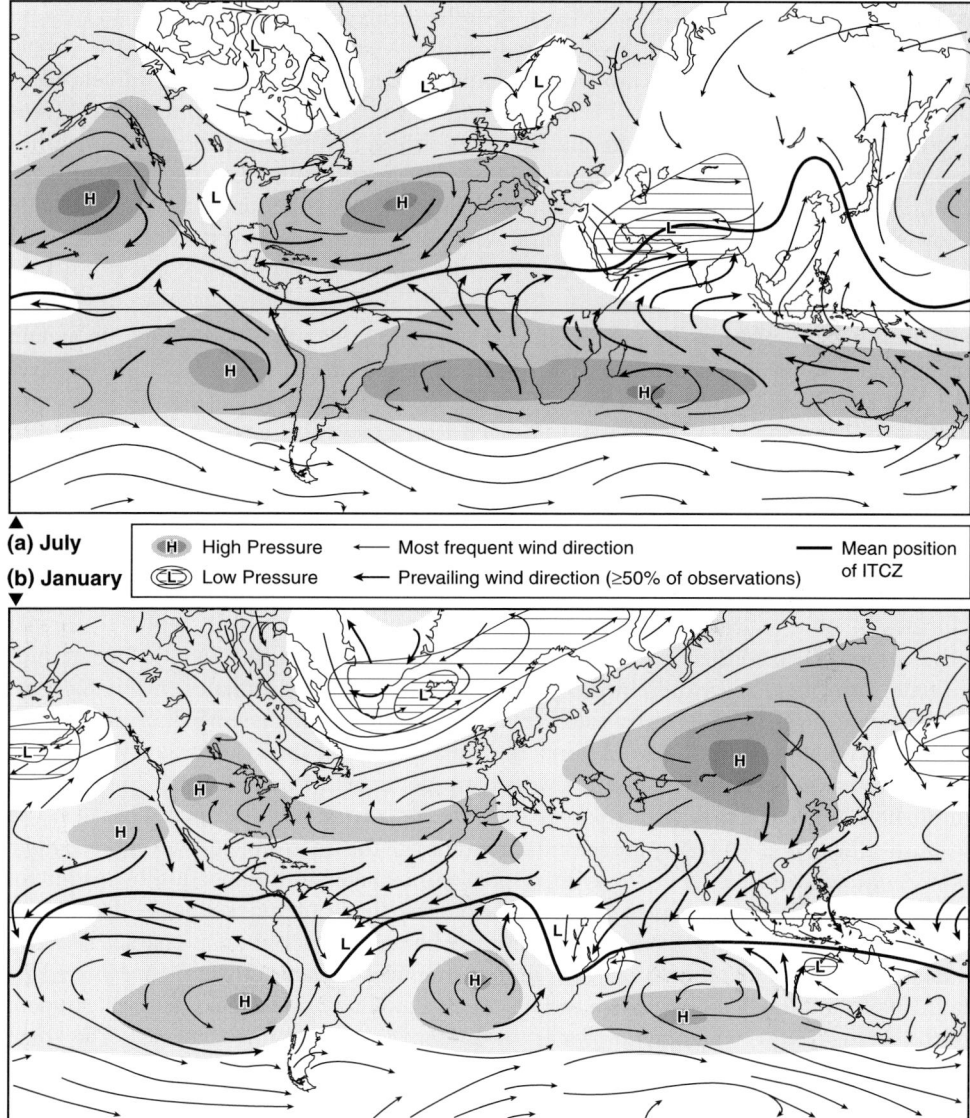

Figure 2. The earth's wind and pressure system, and the position of the Intertropical Convergence Zone (ITCZ) for (a) July and (b) January (re-drawn with permission from Colling et al. (1997)).

Much of Europe lies under a temperate climate. Oceanic polar and sub-polar climates affect the extreme North of the transect only. Europe is strongly influenced by the North Atlantic weather patterns, especially the North Atlantic Oscillation (NAO), at the seasonal to inter-annual time-scales. However, there is no natural boundary between Europe and Asia. Consequently, a marked gradient from Atlantic to continental climate is observed across Europe, with an increase in thermal seasonal contrasts and aridity from West to East. In winter, the sub-tropical Azores anticyclone is centred around 30 °N while the sub-polar low-pressure cell (the Iceland Low) is deep and expanded southward over the warm North Atlantic. Southwesterlies and westerlies bring warmth and moisture to Western Europe, where mild, wet winters prevail from the Mediterranean to the Arctic. Local cyclogenesis bring heavy rains around the Mediterranean. Eastern Europe is influenced by the winter Siberian High responsible for drier and very cold winters. In summer, the Azores sub-tropical High shifts northward and becomes stronger. Westerlies and northwesterlies generate around the Azores anticyclone and flow over Northwestern Europe which remains cool. The Mediterranean region is dry and hot. Eastward, the Siberian High vanishes while a low pressure cell establishes over Central Asia. Summers in Eastern Europe are hot, and relatively wet due to the low atmospheric pressure and to the reactivation of perturbations along the westerly trajectory.

These general patterns are greatly modified by the topography. Three major mountain ranges (the Pyrenees, the Alps, the Caucasus) with peaks exceeding 4000 m.a.s.l. in elevation, act as climatic barriers for atmospheric circulation. Alpine climate dominates above 3000 m.a.s.l. in Southern Europe and at much lower elevation in the smaller ranges of higher latitudes. Another characteristic of the European portion of the PEP III transect is the presence of large inter- and intra-continental seas, the Baltic Sea, the Caspian Sea, the Black Sea and the Mediterranean Sea. These large waterbodies, which act as reservoirs for water recycling and/or as evaporative basins, greatly influence the regional climate.

Africa spans a large range of latitudes from 37 °N to 34.5 °S. The distribution of modern climates is more or less symmetric about the equator as a result of its relatively simple physiography. The northern and southern ends of the continent project into the belt of mid-latitude westerlies, and experience Mediterranean summer-dry climates, receiving most precipitation during winter from westerly cyclonic disturbances. These Mediterranean areas are both flanked by sub-tropical deserts, the Sahara north of the Equator, and the Namib coastal desert in Southwestern Africa, which are dominated by sub-tropical anticyclones throughout the year. These arid zones are separated by a wide belt of tropical climates. The tropical climate is governed by the seasonal migration of the Intertropical Convergence Zone (ITCZ; the meteorological equator) in response to changes in the location of maximum solar heating. This results in northern and southern belts of monsoonal climates with summer rains and winter drought, bracketing a humid equatorial zone characterised by a double rainfall maximum. The zonal temperature and moisture patterns are altered by highlands (the Atlas, Central Sahara and East Africa mountains, and the eastern escarpment of Southern Africa), which act as water towers for the surrounding lowlands. The zonal ("Walker") circulation along the equator, caused by East-West differences in surface and tropospheric temperatures, also greatly influences tropical climates. Rainfall fluctuations in many areas of the continent are statistically linked to the ENSO, but these may be more directly a response to sea-surface temperature (SST) fluctuations in the Indian and Atlantic Oceans which occur in the context of ENSO. The East-West climate asymmetry is linked to topographical features (e.g., the relief of East and Southeast Africa), and to sea-surface conditions (e.g., the cold oceanic Benguela, and the warm Algulhas currents flowing along the western and eastern coasts of Southern Africa, respectively).

Past climate change questions

PEP III is concerned with past climate change along the transect including both north-south and west-east dimensions, and with comparisons with other continental areas (PEPs I and II), polar ice core records and marine records in adjacent seas. The climate change questions posed in PEP III and the methods used depend partly on the time-scale of interest. For example for time stream 1, that focusses on the recent past, instrumental, documentary as well as sedimentary and other proxy methods are available and chronology is more accurate. Key questions concern the nature and past variability of the North Atlantic Oscillation, the nature of the so-called Little Ice Age and Medieval Warm Period and their expression in both high and low latitudes, the cause and effect of past variations in monsoon circulation across Africa from west to east and between hemispheres, evidence for abrupt change in climate especially in dry lands, and the impact of climate variability on human society.

In contrast, time stream 2 issues encompass change taking place over the last two glacial cycles under very different boundary conditions from the present day, both during periods of glaciation and during the period of late- and post-glacial readjustment. It also includes the Eemian interglacial and issues of its similarities and dissimilarities with the Holocene. Key questions include the impact of changes in ice-sheet extent, the influence of changing North Atlantic sea-surface temperatures (SST) and thermohaline circulation on North European climate, the magnitude of climate change from west to east in Europe in relation to changes in the northern ice-sheet, the impact of climate change on the large East African lakes, the relative importance of different orbital forcing regimes in the Mediterranean region, variations in monsoon circulation in glacial and interglacial conditions and the changing impact of Antarctica and the Southern Ocean on the climate of Southern Africa.

The papers in this volume attempt to address many of these questions. The first two papers provide background and context to the PEP III transect: Oldfield and Thompson describe the wide range of records, archives and proxies that are the basis for reconstructing past environmental and climatic variability within the PEP III domain, and Vidal and Arz review evidence for global climate change in the context of ocean variability around the PEP III continental region. Thereafter the volume is constructed around 11 core papers (Chapters 4, 5, 7, 8, 12, 16, 17, 19, 20, 21 and 22) that cover the whole PEP III transect from Southern Africa to Northern Europe. Each of these papers presents a comprehensive synthesis of the evidence for past climate change and its causes on both time-stream 1 and time-stream 2 time-scales. They are augmented, especially for Africa and Mediterranean regions by a number of more specific papers that focus on key sites (e.g., Bar-Matthews and Ayalon, Johnson et al.), specific regions (e.g., Umer et al., Elenga et al., Sowunmi et al., Kallel et al.) or use specific approaches (e.g., Edmunds et al., Ortlieb). There are subsequently two papers on climate modelling, one on data-model comparisons (Bracannot et al.) and the second on new climate model developments, especially for the Holocene (Renssen et al.). The penultimate papers are concerned with the human dimensions of climate change. Robertshaw et al. explore the complex reactions of human society to past climate change and Olago and Odada stress the vulnerability of contemporary African societies in particular to climate change as well as the importance of understanding climate variability over long time-scales in order to manage natural resources sustainably. Finally, the two last papers consider the progress made by the PEP III research community in addressing the key questions posed for the two time-streams and identify future research priorities.

Acknowledgments

This volume is based on the keynote lectures presented at the PEP III conference in Aix-en-Provence in August 2001. We would like to thank all those involved with the organisation of that conference and the conference sponsors, especially IGBP-PAGES, ESF-HOLIVAR, the French Centre National de la recherche scientifique (CNRS), the French Ministère des Affaires Etrangères, the Institut de Recherche pour le Développement (IRD), the Région PACA, the Conseil Général des Bouches du Rhône, the Aix-en-Provence city, the Université Aix-Marseille III, and the Centre Européen de la Recherche et d'Enseignement en Géosciences de l'Environnement (CEREGE, Aix-en-Provence) and the US-NSF, without whom this volume would not have been possible. Each paper was assiduously reviewed by two referees mostly independently of the collective book authorship and we are especially grateful to all those that helped us in this capacity. Our joint editing has benefited enormously from the help of Gail Crick, Hélène Coleman and Heather Binney in the ECRC Office, Elanor McBay in the UCL Geography Drawing Office, and Nicole Page, Jean-Jacques Motte and Christine Vanbesien from the CEREGE. Bill Last and John Smol (DPER Series Editors) and Anna Besse, Judith Terpos and Gert-Jan Geraeds (Kluwer) have provided advise and encouragement throughout the last two years. We are grateful to Cathy Jenks providing the Index. We also would like to thank specifically Frank Oldfield and Keith Alverson in the PAGES International Project Office in Berne for their financial and intellectual support of the PEP III programme. Finally, we hope that this volume is an adequate tribute to the collective effort of those scientists who have supported PEP III activities over the last decade and that it will become a valuable benchmark for those who continue in the future to address questions of climate variability in Africa and Europe.

References

Alverson K., Bradley R. and Pederson T. (eds) 2003. Paleoclimate Global Change and the Future. Springer, Berlin, 221 pp.
Colling A., Dise N., Francis P., Harris N. and Wilson C. 1997. The Dynamic Earth. The Open University, Milton Keynes, 256 pp.
Dodson J.R., Taylor D., Ono Y. and Wang P. 2004. Climate, human and natural systems of the PEP II transect. Quat. Internat. 118-119: 1–203.
Markgraf V. (ed.) 2001. Interhemispheric Climate Linkages. Academic Press, New York, 454 pp.

2. ARCHIVES AND PROXIES ALONG THE PEP III TRANSECT

FRANK OLDFIELD (f.oldfield@btinternet.com)
Department of Geography
University of Liverpool
Liverpool, L69 72T
UK

ROY THOMPSON (roy@ed.ac.uk)
School of Geosciences
The University of Edinburgh
Edinburgh, EH9 3JW
Scotland

Keywords: Palaeoclimate, Palaeoecology, Palaeoenvironment, Palaeo-archives, Palaeo-proxies, Pole-Equator-Pole, Transect, Calibration

Introduction

The PEP III Transect (Gasse and Battarbee, this volume) spans an immense range of environmental and cultural diversity. It includes regions that have been the cradle of Old World civilisations dating back many millennia, it is home to some of the most advanced and favoured societies on the planet, some of the least hospitable environments and some of the most vulnerable and economically impoverished peoples on earth. Documenting and understanding the ways in which climate has varied across the whole length of the Transect and teasing out the past interactions between climate and human welfare pose immense challenges to the scientific community. Moreover, the challenges encompass themes that are of outstanding practical importance for our future as well as part of our compulsive fascination with the past.

The challenges require us to make full use of the instrumental record of recent climate variability and this is the concern of the first part of this chapter. But that alone does not suffice. Valuable though it is, the brief instrumental record fails to encompass the full range of natural climate variability in both time and space that has characterised the last few thousand years. In order to provide a record of that variability, it is necessary to turn to a whole range of 'archives', both documentary and environmental, and to analyse them using methods that can capture datable and decipherable signals of past variability. For this, we rely largely on what are usually termed 'proxies'. After outlining the scope for instrumentally based climate reconstructions across the whole length of the Transect, the present chapter seeks to give a brief introduction to the archives and proxies that provide

R. W. Battarbee et al. (eds) 2004. *Past Climate Variability through Europe and Africa.*
Springer, Dordrecht, The Netherlands.

a basis for reconstructing past environmental and climatic variability within the PEP III domain.

The main archives

Instrumental climate records

Instrumental records are a vital part of any research strategy for palaeoenvironmental reconstructions, especially where our interest focuses on time intervals during which the boundary conditions do not differ too greatly from those over which the records are available. Instrumental data (e.g., Vose et al. (1992)) provide a backbone for palaeoclimatology as they enable us to quantify objectively how the climate has varied and offer the most statistically robust data from which to establish the spatial coherence of the currently dominant modes of climate variability. Furthermore they provide the basis on which diverse proxies can be calibrated to measured climatic properties, as well as forming a bridge that links the longer, proxy records to the period of present day monitoring. Instrumental records cover a whole range of climatic parameters.

For much of the 20th century climate, was viewed as being effectively stationary and only varying slowly over geological time. This static view dominated human decision-making. Recently however, the increasing body of evidence from instrumental records of warming temperatures and of changing patterns of precipitation has led to a new dynamical view of climate change. Instrumental observations of the weather have been aggregated to build up detailed records of climate change as far back as the seventeenth century. In recent years many previously forgotten old ledgers have been unearthed and their valuable contents incorporated into climatic databases. Also during the past decade much progress has been made in improving the fidelity of instrumental climate records by removing systematic biases and by gathering together and analysing daily data rather than relying on monthly averages.

Today the vast observational network (surface and upper-air information from radiosondes), largely established for weather forecasting, coupled with satellite soundings and imagery has led to millions of weather observations being made each day. Data are typically recorded at thousands of synoptic and climatological stations on land; hundreds of aircraft, ships and moored buoys; hundreds of radiosondes and from many satellites. Reanalysis studies, in which daily observations since 1948 have been assimilated into a comprehensive global representation of the daily state of the atmosphere, provide a synthesis of this great wealth of instrumental weather data. These retrospective analyses of weather observations have great potential for providing very valuable data sets to researchers in the climate community. A particularly appealing aspect of reanalysis is that its four-dimensional assimilation and modelling system can transport information from data-rich to data-poor regions providing internally consistent climatic time-series in observationally sparse localities.

The PEP III pole-to-pole transect is rich in instrumental meteorological records and proxy data. It passes through Europe, the region of the world's longest instrumental records (e.g., Camuffo and Jones (2002)), as well as Sub-Saharan Africa a locality which has experienced some of the most pronounced recent climate change of anywhere around the globe (Nicholson and Grist 2001). The instrumental data available for each of the six PEP

III geographical regions of the TS1 (time stream one) science plan (Gasse and Battarbee, this volume) are briefly described.

High latitude, high altitude Europe

Climate reconstructions can be generated for the last three hundred years, even for localities beyond the timberline in sub-arctic Lapland, by using regression analysis to extrapolate the instrumental records from the more populated regions of North and Central Europe. The excellent skill exhibited by the multiple regression methodology relates to the high quality of the European data and to the SW-NE anisotropy of air-temperatures across Europe (Sorvari et al. 2002). High altitude climates, from above the timberline, can similarly be reconstructed from lowland weather series (Agustí-Panareda and Thompson 2002). The high Alps in particular are found to have warmed steadily in recent decades (Schönwiese and Rapp 1997; Böhm et al. 2001). At many localities, permafrost, normally protected from annual and inter-annual temperature fluctuations by its low thermal diffusivity, is now experiencing conditions outside the range of natural climatic variability and is thawing for the first time in millennia (Hoelzle and Haeberli 1995).

Mid latitude Europe

Europe is particularly well endowed with instrumental series (Thompson 1995; 1999). The climatic parameter that provides the longest record is mean air temperature. The renowned Manley (1974) series for Central England starts in 1659. Precipitation series are available from 1697 (Kew), while the longest continuous surface-pressure sequence commences in 1722 at Uppsala in Sweden. The wide geographical coverage of European instrumental records since 1781, means that fluctuations to the regional atmospheric circulation patterns can be established (Luterbacher et al. 1999) and hence climate change hindcast even in the less populated hinterlands of mid latitude Europe. For example, in the Annecy region of the pre-Alps, daily discharge can be reconstructed for rivers back to the 1800s (Foster et al. 2003) and growing season degree-days estimated for marginal agricultural land on an annual basis back to the 1600s (Crook et al. 2002).

The Mediterranean Basin

The Mediterranean climate is characterised by dry summers and wet winters. Mediterranean precipitation records reach back to 1749. Good spatial coverage is provided by over 100 precipitation series spanning the last 150 years. Winter months (NDJF) exhibit a teleconnection seesaw pattern as demonstrated by the correlation coefficients in Table 1. Winters with higher than usual precipitation in the west (Portugal/W. Spain) tend to be dry winters in the east (Israel/N. Egypt), and vice versa. Exactly the same teleconnection pattern is found in the long term precipitation trends (Plate 1) which reveal a 150 year drying trend in the S. and E. Mediterranean (-0.1 mm/month yr^{-1}) which contrasts with an increasing trend in winter precipitation in the N. and W. Mediterranean ($+0.1$ mm/month yr^{-1}). The S. European trend has recently reversed and S. Europe is currently drying in agreement with predictions based on $2xCO_2$ modelling (Ulbrich and Christoph 1999). The consistency of the teleconnection across frequency bands (from annual to centennial) suggests it is not unreasonable to search for the same teleconnection pattern in millennial long proxy records.

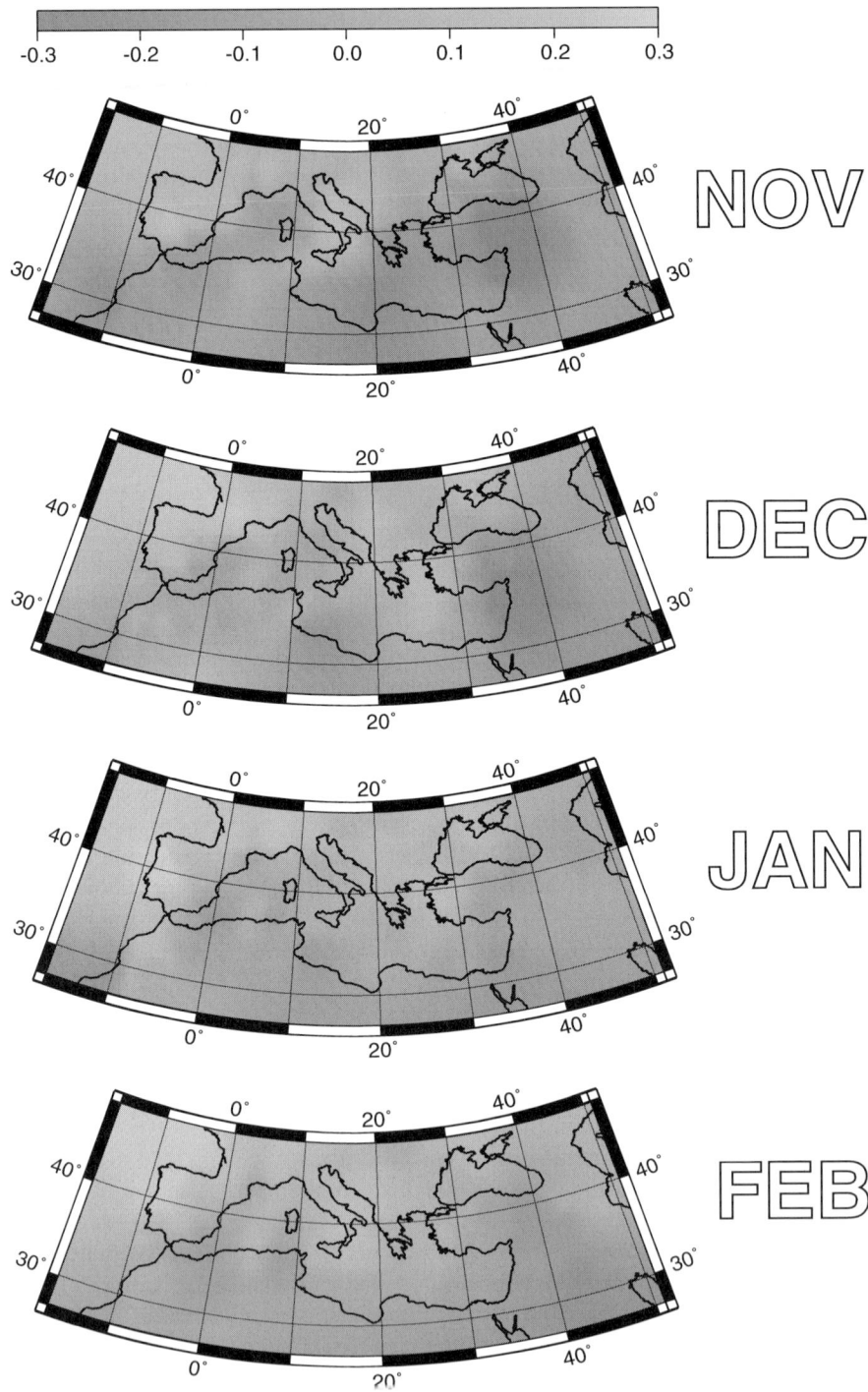

Plate 1. Long-term linear trend in Mediterranean precipitation (mm/month/yr) during the 136 year period (1855 to 1990). Note the drying trend (purple) for all four-winter months in the Near East compared to the increase (blue) for Western Iberia. The null change line runs approximately through Gibraltar-Sicily-Greece-Black Sea. Note how the same east-west teleconnection seesaw is found for precipitation as in the interannual variation of Table 1. (Albers equal area projection). Colour version of this Plate can be found in Appendix, p.627

Table 1. Correlation coefficients (*r*) between the inter-annual precipitation for seven stations in Portugal/Western Spain and nine Eastern Mediterranean stations for the four winter months (NDJF), during the 136-year period (1855 to 1990). Note the block of negative East-West teleconnections in the lower left quadrant, which contrast with the blocks of positive teleconnections in the upper left and lower right quadrants.

	Lis	Coi	Evo	Bad	Por	Val	Bur	Lim	Nic	Bey	Jer	Ism	Sai	Cai	HQ	Alx
Lisbon	1	.70	.88	.82	.69	.70	.66	−.14	−.13	−.02	−.11	−.04	−.04	−.10	−.08	−.16
Coimbra	.70	1	.80	.62	.77	.79	.74	−.20	−.11	−.15	−.18	−.05	−.12	−.12	−.12	−.16
Evora	.88	.80	1	.86	.73	.75	.67	−.13	−.11	−.16	−.23	−.20	.00	−.28	−.25	−.17
Badajoz	.89	.62	.86	1	.56	.70	.66	−.20	−.17	−.19	−.18	−.12	−.16	−.17	−.17	−.17
Porto	.69	.77	.73	.56	1	.69	.63	−.15	−.11	.03	−.12	.03	−.04	−.06	−.14	−.15
Valladolid	.70	.79	.75	.70	.69	1	.81	−.16	−.15	−.03	−.19	−.03	−.11	−.15	−.12	−.15
Burgos	.66	.74	.67	.66	.63	.80	1	−.26	−.20	−.17	−.23	−.12	−.13	−.21	−.26	−.17
Limassol	−.14	−.19	−.13	−.20	−.15	−.16	−.26	1	.48	.15	.35	.11	.06	.17	.63	.19
Nicosia	−.13	−.11	−.11	−.17	−.19	−.15	−.20	.49	1	.15	.37	.10	.14	.08	.39	.49
Beyrouth	−.02	−.15	−.16	−.19	.03	−.03	−.17	.15	.15	1	.31	.59	.41	.58	.22	.24
Jerusalem	−.11	−.18	−.22	−.18	−.12	−.19	−.23	.34	.37	.31	1	.23	.32	.27	.30	.41
Ismailia	−.04	−.05	−.20	−.12	.03	−.03	−.12	.10	.10	.59	.23	1	.42	.85	.17	.19
Port Said	−.04	−.12	.00	−.16	−.04	−.11	−.13	.06	.14	.41	.32	.42	1	.34	.17	.43
Cairo	−.10	−.12	−.28	−.17	−.06	−.15	−.21	.17	.08	.58	.27	.85	.34	1	.15	.19
Cairo HQ	−.08	−.12	−.25	−.17	−.14	−.12	−.26	.63	.40	.22	.30	.17	.17	.15	1	.30
Alexand	−.16	−.16	−.17	−.17	−.14	−.15	−.17	.18	.49	.24	.41	.19	.43	.19	.30	1

Arid and sub-arid belt

Along the African and Arabian segments of the PEP III Transect, instrumental records are much shorter than those from Europe (Nicholson 2001) and tend to be restricted to the coast. Nevertheless in some places records of air temperature and precipitation are available for over 150 years. There have been dramatic climatic changes even over these times, especially in the W. Sahel, with the most pronounced drying of anywhere in the world (Senegal, 1884–1990, JJA, -1.3 mm/month yr^{-1}). At a similar latitude, but further east, through Mali, Niger and Ethiopia the same drying trend is found and averaged around -0.2 to -0.5 mm/month yr^{-1}.

Inter-tropical Africa

20th century instrumental data allow spatial climatic patterns and teleconnections to be well mapped through the tropics. Good coherence is found between various climatic parameters in the tropics. For example, rainy season daytime (maximum) air-temperature and precipitation anomalies in Zimbabwe are closely anti-correlated, due to the effects of clouds (Unganai 1997). Long-term precipitation trends tend to have been low, with the exception of the Angolan coast (Luanda, 1879–1985, MAM, $+0.3$ mm/month yr^{-1}) and

Zimbabwe (1890–1989, DJF, -0.3 mm/month yr^{-1}). Various regions such as Southern Uganda and Central Zimbabwe show statistically significant ENSO signals (Phillips and McIntyre 2000; Unganai 1997). These ENSO signals are however both climatically and geographically limited compared to those from the other PEP transects (Markgraf and Diaz 2001).

Southern Africa

Southern Africa encompasses a particularly wide range of climatic zones, ranging from true desert through Mediterranean and humid sub-tropical to tropical, and a strongly contrasting range of seasonalities all within a small geographic region (500 × 1000 km). Thirty-two precipitation series span the last 100 years. However, they reveal no century long precipitation trends (Richard et al. 2000; 2001). Interannual variability of summer rainfall is linked to sea-surface temperatures in the Indian Ocean (Rocha and Simmonds 1997). Air-temperature data are available from 1857.

Documentary records

Europe especially abounds in documentary records of weather and climate spanning many centuries (e.g., Frenzel et al. (1992), Glaser and Hagedorn (1991)). The special value of documentary evidence lies partly in its spatial and temporal specificity and the increasing skill with which the massive quantities of diverse evidence can be objectively translated into more or less quantifiable descriptors (e.g., Pfister (1992)). These in turn permit analyses with seasonal, even daily resolution as well as interpretation in terms of synoptic patterns, although one must beware of the limitations in comparison to direct instrumental measurements of the weather. For example, a comparison of the spatial coherence of seasonal temperatures, at the inter-annual time scale, between Germany and Switzerland reveals a marked drop in correlation coefficient as the data used changes from instrumental records to documentary evidence. Counter-balancing such limitations are the insights carefully interpreted documentary records can provide into extreme events and their impacts on hydrology, ecosystems and human populations. In most cases, it is extreme events to which human populations are vulnerable, rather than changes in mean annual climate.

Evidence linking changing climatic conditions to contemporary indicators of harvests, human welfare, demographic changes and socio-cultural responses increases the value of documentary evidence, for it can deepen our understanding of the nature of the interaction between environment and society during periods of stress (Brázdil et al. 1999). The richness of the documentary record in providing indicators of human response as well as climate change gives it the added benefit of precluding and indeed transcending interpretations in simple deterministic terms. The full range of applications for documentary reconstructions of past climate has yet to be realised. One novel application is illustrated by the recent work of Bugmann and Pfister (2000) in which Bugmann's forest succession model has been 'driven' by Pfister's reconstructed climate variability over the last five centuries in order to explore the reliability of the model as a predictor of tree line behaviour in the Swiss Alps (Fig. 1).

In the prevailing intellectual climate, those environmental historians whose work links what is essentially a humanistic methodology to areas of application in the natural sciences risk being undervalued by both communities. One of the roles of the leaders of the

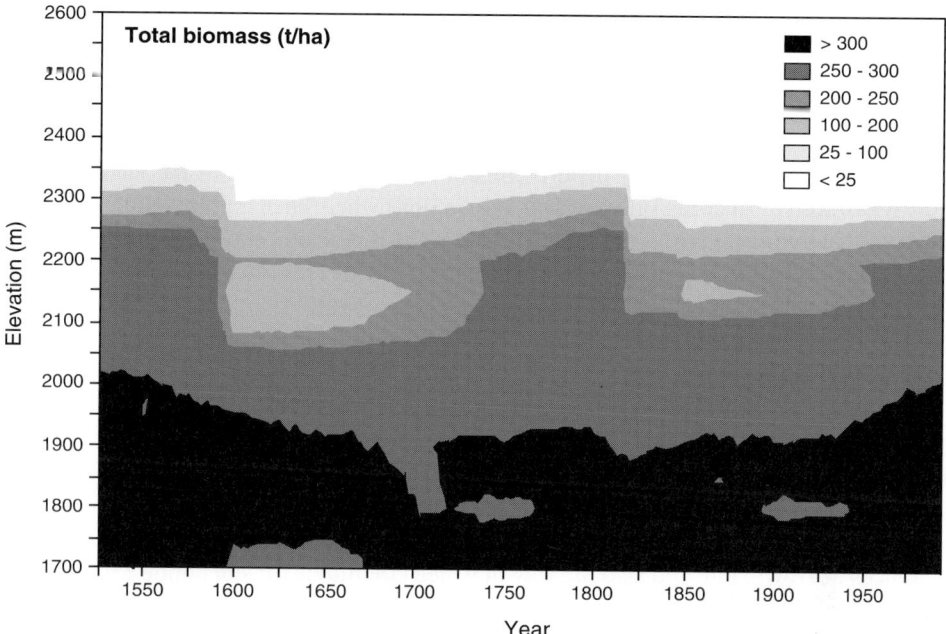

Figure 1. Total biomass (t/ha) on a south-facing slope in the European Alps simulated by a model of forest dynamics as 'forced' by a 470-year compilation of historical climate change from 1525 to 1995. Note the two dramatic drops in the upper-tree line caused by major dieback events following two series of extremely cold summers (1593-97 and 1812-18). (From Bugmann and Pfister (2000)).

palaeoclimate community might be to enhance recognition of the value of document-based reconstructions as well as to encourage those bold and scrupulous enough to bridge the two cultures with skill and success.

Tree rings

One of the most powerful palaeoclimatic tools to emerge over the last few decades has been dendroclimatology, which, by definition, achieves annual resolution and can often be linked to specific characteristics of the growing season (e.g., Lindholm and Eronen (2000)). By now, networks of researchers have developed circum-boreal records distinguished by both rigorous calibration to instrumental records and compelling regional coherence. Spatial cover is also matched by long time series that have served to calibrate radiocarbon dates, provide powerful chronologies in their own right and document high frequency climate variability, including the transient responses to volcanic events (Briffa et al. 1998), throughout the late Holocene. PEP III benefits from the fact that some of the leaders in this research field have spent many years refining and extended the record in Europe and adjacent areas of the former Soviet Union (Briffa 2000; Briffa et al. 2001; Briffa and Matthews 2002). The research field has thus benefited from the creation of major databases that contain evidence which combines precise chronologies with the results obtained by applying a common methodology to comparable archives over a wide area (e.g., Schweingruber et al. (1996)).

In this regard, the community of dendroclimatologists has set an excellent example to the rest of the research community.

Research along the PEP III Transect also includes tree ring, or at least, growth increment studies, from the east Mediterranean (D'Arrigo et al. 2001), the mountains of the Saharan region (Pelfini et al. 2001) and the sub-Saharan Miombo woodland (Trouet et al. 2001). These extend the application of dendroclimatological method beyond the spatial realm where it has been most commonly applied, into regions where such high-resolution proxies are exceptionally rare.

Speleothems

An increasing number of studies exploits the paleoenvironmental signals contained in the calcite precipitated by cave water. Speleothem based studies along the PEP III transect span an amazing range of latitudes and environments from Northern Norway (Lauritzen 1996; 2003) to South Africa (Holmgren et al. 1999; Lee-Thorp et al. 2001), via the Mediterranean region (Aylon et al. 2001; Bar-Matthews et al. 2001; Causse et al. 2001) and Oman (Fleitmann et al. 2001; 2002). The time-span covered is remarkably varied, reflecting differences in growth rate and also the way in which conditions for stalagmite growth have sometimes occupied quite narrow and diverse time intervals, depending on the location of, and water supply to, the site. Thus, even the discontinuities in speleothem development may themselves reflect significant changes in regional climate.

The growing number and wide geographical spread of speleothem records has led to the establishment of a thematic group (SPEP) concerned with this type of archive and led by Stein-Eric Lauritzen (Lauritzen 2001). Palaeoclimatic signatures reside in several speleothem properties ranging from luminescence to geochemical and stable isotope ratios (Lauritzen and Lundberg 1999). One of their major advantages is the scope they offer for dating using the U/Th method (Rosendal et al. 2001). In all cases though, there is a need to understand and, in so far as possible, make quantitative adjustment for all the processes that intervene to modify a climate signature as water passes from the lower atmosphere, through soil and bedrock to the point of calcite precipitation. These processes may be region-, even site- or speleothem-specific. They therefore call for careful study before the sequence of changes that is derived from speleothem analysis can be transformed into climate parameters. The scope speleothems offer, for resolving both high and low frequency variability, gives them special value particularly when viewed alongside other high resolution, precisely datable archives such as tree rings or varved lake sediments.

Corals

Coral records, mainly spanning the last one to two centuries at most, have been studied in favourable localities from the Red Sea (e.g., Klein et al. (1990), Felis et al. (2001)) south-wards to the coast of Madagascar (Zinke et al. 2001). The reliability with which seasonality is reflected in the fluorescent banding of corals coupled with their location in shallow marine environments, makes them uniquely valuable palaeo-archives. The calibration of stable isotope signatures and chemical ratios in corals usually involves comparing them with short instrumental time-series. The correlations that often emerge have encouraged

the use of coral records as indispensable indicators of Sea Surface Temperature (SST) and salinity changes on a seasonal and inter-annual basis. This in turn makes them extremely attractive for documenting the effects of ENSO variability not only in the Pacific region but also along the east coast of tropical Africa. Coral research has important implications for exploring changes to the periodicity of ENSO on decadal to century time-scales, as well as the strength of the Indian Ocean-western Pacific teleconnection and its changes through time.

In addition, increases in the barium/calcium quotient in corals have been shown to reflect increases in fluvial input and associated terrigenous sediment supply to the reef environment as a result of land clearance and soil erosion in recent times (Cole 2001). The sum total of evidence from the coral record therefore provides an elegant illustration of the way in which human-induced changes on land can interact with an overall warming trend, to exacerbate the damaging effects of coral bleaching and associated degradation of reef environments.

Lake sediments

Lake sediment records of almost infinite variety have clearly been among the most popular archives along the whole length of the transect. They defy generalisation. In terms of time spans covered, they range from sites in Northern Fennoscandia where the development of a truly limnic environment postdates the time of isolation from the Baltic as a result of Holocene isostatic rebound (e.g., Hedenstrom and Risberg (1999)), to volcanic or impact crater lakes in central (Creer et al. 1990; Zolitschka 1998) and southern (Allen et al. 1999; 2000; Ramrath et al. 2000; Roberts et al. 2001) Europe, or South Africa (Partridge et al. 1997) respectively, where the sediment record may span several glacial-interglacial cycles. Spatial extent and morphometry are equally varied and we may contrast the small glacial lakes of much of highland Europe (e.g., Birks (2000), Ammann (2000)) with both the major lakes of the African rift system (Johnson 1996) and the once extensive but now dry lakes that formerly covered vast areas of the Southern Sahara region during the early Holocene (Gasse et al. 1987; Kroepelin et al. 2001).

Lake sediments in much of Europe especially, have begun to provide quantitative palaeoclimate proxies through the development of transfer functions linking aspects of lake biology, recoverable from fossil remains, to variations in water and air temperature (e.g., Lotter et al. (1997), Lotter (2003 and Fig. 2)). Other lakes have yielded palaeoclimate records from variations in the stable isotope ratios in ostracods (von Grafenstein et al. 1999), authigenic carbonates (Jones et al. 2002), organic extracts and diatom silica.

Of especial value are lake sediments in which the annual rhythm of sedimentation gives rise to varves, which in turn can be used to provide potentially annual resolution in the palaeo-record as well as to establish a precise and accurate chronology of sediment accumulation (e.g., Zolitschka (1998)). Indeed, variations in the character and thickness of the varves themselves may constitute a quantifiable climate proxy in favourable circumstances and, like other annually resolvable archives, they can give vital information on the incidence and the hydrological and sedimentological impacts of extreme events (e.g., Thorndycraft et al. (1998)). Varves vary greatly in type and origin, but where they can be shown consistently to reflect an annual cycle over a long period of time, their presence

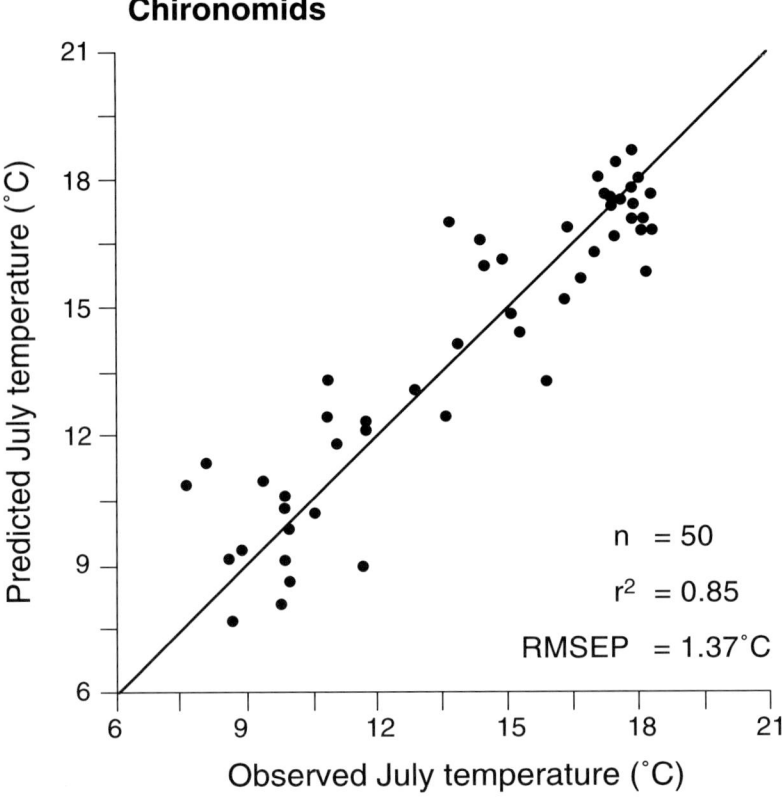

Figure 2. Chironomid data from the Alps illustrates the skill that can now be achieved in climate reconstructions. Here a spatial calibration of July temperatures, based on a training set of lakes along an elevation gradient, gives a cross validation r^2 of 0.85 and a prediction error of 1.37 °C. (From Lotter et al. (1997)).

endows any sediment sequence with added significance, for without them, establishing a reliable chronology of sedimentation can often provide a major and at times, in our present state of knowledge, insurmountable obstacle to making the fullest possible use of palaeolimnological evidence.

Lake basins and their sediments often provide crucial evidence about past lake levels and hence about changes in the balance between precipitation and evaporation. This is especially significant in regions where major fluctuations in available moisture are of crucial significance for human welfare and even survival. In much of Africa, on both sides of the Equator, the evidence for major changes in lake level during the Holocene provides one of the most powerful counters to the polar-centric view of the Holocene as a period of relatively stable climate (Gasse and Van Campo 1994; Verschuren et al. 2000; Bergner et al. 2001). These changes can be identified using both geomorphic and sedimentary evidence; they occur on time-scales ranging from sub-decadal to millennial and their dramatic implications for human societies in the past are becoming increasingly well documented. While climate/vegetation models for N. Africa in the mid-Holocene, with their strengthened summer monsoon, predict pronounced shifts in the climate and in vegetation patterns, the direct effects of orbital changes alone are found to be insufficient to generate the major changes indicated in the palaeoenvironmental reconstructions. Vegetation feedbacks can produce further intensification in monsoonal precipitation during the mid-Holocene especially in W Africa (Ganopolski et al. 1998; Doherty et al. 2000). However, the dramatic and widespread shifts of lake levels in N. Africa are still not fully explained. Soil and ocean feedbacks in fully coupled vegetation-atmosphere-ocean models may provide a means of resolving some of the remaining discrepancies between the modeled and proxy lake-level data.

Peat

Here we consider only those instances where peat stratigraphy and the contained fossil record have been used to provide direct proxies for climate change. This excludes the more numerous instances where peat profiles have provided the context for pollen analytical studies. Ombrotrophic (precipitation-dependent) peatlands are widespread in the cool temperate and boreal parts of the Transect and their direct link with the atmosphere makes them attractive archives for recording past changes in the balance between precipitation and evaporation. It is over a hundred years since peat stratigraphy was first interpreted as a record of climate change in Northern Europe during the Holocene and the terminology to which these early studies gave rise still has wide currency. Plant macrofossils, including *Sphagnum* species with well established habitat requirements relative to the water table, testate amoebae and peat humification measurements have all been shown to reflect variations in surface moisture status at the point of sampling (Charman and Hendon 2001; Barber and Langdon 2001; Barber and Charman 2003). Current work includes several studies designed to quantify this relationship and to test the extent to which the temporal sequences of surface wetness reconstructed from individual cores can be replicated within and between sites (Barber et al. 2000; 2003). This work requires detailed and precise chronological control. Fortunately, ombrotrophic peat can provide the raw material for this, especially if care is taken to eliminate time-transgressive components in the dated material and if the density of radiometric determinations undertaken is close enough to open up the opportunity to establish precise dates by matching onto dendrocalibration 'wiggles' (Van Geel and Mook 1989; Mauquoy et al. 2002).

Groundwater

Groundwater is both an archive and a vital human resource (Edmunds 1999). Its availability as a recorder of past climate in areas where many of the other commonly used archives are absent adds to its attraction. Within the saturated zone of major aquifers in arid and semi-arid regions, and despite the rather low temporal resolution of the record, the combined results from stable isotope and noble gas (Stute and Talma 1998) measurements have provided evidence of outstanding importance for developing quantitative estimates of differences in mean annual temperature between the present day and the last glacial maximum. It is also important in currently arid and semi-arid regions, to combine evidence from ground water with that derived from changing lake levels, for the lakes are often simply the surface expression of water bodies that lie mainly below ground level (Fontes et al. 1993; Edmunds et al. 1999). Interpreting past hydrological changes from the lakes alone thus provides only part of the story.

Much of the evidence already obtained from groundwater studies highlights the fact that in many areas, currently exploited groundwater is 'fossil' and dates from periods of greater moisture availability thousands of years ago. In too many cases, from countries as diverse as Spain, Libya and Northern Nigeria, it is being mined at a rate vastly exceeding the rates of recharge possible under current climatic conditions. So societies find themselves in a dangerous dilemma. While populations and demands on water are growing, inappropriate management is leading to rapidly declining water resources. Human societies that fail to heed the incontrovertible evidence of the extremely slow recharge rates of deep groundwaters do so at their peril.

Records of climate changes in the more recent past can be obtained from profiles of water chemistry within the unsaturated zone (Goni and Edmunds 2001). Conservative elements such as chlorine can be used as tracers that record changes in the degree of evaporative concentration in past groundwater. In this way, the groundwater profiles can preserve a clear record of such events as the Sahel drought of the 1970's and 80's.

Other types of continental archive

A wide range of other archives such as alluvial or colluvial sequences, loess profiles and geomorphic or stratigraphic indicators of past glacier behaviour have all contributed vital information along the Transect. Although, in many cases, they lack the continuity of some of the other archives they can nevertheless often provide essential constraints on inferences derived from other lines of evidence. Many also provide contexts in which palaeo-environmental and archaeological evidence can be directly linked. Special value is inherent to archives in which a direct association can be established between evidence for environmental variability and the archaeological record of societal response.

Special mention should be made of the Kilimanjaro ice core record (Thompson 2001). It forms the only African example of a whole series of ice core records from tropical glaciers ranging from the Andean peaks of South America to the mountains of Central Asia. It will thus form one of the most significant of several emerging links between the PEP Transects. The importance of the record from Kilimanjaro, as from other low latitude ice-cores, is enhanced by the rate at which such archives are disappearing.

Marine sediments

Over the last few decades, the temporal resolution achievable in favourably located marine sediment sequences has made them of increasing interest as a complement to the terrestrially derived evidence that has been the main concern of the PEP Transects. High-resolution records from giant piston cores obtained during IMAGES cruises have proved especially valuable (Cortijo et al. 2000). Much of the marine evidence available over the PEP III Transect takes the form of stable isotope analyses particularly on planktonic foraminifera. From these, along with additional sedimentological, biological and geochemical analyses, reconstructions of past changes in SST, salinity, ventilation, sediment provenance, dust deposition and sea ice extent have been made. These marine proxies provide parallels to the terrestrial record (e.g., Cullen et al. (2000)) as well as evidence for the role that changing ocean circulation has played in modifying and, at times, dramatically amplifying the impact of changes due to external forcing. In the context of the PEP III domain, special interest attaches to past changes in North Atlantic deep-water formation and its impact on ocean circulation (Bond et al. 2001). One of the most important messages inherent in the longer term records from PEP III is the close link on all timescales from decadal to millennial between largely salinity-driven changes in thermohaline circulation in the North Atlantic and dramatic changes in climate regime over a wide area including, but extending well beyond the Atlantic margins of the PEP III Transect (Peterson et al. 2000).

The position of the Mediterranean Sea within the Transect endows its sedimentary record with special significance. One of the most striking features is the presence of sapropel layers (Rossignol-Strick 1985), which are now believed to be linked to periods of increased freshwater discharge into the basin. The coherence of the link between marine evidence for sapropel formation and the terrestrial evidence for moister conditions over much of the Mediterranean region during parts of the early to mid-Holocene is a striking example of the way in which marine and terrestrial archives can be mutually reinforcing. There is scope for exploiting this linkage much more, especially in marine cores with high temporal resolution, where continental 'signatures' such as pollen, or markers of changing terrigenous sediment flux, can be set alongside evidence for changes in the marine realm (e.g., Oldfield et al. (2003a)). In this regard, contexts with rapid sediment accumulation like the western flanks of the Adriatic or the Nile delta may prove especially attractive.

Climate proxies

The dominant concern of the PEP Transects has been with the reconstruction of past *climate*. As a consequence, the interpretation of most of the palaeo-environmental records has been in predominantly climatic terms, sometimes at a rather descriptive, qualitative level, but increasingly in more quantitative ways. This calls for at least two types of interaction between instrumental and non-instrumental data. Wherever proxy climate signatures, whether documentary or environmental, are interpreted in quantitative terms, this requires calibration against instrumental records. This can be achieved by using a spatially distributed training set of observations that span a wide climatic range at the present day (Fig. 2), or by comparing the proxy with an instrumental time series during the period of overlap (Fig. 3).

An elegant example of the use of time series of instrumental data in deciphering the climatic signal preserved by clastic varves is provided by the validation study of Wohlfarth et al. (1998). Multiple regression analysis of the 90-year sequence, from 1860 to 1950, from the Ångermanälven Estuary in North Central Sweden revealed strong correlations between spring/summer precipitation and varve thickness. The clearest climatic relationships were found between the precipitation in the mountains (the source region for the river discharge) and the estuarine varves. Turning to another striking example, dates of the beginning of the grape harvest from wine regions in Central Europe correlate ($r = +0.85$) extremely well with one another (Ladurie and Baulant 1981) and with spring and summer temperatures (Fig. 3). The validation of the grape harvest dates against the thermometer records not only provides reassurance as to the reliability of phenological sources but also a quantitative measure of the dominant climatic variables that control ripening date. Tree-ring widths and

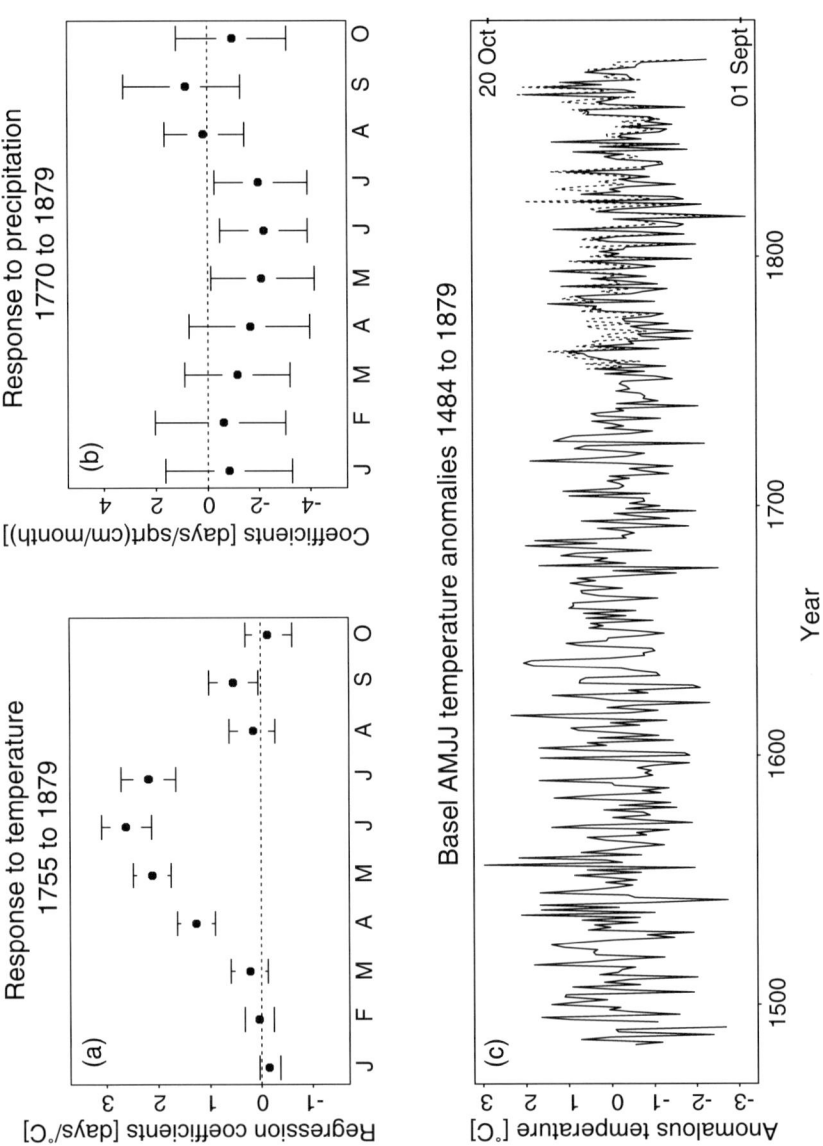

Figure 3. Reconstruction of Basle temperatures, from the fifteenth through nineteenth centuries, based on archival records of grape harvest dates. (a) Temperature anomalies of four months (April through July, AMJJ) are found to display significant relationships with grape harvest date. Warmer months lead to earlier grape harvests (positive correlations). (b) For three months (May through July, MJJ at Paris) wet years are found to result in later harvests (negative correlations). The 95% confidence interval limits in (a) and (b) were derived from bootstrap simulations using the adjusted percentile method. (c) Reconstructed Basle temperature anomalies [left hand scale] from 1484 to 1879 are plotted as a dashed line. The observed temperature anomalies (average anomaly during AMJJ at Basle) reveal a good fit ($r^2 = 0.75$) with the grape harvest dates [right hand scale] during the 124-year period of overlap of the two series. The observed anomalies during the validation period (1755 to 1879) are plotted as a solid line.

late-wood densities provide a third example of calibration of proxy records using the time-series approach. Tree-ring calibrations make a major contribution to most reconstructions of the climate of the past millennium (see e.g., Mann et al. (1999)). These reconstructions are currently of great concern as they demonstrate how the abrupt 20th-century warming has ended a 900-year natural cooling trend. They also strikingly highlight how the 1990s were the warmest decade of the millennium. Consequently they have been a major contributor to the influential 2001 IPCC report being able to reaffirm, in much stronger language than previously, that *'the present climate is changing in ways that cannot be accounted for by natural variability'* (Houghton et al. 2001).

In addition to the need for calibration, developing transcontinental or regional syntheses of past climate change imposes the requirement to harmonise all the many data types, both instrumental and non-instrumental, in terms of their climatic implications and spatial representation. Only through the co-ordination, intercalibration and fusion of all available sources of evidence will it be possible to express the nature of past climate change at a range of spatial and temporal scales and thereby gain an improved understanding of the complex interactions between environmental change and human societies.

The need for quantitative calibration applies to proxy signatures as contrasted, widely used and 'archive-transgressive' as pollen and stable isotopes as well as to a range of much less popular lines of evidence. Among the long list of proxy signatures used, two types merit consideration that is more general.

In the first of these two types the varying frequency of biological remains such as pollen (Guiot et al. 1989; Huntley 1993), diatoms (Bigler and Hall 2002) or chironomids (Brooks et al. 1997), are used to derive statistical relationships between assemblage characteristics in contemporary training sets and climatic parameters. These kinds of statistically based proxy signature have proved highly successful in generating palaeoclimate reconstructions, provided that the range of variability in both the climate space and the biological assemblages has remained reasonably comparable to that of the present day. Uncertainty increases greatly once the record from the past indicates no-analogue situations or presently unrepresented domains of variability (Huntley and Prentice 1993). By contrast, when using the second type of approach to climate reconstruction, for example when using stable isotopes, the interpretation often rests more on an understanding of the processes and the physicochemical laws that control any link between atmospheric properties and the environmental archive.

Increasingly, multiproxy studies are the norm and the climatic inferences derived from a range of proxies can be compared (Birks 2000; Ammann 2000). The multiproxy approach can lend much greater confidence and statistical rigour to the palaeoclimatic inferences derived from a single site or region. It can also open up a rich vein of future research, especially where the climate inferences derived from mutually independent proxies differ significantly. Such frustrating discrepancies are among the best points of departure for improving future understanding.

Chronologies and temporal resolution

Our best estimate suggests that less than 15% of the research presented at the PEP III Aix-en-Provence Conference (August 2001) has the potential for annual resolution and the ascription of a calendar date with an error of less than three percent. An additional few may be resolvable at a decadal level or thereabouts, with a similar accuracy in age ascription. It follows that the vast majority of the work has a time resolution at century level at best and, in many cases, a dating uncertainty equally is great. This unfortunate situation reflects the intractable nature of the problem of chronology for many of the most popular archives, such as lake sediments, which, despite the chronological problems they often pose, can provide proxy records of outstanding interest. Despite the existence of these dating problems, it is abundantly clear that much of the evidence presented has enormous significance for documenting climate variability on all timescales, for providing the raw material from which greater insight into the spatial coherence, persistence and regional expression of major modes of variability can be derived, for developing regional syntheses of past climate change (notable examples of which are contained within this volume), for identifying the key mechanisms of external forcing and internal dynamics and for testing climate models against the ground truth of the past.

This said, future progress in all these respects will depend in part on better chronologies. A greater investment in dating is a prerequisite for improving insights into questions of phasing and spatial coherence, for achieving better harmonisation of data sets between archives, proxies and regions and for the full realisation of many of the main initial goals of the PEP III Transect. The investment must include both a widening of access to dating facilities, especially for those working in less developed countries, and a continuing concern with testing and refining the evidence upon which some standard procedures such as marine reservoir correction of ^{14}C dates are based.

Proxies, ecosystems and people

Using every available line of evidence to reconstruct past climate precludes exploring the effects of climate change on ecosystems without resorting to unacceptably circular argument. Nevertheless, a significant proportion of the work completed along the PEP III Transect does deal explicitly with ecosystem responses, using lines of evidence as diverse as pollen, phytoliths and ^{13}C ratios (Huang et al. 1999) indicative of shifts in the relative importance of plants using different metabolic pathways.

At the stage where the most robust climate proxies have been tested and, in so far as possible, validated, there is a case for using them as independent variables, thus allowing more of the biological records to be interpreted as indicators of ecosystem *responses* to past climate change (Oldfield et al. 2000). It seems inevitable that incorporating this type of approach in the future research agenda of the PEP III scientific community will become attractive, indeed essential, as policy makers become increasingly concerned with understanding the likely response of ecosystems to future climate change. The research encompassed within the PEP III Transect can make a vital contribution to this area of impact studies without compromising its primary aim of palaeoclimate reconstruction.

PEP III spans regions of the world more severely impacted by human activity than almost any others. Evidence for human activities and their effect on terrestrial ecosystems goes back thousands of years, especially in those areas close to the cradle of ancient civilisations. Alongside the reconstruction of climate change, there is the parallel need to document the history of human impacts. Several papers demonstrate how such impacts have often overprinted any decipherable signature of climate change (e.g., Oldfield et al. (2003a, b))

Others illustrate well the difficulty of either disentangling the two types of influence or shedding light on the many ways in which they have interacted in the past (see e.g., Oldfield and Dearing (2003)). Our present biosphere is a no-analogue one created by the combined and interactive influences of human and natural processes. Any suggestion that we may use the past to inform policy for the future has to recognise the crucial importance of both natural and anthropogenic influences. Documenting the history of their interaction in all its richness and complexity is therefore an essential task. At the heart of PEP III has been the need to isolate climate influences, by careful selection of sites and types of archive, as well as by the ingenious use and calibration of proxies. In the longer term, there will also be a parallel and interlocking need to address more explicitly the history of the relationship between natural variability and human activities, not as a one-way street in which human societies merely respond, but as an interactive process. The progress achieved so far within the framework of PEP III encourages us to believe that this challenge too will be met. One of the next steps will be achieved through the European Science Foundation sponsored HOLIVAR initiative which seeks to coordinate and present the full range of evidence for climate variability during the Holocene (www.esf.org/holivar).

Summary

The PEP III Transect is concerned with climate change through Western and Eastern Europe, the Middle-East and Africa. The PEP programme aims to: understand how and why climate has varied in the past; assess how climate change and variability has affected natural ecosystems and human society in the past; and provide a basis both for developing and testing climate models that are needed to forecast climate change in the future. This chapter outlines and illustrates the range of archives and proxies used in the PEP III Transect. In view of the special importance of instrumental records, both for documenting recent climate variability and for calibrating proxies, these are dealt with at some length. Many of the special characteristics of the main archives and proxies are discussed, along with the kinds of information they provide. The major themes of palaeoclimate reconstruction, chronologies and temporal resolution, and past human impact and ecosystem change are considered towards the end of the chapter.

Acknowledgments

We are especially grateful to Sandra Mather for her expert help in producing the Figures.

References

Agustí-Panareda A. and Thompson R. 2002. Reconstructing air temperature at eleven remote alpine and arctic lakes in Europe from 1781 to 1997 AD. J. Paleolim. 28: 7–23.

Allen J.R.M., Watts W.A. and Huntley B. 2000. Weichselian palynostratigraphy, palaeovegetation and palaeoenvironment: the record from Lago Grande di Monticchio, Southern Italy. Qua. Interna. 73/74: 91–110.

Allen J.R.M., Brandt U., Brauer A., Hubberten H.-W., Huntley B., Keller J., Kraml M., Mackensen A., Mingram J., Negendank J.F.W., Nowaczyk N.R., Oberhänsli H., Watts W.A., Wulf S. and

Zolitschka B. 1999. Rapid environmental changes in Southern Europe during the last glacial period. Nature 400: 740–743.

Ammann B. 2000. Biotic responses to rapid climatic changes around the Younger Dryas and minor oscillations on an altitudinal transect in the Swiss Alps. Palaeogeogr. Palaeoclim. Palaeoecol. 159: 191–201.

Aylon A., Vaks A., Gilmour M., Fremkin A., Kaufman A., Matthews A. and Bar-Matthews M. 2001. Pleistocene paleoclimate evidences from speleothem record of a karstic cave located at the desert boundary — Maale-Efraim, Eastern Shomron, Israel. In: Past Climate Variability through Europe and Africa: Abstracts. 56–57. PAGES-PEP III/ESF HOLIVAR International Conference, Aix-en-Provence, August 2001.

Barber K.E., Maddy D., Rose N., Stevenson A.C., Stoneman R. and Thompson R. 2000. Replicated proxy-climate signals over the last 2000 yr from two distant UK peat bogs: new evidence for regional palaeoclimate teleconnections. Quat. Sci. Rev. 19: 481–487.

Barber K.E., Chambers F. and Maddy D. 2003. Holocene palaeoclimates from peat stratigraphy: macrofossil proxy climate records from three oceanic raised bogs in England and Ireland. Quat. Sci. Rev. 22: 521–539.

Barber K.E. and Langdon P. 2001. Testing the palaeoclimatic signal from peat bogs — temperature or precipitation forcing? In: Past Climate Variability through Europe and Africa: Abstracts. 58–59. PAGES-PEP III/ESF HOLIVAR International Conference, Aix-en-Provence, August 2001.

Barber K.E. and Charman D. 2003. Holocene palaeoclimate records from peatlands. In: Mackay A., Battarbee R.W., Birks H.J.B. and Oldfield F. (eds), Global Change in the Holocene. Arnolds, pp. 210–226.

Bar-Matthews M., Almogi-Labin A. and Aylon A. 2001. The east Mediterranean speleothem record vs. the Atlantic and the Indian ocean Monsoonal system. In: Past Climate Variability through Europe and Africa: Abstracts. 58. PAGES-PEP III/ESF HOLIVAR International Conference, Aix-en-Provence, August 2001.

Bergner A.G.N., Trauth M.H. and Bookhagen B. 2001. The magnitude of precipitation/ evaporation changes in the Naivasha basin (Kenya) during the last 150 kyr. In: Past Climate Variability through Europe and Africa: Abstracts. 163–164. PAGES-PEP III/ESF HOLIVAR International Conference, Aix-en-Provence, August 2001.

Bigler C. and Hall R.I. 2002. Diatoms as indicators of climatic and limnological change in Swedish Lapland: a 100-lake calibration set and its validation for palaeoecological reconstructions. J. Paleolim. 27: 97–115.

Birks H.H. (ed.) 2000. Palaeoecosystem reconstructions at Kråkenes lake. J. Paleolim. Special Issue 23:1–114.

Böhm R., Auer I., Brunetti M., Maugeri M., Nanni T. and Schöner W. 2001. Regional temperature variability in the European Alps: 1760–1998 from homogenized instrumental time series. Int. J. Climat. 21: 1779–1801.

Bond G.C., Kromer B., Beer J., Muscheler R., Evans M., Showers W., Hoffmann S., Lotti-Bond R., Hajdas I. and Bonani G. 2001. Persistent solar influence on North Atlantic climate during the Holocene. Science 294: 2130–2136.

Brázdil R., Kotyza O. and Dobrovoln P. 1999. History of Weather and Climate in the Czech Lands, V. Period 1: 500–1599, Masaryk University, Brno.

Briffa K.R. 2000. Annual climate variability in the Holocene: interpreting the message from ancient trees. Quat. Sci. Rev. 19: 87–105.

Briffa K.R., Jones P.D., Schweingruber F.H. and Osborne T.J. 1998. Influence of volcanic eruptions on Northern Hemisphere summer temperature over the last 600 years. Nature 393: 450–455.

Briffa K.R. and Matthews J.A. (eds) 2002. Analysis of dendroclimatological variability and associated natural climates in Eurasia (ADVANCE-10K). Special Issue of The Holocene 12: 639–789.

Briffa K.R., Osborn T.J., Schweingruber F.H., Harris I.C., Jones P.D., Shiyatov S.G. and Vaganov E.A. 2001. Low-frequency temperature variations from a northern tree ring density network. J. Geophys. Res. 106D: 2929.

Brooks S.J., Mayle F.E. and Lowe J.J. 1997. Chironomid-based Late-glacial climate reconstruction for Southeast Scotland. J. Quat. Sci. 12: 161–167.

Bugmann H. and Pfister C. 2000. Impacts of interannual climate variability on past and future forest composition. Reg. Env. Change 1: 112–125.

Camuffo D. and Jones P.D. (eds) 2002. Improved Understanding of Past Climatic Variability from Early Daily European Instrumental Sources. Kluwer Academic Publishers, Dordrecht, 400 pp.

Causse C., Bakalowitz M., Chkir N., Genty D., Karray R., Mellieres M.A., Plagnes V. and Zouari K. 2001. First data on Tunisian speleothems. In: Past Climate Variability through Europe and Africa: Abstracts. 68. PAGES-PEP III/ESF HOLIVAR International Conference, Aix-en-Provence, August, 2001.

Charman D. and Hendon D. 2001. Calibrating peat-based palaeoclimate reconstructions with instrumental climate data. In: Past Climate Variability through Europe and Africa: Abstracts. 69–70. PAGES-PEP III/ESF HOLIVAR International Conference, Aix-en-Provence, August 2001.

Cole J. 2001. Environmental variability in East African coral records: links to human activity and the global tropics. In: Past Climate Variability through Europe and Africa: Abstracts. 26. PAGES-PEP III/ESF HOLIVAR International Conference, Aix-en-Provence, August 2001.

Cortijo E., Labeyrie L., Elliot M., Balbon E. and Tisnerat N. 2000. Rapid climatic variability of the North Atlantic Ocean and global climate: a focus of the IMAGES program. Quat. Sci. Rev. 19: 227–241.

Creer K.M., Thouveny N. and Blunk I. 1990. Climatic and geomagnetic influences on the Lac du Bouchet palaeomagnetic SV record through the last 110 000 years. Phys. Earth Planet. Int. 64: 314–341.

Crook D.S., Siddle D. J., Jones R.T., Dearing J.A., Foster G.C. and Thompson R. 2002. Forestry and flooding in the Annecy Petit Lac Basin, 1730–2000. Environment and History 8: 403–428.

Cullen H.M., deMenocal P.B., Hemming S., Hemming G., Brown F.H., Guilderson T. and Sirocko F. 2000. Climate change and the collapse of the Akkadian Empire: Evidence from the deep-sea. Geology 28: 379–382.

D'Arrigo R., Nuzhet Dalfes H., Cullen H., Jacoby G. and Onol B. 2001. Tree-ring records from Eastern Turkey. In: Past Climate Variability through Europe and Africa: Abstracts. 73. PAGES-PEP III/ESF HOLIVAR International Conference, Aix-en-Provence, August 2001.

Doherty R., Kutzbach J., Foley J. and Pollard D. 2000. Fully coupled climate/dynamical vegetation model simulations over Northern Africa during the mid-Holocene. Clim. Dynam. 16: 561–573.

Edmunds W.M. 1999. Groundwater: a renewable resource? Focus in Sahara and Sahel. Brochure for the GASPAL (Groundwater as Palaeoindicator) project (EC ENRICH, ENV4-CT97-0591).

Edmunds W.M., Fellamn E. and Baba Goni I. 1999. Environmental change, lakes and groundwater in the Sahel of Northen Nigeria. J. Geol. Soc. Lond. 156: 345–355.

Felis T., Patzold J., Al-Moghrabi S.M., Loya Y. and Wefer G. 2001. Northern Red sea corals: Seasonal archive of Middle east climate variability during the Holocene and the last Interglacial. In: Past Climate Variability through Europe and Africa: Abstracts. 80. PAGES-PEP III/ESF HOLIVAR International Conference, Aix-en-Provence, August 2001.

Fleitmann D., Burns S.J., Mudelsee M., Neff U. and Mangini A. 2001. Early- to Late-Holocene Indian Ocean Monsoon variability recorded in stalagmites from Southern Oman. In: Past Climate Variability through Europe and Africa: Abstracts. 82. PAGES-PEP III/ESF HOLIVAR International Conference, Aix-en-Provence, August 2001.

Fleitmann D., Burns S.J., Mudelsee M., Neff U., Kramers J., Mangini A. and Matter A. 2002. Holocene variability in the Indian Ocean Monsoon: A Stalagmite-based high-resolution oxygen isotope record from Southern Oman. PAGES Newsletter 10: 7–8.

Fontes J.C., Gasse F. and Andrews J.N. 1993. Climatic conditions of Holocene groundwater recharge in the Sahel zone of Africa. In: Isotope Techniques in the Study of Past and Current Environmental Changes in the Hydrosphere and the Atmosphere. International Atomic Agency, Vienna.

Foster G.C., Dearing J.A., Jones R.T., Crook D.S., Siddle D.S., Appleby P.G., Thompson R., Nicholson J. and Loizeaux J.-L. 2003. Meteorological and land use controls on geomorphic and fluvial processes in the pre-Alpine environment: an integrated lake-catchment study at the Petit Lac d'Annecy. Hydrological Processes (In press).

Frenzel B., Pfister C. and Gläser B. (Hrsg.) 1992. European climate reconstructed from documentary data: methods and results. 2nd EPC/ESF Workshop, Mainz, 01.-03.03.1990. Stuttgart: Gustav Fischer.

Ganopolski A., Kubatzki C., Claussen M., Brovkin V. and Pethoukhov V. 1998. The influence of vegetation-atmosphere-ocean interaction on climate during the mid-Holocene. Science 280: 1916–1919.

Gasse F. and Battarbee R.W., this volume. Introduction. In: Battarbee R.W., Gasse F., and Stickley C.E. (eds), Past Climate Variability through Europe and Africa. Kluwer Academic Publishers, Dordrecht, the Netherlands, pp. 1–6.

Gasse F., Fontes J.C., Plaziat J.C., Carbonel P., Kaczmarska I., De Deckker P., Soulie-Marsche I., Callot Y. and Dupeuple P.A. 1987. Biological remains, geochemistry and stable isotopes for the reconstruction of environmental and hydrological changes in the Holocene lakes from North Sahara. Palaeogeogr. Paleoclim. Paleoecol. 60: 1–46.

Gasse F. and Van Campo E. 1994. Abrupt post-glacial climate events in the West Asia and North African monsoon domains. Earth Planet. Sci. Lett. 126: 435–456.

Glaser R. and Hagedorn H. 1991. The climate of lower Franconia since 1500. Theor. Appl. Climatol. 43: 101–104.

Goni I.B. and Edmunds W.M. 2001. Evidence of climate change in the S W Chad basin from the geochemistry of unsaturated zone moisture and groundwater. In: Past Climate Variability through Europe and Africa: Abstracts. 89. PAGES-PEP III/ESF HOLIVAR International Conference, Aix-en-Provence, August 2001.

Guiot L., Pons A., de Beaulieu J.-L. and Reille M. 1989. A 140,000 years continental climate reconstruction from two European pollen records. Nature 338: 309–313.

Hedenstrom A. and Risberg J. 1999. Early Holocene shore displacement in Southern Central Sweden as recorded in elevated isolated basins. Boreas 28: 490–504.

Hoelzle M. and Haeberli W. 1995. Simulating the effects of mean annual air temperature changes on permafrost distribution and glacier size. An example from the Upper Engadin, Swiss Alps. Annals of Glaciology 21: 399–405.

Holmgren K., Karlén W., Lauritzen S.E., Lee-Thorp J.A., Partridge T.C., Piketh S., Repinski P., Stevenson C., Svanered O. and Tyson P.D. 1999. A 3000-year high resolution stalagmite-based record of paleoclimate for Northeastern South Africa. The Holocene 9: 295–309.

Houghton J.T., Ding Y., Griggs D.J., Noguer M., van der Linden P.J. and Xiaosu D. (eds) 2001. Climate Change: The Scientific Basis. Contribution of Working Group I to the Third Assessment Report of the Intergovernmental Panel on Climate Change (IPCC). Cambridge University Press, UK.

Huang Y., Freeman K.H., Eglinton T.I. and Street-Perrott F.A. 1999. δ^{13}C analysis of individual lignin phenols in Quaternary lake sediments: a novel proxy for deciphering past terrestrial vegetation changes. Geology 27: 471–474.

Huntley B. 1993. The use of climate response surfaces to reconstruct palaeoclimate from Quaternary pollen and plant macrofossil data. Phil. Trans. Roy. Soc. B. 341: 215–223.

Huntley B. and Prentice I.C. 1993. Holocene vegetation and climates of Europe. In: Wright H.E. Jr., Kutzbach J.E., Webb J. T. III, Ruddiman W.R., Street-Perrott F.A. and Bartlein P.J. (eds), Global Climate since the Last Glacial Maximum. University of Minnesota Press, Minneapolis, 569 pp.

Johnson T.C. 1996. Sedimentary processes and signals of climate change in the large lakes of the East African Rift Valley. In: Johnson T.C. and Odada E.O. (eds), The Limnology, Climatology and Paleoclimatology of the East African Lakes. Gordon and Breach, Amsterdam, pp. 367–412.

Jones R.T., Marshall J.D., Crowley S.F., Bedford A., Richardson N., Bloemendal J. and Oldfield F. 2002. A high resolution multi-proxy Late-glacial record of climate change and intrasystem responses in NW England. J. Quat. Sci. 17: 329–340.

Klein R., Loya Y., Svistzman G., Isdale P.J. and Susic M. 1990. Seasonal rainfall in the Sinai Desert during the late Quaternary inferred from fluorescent bands in fossil corals. Nature 345: 145–147.

Kroepelin S., Hoelzmann P. and Keding B. 2001. Lake evolution and human occupation in the Eastern Sahara during the Holocene. In: Past Climate Variability through Europe and Africa: Abstracts. 106. PAGES-PEP III/ESF HOLIVAR International Conference, Aix-en-Provence, August 2001.

Ladurie E. Le Roy and Baulant M. 1981. Grape harvest from the fifteenth through the nineteenth century. In: Rotberg R.I. and Rabb T.K. (eds), Climate and History: Studies in Interdisciplinary History. Princeton University Press, New Jersey, pp. 259–269.

Lauritzen S.-E. 1996. Calibration of speleothem stable isotopes against historical records: a Holocene temperature curve for North Norway? In: Climatic Change: the Karst Record. Karst Waters Institute Special Publication 2, Charles Town, West Virginia, pp. 78–80.

Lauritzen S.-E. 2001. SPEP: a speleothem transect: contribution from the North Atlantic. In: Past Climate Variability through Europe and Africa: Abstracts. 45. PAGES-PEP III/ESF HOLIVAR International Conference, Aix-en-Provence, August 2001.

Lauritzen S.-E. 2003. Reconstructing Holocene climate records from spelothems. In: Mackay A., Battarbee R.W., Birks H.J.B. and Oldfield F. (eds), Global Change in the Holocene, Arnolds, pp. 227–241.

Lauritzen S.-E. and Lundberg J. (eds) 1999. Speleothems as high-resolution palaeo-climatic archives. Special Issue of The Holocene 9: 643–722.

Lee-Thorp J.A., Holmgren K., Lauritzen S.-E., Linge H., Moberg A., Partridge T.C., Stevenson C. and Tyson P.D. 2001. Rapid climate shifts in the southern African interior throughout the mid to late Holocene. Geophys. Res. Lett. 28: 4507–4510.

Lindholm M. and Eronen M. 2000. A reconstruction of mid-summer temperatures from ring-widths of Scots pine since AD 50 in Northern Fennoscandia. Geograf. Annal. 82: 527–535.

Lotter A.F., Birks H.J.B., Hofman W. and Marchetto A. 1997. Modern diatom, cladocera, chirono-mid, and chrysophyte cyst assemblages as quantitative indicators for the reconstruction of past environmental conditions in the Alps. I. Climate. J. Paleolim. 18: 395–420.

Lotter A.F. 2003. Multi-proxy climatic reconstructions. In: Mackay A., Battarbee R.W., Birks H.J.B. and Oldfield F. (eds), Global Change in the Holocene. Arnolds, pp. 373–383.

Luterbacher J., Schmutz C., Gyalistras D., Xoplaki E. and Wanner H. 1999. Reconstruction of monthly NAO and EU indices back to AD 1675. Geophys. Res. Lett. 26: 2745–2748.

Markgraf V. and Diaz H.F. 2001. The past ENSO record: A synthesis. In: Diaz H.F. and Markgraf V. (eds), El Niño and the Southern Oscillation: Multiscale Variability and Global Impacts. Cambridge University Press, Cambridge, pp. 465–488.

Manley G. 1974. Central England temperatures: monthly means 1659 to 1973. Quart. J. R. Met. Soc. 100: 389–405.

Mann M.E., Bradley R.S. and Hughes M.K. 1999. Northern Hemisphere temperatures during the past millennium: Inferences, uncertainties, and limitations. Geophys. Res. Lett. 26: 759–762.

Mauquoy D., van Geel B., Blaauw M. and van der Plicht 2002. Evidence from northwest European bogs show 'Little Ice Age' climatic changes driven by variations in solar activity. The Holocene 12: 1–6.

Nicholson S.E. 2001. Climatic and environmental change in Africa during the last two centuries. Climate Research 17: 123–144.

Nicholson S.E. and Grist J.P. 2001. A Conceptual model for understanding rainfall variability in the West African Sahel on interannual and interdecadal timescales. Int. J. Climat. 21: 1733–1757.

Oldfield F., Asioli A., Accorsi C.A., Juggins S., Langone L., Rolph T., Trincardi F., Wolff G., Gibbs Z., Vigliotti L., Frignani M. and Branch N. 2003a. A high resolution Late-Holocene palaeo-environmental record from the Central Adriatic Sea. Quat. Sci. Rev. 22: 319–342.

Oldfield F. and Dearing J.A. 2003. The role of human activities. In: Alverson K., Bradley R.S. and Pedersen T.F. (eds), Palaeoclimate, Global Change and the Future. Springer Verlag, Berlin, pp. 143–162.

Oldfield F., Dearing J.A., Gaillard M.-J. and Bugmann H. 2000. Ecosystem processes and human dimensions — The scope and future of HITE (Human Impacts on Terrestrial Ecosystems). PAGES Newsletter 8: 21–23.

Oldfield F., Wake R., Boyle J.F., Jones R., Nolan S., Gibbs Z., Appleby P.G., Fisher E. and Wolff G. 2003b. The Late-Holocene history of Gormire Lake (N.E. England) and its catchment: a multi-proxy reconstruction of past human impact. The Holocene, in press.

Partridge T.C., de Menocal P.B., Lorentz S.A., Paikar M.J. and Vogel J.C. 1997. Orbital forcing of climate over South Africa: a 200 000-year rainfall record from the Pretoria saltpan. Quat. Sci. Rev. 16: 1–9.

Pelfini M., Cremaschi M., Arzuffi L., Di Mauro V., Santilli M. and Zerboni A. 2001. Tree-rings of *Cupressus dupretiana* from the Wadi Tanezzuft area (NW Fezzan, Libyan Sahara). A source of palaeoclimatic information for the middle and late Holocene of the Central Sahara. In: Past Climate Variability through Europe and Africa: Abstracts. 127–128. PAGES-PEP III/ESF HOLIVAR International Conference, Aix-en-Provence, August 2001.

Peterson L.C., Haug G.H., Hughen K.A. and Röhl U. 2000. Rapid changes in the hydrologic cycle of the tropical Atlantic during the last Glacial. Science 290: 1947–1951.

Pfister C. 1992. Monthly temperature and precipitation in Central Europe 1525–1979: quantifying documentary evidence on weather and its effects. In: Bradley R.S. and Jones P.D. (eds), Climate since AD 1500. Routledge, London, pp. 118–142.

Phillips J.G. and McIntyre B. 2000. ENSO and interannual rainfall variability in Uganda: Implications for agricultural management. Int. J. Climat. 20: 171–182.

Ramrath A., Sadori L. and Negendank J.F.W. 2000. Sediments from Lago di Mezzano, Central Italy: a record of Late glacial/Holocene climatic variations and anthropogenic impact. The Holocene 10: 87–95.

Richard Y., Fauchereau N., Poccard I., Rouault M. and Trzaska S. 2001. 20th century droughts in Southern Africa: Spatial and temporal variability, teleconnections with oceanic and atmospheric conditions. Int. J. Climat. 21: 873–885.

Richard Y., Trzaska S., Roucou P. and Rouault M. 2000. Modification of the Southern African rainfall variability /El Niño Southern Oscillation relationship. Clim. Dyn. 16: 886–895.

Roberts N., Reid J., Leng M.J., Kouzugouglu C., Fontugne M., Bertaux J., Woldring H., Bottema S., Black S., Hunt E. and Karabiyikolu M. 2001. The tempo of Holocene climatic change in the eastern Mediterranean region: new high resolution crater lake sediment data from Central Turkey. The Holocene 11: 721–736.

Rocha A. and Simmonds I. 1997. Interannual variability of south-eastern African summer rainfall. Part I: Relationships with air-sea interaction processes. Part II: Modelling the impact of sea-surface temperatures on rainfall and circulation. Int. J. Climat. 17: 235–290.

Rosendahl W., Weigand B. and Kempe S. 2001. Speleothems and Pleistocene climate — new dating results from caves in Germany and their correlation with GRIP ice core record. In: Past Climate Variability through Europe and Africa: Abstracts. 133–134. PAGES-PEP III/ESF HOLIVAR International Conference, Aix-en-Provence, August 2001.

Rossignol-Strick M. 1985. Mediterranean Quaternary sapropels, an immediate response of the African Monsoon to variation of insolation. Palaeogeogr. Palaeoclim. Palaeoecol. 49: 237–263.

Schweingruber F.H. and Briffa K.R. 1996. Tree-ring density networks for climate reconstruction. In: Jones P.D., Bradley R.S. and Jouzel J. (eds), Climate Variations and Forcing Mechanisms of the last 2000 Years. Springer, Berlin, pp. 43–66.

Schönwiese C.-D. and Rapp J. (eds) 1997. Climate Trend Atlas of Europe: Based on Observations 1891–1990. Kluwer, Academic, 224 pp.

Sorvari S., Korhola A. and Thompson R. 2002. Lake diatom response to recent Arctic warming in Finnish Lapland. Global Change Biology 8: 153–163.

Stute M. and Talma S. 1998. Glacial temperatures and moisture transport regimes reconstructed from noble gas and $\delta^{14}O$, Stampriert aquifer, Namibia. In: Isotope Techniques in the Study of Past and Current Environmental Changes in the Hydrosphere and the Atmosphere. Proceedings of the Vienna Symposium 1997. IAEA, Vienna, SM-349-53, pp. 307–328.

Thompson L.G. 2001. Tropical ice core records from the ice fields of Kilimanjaro, ice retreat and asynchronous glaciation. In: Past Climate Variability through Europe and Africa: Abstracts. 24. PAGES-PEP III/ESF HOLIVAR International Conference, Aix-en-Provence, August 2001.

Thompson R. 1995. Complex demodulation and the estimation of the changing continentality of Europe's climate. Int. J. Climatol. 15–2: 175–185.

Thompson R. 1999. A time-series analysis of the changing seasonality of precipitation in the British Isles and neighbouring areas. J. Hydrology 224: 169–183.

Thorndycraft V., Hu Y., Oldfield F., Crooks P.R.J. and Appleby P.G. 1998. Individual flood events detected in the recent sediments of the Petit Lac d'Annecy, Eastern France. The Holocene 8: 741–746.

Trouet V., Haneca K., Coppin P. and Beeckman H. 2001. The value of growth ring series from trees of the miombo woodland in Southern and Eastern Africa as proxy data for climate reconstruction. In: Past Climate Variability through Europe and Africa: Abstracts. 154. PAGES-PEP III/ESF HOLIVAR International Conference, Aix-en-Provence, August 2001.

Ulbrich U. and Christoph M. 1999. A shift of the NAO and increasing storm track activity over Europe due to anthropogenic greenhouse gas forcing. Clim. Dyn. 15: 551–559.

Unganai L.S. 1997. Surface temperature variation over Zimbabwe between 1897 and 1993. Theor. and Appl. Climatol. 56: 89–101.

van Geel B. and Mook W.G. 1989. High resolution ^{14}C dating of organic deposits using natural ^{14}C variations. Radiocarbon 31: 151–155.

Verschuren D., Laird K. and Cumming B.F. 2000. Drought in East Africa during the past 1,100 years. Nature 403: 410–413.

von Grafenstein U., Erlenkeuser H., Brauer A., Jouzel J. and Johnsen S.J. 1999. A Mid-European decadal isotope-climate record from 15,500 to 5000 years BP. Science 284: 1654–1657.

Vose R.S., Schmoyer R.L., Steurer R.L., Peterson T.C., Heim R., Karl T.R. and Eischeid J.K. 1992. The Global Historical Climatology Network: Long-term monthly temperature, precipitation, sea level pressure, and station pressure data. NDP-041. Carbon Dioxide Information Analysis Center, Oak Ridge National Laboratory, Oak Ridge, Tennessee.

Wohlfarth B., Holmquist B., Cata I. and Linderson H. 1998. The climatic significance of clastic varves in the Ångermanälven Estuary, Northern Sweden, AD 1860 to 1950. The Holocene 8: 521–534.

Zinke J., Pfeiffer M., Heiss G., Dullo W.-Chr., Joachimski M. and Eisenhauer A. 2001. A 338 year coral oxygen isotope record off Madagascar: Interannual and interdecadal sea-surface temperature variability. In: Past Climate Variability through Europe and Africa: Abstracts. 163–164. PAGES-PEP III/ESF HOLIVAR International Conference, Aix-en-Provence, August 2001.

Zolitschka B. 1998. A 14000 year sediment yield record from Western Germany based on annually laminated sediments. Geomorphology 22: 1–17.

3. OCEANIC CLIMATE VARIABILITY AT MILLENNIAL TIME SCALES: MODELS OF CLIMATE CONNECTIONS

LAURENCE VIDAL (vidal@cerege.fr)
*Centre Européen de Recherche et d'Enseignement
de Géosciences de l'Environnement (CEREGE)
Europôle de l'Arbois
B.P. 80, F-13454 Aix-en-Provence cedex 4
France*

HELGE ARZ (helge.arz@uni-bremen.de)
*Universität Bremen
Geowissenschaft, 28334 Bremen
Germany*

Keywords: Climate variability, Millennial scale, Climate connections, Hydrological cycle, Tropical/subtropical areas

Introduction

One of the most exciting questions in palaeoclimatology is the study of the complex inter-actions between the different components of the climate system in order to understand how climate changes occur at Milankovitch as well as at millennial and centennial time-scales. The primary objective of this paper is to place the PEP III transect palaeo-data within a global climate context in relation to oceanic climate variability during the last glacial. To take ocean-continent interactions into account is essential to develop our understanding of past climate change.

During recent decades, high resolution palaeoclimatic studies have focused on the last glacial period (60 kyr–10 kyr BP) which has revealed abrupt climate changes on a global scale. Numerous synthesis papers have already been published discussing the relation-ships between fast ocean-atmosphere system reorganisations and millennial scale climate variability at high latitudes (Alley et al. 1999). Indeed the thermohaline circulation is an essential component of the climate system and its stability, but there are still unanswered questions about the respective role of the oceanic circulation, atmospheric circulation and composition, in transferring the climate signal throughout the world. This needs to be further explored.

Recently, several studies have brought to light the role of the tropical/subtropical areas in climate changes. Continental and oceanic low-latitude records have documented the high frequency climate variability during the last glacial period. The inferred continental climate

31

R. W. Battarbee et al. (eds) 2004. *Past Climate Variability through Europe and Africa.*
Springer, Dordrecht, The Netherlands.

conditions in the sub-tropics might provide evidence for the contribution of atmospheric processes associated with rapid climate variability.

Another key question concerns the forcing mechanisms for rapid climate change. One of the most discussed hypotheses relies on ice sheet-ocean-atmosphere interactions driving millennial scale climate variations during the last glacial period (Alley et al. 1999; Ganopolski and Rahmstorf 2001). As already mentioned, ocean-atmosphere coupling is very sensitive to the freshwater flux into the North Atlantic. Modelling of the cryosphere shows that the internal dynamics of the northern hemisphere ice-sheets during periods of ice-building could lead to major reorganisation of the global climate (MacAyeal 1993). Likewise internal oscillations of the ocean-atmosphere system should also be considered as a potential candidate for climate forcing (Winton 1993; Paul and Schulz 2002). On the other hand, studies on rapid climate variability during the Younger Dryas and the Holocene point to an external forcing by revealing a possible connection between solar activity and global climate (Renssen et al. 2000; this volume; Bond et al. 2001). However, knowledge about the importance of the different potential triggers is poor, as climate components have different responses in space and time (and most probably during glacial and interglacial periods).

Therefore, in this paper we concentrate on the interactions between the different climate sub-systems. A review of rapid climate variability during the last glacial period mostly based on marine sediment records with a focus on recent studies based on sub-tropical records is attempted. By compiling these high quality data, we discuss what can be learned about millennial climate changes and connections.

Spatial distribution of climate variability during the last glacial

The North Atlantic area

Detailed studies on marine sequences have shown that the last glacial period was punctuated by quasi-periodic ice-rafted pulses in the North Atlantic, the so-called Heinrich events (HE) (Heinrich 1988). These deposits are the signatures of massive iceberg discharges occurring every 3 to 10 kyr, due to the partial collapse of the circum-Atlantic ice sheets (Maslin et al. (2001), and references therein, Saarnisto and Lunkka (this volume)). Surface water hydrology during these periods revealed a southward shift of the polar front position and a drastic reduction of surface salinity in the North Atlantic (Cortijo et al. 1997). At the same time, air temperatures over Greenland show a marked cooling as indicated from the ice core ($\delta^{18}O$) records (Dansgaard et al. 1993; Bond et al. 1993). Between these cooling periods, a succession of short-lived warm/cold events in Greenland air temperatures are reported, the so-called Dansgaard-Oeschger (D/O) events, with a smaller magnitude in temperature changes compared to those associated with the HE. These short climate swings, related to surges of the surrounding ice sheets coincide with marked changes in the North Atlantic surface water hydrology, although restricted to the northern part of the basin (Elliot et al. 1998; 2001).

Proxies for deep-water hydrology have shown that perturbations of deep-water formation were coupled to millennial scale changes in surface water salinity in the North Atlantic. During the Heinrich events, reduction of deep-water formation and/or changes in convection sites have been clearly shown (Vidal et al. 1997; Zahn et al. 1997; Curry and Oppo 1997) (Fig. 1). However the inter-Heinrich cold intervals (stadials) seem not to be systematically associated with reduction of deep-water ventilation although marked by significant changes in sea surface temperature (SST) in the northern North Atlantic (Elliot et al. 2002). Only few studies report North Atlantic Deep Water (NADW) changes at D/O time scales (Charles et al. 1996; Keigwin and Boyle 1999). One may note that the required temporal resolution in marine sediment records to study this type of climate variability is often not reached and could have contributed to the scarcity of the data.

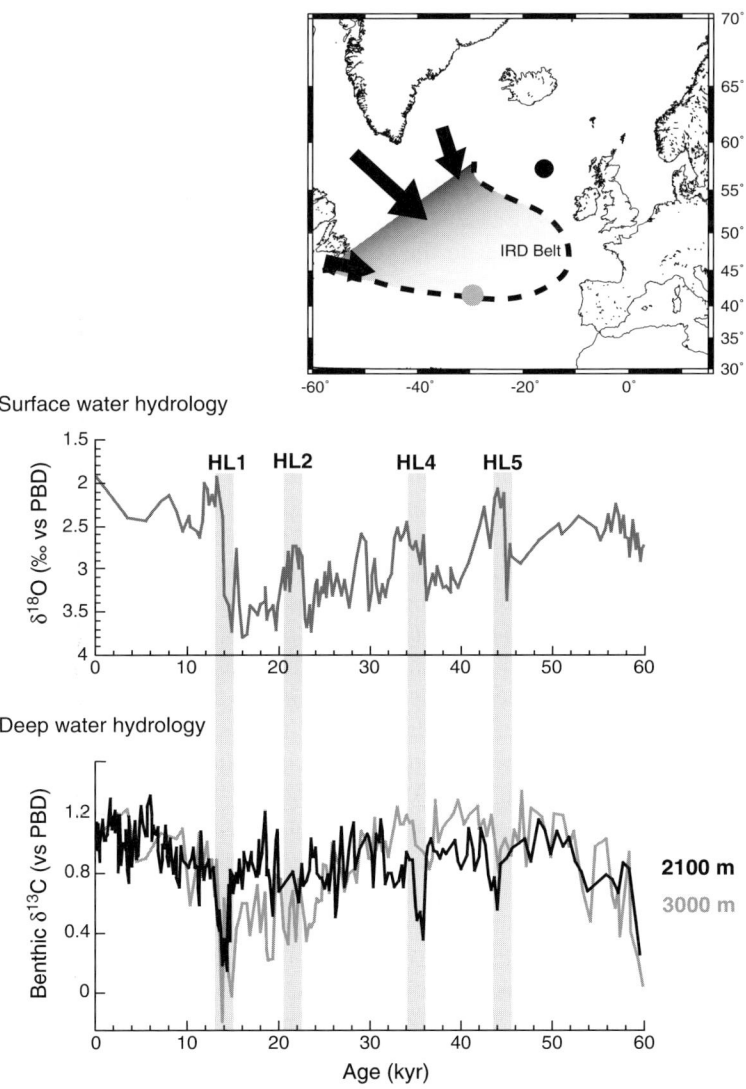

Figure 1. Summary of palaeoclimatic data after Vidal et al. 1997. Planktonic $\delta^{18}O$ record (upper panel) and benthic $\delta^{13}C$ records (lower panel) are shown. The Heinrich events (HE) are indicated. The map shows the IRD belt and the location of the cores.

Thermohaline overturning and NADW production are associated with considerable inter-hemispheric northward heat transport in the Atlantic (Berger and Wefer 1996). Therefore reduction in the NADW production will first act as a positive feedback on the cooling in the North Atlantic area and secondly imprint the climate on a global scale. This is supported by modelling results simulating changes in NADW production in response to freshwater input in the Northern Atlantic Ocean (Manabe and Stouffer 1997; Lohman and Schulz 2000; Ganopolski and Rahmstorf 2001).

Evidence for global imprint of rapid climate variability during the last glacial

It is now clear that rapid climate variability is not confined to the North Atlantic area. Support for a global imprint of climate variability during the last glacial period comes from the polar ice-core data. Analyses of the air bubbles trapped in the ice indicate that atmospheric composition varied on millennial time-scales. Fluctuations of CH_4 concentration parallel the D/O type variability (Chappellaz et al. 1993; Brook et al. 1999) whereas significant variations in CO_2 concentration are closer to the HE type variability (Indermühle et al. 2000).

Marine and terrestrial records showing millennial scale climate change during the last glacial period have also been documented from distant areas (see Voelker et al. (2002) for a review). However the spatial distribution and shape of the signals differ from one location to the other. For example, from the oceanic realm, SST reconstructions from the NE Atlantic (Bard et al. 2000) and from the South Indian Ocean (Bard et al. 1997) as well as upwelling intensity in the Benguela current system (Little et al. 1997) and ice-rafted detritus concentration in deep sea cores from the Southeast Atlantic Ocean (Kanfoush et al. 2000) are modulated by HE-type variability. On the other hand, SSTs from the Mediterranean Sea (Cacho et al. 1999), oxygen minimum zone intensity in the Indian Ocean (Schulz et al. 1998) and in the NE Pacific (Santa Barbara Basin) (Behl and Kennett 1996), productivity and surface salinity in the west equatorial Pacific (de Garidel-Thoron et al. 2001; Stott et al. 2002) mimic the D/O type variability. Continental climate also experienced abrupt climate changes during the last glacial period. Loess sequences in China document changes in wind intensities over Asia (Porter and An 1995). Studies based on lacustrine sediment records provide evidence for strong changes in moisture and precipitation conditions in North America (Grimm et al. 1993), equatorial South America (Baker et al. 2001; Moreno et al. 2001; Lowell et al. 1995) and East Africa (Gasse (2000), for a review on African climate variability). These climate records mostly show variability associated with the Heinrich events. For these type of archives the interpretations can be limited by the lack of record continuity during this period.

Subtropical and tropical climate records
Recently very interesting insights have been gained from glacial marine sequences retrieved on the sub-tropical/tropical continental margin of the Atlantic. From these it has been possible to extract continental climate tracers using the detrital fraction of the sediments to determine changes in input of fine terrigeneous components.

Studies carried out on the north-eastern Brazilian margin (4 °S) suggest that rapid changes of high latitude climate during the last glacial generally coincided with increased rainfall and river runoff in Northeast Brazil (Arz et al. 1998; Behling et al. 2000) (Fig. 2). Taking into account the modern range of seasonal Intertropical Convergence Zone (ITCZ) shifts and the rainfall patterns over North-Eastern Brazil, these studies suggest that during stadial periods/Heinrich events southern Hemisphere trade winds weakened and the ITCZ migrated southward, leading to increased rainfall over North-Eastern Brazil. Millennial-scale hydrological changes during the last glacial are also documented from a high resolution study on the Cariaco Basin sediments in the tropical Atlantic (Peterson et al. 2000). In contrast to the results mentioned before, this area is characterised by increased precipitation during interstadial periods that affected the river discharge into the Cariaco Basin. The authors linked this climate feature to a southward shift of the ITCZ with intensified north-easterly trade winds during cooling periods. Data obtained from sediments from the continental margin off Cameroon (3 °N) provide indirect evidence for rapid fluctuations in rainfall and river discharge volumes in Central Africa at millennial time scales (Schneider et al. 2001). Although preliminary, these results imply that African monsoonal climate has experienced precipitation changes with a pattern characteristic of D/O type cycles with higher precipitation during interstadials.

Therefore it appears that the spatial distribution of climate signals is not uniform as already mentioned by Alley et al. (1999) and Broecker and Hemming (2001) and that no distinct pattern can be distinguished. It seems, however, that climate records that mimic the D/O variability are located in the northern high latitudes of the northern Hemisphere and in tropical/subtropical areas but less apparent in the mid-latitudes of the Atlantic ocean. Climate change that coincides with the Heinrich events variability are found in records from mid latitudes of the northern and southern hemispheres, in the southern high latitudes as well as in tropical areas. This indicates the importance of understanding the mechanisms responsible for the transfer of the climate signals (Fig. 3).

Transfer of the climate signal

The Thermohaline Circulation (THC) link

This topic has been addressed using correlation between methane records of the polar ice cores from both hemispheres (Blunier et al. 1998; Blunier and Brook 2001). Comparison of atmospheric temperature changes in both areas during the last glacial period demonstrate a thermal phase difference. The most prominent cooling periods, linked to the Heinrich events in the Northern Hemisphere are associated with warming phases in the Southern Hemisphere. Similar results were obtained from the comparison between North and South Atlantic sea surface temperature gained from correlations between marine sediments from both areas (Vidal et al. 1999).

Hemispheric asynchrony on millennial time-scales has been attributed to changes in oceanic overturning (Blunier and Brook 2001; Vidal et al. 1999). This behaviour is predicted by coupled ocean-atmosphere circulation models, simulating a reduction of the thermohaline circulation, the so-called "seesaw effect" (Broecker 1998; Stocker 1998; Ganopolski and Rahmstorf 2001). In these models, the warming in the Southern Hemisphere surface waters is interpreted as a compensatory effect, reflecting a weakening of the inflow of southern ocean subsurface waters into the sub-tropical Atlantic. However, all modelling results do not predict a uniform warming of the southern Hemisphere (Manabe and Stouffer

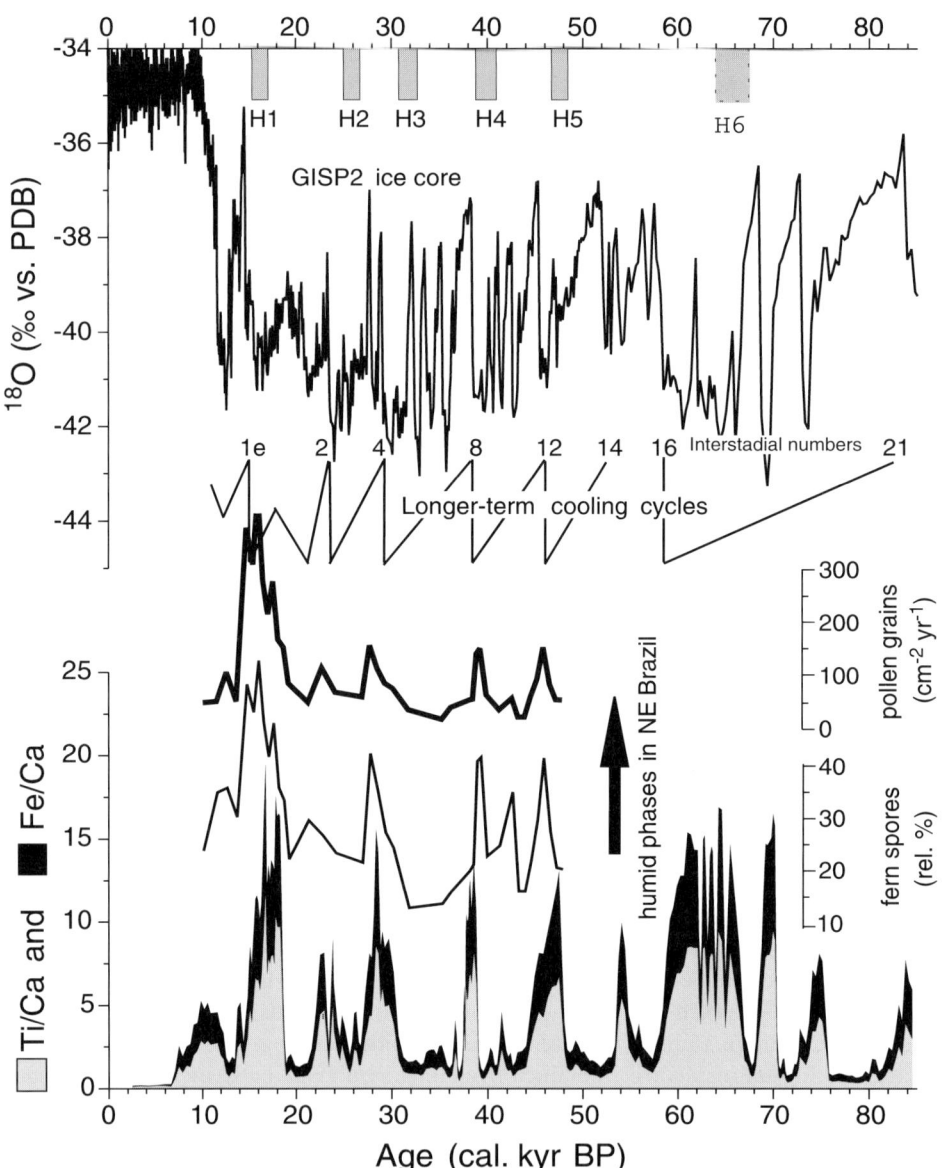

Figure 2. Summary of palaeoclimatic data after Arz et al. (1998) and Behling et al. (2000), showing the total pollen influx, the relative contribution of fern spores and the XRF Ti-Ca and Fe-Ca ratios from sediment cores GeoB 3104-1/3912-1. More humid periods in NE Brazil line up with the Heinrich Events as recorded in the Greenland ice cores.

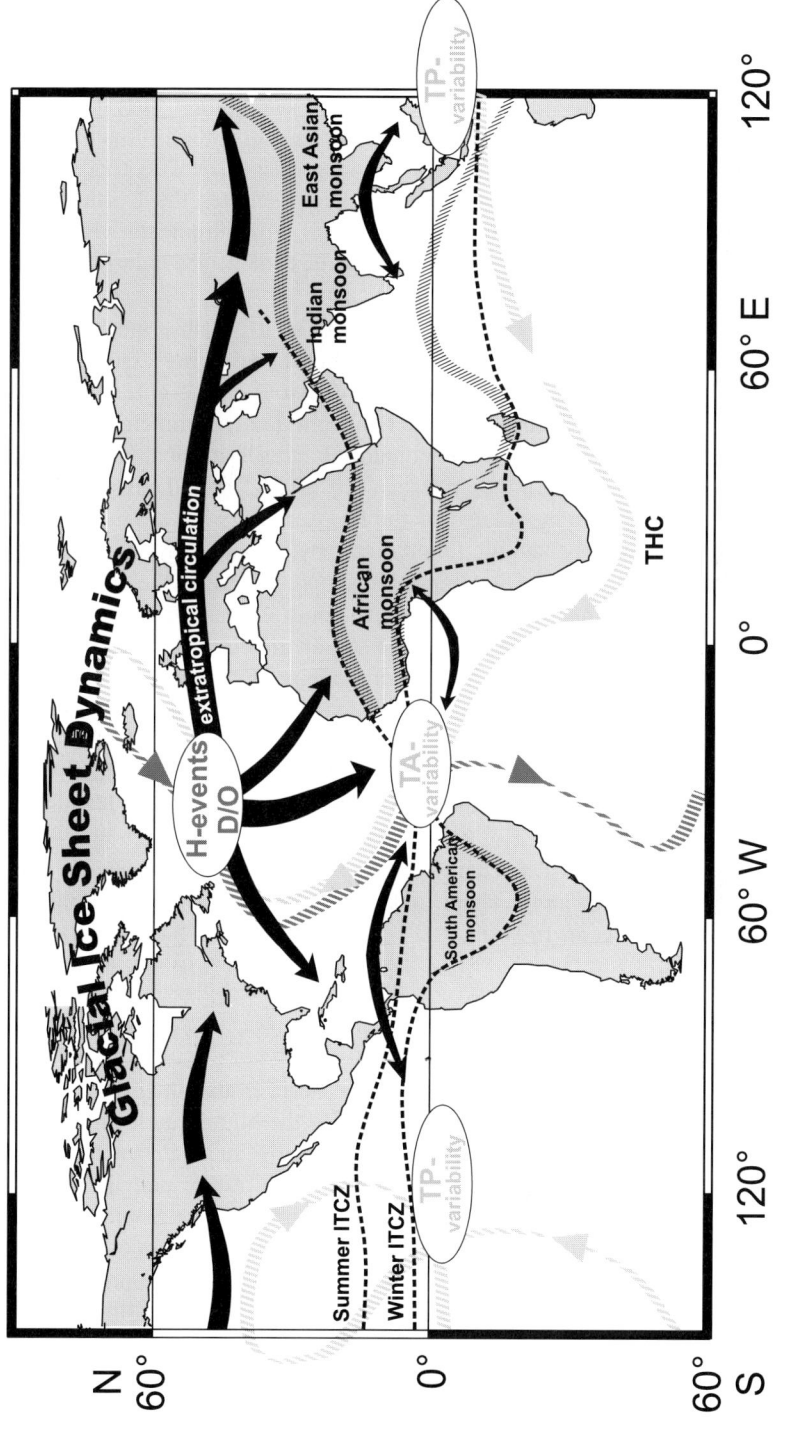

Figure 3. Schematic representation of different modes of climate connections: thermohaline circulation (THC) (grey arrows); atmospheric circulation (including hydrological cycle) (black arrows); areas with potential effects on climate variability in the northern Atlantic (H-events, D/O), tropical Atlantic (TA variability) and tropical Pacific (TP variability); ITCZ positions (winter and summer) (dotted black lines), and regions influenced by monsoonal climates (black rectangles).

1997; Rind et al. 2001) and differences are observed in surface water temperature response for the Pacific and Southern Ocean.

The operation of bipolar seesaw also affects atmospheric composition. Marchal et al. (1998) simulated the evolution of the atmospheric CO_2 concentration in response to a freshwater-induced collapse of the THC using a coupled ocean circulation-biogeochemical model. In their simulation, a THC shut-down leads to an increase of the CO_2 concentration due to a warming of the Southern Ocean which over-compensates the North Atlantic cooling. This is in agreement with Antarctic ice core CO_2 measurements which show an increase during the most prominent HE events (Stauffer et al. 1998; Indermühle et al. 2000).

Most of the observed data are in favour of the THC as a possible mechanism for propagating the climate signal at least linked to the climate variability associated with Heinrich events (Fig. 3). However, the out-of-phase behaviour of the climate response to changes in oceanic overturning seems not to be hemispherically symmetric, but rather characterised by a non-uniform response of the equatorial and sub-tropical latitudes. This is suggested from a well-dated southern Indian Ocean (20 °S) SST record covering the last 40 kyr (Bard et al. 1997). SSTs are modulated by the HE variability but are in phase with North Atlantic climate changes; a behaviour not predicted by the seesaw mode. It has to be noted that only very few high resolution Southern hemisphere records of the glacial period are available (Cortese and Abelmann 2002). The north-south thermal "antiphasing" may be limited in space (i.e., in the Atlantic area) and the seesaw model may be too simple to account for the real longitudinal and latitudinal heterogeneity of millennial-scale climate change (Steig 2001). New results providing a record of sea-surface temperature in the mid-latitude from the Indian Ocean (between 30 °S and 50 °S) for the last deglaciation support this feature (Stenni et al. 2001). The study shows that the southern hemisphere Indian Ocean is neither in phase with the North nor with the high latitudes of the Southern Hemisphere during this period.

By modifying heat transport, NADW changes have probably had most effect on atmospheric circulation that could imprint the shape and the phasing of oceanically-influenced climate signals (Mikolajewicz et al. 1997).

The atmospheric link

Certainly, high frequency climate variability as discussed earlier also involves strong atmospheric teleconnective processes either as a response to the high latitude changes and THC variations, operating as feedback and amplifier mechanisms (Street-Perrott and Perrott 1990; Alley et al. 1999; Pierrehumbert 1999; Peterson et al. 2000), or perhaps acting as a trigger originating from, for example, the low latitude (tropical/sub-tropical) climate systems (e.g., Clement et al. (1999), McIntyre and Molfino (1996), Kudrass et al. (2001)).

One such process is likely to be the extra-tropical circulation, i.e., the northern hemisphere polar and mid-latitude circulation, exporting North Atlantic variability to remote areas over the Northeast Pacific, and American and Eurasian continents (Behl and Kennett 1996; Kennett and Ingram 1995; Benson 1999; Mayewski et al. 1994; Cacho et al. 2000; Thouveny et al. 1994; Porter and An 1995; An 2000). From low latitude proxy data, however, we learn that major reorganizations on the millennial time-scale took place also in the tropical/sub-tropical climate system involving changes especially in the hydrological cycle and therefore acting at least as important amplifier mechanisms through, for example, changes in water vapour distribution and transport (Street-Perrott and Perrott 1990; Arz et al. 1998; Behling et al. 2000; Peterson et al. 2000; Schneider et al. 2001). A key role of the tropics is also expected when looking at the ice-core methane records, which suggest at least 50% contribution of tropical sources to the millennial-scale glacial shifts in the global methane budget of the atmosphere (Brook et al. 1996). Perhaps the most intriguing conclusion drawn from the low-latitude palaeorecords is that high latitude stadial events were accompanied by a southward shift of the ITCZ over the western tropical Atlantic leading to variable precipitation patterns over the northern and north-western part of the South American continent, and that simultaneously the African and Asian Monsoon weakened and Central Africa experienced less precipitation. The question arises whether the ITCZ displacements and monsoonal variability are seen as a direct response to the strengthened high-pressure cell over the North Atlantic due to generally lower SSTs during stadials (Bond et al. 1993; Elliot et al. 2001) or if they rather represent a consequence of the slowing down of the THC, which may have initiated a surface warming of the tropical/sub-tropical ocean (Arz et al. 1999; Rühlemann et al. 1999; Manabe and Stouffer 1997) leading to a southward displacement of the thermal equator (Fig. 3). Although these processes can hardly be distinguished at present, it seems unlikely that thermohaline overturning could control these fast shifts in the tropical hydrological cycle only, as there is no direct evidence for changes in NADW production on D/O time-scales. Moreover, the spatial distribution of the D/O-type climate records is especially well-expressed in tropical and sub-tropical monsoonal areas, locations with strong atmospheric responses to changes in the North Atlantic (Alley et al. 1999). Modelling studies, for example, demonstrate such possible climate connections between the North Atlantic and areas under monsoonal circulation (Hostetler et al. 1999).

An alternative hypothesis is proposed by recent modelling work based on ENSO climate variability. These studies demonstrate possible inter-basin climate connections, which involve the tropical areas as possible triggers for climate change in the northern latitudes through changes in the hydrological cycle (Fig. 3) (Cane 1998; Clement and Cane 1999; Schmittner et al. 2000). These hypotheses, however, need to be tested further.

Altogether these results suggest that the effects of ocean-atmosphere reorganizations were different during HE and D/O events. It seems that the oceanic link and atmospheric processes were operating during the HE, whereas predominantly atmospheric processes, probably linked to North Atlantic changes, modulate the D/O scale climate variability in and outside the Atlantic area.

Conclusions and outlook

A review of high-resolution palaeoclimate records of the last glacial period help to provide evidence for climate modes associated with millennial-scale climate variability. The oceanic link for transferring climate signals linked to Heinrich events seems probable but is probably limited in space, to the Atlantic area. On the other hand, atmospheric processes are more likely to explain the distribution of climate events related to the D/O type variability. It seems that the tropical/sub-tropical areas are sensitive to rapid climate change and may play an amplifier role. In other areas (such as the mid-latitudes) the climate signals (oceanic and continental) could be damped down by the oceanic inertia.

To constrain atmospheric processes associated with rapid climate changes during the last glacial, studies of land-ocean-climate interactions need to be encouraged. The value of reconstructing continental climate conditions in sub-tropical areas has been well demonstrated. More work is needed to determine how rapidly changes in the position of the ITCZ take place and to determine the influence of oceanic hydrology on continental climate. High resolution analysis of lacustrine and continental-margin marine sediments should consequently be promoted. The next step should be a focus on particularly interesting areas, such as the East Pacific and the continental climate over Central America, in order to estimate inter-basin moisture transfer (between the tropical Atlantic and tropical Pacific) which is an essential component of high latitude climate (Broecker and Denton 1989).

Summary

The last glacial period has been punctuated by abrupt climate swings occuring on millennial time scale, first revealed in marine cores from the North Atlantic and subsequently documented on a global scale. Beside the question of the forcing mechanisms triggering these climate changes, it is important to understand how climate signals are transferred throughout the globe and which climate components are involved. In this paper we examine and review paleoclimate proxies from several archives characterising the different components of the climate system. A special emphasis is placed on recent results obtained from tropical/subtropical areas that help to determine oceanic and continental climate conditions. The knowledge of ocean-continent interactions is essential to estimate the atmospheric processes (in addition to the ocean) that were operating during these rapid climate events.

Acknowledgments

L.V. thanks the PAGES PEP III meeting scientific committee for the invitation to speak at the Aix-en-Provence conference. We are grateful to the two anonymous referees for reviews of the manuscript.

References

Alley R.B., Clark P.U., Keigwin L.D. and Webb R.S. 1999. Making sense of millennial scale climate change. In: Clark P.U., Webb R.S. and Keigwin L.D. (eds), Mechanisms of Global Climate Change at Millennial Time Scale. Geophysical Monograph 112, pp. 385–394.
An Z. 2000. The history and variability of the East Asian palaeomonsoon climate. Quat. Sci. Rev. 19: 171–187.
Arz H.W., Pätzold J. and Wefer G. 1998. Correlated millennial scale changes in surface hydrography and terrigeneous sediment yield inferred from last glacial marine deposits off Northeastern Brazil. Quat. Res. 50: 157–166.
Arz H.W., Pätzold J. and Wefer G. 1999. The deglacial history of the western tropical Atlantic as inferred from high resolution stable isotope records off Northeastern Brazil. Ear. Planet. Sci. Lett. 167: 105–117.

Baker P.A., Rigsby C.A., Seltzer G.O., Fritz S.C., Lowenstein T.K., Bacher N.P. and Veliz C. 2001. Tropical climate changes at millennial and orbital timescales on the Bolivian Altiplano. Nature 409: 698–701.

Bard E., Rostek F. and Sonzogni C. 1997. Interhemispheric synchrony of the last deglaciation inferred from alkenone palaeothermometry. Nature 385: 707–710.

Bard E., Rostek F., Turon J.-L. and Gendreau S. 2000. Hydrological impact of Heinrich events in the subtropical Northeast Atlantic. Science 289: 1321–1324.

Behl R.J. and Kennett J.P. 1996. Brief interstadial events in the Santa Barbara basin, NE Pacific during the past 60 kyr. Nature 379: 243–246.

Behling H., Arz H.W., Pätzold J. and Wefer G. 2000. Late Quaternary vegetational and climate dynamics in Northeastern Brazil, inferences from marine core GeoB 3104-1. Quat. Sci. Rev. 19: 981–994.

Benson L. 1999. Records of millennial-scale climate change from the Great Basin of the Western United States. In: Clark P.U., Webb R.S. and Keigwin L.D. (eds), Mechanisms of Global Climate Change at Millennial Time Scale. Geophysical Monograph 112, pp. 203–225.

Berger W. and Wefer G. 1996. Expeditions into the past: Palaeoceanographic studies in the South Atlantic. In: Wefer G., Berger W., Siedler G. and Webb D.J. (eds), The South Atlantic: Present and Past Circulation. Springer-Verlag, Berlin, pp. 363–410.

Blunier T., Chappellaz J., Schwander J., Dällenbach J., Stauffer B., Stocker T.F., Raynaud D., Jouzel J., Clausen H.B., Hammer C.U. and Johnsen S.J. 1998. Asyn-chrony of Antarctic and Greenland climate during the last glacial period. Nature 394: 739–743.

Blunier T. and Brook E.J. 2001. Timing of millennial-scale climate change in Antarctica and Greenland during the last glacial period. Science 291: 109–112.

Bond G., Broecker W.S., Johnsen S., McManus J., Labeyrie L., Jouzel J. and Bonani G. 1993. Correlations between climate records from North Atlantic sediments and Greenland ice. Nature 365: 143–147.

Bond G., Kromer B., Beer J., Muscheler R., Evans M.N., Showers W., Hoffmann S., Lotti-Bond R., Hakdas I. and Bonani G. 2001. Persistent solar influence on North Atlantic climate during the Holocene. Science 294: 2130–2136.

Broecker W.S. and Denton G. 1989. The role of ocean-atmosphere reorganizations in glacial cycles. Geoch. et Cosmoch. Acta. 53: 2465–2501.

Broecker W.S. 1998. Palaeocean circulation during the last deglaciation: a bipolar seesaw? Palaeoceanography 13:119–121.

Broecker W.S. and Hemming S. 2001. Climate swings come into focus. Science 294: 2308–2309.

Brook E.J., Sowers T. and Orchardo J. 1996. Rapid variations in atmospheric methane concentration during the past 110,000 years. Science 273: 1087–1091.

Brook E.J. and Harder S. 1999. Atmospheric methane and millennial-scale climate change. In: Clark P.U., Webb R.S. and Keigwin L.D. (eds), Mechanisms of Global Climate Change at Millennial Time Scale. Geophysical Monograph 112, pp. 165–175.

Cacho I., Grimalt J.O., Pelejero C., Canals M., Sierro F.J., Abel Flores J. and Shackleton N. 1999. Dansgaard-Oeschger and Heinrich event imprints in Alboran Sea palaeotemperatures. Palaeoceanography 14: 698–705.

Cacho I., Grimalt J.O., Sierro F., Shackleton N. and Canals M. 2000. Evidence for enhanced Mediterranean thermohaline circulation during rapid climatic coolings. Ear. Planet. Sci. Lett. 183: 417–429.

Cane M. 1998. A role for the tropical Pacific. Science 282: 59–60.

Chappellaz J., Blunier T., Raynaud D., Barnola J.-M., Schwander J. and Stauffer B. 1993. Synchronous changes in atmospheric CH_4 and Greenland climate between 40 and 8 kyr BP. Nature 366: 443–445.

Charles C.D., Lynch-Stieglitz J., Ninnemann U.S. and Fairbanks R.G. 1996. Climate connections between the hemisphere revealed by deep sea sediment core/ice core correlations. Ear. Planet. Sci. Lett. 142: 19–27.

Clement A.C. and Cane M. 1999. A role for the tropical Pacific coupled Ocean-Atmosphere system on Milankovitch and millennial timescales. Part I: A modelling study of tropical Pacific variability. In: Clark P.U., Webb R.S. and Keigwin L.D. (eds), Mechanisms of Global Climate Change at Millennial Time Scale. Geophysical Monograph 112, pp. 363–371.

Clement A.C., Seager R. and Cane M. 1999. Orbital controls on the El Nino/Southern Oscillation and the tropical climate. Palaeoceanography 14: 441–456.

Cortese G. and Abelmann A. 2002. Radiolarian-based palaeotemperatures during the last 160 kyr at ODP Site 1089 (Southern Ocean, Atlantic sector). Palaeogeogr. Palaeoclim. Palaeoecol. 182: 259–286.

Cortijo E., Labeyrie L., Vidal L., Vautravers M., Chapman M., Duplessy J.-C., Elliot M., Arnold M., Turon J.-L. and Auffret G. 1997. Changes in the sea surface hydrology associated with Heinrich event 4 in the North Atlantic Ocean between 40 deg. and 60 deg. N. Ear. Planet. Sci. Lett. 146: 27–45.

Curry W.B. and Oppo D.W. 1997. Synchronous, high frequency oscillations in tropical sea surface temperatures and North Atlantic deep water production during the last glacial cycle. Palaeoceanography 12: 1–14.

Dansgaard W., Johnson S.J., Clausen H.B., Dahl-Jensen D., Gundenstrup N.S., Hammer C.U., Hvidberg C.S., Steffensen J.P., Sveinbjörnsdottir A.E., Jouzel J. and Bond G. 1993. Evidence for general instability of past climate from a 250-kyr Ice-core record. Nature 364: 218–220.

de Garidel-Thoron T., Beaufort L., Linsley B.K. and Dannenmann S. 2001. Millennial-scale dynamics of the east Asian winter monsoon during the last 200,000 years. Palaeoceanography 16: 491–502.

Elliot M., Labeyrie L., Bond G., Cortijo E., Turon J.-L., Tisnerat N. and Duplessy J.-C. 1998. Millennium timescale iceberg discharges in the Irminger Basin during the last glacial period: relationship with the Heinrich events and environmental settings. Palaeoceanography 13: 433–446.

Elliot M., Labeyrie L., Dokken T. and Manthé S. 2001. Coherent patterns of ice-rafted debris deposits in the Nordic regions during the last glacial (10–60 kyr). Ear. Planet. Sci. Lett. 194: 151–163.

Elliot M., Labeyrie L. and Duplessy J.-C. 2002. Changes in north Atlantic deep-water formation associated with the Dansgaard-Oeschger temperature oscillations (10–60 kyr). Quat. Sci. Rev. 21: 1153–1165.

Ganopolski A. and Rahmstorf S. 2001. Rapid changes of glacial climate simulated in a coupled climate model. Nature 409: 153–158.

Gasse F. 2000. Hydrological changes in the African tropics since the Last Glacial Maximum. Quat. Sci. Rev. 19: 189–211.

Grimm E.C., Jacobson G.L., Watts W., Hansen B.C.S. and Maasch K. 1993. A 50,000-year record of climate oscillations from Florida and its temporal correlation with Heinrich events. Science 261: 198–200.

Heinrich H. 1988. Origin and consequences of cyclic ice rafting in the northeast Atlantic Ocean during the past 130,000 years. Quat. Res. 29: 142–152.

Hostetler S.W., Clark P.U., Bartlein P.J., Mix A.C. and Pisias N.J. 1999. Atmospheric transmission of North Atlantic Heinrich events. J. Geophys. Res. 104: 3947–3952.

Indermühle A., Monnin E., Stauffer B. and Stocker T. 2000. Atmospheric CO_2 concentration from 60 to 20 kyr BP from the Taylor Dome ice core, Antarctica. Geophys. Res. Lett. 27: 735–738.

Kanfoush S.L., Hodell D.A., Charjes C.D., Guilderson T.P., Mortyn P.G. and Ninnemann U.S. 2000. Millennial-scale instability of the Antarctic Ice sheet during the last glaciation. Science 288: 1815–1818.

Keigwin L.D. and Boyle E. 1999. Surface and deep ocean variability in the northern Sargasso Sea during marine isotope stage 3. Palaeoceanography 14: 164–170.

Kennett J.P. and Ingram B.L. 1995. A 20,000-year record of ocean circulation and climate change from the Santa Barbara basin. Nature 377: 510–514.

Kudrass H.R., Hofmann A., Doose H., Emeis K. and Erlenkeuser H. 2001. Modulation and amplification of climatic changes in the Northern Hemisphere by the Indian summer monsoon during the past 80 k.y. Geology 29: 63–66.

Little M.G., Schneider R.R., Kroon D., Price B., Summerhayes C.P. and Segl M. 1997. Trade wind forcing of upwelling, seasonality, and Heinrich events as a response to sub-Milankovitch climate variability. Palaeoceanography 12: 568–576.

Lohman G. and Schulz M. 2002. Reconciling Bölling warmth with peak deglacial meltwater discharge. Palaeoceanography 15: 537–540.

Lowell T.V., Heusser C.J., Andersen B.G., Moreno P.I., Hauser A., Heusser L.E., Schluchter G., Marchant D.R. and Denton G. 1995. Interhemispheric correlation of Late Pleistocene glacial events. Science 269: 1541–1549.

MacAyeal D.R. 1993. Binge/purge oscillations of the Laurentide ice sheet as a cause of the North Atlantic's Heinrich events. Palaeoceanography 8: 775–784.

Manabe S. and Stouffer R.J. 1997. Coupled ocean-atmosphere model response to freshwater input: Comparison to Younger Dryas event. Palaeoceanography 12: 321–336.

Marchal O., Stocker T.F. and Joos F. 1998. Impact of oceanic reorganizations on the ocean carbon cycle and atmospheric carbon dioxide content. Palaeoceanography 13: 225–244.

Maslin M., Seidov D. and Lowe J. 2001. Synthesis of the nature and causes of rapid climate transitions during the Quaternary. In: Seidov D., Haupt B.J. and Maslin M. (eds), The Oceans and Rapid Climate Changes: Past, Present, and Future. Geophysical Monograph 126, pp. 9–52.

Mayewski P.A., Meeker L.D., Whitlow S., Twickler M.S., Morrison M.C., Bloomfield P., Bond G.C., Alley R.B., Gow A.J., Grootes P.M., Meese D.A., Ram M., Taylor K.C. and Wumkes W. 1994. Changes in atmospheric circulation and ocean ice cover over the North-Atlantic during the last 41,000 Years. Science 263: 1747–1751.

McIntyre A. and Molfino B. 1996. Forcing of Atlantic equatorial and subpolar millennial cycles by precession. Science 274: 1867–1870.

Mikolajewicz U., Crowley T.J., Schiller A. and Voss R. 1997. Modelling teleconnections between the North Atlantic and North Pacific during the Younger Dryas. Nature 387: 384–387.

Moreno P.I., Jacobson G.L., Lowell T.V. and Denton G. 2001. Interhemispheric climate links revealed by late-glacial cooling episode in Southern Chile. Nature 409: 804–808.

Paul A. and Schulz M. 2002. Holocene climate variability on centennial to millennial time scales: 2. Internal feedbacks and external forcing as possible causes. In: Wefer G., Berger W.H., Behre K.E. and Jansen E. (eds), Climate Development and History of the North Atlantic Realm. Springer Verlag, Berlin, pp. 55–73.

Peterson L.C., Haug G.H., Hughen K.A. and Röhl U. 2000. Rapid changes in the hydrologic cycle of the tropical Atlantic during the last Glacial. Science 290: 1947–1951.

Pierrehumbert R.T. 1999. Huascaran $d^{18}O$ as an indicator of tropical climate during the Last Glacial Maximum. Geophys. Res. Lett. 26: 1345–1348.

Porter S.C. and An Z. 1995. Correlation between climate events in the North Atlantic and China during the last glaciation. Nature 375: 305–308.

Renssen H., van Geel B., van der Plicht J. and Magny M. 2000. Reduced solar activity as trigger for the start of the Younger Dryas? Quat. Int. 68-71: 373–383.

Rind D., Russell G.L., Schmidt G.A., Sheth S., Collins D., Demenocal P. and Teller J. 2001. Effects of glacial meltwater in the GISS Coupled Atmosphere-Ocean Model: Part II: A bi-polar seesaw in Atlantic Deep Water production. J. Geophys. Res. 106: 27355–27366.

Rühlemann C., Mulitza S., Müller P.J., Wefer G. and Zahn R. 1999. Warming of the tropical Atlantic Ocean and slowdown of thermohaline circulation during the last deglaciation. Nature 402: 511–514.

Saarnisto M. and Lunkka J.P., this volume. Climate variability during the last interglacial-glacial cycle in NW Eurasia. In: Battarbee R.W., Gasse F. and Stickley C.E. (eds), Past Climate Variability through Europe and Africa. Kluwer, Dordrecht. pp. 443–464.

Schmittner A., Appenzeller C. and Stocker T.F. 2000. Enhanced Atlantic freshwater export during the El Nino. Geophys. Res. Lett. 27: 1163–1166.

Schneider R.R., Adegbie A.T., Röhl U. and Wefer G. 2001. Glacial millennial-scale fluctuations in African precipitation recorded in terrigenous sediment supply and freshwater signals offshore Cameroon. Abstract vol. of PEP III poster session, Aix-en-Provence, pp. 143.

Schulz H., von Rad U. and Erlenkeuser H. 1998. Correlation between Arabian Sea and Greenland climate oscillations of the past 110,000 years. Nature 393: 54–57.

Stauffer B., Blunier T., Dällenbach A., Indermühle A., Schwander J., Stocket T.F., Tschumi J., Chappellaz J., Raynaud D., Hammer C.U. and Clausen H.B. 1998. Atmospheric CO_2 concentration and millennial-scale climate change during the last glacial period. Nature 392: 59–62.

Steig E.J. 2001. No two latitudes alike. Science 293: 2015–2016.

Stenni B., Masson-Delmotte V., Johnsen S., Jouzel J., Longinelli A., Monnin E., Röthlisberger R. and Selmo E. 2001. An oceanic cold reversal during the last deglaciation. Nature 293: 2074–2077.

Stoker T.F. 1998. The seesaw effect. Science 282: 61–62.

Stott L., Poulsen C., Lund S. and Thunell R. 2002. Super ENSO and global climate oscillations at millennial time scales. Science 297: 222–226.

Street-Perrott A. and Perrott A. 1990. Abrupt climate fluctuations in the tropics: the influence of Atlantic Ocean circulation. Nature 343: 607–612.

Thouveny N., de Beaulieu J.-L., Bonifay E., Créer K.M., Guiot J., Icole M., Johnsen S., Jouzel J., Reille M., Williams T. and Williamson D. 1994. Climate variations in Europe over the past 140 kyr deduced from rock magnetis. Nature 371: 503–506.

Vidal L., Labeyrie L., Cortijo E., Arnold M., Duplessy J.-C., Michel E., Becqué S. and van Weering T.C.E. 1997. Evidence for changes in the North Atlantic Deep Water linked to meltwater surges during the Heinrich events. Ear. Planet. Sci. Lett. 146: 13–27.

Vidal L., Schneider R.R., Marchal O., Bickert T., Stocker T.F. and Wefer G. 1999. Link between the North and South Atlantic during the Heinrich events of the last glacial period. Clim. Dyn. 15: 909–919.

Voelker A.H.L. and workshop participants 2002. Global distribution of centennial-scale records for Marine Isotope Stage (MIS) 3: a database. Quat. Sci. Rev. 21: 1185–1212.

Winton M. 1993. Deep decoupling oscillations of the oceanic thermohaline circulation. In: Peltier W.R. (ed.), Ice in the Climate System. Springer Verlag, Berlin, pp. 417–432.

Zahn R., Schöenfeld J., Kudrass H.R., Park M.-H., Erlenkeuser H. and Grootes P. 1997. Thermohaline instability in the North Atlantic during meltwater events: stable isotope and ice rafted detritus records from core SO75-26KL, Portuguese margin. Palaeoceanography 12: 696–710.

4. BETWEEN AGULHAS AND BENGUELA: RESPONSES OF SOUTHERN AFRICAN CLIMATES OF THE LATE PLEISTOCENE TO CURRENT FLUXES, ORBITAL PRECESSION AND THE EXTENT OF THE CIRCUM-ANTARCTIC VORTEX

TIM C. PARTRIDGE (tcp@iafrica.com)
Climatology Research Group
University of the Witwatersrand
Private Bag 3, WITS 2050
Johannesburg
South Africa

LOUIS SCOTT (Scottl.sci@mail.uovs.ac.za)
Department of Plant Sciences
University of the Free State
P.O. Box 339, Bloemfontein 9300
South Africa

RALPH R. SCHNEIDER (rschneid@uni-bremen.de)
Fachbereich Geowissenschaften
Universität Bremen
Klagenfurter Strasse
28359 Bremen
Germany

Keywords: Miombo, Fynbos, Tswaing Crater, Rainfall proxy, Heinrich Events, Dune activity, Antarctic Cold Reversal, Benguela upwelling, Kalahari, Palaeo-lakes

Introduction

Terrestrial records for the Middle Pleistocene of Southern Africa are sparse and mostly fragmentary because the landscape was largely devoid of suitable depositories. Fortunately, one continuous lacustrine sequence, that of the Tswaing impact crater near Pretoria, South Africa (Fig. 1), spans the period from ~200 kyr to the present. Recent analyses of marine cores off the west coast of Southern Africa have, in addition, vouchsafed important proxy evidence of palaeo-wind regimes, periods of aridity and terrestrial flora. A generalised, but largely consistent, picture of palaeoenvironmental fluctuations during Isotope Stage 6 can therefore be drawn.

R. W. Battarbee et al. (eds) 2004. *Past Climate Variability through Europe and Africa.*
Springer, Dordrecht, The Netherlands.

Figure 1. Locality plan for Southern Africa.

Isotope Stage 5 is scarcely better documented in terrestrial settings. A scattering of archaeological sites spans this period, but dating is problematical and thus precludes all but the most general conclusions in respect of forcings and responses. This was an interval which saw the emergence in South Africa of some of the world's first "modern" humans with their Middle Stone Age tool-kits. In contrast to scarce continental climate archives, marine sedimentary records of pollen associations and abundances, as well as grain-size distributions of the terrigenous silt fraction, from off the south-west African coast, provide new insights into climate changes over the western part of Southern Africa during the penultimate interglacial.

Palaeo-environmental signatures were better defined both regionally and temporally after the onset of the Last Glacial. Particularly striking is the evidence for a decline in the dominance of precessional forcing in the summer rainfall region after about 50 kyr; instead, climatic changes appear to have been driven to an increasing extent by stadial/interstadial cycles. It is not evident, however, that events in the northern hemisphere pre-dated those in the southern mid-latitudes of Africa. Among the most important of the latter were discrete, and temporally restricted, episodes of dune mobility and one or more intervals during which mega-lakes were widespread. As temperatures declined and climatic variability increased (as a result of millennial-scale variations in the extent of Antarctic sea-ice, and consequent fluctuations in the extent and intensity of the Antarctic vortex), so human populations adapted their technologies and territorial ranges. The advent of the Late Stone Age of Southern Africa, with its fundamentally different traditions of tool making, was one notable response.

Conditions during the Last Glacial Maximum can be reconstructed with relative confidence. Fairly uniform temperature declines over a broad latitudinal band brought widespread desiccation, and many lakes, for example, the Tswaing Crater became brackish or dried out completely (contrary to earlier belief). During this and the subsequent deglacial period, events in Antarctica were pre-eminent in forcing changes in climate over the subcontinent and generally led oceanic responses to variations in the thermohaline circulation. As temperatures rose towards Holocene values the brief but pervasive cooling of the Younger Dryas can be identified in sequences where resolution and dating control are sufficiently good.

The primacy of precessional forcing during isotope stage 6

The record of the Tswaing Crater

That the influence of orbital precession was particularly strong during Isotope Stage 6 is well known (Berger 1978; Barker et al., this volume). While the effects of changing receipts of insolation on regional temperature regimes have received considerable attention, responses in precipitation have been less easy to track. A core recovered from the 90 m lacustrine infilling of the Tswaing impact crater, 40 km north of Pretoria (Fig. 1), has yielded several proxy environmental records spanning the last 200 kyr (Partridge et al. 1999a). The most useful of these permitted the estimation of rainfall variations in this summer rainfall area from principal component analysis of sedimentological parameters, chiefly clastic particle size (Partridge et al. 1997). The clastic component of the sediments has been derived from weathering and soil formation on the granitic rocks of the crater walls. Since a close relationship was found to exist between the fine particle fraction (-20μm) of pristine granitic soils and rainfall across the east-west climatic gradient which exists presently over Southern Africa ($r = 0.88$ for mean annual precipitation ranging from 440 to 1500 mm (Partridge et al. 1997)), a transfer function could be developed linking variations in the -20μm clastic fraction of the lake core to past rainfall. Age control was provided for the last 43 kyr by high-precision calibrated radiocarbon dating, from which a fairly uniform sediment accumulation rate was apparent. Below the 20 metres covered by these determinations the only date obtained was from fission-track measurements on impact glasses underlying the lacustrine sequence. The precision of the age estimate of $\sim$200 kyr for this level, derived from measurements carried out on 421 individual glass particles, is about 20%, but it gives a mean accumulation rate almost identical to that for the last 43 kyr.

The rainfall proxy derived from the sediments showed regular, well defined variations with a periodicity of about 23 kyr (Fig. 2). Power spectral analysis confirmed this precessional signal; fine tuning to the summer insolation curve at 30 °S (Berger 1978) permitted the interpolation of ages between the dated upper and lower parts of the sequence. The tuning process adjusted ages based on mean sediment accumulation rate by an average value of 3.3 kyr, with a maximum single adjustment of 8.8 kyr (Partridge et al. 1997).

From Figure 2 it is apparent that changes in inferred precipitation track precessional insolation fluctuations almost perfectly, both in frequency and amplitude, during the period between $\sim$200 kyr and 60 kyr. Of significance is the fact that a coefficient expressing the sensitivity of past precipitation to changes in insolation can be derived from the proxy data.

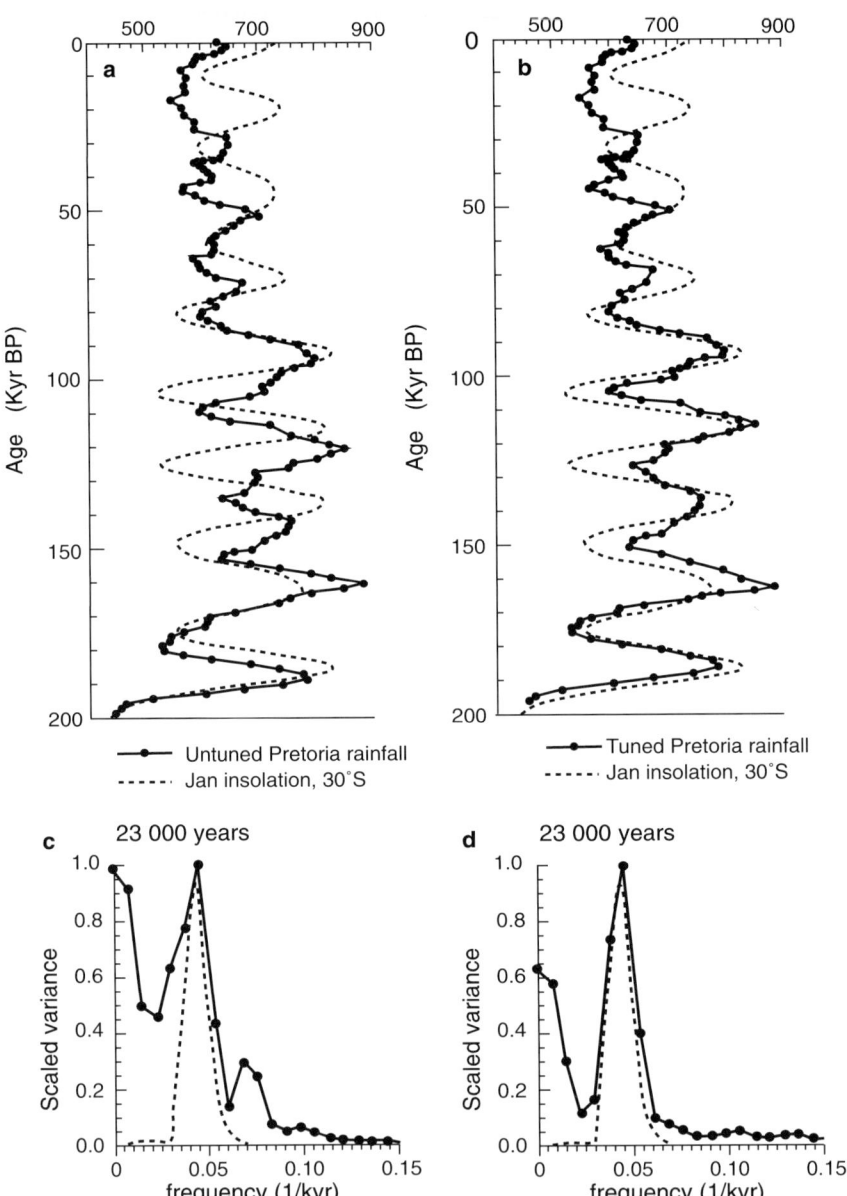

Figure 2. The Tswaing proxy rainfall time-series derived from a granulometric transfer function. (a) Age model derived from raw dated using calibrated radiocarbon ages and 200 kyr estimate for the base of the lacustrine succession; (b) Age model obtained by tuning the time-series to the precessional insolation curve. Note the dominance of precessional variability in both series; (c) The dominant period of variation in the untuned rainfall time series is 23 kyr; (d) The relative proportion of total variance at this frequency (46%) remained essentially unchanged after the tuning process (from Partridge et al. (1997)).

Rainfall at the site is estimated by the proxy to have varied from a low of 535 mm to a high of 900 mm (a change of 68%), relative to a range from 920–1060 W/m^2 in summer insolation, giving a sensitivity coefficient of 4.5; this should be compared with values determined in climate model experiments of 5.0 for the North African monsoon and 3.5 for the entire northern subtropics (Prell and Kutzbach 1987). Of note, too, is the symmetrically inverse (anti-phase) relationship that exists between the Tswaing record and a proxy for the strength of the North African monsoon derived from marine microfaunal variations by McIntyre et al. (1989) at a site located at 20 °N off the coast of West Africa (Partridge et al. 1997). Supporting evidence for contrasting responses between the two hemispheres comes from the Mozambique Channel (Van Campo et al. 1990) and Lake Tritrivakely in Madagascar (Gasse and Van Campo 1998; 2001).

The foregoing evidence confirms that, when precessional forcing is strong (as was the case during Isotope Stage 6), summer rainfall in the mid-latitudes of Africa fluctuated in direct relation, and with a relatively high degree of sensitivity, to changes in receipts of solar radiation. Signal noise is relatively low in the Tswaing sediment record, and the effects of other forcings (e.g., changes in intensity of the thermohaline circulation, as expressed in current fluxes marginal to Southern Africa) appear to have been minimal, at least within the more humid region of the subcontinent. However, the pollen record from Tswaing suggests that temperature fluctuations between ∼200 kyr and 60 kyr were independent of sediment-derived moisture fluctuations (Partridge et al. 1999a). Microscopic charcoal indicates changes in fire regime during this period that could have been linked to the presence of people, as no correlation to palaeoclimate parameters is seen (Scott 2002a).

Marine core records

The terrigenous silt fraction of sediments recovered from Walvis Ridge, SE Atlantic Ocean, reveals a history of south-western African climate of the last 300 kyr (Stuut et al. (2002), Fig. 3). Grain-size distributions can be explained by three end-members. The two coarsest end-members are interpreted as aeolian dust, and the third end-member as hemipelagic mud originating from fluvial supply. The ratio of the two aeolian end-members reflects the wind-transported grain-size and is linked to the intensity of the SE trades (or South Atlantic tropical easterlies). In general, trade winds were intensified during glacials compared to interglacials. Changes in the ratio of the two aeolian end-members to the hemipelagic one are interpreted as reflecting variations in south-western African aridity. The results suggest that late Quaternary climates of this region were relatively arid during the interglacial stages and relatively humid during the glacial stages. This is in conflict with results from palynological investigations in sediment cores from the continental margin off Namibia (Shi and Dupont 1997; Shi et al. 1998; 2000; 2001), which imply glacial extension of grassland and semi-desert or desert vegetation types, as well as shrinkage of the Miombo woodlands, and thus more arid conditions. However, higher influxes of dry vegetation pollen types under glacial climates may have been related to stronger trade wind intensity rather than to an areal extension of semi-desert and desert vegetation within the coastal belt of South Western Africa. The northward progression of Fynbos, including Restionaceae and Asteraceae vegetation in glacial climates, viewed in conjunction with more humid conditions as indicated by the grain-size records, could then imply a northward extension of the winter rainfall belt along the western lowlands. Thus, unless these vegetation changes

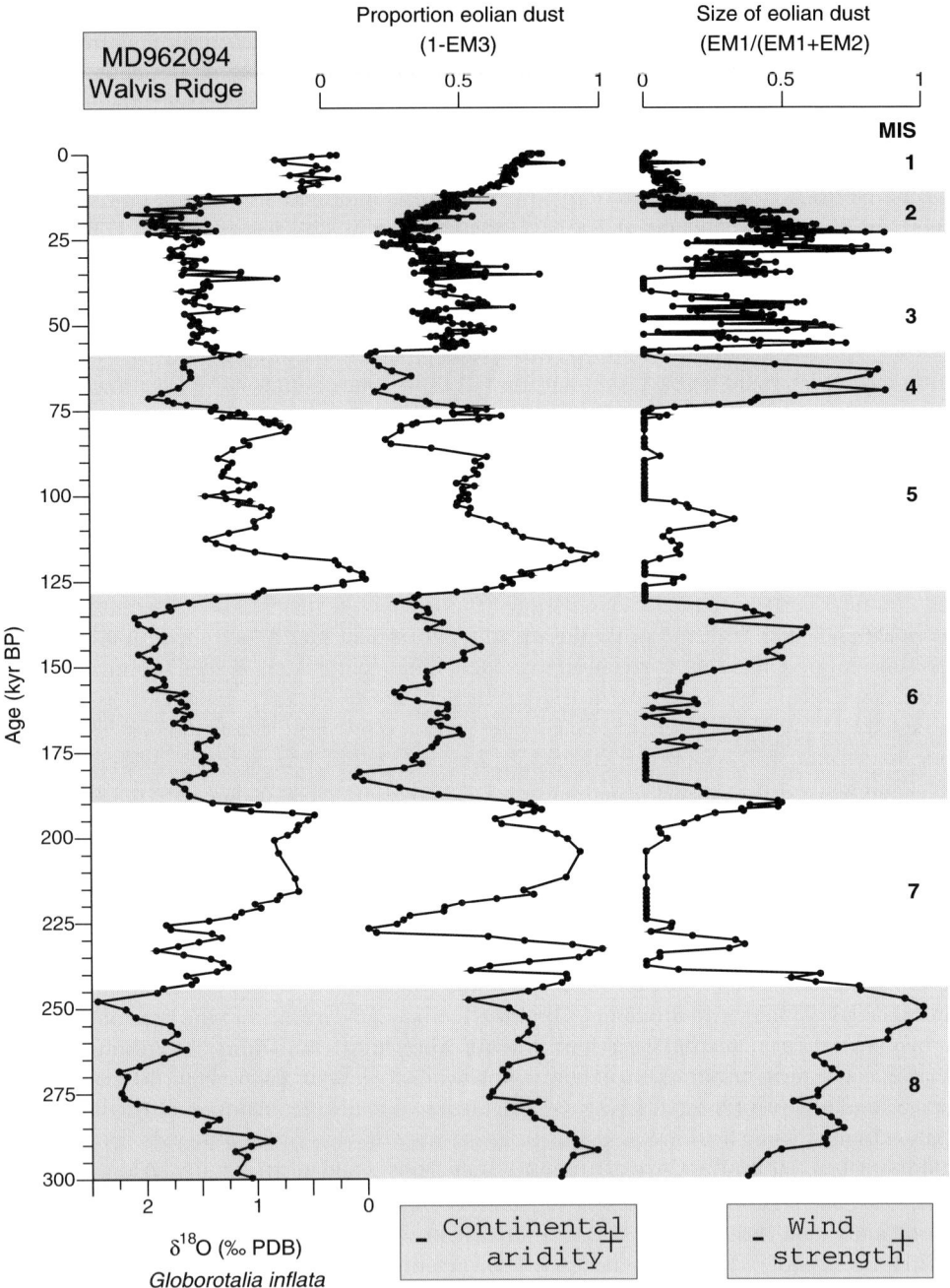

Figure 3. Time series of variations in median grain size and end member contributions of the terrigenous sediment fraction in marine core MD 962094 compared with global climate. (a) *Globorotalia inflata* $\delta^{18}O$; (b) Median grain size; (c) End member contributions of the three end members (from Stuut et al. (2002)).

were brought about by simple lowering of vegetation belts, they may have been induced by meridional shifts in the atmospheric circulation system and a northward displacement of the zone of westerlies, causing increased winter rainfall in parts of Western South Africa and Namibia. At the same time, equator-ward shifts of the southern ocean frontal systems and Antarctic sea-ice margin increased the meridional pressure gradient, strengthening atmospheric circulation and trade-wind intensity during glacial periods. The west coast winter rainfall region may thus have undergone changes different from those elsewhere over South Africa. Evidence for increased onshore precipitation over parts of the Western Cape during the Last Glacial Maximum has been assembled by Cartwright and Parkington (1997), Cowling et al. (1999) and Parkington et al. (2000) (see section 5 below). However, terrestrial proxies for the Namib desert to the north point strongly to greater aridity at that time (Vogel 1987), and it is possible that problems with the interpretation of the marine sediment data remain unresolved.

Precessional periodicity, although very pronounced in the Tswaing record from Eastern South Africa, is strongly subdued compared to the glacial-interglacial 100 kyr cycle in the grain-size and pollen records off Western South Africa. Interestingly, glacial Stage 6 is the period that is characterised by the strongest precessional component over the last 300 kyr, both in the terrigenous proxy records for Western South Africa and at Tswaing.

Glimpses of the Eemian

Although a number of terrestrial sites in Southern Africa span this interval, only four are of sufficient significance to warrant mention in this review. At Tswaing the lake shallowed and became more concentrated between 140 and 125 kyr, but was still relatively deep. The Eemian itself (Isotope Stage 5e) coincided with a precessional low during which proxy data suggest that rainfall was similar to that of today (ranging perhaps to 10% higher). By comparison rainfall during the subsequent 5d interval was about 35% higher. Unfortunately no Eemian pollen were preserved at Tswaing (Scott 1999a), but there is good diatom preservation (Metcalfe 1999) and the sediments appear to be annually laminated, suggesting that high resolution proxies could be derived. The entire interval coincides with a δ^{18}O anomaly in lake carbonates of at least 5‰, the implications of which remain to be determined (N.J. Shackleton, pers. comm.).

At Border Cave closer to the Indian Ocean seaboard (Fig. 1), micro-mammalian evidence indicates elevated temperatures and a southward shift of the Miombo boundary of perhaps 3° to beyond 27 °S (Avery 1992). Pollen evidence from a core offshore of Angola places this boundary, inland of the Atlantic coast, around 17 °S at this time (Shi and Dupont 1997). Rainfall at Border Cave of around double today's 800 mm is implied. Although early dates for this site, derived from amino-acid racemization, remain controversial, sediment accumulation rates during the latter part of the sequence are well calibrated by radiocarbon (Butzer et al. 1978). Careful extrapolation and the use of micro-mammalian indicators leave little doubt concerning the position of the preceding interglacial temperature optimum but are of little use in resolving the timing of responses with any degree of precision.

Another important site at Florisbad in the Free State (Fig. 1) has a somewhat firmer chronology based on ESR dating of tooth enamel from antelopes occurring in key levels (Grun et al. 1996). An Eemian fauna associated with a Middle Stone Age industry contains moisture-loving species such as hippopotamus and lechwe, while hyena copro-lites have yielded pollen indicating a significantly more grassy regional environment than characterises this area today (Partridge and Scott 2000).

Finally, the cave site of Klasies River Mouth on the coast of the southern Cape (Fig. 1) has produced fragments of anatomically modern humans in association with Middle Stone Age tools. The horizon in question directly overlies an elevated beach gravel and is capped by speleothems that have uranium series dates a little in excess of 100 kyr (Deacon and Lancaster 1988). The occupation therefore postdates (probably by a fairly small margin) the peak high sea level of Isotope Stage 5. The molluscan fauna from the beach gravel, and others from Eemian deposits in estuaries of the south eastern (Agulhas) coast, indicate significantly elevated inshore temperatures (Maud 2000).

The available terrestrial evidence, although sparse and not particularly well dated, suggests that, throughout the Eemian, Southern Africa remained strongly influenced by orbital precession. By contrast, the Holocene interglacial was a period of significantly lower precessional amplitude and was characterised by frequent medium and small oscillations in key climatic indicators that may relate to fluctuations in solar output and other forcings.

For the Eemian (marine Isotope Stage 5e) marine sediments off Namibia indicate a sig-nificant decrease in pollen influx typical of grassland, desert and semi-desert environments, paralleled by a slight increase in dry forest pollen (Shi et al. (2001), Fig. 4), and a shift in grain-size distribution indicative of drier conditions in the coastal belt but weaker trade winds than during the glacial periods. The precessional signal is not as pronounced as in glacial Stage 6. The marine records do not imply any significant differences in continental climate conditions between the Eemian and Holocene.

The early Last Glacial

The onset of the Last Glacial is well reflected in the Tswaing core. Below 45 m (ca. 97 kyr) the diatoms indicate low salinity and moderate alkalinity within the lake (Metcalfe 1999). Above 41 m (ca. 94 kyr) there was a pronounced increase in both (probably ~5–30%) and anoxic conditions probably developed. The lake changed fundamentally from a freshwater system to something akin to the modern, highly alkaline and saline system, a change which was not reversed through the remainder of its history. The proxy rainfall record mirrors this important transition, with precipitation falling markedly after about 90 kyr. Pollen and diatoms at Tswaing (Scott 1999a; Partridge et al. 1999a) and pollen at Florisbad (Partridge and Scott 2000) confirm that moisture levels did not remain constant.

The strong influence of orbital precession, evident in the early part of the Tswaing record, begins to disappear after about 60 kyr. Prior to this time estimated fluctuations in rainfall accord only partially with temperature changes evident in other (mainly marine) data series, but the influence of radiative forcing, especially at lower latitudes, clearly remained paramount. As the amplitude of the precessional signal lessened, so other influences are reflected to an increasing extent, both at Tswaing and elsewhere. In particular, changes at timescales matching the stadial/interstadial Bond Cycles and briefer Heinrich Events, so well documented in the millennial oceanic records of the North Atlantic, begin to make their appearance. The ensuing discussion is from Partridge (2002).

In Figure 5 the largest Heinrich Events are shown as horizontal bars in relation to both the Tswaing proxy rainfall record and the original time-series generated by principal component

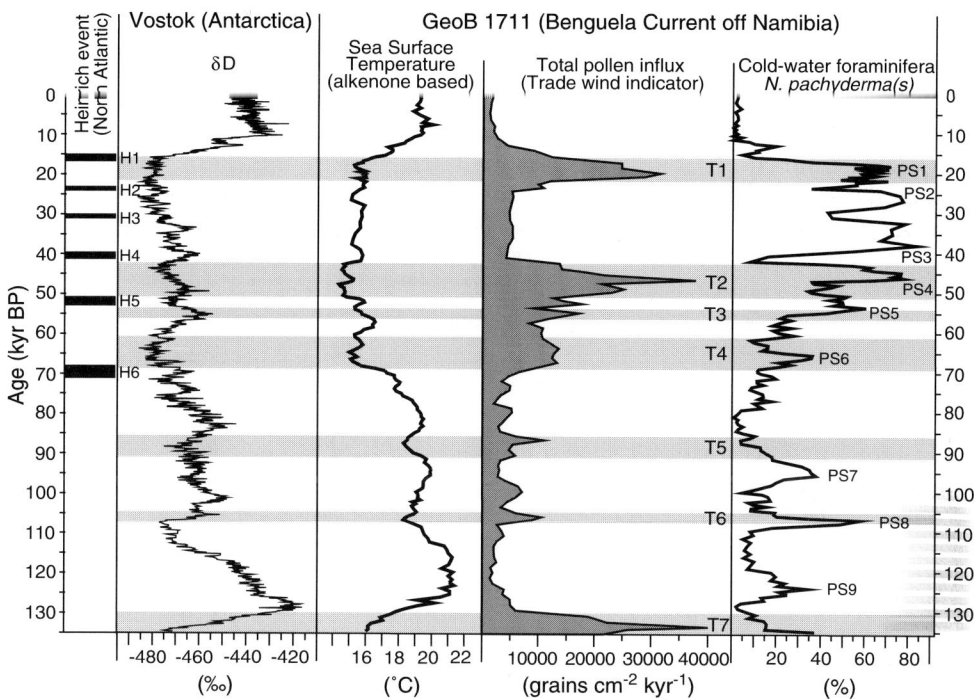

Figure 4. Proxy records from marine core GeoB1711-4: pollen influx of vegetation groups and cumulative pollen percentages compared with oxygen isotopes from *Cibicidoides wuellerstorfi* (in parts per mille vs. PDB) and SSTs derived from alkenones. Solid arrows show oxygen isotope stratigraphic control points. Trade wind enhancements are numbered T1 to T6 and shown by shaded bands (from Shi et al. (2001)).

analysis. It must be reiterated that this part of the record has excellent chronological control from calibrated radiocarbon dates (Partridge et al. 1997). The original time series is included in Figure 5 to ensure that apparent changes reflected in the proxy record are not an artefact of any smoothing involved in the generation of this record.

It is clear that Heinrich Events 4 and 5 significantly post-date the onset of more arid conditions, as reflected by marked excursions in the proxy rainfall series. The drying phases associated with some of the later Heinrich Events are less pronounced, but the commencement of each appears to pre-date the onset of ice-rafting as determined from sediment cores in the North Atlantic. The time of commencement of each drying event is shown in Figure 5. The date of onset of the drying cycle associated with H5 was extrapolated on the basis of the sediment accumulation rate defined by the two earliest radiocarbon dates for the lacustrine sequence. The resulting age estimate precedes the earliest date by 3.75 kyr. The results of the comparison are summarized in Table 1.

Also shown in Figure 5 are the principal periods of dune building in the Kalahari, several hundred km west of the Tswaing crater. Stokes and his co-workers (Stokes et al. 1997a; 1997b; 1998; 1999), have identified these periods on the basis of a large number of luminescence dates. Additional age determinations in Northern Namibia by Thomas and his co-workers (Thomas et al. 2000), have added to this tally and have defined an additional

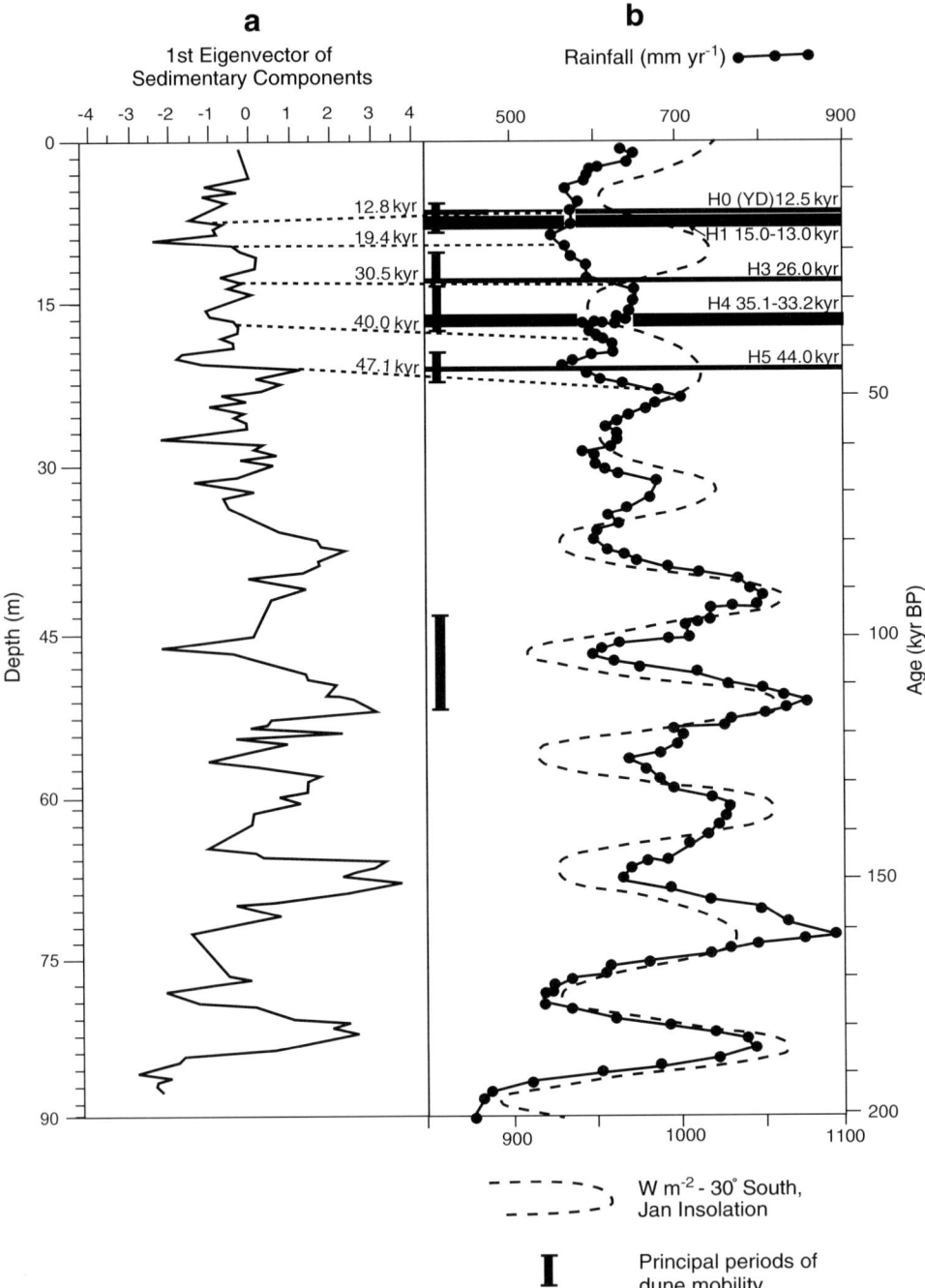

Figure 5. (a) First eigenvector of sedimentological components of Tswaing borehole core, with onset of arid episodes indicated; (b) Proxy rainfall time-series derived from (a), with principal Heinrich Events and periods of dune mobility in the Kalahari superimposed (after Partridge (2002)).

Table 1. Results of analysis of Tswaing rainfall series

Heinrich Event	Age[a] (kyr BP)	Depth of onset of drying at Tswaing (a) and corresponding date (b)		Lead at Tswaing (kyr)
		(a) - m	(b) - kyr	
HO	12.5	7.5	12.8	0.3
H1	15.0–13.0	9.0	19.4	4.4
H3	26.0	11.5	30.5	4.5
H4	35.1–33.2	17.0	40.0	4.9
H5	44.0	21.5	47.1	3.1

[a]after Broecker et al. 1992 and Bond et al. 1992.

period of dune mobility. These results show that arid periods were relatively short in relation to the intervening more humid spells (such as that of today), during which dunes remained stabilised by vegetation. The most important intervals of dune activity were:

16–10 kyr	— SW Kalahari
26–20 kyr	— SW, N and E Kalahari
36–29 kyr	— N Kalahari
46–41 kyr	— N and E Kalahari
115–95 kyr	— N and E Kalahari

As has been observed elsewhere, the Kalahari Sand has been in existence for very much longer than is indicated by these dates (Partridge 1995). Moreover, it is probable that some intervals of dune activity have remained undocumented as a result of sampling difficulties. The near perfect synchrony of the beginning of several such intervals with the onset of arid phases at Tswaing, and their anticipation of Heinrich Events on at least four occasions, is nonetheless remarkable and is indicative of well developed global atmospheric links. It should be noted that the earliest phase of dune building thusfar identified in the Kalahari (115–95 kyr) corresponds closely with an arid cycle (and associated precessional low) in the Tswaing record, and pre-dates the earliest of the Heinrich Events by a substantial margin.

From the foregoing analysis it is clear that episodes of aridity at Tswaing pre-date most corresponding Heinrich Events by between 3.1 and 4.9 kyr; in the case of the Younger Dryas the lead is a mere 300 yr, confirming that this event was much more sharply defined than those which preceded it. This pattern of dry events with relatively abrupt onsets contrasts with the more gradual and prolonged cooling cycles evident within North Atlantic oceanic temperature records (Bond et al. 1993; Bond and Lotti 1995). These cycles lasted 10–15 kyr and are linked to the succession of interstadial Dansgaard-Oeschger cycles that characterise the interval between Isotope Stage 5a and the Holocene. Each cycle was terminated by a period of major iceberg discharge, which was followed by a large and rapid increase in SSTs to near interglacial values. It seems improbable that a mechanism exists through which the passing of a particular threshold during slow cooling in the North Atlantic could trigger an abrupt response in the opposite hemisphere. More probable sources of clues to explain the apparent southern hemisphere lead are the ice core records from Antarctica. When the Antarctic records are examined together (e.g., Stocker (2000), Fig. 5), it is evident that, almost without exception, the onset of each arid phase in the Tswaing record coincides with the end of a period of declining Antarctic temperatures lasting $\sim$2–5 kyr. The most recent of these, the Antarctic Cold Reversal, ended some 300 yr before the onset of HO (the Younger Dryas) (Stocker 2000).

The evidence presented above appears sufficiently robust to posit that atmospheric mechanisms link these events. Cooling over Antarctica would be accompanied by growth of sea-ice which would, in turn, shift the position of the Antarctic Convergence and increase the extent and intensity of the circumpolar vortex. Mayewski et al. (1996) have drawn attention to several intervals during the last deglaciation when dust and especially sea-salt increased substantially in the Taylor Dome ice core. These correlate with climatic reversals, indicating that, during episodes of cooling in Antarctica, windiness around the ice margin increased significantly. The accompanying propagation of the westerlies equator-wards brings major responses in southern African climates (Tyson 1986; Tyson and Partridge 2000). Rainfall, especially in the western interior, is reduced as the zone of Hadley cell subsistence shifts and is enlarged, and a higher (but nonetheless small) percentage occurs in the winter season. The extension of the vortex also affects the strength of the South Atlantic tropical easterlies (the south-east trade winds). Stronger trades increase Benguela upwelling along the west coast of Southern Africa and carry more heat and moisture across the equator into the northern hemisphere (Leuscher and Sirocko 2000). This has been invoked as a factor in the enhanced build-up of the northern hemisphere ice sheets (Ruddiman and McIntyre 1981; Imbrie et al. 1989) and can also explain the $\sim$3 kyr lead in responses in the South Atlantic over those in the North Atlantic (Little et al. 1997). The mean lead evident from the Tswaing record, based on the ages presented in Table 1, is 3.4 kyr, but increases to 4.2 kyr if the anomalously short lead time associated with the Younger Dryas is ignored. These disparities are substantially larger than the error limits of the radiocarbon dates.

During the past 50 kyr the onset of aridity, and sometimes of dune building in the interior of Southern Africa, thus preceded most major discharges of ice into the North Atlantic by a significant margin; that these events were linked is, however, clearly demonstrable. The reduction in thermohaline circulation following North Atlantic cooling, and its complete cessation during strong Heinrich Events (Stocker 2000), restricted heat transfer from the western Indian Ocean into the eastern Atlantic, via the Agulhas Current (Zonneveld et al. 1997), thereby amplifying and prolonging arid interludes within the Kalahari. These changes in the oceanic conveyor must therefore be categorised as positive feedbacks, rather than the primary cause of regional cooling and aridification. Fluctuations in the extent of Antarctic sea-ice, perhaps driven by non-linear responses to orbital precession as envisaged by Curry and Oppo (1997), appear to have set in train this repeated concatenation of events. Such irregular responses seem to be more prevalent when the strength of the precessional signal is reduced.

The above conclusions suggest that global climate changes during the past 50 kyr were neither as synchronous, nor as symmetrical, as many workers have proposed, nor were they necessarily forced from the northern hemisphere. This accords with the recent finding that the penultimate deglaciation was triggered by events in the tropical oceans or, most plausibly, in the southern hemisphere (Henderson and Slowey 2000).

The history of dune activity and the aerial distribution of linear dune ridges are of use not only in flagging the duration of major modal changes in regional climate, but in the reconstruction of the regional patterns of atmospheric circulation that accompanied them.

Dune alignment can be used to reconstruct past wind regimes: reactivation of dunes in the most arid south-western area of the Kalahari would have necessitated a 50% reduction in rainfall (at present temperatures) or a 20% increase in the time when the wind exceeded the transport threshold (Lancaster 2000). Far greater changes would have been required to restore dune mobility in more humid areas to the north. Based on alignments, the NE Kalahari dunes formed when the semi-permanent South African anticyclone was 2° north of its present position. These dunes were most recently reactivated between 46–41 kyr, 36–29 kyr and 26–20 kyr, based on OSL dates (Stokes et al. 1999). Pollen in a nearshore Atlantic core near 12 °S indicates a northward shift of desert conditions along the western seaboard to 12°–13 °S during the latter interval (Shi and Dupont 1997). Pollen from Tswaing suggest that drier warm savanna developed ca. 62–60 kyr and ca. 44–39 kyr (Scott 1999b; Partridge et al. 1999a). The dune building interval from 46–41 kyr seems to correspond with more grassy vegetation in Southern Botswana after 47 kyr, as indicated by stable carbon isotope ratios in a stalagmite from Lobatse Cave (Holmgren et al. 1995).

Continental climate indicators from marine sediments offshore of Namibia in general support the findings from the Tswaing record. The pattern of grain-size variations indicative of trade wind strength and arid/humid cycles shows a clear trend from precession- and obliquity-related variance towards frequent millennial changes between 80 and 20 kyr. In terms of this evidence changes in wind strength seem to have been more important than changes in precipitation (Stuut et al. (2002), Fig. 3). Although absolute dating of the relevant sediment core is far from satisfactory for the mid-glacial period (MIS 3), the high, frequent oscillations in aeolian input of fine and coarse silt, indicative of changes in wind strength, are probably connected to the occurrence of Bond Cycles and Heinrich Events in the northern hemisphere, as well as to millennial scale oscillations in upwelling conditions off Namibia, as implied by the strong variance in planktonic foraminifera species associations in continent-margin sediments (Little et al. 1997). However, the absence of high-resolution absolute age control for these marine sediment records makes it difficult to decipher the phasing of grain-size variations with respect to high frequent climate oscillations in the high latitudes of both hemispheres, as well as to the continental records of dune formation and precipitation changes in South Africa.

The only existing pollen record from marine sediments offshore of Namibia that covers the entire Last Glacial (Shi et al. (2001), Fig. 4) has a yet lower temporal resolution. Nonetheless, changes in pollen influx patterns and vegetation associations strongly suggest that cold and dry periods on land were connected generally to cooler surface ocean temperatures and to enhanced coastal upwelling in Namibian waters. In sum, the terrestrial and marine evidence presented above suggests that the southern limit of mid-latitude aridity in the west of Southern Africa (the Kalahari-Miombo boundary) shifted northwards during cold, arid spikes within the Last Glacial, the largest shift being associated with Heinrich Event 2 during the Last Glacial Maximum. Thus, between the Eemian and the LGM, this important ecotone migrated latitudinally through fully 5 degrees.

As indicated previously, episodes of aridity during Isotope Stage 3 were relatively short-lived. By contrast, available moisture during some intervening periods must have been significantly in excess of present values. In the Kalahari region of Botswana, palaeo-shorelines indicate that mega-lakes were present in areas marginal to the present inland delta of the Okavango River and between it and the Zambezi River. These features are juxtaposed with dune ridges mobilised during earlier arid intervals. Further to the south, in the semi-arid Karoo region of South Africa, presently ephemeral playas contained permanent water bodies for extended periods, while on the eastern coastal plain of the subcontinent, which today hosts scattered small lakes, large fresh-water systems were common.

The major palaeo-basins linked to the Okavango and Zambezi rivers may have been controlled, in part, by neotectonic adjustments along a series of faults which extend from south-west to north-east between Namibia and the Zambezi valley (Thomas and Shaw 1991). These probably played an important role in providing conduits for inflows from both river systems and in creating topographic thresholds for different phases of lake formation. However, the accordance of crest elevations of strandlines and offshore barrier bars argues for the dominance of climatic factors and for the hydrological unity of the system, at least when it was at its maximum extent; in fact, little subsequent deformation of these features has occurred as a result of tectonic movements. Hence the 940–945 m crest level present in the Ngami Basin to the south-west of the delta has a counterpart in the Mababe Basin to the north-east, as well as in the Makgadikgadi Basin some 100 km to the south-east (Thomas and Shaw 1991). The entire system covered an area in excess of 80,000 km^2, thus exceeding in area today's largest lake (Lake Victoria in Uganda). Named Lake Palaeo-Makgadikgadi by Grey and Cook (1977), the age of this feature, and those of smaller lakes which were preserved later in discrete sub-basins, have remained elusive because the scores of radiocarbon ages for their shorelines were obtained from pedogenic calcretes. Such secondary accumulations have been shown to be polygenetic and almost invariably post-date the deposits in which they formed, often by a substantial margin (Partridge and Scott 2000). All that may be concluded from the available data is that Lake Palaeo-Makgadikgadi probably formed sometime during Isotope Stage 3; water budget analyses indicate that an inflow of some 50 km^{3}yr^{-1} would have been required to sustain such a lake, which implies major input from the Zambezi River (Thomas and Shaw 1991). A somewhat lower stage at 936 m (the Lake Thamalakane Stage), evident in basins peripheral to the Okavango Delta, was characterised by multiple transgressions, suggesting that this threshold was reached repeatedly over a considerable period of time. Water from Lake Thamalakane overflowed via the Boteti River into the Makgadikgadi Basin to feed the 920 m lake which it enclosed at that time. Diatomaceous earths associated with this level have yielded OSL dates of 32–37 kyr (Shaw et al. 1997). It is therefore likely that Lake Thamalakane filled (perhaps repeatedly) during Isotope Stage 3. At the Tswaing Crater, about 1300 km to the south-east, an increase in forest pollen and sedimentological evidence for higher rainfall after ca. 57 kyr (Scott 1999a; Partridge et al. 1999a) reflect an earlier moist interval when lakes may have existed in the Kalahari.

Presently ephemeral playas, such as the Alexandersfontein Basin in the western interior of South Africa, have analogously elevated shorelines indicating that they formerly contained relative deep and extensive water bodies. Calcrete ^{14}C dates which were previously held to document a high lake stand of Alexandersfontein between 21 kyr and 16 kyr (Butzer et al. 1973) must be discarded for the reasons given previously; all that may be concluded is that this lake phase probably belongs within Isotope Stage 3. Better control is, however, available for palaeo-lake deposits on the eastern coastal plain of South Africa. Extensive occurrences of diatomite around Mbazwana point not only to the former presence of large, freshwater coastal lakes, but contain peat horizons which date them to between 36 and 23 kyr. The lower diatomaceous levels contain deep water species indicative of fresh conditions, but are replaced by forms characteristic of shallow, more saline environments towards the top of the sequence (Maud et al. 1993).

That terrestrial temperatures during at least the latter part of Isotope Stage 3 were significantly lower than those of today is indicated by analyses of nitrogen and noble gases in an artesian aquifer at Uitenhage in the Eastern Cape (Heaton et al. 1986) and from ^{18}O values in stalagmite from the nearby Cango Cave (Talma and Vogel 1992). Confirmation of an average temperature depression, after about 40 kyr, of around 4 °C has been provided by noble gases in the Stampriet aquifer, some 900 km to the north-west (Stute and Talma 1998). Fynbos pollen in Tswaing indicate a depression of vegetation zones of ca. 100 m around 38–35 kyr (Scott 1999a; Partridge et al. 1999a). It is thus clear that, while rainfall oscillated significantly during Isotope Stage 3 in response both to fluctuations in the extent of the circum-Antarctic vortex and changes in the thermohaline circulation around Southern Africa, temperatures remained relatively constant over considerable areas.

The Last Glacial Maximum: human responses to widespread cold and aridity

Palaeoenvironmental conditions in Southern Africa during the Last Glacial Maximum have been synthesised using the full array of proxy records stored in the Palaeoclimates of the Southern Hemisphere (PASH) database relating to the subcontinent (Partridge et al. (1999b), based on the work, *inter alia*, of Avery (1982), Bonnefille et al. (1990), Deacon and Lancaster (1988), Heaton et al. (1986), Lancaster (1979, 1981, 1988), Lee-Thorp and Beaumont (1995), Livingstone (1971), Mahaney (1990), Maley (1993), Partridge (1995), Partridge et al. (1990, 1997), Rosqvist (1990), Runge and Runge (1995), Scott (1982, 1987, 1994), Scott and Thackeray (1987), Stager (1988), Stager et al. (1986), Stute and Talma (1998), Talma and Vogel (1992), Thackeray (1987), Thackeray and Lee-Thorp (1992), Vincens (1993), and Vogel (1987)). Of these a few have been decisive in defining temperatures and the regional distribution of aridity.

As discussed previously, isotopic analyses of artesian groundwaters, together with data from the Cango speleothem, confirm temperature lowering of 5°–6 °C over much of Southern Africa south of 24°. A clear south-to-north meridional temperature gradient is nonetheless evident in the reconstructed palaeotemperature field (Fig. 6). Rainfall appears to have been lower over the entire subcontinent, but regional variations interpreted from lake levels, pollen, dune activity, micro-mammals and other proxies were more disparate, ranging from less than 30% over the Kalahari to more than 70% of present values further east (Fig. 6). There is also some evidence to suggest that specific localities within the western Cape may have received more rain than today through their greater exposure to storm-track belts as the circum-Antarctic vortex expanded (Parkington et al. 2000). Localised orographic effects would also have been important here. Lake level data, which were thought previously to indicate high stands during the LGM, must now be reviewed because of dating difficulties. In fact, all southern African lakes with reliable chronologies were apparently low during this period; many became brack or dried out entirely. Evaporation was therefore not sufficiently low to compensate for the reduced precipitation (indeed, increased dune activity and dust deposition indicate significantly higher wind speeds, which would have largely offset reductions in evaporation due to lowered temperatures). Reconstructions of biomes during the LGM indicate that beyond 13 °S the western half of Southern Africa

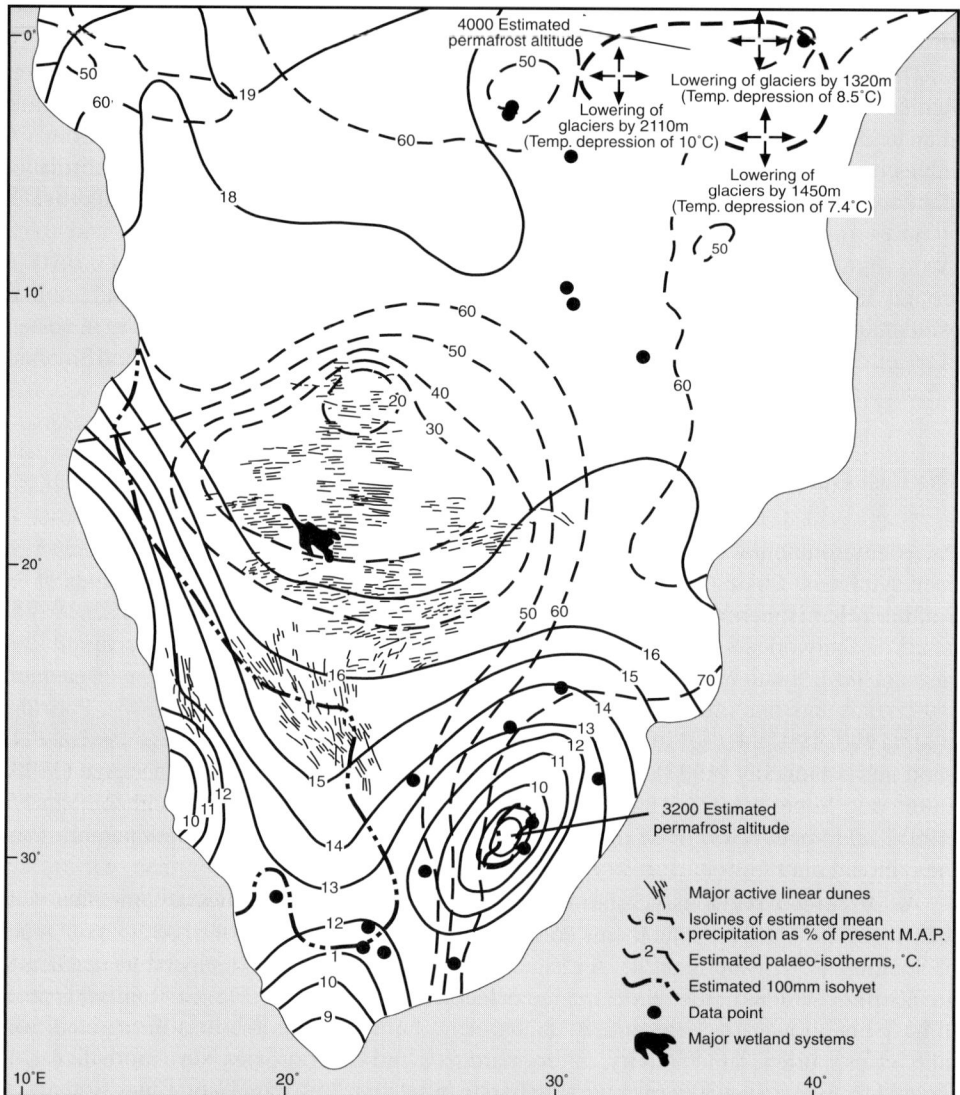

Figure 6. Palaeoclimatic reconstruction of rainfall and temperature conditions at the time of the Last Glacial Maximum at 21 000–18 000 BP (after Partridge et al. (1999b)).

was desert, with xerophytic woodland occupying much of the remainder; the upland areas carried steppe vegetation (Partridge et al. 1999b).

As is the case today, Southern Africa during the LGM seems to have been subject to two seasonally distinct climate systems: the winter rainfall system of the south-west and the summer rainfall system of the remainder of the subcontinent. C_4 grasses dominate the summer rainfall region, while C_3 grasses are widespread in the winter rainfall Fynbos (macchia) biome (Vogel et al. 1978). Isotopic data from both the Cango speleothem and antelope tooth enamel in archaeological sites suggest that, in contrast to evidence for the expansion of C_4 plant communities in the tropics during the LGM, no parallel trend was apparent near the southerly limit of C_4 habitats in the southern Cape (Lee-Thorp and Talma 2000; Scott and Vogel 2000; Scott 2002b). There is, in fact, evidence that a modest increase in C_3 grasses occurred in this winter rainfall area, indicating that the seasonality of precipitation was the dominant factor then as it is now, and that the glacial temperature depression was sufficiently large to suppress the advantage of reduced CO_2 for C_4 grasses.

There is no evidence that, within either of the two climate systems, rainfall seasonality was reversed at this or at any other time during the Last Glacial. What is clear, however, is that expansion of the circumpolar vortex and the influence of the westerlies would tend to extend the winter rainfall belt and the range of C_3 grasses. And, while there is some evidence to suggest that semi-arid areas in the western parts of the sub-continent to the north of the extended winter rainfall region received more winter rainfall during dry phases, their total winter precipitation receipts continued to represent a negligible proportion of the annual total (Tyson and Partridge 2000). Important supporting evidence has been gleaned from the Cango speleothem, which comes from a transitional area subject to year round rainfall, in which some 40% of grass is of the C_4 type. Fluctuations in its ^{13}C values indicate an overwhelming dominance of C_3 plants during glacial times whereas, during the Holocene, evidence for an increasing C_4 component suggests that grasses were responding to a higher proportion of summer rainfall (Lee-Thorp and Talma 2000; Scott and Vogel 2000). As mentioned previously, there is evidence from archaeological charcoal assemblages that soil moisture in the Western Cape was relatively high during the Last Glacial Maximum, contrasting with the situation in the eastern Cape (Cowling et al. 2000).

The advent of the LGM brought with it major cultural responses throughout Southern Africa. Anatomically modern humans had left remains of their Middle Stone Age (MSA) tool-kits in numerous and widely dispersed archaeological sites during Isotope Stages 5, 4 and 3; increased visibility appears to have coincided with periods of elevated moisture availability. With few exceptions MSA technologies persisted across Southern Africa until 20–25 kyr (Mitchell 2000). The relatively prolonged period of desiccation that was initiated before the LGM, but persisted to include both it and the briefer Heinrich Event 2 embedded within it (Fig. 4), clearly had a decisive impact both on demography and technological innovation. At those sites which remained occupied during the cold and aridity which spread over most of Southern Africa, the microlithic Robberg Industrial Complex, characterised by the use of bladelets, ushered in the Later Stone Age. Not surprisingly, the greatest density of sites is found in the more humid areas of the Cape Fold Mountains, the Drakensberg, KwaZulu-Natal and Mpumalanga (Mitchell 2000). Subsistence strategies appear to have changed to encompass more plant foods, with hunting increasingly adapted to the exploitation of grazing antelopes and to the opportunities afforded by local environmental circumstances. The Robberg Complex remained dominant until the early Holocene, although it was succeeded locally by variants of the non-microlithic Oakhurst Industrial Complex after about 12 kyr.

Deglaciation: the interaction between Antarctic and thermohaline forcings

During the deglacial period southern African climates responded most obviously to changes originating in Antarctica and to the fluctuating heat budget of the western Indian Ocean. The latter is the source of much of the precipitation falling over the summer rainfall region. Less than 30% of the subcontinent receives winter rainfall brought by temperate low pressure systems originating within the South Atlantic, although the area of Atlantic influence is extended northwards along the west coast by cold Benguela upwelling. Changes in the intensity of the Benguela system and in the strength of the warm Agulhas current along the south eastern margin of Southern Africa are modulated by fluctuations within the global thermohaline circulation.

The extent of Antarctic sea-ice controls the position of the Antarctic Convergence, which influences the diameter of the circumpolar vortex, while the thermal gradient across Antarctica determines its intensity. These factors also influence the strength of the South Atlantic tropical easterlies (trade winds). Strong easterlies increase Benguela upwelling as well as heat and moisture movement into the northern hemisphere, as observed previously. An enlarged circumpolar vortex extends northwards the influence of the mid-latitude westerlies, simultaneously shifting and enlarging the zone of Hadley cell subsidence over the western hinterland of Southern Africa. Northward movement of the ITCZ diminishes the influence of precessional forcing via the tropical easterlies which convey moisture from the western Indian Ocean. Such changes in latitudinal gradients influence the timing and synchroneity of climatic responses and, to a lesser degree, the extent of zones of contrasting rainfall seasonality.

Sonzogni et al. (1998) have shown that the break in synchroneity between events in the Antarctic and those in the western Indian Ocean (and the North Atlantic) coincides with the Antarctic Convergence. To the north of this responses were linked to fluxes in the thermohaline circulation, with significant lags evident in the distal ocean basins of the southern hemisphere. Thus phases of warming in Antarctica led those in the western Indian Ocean (and the northern hemisphere) by between 2 and 5 kyr. In particular, deglacial warming in Antarctica began around 20–18 cal. kyr, but became evident around 20 °S in Indian Ocean cores only around 15 cal. kyr, in phase with northern hemisphere time series (Sonzogni et al. 1998). By contrast, a clear warming trend is evident in the Cango Cave record after 20 cal. kyr (Lee-Thorp and Talma 2000), showing that responses occurred earliest in southern continental settings. Comparable evidence comes from the pollen record offshore of Namibia spanning the Last Glacial (Shi et al. (2001), Fig. 4). Highest influxes of grassland, desert and semi-desert pollen occurred during the coldest periods in Antarctica, as indicated by the VOSTOK ice-core deuterium record, and do not seem to be associated with pronounced cold stadials in the glacial Bond Cycles typically observed in northern hemisphere climate records (Fig. 4). This suggests that changes in humidity over Western South Africa during the early and mid Last Glacial were likewise governed by latitudinal movements of circum-Antarctic frontal systems and the belt of westerly winds delivering the moisture to the winter rainfall region of South Africa. In addition, a high resolution marine pollen record covering the last 20 kyr strongly supports the hypothesis that continental climate in South-West Africa has varied in concert with Antarctic Last Glacial to Holocene climate evolution and was not intimately connected to the path of deglaciation typical of the northern hemisphere (Shi et al. 1998; 2000).

Despite these obvious differences, Younger Dryas cooling is evident in nearshore and marine records where resolution is sufficient. Core MD 79257 at 20 °S in the western Indian Ocean (Sonzogni et al. 1998) and mollusc stable isotope data for the Atlantic Ocean seaboard (Cohen et al. 1992) contain clear and well-dated Younger Dryas signals. The terrestrial pollen record from Wonderkrater (Scott and Thackeray 1987; Scott et al. 1995; Thackeray 1990) shows marked temperature oscillations near the time of the Younger Dryas, but correlation will depend on the outcome of new dating that is in progress (Holmgren et al. 2003). Thereafter, warming resumed, with the Holocene temperature range being attained in most time series at, or shortly after, 10 kyr. Regional variations in both available moisture and temperature during the Holocene, as documented in several high resolution records, were considerable although generally of smaller amplitude than those of glacial times (see Scott and Lee-Thorp (2003)).

Of note is the fact that the Antarctic Cold Reversal, centred on 13.5 kyr does not appear to be restricted to ice-cores from the Antarctic continent. Contemporaneous cooling is evident in the speleothem record from Cold Air Cave in Northern South Africa (Holmgren et al. 2003). A similar cooling trend is recorded also in the Huascaran ice-core from equatorial South America (Thompson et al. 1995), supporting the notion that the influence of this event was felt as far north as the equator, rather than being limited to southern polar latitudes.

Summary

This review of the late Pleistocene climatic history of Southern Africa serves to highlight a number of issues. The dominant influence of orbital precession in modulating continental humidity during Isotope Stage 6, and in some areas until after the end of Isotope Stage 4, is striking. Overprinting of precession by fluctuations in ice-volume, deep-water formation/thermohaline activity, and atmospheric circulation over and around Antarctica only became important after precessional forcing weakened during Isotope Stage 3. While influences from adjoining areas of the Atlantic and Indian oceans were clearly important in producing differing receipts of rainfall from region to region, events in Antarctica during the past 50 kyr seem to have dominated major responses, especially those governing the timing and distribution of major intervals of aridity. And there is evidence to suggest that the forcing of some of these events was hemispherically asymmetrical, with responses in Southern Africa preceding those in the North Atlantic by 3–4 kyr. A feature of the record of continental desiccation in Southern Africa is its spikiness: episodes of dune remobilisation were relatively short and areally restricted. Glacial-interglacial shifts in major ecotones were relatively large; for example, the Miombo/desert boundary in the western hinterland shifted through about 5 degrees of latitude between Isotope Stage 5e and Isotope Stage 2.

The primary conclusion of this synthesis is that events in Antarctica appear to create important thresholds, the crossing of which drives major climatic changes of regional hemispheric, or perhaps even global, significance. The timing and regional manifestations of these responses must become a major focus of future palaeoclimatic research in Southern Africa. Many more long, well-dated and highly resolved records are needed for this purpose. We venture to suggest that, as the effects of global warming increase, it may be the capacity of global environmental models to reflect responses in Antarctica, and the way in which these will, in turn, impact upon other areas of the globe, that proves decisive in helping to predict some hitherto unexpected effects of rising temperature.

Acknowledgments

Support for this research from the Water Research Commission and the National Research Foundation, South Africa, is acknowledged with gratitude. R.R.S. acknowledges financial support by the German Science Foundation through grant SFB 261 at Bremen University and support for IMAGES Calypso coring off South Africa.

References

Avery D.M. 1982. Micromammals as palaeoenvironmental indicators and an interpretation of the Late Quaternary in the southern Cape Province, South Africa. Ann. S. Afr. Mus. 85: 183–377.

Avery D.M. 1992. The environment of early modern man at Border Cave, South Africa: micromammalian evidence. Palaeogeogr. Palaeoclimatol. Palaeoecol. 91: 71–87.

Barker P.A., Talbot M.R., Street-Perrott F.A., Marret J., Scourse J. and Odada E., this volume. Late Quaternary climatic variability in intertropical Africa. In: Battarbee R.W., Gasse F. and Stickley C.E. (eds), Past Climate Variability through Europe and Africa. Kluwer Academic Publishers, Dordrecht, the Netherlands, pp. 117–138.

Berger A.L. 1978. Long-term variations of daily insolation and Quaternary climate changes. J. Atmos. Sci. 35: 2362–2367.

Bond G.C., Broecker W.S., Johnsen S., McManus J., Labeyrie L., Jouzel J. and Bonani G. 1993. Correlations between climate records from North Atlantic sediments and Greenland ice. Nature 365: 143–147.

Bond G.C., Heinrich H., Broecker W.S., Labeyrie L., McManus J., Andrews J., Houn S., Jantschik R., Clasen S., Simet C., Tedesco K., Klas M., Bonani G. and Ivy S. 1992. Evidence for massive discharges of icebergs in the North Atlantic ocean during the last glacial period. Nature 360: 245–249.

Bond G.C. and Lotti R. 1995. Iceberg discharges in the North Atlantic on millennial time scales during the last glaciation. Science 278: 1257–1266.

Bonnefille R., Roeland J.C. and Guiot J. 1990. Temperatures and rainfall estimates for the past 40 000 years in equatorial Africa. Nature 346: 347–349.

Broecker W.S., Bond G.C., Klas M., Clark E. and McManus J. 1992. Origin of the northern Atlantic's Heinrich events. Clim. Dyn. 6: 265–273.

Butzer K.W., Beaumont P.B. and Vogel J.C. 1978. Lithostratigraphy of Border Cave, Kwazulu, South Africa: a Middle Stone Age sequence beginning c. 195 000 BP. J. Archaeol. Sci. 5: 317–341.

Butzer K.W., Fock G.J., Stuckenrath R. and Zilch A. 1973. Paleohydrology of Late Pleistocene lake Alexandersfontein, Kimberley, South Africa. Nature 243: 328–330.

Cartwright C. and Parkington J.E. 1997. The wood charcoal assemblages from Elands Bay Cave, south western Cape: principles, procedures and preliminary interpretation. S. Afr. Archaeol. Bull. 52: 59–72.

Cohen A.L., Parkington J.E., Brundrit G.B. and van der Merwe N.J. 1992. Holocene marine climate record in mollusc shells from the southwest African coast. J. Quat. Res. 38: 379–385.

Cowling R.M., Cartwright C.R., Parkington J.E. and Allsop J.C. 1999. Fossil wood charcoal assemblages from Elands Bay Cave, South Africa: implications for Late Quaternary vegetation and climates in the winter rainfall fynbos biome. J. Biogeogr. 26: 367–378.

Curry W.B. and Oppo D.W. 1997. Synchronous high-frequency oscillations in tropical sea-surface temperatures and North Atlantic deep water production during the last glacial cycle. Paleoceanography 12: 1–14.

Deacon J. and Lancaster I.N. 1988. Late Quaternary Palaeoenvironments of Southern Africa. Clarendon Press, Oxford, 220 pp.

Gasse F. and Van Campo E. 1998. A 40 000 yr pollen and diatom record from Lake Tritrivakely, Madagascar, in the southern tropics. Quat. Res. 49: 299 311.

Gasse F. and Van Campo E. 2001. Late Quaternary environmental changes from a pollen and diatom record in the southern tropics (Lake Tritrivakely, Madagascar), Palaeogeogr. Palaeoclimatol. Palaeoecol. 167: 287–308.

Grey D.R.C. and Cooke H.J. 1977. Some problems in the Quaternary evolution of the landforms of Northern Botswana. Catena 4: 123–133.

Grün R., Brink J.S., Spooner N.A., Taylor L., Stringer C.B., Franciscus R.G. and Murray A.S. 1996. Direct dating of Florisbad hominid. Nature 382: 500–501.

Heaton T.H.E., Talma A.S. and Vogel J.C. 1986. Dissolved gas palaeotemperatures and ^{18}O variations derived from groundwater near Uitenhage, South Africa. Quat. Res. 25: 79–88.

Henderson G.M. and Slowey N.C. 2000. Evidence from U-Th dating against Northern Hemisphere forcing of the penultimate deglaciation. Nature 404: 61–66.

Holmgren K., Karlén W. and Shaw P.A. 1995. Paleoclimatic significance of the stable isotope composition and petrology of a Late Pleistocene stalagmite from Botswana. Quat. Res. 43: 320–328.

Holmgren K., Lee-Thorp J.A., Cooper G., Lundblad K., Partridge T.C., Scott L., Sithaldeen R., Talma A.S. and Tyson P.D. Persistent millennial-scale variability of the past 25 thousand years in Southern Africa. Quat. Sci. Rev. (in press). Still in press?

Imbrie J., McIntyre A. and Mix A. 1989. Oceanic response to orbital forcing in the late Quaternary: observational and experimental strategies. In: Berger A. (ed.), Climate and Geo-Sciences. Kluver Academic Publishers, Norwell, pp. 121–164.

Lancaster I.N. 1979. Quaternary environments of the arid zone of Southern Africa. Department of Geography and Environmental Studies Occasional Paper, 22.

Lancaster I.N. 1981. Palaeoenvironmental implications of fixed dune systems in Southern Africa. Palaeogeogr. Palaeoclimatol. Palaeoecol. 33: 327–346.

Lancaster I.N. 1988. Development of linear dunes in the South Western Kalahari, Southern Africa. J. Arid Envts. 14: 233–344.

Lancaster I.N. 2000. Eolian deposits. In: Partridge T.C. and Maud R.R. (eds), The Cenozoic of Southern Africa. Oxford University Press, New York, pp. 73–87.

Lee-Thorp J.A. and Beaumont P.B. 1995. Vegetation and seasonality shifts during the Late Quaternary deduced from the $^{13}C/^{12}C$ ratios of grazers at Equus Cave, South Africa. Quat. Res. 43: 426–432.

Lee-Thorp J.A. and Talma A.S. 2000. Stable light isotopes and environments in the Southern African Quaternary and Late Pliocene. In: Partridge T.C. and Maud R.R. (eds), The Cenozoic of Southern Africa. Oxford University Press, New York, pp. 236–251.

Leuscher D.C. and Sirocko F. 2000. The low-latitude monsoon climate during Dansgaard-Oeschger cycles and Heinrich Events, Quat. Sci. Rev. 19: 243–254.

Little M.G., Schneider R.R., Kroon D., Price N.B., Summerhays C. and Segl M. 1997. Trade wind forcing of upwelling seasonality and Heinrich events as a response to sub-Milankovitch climate variability. Paleoceanography 12: 568–576.

Livingstone D.A. 1971. A 22 000-year pollen record from the plateau of Zambia. Limnol. Oceanogr. 16: 349–356.

Mahaney W.C. 1990. Ice on the Equator. Caston, Elliston Bay. 386 pp.

Maley J. 1993. The climatic and vegetational history of the equatorial regions of Africa during the Upper Quaternary. In: Shaw T., Sinclair P., Barsey A. and Okpoko A. (eds), The Archaeology of Africa. Routledge, London, pp. 43–52.

Maud R.R. 2000. Estuarine deposits. In: Partridge T.C. and Maud R.R. (eds), The Cenozoic of Southern Africa. Oxford University Press, New York, pp. 162–172.

Maud R.R., Partridge T.C., Alhonen P., Donner J. and Vogel J.C. 1993. A preliminary assessment of the environmental conditions represented by the Mbazwana diatomite (Zululand coast). Conference Abstract, 11th Conference of S. African Society for Quaternary Research, Kimberley, July 1993.

Mayewski P.A., Twickler M.S., Whitlow S.T., Meeker L.D., Yang Q., Thomas J., Kreutz K., Grootes P.M., Steig E.J., Waddington E.D., Saltzman E.S., Whung P.-Y. and Taylor K.C. 1996. Climate change during the last deglaciation in Antarctica. Science 272: 1636–1638.

McIntyre A., Ruddiman W.F., Karlin K. and Mix A.C. 1989. Surface water response of the equatorial Atlantic Ocean to orbital forcing. Paleoceanography 4: 19–55.

Metcalfe S.E. 1999. Diatoms from the Pretoria Saltpan — a record of lake evolution and environmental change. In: Partridge T.C. (ed.), Tswaing: Investigations into the Origin, Age and Palaeo-Environments of the Pretoria Saltpan, Mem. 85, Council for Geoscience, Pretoria, pp. 172–192.

Mitchell P.J. 2000. The Quaternary archaeological record in Southern Africa. In: Partridge T.C. and Maud R.R. (eds), The Cenozoic of Southern Africa. Oxford University Press, New York, pp. 357–370.

Parkington J.E., Cartwright C., Cowling R.M., Baxter A. and Meadows M.E. 2000. Palaeovegetation at the Last Glacial Maximum in the western Cape, South Africa: wood, charcoal and pollen evidence from Elands Bay Cave. S. Afr. J. Sci. 96: 543–546.

Partridge T.C. 1995. Palaeoclimates of the arid and semi-arid zones of Southern Africa during the last climatic cycle. Mem. Geol. Soc. France, 167: 73–83.

Partridge T.C. 2002. Were Heinrich Events forced from the southern hemisphere? S. Afr. J. Sci. 98: 43–46.

Partridge T.C., Avery D.M., Botha G.A., Brink J.S., Deacon J., Herbert R.S., Maud R.R., Scott L., Talma A.S. and Vogel J.C. 1990. Late Pleistocene and Holocene climatic change in Southern Africa. S. Afr. J. Sci. 86: 302–306.

Partridge T.C., deMenocal P.B., Lorentz S.A., Paiker M.J. and Vogel J.C. 1997. Orbital forcing of climate over South Africa: a 200 000-year rainfall record from the Pretoria Saltpan. Quat. Sci. Rev. 16: 1–9.

Partridge T.C., Metcalfe S.E. and Scott L. 1999a. Conclusions and implications for a model of regional palaeoclimates during the last two glacial cycles. In: Partridge T.C. (ed.), Tswaing: Investigations into the Origin, Age and Palaeoenvironments of the Pretoria Saltpan. Mem. 85, South African Council for Geoscience, pp. 193–198.

Partridge T.C., Scott L. and Hamilton J.E. 1999b. Synthetic reconstructions of southern African environments during the Last Glacial Maximum (21–18 kyr) and the Holocene Altithermal (8–6 kyr). Quat. Int. 57/58: 207–214.

Partridge T.C. and Scott L. 2000. Lakes and pans. In: Partridge T.C. and Maud R.R. (eds), The Cenozoic of Southern Africa. Oxford University Press, New York, pp. 145–161.

Prell W.L. and Kutzbach J.E. 1987. Monsoon variability over the past 150,000 years. J. Geophys. Res. 92: 8411–8425.

Rosqvist G. 1990. Quaternary glaciations in Africa. Quat. Sci. Rev. 9: 281–297.

Ruddiman W.F. and McIntyre A. 1981. Oceanic mechanisms for amplification of the 23,000-year ice-volume cycle. Science 212: 617–627.

Runge J. and Runge F. 1995. Late Quaternary palaeoenvironmental conditions in Eastern Zaire (Kivu) deduced from remote sensing morpho-pedological and sedimentological studies (Phytoliths, Pollen, C-14 data). Proc. 2nd Internat. Palynol. Conf. Tervuren, Belgium, March 1995.

Scott L. 1982. A Late Quaternary pollen record from the Transvaal Bushveld. Quat. Res. 17: 339–370.

Scott L. 1987. Pollen analysis of Hyena coprolites from Equus Cave, Taung, Southern Kalahari (South Africa). Quat. Res. 28: 144–156.

Scott L. 1994. Palynology of late Pleistocene hyrax middens, south-western Cape Province, South Africa: a preliminary report, Hist. Biol. 9: 71–81.

Scott L. 1999a. Palynological analysis of the Pretoria Saltpan (Tswaing Crater) sediments and veg-
      etation history in the bushveld savanna biome, South Africa. In: Partridge T.C. (ed.), Tswaing:
      Investigations into the Origin, Age and Palaeoenvironments of the Pretoria Saltpan, Mem. 85,
      South African Council for Geoscience: 143–166.

Scott L. 1999b. The vegetation history and climate in the Savanna Biome, South Africa, since
      190,000 Ka: a comparison of pollen data from the Tswaing Crater (the Pretoria Saltpan) and
      Wonderkrater. Quat. Internat. 57-58: 215–223.

Scott L. 2002a. Microscopic charcoal in sediments: Quaternary fire history of the grassland and
      savanna regions in South Africa, J. Quat. Sci. 17: 77–86.

Scott L. 2002b. Grassland development under glacial and interglacial conditions in Southern Africa:
      review of pollen, phytolith and isotope evidence. Palaeogeogr. Palaeoclimatol. Palaeoecol. 177:
      47–57.

Scott L. and Lee-Thorp J.A., this volume. Holocene climatic trends and rhythms in Southern Africa.
      In: Battarbee R.W., Gasse F. and Stickley C.E. (eds), Past Climate Variability through Europe
      and Africa. Kluwer Academic Publishers, Dordrecht, the Netherlands, pp. 69–91.

Scott L., Steenkamp M. and Beaumont P.B. 1995. Palaeoenvironmental conditions in South Africa
      at the Pleistocene-Holocene transition. Quat. Sci. Rev. 14: 937–994.

Scott L. and Thackeray J.F. 1987. Multivariate analysis of late Pleistocene and Holocene pollen
      spectra from Wonderkrater, Transvaal, South Africa. S. Afr. J. Sci. 83: 93–98.

Scott L. and Vogel J.C. 2000. Evidence for environmental conditions during the last 20 000 years in
      Southern African from ^{13}C in fossil hyrax dung. Glob. Planet. Chan. 26: 207–215.

Shaw P.A., Stokes S., Thomas D.S.G., Davies F.B.M. and Holmgren K. 1997. Palaeoecology and age
      of a Quaternary high lake level in the Makgadikgadi basin of the Middle Kalahari, Botswana. S.
      Afr. J. Sci. 93: 273–276.

Shi N. and Dupont L.M. 1997. Vegetation and climatic history of Southwest Africa: a marine
      palynological record of the last 300,000 years. Veg. Hist. Archaeobot. 6: 117–131.

Shi N., Dupont L., Beug H.J. and Schneider R.R. 1998. Vegetation and climate changes during the
      last 21,000 yr in SW Africa — Evidence from a high resolution marine palynological record. Veg.
      Hist. Archaeobot. 7: 127–140.

Shi N., Dupont L., Beug H.J. and Schneider R.R. 2000. Correlation between vegetation in
      Southwestern Africa and oceanic upwelling in the past 21,000 years. Quat. Res. 54: 72–80.

Shi N., Schneider R.R., Beug H.J. and Dupont L.M. 2001. Southeast trade wind variations during the
      last 135 kyr: evidence from pollen spectra in eastern South Atlantic sediments. Ear. Planet. Sci.
      Lett. 5783: 1–11.

Sonzogni C., Bard E. and Rostek E. 1998. Tropical sea-surface temperatures during the Last Glacial
      period: a view based on alkenones in Indian Ocean sediments. Quat. Sci. Rev. 17: 1185–1201.

Stager J.C. 1988. Environmental changes at Lake Cheshi, Zambia, since 40,000 years BP. Quat. Res.
      29: 54–65.

Stager J.C., Reintha P.N. and Livingstone D.A. 1986. A 25,000-year history for Lake Victoria, East
      Africa, and some comment on its significance for the evolution of cichlid fishes. Freshwat. Biol.
      16: 15–19.

Stocker T.F. 2000. Past and future reorganizations in the climate system. Quat. Sci. Rev. 19: 301–319.

Stokes S., Haynes G., Thomas D.S.G., Higginson M. and Malifa M. 1998. Punctuated aridity in
      Southern Africa during the last glacial cycle: the chronology of linear dune construction in the
      Northeastern Kalahari. Palaeogeogr. Palaeoclimatol. Palaeoecol. 137: 305–322.

Stokes S., Thomas D.S.G. and Shaw P.A. 1997a. New chronological evidence for the nature and
      timing of linear dune development in the Southwest Kalahari Desert. Geomorphology 20: 21–93.

Stokes S., Thomas D.S.G. and Washington R. 1997b. Multiple episodes of aridity in Southern Africa
      since the last interglacial. Nature 261: 385–390.

Stokes S., Washington R. and Preston A. 1999. Late Quaternary evolution of the central Southern Kalahari: environmental responses to changing climatic conditions. In: Andrews P. and Banham P. (eds), Late Cenozoic Environments and Hominid Evolution: a Tribute to Bill Bishop. The Geological Society, London, pp. 247–268.

Stute M. and Talma A.S. 1998. Glacial temperatures and moisture transport regimes reconstructed from noble gases and $\delta^{18}O$, Stampriet aquifer, Namibia. In: Isotope Techniques in the Study of Environmental Change. Proc. Symp. Internat. Atomic Energy Agency, Vienna, pp. 307–328.

Stuut J.-B.W., Prins M.A., Schneider R.R., Weltje G.J., Jansen J.H.F. and Postma G. 2002. A 300 kyr record of aridity and wind strength in Southwestern Africa: evidence from grain-size distributions in sediments on Walvis Ridge, SE Atlantic. Mar. Geol. 180: 221–233.

Talma A.S. and Vogel J.C. 1992. Late Quaternary palaeotemperatures derived from a speleothem from Cango Caves, Cape Province, South Africa. Quat. Res. 37: 203–213.

Thackeray J.F. 1987. Late Quaternary environmental changes inferred from small mammalian fauna, Southern Africa. Clim. Change 10: 285–305.

Thackeray J.F. 1990. Temperature indices from late Quaternary sequences in South Africa: comparisons with the Vostok core, S. Afr. Geog. J. 72: 47–49.

Thackeray J.F. and Lee-Thorp J.A. 1992. Isotopic analysis of equid teeth from Wonderwerk Cave, northern Cape Province, South Africa. Palaeogeogr. Palaeoclimatol. Palaeoecol. 99: 141–150.

Thomas D.S.G., O'Connor P.W., Bateman M.D., Shaw P.A., Stokes S. and Nash D.J. 2000. Dune activity as a record of late Quaternary aridity in the Northern Kalahari: new evidence from Northern Namibia interpreted in the context of regional arid and humid chronologies. Palaeogeogr. Palaeoclimatol. Palaeoecol. 156: 243–259.

Thomas D.S.G. and Shaw P.A. 1991. The Kalahari Environment. Cambridge University Press, Cambridge. 284 pp.

Thompson L.G., Mosley-Thompson E., Davis M.E., Lin P.-N., Henderson K.A., Cole-Dai J., Bolzan J.F. and Liu K.B. 1995. Late glacial stage and Holocene tropical ice core records from Huascaran, Peru. Science 269: 46–50.

Tyson P.D. 1986. Climatic Change and Variability in Southern Africa. Oxford University Press, Cape Town. 256 pp.

Tyson P.D. and Partridge T.C. 2000. Evolution of Cenozoic climates. In: Partridge T.C. and Maud R.R. (eds), The Cenozoic of Southern Africa. Oxford University Press, New York, pp. 371–387.

Van Campo E., Duplessy J.C., Prell W.L., Barratt N. and Sabatier R. 1990. Comparison of terrestrial and marine temperature estimates for the past 135 kyr off Southeast Africa: a test for GCM simulations of palaeoclimate. Nature. 348: 209–212.

Vincens A. 1993. Nouvelle sequence pollinique du Lac Tanganyika: 30,000 ans d'histoire botanique et climatique du Bassin Nord. Rev. Palaeobot. Palynol. 78: 381–394.

Vogel J.C. 1987. Chronological framework for palaeoclimatic events in the Namib. NPRL Research Report CFIS 145, Council for Scientific and Industrial Research, Pretoria.

Vogel J.C., Fuls A. and Ellis R.P. 1978. The geographical distribution of Kranz grasses in South Africa. S. Afr. J. Sci. 74: 209–215.

Zonneveld K.A.F., Ganssen G., Troelstra S., Versteegh G.J.M. and Visscher H. 1997. Mechanisms forcing abrupt fluctuations of the Indian Ocean summer monsoon during the last deglaciation. Quat. Sci. Rev. 16: 187–201.

5. HOLOCENE CLIMATIC TRENDS AND RHYTHMS IN SOUTHERN AFRICA

LOUIS SCOTT (Scottl.sci@mail.uovs.ac.za)
Department of Plant Sciences
University of the Free State
P.O. Box 339, Bloemfontein 9300
South Africa

JULIA A. LEE-THORP (jlt@science.uct.ac.za)
Department of Archaeology and Quaternary Research Centre
University of Cape Town
Private Bag Rondebosch 7701
South Africa

Keywords: Aridity, Rainfall seasonality, Grassland savanna, Isotopes, Pollen, Human response, Holocene, Southern Africa

Introduction

Southern Africa's unique mid-latitude oceanic position invites broad comparisons with palaeoclimatic records across the PEP III transect including long-distance thermohaline circulation teleconnections. Atmospheric and oceanic circulation systems around Southern Africa (Fig. 1) are linked (Lutjeharms et al. 2001). They interact to influence distribution of biomes including prominence of C_4 and C_3 grasses in the summer-rain and winter rain regions respectively (Vogel et al. 1978; Cowling et al. 1997) (Fig. 2). Climate is dominated by two systems, the westerlies and easterlies (Fig. 1) and shifts in these systems undoubtedly affected the climate history during the Holocene. The generally moderating effect of the oceans, in particular the warm Agulhas western boundary current on the east coast (Lutjeharms et al. 2001), and the semi-arid nature of the region suggests that moisture rather than temperature changes is the more important climate parameter, at least on the Holocene time scale. Furthermore, the cold Benguela upwelling zone and its associated atmospheric circulation system in the South Atlantic is strongly linked to coastal aridity on the west coast.

Asymmetry in the continental configuration of Africa introduces a number of contrasts with the north of the continent, in climate patterns, biome distributions and orbital parameters. In contrast to equivalent regions in Northern Africa that border Eurasia, Southern Africa lacks the influence of other continents. Symmetry of biomes on both sides of the equator in Sub-Saharan Africa is strong near the equator but weakens towards the northern

R. W. Battarbee et al. (eds) 2004. *Past Climate Variability through Europe and Africa.*
Springer, Dordrecht, The Netherlands.

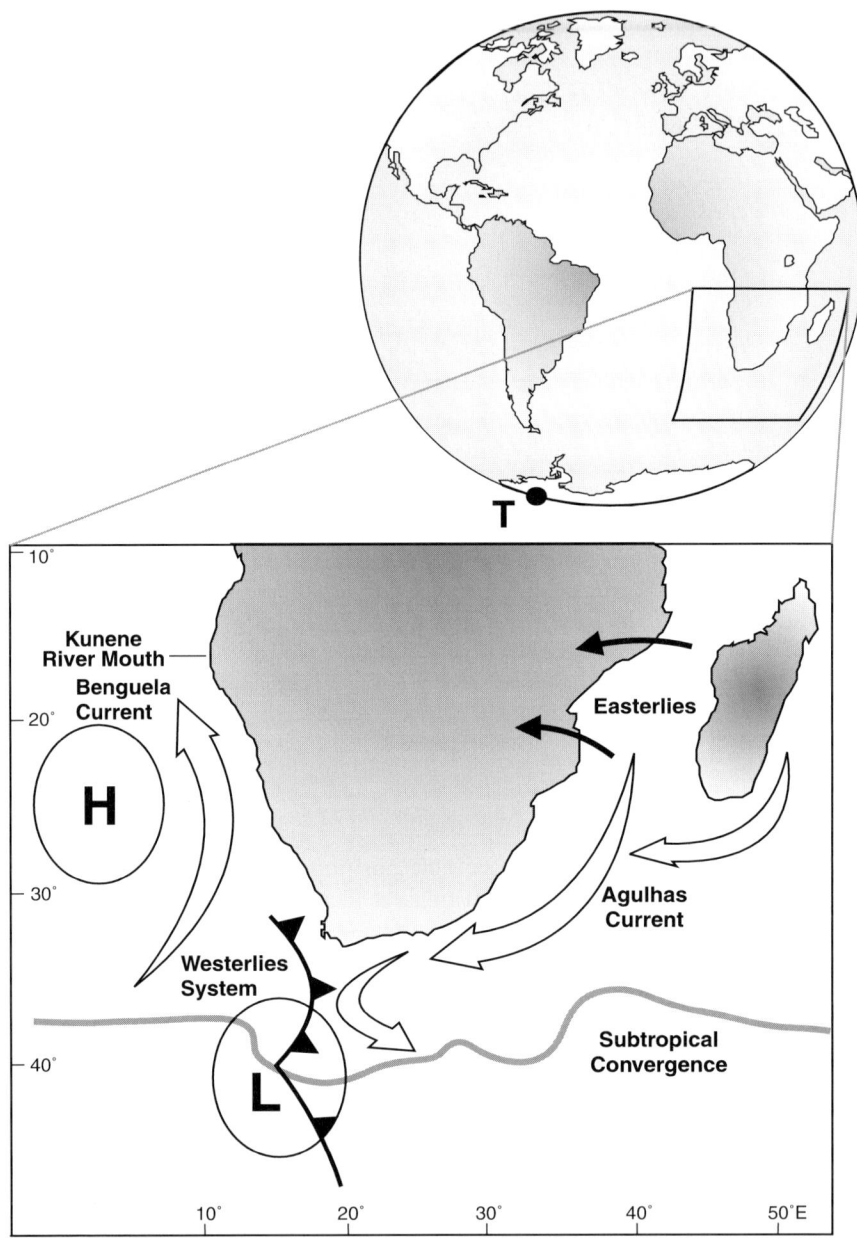

Figure 1. Map showing the global oceanic position of Southern Africa. The climate of the summer rain region is under influence of easterly winds and the Agulhas current, and the western winter rain and arid regions under the influence of westerlies frontal systems and the Benguela upwelling system respectively. L and H – areas of low and high pressure, T = Taylor Dome.

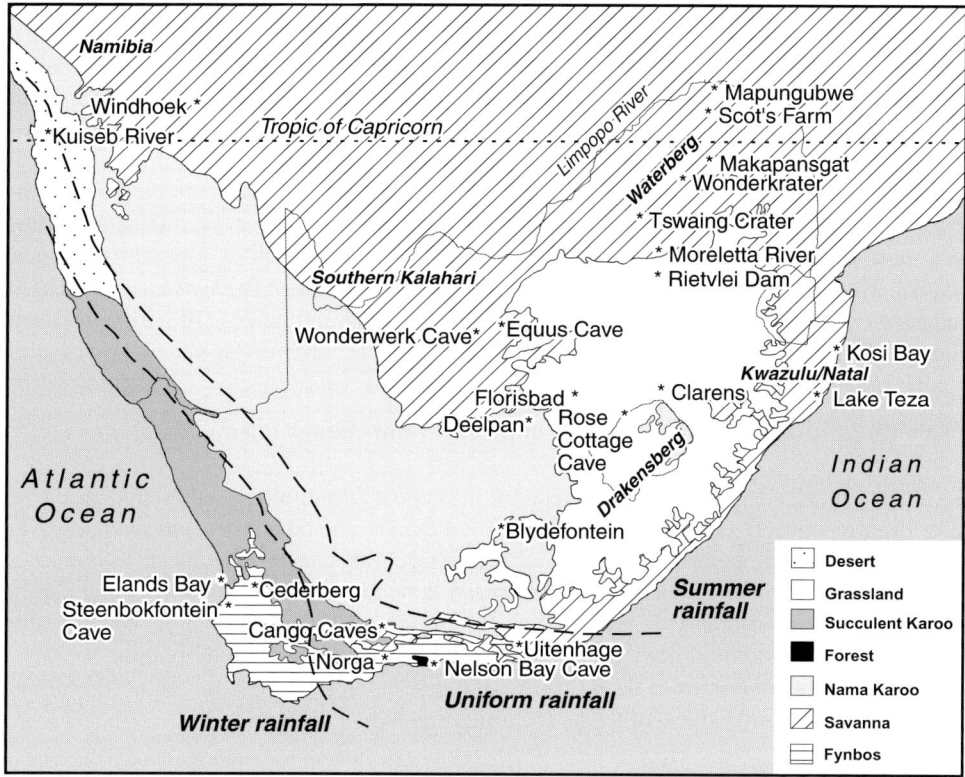

Figure 2. Map showing the approximate positions of major biomes in Southern Africa and the location of sites mentioned in the text. The dashed lines indicate the intermediate zone between summer and winter rainfall regions.

and southern extremes of the continent. The northern parts of Southern Africa with sub-tropical and tropical grasslands and savannas have equivalents elsewhere in Africa but the Southern African Desert, Succulent Karoo, Nama Karoo, and Fynbos biomes (Fig. 2) are unequaled (Cowling et al. 1997). The southern tip of Africa, which receives winter rain, has a Mediterranean type of environment (the Fynbos Biome) but it has Gondwanaland affinities unlike its northern African counterpart, which shows a Eurasian structure.

Biome and climate patterns on both sides of the equator probably fluctuated asymmetrically in the past following orbital forcing with long-term moisture development in Southern Africa out of phase with the northern tropical parts of the continent. (Street-Perrott and Perrott 1993). Lake Malawi is the northernmost site that conforms to the southern pattern (Finney et al. 1996). The 8.2 kyr (1000 years) drought event that occurred in Northern Africa near the peak of the precession cycle (Gasse 2000) is an example of a rapid event for which inter-hemispheric connections should be investigated further in Southern Africa in order to understand possible connections and anomalies in the broad pattern.

Continuous lake sequences are unfortunately very rare in Southern Africa; therefore a diverse range of archives has been used in palaeoclimate reconstruction. Several factors inhibit accumulation of good-quality, high-resolution records, and this is a particular disadvantage for determination of Holocene climates where greater resolution is required. The underlying uplifted geological structure and ancient erosional regime of the interior and a lack of recent volcanics do not favour formation of basins like small craters and marshes. Hence, organic-rich accumulations are rare; furthermore evaporation and desiccation in the region have lowered the quality of deposits where they do occur. Continental data are spatially patchy and obtained largely from relatively coarsely dated, and often pulsed, pollen sequences and faunal abundance, isotope and other biological data from archaeological sites, and more rarely from aquifers. Some continuous stalagmite records are capable of providing information on long-term trends and high frequency variability with good chronological control. They can address key climatological questions, such as if the Little Ice Age (LIA) and Medieval Warm Period (MWP) were manifested on the southern tip of Africa, and how climate changes affected human settlements and biota.

Evidence for millennial scale climate and vegetation change during the Holocene

By pairing more abundant but lower resolution records from pollen, sediments, etc., with rarer high-resolution, chronologically-controlled data from stalagmites and well-resolved archaeological sites, palaeoenvironmental reconstructions can be optimised. Evidence of Holocene environmental conditions is presented here according to time-slices that can be broadly correlated over a wide area of the sub-continent although they show considerable internal variability and do not by any means represent uniform climatic phases. As far as possible the information from each time-slice is presented starting from the northernmost region and ending with the south. Site localities are shown in Figure 2.

11–7.5 kyr

In general, relative aridity is indicated for this period, excepting in the southwestern winter-rainfall region, which was relatively mesic.

A pollen sequence from Atlantic marine sediments offshore of the Kunene River Mouth (Fig. 1) suggests relative aridity in the area bordering Namibia and Angola before 7.5 kyr (Shi et al. 1998). Geomorphological data from Northern Namibia also suggest dry conditions although not as dry as the preceding terminal Pleistocene (Eitel et al. 2001). Currently submerged speleothems from Aikab, representing lower water tables than at present, may suggest aridity ca. 10 kyr (Brook et al. 1999).

In the central region south of the Limpopo River (Fig. 2), pollen sequences at Won-derkrater (Fig. 3), Rietvlei, Tate Vondo, and Tswaing Crater show an extension of Kalahari Thornveld elements between ca. 10 and 7.3 kyr, indicating dryness together with rising temperatures (Scott 1993; 1999). In the Southern Kalahari, micromammal data at Wonder-werk Cave (Avery 1981) suggest arid, open conditions. Carbon isotope data from grazing herbivores at Equus and Wonderwerk Caves show that C_4 grasses remained dominant in this region throughout the Holocene (Lee-Thorp and Beaumont 1995), indicating no changes in rainfall seasonality.

In the eastern highlands of the Free State and Lesotho, rapid, large temperature shifts of several degrees Celsius magnitude are indicated by carbon isotope data from large-bodied grazers in the well-dated Rose Cottage Cave sequence, reflecting changes in the proportions

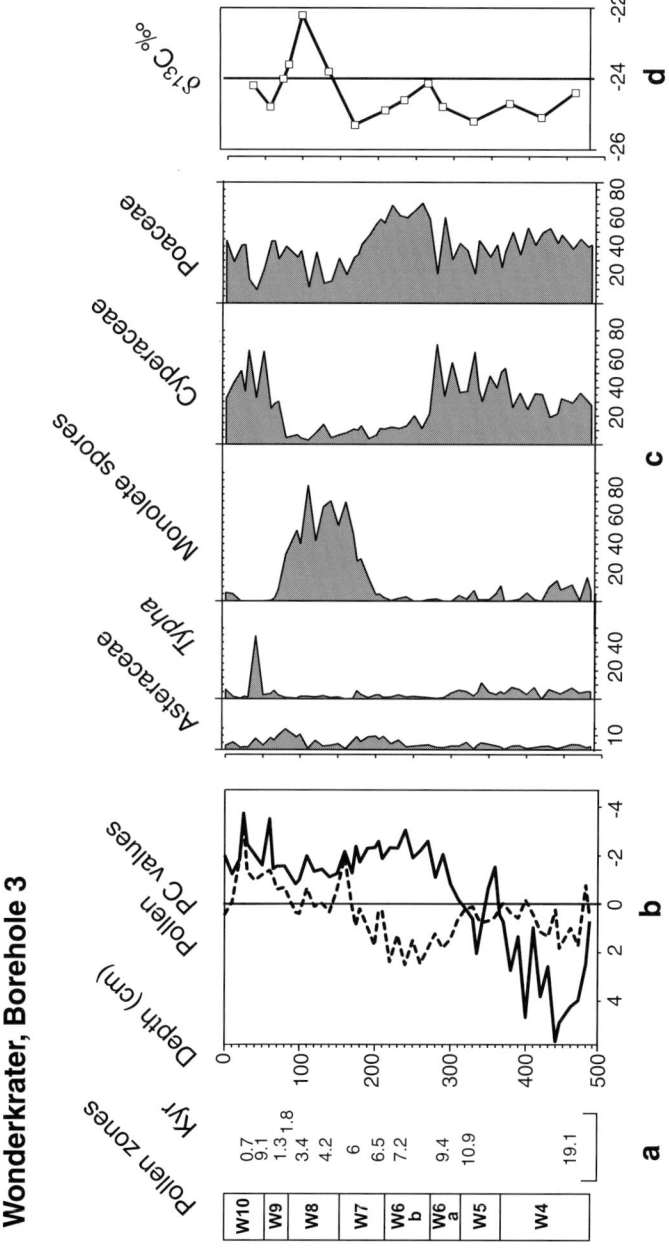

Figure 3. (a) Selected calibrations of radiocarbon dates (Scott 1982a; Scott et al. 2003), (b) temperature (solid) and moisture (dashed) indications for the Wonderkrater Borehole 3 based on principal components analysis of indicator pollen types (excluding taxa in c) (Scott 1999), (c) percentages of prominent pollen types (Scott 1982a), and (d) $\delta^{13}C$ values of the organics in the spring deposits.

of C_3 and C_4 grasses (Smith 1997; Smith et al. 2002). In this highland area the proportion of C_3 grasses increases with altitude and decreasing temperatures, while that of C_4 grasses increases with decreasing altitude and warmth. Cool conditions with C_3 grasses pertained at 10, 9.5 and 8 kyr, interspersed with warmer conditions (more C_4 grass) at 9.9, 9, and finally after 7.6 kyr (Fig. 4a). Indices from charcoal abundance data of species adapted to different temperatures, from the same sites (Esterhuysen et al. 1999), also show rapid early Holocene variation but suggest that associations between temperature and moisture conditions were not fixed. Gas solubility data in the Uitenhage aquifer (Fig. 4b) shows broadly similar temperature shifts (Heaton et al. 1986) but the record lacks the resolution to resolve even millennial scale shifts at this time. Dominance of Asteraceae pollen at Blydefontein and prominent *Euryops* (Asteraceae) charcoal in the Southern Drakensberg indicate aridity from 11 to 8.7 kyr and again at ~7.1 kyr (Tusenius 1989; Bousman 1991).

Vegetation typical of the current arid Succulent Karoo biome (Fig. 2) developed at Eksteenfontein between ca. 11 and 9.4 kyr (Scott et al. 1995). Along the south coast relatively high rainfall occurring in both winter and summer currently promotes coastal forests or thicket; both C_3 and C_4 grasses are present where grass occurs (Cowling 1983). Relatively more fynbos and less forest at ~8.7 to 7.7 kyr is suggested by pollen at the currently forested site of Groenvlei (Martin 1968). At the coastal archaeological site of Nelsons Bay Cave, however, consistently low nitrogen isotope ratios ($^{15}N/^{14}N$) from grazing fauna through the Holocene, suggest that little change occurred in rainfall receipts in the region (Sealy 1997).

In contrast, the southwestern winter rainfall region experienced relatively mesic conditions during the early Holocene. Archaeological evidence shows a higher abundance of mesic woody elements in charcoal assemblages until ca. 8.5 kyr (Cartwright and Parkington 1997; Cowling et al. 1999) and lower $\delta^{18}O$ in ostrich eggshell at coastal Elands Bay Cave (Lee-Thorp and Talma 2000). Pollen in a peat deposit in the adjacent Cederberg mountain range, however, shows little change during this period (Meadows and Sugden 1991). A hyrax dung pollen sequence slightly further northwest in the Cederberg (Pakhuis Pass) closer to the Nama Karoo boundary, suggests slightly more mesic conditions with prominent *Dodonaea* and other woody elements (Scott 1994).

7.5–5.1 kyr

Numerous proxies show that the temperature optimum beginning shortly before 7.5 kyr (Scott 1982a; 1993; Lee-Thorp and Talma 2000) also showed moisture improvement, at least in the summer rainfall region.

On the western side of the subcontinent, desert pollen declined in offshore (windblown) sediments near the Kunene River Mouth at 7 kyr in favour of more mesic afro-montane elements (Shi et al. 1998). In Central Namibia near Windhoek, a warmer, moister phase with grass and woodland pollen is indicated between 7.9 and 6.4 kyr (Scott et al. 1991). Dates for submerged speleothems in Northern Namibia, however, indicate low water tables (Brook et al. 1999), seeming to contradict the pollen indicators.

In the savanna region of the northern interior pollen data from Wonderkrater (Fig. 3) suggest generally rising temperatures between ~9 and 6 kyr, with a delayed moisture increase ~6.5 kyr (Scott 1982a). Higher humate levels (recorded in the grey scale) and lower

a

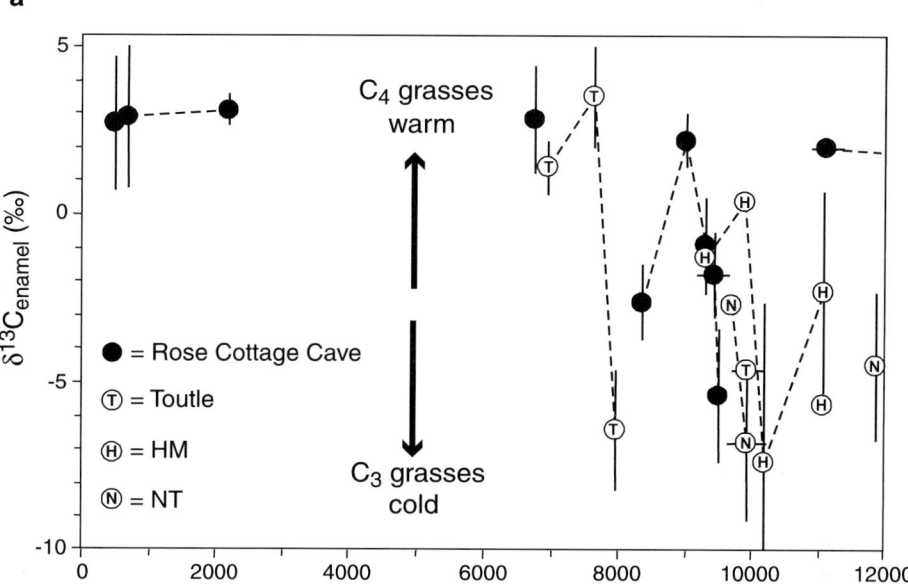

b

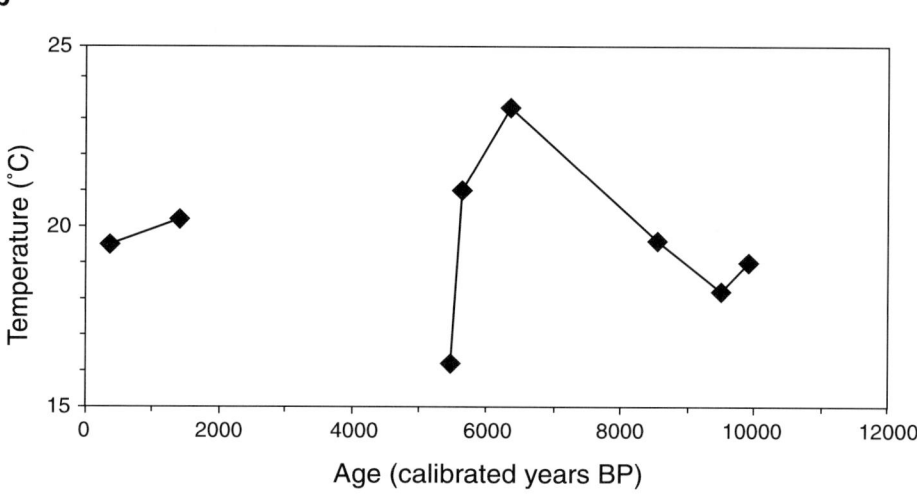

Figure 4. (a) Proportions of C_4 grasses indicated by $\delta^{13}C$ in grazer tooth enamel from Rose Cottage Cave and three nearby sites in the eastern grassland that provide indications of temperature shifts (Smith et al. 2002). (b) Temperature curve derived from dissolved gases for the Uitenhage Aquifer, shown with the $\delta^{18}O$ series (based on data from Heaton et al. (1986)).

$\delta^{13}C$ in the nearby Makapansgat Valley T7 stalagmite (Fig. 5) indicate wetter, warmer conditions with enhanced C_3 vegetation density (woody, shrubby and herbal forms) at 6.4 kyr to 5.1 kyr (Lee-Thorp et al. 2001). In the Makapansgat stalagmites $\delta^{18}O$ values are interpreted as reflecting primarily isotopic composition of rainfall, and hence, hydrological conditions (Holmgren et al. 1999; Lee-Thorp et al. 2001). Following this interpretation, higher $\delta^{18}O$ values in this period (6.4–5.1 kyr) indicate moister, warmer conditions. The highly resolved T7 sequence, recently augmented by similar results for another stalagmite from the same site (Holmgren et al. 2003), shows that climate variability occurred on short time-scales including interruptions of the "optimum", which are blurred in all of the less resolved records. Warm conditions were continuous until 5.7 kyr, where after 2 reversals occurred before 5.1 kyr (Lee-Thorp et al. 2001).

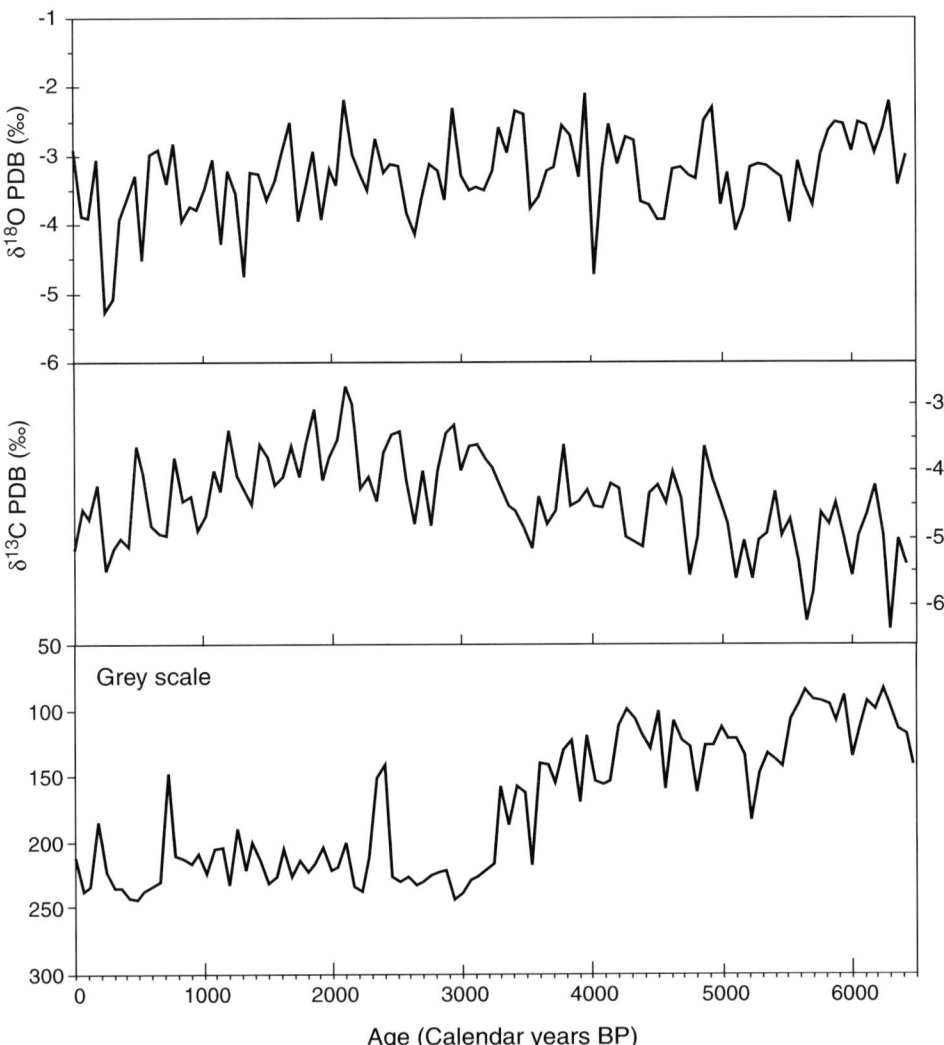

Figure 5. $\delta^{13}C$, $\delta^{18}O$ and grey scale series for the T7 stalagmite, Cold Air Cave in the Makapansgat Valley, plotted against age after interpolation to an average age interval of 12 years (from Lee-Thorp et al. (2001)).

In the highveld grassland to the south, high $\delta^{13}C$ in grazers at Rose Cottage Cave and other sites show that C_4 grass was as abundant as today by ~7.6 kyr. Warmer temperatures supporting C_4 grasses prevailed from this period onwards, or at least for those periods represented in the archaeological sites (Smith 1997; Smith et al. 2002). Earlier interpretations of sediments at Florisbad and surroundings (Butzer 1984; Visser and Joubert 1991) suggested semi-arid conditions despite higher water levels in pans. Recent pollen data suggests that high seasonal evaporation rates may account for the apparent contradiction, and that wetness increased after 7.2 kyr at Florisbad (Scott and Nyakale 2002). As today, frost on the 'highveld' continued to prevent the spread of trees at Florisbad, but at ~7.2 kyr a brief southward expansion of savanna, and hence less frost, is indicated in the Rietvlei Dam pollen sequence (Scott and Vogel 1983). Savanna expansion is also suggested by limited pollen evidence from pan sediment sites associated with stone tools belonging to the Oakhurst Complex (Horowitz et al. 1978) nearly 100 Km to the west of Florisbad nearer the boundary of the Savanna Biome.

From the eastern limits of the Fynbos Biome, the dissolved gas temperature record from the Uitenhage aquifer (Heaton et al. 1986) shows a temperature increase of ~4 °C at 6.3 kyr, above the 8.5 kyr temperature determination, followed by marked cooling at ~5.5 kyr (Fig. 4b). This pattern is in broad agreement with the higher resolution Makapansgat and Cango Cave stalagmite $\delta^{18}O$ records (Fig. 5). Although it is not possible to determine the amplitude of temperature depression from the Makapansgat stalagmites, for Cango a mean annual temperature depression of ~1.5 °C occurs at 5 kyr (Talma and Vogel 1992; Lee-Thorp and Talma 2000). On the coast, $\delta^{18}O$ in marine molluscs from Nelson's Bay Cave indicate warmer sea surface temperatures (SST) at ca. 6.8 kyr (Cohen and Tyson 1995). Higher SST was interpreted to be a result of intensified westerlies and their associated frontal disturbances by Cohen and Tyson (1995), but other influences, for instance slightly higher Agulhas current influence or temperatures could also be considered. The $\delta^{13}C$ record from large grazing herbivores indicates a short episode with fewer C_4 grasses at ~5.5 kyr (Sealy 1997), suggesting greater influence of winter rainfall or possibly also lower temperatures, which is apparently not reflected in the marine mollusc $\delta^{18}O$ record. Unfortunately, dating resolution in the archaeological site and especially the aquifer is not high enough to determine whether the 5.5 kyr and 5 kyr "events" are the same or not.

Numerous coastal records show that sea-levels were about 2 m higher in the middle Holocene. These levels are recorded at Groenvlei (Martin 1968), the Verlorenvlei and along the western Cape coast near Elands Bay (Fig. 2) (Miller et al. 1993; Meadows et al. 1996), and at Lake Teza and surrounding coast line, before 5.6 kyr (Ramsay 1995; Scott and Steenkamp 1996). The change in pollen trapping between marine and freshwater marshy situations complicates estimates of moisture conditions during this phase. A greater abundance of grasses occurred until shortly after 5 kyr at Verlorenvlei after which relatively dry conditions may be indicated by Asteraceae pollen (Meadows et al. 1996).

A climate sequence for the Western Cape winter rainfall region is difficult to establish owing to a lack of indicators during this period, at least partly due to the rarity of human occupation sites.

5.1–2 kyr

Various proxy data sequences show that an overall aridification trend began across the subcontinent soon after 5 kyr. There are, however, some contradictions in these records, some of which may simply be due to poor chronological resolution that confuses different episodes, others too subtle, imperfectly understood, differences in the perspectives offered by different proxies. For instance, $\delta^{13}C$ in the geographically widely separated Makapansgat and Cango stalagmites (Fig. 2) show decided increases in C$_4$ grass after 4 kyr, reaching peak abundances at 2–2.4 kyr (Figs. 5 and 6). Pollen data, however, suggest increasing abundance of Asteraceae (Fig. 3), a C$_3$ shrub characteristic of summer rain areas incorporating also some winter rain, as typically occurs in the dry Karoo region today.

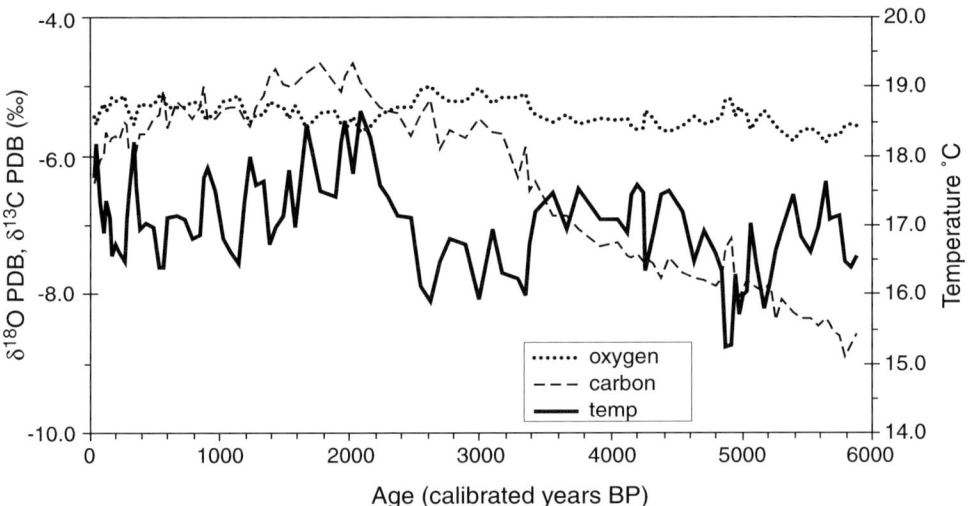

Figure 6. $\delta^{13}C$, $\delta^{18}O$ and the derived temperature sequence of the Cango Caves stalagmite (from Talma and Vogel (1992), Lee-Thorp and Talma (2000)).

A variable, modestly moist/warm period is indicated by the Makapansgat stalagmite $\delta^{18}O$ and grey series (showing relatively high humate levels) for the interior summer rainfall region from 4.3 to 3.2 kyr (Fig. 5) (Lee-Thorp et al. 2001). Termination of this event corresponds with a vegetation change evident in the $\delta^{13}C$ and grey index; lower $\delta^{18}O$ values thereafter suggest increasing aridity. Two relatively modest $\delta^{18}O$ minima corresponding to drying and/or cooling are clustered at 3.1 and 2.6 kyr. The Cango Cave $\delta^{18}O$ sequence (Talma and Vogel 1992) indicates variably lower temperatures between 3.3 and 2.5 kyr, while principal component analysis of selected pollen at Wonderkrater (excluding Asteraceae), and enhanced Ericaceae pollen in the Blydefontein Shelter deposits ~3.1 kyr also suggest cooler conditions (Scott and Thackeray 1987; Scott 1999; C.B. Bousman and L. Scott, unpublished). Two brief episodes of higher $\delta^{18}O$ in the Makapansgat stalagmite, suggesting moister conditions, at 2.1 and 2.9 kyr correspond to two distinct peaks of swamp forest growth at Scot's Farm on the northern foot of the Soutpansberg (Scott 1982b).

The values of $\delta^{13}C$ in the Makapansgat stalagmites peak at ~2.2 kyr, indicating a stepwise trend towards higher C$_4$ grass abundance (Fig. 5). This matches the pattern observed in the Cango stalagmite in the southern Karoo/fynbos transition (Fig. 6) (Talma and Vogel 1992), suggesting those conditions supported more C$_4$ grasses. In contrast, the Wonderkrater spring pollen sequence (Fig. 3) shows a strong local representation of fern spores after 4.1 kyr, and little Poaceae (grass) pollen. Asteraceae remained relatively prominent but since no other indications of aridity are observed, their presence could reflect cooler conditions similar to adjacent high-lying surroundings like the Waterberg (Scott 1982a). Development of cooler conditions is supported by independent PCA that exclude Asteraceae pollen (Scott and Thackeray 1987, Scott 1999) (Fig. 3). Sediment δ^{13}C values at Wonderkrater suggest that, apart from locally abundant C$_3$ *Phragmites* (Vogel et al. 1978) the remaining grasses must have included C$_4$ types (Fig. 3d).

Pollen data from Florisbad suggests relatively mesic, or grassy, conditions in the central highlands after 4.9 kyr (Scott and Nyakale 2002), also observed at Deelpan and Blydefontein (Scott 1993). Shrubby Asteraceae pollen became gradually more prevalent between 4.4 and 2 kyr at a number of highveld grassland and Karoo sites suggesting karroid vegetation but the pattern is not replicated at Florisbad, where increased Cyperaceae pollen by 3.9 kyr at the site indicates local wetness. In Equus Cave in the Southern Kalahari, ostrich eggshell δ^{15}N and δ^{18}O eggshell apparently do not suggest dry conditions from 5–2 kyr (Johnson et al. 1997). Although these data are few, they are supported by relatively low δ^{15}N values from grazing fauna at Wonderwerk Cave, and a micro-mammmal-based moisture index, suggesting relatively moist conditions for this region between ~4.2 and 2 kyr (Thackeray and Lee-Thorp 1992).

On the east coast, *Podocarpus* pollen fluctuations in lake and swamp deposits suggest extensive mesic forests during the middle Holocene that began to diminish in size after 3.1 kyr (Scott and Steenkamp 1996; Mazus 2000). Vessel diameter of fossil wood charcoal in the higher altitude Drakensberg mountain region indicates relatively wetter conditions ca. 2.4 kyr (February 1994).

Marked climatic change is indicated in the southwestern Cape from 4.7 kyr. A conspicuous arid episode is indicated by high δ^{18}O ostrich eggshell values at Elands Bay Cave between 4.7 and 3 kyr (Lee-Thorp and Talma 2000). At the same time, increased Asteraceae shrubs are recorded in charcoal assemblages from this site, indicating general and xeric thicket but no mesic thicket like *Dodonaea* (Cartwright and Parkington 1997; Cowling et al. 1999). Meadows et al. (1996) note increasing Asteraceae at Verlorenvlei (adjacent to Elands Bay). Sea surface temperatures off the west coast were cooler ca. 3 kyr (Cohen et al. 1992; Jerardino 1995), suggesting increased upwelling intensity, or reduced Agulhas water influence, or possibly influx of cooler waters from the Southern Ocean. The Elands Bay ostrich eggshell series suggests the return of more mesic conditions from 2.7–1.2 kyr (Lee-Thorp and Talma 2000), while pollen from hyrax dung middens in the Cederberg show more Cyperaceae, "renosterbos" (*Stoebe/Elytropappus* type) and declining *Dodonaea* thicket ca. 2 kyr (Scott 1994).

In contrast to the Western Cape, in the Southern Cape region, increased forest pollen in a peat sequence at Norga points to relatively moist conditions at 3.1 kyr (Scholtz 1986).

2 kyr - present

There is good evidence for a number of environmental shifts during the last 2000 years, but the most noteworthy event may be the climatic fluctuation during the Little Ice Age (LIA) period. Evidence for a lower sea level on the southern coast during its earlier phase (1520 or 1570 AD) may be a reflection of widespread anomalous conditions during this phase (Marker 1997).

High numbers of grass pollen in hyrax dung from the Namib Desert (Kuiseb River) suggest relatively moist but fluctuating conditions ~2 kyr, while $\delta^{13}C$ analysis points to slightly higher C_4 dietary contributions (Scott and Vogel 2000). Lake-deposits with pollen from Otjikoto further to the north in Namibia also seem to suggest moist conditions around this time although there are some uncertainties about the dates (Scott et al. 1991). C_4 grass contribution to diets of hyraxes in the Kuiseb River decreased soon after 2 kyr and by ca. 0.9 kyr very little grass pollen is recorded in the dung sequence. The results show some agreement with Makapansgat especially around the time of the LIA but the temperature interpretations based on PCA curves of frost sensitive forms in the Kuiseb do not seem to correspond well with regional interpretations extrapolated from Makapansgat and Cango. Indeed sensitivity to temperature seems to vary widely across the subcontinent.

The highest Asteraceae values occur in the Wonderkrater pollen sequence at ~2 kyr, followed by a decline (Scott 1982a). This evidence continues the apparent contradictory pattern between the pollen and Makapansgat $\delta^{13}C$ values, which indicate high proportions of C_4 grasses in the vegetation overburden. Both $\delta^{13}C$ and $\delta^{18}O$ in the Makapansgat stalagmites show a decreasing trend after 2 kyr, indicating a decline in C_4 grasses in step with episodic sharp $\delta^{18}O$ minima, suggesting that rain formed at markedly higher, cooler altitudes, and in turn, that stormier and cooler conditions pertained (Holmgren et al. 1999). The series of oscillations culminated in a marked $\delta^{18}O$ minimum at AD 1750, which seems to represent the regional manifestation of the LIA peak (Fig. 5). Corresponding variations in pollen data include a decline of frost sensitive arboreal species in the bushveld at Moreletta River (Pretoria) near the northern highveld boundary (Fig. 2) (Scott 1984) while the detailed pollen diagram of Wonderkrater (Scott 1982a) suggests a similar decline in trees.

In the eastern summer-rain region in coastal KwaZulu-Natal, *Podocarpus* pollen in lake and swamp deposits began to retreat northward to the Kosi Bay area after by 1.3 kyr (Mazus 2000) suggesting increasing dryness.

At Florisbad, a phase with summer rain and strong evaporation indicated by prominent grass and Chenopodiaceae pollen at ~2 kyr (Scott and Nyakale 2002) is followed by moister conditions by 1.5 kyr with higher grass and Cyperaceae pollen. Near Florisbad, in similar grassland to that surrounding the site, the spring sequence at Deelpan (Meriba I), shows a marked increase in halophytes indicating strong evaporation and warm temperatures prior to ca. 0.7 kyr (Scott and Brink 1992). Both $\delta^{15}N$ and $\delta^{18}O$ from ostrich eggshells in Equus Cave suggest warm and arid, but fluctuating, conditions from 1.5 kyr (Johnson et al. 1997).

In the Karoo, dry conditions are indicated by presence of Asteraceae pollen and isotope ratios in ostrich eggshell and soil humates from Blydefontein ca. 2 kyr (Scott and Bousman 1990; Bousman 1991). After 2 kyr, grass pollen shows a recovery, confirmed by isotope ratios in ostrich eggshell and soil organics at Blydefontein (Bousman 1991). Pollen in hyrax dung suggests that grassy vegetation occurred at Blydefontein ca. 900–1500 AD (Bousman and Scott 1994). Karoo shrubs increased at the cost of grass cover since ca. 1700 AD, possibly as a result of generally decreasing summer rainfall receipts.

Slightly further south, a decline towards present levels of C_4 grass is registered in the Cango stalagmite after 2 kyr. Abundance of a typical summer rainfall, frost resistant tree species, *Acacia karroo*, increased in nearby Boomplaas Cave at 2 kyr (Deacon et al. 1984; Scholtz 1986). Because of its position at the transition between the winter and summer rain regimes, this area is highly sensitive to a decline in summer rainfall.

Dietary δ^{13}C values in hyrax dung from Pakhuis Pass (Cederberg) decline sharply ca. 1–0.9 kyr (Scott and Vogel 2000), suggesting cooler winter-rain conditions. Following a moister episode, δ^{18}O values of ostrich eggshell in Elands Bay Cave register arid and/or more evaporative conditions for most of the last 1000 years (Lee-Thorp and Talma 2000).

Evidence for sub-millennial variability

Long continuous records capable of yielding information on short term variability are extremely scarce. Spectral analysis of the Makapansgat δ^{18}O series in the T7 stalagmite shows significant periodic components in this rainfall-linked proxy at 57 years (with associated peaks at 53 and 65 years) reminiscent of the North Atlantic Oscillation, while the δ^{13}C and grey scale parameters peak at 540 and 330 years respectively (Lee-Thorp et al. 2001). When these records are subjected to wavelet analysis, however, it can be seen that dominant periodic components shifted over time (Tyson et al. 2002). Wavelet analysis of the δ^{18}O series shows periodicity at 500 years, and a cluster near 100 years between AD 0–1000, which shifts to 70–80 years after AD 1000 (Tyson et al. 2002).

Pollen data did not produce the same high resolution or continuity as the stalagmites but a 20th century pollen record from hyrax dung at Clarens suggests that this material is capable of recording decadal variations (Carrion et al. 1999). Quasi-centennial scale moisture cycles can be observed in some pollen records, for example, in the grass pollen abundance in desert hyrax dung accumulations from the Kuiseb River (Scott 1996).

Mechanism of environmental change on different temporal scales

In order to understand the mechanism of climatic change over Southern Africa during the Holocene we need to establish the rapidity of shifts in moisture and temperature patterns, whether shifts appeared simultaneously over large parts of the continent, and how these changes relate to events elsewhere. Spatially separated biomes can be expected to respond in different ways to forcing of seasonal and moisture regimes, as well as to temperature. The winter rainfall area stands in contrast with the northern, central and eastern parts in view of more mesic conditions on the southwestern coastal region during the early Holocene (Cartwright and Parkington 1997; Cowling et al. 1999), and drier conditions in the mid-Holocene when conditions in the interior were more mesic.

On a broader scale the rapid event of ca. 8.2 kyr widely observed in North Atlantic sediments and also in the northern and equatorial parts of Africa (Gasse 2000; Stager and Mayewski 1997), can at present only be tentatively linked with any anomaly in the Southern Hemisphere. A climate transition does seem to have occurred at 8 kyr, with a marked temperature depression observed on the eastern highlands (Smith et al. 2002), but whether or not this shift is linked to the 8.2 kyr "event", or rather forms part of large millennial-scale climate shifts is uncertain. A mechanism which could link the 8.2 kyr event in the North Atlantic regions with a marked temperature depression in the Southern African highlands is in any case, not clear. The general moisture pattern seems to show the expected anti-phase response of monsoon-like (ITCZ) moisture distribution to the precessional cycle (Street-Perrott and Perrott 1993).

Two alternative explanations have been proposed for the development of moister conditions during the middle Holocene in the central interior of Southern Africa during the middle Holocene (Scott 1993). In one, higher incidence of summer rain spread gradually from the savanna in the north to the Karoo in the south, based on the first appearance of early Holocene pollen moisture indicators. The second explanation points to moist pulses as indicated by geomorphic evidence (Butzer 1984). The new Florisbad data support the 'pulses' proposal (Scott and Nyakale 2002), but the later arrival of wetness at Blydefontein to the south perhaps supports the gradual spreading of moisture (Scott 1993). In view of the high amplitude, rapid changes demonstrated in the Makapansgat stalagmites, "gradualism" from pollen diagrams, or "pulses" from geomorphic evidence may merely be related to poor chronological control and resolution. The stalagmite data provide the strongest evidence for a pattern of extremely rapid changes — on the order of a couple of decades or less in many cases (Lee-Thorp et al. 2001). This rapidity and scale suggests that apparently long-term climate or environmental states noted from archaeological, geomorphological, and palynological evidence may be crude averages of a variety of conditions.

Periodic elements evident in the Makapansgat data differ for all three variables of $\delta^{18}O$, $\delta^{13}C$ and grey scale, suggesting separate influences on their respective time evolution (Lee-Thorp et al. 2001). Different processes associated with moisture cycles on centennial scales and millennial scales (moisture, seasonality, or temperature-dependent evapo-transpiration rates) are likely superimposed on each other. The long-term pattern seems to be reflected in the general $\delta^{13}C$ curves from Makapansgat and Cango, and to a lesser degree in the $\delta^{18}O$ curves, while strong shorter-term oscillations superimposed on these trends are observed. Given the constraints of chronological control, there is no discernible lag between the patterns of the two $\delta^{13}C$ curves, both reaching the highest values at $\sim$2.2 kyr (Figs. 5 and 6) (Talma and Vogel 1992; Lee-Thorp and Talma 2000). This observation argues for coincident climate responses from north to south. The rise and recent partial decline in C_4 grasses during the last 3 kyr may also be a seasonal feedback of incoming solar radiation, induced by orbital forcing in the precessional cycle.

The evidence from the middle Holocene (ca. 6 kyr) suggests a scenario for dominant influence from moist easterlies and reduced influence from the westerlies, and accords well with warmer temperatures and a reduced Antarctic circumpolar vortex recorded in the Taylor Dome ice core as higher $\delta^{18}O$ (Fig. 7) and lower sodium concentrations respectively (Steig et al. 2000). The regional vegetation pattern in the interior of Southern Africa over the late Holocene between 5 and 2 kyr likely responded to generally rising summer seasonality, promoting C_4 grasses, as reflected in stable carbon isotopes. The near-simultaneous general increase in shrubby Asteraceous vegetation seems at first glance to contradict this scenario; but higher Asteraceae abundance can, apart from lower summer rain, also be related to cooling. Hence, in combination the data may reflect interplay between cycles of drying, cooling and more seasonally restricted rainfall. Tree-cover also declined markedly according to pollen evidence from Wonderkrater and as implied by the stalagmite data. This development might be related to increased incidence of frost (Scott 1982a; Scott and Thackeray 1987; Lee-Thorp et al. 2001), or fire, but fire disturbance at Wonderkrater is not suggested until recently (Scott 2002). In the case of the karroid Cango region where no savanna trees are present, a similar argument can be advanced if mountain scrub is considered instead. An increase in charcoal of a typically summer rainfall and relatively frost resistant species, *Acacia karroo*, in nearby Boomplaas Cave at 2 kyr (Deacon et al.

1984; Scholtz 1986) may reflect a replacement of mountain scrub as firewood, by valley material. More open Asteraceous vegetation in the Western Cape (Scott 1994; Cartwright and Parkington 1997; Cowling et al. 1999) would support a regional pattern of more open thicket.

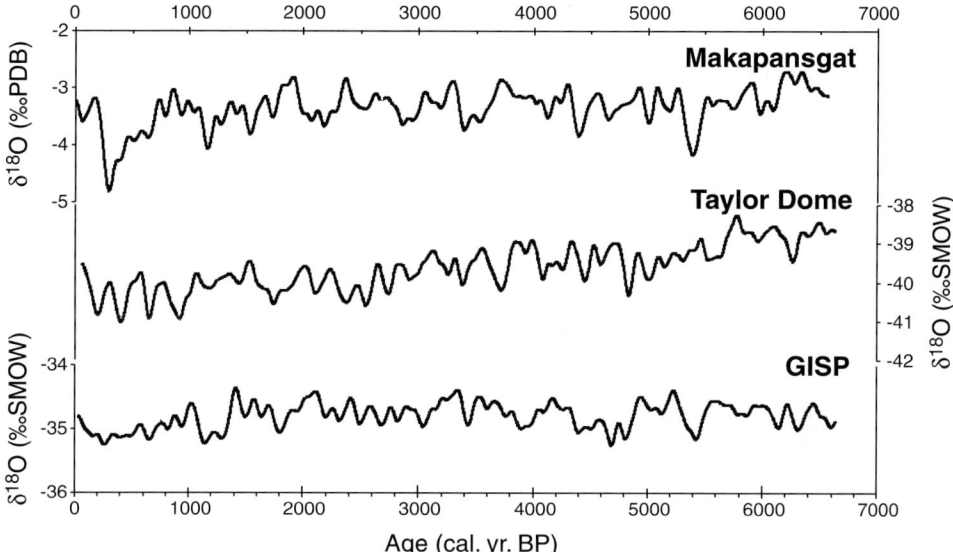

Figure 7. Makapansgat stalagmite $\delta^{18}O$ compared with $\delta^{18}O$ values in the Taylor Dome (Antarctica) and GISP II (Greenland) ice core (Grootes et al. 1993; Steig et al. 2000), all plotted with a Gaussian low-pass filter (~100 year running mean). (From Lee-Thorp et al. (2001)).

A comparison of the Makapansgat $\delta^{18}O$ records with the near-coastal Antarctic Taylor Dome ice core $\delta^{18}O$, sodium (Na) and other trace element concentration records (Steig et al. 2000) indicates strong correspondences despite the large distance between them (Fig. 1). Warming (^{18}O enrichment) at Taylor Dome prior to ~5000 years is matched at Makapansgat (Fig. 7). A weakened Antarctic circumpolar vortex is indicated by the lowest sodium concentrations for the entire Holocene in Taylor Dome at this time (Steig et al. 2000). Concurrent reduction in wind-driven mixing in Lake Victoria on the equator was used to infer reduced influence of the circumpolar vortex far to the north (Stager and Mayewski 1997). Warming also occurs in Taylor Dome (indicated as higher $\delta^{18}O$) at about the same time as higher $\delta^{18}O$ at Makapansgat during a variable, modest warm/humid phase at 4.3–3.2 kyr. A shift towards lower $\delta^{18}O$ in Taylor Dome occurs at ~3 kyr as seen in Makapansgat, while both records show declining $\delta^{18}O$ trends thereafter, observed also in the Greenland GISP 2 ice core (Grootes et al. 1993) (Fig. 7). The drier, cooler conditions inferred from the Makapansgat stalagmite record are reflected in various more poorly resolved proxy data across Southern Africa (Tyson and Lindesay 1992). Moreover higher Na levels in Taylor Dome indicate enhanced Antarctic circumpolar atmospheric circulation at this time. Elsewhere in Africa, trends towards aridification are observed in African lake levels including Lake Victoria, and Lake Malawi fell shortly after 1.5 kyr (Stager and Mayewski 1997; Johnson et al. 2001).

Demographic responses

Some correlation between major cultural manifestations and climatic or environmental changes can be drawn, but the nature of these relationships is complex and difficult to determine, and very often several explanations for cultural shifts, including social causes, must be considered. Farmers who arrived shortly after 2 kyr with their crops and domestic animals and their attendant requirements for water and food probably had fewer mechanisms for dealing with variable conditions than the highly flexible hunter-gatherer societies before them.

Deacon (1974) first observed that archaeological visibility of Later Stone Age people in Southern Africa might be related in some way to climatic conditions as illustrated by a comparison of frequency of radiocarbon dates associated with archaeological finds in the interior. In the earlier part of the Holocene, Oakhurst Complex stone tools of the Later Stone Age appeared over wide parts of Southern Africa (Mitchell 2000). Development of fluctuating and often arid conditions during this phase presented challenges for hunter-gatherers, which may be at least partly reflected in the low density of sites occupied, and the nature of resource exploitation and of stone tool manufacture. The lowest frequency of dates for the sub-continent as a whole occurs at ca. 8.7 kyr during the early Holocene dry phase. A recent update of the pattern shows a more complex picture but supports the pattern (Mitchell 2000). One of the shifts was in hunting strategy: small solitary bovids were taken instead of the large, mobile grazers typical of the Pleistocene (Deacon et al. 1984). Assemblages of the Oakhurst Complex disappeared over a wide region in South Africa ca. 8.7 kyr.

Expansion of microlithic Wilton-type industries after ca. 7.5 kyr (Mitchell 2000) took place in conditions of increasing moisture availability. Deacon's (1974) radiocarbon frequencies show that Later Stone Age sites, and hence archaeological 'visibility' of hunter-gatherers in the interior landscape, peaked at ~3.1 kyr. In the Eastern Cape, population densities begin to rise after 6.8 kyr, the range of sites occupied had increased after ~5 kyr, and between 5 and 2 kyr resource exploitation intensified, including a concentration on riverine resources (Hall 2000). This evidence, together, suggests a significant increase in population density of hunter gatherers. It appears that the slightly cooler conditions pertaining by this time were favourable for people over wide parts of the subcontinent.

The pattern differs in the Western Cape where archaeological visibility of coastal hunter-gatherers was low between ~7.8 and 4.2 kyr. The possibility has long been entertained that marked regional aridity rendered the Western Cape region unattractive to hunter-gatherers (Parkington et al. 1987), although other factors including changes in rocky shorelines associated with a higher sea-level at this time, might also have made it less attractive (Miller et al. 1993; Jerardino and Yates 1996). New $\delta^{18}O$ mollusc data from the Steenbokfontein Cave, where human occupation was more continuous, suggest that sea-surface temperatures were not different from those before or after the occupation 'hiatus', implying that coastal Benguela upwelling, usually associated with coastal aridity, was not unusually high at this time (A. Jerardino, pers. comm.). The period between 4 and 2 kyr, is known as the "megamidden" period because of the accumulation of enormous shell middens along the coast (Parkington et al. 1987). A very high proportion of the human burials found in coastal caves, shelters or simply along the dune cordon, have been assigned to this period based on their radiocarbon dates (Parkington et al. (1987), Lee-Thorp et al. (1989), Sealy, pers. comm.).

Agriculture arrived relatively late in Southern Africa, shortly after 2 kyr, and therefore its impact has generally not been as prolonged as in other African regions to the north. The Iron Age farmers who penetrated to the eastern and central regions of the subcontinent manufactured iron implements and brought the C_4 crops sorghum and millet, as well as domestic sheep and later, cattle. These cereals require summer rain. Climate and environmental patterns over the last 2 kyr may be compared with a relative wealth of archaeological data from areas occupied by these semi-sedentary farmers. Agriculture based on C_4 crops was never possible in the drier, westerly winter rainfall regions. Transhumant Khoisan pastoralists with sheep and later cattle entered the region by about 2 kyr (Sealy and Yates 1994), probably when conditions were relatively moister and grassier (Scott et al. 1991; Scott 1996; Bousman 1998).

A marked increase in population size and social complexity amongst the farming and iron-producing groups took place in and around the Shashe-Limpopo region beginning about AD 800. Based on the temporal distribution of radiocarbon dates for Iron Age groups (Vogel 1995), Huffman (1996) proposed that populations expanded during warmer, moister phases when crop growing and cattle farming flourished and that they declined or moved away during drier periods. Social complexity increased rapidly leading to the development of stratified societies and large towns most apparent between AD 1000–1270 at the K2 and Mapungubwe sites (Vogel 1995; Huffman 1996). Since this area currently experiences very low (<330 mm/yr) rainfall with a high co-efficient of variation less suitable for maintenance of large domestic herds or cultivation, it is widely considered that the scale of settlement, and development of wealth, was made possible by more favourable conditions existing at the time (Huffman 1996; Tyson et al. 2000). The recent data from Makapansgat and Wonderkrater to the southeast confirm that between AD 1000 and 1300 regional conditions were at times slightly moister and warmer than today, although strictly speaking they fall within a general trend of continental drying (Scott and Thackeray 1987; Holmgren et al. 1999; Lee-Thorp et al. 2001). Nevertheless, even slight increases in moisture, or perhaps greater predictability, would have promoted a suitable environment for herding of cattle and sheep, because of the predominance of 'sweet' grasses (grasses which remain nutritious throughout the year). An 'oversupply' of rain, although this might seem to be advantageous in a generally arid region, brings disadvantages in the form of less nutritious grasses, and worse, diseases such as malaria and tsetse fly.

The subsequent collapse of Mapungubwe at AD 1280, apparent dispersal of the population, and subsequent rise to power of Great Zimbabwe further north, has given rise to speculation about a connection to worsening conditions associated with emergence of "Little Ice Age" conditions (Huffman 1996). This scenario now seems unlikely for several reasons. Firstly, the firm Makapansgat chronology shows that marked dry, cool "LIA" conditions emerged only later, about AD 1600 (Lee-Thorp et al. 2001). More recent excavations do not suggest a population disappearance, merely the demise of Mapungubwe itself, and some sites post-dating AD 1600 appear in low-lying seasonally inundated areas, suggesting that they were clustered close to remaining pans and water sources (S. Hall, pers. comm.). Finally, new nitrogen isotope data from a number of sites suggest that no climate deterioration took place simultaneously with the collapse of Mapungubwe, rather drier conditions emerged much later (J.M. Smith, pers. comm).

One other relationship remains to be investigated, that is, the effect of intense farming on a landscape which is often marginal for dryland agriculture and introduced non-indigenous animals. Pollen preserved in cow dung from Iron Age "kraals" (cattle enclosures) over the last 1500 years suggests that environments were grassier and more open than the savanna environments of today (Carrion et al. 2000). The openness was probably due to local clearing of wood near settlements for domestic and manufacturing (iron smelting) fuel, while the regional woodland was naturally more open than at present. A study of microscopic charcoal as an indicator of fire in swamp and lake sediments of the savanna, does not, with the exception of the Wonderkrater site, suggest that the arrival of farmers coincided with higher fire incidences (Scott 2002). Controlled use of fire and protection of grassland for grazing and removal of fuel probably account for high grass pollen and low microscopic charcoal counts.

Conclusion

Large parts of the subcontinent remain unexplored with regard to palaeoenvironments and records are scattered and often incomplete. A great deal more research is therefore needed to improve our current understanding of Holocene change. Correspondence between rare available high resolution continental climate data with a near-coastal Antarctic ice core suggest that the climate of the subcontinent continued to be strongly influenced by shifts in intensity and extent of frontal systems associated with the circumpolar vortex. This major forcing probably holds an important key to synchroneity with global climate trends and events observed for Southern African climates.

Southward spreading of summer-rain moisture related to the ITCZ during the middle and late Holocene was characterised by strong centennial scale variability over different regions of the interior. On a millennial scale the link between moisture and temperature conditions seems less certain, but at smaller centennial scale variability this link is observed. A picture is emerging that shows that biota and human settlements across the subcontinent were strongly influenced by the relatively mild (compared to Glacial period, Partridge et al. (2004)) climate fluctuations which occurred during the Holocene.

Summary

Although palaeoclimate records across Southern Africa are patchy in time and space, it seems clear that Holocene climates and environments varied on scales that carried significant implications for distribution of fauna and flora, and for human settlement. Large millennial-scale temperature shifts of up to several degrees Celsius are indicated by changes in proportions of C_3 and C_4 grasses in the sensitive eastern highlands until $\sim$7 kyr, and by dissolved gases in the Uitenhage aquifer. In the north, Makapansgat stalagmite records indicate shifts of similar duration and scale in addition to smaller, more rapid shifts. Visibility of prehistoric human populations was lowest between 8 and 9 kyr. The warming trend from $\sim$7 kyr was accompanied by spreading summer moisture, culminating in a warm, wet phase from 6.5–5.1 kyr according to stalagmite and pollen data. The Western Cape, in contrast, seems to have been relatively arid. The suggestion is that the easterlies circulation system dominated at this time and the westerlies system was reduced, concurrent with a reduced circumpolar vortex recorded in coastal Antarctica. Most records after 5 kyr reflect an overall but variable drying trend. High proportions of Asteraceae pollen are recorded in many sites from ~4.4–2 kyr. Stalagmite sequences record increasing development of C_4 grasslands, peaking just before 2 kyr. A series of ~1 °C temperature fluctuations associated with drier, stormier and cooler conditions culminated at 1750 AD. Climate variability as determined from the Makapansgat stalagmites also shows periodic components at centennial and multi-decadal scales. Shifts in intensity and extent of frontal systems associated with the Antarctic circumpolar vortex likely hold an important key to synchroneity with global climate trends and events.

Acknowledgments

We thank our colleagues in the Southern African Society for Quaternary Research and the Quaternary Research Centre for their contributions. Tim Partridge suggested the title of this paper. Anne Westoby did the original drafting of Figure 1. Jeannette Smith, Siep Talma and Karin Homgren provided data for Figures 4, 6 and 7, respectively. Two referees, Raymonde Bonnefille and Lydie du Pont, provided constructive critiques. This work was funded by the National Research Foundation, the University of the Free State (GUN 2053236), the University of Cape Town (GUN 2047166) and the Water Research Commission of South Africa. We thank Françoise Gasse and Rick Battarbee for their efforts in leading the PEP III project and for organizing an excellent conference in Aix-en-Provence (August 2001).

References

Avery D.M. 1981. Holocene micromammalian faunas from the Northern Cape, South Africa. S. Afr. J. Sci. 77: 265–73.

Bousman C.B. 1991. Holocene palaeoecocology and Later Stone Age hunter gatherer adaptations in the South African interior plateau. Ph.D. thesis, Southern Methodist University, Dallas, Texas, 371 pp.

Bousman C.B. 1998. The chronological evidence for the introduction of domestic stock into Southern Africa. African Archaeological Review 15: 133–150.

Bousman C.B. and Scott L. 1994. Climate or overgrazing?: the palynological evidence for vegetation change in the eastern Karoo. S. Afr. J. Sci.: 575–578.

Brook G.A., Marais E., Cowart J.B. 1999. Evidence of wetter and drier conditions in Namibia from tufas and submerged speleothems. Cimbebasia 15: 29–39.

Butzer K.W. 1984. Archaeology and Quaternary environment in the interior of Southern Africa. In: Klein R.G. (ed.), Southern African Prehistory and Palaeoenvironments. Balkema, Rotterdam, pp. 1–64.

Carrion J.S., Scott L. and Vogel J.C. 1999. Twentieth century changes in montane vegetation in the Eastern Free State, South Africa, derived from palynology of hyrax middens. J. Quat. Sci. 14: 1–16.

Carrion J.S., Scott L., Huffman T. and Dryer C. 2000. Pollen analysis of Iron Age cow dung in Southern Africa. Veg. Hist. Archaeobot. 9: 239–249.

Cartwright C. and Parkington J. 1997. The wood charcoal assemblages from Elands Bay Cave, Soutwestern Cape: principles, procedures and preliminary interpretation. S. Afr. Archaeol. Bull. 52: 59–72.

Cohen A.L., Parkington J.E., Brundrit G.B. and van der Merwe N.J. 1992. A Holocene sea surface temperature record in mollusc shells from the southwest African coast. Quat. Res. 38: 379–385.

Cohen A. and Tyson P.D. 1995. Sea surface temperatures during the Holocene on the south coast of Africa. The Holocene 5: 304–312.

Cowling R.M. 1983. The occurrence of C_3 and C_4 grasses in fynbos and allied shrublands in the southeastern Cape, South Africa. Oecologia 58: 121–127.

Cowling R.M., Cartwright C.R., Parkington J.E. and Allsop J.C. 1999. Fossil wood charcoal assemblages from Elands Bay Cave, South Africa: implications for Late Quaternary vegetation and climates in the winter-rainfall fynbos biome. J. Biogeogr. 26: 367–378.

Cowling R.M., Richardson D.M. and Pierce S.M. 1997. Vegetation of Southern Africa. Cambridge University Press, Cambridge, 615 pp.

Deacon J. 1974. Patterning in the radiocarbon dates for the Wilton/Smithfield Complex in Southern Africa. S. Afr. Archaeol. Bull. 29: 3–18.

Deacon H.J., Deacon J., Scholtz A., Thackeray J.F., Brink J.S. and Vogel J.C. 1984. Correlation of palaeoenvironmental data from the late Pleistocene and Holocene deposits at Boomplaas Cave, Southern Cape. In: Vogel J.C. (ed.), Late Cainozoic Palaeoclimates of the Southern Hemisphere. Balkema, Rotterdam, pp. 339–352.

Eitel B., Blümel W.D., Hüser K. and Mauz B. 2001. Dust and loessic alluvial deposits in Northwestern Namibia (Damaraland, Kaokoveld): sedimentological and palaeoclimatic evidence based on luminescence data. Quat. Int. 57: 57–65.

Esterhuysen A.B., Mitchell P. and Thackeray J.F. 1999. Climatic change across the Pleistocene/Holocene boundary in the Caledon River Southern Africa: Results of a factor analysis of charcoal assemblages. S. Afr. Field Archaeol. 8: 28–34.

February E. 1994. Rainfall reconstruction using wood charcoal from two archaeological sites in South Africa, Quat. Res. 42: 100–107.

Finney B.P., Scholz C.A., Johnson T.C., Trumbore S. and Southon J. 1996. Late Quaternary lake-level changes in Lake Malawi. In: Johnson T.C. and Odada E. (eds), Climatology and Palaeoclimatology of East African Lakes. Gordon and Beach, Toronto, pp. 495–508.

Gasse F. 2000. Hydrological changes in the African tropics since the Last Glacial Maximum. Quat. Sci. Rev. 19: 189–211.

Grootes P.M., Stuiver M., White J.W.C., Johnsen S. and Jouzel J. 1993. Comparison of oxygen isotope records from the GISP2 and GRIP Greenland ice cores. Nature 366: 552–554.

Hall S. 2000. Burial sequences in the Later Stone Age of the Eastern Cape Province, South Africa. S. Afr. Archaeol. Bull. 55: 137–146.

Heaton T.H.E., Talma A.S. and Vogel J.C. 1986. Dissolved gas palaeotemperatures and ^{18}O variations derived from groundwater near Uitenhage, South Africa. Quat. Res. 25: 79–88.

Holmgren K., Karlén W., Lauritzen S.E., Lee-Thorp J.A., Partridge T.C., Piketh S., Repinski P., Stevenson J., Svanered O. and Tyson P.D. 1999. A 3000-year high-resolution stalagmite-based record of palaeoclimate for North-Eastern South Africa. The Holocene 9: 295–309.

Holmgren K., Lee-Thorp J.A., Cooper G.R.J., Lundblad K., Partridge T.C., Scott L., Sithaldeen R., Talma A.S. and Tyson P.D. 2003. Persistent millennial-scale variability over the past 25,000 years in Southern Africa. Quat. Sci. Rev. 22: 2311–2326.

Horowitz A., Sampson C.G., Scott L. and Vogel J.C. 1978. Analysis of the Voigtspost site, O.F.S., South Africa. S. Afr. Archaeol. Bull. 33: 152–9.

Huffman T. 1996. Archaeological evidence for climatic change during the last 2000 years in Southern Africa. Quat. Int. 33: 55–60.

Jerardino A. 1995. Late Holocene neoglacial episodes in Southern South America and Southern Africa: a comparison. The Holocene 5: 361–368.

Jerardino A. and Yates R. 1996. Preliminary results from excavations at Steenboktontein Cave: implications for past and future research. S. Afr. Archaeol. Bull. 51: 7–16.

Johnson B.J., Miller G.H., Fogel M.L. and Beaumont P.B. 1997. The determination of Late Quaternary palaeoenvironments at Equus Cave, South Africa, using stable isotopes and amino acid
racemization in ostrich eggshell. Palaeogeogr. Palaeoclim. Palaeoecol. 136: 121–137.

Johnson T.C., Barry Y., Chan Y. and Wilkinson P. 2001. Decadal record of climate variability spanning
the past 700 yr in the Southern Tropics of East Africa. Geology 29: 83–86.

Lee-Thorp J.A., Sealy J.C. and van der Merwe N.J. 1989. Stable carbon isotope ratio differences
between bone collagen and bone apatite, and their relationship to diet. J. Arch. Sci. 16: 585–599.

Lee-Thorp J.A. and Beaumont P.B. 1995. Vegetation and seasonality shifts during the Late Quaternary
deduced from $^{13}C/^{12}C$ ratios of grazers at Equus Cave, South Africa. Quat. Res. 43: 426–432.

Lee-Thorp J.A. and Talma A.S. 2000. Stable light isotopes and past environments in the Southern
African Quaternary. In: Partridge T.C. and Maud R.R. (eds), The Cenozoic of Southern Africa.
Oxford Monographs on Geology and Geophysics 40, Oxford University Press, New York, pp.
236–251.

Lee-Thorp J.A., Holmgren K., Lauritzen S.-E., Linge H., Moberg A., Partridge T.C., Stevenson C.
and Tyson P.D. 2001. Rapid climate shifts in the Southern African interior throughout the mid to
late Holocene, Geophys. Res. Lett. 28 (23): 4507–4510.

Lutjeharms J.R.E., Monteiro P.M.S., Tyson P.D. and Obura D. 2001. The oceans around Southern
Africa and regional effects of climate change. S. Afr. J. Sci. 97: 119–130.

Marker M.E. 1997. Evidence for a Holocene low sea-level at Knysna. S. Afr. Geogr. J. Special Edition:
106–107.

Martin A.R.H. 1968. Pollen analysis of Groenvlei lake sediments, Knysna (South Africa). Rev.
Palaeobot. Palynol. 7: 107–144.

Mazus H. 2000. Clues on the history of Podocarpus forest in Maputaland, South Africa, during the
Quaternary, based on pollen analysis. African Geoscience Rev. 7: 75–82.

Meadows M.E., Baxter A.J. and Parkington J.E. 1996. Late Holocene environments at Verlorenvlei,
Western Cape Province, South Africa. Quat. Int. 33: 81–95.

Meadows M.E. and Sugden J.M. 1991. A Vegetation history of the last 14 000 years on the Cederberg,
South-Western Cape Province. S. Afr. J. Sci. 87: 34–43.

Miller D.E., Yates R.J., Parkington J.E. and Vogel J.C. 1993. Radiocarbon-dated evidence relating to
a mid-Holocene relative high sea level on the South-Western Cape coast, South Africa. S. Afr. J.
Sci. 89: 35–44.

Mitchell P. 2000. The Quaternary Archaeological record in Southern Africa. In: Partridge T.C.
and Maud R.R. (eds), The Cenozoic of Southern Africa. Oxford Monographs on Geology and
Geophysics 40, Oxford University Press, New York, pp. 357–370.

Partridge T.C., Scott L. and Schneider R.R. 2004., this volume. Between Agulhas and Benguela:
responses of Southern African climates of the Late Pleistocene to current fluxes, orbital precession
and extent of the Circum-Antarctic vortex. In: Battarbee R.W., Gasse F. and Stickley C.E. (eds),
Past Climate Variability through Europe and Africa. Kluwer Academic Publishers, Dordrecht, the
Netherlands, pp. 45–68.

Parkington J.E., Poggenpoel C., Buchanan W., Robey T., Manhire A. and Sealy J.C. 1987. Holocene
coastal settlement patterns in the Western Cape. In: Bailey G.N. and Parkignton J.E. (eds), The
Archaeology of Prehistoric Coastlines. Cambridge University Press, Cambridge, pp. 22–41.

Ramsay P.J. 1995. 9000 years of sea-level change along the Southern African coastline. Quat. Int.
31: 71–75.

Scholtz A. 1986. Palynological and palaeobotanical studies in the Southern Cape. M.Sc.-thesis,
University of Stellenbosch, Stellenbosch, 280 pp.

Scott L. 1982a. A Late Quaternary pollen record from the Transvaal bushveld, South Africa. Quat.
Res. 17: 339–370.

Scott L. 1982b. A 5000-year old pollen record from spring deposits in the bushveld at the north of
the Soutpansberg, South Africa. Palaeoecol. Afr. 14: 45–55.

Scott L. 1984. Reconstruction of Late Quaternary palaeoenvironments in the Transvaal region, South Africa, on the basis of palynological evidence. In: Vogel J.C. (ed.), Late Cainozoic Palaeoclimates of the Southern Hemisphere. Balkema, Rotterdam, pp. 317–327.

Scott L. 1993. Palynological evidence for Late Quaternary warming episodes in Southern Africa, Palaeogeogr. Palaeoclim. Palaeoecol. 101: 229–235.

Scott L. 1994. Palynology of late Pleistocene hyrax middens, South-Western Cape Province, South Africa: a preliminary report. Hist. Biol. 9: 71–81.

Scott L. 1996. Palynology of hyrax middens: 2000 years of palaoenvironmental history in Namibia Quat. Int. 33: 73–79.

Scott L. 1999. The vegetation history and climate in the Savanna Biome, South Africa, since 190 000 KA: A comparison of pollen data from the Tswaing Crater (the Pretoria Saltpan) and Wonderkrater. Quat. Int. 57-58: 215–223.

Scott L. 2002. Microscopic charcoal in sediments: Quaternary fire history of the grassland and savanna regions in South Africa. J. Quat. Sci. 17: 77–86.

Scott L. and Bousman C.B. 1990. Palynological analysis of hyrax middens from Southern Africa. Palaeogeogr. Palaeoclim. Palaeoecol. 79: 367–379.

Scott L. and Brink J.S. 1992. Quaternary palaeoenvironments of pans in Central South Africa: palynological and palaeontological evidence. S. Afr. Geographer 19: 22–34.

Scott L., Cooremans B., de Wet J.S. and Vogel J.C. 1991. Holocene environmental changes in Namibia inferred from pollen analysis of swamp and lake deposits. The Holocene 1: 8–13.

Scott L., Holmgren K., Talma A.S., Woodborne S. and Vogel J.C. 2003. Age interpretation of the Wonderkrater spring sediments and vegetation change in the savanna biome, Limpopo Province, South Africa. S.A.J. Sci. 99: 484–488.

Scott L. and Nyakale M. 2002. Pollen indications of Holocene palaeoenvironments at Florisbad in the Central Free State, South Africa. The Holocene 14: 497–503.

Scott L. and Steenkamp M. 1996. Environmental history and recent human disturbance at coastal Lake Teza, Kwazulu/Natal. S. Afr. J. Sci. 92: 348–350.

Scott L., Steenkamp M. and Beaumont P.B. 1995. Palaeoenvironmental conditions in South Africa at the Pleistocene-Holocene transition. Quat. Sci. Rev. 14: 937–94.

Scott L. and Thackeray J.F. 1987. Multivariate analysis of Late Pleistocene and Holocene pollen spectra from Wonderkrater, Transvaal, S. Africa. S. Afr. J. Sci. 83: 93–98.

Scott L. and Vogel J.C. 1983. Late Quaternary pollen profile from the Transvaal highveld, South Africa. S. Afr. J. Sci. 79: 266–272.

Scott L. and Vogel J.C. 2000. Evidence for environmental conditions during the last 20,000 years in Southern Africa from ^{13}C in fossil hyrax dung. Global Planet. Change 26: 207–215.

Sealy J.C. 1997. Seasonality of rainfall around the Last Glacial Maximum as reconstructed from carbon isotope analyses of animal bones from Nelson Bay Cave. S. Afr. J. Sci. 92: 441–444.

Sealy J.C. and Yates R. 1994. The chronology and introduction of pastoralism to the Cape, South Africa. Antquity 68: 58–67.

Shi N., Dupont L.M., Beug H.-J., Schneider R. 1998. Vegetation and climate changes during the last 21 000 years in SW Africa based on a marine pollen record. Veg. Hist. Archaeobot. 7: 127–140.

Smith J.M. 1997. Stable isotope analysis of fauna and soils from sites in the Eastern Free State and Western Lesotho, Southern Africa: a palaeoenvironmental interpretation. M.Sc. thesis, UCT, Cape Town.

Smith J.M., Lee-Thorp J.A. and Sealy J.C. 2002. Stable carbon and oxygen isotopic evidence for late Pleistocene and early — middle Holocene climatic fluctuations in the Caledon River Valley, Southern Africa. J. Quat. Sci. 17: 683–695.

Stager J.C. and Mayewski P.A. 1997. Abrupt Early to Mid-Holocene climatic transition registered at the Equator and poles. Science 276: 1834–1836.

Steig E.J., Morse D.L., Waddington E.D., Stuiver M., Grootes P.M., Mayewski P.A., Twickler M.S. and Whitlow S.I. 2000. Wisconsinian and Holocene climate history from an ice core at Taylor Dome, Western Ross Embayment, Antarctica. Geogr. Ann. A 82: 213–235.

Street-Perrott A.F. and Perrott R.A. 1993. Holocene vegetation, lake levels, and climate of Africa. In: Wright H.E., Kutzbach J.E., Webb T., Ruddiman W.E., Street-Perrott F.A. and Bartlein P.J. (eds), Global Climates since the Last Glacial Maximum. University of Minnesota Press, Minneapolis, pp. 318–356.

Talma A.S. and Vogel J.C. 1992. Late Quaternary paleotemperatures derived from a speleothem from Cango Caves, Cape Province, South Africa. Quat. Res. 37, 203–213.

Thackeray J.F. and Lee-Thorp J.A. 1992. Isotopic analysis of Equid teeth from Wonderwerk Cave, northern Cape Province, South Africa. Palaeogeogr. Palaeoclim. Palaeoecol. 99: 141–150.

Tusenius M.L. 1989. Charcoal analytical studies in the North-Eastern Cape, South Africa. S. Afr. Archaeol. Soc. Goodwin Series 6: 77–83.

Tyson P.D., Cooper G.R.J. and McCarthy T.S. 2002. Millennial to multi-decadal variability in the climate of Southern Africa. Int. J. Climatol. 22: 1105–1117.

Tyson P.D. and Lindesay J.A. 1992. The climate of the last 2000 years in Southern Africa. The Holocene 2, 3: 271–278.

Tyson P.D., Karlen W., Holmgren K. and Heiss G.A. 2000. The Little Ice Age and medieval warming in South Africa. S. Afr. J. Sci. 96: 121–126.

Visser J.N.J. and Joubert A. 1991. Cyclicity in the late Pleistocene to Holocene and lacustrine deposits at Florisbad, Orange Free State. S. Afr. J. Geol. 94: 123–131.

Vogel J.C., Fuls A. and Ellis R.P. 1978. The geographical distribution of Kranz Grasses in South Africa. S. Afr. J. Sci. 74: 209–215.

Vogel J.C. 1995. The temporal distribution of radiocarbon dates for the Iron Age in Southern Africa. S. Afr. Archaeol. Bull. 50: 106–109.

6. DIATOM PRODUCTIVITY IN NORTHERN LAKE MALAWI DURING THE PAST 25,000 YEARS: IMPLICATIONS FOR THE POSITION OF THE INTERTROPICAL CONVERGENCE ZONE AT MILLENNIAL AND SHORTER TIME SCALES

THOMAS C. JOHNSON (tcj@d.umn.edu)
Large Lakes Observatory
University of Minnesota Duluth
Duluth, MN 55812
USA

ERIK T. BROWN (etbrown@d.umn.edu)
Large Lakes Observatory
University of Minnesota Duluth
Duluth, MN 55812
USA

JAMES MCMANUS (jmcmanus@coas.oregonstate.edu)
Large Lakes Observatory
University of Minnesota Duluth
Duluth, MN 55812
USA

Keywords: Africa, Palaeoclimate, Rift Valley, Tropical lakes, Little Ice Age, Intertropical Convergence Zone, De-glaciation

Introduction

The large lakes of the East African Rift Valley provide a magnificent array of sedimentary basins that are actively recording the modern climate dynamics of tropical East Africa. The basins are perhaps 10 million or more years old (Cohen et al. 1993), and their deepest reaches have archived continuous records of past climate change that are unparalleled anywhere else in the tropics in terms of longevity coupled with resolution (Johnson 1993). These lakes are highly sensitive to climate variability, and their sediments carry rich signals of past environmental dynamics, inscribed in the assemblages of microfossils, the abundance and isotopic composition of endogenic minerals, bulk sediment texture and structure, and organic geochemistry.

The International Decade for the East African Lakes (IDEAL) is a programme that has helped to steer and coordinate palaeoclimate studies on the rift lakes since the programme's inception in 1993 (Johnson 1993). The IDEAL community of scientists from

R. W. Battarbee et al. (eds) 2004. *Past Climate Variability through Europe and Africa.*
Springer, Dordrecht, The Netherlands.

North America, Europe, and Africa is investigating not only the climatic history of the large rift lakes, but modern processes within their basins as well. The "decade" of IDEAL officially started in 1995 with its first field programme on Lake Victoria (Lehman 1998). Since then IDEAL scientists have continued to carry on investigations on Lakes Edward, Malawi and Tanganyika, as well as on smaller lakes in East Africa.

The IDEAL palaeoclimate community is presently turning its focus to Lake Malawi, where a major drilling programme is planned for the last two months of 2004. Lake Malawi is the southern-most of the rift lakes, and it appears to be exceptionally promising as a target for long, continuous records of past climate change. It has responded somewhat differently from lakes farther north in Africa to global climate forcing associated with orbital (Milankovitch) cycles of insolation. In contrast to Lake Tanganyika and the lakes of North Africa, Lake Malawi experienced relatively dry conditions during much of the early Holocene (Finney et al. 1996; Johnson 1996). The history of Lake Malawi derived from drilling operations therefore may provide important insight into how the southern tropics of Africa have responded differently from the northern tropics on time-scales ranging from millennia to decades.

This paper presents recent data derived from piston cores recovered from the north basin of Lake Malawi. It focuses on the abundance and mass accumulation rate of biogenic silica (BSi MAR) measured in the cores, supplemented by some trace-element data and the species composition of the diatom assemblages. These results imply linkages between the climate dynamics of tropical Southern Africa and extratropical regions as far away as Greenland.

Lake Malawi

Lake Malawi extends from $\sim$10 °S to 15 °S and is the fifth largest lake in the world, having a volume of 6140 km^3, a length of about 600 km, and a maximum depth of 700 m (Fig. 1) (Eccles 1974). The lake's structure consists of four half graben, bounded by major border faults (Rosendahl 1987). The interplay among physical dynamics, chemistry, and biology of the lake produces anoxic conditions in the water column below $\sim$200 m depth (Eccles 1974; Halfman 1996). The anoxic nature of the deep basins, coupled with high sediment accumulation rates, provides a sedimentary environment well suited for high-resolution studies of palaeoclimate.

The climate over Malawi exhibits sufficient seasonality in rainfall and wind regime to affect the physical dynamics of the lake. The chemocline in Lake Malawi is about twice as deep as in Lake Tanganyika due to the cooler winter conditions over Malawi compared to the more equatorial setting of Tanganyika. Rainfall is strongly influenced by the seasonal migration of the Inter-Tropical Convergence Zone (ITCZ) to the north and south of the equator (Fig. 2). Lakes located near the equator experience two rainy seasons per year, but at the more southerly latitude of Lake Malawi, the rainy season occurs only once during the months of austral summer. Orographic effects also influence Malawi's climate (Nicholson 1996); within the basin, annual rainfall ranges from 750 mm in the lowlands at the southern end of the lake to >2400 mm just north of the lake. When the ITCZ migrates northwards during austral winter, Southern Africa is dominated by a high pressure cell that suppresses convection of moisture off the adjacent Southwest Indian Ocean (Lindesay 1998). The pattern of climate variability throughout the region has been attributed to the influence of

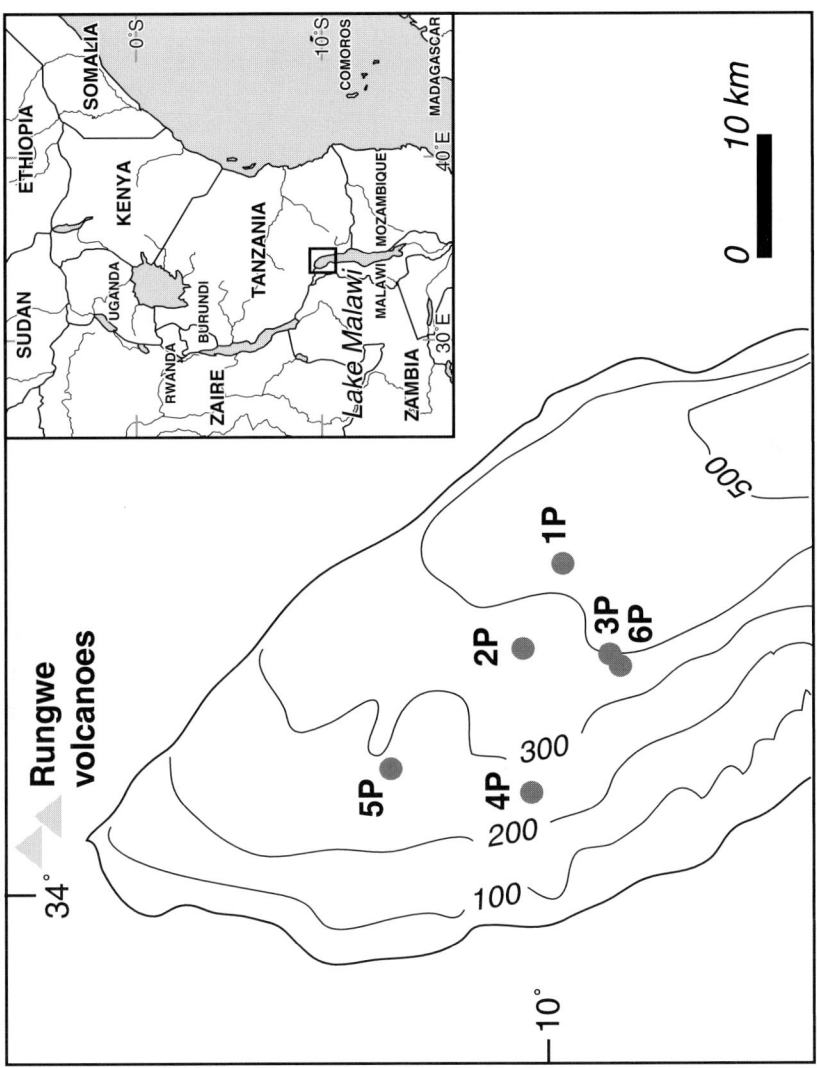

Figure 1. Bathymetric map of northern Lake Malawi, showing core locations (grey circles) and location of the Rungwe volcanic field. Contours in metres.

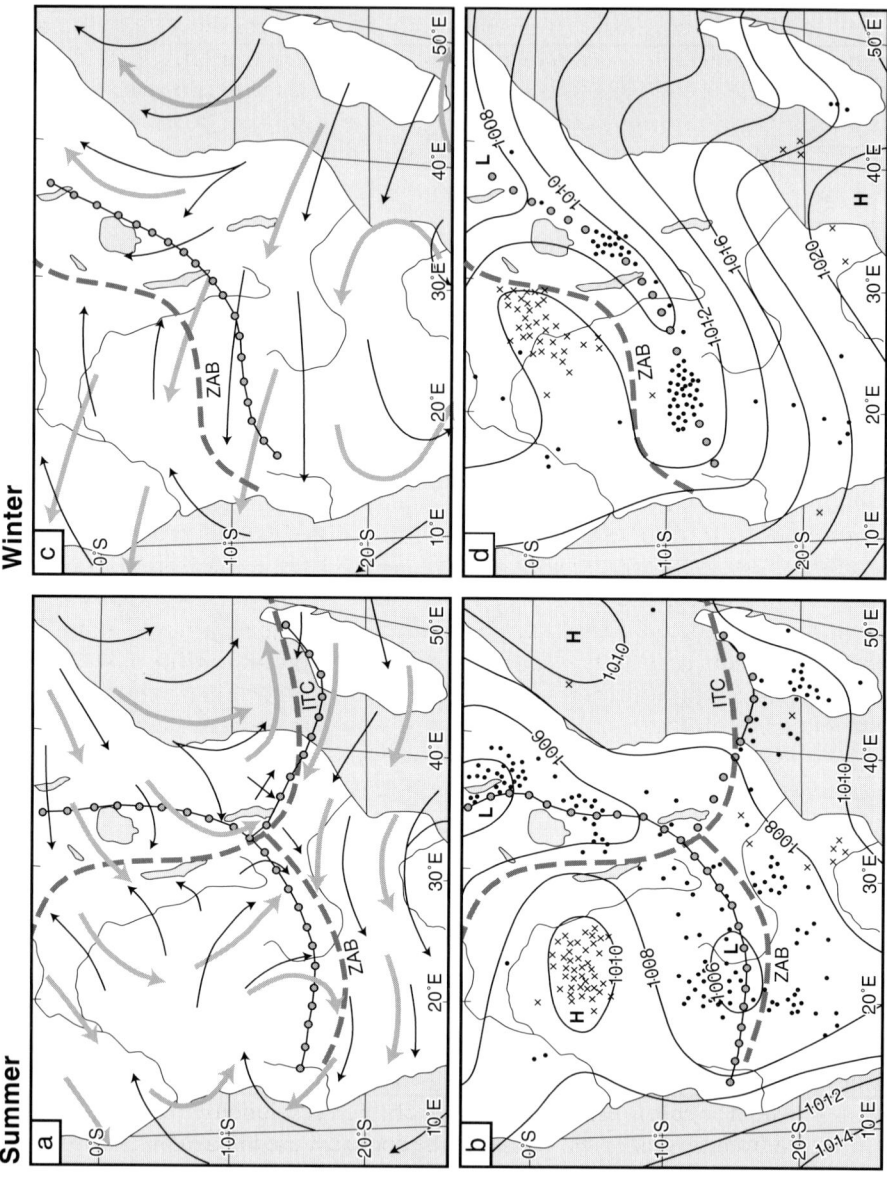

Figure 2. Location of the Intertropical Convergence Zone (ITCZ) and Zaire Air Boundary (ZAB) over East Africa during austral summer and winter. Double arrows indicate winds at 3 km and single arrows indicate lower level winds (a and c). X's indicate high pressure regions and dots indicate regions of individual low pressure systems (Modified from Tyson (1986)).

sea surface temperature (SST) patterns in the Atlantic and Indian Oceans (Nicholson 1996) and, more recently, to the combined effects of the equatorial Pacific (ENSO) and Atlantic SST anomalies (Camberlin et al. 2001).

The pattern of sedimentation in Lake Malawi is fairly typical of lacustrine rift basins. Sand interbedded with mud dominates the nearshore facies to a depth of about 100 m in most regions of the lake, and diatomaceous mud or turbidites prevail farther offshore (Johnson and Ng'ang'a 1990; Wells et al. 1994). Ferruginous muddy sands with laminated nodules of goethite and manganite are commonly observed in a depth range of 80 to 160 m. Here winnowing by currents associated with large storm waves, coupled with the redox reactions associated with the chemocline, yield an unusual sediment composition and texture (Owen et al. 1996).

Laminated sediments are found in many regions of the lake. ^{210}Pb dating of cores confirms that these laminations are varves, representing annual variations in sediment deposition. The varves consist of light/dark couplets of diatom ooze interbedded with dark layers of more terrestrially derived material, including clay, silt and microscopic plant debris and charcoal (Pilskaln and Johnson 1991; Johnson et al. 2000). The varves are associated with seasonal changes in lake mixing conditions and rainfall, with greater diatom productivity during the dry, windy season creating the light, diatomaceous layer and greater terrestrial input during the rainy season creating the dark, clastic layer. This contention is consistent with time-series sediment trap studies that demonstrate strong seasonal variations in the flux of diatoms to the lake floor during the windy season (Francois et al. 1996).

Recent sediments recovered from the south and central parts of the lake are not varved, even in depths where the water is anoxic. This reflects major differences in the sedimentary environments of the northern and south/central basins. Most of the major rivers flow into the northern basin (Scholz et al. 1993; Johnson et al. 1995), so the seasonality of sediment influx is greater there than elsewhere in the lake. Sediment deposition in the south/central basin does not faithfully record the lake's seasonality, with sediments arriving there after several cycles of deposition, resuspension and redeposition. These processes would tend to mix sedimentary components that were delivered to the offshore basins in different seasons and create homogenous sediment, despite the lack of bioturbation in the anoxic waters of the hypolimnion.

Methods

We recovered six piston cores and six multi-cores from the north basin of Lake Malawi in February 1998 (Fig. 1). We knew that short cores collected from the region in 1993 were varved throughout their ca. 60 cm lengths. Our hope was to recover cores that were continuously varved over a ten-metre interval, spanning the entire Holocene. By taking a suite of cores, we would be able to establish an extremely high-resolution chronology, based on a stacked record from several cores, similar to the approach used in dendrochronology.

The cores were collected aboard the R/V Usipa, a 15 m research catamaran. We used the University of Minnesota's Limnological Research Center (LRC) Kullenberg piston corer and an Ocean Instruments Multi-Corer. Core sites were selected based on seismic reflection profiles previously obtained in the area by Duke University's Project PROBE and Project SEPRO (Johnson and Ng'ang'a 1990; Scott et al. 1991; Johnson et al. 1995). Core positions were established by GPS navigation.

The cores were air-freighted back to the LRC of the University of Minnesota, where they were scanned by a Geotek multi-sensor instrument for magnetic susceptibility, gamma-ray bulk density and sound velocity. The cores were then split, visually described, and sampled at 10 cm intervals for water content and for coulometric analysis of organic and inorganic carbon (Barry et al. 2002). The cores were initially analysed at 10 cm intervals for the abundance of biogenic silica (BSi), using a time series digestion (DeMaster 1979) in 0.5N NaOH at 85 °C. Two cores, M98-1P and 2P, were subsequently sampled at 1 cm intervals to generate high-resolution profiles of % BSi. The analytical procedure for BSi was modified to entail a single determination of dissolved silica in the digestion solution after the sediment digested for 42.5 minutes. The timing of the extraction was determined after examining the results of 168 analyses of Lake Malawi sediments using DeMaster's (1979) more labour-intensive, time-series procedure. This new, more efficient method produced results that agree well with the DeMaster technique (standard deviation of only 2.2% BSi).

Core M98-2P was also sampled at 10 cm intervals for analysis of certain trace elements, including niobium and titanium, and for total phosphorus (TP) and inorganic phosphorus (IP). Nb and Ti were determined by inductively coupled plasma-mass spectrometry on total sediment digests. Based on repeated analyses of standard reference materials (National Institute of Standards and Technology (NIST) and the Geological Survey of Canada (GSC)) we estimate our accuracy and precision for the ratio of Nb/Ti to be better than 4%. Phosphorus (TP and IP) was analysed in the sediments following the procedure outlined in (Aspila et al. 1976). With this approach we obtained recoveries on NIST reference materials of 94% or better for total P. The precision of the P analyses, based on duplicate and triplicate determinations, were typically better than 5%.

The chronology in the cores is based primarily on radiocarbon dates supported by stratigraphic correlations among the cores and, in the youngest sediments, varve counts supplemented by ^{210}Pb dates (Johnson et al. 2001; Barry et al. 2002). Bulk organic carbon was dated in some samples and in others, pollen extracts (the organic residue from standard pollen preparation procedures) were dated. There was no significant difference between dates of these different components when compared using replicate samples (Barry 2001). The ages were corrected by subtracting 450 years from the reported radiocarbon age, based on four radiocarbon dates recovered on both bulk organic matter and pollen extracts in a multi-core from the north basin, M98-11MC, immediately adjacent to core site M98-1P (Barry 2001). These dates were compared to the chronology of the multi-core established by ^{210}Pb and varve counts (Johnson et al. 2001). The 450-year offset is believed to be due primarily to old, reworked organic matter mixed in with the recently formed organic matter that is entrained in the sediment falling to the lake floor, rather than a reservoir effect. The corrected radiocarbon dates were converted to calendar ages using the programme, CALIB version 4.3 for Macintosh, provided by the University of Washington (Stuiver and Reimer 1993).

Results

General stratigraphy

The piston cores unfortunately were not varved continuously throughout their lengths. Most of them consist of five stratigraphic units that could be correlated visually among the cores

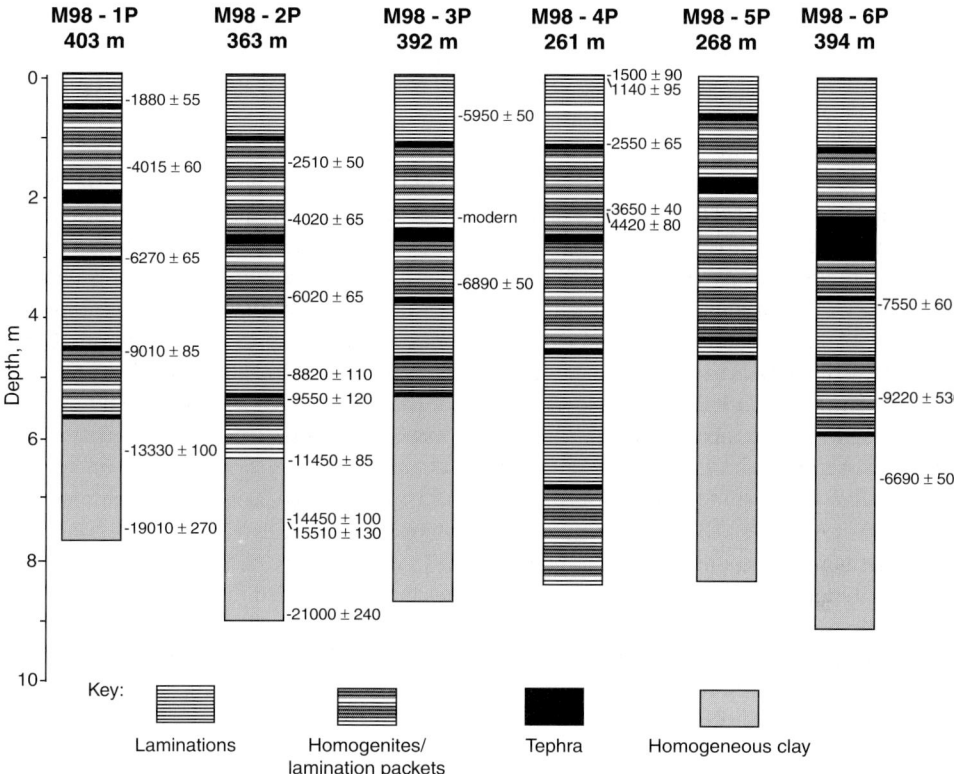

Figure 3. General stratigraphy and radiocarbon dates in 6 piston cores recovered from Lake Malawi in 1998. Core locations are shown in Figure 1.

(Fig. 3) (Barry et al. 2002):

• an upper varved unit about one metre thick (Unit I)

• an upper mixed unit of alternating varves and homogenites, 2–3 m thick (Unit II)

• a lower varved unit, 1–2 m thick (Unit III)

• a lower mixed unit of alternating varves and homogenites, 1–2 m thick (Unit IV)

• a homogeneous silty clay (Unit V)

The varves are typically 0.7–1.7 mm thick, consisting of diatom ooze alternating with silty-clay representing, respectively, the dry windy season and the rainy season. In Units II and IV, individual varve packets are typically 2–3 cm thick and the intervening homogenites, which are diatomaceous silty-clays, are of comparable thickness. The basal contact of individual homogenites are typically erosional, cross-cutting the underlying varves. The homogenites are interpreted to be muddy turbidites, derived from debris flows which perhaps were triggered by earthquakes or storm surges up-slope from the core sites. The upper 50 cm of

Unit I contains three homogenites, which can be correlated among cores M98-4PG, 10MC and 11MC. They are 3–4 cm thick at the shallower core site (4PG) and 1–2 cm thick at the deeper sites of the two multi-cores. Varve counts from the top of each core to the top of the third homogenite are 252, 252 and 236, respectively. The close agreement among these counts suggests that the turbidity currents that form these homogenites do not erode many of the underlying varves. Unit V is a stiff brown clay that is distinctly different from the overlying units in its absence of varves. There are subtle colour changes throughout the unit, at times with sharp contacts. Interspersed among all of the units are a few, thin layers of tephra, ranging from a few centimetres to only millimetres thick. The tephra is derived from the Rungwe volcano field to the north of the lake.

Several radiocarbon dates of organic matter were obtained in the cores (Fig. 3). They indicate that Units I-IV are Holocene in age and Unit V is Late Pleistocene, extending back to about 21,000 ^{14}C yr BP or about 25,000 cal yr BP (Barry 2001). Variation in apparent sediment age at the top of the cores indicates some over-penetration, and perhaps also variable old-carbon contamination between sites.

Most of the detailed analyses on the north basin cores thus far have been focused on cores M98-1P and 2P. These cores have representative stratigraphy for the north basin and have been radiocarbon-dated most extensively.

Biogenic silica

There is pronounced structure in biogenic silica profiles spanning the last 25,000 years (Fig. 4). The profiles correlate well with one another, indicating that the cores record basin-wide phenomena. For example, they all show a %BSi minimum, corresponding with the upper part of Unit IV. However, these correlatable signals are restricted to the north basin of the lake. Cores recovered farther to the south, near the centre of the lake, show little variation in biogenic silica over the past several thousand years. The high-resolution profiles of %BSi in north-basin cores M98-1P and 2P, recovered 10 km apart in 363 m and 403 m water depth, respectively, also show remarkable similarity, not only in their gross structure but also in the presence of exceptionally high or low values (Fig. 5). We assume that these exceptional peaks and troughs are isochronous horizons in the cores. We used ten of these BSi horizons and four ash layers to refine the chronology in M98-1P and 2P in the following manner. Ages for the stratigraphic tie points were established independently in each core by linear interpolation between corrected, calibrated dates in the respective core. These independently-derived dates were then compared in M98-1P and M98-2P and a single, "correlated age" was assigned to the stratigraphic horizon in both cores. In most instances the age determined in M98-2P was used because this core has more dates, especially prior to 10 kyr (Table 1).

The biogenic silica profiles of M98-1P and 2P were re-plotted in terms of mass accumulation rate of biogenic silica (BSi MAR), using the relationship:

$$BSi\,MAR = MAR_{sed} * \% \,BSi/100 \qquad (1)$$

and:

$$MAR_{sed} = LSR^{*}(1 - \Phi)^{*}\rho, \qquad (2)$$

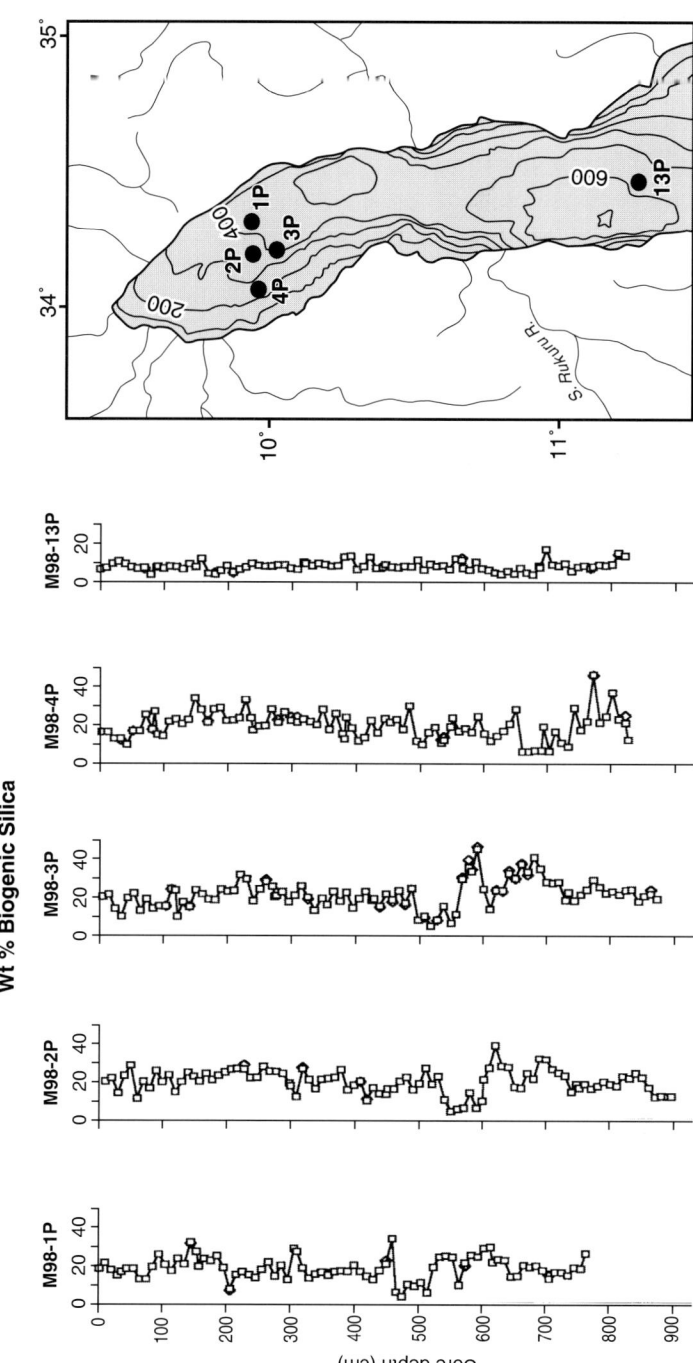

Wt % Biogenic Silica

Figure 4. Weight per cent biogenic silica in five piston cores from the northern half of Lake Malawi. Note the lesser abundance and variability in %BSi in core M98-13P from the central basin.

Table 1. Stratigraphic marker horizons used in the two cores to establish a consistent age model for the two cores.

Marker Horizon	Type	Age in 1P (Cal ybp)	Age in 2P (Cal ybp)	Correlated Age
A2	ash	1044	960	1000
A3	coarse ash	4410	4375	4400
G	BSi low	5393	5473	5473
A4	ash	8633	9084	9084
d	BSi peak	9598	10275	10275
D	BSi low	10023	10621	10621
E	BSi low	11337	11634	11634
F	BSi low	13369	13111	13111
A5	ash	14838	15409	15200
I	BSi low	16367	17285	17285
B	BSi low	19393	19105	19105
b	BSi peak	21855	21195	21195
A	BSi low	22163	21451	21451
a	BSi peak	22573	21877	21877

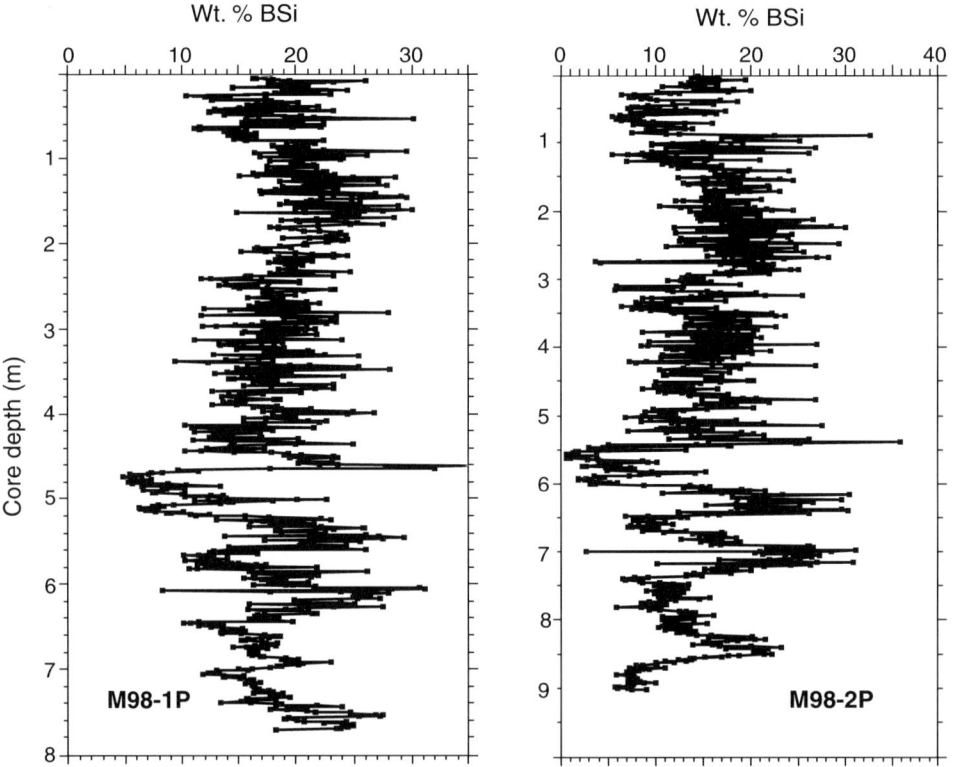

Figure 5. High-resolution profiles of weight per cent biogenic silica in cores M98-1P and M98-2P. Certain peaks and valleys, as well as extreme values, in the two cores are readily correlated.

where: MAR_{sed} is the mass accumulation rate of the dry bulk sediment

Φ is sediment porosity, derived from water content

ρ is dry sediment density and assumed equal to 2.54 g/cm^3

LSR = linear sediment accumulation rate interpolated between any two dated horizons.

The BSi MAR profiles of the two cores are nearly identical and, at the millennial scale, resemble the Greenland ice core records of $\delta^{18}O$ (e.g., the GRIP record) (Fig. 6). Both BSi MAR profiles have lower mean values in the Pleistocene than in the Holocene, and the transition from Pleistocene to Holocene values occurs in two abrupt steps, separated by an interval of Pleistocene-like conditions. However the timing of the glacial-interglacial transition is different from the GRIP record. The profiles in both cores exhibit an abrupt shift to higher values beginning at about 13 kyr BP followed by a return to lower values about a thousand years later. The BSi MAR rises abruptly once again around 10.3 kyr, to Holocene values. The initial rise to Holocene-like values at 13 kyr occurred just when the Northern Hemisphere plunged back into glacial conditions at the beginning of the Younger Dryas. We initially tried to force the Malawi age model to conform to the GRIP record, with the abrupt rise in BSi MAR dated at the beginning of the Bolling-Allerod at 15 kyr. But this is impossible because there is no conceivable mechanism that would yield radiocarbon ages in the Malawi cores that would consistently be 1600 years too young. Interestingly, all of the subdued peaks and valleys of the Pleistocene portion of the BSi MAR profile are anti-phased with the GRIP record (Fig. 6). When BSi MAR values increase (i.e., towards Holocene values), the GRIP record shifts towards colder inferred temperatures. The Malawi BSi MAR record does not relate as closely to the Antarctic ice core records at Byrd or Vostok (Blunier and Brook 2001), nor to the equatorial Atlantic sea surface temperature record (Vidal et al. 1999) as to the Greenland records.

The inverse correlation between the Malawi BSi and GRIP records breaks down after 11 kyr, after which time the Malawi record exhibits considerable change in the Holocene while the GRIP record remains remarkably stable. The BSi MAR was relatively low around 9.2 kyr, 5.5 kyr and 1 kyr, and it was at its highest around 3–4.5 kyr. The climatic tie to the high latitudes of the north appears to have weakened considerably with the demise of the continental ice sheets.

The variance in BSi MAR in the Holocene sediments is far greater than in the Pleistocene (Fig. 6), suggesting more inter-decadal variability in the ecosystem, and perhaps in tropical African climate, in interglacial than in glacial times.

Diatom assemblages

Gasse et al. (2002) record the species abundances of diatoms in M98-2P, which show substantial variability over the past 25,000 years. One of the major conclusions from this study that will be shown to be relevant to the BSi MAR profiles is based on the relative abundance of periphytic diatoms. It varies from more than 20% of the diatom assemblage for much of the period before 15.5 kyr to less than 10% for most of the time since 15 kyr. Periods of high abundance of periphytic diatoms are interpreted to reflect low stands in lake level which prevailed between about 23 and 15.5 kyr, and briefly around 10.5 kyr (Fig. 7) (Gasse et al. 2002). Fluctuations are also observed in the relative abundance of major planktonic

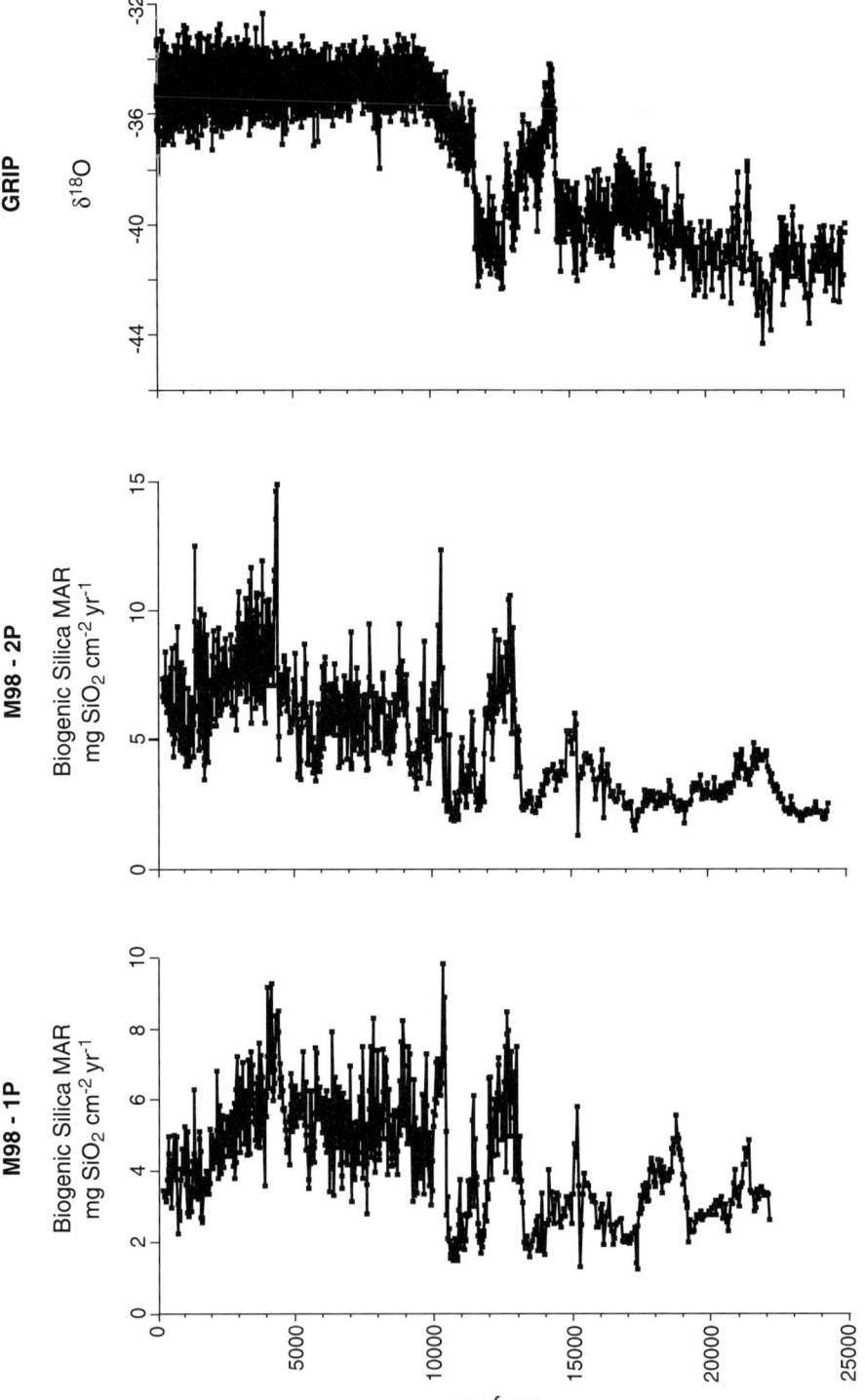

Figure 6. The mass accumulation rate of biogenic silica in cores M98-1P and 2P (left, centre) compared to the Greenland ice core GRIP oxygen isotope record (right).

diatoms (*Aulacoseira nyassensis, Stephanodiscus* spp. and *Cyclostephanos* spp.) that are inferred by Gasse et al. (2002) to indicate different degrees of nutrient upwelling, but the impact of differential preservation of different genera of diatoms cannot be ruled out as the primary cause of these fluctuations.

Nb/Ti *ratios in bulk sediments*

The abundance of certain trace elements was determined in bulk sediment by inductively coupled plasma-mass spectrometry (ICP-MS). Here we use the Nb/Ti ratio as a measure of the relative abundance of volcaniclastic debris in the sediments. Titanium is a component of most aluminosilicate minerals derived from the drainage basin, whereas niobium is an incompatible element that is not readily incorporated into silicate minerals when they are formed. Consequently niobium is concentrated in magma as silicate minerals crystallize, and ultimately becomes incorporated into the glass residue, including tephra, derived from volcanic eruptions. Discrete tephra layers in the piston cores from northern Lake Malawi have Nb/Ti ratios of .04–.06, whereas the background sediment has ratio values typically in the range of 0.01 to 0.02. While tephra has been introduced directly to the lake at times of volcanic eruption, additional volcaniclastic debris enters the lake after it has been chemically weathered and eroded from the drainage basin. The weathered residue of volcanic ash consists of new clay minerals and degraded oxides and hydroxides that still carry the high Nb/Ti ratio in its bulk chemistry. Except for the discrete tephra layers in the cores, we rarely observe volcanic glass shards in the sediments. Nevertheless a profile of Nb/Ti in M98-2P exhibits considerable structure that we attribute to greater or lesser abundance of volcaniclastic material.

A remarkable coincidence occurs between the peaks of BSi MAR and Nb/Ti in M98-2P (Fig. 7). The volcanic ash in Lake Malawi is probably derived from just one area, the Rungwe volcanoes to the north of the lake (Fig. 1). These are late Tertiary to Holocene (with eruptions occurring as recently as the early 19th century) basalts and phonolytic trachytes (Harkin 1955). Discounting the sharp peaks associated with the discrete ash layers, the Nb/Ti ratio appears to have higher average values in the Pleistocene than in the Holocene sediments (Fig. 7).

Phosphorus

While the BSi MAR is interpreted to reflect past diatom productivity in the overlying waters, diatom productivity may not reflect primary production as a whole. It is conceivable that primary productivity by some other group of phytoplankton, such as green or blue-green algae, had already increased by 15.5 kyr, when the periphytic diatom record indicates that lake level rose. To test this, we examined the mass accumulation rate of phosphorus, another potential proxy of palaeoproductivity in marine and lake sediments, in core M98-2P. Phosphorus in sediments is associated with organic matter (organic phosphorus) and in minerals such as apatite and vivianite (inorganic phosphorus). With lower values in the Pleistocene and higher values in the Holocene, the total phosphorus MAR mimics the BSi MAR in M98-2P, at least on a glacial-interglacial scale (Fig. 7). However the TP MAR profile does not show nearly as much structure as the BSi MAR, and in particular

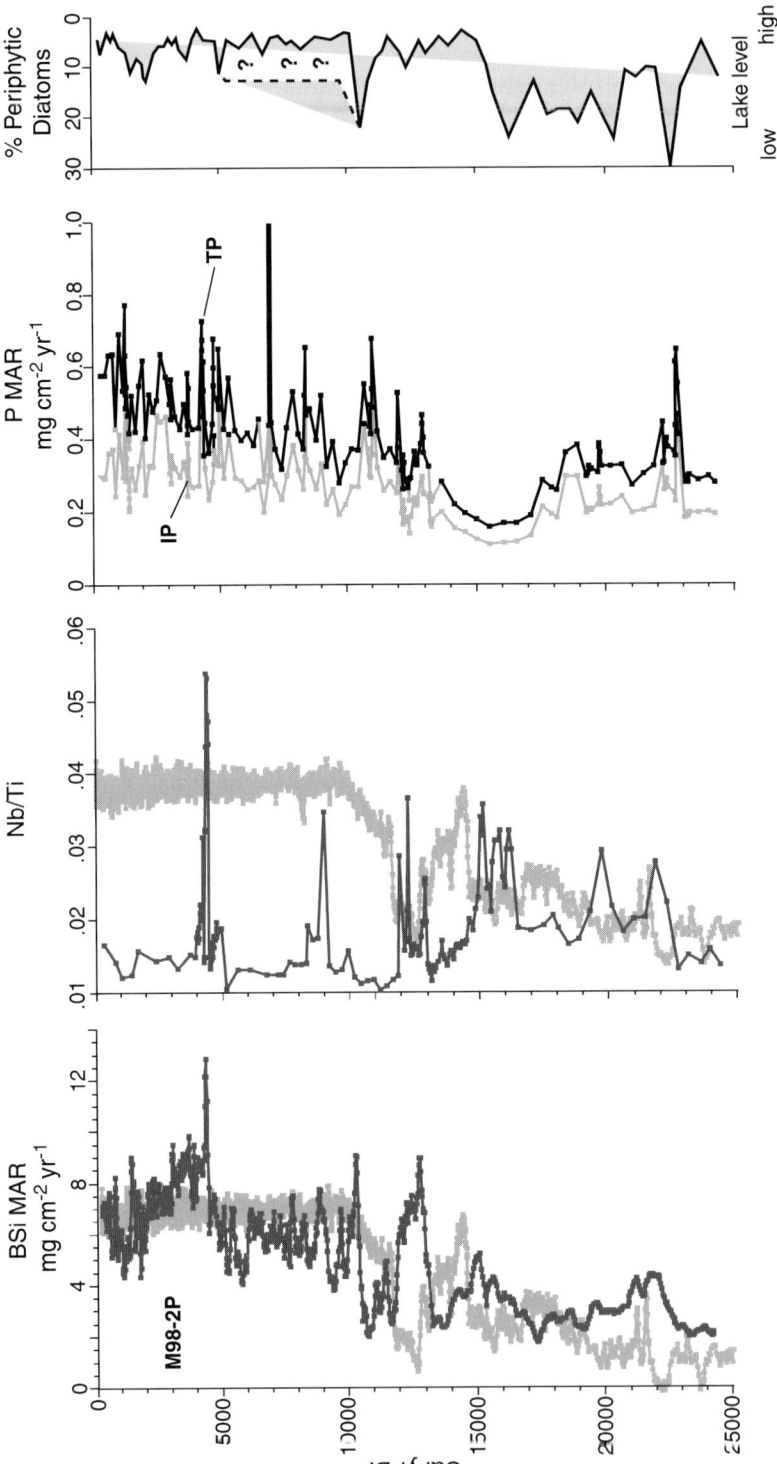

Figure 7. Far left: The smoothed profile of BSi MAR in M98-2P (dark curve) with the smoothed GRIP oxygen isotope record superimposed (light curve). Left centre: The ratio of niobium to titanium in M98-2P (dark curve) with the smoothed GRIP oxygen isotope record superimposed (light curve). Right centre: The mass accumulation rate of total phosphorus (TP) and inorganic phosphorus (IP) in M98-2P. The difference between TP and IP is the abundance of organic phosphorus in the core. Far right: the relative abundance of periphytic diatoms in M98-2P (from Gasse et al. (2002)). The question marks and dashed line in the diatom profile refer to the early Holocene lowstand of the lake not revealed by the diatom assemblage but inferred from the endogenic calcite found in sediments of this age in the southern part of the lake (see text).

does not show the pronounced two-step transition from glacial to interglacial conditions. Nevertheless, the rise in TP MAR to Holocene values accompanies the more abrupt rise in BSi MAR at 13 kyr, and not when diatom floras indicate lake level rose, at 15.5 kyr.

Discussion

Spatial differences in the biogenic silica profiles (i.e., between the northern and central basins of the lake, Fig. 4) are explained by the upwelling dynamics of a long rift lake. Winds are channeled along the axis of the rift valley, thereby establishing a preponderance of either northerly or southerly winds over the lake. The former will promote upwelling at the north end and the latter the opposite. Under such circumstances, upwelling will rarely occur in the central basin.

Upwelling can also be strongly influenced by internal seiches. Plisnier (2000) documented this clearly for Lake Tanganyika, where upwelling is well established at the south end of the lake during the season of intense southerly winds. At the end of the windy season, the thermocline is depressed in the north basin due to prevailing winds out of the south. As the winds subside, the thermocline rebounds, causing an internal wave that oscillates back and forth through the basin with a period of about a month. The crest of this internal wave breaks periodically into the photic zone at the north end of the lake, creating a pulse of high productivity (Plisnier 2000). The same happens at the south end of the lake, but not in the middle, where the internal seiche may not shoal and break. So primary production dynamics appear to be amplified at the ends of Lake Tanganyika and much less variable near the centre of the lake. This is probably the case in most long, narrow, rift lakes, including Malawi. The pattern of upwelling associated with axial winds and perhaps breaking internal waves would explain the differences observed in biogenic silica profiles from the center of the lake versus the north basin.

We envisage two possible mechanisms to explain the temporal variations in BSi MAR in northern Lake Malawi. First, the supply of silica to the lake is dependent on the weathering rate in the catchment: warm, wet conditions promote chemical weathering of bedrock and enhance the supply of dissolved silica to the lake. The concentration of dissolved silica remains similar over a variety of runoff regimes in 13 major rivers flowing into Lake Malawi (Hecky et al. 2003). There appears to be a buffering effect that maintains relatively constant dissolved silica concentrations in many rivers (Burton and Liss 1973; Lesack et al. 1984). For this reason, the river loading of dissolved silica to Lake Malawi can be assumed proportional to river discharge; high under moist conditions and low under dry conditions. Second, an increase in the frequency of northerly winds over the lake would increase upwelling and diatom productivity in the north basin relative to the south basin. This would increase the supply of recycled biogenic silica and other nutrients in the north basin at the expense of the southern reaches of the lake. At present, the winds over Lake Malawi prevail out of the south throughout most of the year, resulting in the highest productivity being situated in the southern arms of the lake (Patterson and Kachinjika 1995; Bootsma and Hecky 1998). The internal wave dynamics described in the previous paragraph may contribute to higher productivity at each end of the lake regardless of which way the wind blows. But upwelling and associated high productivity will be far longer in response to the prolonged offshore wind at the upwind end of the lake than it will be to the relatively short-lived breaking of internal waves at the downwind end.

The first mechanism undoubtedly explains the difference between glacial and inter-glacial BSi MAR's. The relatively low BSi MAR in Pleistocene sediments is attributed to the cool, dry conditions that were pervasive throughout tropical Africa during the last glacial maximum (LGM) (Gasse 2000). The relative abundance of periphytic diatoms indicates that the Malawi basin was relatively dry during the late Pleistocene. The magnitude of this lowstand is not yet known, but was probably between 100 m and 200 m lower than present, based on seismic-reflection profiles and sediment-core stratigraphy (Johnson and Ng'ang'a 1990). Substantial drops in periphytic diatom abundance occur at about 22 kyr and 16 kyr, presumably associated with rises in lake level, and are followed by modest rises in BSi MAR. Other than these possible ties at 22 and 16 kyr, Pleistocene lake level, as inferred from the diatom record, and fluctuations in BSi MAR do not appear to be tightly coupled.

The second mechanism (more frequent northerly winds) can explain the anti-phase relationship between the Pleistocene BSi MAR and the GRIP record depicted in Figure 7. The Nb/Ti ratio in the sediments of core M98-2P support this hypothesis. Discounting the sharp peaks associated with discrete ash layers in the Holocene sediments, the Nb/Ti ratio appears to have higher average values in the Pleistocene than in the Holocene sediments (Fig. 7). Northerly winds blowing over the cool, dry volcanic landscape would have trans-ported ash or its weathered residue to the lake. Times of higher Nb/Ti in the Pleistocene sediments reflect more frequent or more intense north winds over the basin. These winds would have caused upwelling at the north end of the lake, promoting high diatom produc-tivity. Northern Lake Malawi is exposed to northerly winds in austral summer, when the intertropical convergence zone (ITCZ) has migrated to its southern position (Fig. 2) (Tyson 1986). The coincidence of high Nb/Ti ratios and high BSi MAR in the late Pleistocene sediments of northern Lake Malawi, with the periods of relatively cool conditions over Greenland, suggest that the ITCZ either migrated farther southward, or remained at its southern terminus longer, at these times. Given the uncertainty in the chronology of the Malawi cores, we cannot determine whether the high latitudes are impacting the behavior of the ITCZ, or vice versa.

The first dramatic increase in the diatom production of northern Lake Malawi to Holocene-like conditions started at 13 kyr, coincident with the onset of the Younger Dryas in the Northern Hemisphere. The benthic diatom record suggests that lake level had already risen substantially, more than 2000 years previously. The region's climate was moister than during the Pleistocene, and the biology in the north basin was poised to respond. We hypothesise that this biological response awaited the onset of northerly winds associated with the Younger Dryas, and perhaps the termination of the Antarctic Cold Reversal (ACR), which marked the resumption of warming to Holocene conditions in high southern latitudes.

It is not surprising that the phosphorus MAR does not faithfully track the BSi record in the Holocene (Fig. 7). Changes in soil development associated with varying climate or volcanic sources of P and BSi will influence the nutrient source function to the lake (Chadwick et al. 1999; Filipelli and Souch 1999). Furthermore, the lake's internal phos-phorus cycle is likely to be tightly coupled to the redox behavior of iron, especially where the chemocline, marking the boundary between oxic and anoxic waters, intersects the lake floor. The chemocline is presently at about 200 m depth in Lake Malawi, but it has migrated vertically by more than 100 m during the Holocene (Finney and Johnson 1991; Brown et al. 2000). This dynamic behavior of the chemocline very likely affects the geochemical mass balance of phosphorus in Lake Malawi in ways that will impact its delivery to sediments in the deep, anoxic basins offshore.

While the diatom record implies that lake level has been relatively high, i.e., near present-day level for most of the past 10,000 years, the BSi MAR record exhibits relatively low values in the early Holocene compared to the late Holocene, with a sharp break at about 4.5 kyr. Piston cores recovered from water depths less than 265 m in southern Lake Malawi contain endogenic calcite in sediments that date between 11.2 kyr and 5 kyr (Finney and Johnson 1991; Finney et al. 1996; Ricketts and Johnson 1996). The carbonate record is interpreted to reflect a time when the lake was a closed basin, i.e., with a water level that was lower than the present outlet. It appears that the benthic diatom record is not sufficiently sensitive to register this intermediate early Holocene lake level. The slightly depressed BSi MAR in the north basin of the lake may be attributed to these relatively dry conditions.

The abrupt rise in BSi MAR at 4.5 kyr may have marked the onset of more frequent northerly winds in the basin once again. The relatively low Nb/Ti ratios throughout the Holocene show a slight rise just prior to 4.5 kyr, then a brief excursion to very high values associated with a volcanic ash layer in the core, and then a return to the slightly elevated values of the late Holocene.

Alternatively, a shift to wetter conditions around 4.5 kyr may be the explanation for the sudden rise in BSi MAR. The periphytic diatom record, while somewhat insensitive to the subtleties of the Holocene lake level, is at least consistent with a highstand culminating shortly after 4.5 kyr (Fig. 7). This rise in lake level would be just the opposite of many lakes in Northern Africa (e.g., Abhe, Ziway Shala, Turkana) that experienced an abrupt shift to lower levels at around the same time. This behavior of Lake Malawi, coupled with its status as a closed basin during the early Holocene when the lakes in equatorial and Northern Africa were high (Street-Perrott and Roberts 1983; Johnson 1996; Gasse 2000), once again suggests antiphase behaviour between Malawi and lakes to the north. The very high diatom productivity which was initiated around 4.5 kyr subsided by 1.5 kyr.

During the last millennium, the BSi MAR appears to have maintained a subtle link to the climate of the Northern Hemisphere. Diatom productivity was high during the Little Ice Age (LIA) (Fig. 8) (Johnson et al. 2001) and relatively low around 1 kyr, the time of the Medieval Warm Period (MWP) (Fig. 5). These trends are consistent with the observations of millennial-scale variability observed in the late Pleistocene, and suggest again that north winds were more prevalent over the Malawi basin when the climate of the Northern Hemisphere was relatively cool. Johnson et al. (2001) attributed the high diatom productivity in the north basin during the LIA to a lowstand of the lake at that time and the influence that the lowstand would have on the geochemical mass balance of silica in the lake system. Owen et al. (1990) independently concluded that the lake level was low around the time of the LIA, and Johnson et al. (2001) suggested that the BSi record from northern Lake Malawi could be used to refine the timing of that lowstand. Now we lean more towards the north-wind hypothesis to explain the observed decade-scale variability in diatom productivity over the past few centuries. The evidence for a lowstand in the cores examined by Owen et al. (1990) remains unchallenged, but how this may or may not tie into the biogenic silica record in the north basin, or the position of the ITCZ, remains open to further examination.

If the ITCZ indeed migrated farther south or remained at its southern terminus longer, this would imply that Malawi should experience longer summer rains, and therefore have

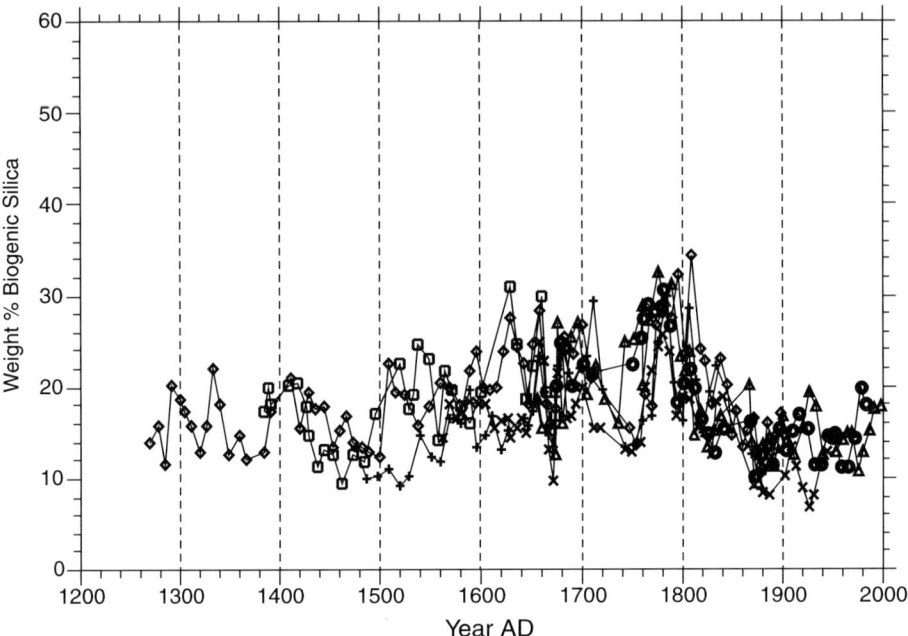

Figure 8. High resolution record of biogenic silica abundance in six short cores from the north basin. The chronology is based on varve counts and ^{210}Pb dating (Modified from Johnson et al. (2001)). The sediment accumulation rates in these cores is relatively constant so the per cent BSi profiles would be nearly identical to profiles of BSi MAR in the same cores.

been at a highstand during the LIA. However the vertical convection associated with the ITCZ could be weakened under such circumstances, leading to less rainfall. Position and intensity of the ITCZ will impact the amount of rain that will fall on the Malawi basin.

Farther to the south in Africa, regions not ordinarily exposed to long summer rains associated with the ITCZ may indeed experience more rainfall when the ITCZ migrates farther southward, regardless of the convection intensity. A high-resolution record of past climate has been generated from speleothems in Cold Air Cave at Makapansgat Valley in the Northern Province of South Africa, at about 24 °S Latitude (Holmgren et al. 1999), about 1500 km to the south of Lake Malawi's north basin. The chronology of this record has been revised in Lee-Thorp et al. (2001). Holmgren et al. (1999) and Lee-Thorp et al. (2001) attribute enrichment in both $\delta^{18}O$ and $\delta^{13}C$, as well as dark coloration of the stalagmite growth layers, to relatively wet conditions. There is not a direct correlation between rainfall over Malawi and the Makapansgat Valley (Nicholson and Entekhabi 1987), but there could conceivably be a relationship between north winds over Malawi and high rainfall in the Northern Province of South Africa. A comparison of the high-resolution record of biogenic-silica deposition in northern Lake Malawi over the past 700 years (Johnson et al. 2001) with the Cold Air Cave record shows all periods of high diatom productivity (attributed to north winds) to correspond with wet conditions in the Makapansgat Valley (Fig. 9). However the relationship is not perfect: clearly the tie between the climate of the Malawi basin and Makapansgat Valley is not a simple or consistent one. On a longer time-scale

the relationship between the records of Cold Air Cave and northern Lake Malawi is less impressive (Fig. 10).

There is renewed interest in the impact of solar variability on climate, based in part on the results of climate models (e.g., Tett et al. (1999), Crowley (2000)). This interest is further enhanced by a few correlations between palaeoclimate records and the history of solar activity derived from the abundance of the cosmogenic radionuclides ^{10}Be and ^{14}C in tree rings and ice cores. A compelling argument is made by Verschuren (2003) and Verschuren et al. (2000) for the influence of solar activity on the hydrological budget of Lake Naivasha in Kenya. There, periods of low solar activity (the Maunder, Sporer and Wolf minima) coincide with high effective moisture. Tyson et al. (2000) suggest a possible link between changes in solar irradiance and temperature at Makapansgat Valley based on the δ^{18}O record in the Cold Air Cave stalagmite and the δ^{14}C and δ^{10}Be records in ice cores. However the Lake Malawi high-resolution record spanning the past 700 years shows no consistent trend for Malawi with regard to times of solar irradiance maxima or minima (Fig. 9). Periods of maximum diatom productivity in northern Lake Malawi fall during times of high, low and sometimes transitional values of solar activity (Fig. 9).

Summary

The sediments of northern Lake Malawi carry high-resolution signals of past climate change that can be linked to diatom productivity and, we hypothesise, to the hydrological budget and the recurrence of north winds over the lake basin. Diatom productivity was relatively low during the late Pleistocene compared to the Holocene, due to relatively dry conditions in tropical Africa during the last glacial period. Millennial scale perturbations in the mass accumulation rate of diatoms, especially prior to 11 kyr, are attributed to the frequency or strength of northerly winds blowing over the lake. A prevalence of northerly winds would occur when the ITCZ is at the southern terminus of its seasonal migration to the south during austral summer. If it migrated farther south, or maintained its southern position for a longer duration than normal, winds would more likely be channeled over the lake basin out of the north than out of the south. Under these circumstances, diatom productivity in the north basin of the lake would be enhanced. Prior to 11 kyr, diatom productivity was higher when the oxygen-isotope records in Greenland ice cores indicate relatively cool conditions in the high northern latitudes. The Lake Malawi record shows far more variability in the Holocene than do the Greenland ice-core records, and a subtle link between diatom productivity in Malawi and the climate of the Northern Hemisphere apparently continues to exist in a manner consistent with the late Pleistocene. Diatom production was relatively high during the Little Ice Age and relatively low during the Medieval Warm Period.

There is not a consistent link between the high-resolution records of Lake Malawi and Lake Naivasha (Kenya), or between the record of Lake Malawi and the history of solar irradiance inferred from δ^{14}C and δ^{10}Be records in ice cores and tree rings. There appears to be a correlation between high diatom productivity in northern Lake Malawi and high rainfall in the vicinity of Cold Air Cave, South Africa, at least on a decadal time scale spanning the past seven centuries. The link is not perfect, however, and the two phenomena are even less coupled when compared over the past six millennia. The climate of Africa is complex, and since establishing clear, consistent relationships between locales as widespread as South

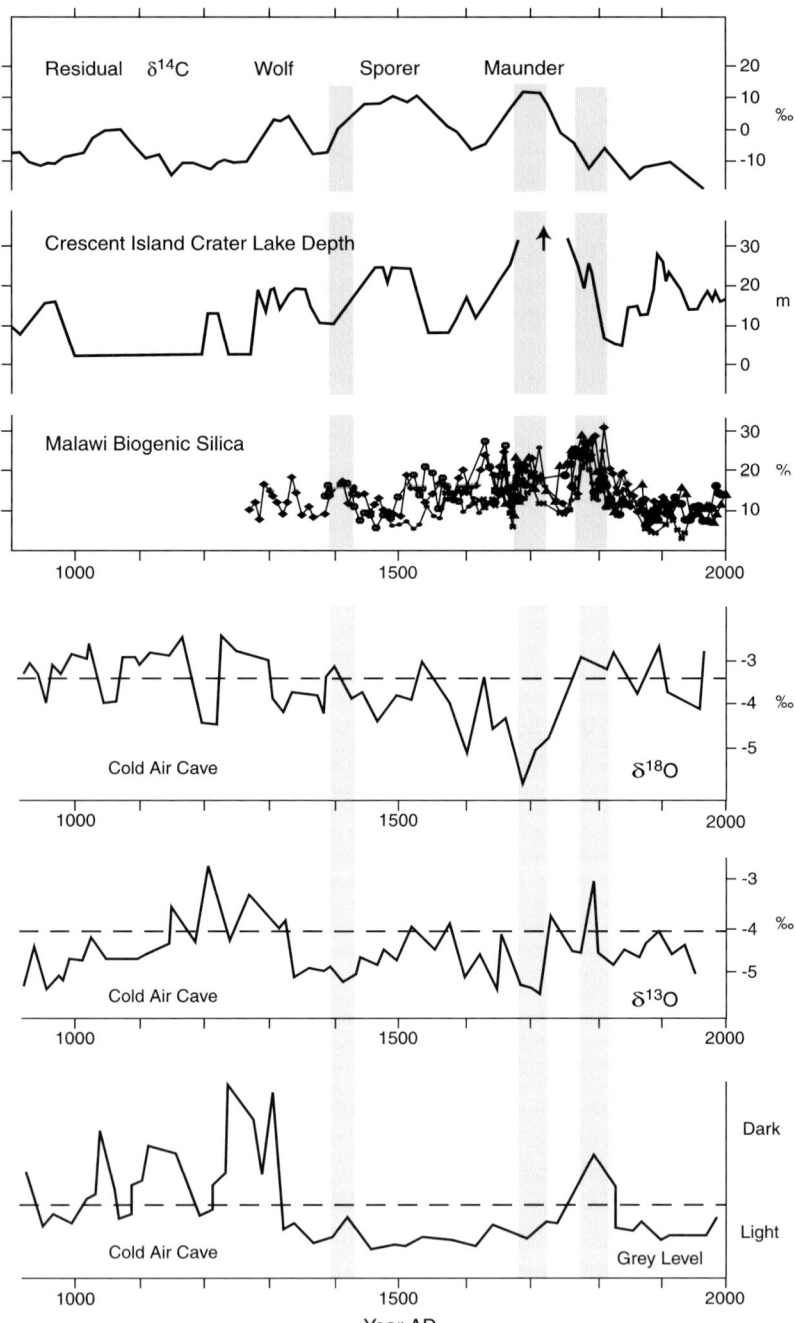

Figure 9. Summary diagram illustrating the high resolution record of diatom abundance in northern Lake Malawi, compared to lake level at Crescent Island Crater in Lake Naivasha, Kenya (Vershuren et al. 2000), solar activity as estimated by the residual activity of ^{14}C measured in tree rings and ice cores (Stuiver and Reimer 1993 in Verschuren et al. 2000), and the results from Cold Air Cave, Northern South Africa (Lee-Thorp et al. 2001; Tyson et al. 2000).

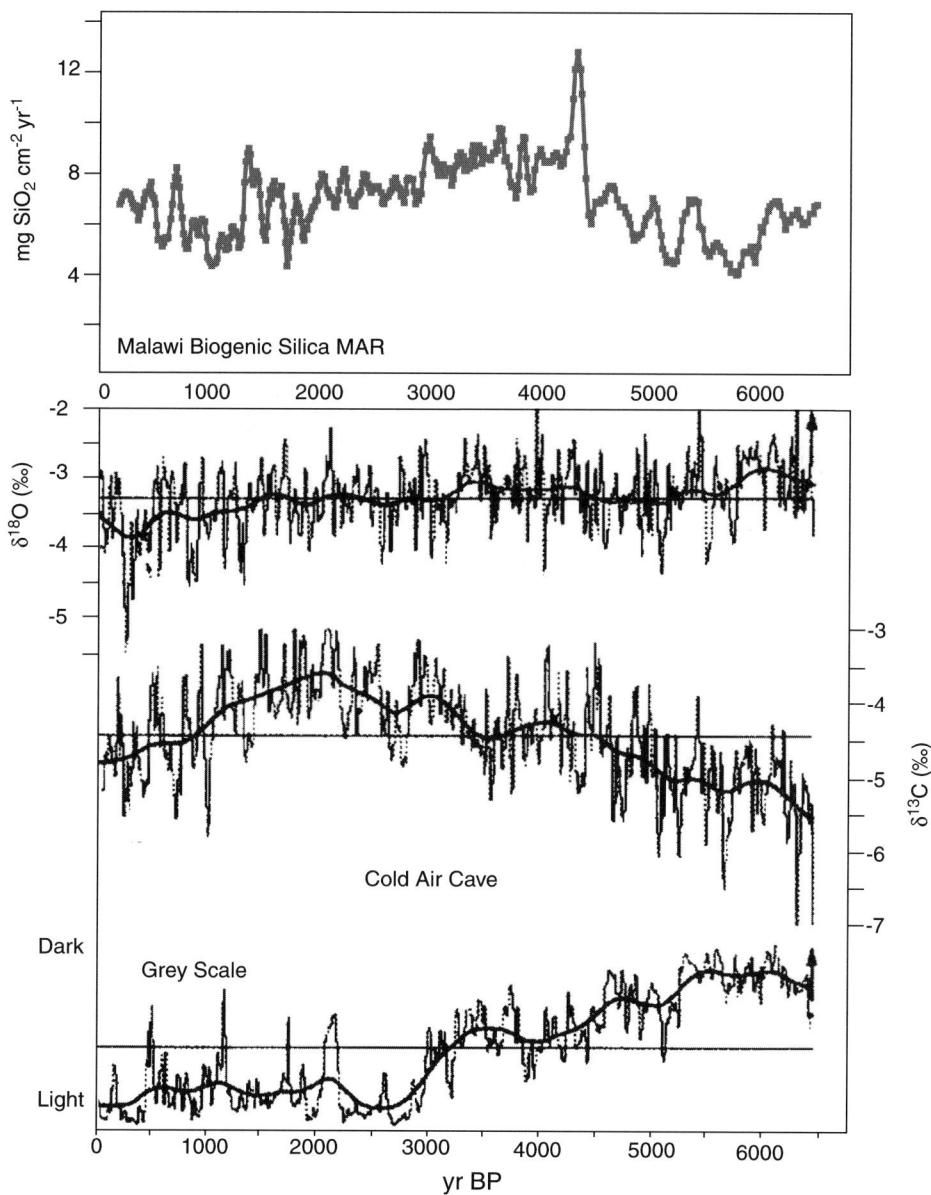

Figure 10. Comparison of the Malawi BSi MAR record for the past 6500 years with the Cold Air Cave record (Lee-Thorp et al. 2001), showing no consistent relationship.

Africa and Kenya is not possible even with the modern climate records, we should find it no more simple with our high-resolution records of past climate change.

Acknowledgments

We thank the Government of Malawi for permission to carry out field work in their country and for providing logistical support. The SADCC-GEF Lake Malawi Biodiversity Project provided access to their field station at Senga Bay and generously provided ship time aboard the R/V Usipa, under the command of Captain Mark Day. Françoise Gasse, Phil Barker and Mike Talbot have provided much useful feedback on an earlier manuscript on the Lake Malawi records, and their input has had considerable influence on the interpretations presented here. Mike Talbot and Dirk Verschuren provided valuable reviews prior to the final draft. We thank Yvonne Chan, Sarah Grosshuesch and Jason Agnich for overseeing the sediment analyses at the Large Lakes Observatory. This material is based upon work supported by the National Science Foundation under Grant No. ATM-9709291. This is publication number 132 of the International Decade for the East African Lakes (IDEAL).

References

Aspila K.I., Agemian H. and Chau A.S.Y. 1976. A semi-automated method for the determination of inorganic, organic, and total phosphate in sediments. Analyst 101: 187–197.
Barry S.L. 2001. Stratigraphic Correlation and High-Resolution Geochronology of Varved Sediments from Lake Malawi, East Africa. M.S.Thesis, University of Minnesota, Duluth, 125 pp.
Barry S.L., Filippi M.L., Talbot M.R. and Johnson T.C. 2002. A 20,000-yr sedimentological record from Lake Malawi, East Africa: the Late-Pleistocene / Holocene transition in the southern tropics. In: Odada E.O. and Olago D.O. (eds), The East African Great Lakes: Limnology, Palaeoclimatology and Biodiversity. Kluwer Academic Publishers, Dordrecht, pp. 369–392.
Blunier T. and Brook E.J. 2001. Timing of millennial-scale climate change in Antarctica and Greenland during the last glacial period. Science 291: 109–112.
Bootsma H.A. and Hecky R.E. 1998. Water Quality Report: Draft Document.
Brown E.T., Le Callonnec L. and German C.R. 2000. Geochemical cycling of redox-sensitive metals in sediments from Lake Malawi: A diagnostic paleotracer for episodic changes in mixing depth. Geochim. Cosmochim. Acta 64: 3515–3523.
Camberlin P., Janicot S. and Poccard I. 2001. Seasonality and atmospheric dynamics of the teleconnection between African rainfall and tropical sea-surface temperature: Atlantic vs. ENSO. Int. J. Climatol. 21: 973–1005.
Chadwick O.A., Derry L.A., Vitousek P.M., Huebert B.J. and Hedin L.O. 1999. Changing sources of nutrients during four million years of ecosystem development. Nature 397: 491–497.
Cohen A.S., Soreghan M. and Scholz C.A. 1993. Estimating the age of formation of lakes: an example from Lake Tanganyika, East African rift system. Geology 21: 511–514.
Crowley T.J. 2000. Causes of climate change over the past 1000 years. Science 289: 270–277.
DeMaster D.J. 1979. The marine budgets of silica and Si^{32}. Ph.D. Dissertation, Yale University, New Haven, 308 pp.
Eccles D.H. 1974. An outline of the physical limnology of Lake Malawi (Lake Nyasa). Limnol. Oceanogr. 19: 730–742.
Filipelli G.M. and Souch C. 1999. Effects of climate and landscape development on the terrestrial phosphorus cycle. Geology 27: 171–174.

Finney B.P. and Johnson T.C. 1991. Sedimentation in Lake Malawi (East Africa) during the past 10,000 years: a continuous paleoclimate record from the southern tropics. Palaeogeog. Palaeoclimat. Palaeoecol. 85: 351–366.

Finney B.P., Scholz C.A., Johnson T.C., Trumbore S. and Southon J. 1996. Late Quaternary lake level changes of Lake Malawi. In: Johnson T.C. and Odada E.O. (eds), The Limnology, Climatology, and Palaeoclimatology of the East African Lakes. Gordon and Breach, Toronto, pp. 495–508.

Francois R., Pilskaln C.H. and Altabet M.A. 1996. Seasonal variation in the nitrogen isotopic composition of sediment trap materials collected in Lake Malawi. In: Johnson T.C. and Odada E.O. (eds), The Limnology, Climatology, and Paleoclimatology of the East African Lakes. Gordon and Breach, Amsterdam, pp. 241–250.

Gasse F. 2000. Hydrological changes in the African tropics since the Last Glacial Maximum. Quat. Sci. Rev. 19: 189–211.

Gasse F., Barker P. and Johnson T.C. 2002. A 24,600 yr diatom record from the northern basin of Lake Malawi. In: Odada E.O. and Olago D.O. (eds), The East African Great Lakes: Limnology, Palaeoclimatology and Biodiversity. Kluwer Academic Publishers, Dordrecht, pp. 393–414.

Halfman J.D. 1996. CTD - transmissometer profiles from Lakes Malawi and Turkana. In: Johnson T.C. and Odada E.O. (eds), The Limnology, Climatology and Paleoclimatology of the East African Lakes. Gordon and Breach, Amsterdam, pp. 169–182.

Harkin D.A. 1955. The Sarabwe lava flow, Kiejo, Rungwe district. Tanganyika Notes 40: 20–23.

Hecky R.E., Bootsma H.A. and Kingdon M.L. 2003. Impact of lend use on sediment and nutrient yields to Lake Malawi/Nyasa (Africa). Journal of Great Lakes Research 29: 139–158

Holmgren K., Karlen W., Lauritzen S.E., Lee-Thorp J., Partridge T.C., Piketh S., Repinski P., Stevenson C., Svanered O. and Tyson P.D. 1999. A 3000-year high-resolution stalagmite-based record of palaeoclimate for Northeastern South Africa. The Holocene 9: 295–309.

Johnson T.C. 1993. IDEAL Science and Implementation Plan. PAGES Technical Report 93-1: 1–37.

Johnson T.C. 1996. Sedimentary processes and signals of past climatic change in the large lakes of the East African Rift Valley. In: Johnson T.C. and Odada E.O. (eds), The Limnology, Climatology and Paleoclimatology of the East African Lakes. Gordon and Breach, Amsterdam, pp. 367–412.

Johnson T.C. and Ng'ang'a P. 1990. Reflections on a rift lake. In: Katz B.J. (ed.), Lacustrine Basin Exploration: Case Studies and Modern Analogs. Amer. Assoc. Petrol. Geol., Tulsa, pp. 113–136.

Johnson T.C., Wells J.D. and Scholz C.A. 1995. Deltaic sedimentation in a modern rift lake. Geol. Soc. Amer. Bull. 107: 812–829.

Johnson T.C., Kelts K. and Odada E.O. 2000. The Holocene history of Lake Victoria. Ambio 29: 2–11.

Johnson T.C., Barry S.L., Chan Y. and Wilkinson P. 2001. Decadal record of climate variability spanning the last 700 years in the southern tropics of East Africa. Geology 29: 83–86.

Lehman D. 1998. History and ontogeny of IDEAL. In: Lehman J.T. (ed.), Environmental Change and Response in East African Lakes. Kluwer Academic Publishers, Dordrecht, Germany, pp. 1–6.

Lindesay J.A. 1998. Present climates of Southern Africa. In: Hobbs J.E., Lindesay J.A. and Bridgman H.A. (eds), Climates of the Southern Continents: Present, Past and Future. John Wiley and Sons, New York, pp. 5–62.

Nicholson S.A. and Entekhabi D. 1987. Rainfall variability in equatorial and Southern Africa: Relationships with sea surface temperatures along the southwestern coast of Africa. J. Clim. Appl. Met. 26: 561–578.

Nicholson S.E. 1996. A review of climate dynamics and climate variability in Eastern Africa. In: Johnson T.C. and Odada E.O. (eds), The Limnology, Climatology, and Paleoclimatology of the East African Lakes. Gordon and Breach, Toronto, pp. 25–56.

Owen R.B., Renaut R.W. and Williams T.M. 1996. Characteristics and origins of laminated ferromanganese nodules from Lake Malawi, Central Africa. In: Johnson T.C. and Odada E.O. (eds), The

Limnology, Climatology, and Paleoclimatology of the East African Lakes. Gordon and Breach, Toronto, pp. 461–474.

Owen R.B., Crossley R., Johnson T.C., Tweddle D., Kornfleld I., Davidson S., Eccles D.H. and Engstrom D.E. 1990. Major low levels of Lake Malawi and implications for speciation rates in cichlid fishes. Proc. Royal Soc. London. B. 240: 519–553.

Patterson G. and Kachinjika O. 1995. Limnology and phytoplankton ecology. In: Menz A. (ed.), The Fishery Potential and Productivity of the Pelagic Zone of Lake Malawi/Niassa. Natural Resources Institute, Overseas Development Administration, Chatham, pp. 1–68.

Pilskaln C. and Johnson T.C. 1991. Seasonal signals in Lake Malawi sediments. Limnol. Oceanogr. 36: 544–557.

Plisnier P.-D. 2000. Climate, limnology and fisheries changes in Lake Tanganyika. In: Odada E. (ed.), Proc. 2nd IDEAL Symposium, Club Makakola, Malawi, p. 25.

Ricketts R.D. and Johnson T.C. 1996. Early Holocene changes in lake level and productivity in Lake Malawi as interpreted from oxygen and carbon isotopic measurements of authigenic carbonates. In: Johnson T.C. and Odada E.O. (eds), The Limnology, Climatology and Paleoclimatology of the East African lakes. Gordon and Breach, Amsterdam, pp. 475–493.

Rosendahl B.R. 1987. Architecture of continental rifts with special reference to East Africa. Ann. Rev. Earth Planet. Sci. 15: 445–504.

Scholz C.A., Johnson T.C. and McGill J.W. 1993. Deltaic sedimentation in a rift valley lake: new seismic reflection data from Lake Malawi (Nyasa), East Africa. Geology 21: 395–398.

Scott D.L., Ng'ang'a P., Johnson T.C. and Rosendahl B.R. 1991. High-resolution character of Lake Malawi (Nyasa), East Africa, and its relationship to sedimentary processes. Int. Assoc. Sedimentol. Special Publication 13: 129–145.

Street-Perrott F.A. and Roberts N. 1983. Fluctuations in closed-basin lakes as an indicator of past atmospheric circulation patterns. In: Street-Perrott A., Beran M. and Ratcliffe R.A.S. (eds), Variations In the Global Water Budget. Reidel, Dordrecht pp. 331–345.

Stuiver M. and Reimer P.J. 1993. Long-term C14 variations. Radiocarbon 35: 215–230.

Tett S.F.B., Stott P.A., Allen M.R., Ingram W.J. and Mitchell J.F.B. 1999. Causes of twentieth-century temperature change near the Earth's surface. Nature 399: 569–572.

Tyson P.D. 1986. Climatic Change and Variability in Southern Africa. Oxford University Press, Oxford, 220 pp.

Verschuren D., this volume. Decadal and century-scale climate variability in tropical Africa during the past 2000 years. In: Battarbee R.W., Gasse F. and Stickley C.E. (eds), Past Climate Variability through Europe and Africa. Kluwer Academic Publishers, Dordrecht, the Netherlands, pp. 139–158.

Verschuren D., Laird K.R. and Gumming B.F. 2000. Rainfall and drought in equatorial East Africa during the past 1,100 years. Nature 403: 410–414.

Vidal L., Schneider R.R., Marchal O., Bickers T., Stocker T.F. and Wefer G. 1999. Link between the North and South Atlantic during the Heinrich events of the last glacial period. Clim. Dynam. 15: 909–919.

Wells J.T., Scholz C.A. and Johnson T.C. 1994. Highstand deltas in Lake Malawi, East Africa: Environments of deposition and processes of sedimentation. In: Lomondo A.J., Schreiber B.C. and Harris P.M. (eds), Lacustrine Reservoirs and Depositional Systems. Soc. Sediment. Geol., Tulsa, pp. 1–36.

7. LATE QUATERNARY CLIMATIC VARIABILITY IN INTERTROPICAL AFRICA

PHILIP A. BARKER (p.barker@lancs.ac.uk)
Department of Geography
Lancaster University
Lancaster, LA1 4YB, UK

MICHAEL R. TALBOT (michael.talbot@geol.uib.no)
Geological Institute
University of Bergen
5007 Bergen, Norway

F. ALAYNE STREET-PERROTT
(f.a.street-perrott@swansea.ac.uk)
Department of Geography
University of Wales Swansea
Swansea, SA2 8PP, UK

FABIENNE MARRET (f.marret@bangor.ac.uk)
School of Ocean Sciences
University of Wales Bangor
Menai Bridge
Anglesey, LL59 5EY, UK

JAMES SCOURSE (j.scourse@bangor.ac.uk)
School of Ocean Sciences
University of Wales Bangor
Menai Bridge
Anglesey, LL59 5EY, UK

ERIC O. ODADA (pass@uonbi.ac.ke)
Department of Geology
University of Nairobi
PO Box 30197, Nairobi, Kenya

Keywords: Tropical Africa, Palaeoclimate, Lake-levels, Insolation, Carbon, Oxygen isotopes, Diatoms, Congo Fan, Late Quaternary

117

R. W. Battarbee et al. (eds) 2004. *Past Climate Variability through Europe and Africa.*
Springer, Dordrecht, The Netherlands.

Introduction

Tropical Africa is at the geographical heart of the PEP III transect and forms part of the heat engine which drives the meridional circulation of the atmosphere. The tropics are therefore central to studies of climate change not only in the equatorial belt but also in sub-tropical regions (Yin and Battisti 2001), and may even lead some high latitude climate changes (Henderson and Slowey 2000). Despite this pivotal role, the tropics have historically been the poor relation of temperate regions in palaeoenvironmental research. Here we synthesise new palaeoclimate information derived primarily from forest and savanna regions of tropical Africa. These recent studies have helped to close some empirical gaps in data coverage within the region and at the same time have led to a new conceptual understanding of past tropical climates. The latter has been achieved through the rigorous application of classical methods as well as the development of several new proxies of environmental change.

We focus our discussion on several key palaeoclimate issues that fall within the scope of PAGES time stream 2 (cf. Gasse and Battarbee (this volume)). Although this is defined as spanning the last two glacial cycles, we will place greatest emphasis on the events since the penultimate glacial (marine isotope stage 6, hereafter MIS 6) and especially on the last 30,000 calendar years BP (abbreviated below to kyr) (broadly MIS 1-3). At these long time-scales, orbital-forcing factors, modified by earth-surface feedback mechanisms, ought to be dominant and be recognisable in the biotic and hydrological systems. We will then examine the evidence for abrupt events at century to millennial scales, which are prominent features in many climate reconstructions since the Last Glacial Maximum (LGM). Climate forcing at these time-scales is tightly coupled to oceanic changes and so we will extend our discussion to a number of offshore sites. Numerous terrestrial archives are now being explored including some excellent high-resolution speleothem records (Holmgren et al. 2001), but to limit our study we will confine our discussion to lakes and the major river systems (Fig. 1).

Few proxies are directly related to the fundamental climate variables of precipitation and temperature. Even in the tropics where shifts from wet to dry conditions commonly outweigh temperature changes, it is difficult to tease these variables apart. Moreover, the part played by atmospheric gases such as CO_2 in terrestrial environmental change is now also recognised (Street-Perrott 1994; Jolly and Haxeltine 1997; Street-Perrott et al. 1997). Identifying the specific contribution of the different variables is of primary concern to tropical palaeoenvironmental studies and has promoted the development of several novel approaches to environmental reconstruction.

Climate forcing factors and contemporary processes

At the glacial-interglacial time-scale the major forcing factors controlling the tropical African climate include orbital configuration, the volume of the Northern Hemisphere ice sheets, surface boundary conditions, oceanic properties and atmospheric transparency (Street-Perrott et al. 1990; Kutzbach and Liu 1997; Texier et al. 2000). The dominant paradigm for the last two decades has been that much of the hydrological variability in the region stems from the control exerted by the 19–23 kyr Milankovitch precessional cycle (Kutzbach and Street-Perrott 1985; Kutzbach and Guetter 1986). Insolation controls the strength of the monsoon circulations through the differential heat capacity of land and

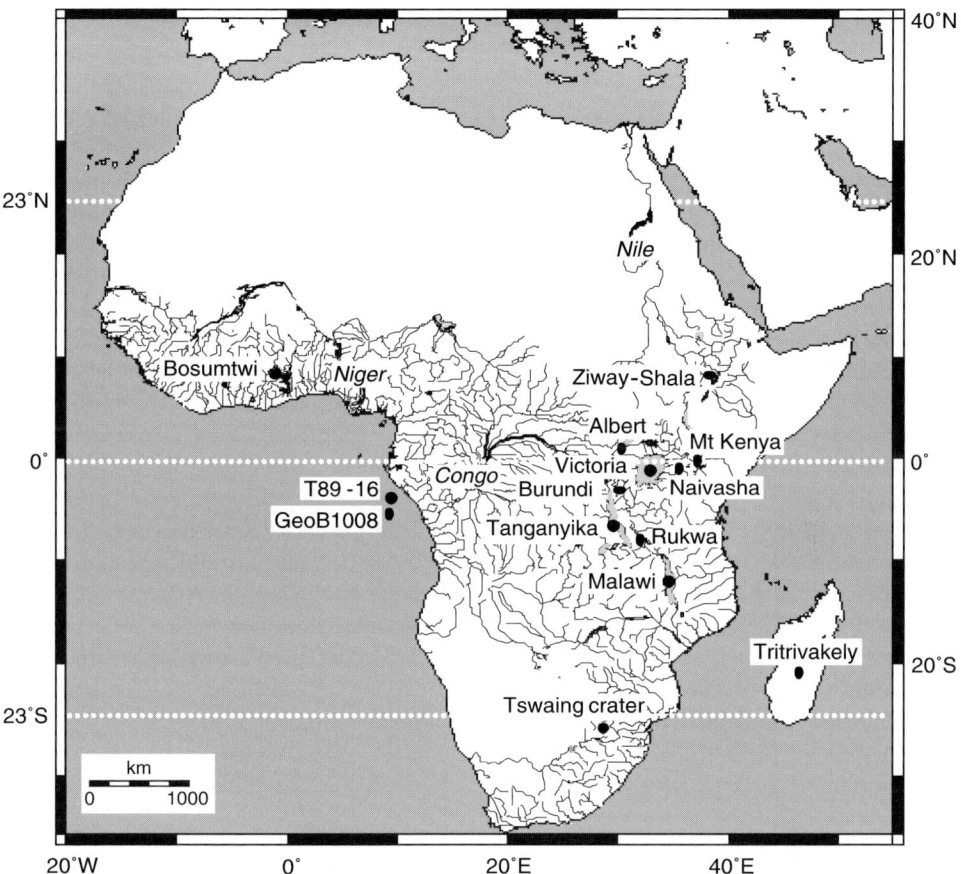

Figure 1. Location of the lakes, rivers and marine cores mentioned in the text. The base map was drawn online at www.aquarius.geomar.de.

oceans, and fuels the large-scale convective overturning in the tropics. A sensitivity analysis for the northern tropics suggests that a 10% increase in insolation would create an average 35% (range 25–50%) increase in precipitation (Prell and Kutzbach 1992) and a similar figure of 45% has been calculated for the southern tropics (Partridge et al. 1997). Lake level fluctuations across the northern and equatorial tropics provide strong support for orbital forcing as a dominant climate mechanism from the late-glacial to the mid-Holocene (Kutzbach and Street-Perrott 1985; Street-Perrott et al. 1990). It is less clear to what extent the Northern and Southern hemisphere tropics responded in antiphase as the theory would predict.

In addition to orbital configuration, atmospheric CO_2 and CH_4 are essential in forcing climate changes, especially through the amplification of insolation changes (Shackleton 2000). High-resolution gas records from polar ice cores have confirmed the magnitude of global average changes in these variables between the LGM and the Holocene (Indermühle et al. 1999). The impact of trace-gas levels on tropical ecosystems and conversely, the role of tropical wetlands in regulating the global carbon dioxide and methane cycles, cannot be ignored. The East African great lakes are major methane producers (Craig 1974; Tietze et al. 1980; Adams and Ochola 2002), but of even greater significance in terms of their global climatic impact are the vast wetlands that develop during periods of enhanced humidity (Petit-Maire 1990; Petit-Maire et al. 1991; Street-Perrott 1992; 1994).

Nicholson (2000) suggests that the major glacial-interglacial shifts in African climate are caused by climate dynamics similar to those operating today, even if the boundary conditions differ. This general pattern is contradicted by events within the last millennium that show a negative relationship between high latitude temperatures and East African rainfall (Verschuren et al. 2000), demonstrating that complex and poorly understood climate processes operate at sub-millennial time-scales. Nevertheless, on the historical time-scale, major climate shifts can be traced across the continent and are coupled with large-scale features of the general circulation (Nicholson 2000; Verschuren, this volume). Precipitation is associated with the proximity of atmospheric convergence zones. Classical climatological theory implies that the double passage of the Intertropical Convergence Zone (ITCZ) explains the bimodal rains of the equatorial region (Nicholson and Flohn 1980). In addition, a marked west-east rainfall gradient is maintained by the moist southwesterly air flows from the Atlantic. The broadly North-South Congo Air Boundary in the central-southern part of the continent marks the convergence of low level flows from the Atlantic and Indian Oceans. Both northern and southern parts of the tropical region receive rainfall carried by monsoon circulation. Synoptic patterns for the intertropical region are strongly affected by topography and localised factors such as lake effects or coastal influences rendering the development of coherent syntheses difficult (Nicholson 2000).

From marine isotope stage 6 to the LGM

A key aim of scientists working within this sector of the PEP III transect is to evaluate the role of the precession cycle in forcing the monsoon circulations and whether their response was asymmetrical between the hemispheres (Kutzbach and Street-Perrott 1985; Partridge et al. 1997). Until recently, no records of sufficient duration and resolution were available to examine this prediction through several precession cycles, but within the last 5 years several records have been published that contain relatively continuous sequences extending back to the last interglacial (MIS 5e or Eemian). The marine record GeoB1008-3 (Fig. 2) off the Congo river mouth (6.5 °S, although the Congo basin extends to ~6 °N) is one of the most complete to encompass this period (Schneider et al. 1996). Glacial and interglacial stages are well distinguished by sea-surface temperatures (SST), with the last interglacial in particular being warmer than the present day. Figure 2 also shows that the northern hemisphere precession 19–23 kyr cycle is an important component of the alkenone-derived SST curve. The pollen record corresponds closely to the alkenone curve with the lowland rainforest pollen sum peaking during MIS 5e, 5c and 5a (Jahns 1996). Peaks of the Afro-montane coniferous tree *Podocarpus* mark cooler intervals within stage 5 (sub-stages 5d and 5c). Herbaceous pollen indicative of vegetation that is more open is most abundant during the glacial maxima of stage 6 and 2.

The study of new sections from Lake Naivasha (0.7 °S) and the application of ^{40}Ar/^{39}Ar dating have extended the 30 kyr-long lake level record from Central Kenya (Richardson

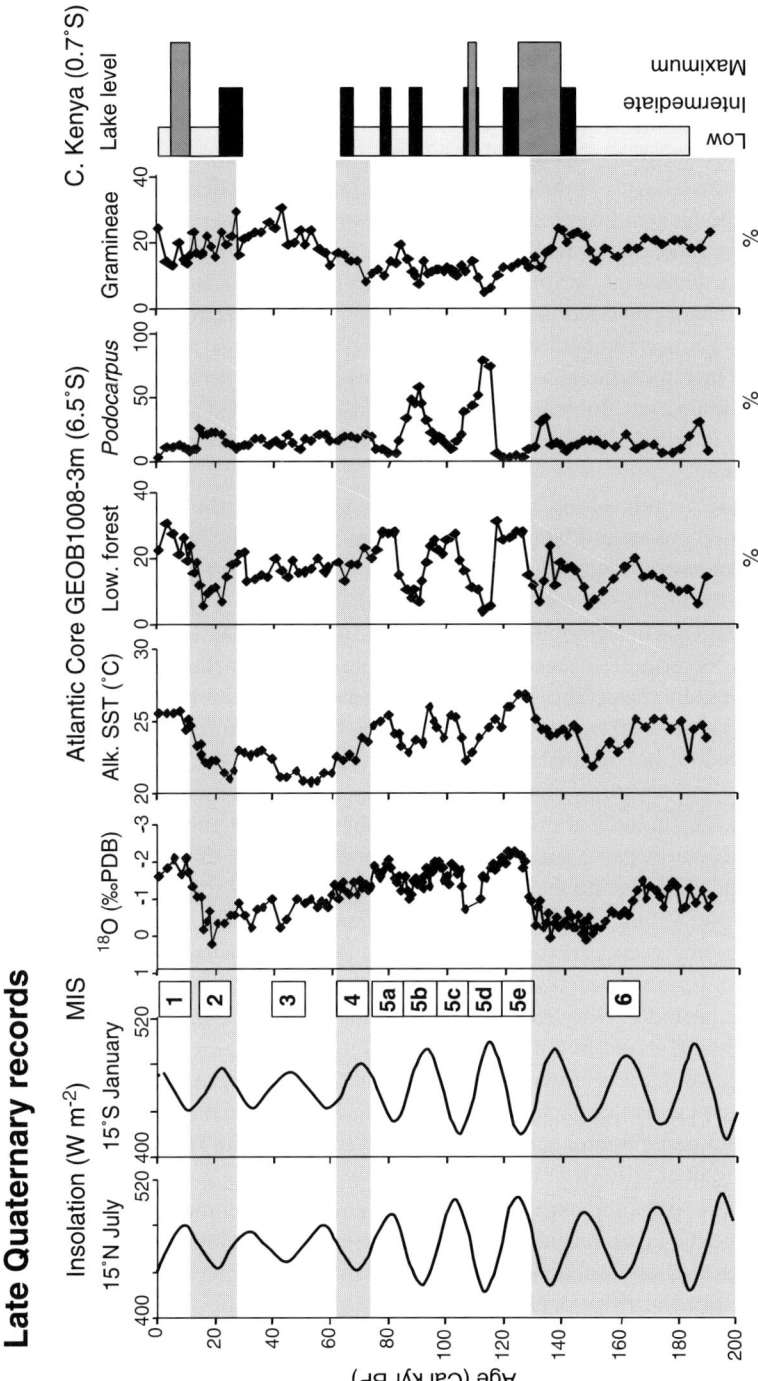

Figure 2. Late Quaternary records of δ^{18}O, alkenone derived SST and pollen percentages (lowland forest, the Afromontane conifer *Podocarpus* and Gramineae) from marine core GeoB1008-3 (Schneider et al. 1996; Jahns et al. 1996) compared to relative lake level changes from Central Kenya (primarily from Lake Naivasha with additional data from Nakuru and Elmenteita) (Richardson and Dussinger 1986; Trauth et al. 2001). Summer insolation curves for 15 °N and 15 °S are from Berger and Loutre (1991). Marine isotope stages (MIS) 1-6 are labelled and the colder stages 2, 4 and 6 are shaded grey.

and Dussinger 1986) back to 175 kyr (Trauth et al. 2001), making this the best dated long record from equatorial Africa. According to Trauth et al. (2001) the chronology of high stands identified during this period mainly follows the periodicity of the 19–23 kyr cycle generated by orbital precession (Fig. 2). High stands centred on 135 kyr, 110 kyr, 90 kyr and 66 kyr correspond to maxima of March insolation (March being the beginning of the period of equatorial long rains at the present day). The first of these events was coeval with a highstand in the Magadi-Natron basin (Kenya-Tanzania border) marked by stroma-tolites 47 m above the present lake and dated by U/Th to 130–140 kyr (Hillaire-Marcel and Casanova 1987). The Lake Naivasha highstand began significantly earlier (~146 kyr) than the onset of the last interglacial (termination 2 dated by SPECMAP to 127 ± 6 kyr; Imbrie et al. 1984) and therefore preceded the warming of the Northern Hemisphere predicted by orbital calculations (Karner and Muller 2000). Detailed comparison of the structure of wet episodes during the last interglacial with those of the Holocene is now needed to ascertain whether millennial-scale hydrological instability was also a feature of the Eemian (MIS 5e). Moreover, the Kenyan lake level changes must be tested against other records to establish their regional extent.

South of the equator, three long lake sediment records show the signature of orbital forcing. The combined pollen and lake level study from Tritrivakely, Madagascar, provides a continuous palaeoenvironmental record beginning in the last interglacial (Gasse and Van Campo 1998; 2001). The chronology is problematic before 41 kyr (the limit of the radiocarbon series) and has been modelled by extrapolation of the ^{14}C dates and tuning to the Vostok δ^{18}O record. Nevertheless, recognisable structure emerges in the pollen record where warm phases indicated by the establishment of wooded grassland occur at around 125 kyr, 100 kyr, 83 kyr, 60 kyr and 10 kyr ± 5 kyr. Lake high stands (before 143 kyr and around 115 kyr if the age model is correct) and low stands (around 125 kyr and 105 kyr) match pollen-inferred cold and warming phases, respectively. As in other records the separation of climate variables is difficult. Summer rain during phases of high summer insolation was not heavy enough to compensate for the large evaporation losses during warm summers and dry winters, especially during the cold LGM, which was drier than today.

A new core recently obtained from a low sedimentation site on the Kyrvala Island ridge in Lake Tanganyika preserves a record of the last ca. 80 kyr in just over 9 m of sediment (Scholz et al., 2003). Evidence of shallow water conditions with possible exposure of the core site indicates a period of very low lake level prior to 80 kyr ago. Above this section the sediments are characterised by large variations in organic carbon (TOC) content. The uncertain chronology of the pre-radiocarbon section of the core at present precludes direct comparison with regional insolation, but Scholz et al. (2003) suggest that some of the shorter periods of enhanced organic matter accumulation may coincide with Heinrich events in the North Atlantic (Bond et al. 1993).

Orbital forcing can be used to explain many of the major environmental changes shown in these records since the last interglacial. However, the sedimentological studies from the Tswaing Crater (Pretoria Salt Pan) demonstrate how orbital mechanisms may be overridden by other forcing factors. Partridge et al. (1997, this volume) have shown, using a grain-size record calibrated against rainfall, that austral summer insolation correlates with wet events at this site from 200 kyr to the LGM. However, the relationship breaks down at the LGM, where the phase of the precession cycle predicts wetness, yet the grain-size and diatom data (Metcalfe 1999) indicate aridity. It seems that when orbital eccentricity is small, direct insolation effects can be overridden by higher latitude changes during full glacial periods (deMenocal et al. 1993).

On Mt Kenya the response of vegetation to multi-faceted environmental change has been recognised through compound-specific carbon isotope determinations of organic matter (Street-Perrott et al. 1997; Huang et al. 1999). Studies at Sacred Lake have shown that bulk $\delta^{13}C$ values were up to 17 per mille higher in glacial times than today and that the composition of the bulk organic matter (TOC) was dominated by plants (including algae) possessing CO_2 concentrating mechanisms (Street-Perrott et al. 1997). The inference from isotopic values of terrestrial biomarkers that C_4 grasses were present has been confirmed by grass cuticle analysis (Wooller et al. 2000; Ficken et al. 2002) and further evidence for reduced CO_2 is shown by increased stomatal density of grasses at the LGM (Wooller and Agnew 2002). The shift toward grasses and sedges utilising the more efficient C_4 photosynthetic pathway would be consistent with lower CO_2, greater aridity and increased fire frequency. All three variables probably contributed to the lowering of vegetation belts and would tend to moderate the more extreme estimates of temperature lowering during the LGM in the tropics based on the simplistic application of modern lapse rates. Time-series analysis reveals that the precessional cycle and its harmonics dominate the frequency spectrum of the Sacred Lake $\delta^{13}C$ values (Olago et al. 2000), a further demonstration of the tight coupling between the tropical monsoons, earth surface processes and orbital forcing.

These long Late Quaternary palaeoenvironmental records complement the much larger number of studies from tropical Africa that provide continuous data from before the LGM to the present. The number of sites now investigated has enabled the regional significance of localised records to be established and enable key periods to be reconstructed with confidence (Pinot et al. 1999). A wide range of proxies from different archives shows that intertropical Africa was both drier and cooler than the present day at the LGM (Gasse 2000). Quantitative estimates from pollen transfer functions confirm a reduction of 32% in rainfall (Bonnefille and Chalié 2000) and a temperature fall of 2–5 °C in the Burundi highlands (Bonnefille et al. 1990) (Fig. 3). Few pollen records are available from the lowland tropics although offshore pollen data confirm a reduction in rainforest elements of the flora (Marret et al. 2001).

Lake-level data agree with this picture of conditions at least as dry as present at the LGM. New records from Lake Malawi and Lake Rukwa extend the pattern of negative water balance established in northern and equatorial Africa to the southeast quadrant of this study region (Fig. 3). Diatom data from Lake Rukwa indicate that the basin was occupied by a swamp-like environment in clear contrast to the high water levels of the early Holocene (Barker et al. 2002). Lake Malawi has previously been regarded as an outlier by indicating water level fluctuations in antiphase with most other tropical African lakes. However, recent work on well-dated deepwater (363 m) cores from the Northern Basin has now shown that Malawi was low at the LGM (Johnson et al. 2002; this volume; Gasse et al. 2002). The evidence for this interpretation comes from the abundance of diatom periphyton during the LGM indicating that the coring site was much closer to the lake margin than today. This assertion is supported by other data from these cores including evidence for a reduced silica supply to the lake (Johnson et al. 2002; this volume) and increased bulk $\delta^{13}C$ values suggesting less C_3 vegetation in the catchment and/or a higher contribution of algal detritus. The contrast with earlier studies arises from the more robust chronology and greater stratigraphical continuity of the northern basin cores.

Precipitation and palaeohydrology of south East Africa
Diatom periphyton and pollen-precipitation records

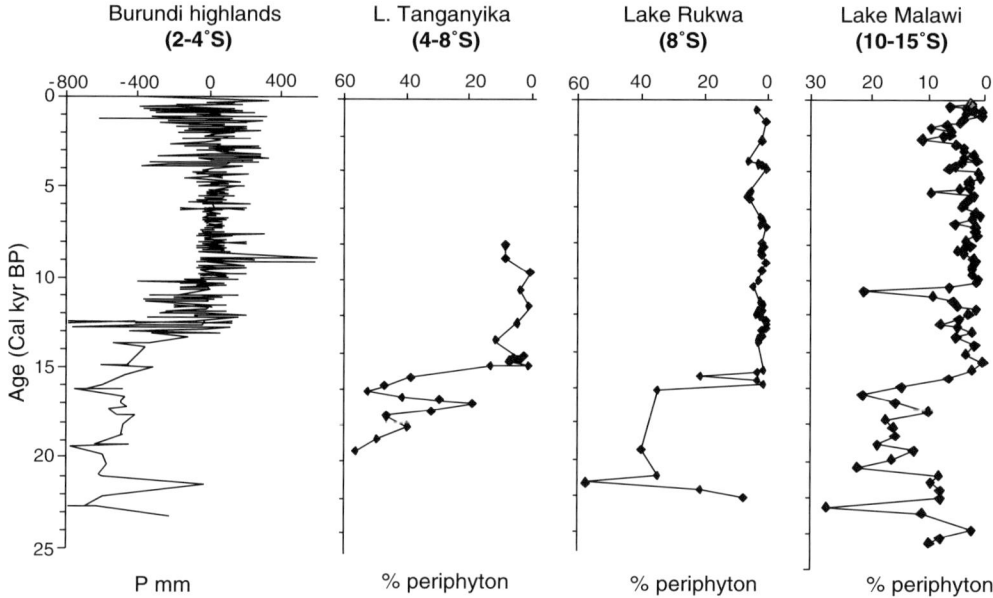

Figure 3. Precipitation and palaeohydrological records of the LGM from East Africa south of the equator. Precipitation in Burundi was calculated using pollen transfer functions by Bonnefille and Chalié (2000). Diatom periphyton is used as a surrogate for relative lake level change (note reversed scales). Data from Gasse et al. (1989), Barker et al. (2002) and Gasse et al. (2002).

The mechanisms responsible for relative dry glacial climates are revealed in the comparison of General Circulation Model (GCM) simulations by PMIP (Paleoclimate Modelling Intercomparison Project) (Pinot et al. 1999). All those model runs that used computed SSTs simulate aridity across this region of the PEP III transect; however, those based on empirical (and higher) SST estimates by CLIMAP (Climate: Long-range Investigation, Mapping, and Prediction) suggested wetter conditions than today, particularly over East Africa (Kutzbach and Guetter 1986). It is clear that at the LGM the tropical African climate was strongly affected by the cooling of the surrounding oceans, which in turn fluctuated in line with the development of the polar ice sheets, as well as the reduction in atmospheric greenhouse gases (Kutzbach and Street-Perrott 1985; Indermühle et al. 1999).

Intertropical Africa since the LGM

Insolation in the Northern Hemisphere tropics rose from a minimum at 22 kyr to a maximum at 12 kyr (Berger and Loutre 1991). Insolation alone would lead to a gradual enhancement of monsoon rainfall by 35–45% in northern intertropical Africa (Prell and Kutzbach 1992).

Such a smooth, progressive transition is not shown by the data (Kutzbach and Street-Perrott 1985), rather the evidence reveals abrupt changes and irregular climate fluctuations. Recovery from the dry glacial maximum began early in many sites, especially in the southern-most part of the study region. At Lake Tritrivakely in Madagascar the lake began to rise after 17.5–17 kyr. In Lake Malawi a transitional phase can be inferred from the diatom data beginning as early as 17.3 kyr, whereas the Lake Tanganyika diatom record suggests an initial P-E increase between 21 and 15 kyr (Gasse et al. 1989). Lake Rukwa was saline between ca. 16 and 15 kyr but slightly deeper than at the LGM, which compares to a short-lived transgression in Lakes Albert (Beuning et al. 1997b) and Victoria (Talbot and Livingstone 1989; Johnson 1996). In these lakes, an early post-LGM humid period is recorded by lake sediments bracketed by two palaeosols (Talbot and Livingstone 1989). The brief wet phase thus took place between about 16 kyr and 15 kyr, but pre-dated the major transgression that began after 15 kyr (Talbot and Laerdal 2000). In the Congo Fan record, the first increase in Congo discharge after the LGM is registered between 16.5 kyr and 16 kyr, followed by a return to lower values (Marret et al. 2001) (Fig. 4).

In contrast to these relatively short and low-amplitude early transgressions, the lake level rise ca. 15 kyr was spectacular. Many equatorial lakes such as Victoria (Talbot and Laerdal 2000), Magadi (Roberts et al. 1993), Manyara (Holdship 1976) and Tanganyika (Gasse et al. 1989) rose substantially at this time. The abrupt onset of the so-called African humid period is thought to represent a non-linear response to insolation changes triggered (at least in North Africa at $20\,°N$) when insolation reached $470\,W\,m^{-2}$ (or 4.2% above modern values; Berger and Loutre (1991)). The amplitude of the hydrological response to insolation forcing requires positive feedback from vegetation, surface water and sea-surface temperature changes (deMenocal et al. 2000).

The Congo Fan record (Marret et al. 2001) is a multi-proxy dataset consisting of oxygen isotopic, biomarker (long chain alkane/alkenone) and palynological data (Fig. 4). High Congo palaeo-discharge events are recorded by rapid increases in the fluxes of pollen, *Pediastrum* (a freshwater green alga) and charred grass cuticle fluxes, high ratios of long chain terrestrial n-alkanes to marine alkenones, an increase in the alkenone (U^K_{37}) SST index, decreased $\delta^{18}O$ ratios and an increase in the sediment accumulation rate of terrigenous material. These changes reflect flushes of terrigenous matter, including pene-contemporaneous pollen, to the fan and reduced salinity in the Congo plume. Contemporaneous increases in dinoflagellate cyst flux and changes in cyst-assemblage composition indicate that these events stimulated river-induced upwelling and associated productivity. A major discharge pulse at around 13.5 kyr marks the most significant event in this 30 kyr record. The rise in lowland rainforest taxa, which continues into the Holocene, begins at this point, coincident with a decline in dry grassland taxa. The surface-water salinity record off the Niger Delta closely matches the evidence for a major increase in discharge from the Congo. Here, isotopic evidence from planktonic foraminifera indicates an abrupt decline in salinity beginning at ca. 13.5 kyr due to a dramatic rise in freshwater outflow from the Niger (Pastouret et al. 1978).

The major part of the PEP III transect demonstrates the extent and influence of the Younger Dryas cold event (ca. 12.5–11.5 kyr) on surface systems. The influence of this episode in tropical Africa is profound and manifested through an abrupt switch to relative drought. Substantial regressions lasting about a thousand years are shown in Lake Bosumtwi (Talbot and Johannessen 1992), Lake Magadi (Roberts et al. 1993), the Ziway-Shala basin

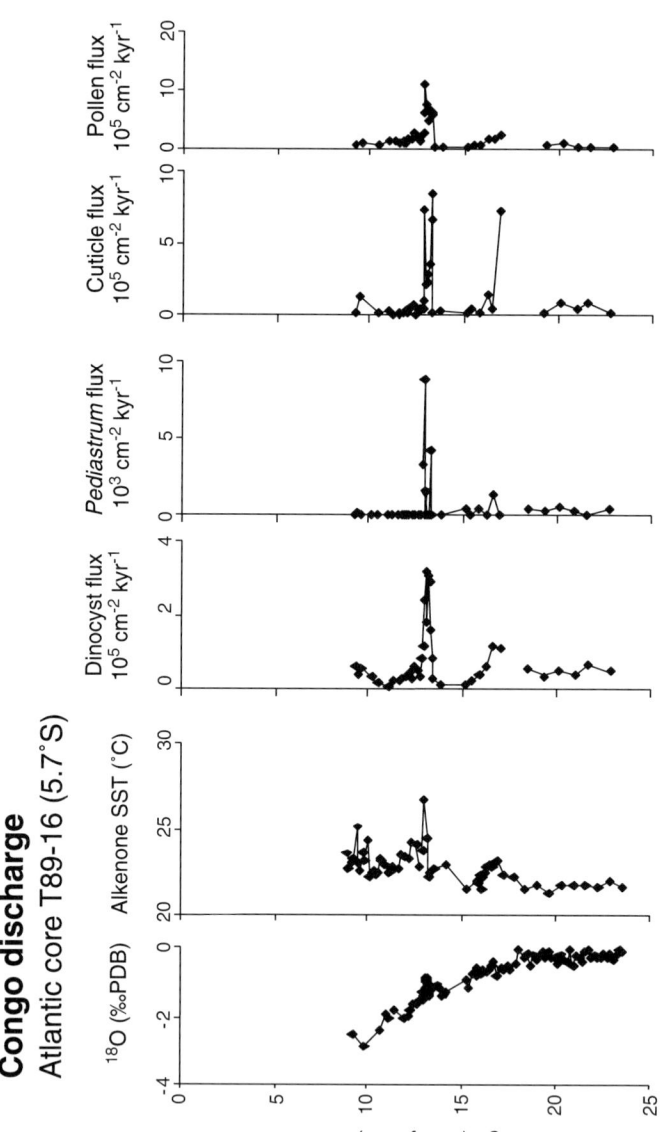

Figure 4. Evidence for changes in Congo discharge shown by fluxes of pollen, green algae (*Pediastrum*) and dinoflagellates recovered from marine core T89-16 (Marret et al. 2001).

(Gillespie et al. 1983), Lake Tanganyika (Gasse et al. 1989), Lake Victoria (Johnson et al. 2000), and Lake Albert (Beuning et al. 1997b). Further evidence for drought during the Younger Dryas comes from the cessation of peat growth in the Aberdare Mountains (Street-Perrott and Perrott 1990) and in a shift to grassland pollen in the Burundi highlands (Bonnefille et al. 1995). The response of lakes in the south-eastern tropics of Africa is less clear. The percentage of benthic diatoms in sediments from the northern basin of Lake Malawi is relatively high during the Younger Dryas although the increase begins 800 years before the inferred cooling in Greenland (Gasse et al. 2002). These changes are chronologically equivalent to the Antarctic Cold Reversal (Jouzel et al. 1995) and may reflect the southern de-glaciation rather than the break-up of the northern ice sheets (Johnson et al. 2002). Direct correlation between the Greenland and Congo Fan records is currently precluded by uncertainties in the ^{14}C age model for the latter record which is based on bulk carbonate (Marret et al. 2001). However, there is a clear "twin peak" palaeo-discharge event in the Congo record in which palaeo-discharge highs centred on 13.5 kyr and 13 kyr are separated by a drier phase with lower discharge. The current age model would suggest that these events led the Younger Dryas but it is equally plausible that the drier phase was coincident with the Younger Dryas.

Holocene events

Following the dry conditions of the Younger Dryas, the African humid phase became re-established in the intertropical region. Many of the major lakes overflowed and produced cascading systems that linked the lakes to the African continent's major drainage systems. New insights into the chronology of isolation and connection of the White Nile and its headwater lakes are provided by Sr-isotope analysis which can provide a sensitive tracer of water source as Sr isotopes are not fractionated in the hydrological cycle (Talbot et al. 2000; Talbot and Brendeland 2001). These studies indicate that flow from Lake Victoria via Lake Albert into the main Nile became re-established between 15 kyr and 14 kyr. The ensuing period of elevated discharge in the Nile (Adamson et al. 1980; Rossignol-Strick et al. 1982) coincides with the evidence for increased outflow from the Congo and Niger (see above), confirming a general increase in precipitation in the interior of the African continent at this time.

The general African humid phase came to an abrupt end ca. 5.5 kyr. According to deMenocal (2000), in at least the Northern Hemisphere, the mechanisms responsible may be similar to those that led to its onset. Insolation declined gradually but crossed the critical threshold of 4.2% greater than the present day at about this time triggering vegetation-albedo and surface-albedo feedback mechanisms and amplifying the orbital forcing (Claussen et al. 1999; Texier et al. 2000). Lake levels at many sites in tropical Africa show a series of century to millennial-scale arid events throughout the Holocene (Gasse and Van Campo 1994) (Fig. 5). These abrupt events have strong links to changes in Atlantic SSTs and the thermohaline circulation that influenced both tropical and extra-tropical regions of Africa (Street-Perrott and Perrott 1990; Lamb et al. 1995; Gasse 2000). However, the coincidence of these periods with similar short-lived events in Tibet also implies a connection to the Indian Ocean monsoon (Gasse and Van Campo 1994). Better ways of describing and explaining the abrupt hydrological changes within the Holocene are important if we are to understand these decadal-scale transitions.

High resolution Holocene lake records

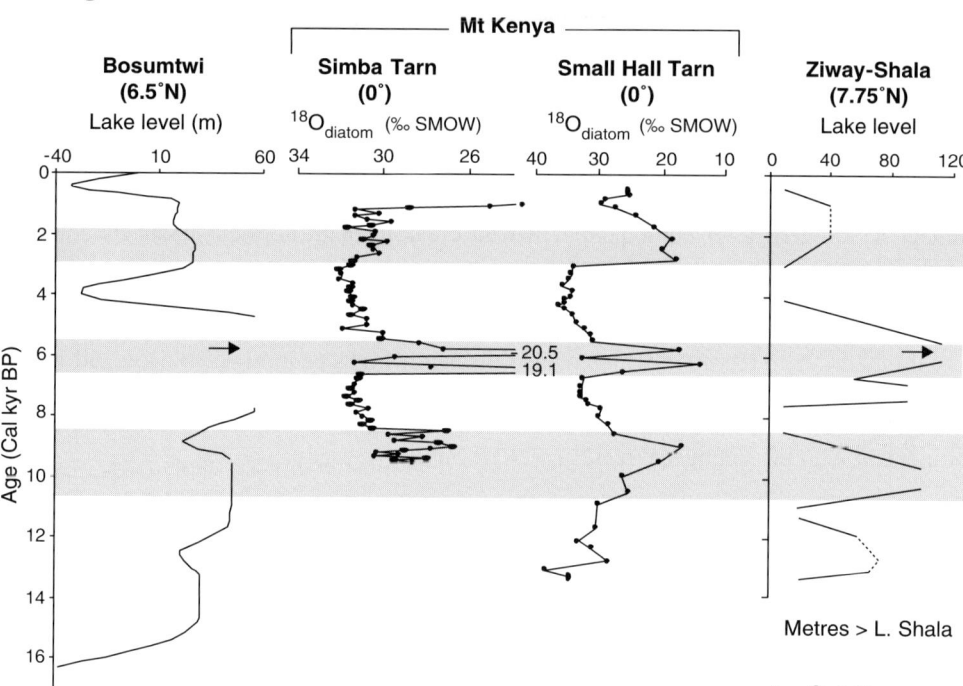

Figure 5. Comparison of high resolution Holocene lake level records from Bosumtwi (Talbot and Johansson 1992) and Ziway-Shala (Gillespie et al. 1983) with oxygen isotope records from high altitude tarns on Mt Kenya (Barker et al. 2001b). Note reversed scales. The grey shading represents periods of high P-E on Mt Kenya.

A recent study of oxygen isotope variations in diatom silica from alpine tarns on Mt Kenya has revealed a series of drier intervals corresponding to these abrupt Holocene events (Barker et al. 2001b). Whereas the levels of lakes at lower elevations record dry events lasting several centuries, the equivalent periods on Mt Kenya were apparently millennial in duration (Fig. 5). It seems that at extreme altitude lake- water isotopes are very sensitive to variation in convective intensity, cloud height and changes in evaporation. Therefore, high altitude lake isotope records give great precision in the delimitation of extreme wet periods but are less sensitive to drought; an inverse response to that of lake level, where the record is truncated once a drainage sill is reached.

Defining the duration and chronology of centennial-to-millennial scale climatic events is difficult given the imprecision of most ^{14}C chronologies and the heterogeneous nature of palaeoenvironmental archives. Nevertheless, two major dry periods can be identified after the Younger Dryas event. The first (8.4–8 kyr) coincides with the 8.2 kyr event found in the Greenland ice sheet (Alley et al. 1997), European lakes (von Grafenstein et al. 1998; Barber et al., this volume) and oceanic records (Barber et al. 1999). In Lake Victoria diatom production declined between 9.8 kyr and 7.5 kyr (Johnson et al. 2000), a period much longer than the duration of the 8.2 kyr event. However, shorter regressions ca. 8.4–8 kyr (Gasse and van Campo 1994; Gasse 2000) were found in Lake Abhé (Gasse 1977), the Ziway-Shala system (Gillespie et al. 1983), and Lake Bosumtwi (Talbot et al. 1984).

The second widespread regression was centred on 4.2–4 kyr (Gasse 2000). This period is important because of its relation to the collapse of the Old Kingdom in the Nile valley (Hassan 1997). Sedimentological and Sr-isotopic evidence for a regression and a period of probable isolation of Lake Albert at this time (Talbot and Brendeland 2001) suggests that the collapse must in part have been due to a drastic decline in the base-flow of the White Nile. In East Africa this event corresponds to a decline in moist rainforest and its replacement by dry forest types (Street-Perrott and Perrott 1993). In west equatorial Africa, this period is also recorded in a number of terrestrial and marine sites (Maley 1997; Marret et al. 1999).

Conclusions

Deciphering the imprint of environmental change has resulted in the application of novel proxies in palaeolimnological studies of tropical Africa. New techniques have been used to overcome limitations and exploit opportunities. Stable isotope proxies have played an important part in revealing the palaeoclimatic time-line summarised above. As in other parts of the PEP III transect, oxygen isotopes have been used to trace the hydrological cycle using sedimentary carbonate (e.g., Lamb et al. (2000), Ricketts and Johnson (1996)). Unfortunately carbonate is absent in many key sites in tropical Africa and other host materials have had to be investigated. Diatom silica and sponge spicules offer the opportunity to exploit lake sediment rich in biogenic silica; these have significant potential if methodological and interpretative difficulties can be addressed (Rietti-Shati et al. 1998; Barker et al. 2001b). Studies of oxygen isotopes in plant cellulose can also be used as in Lake Victoria (Beuning et al. 1997a) and in peat bogs of the Burundi highlands (Aucour et al. 1996). In all these studies interpretation of $\delta^{18}O$ has to be coupled with studies of the modern environment.

Limnological nutrient cycles may well be coupled to broader atmospheric changes and in part reflect climate changes. For example, in shallow lakes the relative importance of autochthonous carbon has created a sedimentary archive of shifts in the regional biota that reflect variations in climate, fire and atmospheric CO_2 (Street-Perrott et al. 1997; Wooller et al. 2000; 2002; Barker et al. 2001a; Ficken et al. 2002). In deep lakes the connection between bulk $\delta^{13}C$ values and atmospheric processes is more indirect, due to the dominance of aquatic productivity and the lower amplitude of change in $\delta^{13}C$. The glacial-interglacial range of $\delta^{13}C$ in Lake Bosumtwi is 26 per mille and in Sacred Lake is 17 per mille, whereas equivalent values for Lakes Victoria, Tanganyika and Malawi are 7, 9 and 4 per mille respectively (Fig. 6). Additional insights into the relative roles of climatic versus atmospheric CO_2 variations in forcing tropical vegetation change may be provided by coupled morphological-carbon isotope studies of charred grass cuticle preserved in the $>180\,\mu m$ fraction of lake and bog sediments. Many African grasses are ecologically sensitive and studies of palaeo-grassland assemblages have the potential to yield refined palaeoenvironmental reconstructions (Collatz et al. 1998; Wooller et al. 2000; 2002; Ficken et al. 2002; Beuning and Scott 2002). Nitrogen isotopes are also important in reconstructing nutrient cycling although the $\delta^{15}N$ signal is not a direct proxy of regional changes. A comparison of $\delta^{15}N$ values in lakes Victoria, Tanganyika and Bosumtwi illustrates the site-specific character of this isotope signal (Fig. 7). Nitrogen isotopes contribute to the functional understanding of lakes (e.g., N-fixation and limitation, NH_3 volatilization under high pH, flushing rates of soil N, dissolved inorganic N abundance) but regional synergy is harder to demonstrate.

The development, refinement and application of new methods of palaeoenvironmental reconstruction in the last decade have produced considerable progress in the testing of forcing factors controlling the African climate system and the identification of teleconnections between tropical Africa and other regions. Meridional linkages have been proposed between lake levels and the records from the polar ice cores and the surrounding oceans (Street-Perrott and Perrott 1990; Stager and Mayewski 1997; Johnson et al. 2002). Teleconnections have often assumed a passive role for the terrestrial tropics and have viewed the records in responsive mode. It is interesting to reverse this argument and to consider the extent to which changes in the terrestrial tropics and the surrounding oceans have generated some of these higher latitude changes (cf., Leuschner and Sirocko 2000). For example, Lake Victoria and lakes on Mt Kenya show a dry event beginning before the 8.2 kyr cold spell recorded in the Greenland ice sheet (Alley et al. 1997) and at glacial termination two tropical lakes rose before significant insolation changes at 65 °N (Trauth et al. 2001). The mechanisms by which the tropics could lead high latitude climate change at the millennial scale will differ with respect to particular events. One particularly important process involves methane emission from tropical wetlands formed during the initial phase of rising P-E that contributes to warming that leads Northern Hemisphere deglaciation (Petit-Maire 1990; Petit-Maire et al. 1991; Street-Perrott 1992; 1994). Other mechanisms that could drive or amplify changes at higher latitudes include changes in latent heat transport (Street and Grove 1976), albedo (Street-Perrott et al. 1990), SST (Bard et al. 1997; Henderson and Slowey 2000) and ENSO (Beaufort et al. 2001). The tropics are therefore central to our understanding of global climate change and further work is essential to understand the part played by different processes on specific events of various durations.

This synthesis has attempted to demonstrate that palaeoenvironmental records from tropical Africa are both coherent between the continent and its surrounding oceans, and linked to meridional changes along the PEP III transect. Terrestrial and oceanic data indicate that insolation forcing is integral to our understanding of African climates on multi-millennial time-scales, although it may be overridden during periods of full glaciation. The palaeoenvironmental record from tropical Africa also reveals rapid swings between wet and dry climates in just a few decades that have enormous implications for human societies, yet these are poorly understood and remain an important target for future research (cf. Olago and Odada (this volume)). In the tropics, as elsewhere, the prediction of future changes will be achieved by the continued development of new ways of revealing past climate signatures and mechanisms.

Summary

Classical techniques and newly developed environmental proxies inform this synthesis of recently published palaeoclimate data from intertropical Africa. Particular advances have been made in the application of oxygen isotope techniques to climate reconstruction from silica and organic hosts, as well as using carbon isotopes and compound specific analysis

Carbon cycle

Bulk ¹³C values (‰PDB) of lacustrine organic matter

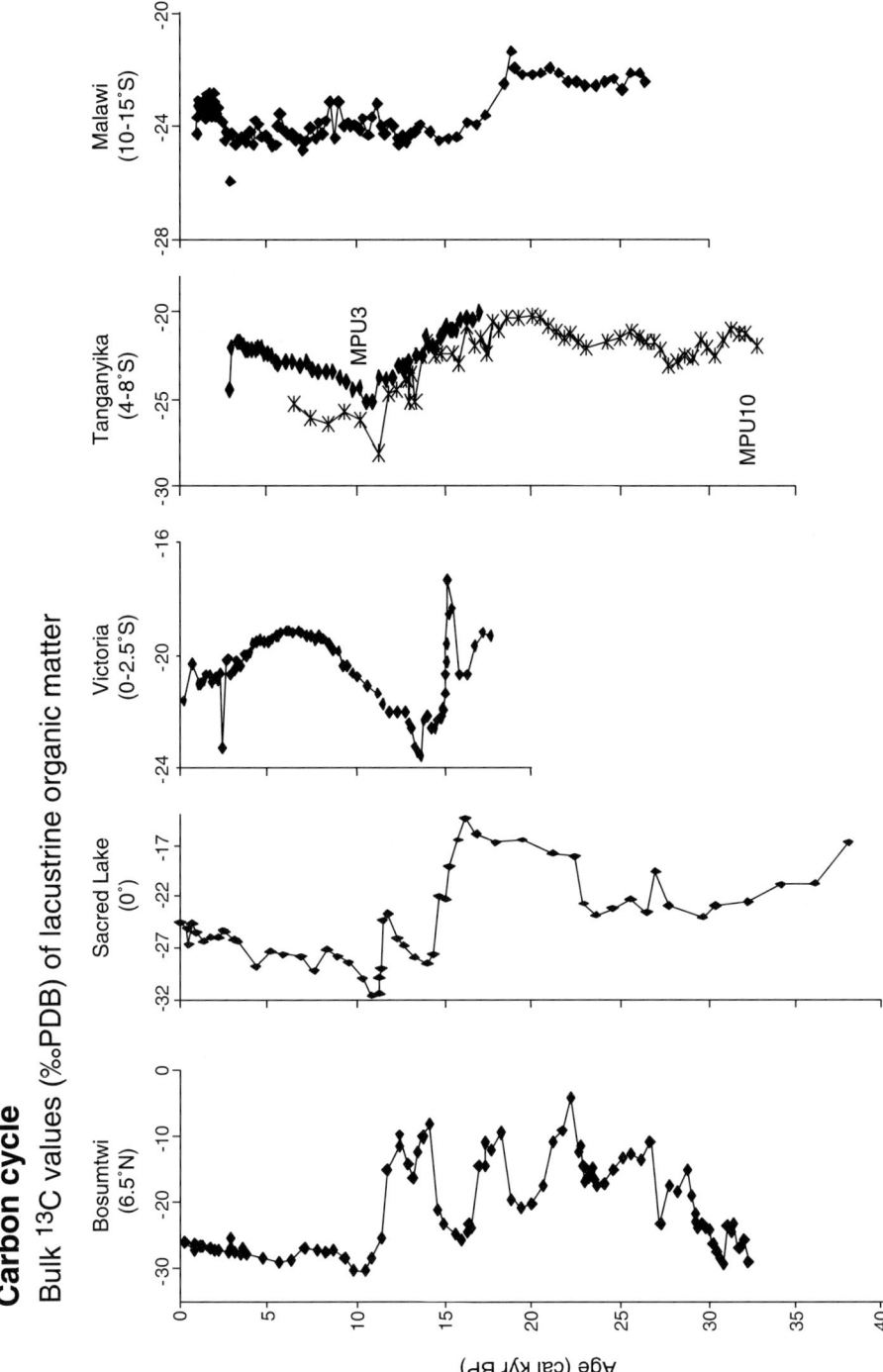

Figure 6. Changes in the carbon cycle recorded by bulk $\delta^{13}C$ values in crater lakes Bosumtwi (Talbot and Johannessen 1992) and Sacred Lake (Street-Perrott et al. 1997), together with records from Lake Victoria (Talbot and Laerdal 2000), Lake Tanganyika (Talbot and Jensen, unpub.) and Lake Malawi (Filippi and Talbot, unpub.).

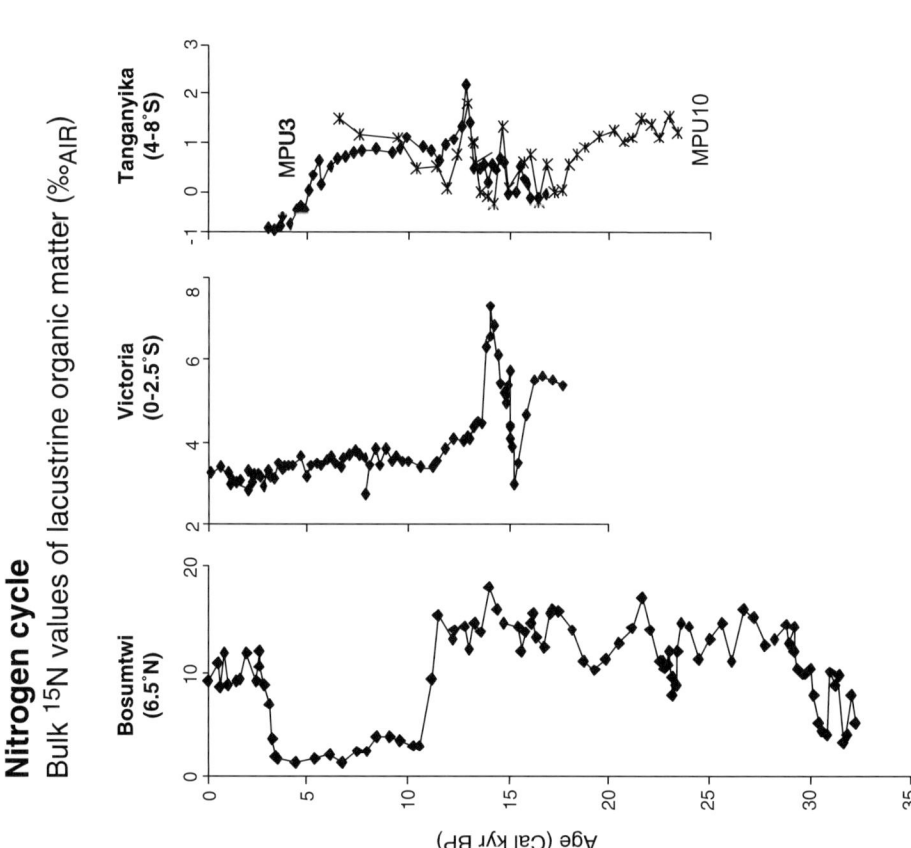

Figure 7. The nitrogen cycle in Lakes Bosumtwi (Talbot and Johannessen 1992), Victoria (Talbot and Laerdal 2000) and Tanganyika (Talbot and Jensen, unpub).

of carbon cycle dynamics. Nitrogen isotopes are closely linked to localised ecosystem processes and show less regional correlation. The close interaction between continental African climate and conditions in the surrounding oceans is clear in the examples discussed. We have also considered lacustrine and offshore sequences that give an insight into tropical African climate from marine isotope stage 6, the more numerous records available for the last 25 kyr, and abrupt climate shifts revealed by high-resolution records since the LGM. Long lake-level records from East Africa and marine cores from the Congo fan show the importance of orbital forcing in generating multi-millennial scale climate change. Nevertheless, recent studies from the southern African tropics indicate that the effects of insolation can be overridden under full glacial conditions. Forcing factors promoting millennial scale changes in the tropics are explored and the role of the tropics in driving changes at higher latitudes is briefly discussed. We note that at glacial termination two, tropical lake levels rose before insolation increased, and that methane emissions from greatly expanded tropical lakes during the deglacial periods is an important feedback process in global climate readjustment.

Acknowledgements

The authors are grateful to the scientists who provided data for this synthesis especially R. Bonnefille, F. Chalié, F. Gasse, M. Trauth and E. Van Campo. The original studies were supported by numerous programmes and funding bodies including NSF (IDEAL), EU (RUKWA), NERC (Mt Kenya, Congo Fan), NFR (Victoria, Albert, Tanganyika, Malawi, the Nile).

References

Adams D.D. and Ochola S.O. 2002. A review of sediment gas cycling in lakes with reference to Lake Victoria and sediment gas measurements in Lake Tanganyika. In: Odada E.O. and Olago D.O. (eds), The East African Great Lakes: Limnology, Palaeoclimatology and Biodiversity. Advances in Global Change Research 12. Kluwer Academic Publishers, Dordrecht, pp. 277–305.

Adamson D.A., Gasse F., Street F.A. and Williams M.A.J. 1980. Late Quaternary history of the Nile. Nature 287: 50–55.

Alley R.B., Mayewski P.A., Sowers T., Stuiver M., Taylor K.C. and Clark P.U. 1997. Holocene climatic instability: A prominent, widespread event 8200 yr ago. Geology 25: 483–486.

Aucour A.M., HillaireMarcel C. and Bonnefille R. 1996. Oxygen isotopes in cellulose from modern and Quaternary intertropical peatbogs: Implications for palaeohydrology. Chem. Geol. 129: 341–359.

Barber D.C., Dyke A., Hillaire-Marcel C., Jennings A.E., Andrews J.T., Kerwin M.W., Bilodeau G., McNeely R., Southon J., Morehead M.D. and Gagnon J.M. 1999. Forcing of the cold event of 8200 years ago by catastrophic drainage of Laurentide lakes. Nature 400: 344–348.

Barber K., Zolitschka B., Tarasov P. and Lotter A.F., this volume. Atlantic to Urals — the Holocene climatic record of Mid-Latitude Europe. In: Battarbee R.W., Gasse F. and Stickley C.E. (eds), Past Climate Variability through Europe and Africa. Kluwer Academic Publishers, Dordrecht, the Netherlands, pp. 417–442.

Bard E., Rostek F. and Sonzogni C. 1997. Interhemispheric synchrony of the last deglaciation inferred from alkenone palaeothermometry. Nature 385: 707–710.

Barker P., Perrott R.A., Street-Perrott F.A. and Huang Y. 2001a. Evolution of the carbon cycle in Lake Kimilili, Mt. Elgon since 14,000 cal yr BP: a multi-proxy study. Palaeoecol. Afr. 27: 77–94.

Barker P.A., Street-Perrott F.A., Leng M.J., Greenwood P.B., Swain D.L., Perrott R.A., Telford R.J. and Ficken K.J. 2001b. A 14,000-year oxygen isotope record from diatom silica in two alpine lakes on Mt. Kenya. Science 292: 2307–2310.

Barker P., Telford R., Gasse F. and Thevenon R. 2002. Late Pleistocene and Holocene palaeohydrology of Lake Rukwa, Tanzania, inferred from diatom analysis. Palaeogeogr. Palaeoclimatol. Palaeoecol. 187: 295–305.

Beaufort L., de Garidel-Thoron T., Mix A.C. and Pisias N.G. 2001. ENSO-like forcing on oceanic primary production during the Late Pleistocene. Science 293: 2440–2444.

Berger A. and Loutre M.F. 1991. Insolation values for the climate of the last 10 million years. Quat. Sci. Rev. 10: 297–317.

Beuning K.R.M., Kelts K., Ito E. and Johnson T.C. 1997a. Paleohydrology of Lake Victoria, East Africa, inferred from oxygen isotope ratios in sediment cellulose. Geology 25: 1083–1086.

Beuning K.R.M. and Scott J.E. 2002. Effects of charring on the carbon isotopic composition of grass (Poaceae) epidermis. Palaeogeogr. Palaeoclimatol. Palaeoecol. 177: 169–181.

Beuning K.R.M., Talbot M.R. and Kelts K. 1997b. A revised 30,000-year paleoclimatic and paleohydrologic history of Lake Albert, East Africa. Palaeogeogr. Palaeoclimatol. Palaeoecol. 136: 259–279.

Bond G., Broecker W., Johnsen S., McManus J., Labeyrie L., Jouzel J. and Bonani G. 1993. Correlations between climate records from North-Atlantic sediments and Greenland ice. Nature 365: 143–147.

Bonnefille R. and Chalié F. 2000. Pollen-inferred precipitation time-series from equatorial mountains, Africa, the last 40 kyr BP. Glob. Planet. Chan. 26: 25–50.

Bonnefille R., Riollet G., Buchet G., Icole M., Lafont R. and Arnold M. 1995. Glacial/interglacial record from intertropical Africa, high resolution pollen and carbon data at Rusaka, Burundi. Quat. Sci. Rev. 14: 917–936.

Bonnefille R., Roeland J.C. and Guiot J. 1990. Temperature and rainfall estimates for the past ca. 40,000 years in equatorial Africa. Nature 346: 347–349.

Claussen M., Kubatzki C., Brovkin V., Ganopolski A., Hoelzmann P. and Pachur H.J. 1999. Simulation of an abrupt change in Saharan vegetation in the mid-Holocene. Geophys. Res. Lett. 26: 2037–2040.

Collatz G.J., Berry J.A. and Clark J.S. 1998. Effects of climate and atmospheric CO_2 partial pressure on the global distribution of C_4 grasses: present, past, and future. Oecologia 144: 441–454.

Craig H. 1974. Lake Tanganyika Geochemical and Hydrographic Study: 1973 Expedition. SIO Reference Series. Scripps Institution of Oceanography, 83 pp.

deMenocal P., Ortiz J., Guilderson T., Adkins J., Sarnthein M., Baker L. and Yarusinsky M. 2000. Abrupt onset and termination of the African Humid Period: rapid climate responses to gradual insolation forcing. Quat. Sci. Rev. 19: 347–361.

deMenocal P.B., Ruddiman W.F. and Pokras E.M. 1993. Influences of high-latitude and low-latitude processes on African terrestrial climate — Pleistocene eolian records from equatorial Atlantic-Ocean Drilling Program Site-663. Paleoceanography 8: 209–242.

Ficken K.J., Wooller M.J., Swain D.L., Street-Perrott F.A. and Eglinton G. 2002. Reconstruction of a subalpine grass-dominated ecosystem, Lake Rutundu, Mount Kenya: a novel multi-proxy approach. Palaeogeogr. Palaeoclimatol. Palaeoecol. 177: 137–149.

Gasse F. 1977. Evolution of Lake Abhé (Ethiopia and TFAI) from 70,000 BP. Nature 265: 42–45.

Gasse F. 2000. Hydrological changes in the African tropics since the Last Glacial Maximum. Quat. Sci. Rev. 19: 189–211.

Gasse F., Barker P.A. and Johnson T.C. 2002. A 24 000 yr diatom record from the northern basin of Lake Malawi. In: Odada E.O. and Olago D.O. (eds), The East African Great Lakes: Limnology,

Paleolimnology and Biodiversity. Advances in Global Change Research 12. Kluwer Academic Publishers, Dordrecht. pp. 393–414.

Gasse F., Lédée V., Massault M. and Fontes J.C. 1989. Water-level fluctuations of Lake Tanganyika in phase with oceanic changes during the last glaciation and deglaciation. Nature 342: 57–59.

Gasse F. and Battarbee R.W., this volume. Introduction. In: Battarbee R.W., Gasse F. and Stickley C.E. (eds), Past Climate Variability through Europe and Africa. Kluwer Academic Publishers, Dordrecht, the Netherlands, pp. 1–6.

Gasse F. and Van Campo E. 1994. Abrupt post-glacial climatic events in West Asia and North African monsoon domains. Earth Planet. Sci. Lett. 126: 435–456.

Gasse F. and Van Campo E. 1998. A 40 000-yr pollen and diatom record from Lake Tritrivakely, Madagascar, in the southern tropics. Quat. Res. 49: 299–311.

Gasse F. and Van Campo E. 2001. Late Quaternary environmental changes from a pollen and diatom record in the southern tropics (Lake Tritrivakely, Madagascar). Palaeogeogr. Palaeoclimatol. Palaeoecol. 167: 287–308.

Gillespie R., Street-Perrott F.A. and Switsur R. 1983. Post-glacial arid episodes in Ethiopia have implications for climate prediction. Nature 306: 680–683.

Hassan F.A. 1997. Nile floods and political disorder in early Egypt. In: Dalfes H.N., Kukula G. and Weiss H. (eds), Third Millennium BC Abrupt Climate Change and the Old World Collapse. Springer-Verlag, Berlin. pp. 1–23.

Henderson G.M. and Slowey N.C. 2000. Evidence from U-Th dating against Northern Hemisphere forcing of the penultimate deglaciation. Nature 404: 61–66.

Hillaire-Marcel C. and Casanova J. 1987. Isotopic hydrology and palaeohydrology of the Magadi (Kenya) -Natron (Tanzania) basin during the late Quaternary. Palaeogeogr. Palaeoclimatol. Palaeoecol. 58: 155–181.

Holdship S.A. 1976. The palaeolimnology of Lake Manyara, Tanzania: a diatom analysis of a 56 meter sediment core. Unpublished Ph.D. thesis, Duke University.

Holmgren K., Tyson P.D., Moberg A. and Svanered O. 2001. A preliminary 3000-year regional temperature reconstruction for South Africa. S. Afr. J. Sci. 97: 49–51.

Huang Y., Freeman K.H., Eglinton T.I. and Street-Perrott F.A. 1999. $\delta^{13}C$ analyses of individual lignin phenols in Quaternary lake sediments: a novel proxy for deciphering past vegetation changes. Geology 27: 471–474.

Imbrie J., Hays J.D., Martinson D.G., McIntyre A., Mix A.C., Morley J.J., Pisias N.G., Prell W.L. and Shackleton N.J. 1984. The orbital theory of Pleistocene climate: support from a revised chronology of the marine $\delta^{18}O$ record. In: Berger A., Imbrie J., Hays J.D., Kukla G. and Saltzman B. (eds), Milankovitch and Climate. Reidel, Dordrecht. pp. 269–306.

Indermühle A., Stocker T.F., Joos F., Fischer H., Smith H.J., Wahlen M., Deck B., Mastroianni D., Tschumi J., Blunier T., Meyer R. and Stauffer B. 1999. Holocene carbon-cycle dynamics based on CO_2 trapped in ice at Taylor Dome, Antarctica. Nature 398: 121–126.

Jahns S. 1996. Vegetation history and climate changes in West Equatorial Africa during the Late Pleistocene and Holocene, based on a marine pollen diagram from the Congo fan. Veg. Hist. and Archaeobot. 5: 207–213.

Johnson T.C. 1996. Sedimentary processes and signals of past climatic change in the large lakes of the East African Rift valley. In: Johnson T.C. and Odada E.O. (eds), The Limnology, Climatology and Paleoclimatology of the East African Lakes. Gordon and Breach, Amsterdam. pp. 367–412.

Johnson T.C., Brown E., McManus J., Barry S., Barker P. and Gasse F. 2002. A high resolution paleoclimate record spanning the past 25 000 years in Southern East Africa. Science 296: 113–132.

Johnson T.C., Brown E.T. and McManus J., this volume. Diatom productivity in Northern Lake Malawi during the past 25,000 years: implications for the position of the imtertropical convergence zone at millennial and shorter time scales. In: Battarbee R.W., Gasse F. and Stickley C.E. (eds),

Past Climate Variability through Europe and Africa. Kluwer Academic Publishers, Dordrecht, the Netherlands, pp. 93–116.

Johnson T.C., Kelts K. and Odada E. 2000. The Holocene history of Lake Victoria. Ambio 29: 2–11.

Jolly D. and Haxeltine A. 1997. Effect of Low glacial atmospheric CO_2 on Tropical African montane vegetation. Science 276: 786–788.

Jouzel J., Vaikmae R., Petit J.R., Martin M., Duclos Y., Stievenard M., Lorius C., Toots M., Melieres M.A., Burckle L.H., Barkov N.I. and Kotlyakov V.M. 1995. The 2-step shape and Timing of the Last Deglaciation in Antarctica. Clim. Dyn. 11: 151–161.

Karner D.B. and Muller R.A. 2000. A causality problem for Milankovitch. Science 288: 2143–2144.

Kutzbach J.E. and Guetter P.J. 1986. The influence of changing orbital parameters and surface boundary conditions on climatic simulations for the past 18 000 years. J. Atmos. Sci. 43: 1726–1759.

Kutzbach J.E. and Liu Z. 1997. Response of the African monsoon to orbital forcing and ocean feedbacks in the middle Holocene. Science 278: 440–443.

Kutzbach J.E. and Street-Perrott F.A. 1985. Milankovitch forcing of fluctuations in the level of tropical lakes from 18 to 0 kyr BP. Nature 317: 130–134.

Lamb A.L., Leng M.J., Lamb H.F. and Mohammed M.U. 2000. A 9000-year oxygen and carbon isotope record of hydrological change in a small Ethiopian crater lake. Holocene 10: 167–177.

Lamb H.F., Gasse F., Benkaddour A., Elhamouti N., Vanderkaars S., Perkins W.T., Pearce N.J. and Roberts C.N. 1995. Relation between century-scale Holocene arid intervals in tropical and temperate zones. Nature 373: 134–137.

Leuschner D.C. and Sirocko F. 2000. The low-latitude monsoon climate during Dansgaard-Oeschger and Heinrich Events. Quat. Sci. Rev. 19: 243–254.

Maley J. 1997. Middle to Late Holocene changes in tropical Africa and other continents: paleomonsoon and sea-surface temperature variations. In: Nüzhet-Dalfes H., Kukla G. and Weiss H. (eds), Third Millennium BC Climate Change and Old World Collapse. NATO ASI Series, Global Environmental Change. Springer-Verlag, Berlin. pp. 611–640.

Marret F., Scourse J., Jansen J.H.F. and Schneider R. 1999. Changements climatiques et paléocéanographiques en Afrique centrale atlantique au cours de la dernière déglaciation: contribution palynologique. C. r. Acad. Sci., Paris série IIa 329: 721–726.

Marret F., Scourse J., Versteegh G., Jansen J.H.F. and Schneider R. 2001. Integrated marine and terrestrial evidence for abrupt Congo River palaeodischarge fluctuations during the last deglaciation. J. Quat. Sci. 16: 761–766.

Metcalfe S.E. 1999. Diatoms from the Pretoria Saltpan — a record of lake evolution and environmental change. In: Partridge T.C. (ed.), Investigations into the Origin, Age and Palaeoenvironments of the Pretoria Saltpan. Geol. Surv. S. Afr. Mem. pp. 172–192.

Nicholson S.E. 2000. The nature of rainfall variability over Africa on time scales of decades to millennia. Glob. Planet. Chan. 26: 137–158.

Nicholson S.E. and Flohn H. 1980. African environmental and climatic changes and the general atmospheric circulation in the Late Pleistocene and Holocene. Clim. Chan. 2: 313–348.

Olago D. and Odada E. This voulme. Palaeoclimate research in Africa: relevance to sustainable development, environmental management and significance for the future. In: Battarbee R.W., Gasse F. and Stickley C.E. (eds), Past Climate Variability through Europe and Africa. Kluwer Academic Publishers, Dordrecht, the Netherlands, pp. 551–565.

Olago D.O., Street-Perrott F.A., Perrott R.A., Ivanovich M., Harkness D.D. and Odada E.O. 2000. Long-term temporal characteristics of palaeomonsoon dynamics in equatorial Africa. Glob. Planet. Chan. 26: 159–171.

Partridge T.C., deMenocal P.B., Lorentz S.A., Paiker M.J. and Vogel J.C. 1997. Orbital forcing of climate over South Africa: A 200,000-year rainfall record from the Pretoria Saltpan. Quat. Sci. Rev. 16: 1125–1133.

Partridge T.C., Scott L. and Schneider R.R., this volume. Between Agulhas and Benguela: responses of Southern African climates of the Late Pleistocene to current fluxes, orbital precession and extent of the Circum-Antarctic vortex. In: Battarbee R.W., Gasse F. and Stickley C.E. (eds), Past Climate Variability through Europe and Africa. Kluwer Academic Publishers, Dordrecht, the Netherlands, pp. 45–68.

Pastouret L., Charnley H., Delibrias G., Duplessy J.C. and Thiede J. 1978. Late Quaternary climatic changes in western tropical Africa deduced from deep-sea sedimentation off the Niger Delta. Oceanol. Acta 1: 217–32.

Petit-Maire N. 1990. Will greenhouse warming green the Sahara? Episodes 13: 103–107.

Petit-Maire N., Fontugne M. and Rouland C. 1991. Atmospheric methane ratio and environmental changes in the Sahara and Sahel during the last 130 kyrs. Palaeogeogr. Palaeoclimatol. Palaeoecol. 86: 197–204.

Pinot S., Ramstein G., Harrison S.P., Prentice I.C., Guiot J., Stute M. and Joussaume S. 1999. Tropical paleoclimates at the Last Glacial Maximum: comparison of Paleoclimate Modeling Intercomparison Project (PMIP) simulations and paleodata. Clim. Dyn. 15: 857–874.

Prell W.L. and Kutzbach J.E. 1992. Sensitivity of the Indian monsoon to forcing parameters and implications for its evolution. Nature 360: 647–652.

Richardson J.L. and Dussinger R.A. 1986. Paleolimnology of mid-elevation lakes in the Kenya Rift Valley. Hydrobiologia 143: 167–174.

Ricketts R.D. and Johnson T.C. 1996. Climate-change in the Turkana basin as deduced from a 4 000 year long $\delta^{18}O$ record. Ear. Planet. Sci. Lett. 142: 7–17.

Rietti-Shati M., Shemesh A. and Karlén W. 1998. A 3 000-year climatic record from biogenic silica oxygen isotopes in an equatorial high-altitude lake. Science 281: 980–982.

Roberts N., Taieb M., Barker P., Damnati B., Icole M. and Williamson D. 1993. Timing of the Younger Dryas event in East Africa from lake-level changes. Nature 366: 146–148.

Rossignol-Strick M., Paterne M., Bassinot F.C., Emeis K.-C. and DeLange G.J. 1982. An unusual mid-Pleistocene monsoon period over Africa and Asia. Nature 392: 269–272.

Schneider R.R., Müller P.J., Rulhand G., Meinecke G., Schmidt H. and Wefer G. 1996. Late Quaternary surface temperatures and productivity in the east-equatorial South Atlantic: response to changes in trade/monsoon wind forcing and surface water advection. In: Wefer G., Berger W.H., Siedler G. and Webb D.J. (eds), The South Atlantic: Present and Past Circulation. Springer-Verlag, Berlin. pp. 527–551.

Scholz C.A., King J.W., Ellis G.S., Swart P.K., Stager J.C. and Colman S.M. 2003. Paleolimnology of Lake Tanganyika, East Africa, over the past 100 kyr. J. Paleolimnol. 30: 139–150.

Shackleton N.J. 2000. The 100 000-year ice-age cycle identified and found to lag temperature, carbon dioxide, and orbital eccentricity. Science 289: 1897–1902.

Stager J.C. and Mayewski P.A. 1997. Abrupt early to mid-Holocene climatic transition registered at the equator and the poles. Science 276: 1834–1836.

Street F.A. and Grove A.T. 1976. Environmental and climatic implications of late Quaternary lake-level fluctuations in Africa. Nature 261: 385–390.

Street-Perrott F.A. 1992. Tropical wetland sources. Nature 355: 23–24.

Street-Perrott F.A. 1994. Palaeo-perspectives: Changes in terrestrial ecosystems. Ambio 23: 37–43.

Street-Perrott F.A., Huang Y., Perrott R.A., Eglinton G., Barker P., Ben, Khelifa L., Harkness D.A. and Olago D. 1997. Impact of lower atmospheric CO_2 on tropical mountain ecosystems: carbon-isotope evidence. Science 278: 1422–1426.

Street-Perrott F.A., Mitchell J.F.B., Marchand D.S. and Brunner J.S. 1990. Milankovitch and albedo forcing of the tropical monsoons: a comparison of geological evidence and numerical simulations for 9 000 yr BP. Trans. r. Soc. Edinb. Ear. Sci. 81: 407–427.

Street-Perrott F.A. and Perrott R.A. 1990. Abrupt climate fluctuations in the tropics: the influence of Atlantic Ocean circulation. Nature 343: 607–611.

Street-Perrott F.A. and Perrott R.A. 1993. Africa. In: Wright H.E., Kutzbach J.E., Webb III T., Ruddiman W.F., Street-Perrott F.A. and Bartlein P.J. (eds), Global Climates since the Last Glacial Maximum. University of Minnesota Press, Minneapolis.

Talbot M.R. and Brendeland K.I. 2001. Strontium isotopes as palaeohydrological tracers in the White Nile headwater lakes, East Africa. EOS Transactions, AGU Fall Meeting Supplement 82, Abstract PP21C-05.

Talbot M.R. and Johannessen T. 1992. A high-resolution paleoclimatic record for the last 27 500 years in Tropical West Africa from the carbon and nitrogen isotopic composition of lacustrine organic matter. Ear. Plan. Sci. Lett. 110: 23–37.

Talbot M.R. and Laerdal T. 2000. The Late Pleistocene-Holocene palaeolimnology of Lake Victoria, East Africa, based upon elemental and isotopic analyses of sedimentary organic matter. J. Paleolim. 23: 141–164.

Talbot M.R. and Livingstone D.A. 1989. Hydrogen index and carbon isotopes of lacustrine organic matter as lake level indicators. Palaeogeogr. Palaeoclimatol. Palaeoecol. 70: 121–137.

Talbot M.R., Williams M.A.J. and Adamson D.A. 2000. Strontium isotope evidence for late Pleistocene reestablishment of an integrated Nile drainage network. Geology 28: 343–346.

Talbot M.R., Livingstone D.A., Palmer D.G., Maley J., Melack J.M., Delibrias G. and Gulliksen J. 1984. Preliminary results from sediment cores from Lake Bosumtwi, Ghana. Palaeoecology of Africa 16: 173–192.

Texier D., deNoblet N. and Braconnot P. 2000. Sensitivity of the African and Asian monsoons to mid-Holocene insolation and data-inferred surface changes. J. Clim. 13: 164–181.

Tietze K., Geyh M., Müller H., Schröder L., Stahl W. and Wehner H. 1980. The genesis of methane in Lake Kivu (Central Africa). Geolog. Rundschau. 69: 452–472.

Trauth M.H., Deino A. and Strecker M.R. 2001. Response of the East African climate to orbital forcing during the last interglacial (130–117 ka) and the early last glacial (117–60 ka). Geology 29: 499–502.

Verschuren D., this volume. Decadal and century-scale climate variability in tropical Africa during the past 2000 years. In: Battarbee R.W., Gasse F. and Stickley C.E. (eds), Past Climate Variability through Europe and Africa. Kluwer Academic Publishers, Dordrecht, the Netherlands, pp. 139–158.

Verschuren D., Laird K.R. and Cumming B.F. 2000. Rainfall and drought in equatorial East Africa during the past 1 100 years. Nature 403: 410–414.

von Grafenstein U., Erlenkeuser H., Muller J., Jouzel J. and Johnsen S. 1998. The cold event 8 200 years ago documented in oxygen isotope records of precipitation in Europe and Greenland. Clim. Dyn. 14: 73–81.

Wooller M.J. and Agnew A.D.Q. 2002. Changes in graminoid stomatal morphology over the last glacial- interglacial transition: evidence from Mount Kenya, East Africa. Palaeogeogr. Palaeoclim. Palaeoecol. 177: 123–136.

Wooller M.J., Street-Perrott F.A. and Agnew A.D.Q. 2000. Late Quaternary fires and grassland palaeoecology of Mount Kenya, East Africa: evidence from charred grass cuticles in lake sediments. Palaeogeogr. Palaeoclim. Palaeoecol. 164: 207–230.

Yin J.H. and Battisti D.S. 2001. The importance of tropical sea surface temperature patterns in simulations of last glacial maximum climate. J. Clim. 14: 565–581.

8. DECADAL AND CENTURY-SCALE CLIMATE VARIABILITY IN TROPICAL AFRICA DURING THE PAST 2000 YEARS

D. VERSCHUREN (dirk.verschuren@UGent.be)
Research group Limnology
Department of Biology
Ghent University
Ledeganckstraat 35
B-9000 Gent
Belgium

Keywords: Little Ice Age, Medieval Warm Period, Solar climate forcing, Africa, Regional climate linkages

Introduction

Holocene climate in high-latitude regions of the world has been relatively stable compared to glacial climates. In contrast, tropical Africa and other low-latitude continental regions were marked by a succession of millennium-scale wet and dry episodes, separated by rather abrupt transitions (Gasse and Van Campo 1994; Lamb et al. 1995; Gasse 2000). These continent-wide fluctuations in the balance of rainfall and evaporation must somehow have resulted from large-scale variation in the position or intensity of large-scale tropical monsoon systems, but their relationship to Holocene climate variability in extra-tropical regions (e.g., Bond et al. (2001)) and the likely mechanisms of external climate forcing are only just beginning to be revealed (Gupta et al. 2003; Hoelzmann et al., this volume).

Compared to this marked hydrological instability of African climate during the early and middle Holocene, the last 2000 years have commonly been thought of as rather stable and uneventful. This idea can be traced back to the first reviews of late-Quaternary vegetation and lake-level change in Africa (Butzer et al. 1972; Livingstone 1975; 1980; Hamilton 1982), which in their focus on the prominent late-Glacial and early-Holocene events found little worth mentioning in the late Holocene. Low time resolution and poor age control meant that the last 2000 years were typically represented by just a few data points floating on an interpolated section of the time line. In addition, evidence for 20th-century landscape disturbance was often missing because soft surface muds had not been recovered, or had been discarded. This lack of a reference frame for signatures of pre-modern human impact, together with the assumption of relative climatic stability, helped perpetuate among palaeoecologists, archaeologists, and geomorphologists the paradigm that most evidence for vegetation and landscape change in tropical Africa younger than 2000 years is due to human activity (Taylor 1990; Jolly et al. 1997; Eriksson 1998).

R. W. Battarbee et al. (eds) 2004. *Past Climate Variability through Europe and Africa.*
Springer, Dordrecht, The Netherlands.

Yet already 20 years ago, Crossley et al. (1984) assembled dated geomorphological and archaeological evidence from the fluctuating shorelines of lakes Chilwa and Malawi which suggested that climatic conditions during the last millennium in south-eastern tropical Africa must have been quite unstable, involving moisture-balance fluctuations large enough to seriously affect the living conditions of indigenous peoples. Early geomorphological and sedimentary data from Lake Bosumtwi in Ghana (Talbot and Delibrias 1980) likewise indicated that humid West Africa experienced at least one episode of severe drought within the last millennium comparable in magnitude to the major droughts of the middle Holocene. Possible synchroneity with similar events reported from Lake Chad in the Sahel (Maley 1973) and Lake Abhé near the Red Sea (Gasse 1977) clearly suggested that the last 2000 years of Africa's climate history were punctuated by major, century-scale climatic anomalies of possibly continent-wide extent. Recognising the ambiguity of some disturbance indicators in African pollen records (Taylor et al. 2000), modern palaeoecological studies attempt to distinguish between the effects of climate change and human activities on African vegetation, and these now yield clear signatures of climate-driven vegetation change during the last millennium, both in semi-arid Central Kenya (Lamb et al. 2003) and the humid forests of Western Uganda (Marchant and Taylor 1998).

Given the reality of African climatic instability at all time scales (Nicholson 2000), one of the great challenges of tropical palaeoclimatology today is to reconstruct Africa's climate history with a time resolution and precision sufficient to elucidate the causes and pace-makers of decadal to century-scale rainfall variability, to help evaluate how future interaction of anthropogenic climate forcing with natural climate dynamics may change the frequency and intensity of severe, socio-economically disruptive drought.

This chapter presents a brief review of the available evidence for climate change in tropical Africa during the past 2000 years, in an attempt to identify patterns of spatial coherence which may reveal links with the climate history of extra-tropical regions, and hint at possible causal mechanisms. It will be evident that the limited age control of much of the older published work hampers evaluation of the synchroneity of climate events between sites: only a handful of the available climate-proxy records extend beyond the last 300–400 years with better than century-scale resolution and age control.

High-resolution climate-proxy archives in tropical Africa

Documentary records

For all practical purposes, in tropical Africa the 'historical' period can be said to have started in the mid-19th century with state-sponsored European exploration, and then colonisation, of the continent's interior. Instrumental weather records are mostly limited to the last 120 years, and Africa's long cultural history notwithstanding, documentary records of weather-related information extending to before AD 1800 are rare (Fig. 1; see Nicholson (2001a, b) for the period from AD 1800). Old chronicles that do exist are often discontinuous, deal with the societal impact of anomalous weather rather than the weather events themselves, and are difficult to normalise with regard to long-term climate variability because of interference from various factors other than climate. One of the least tractable of these factors is the reference period for a particular observation. Often this is only a decade or so, and rarely longer than the observer's own life-time experience. Consequently, a few

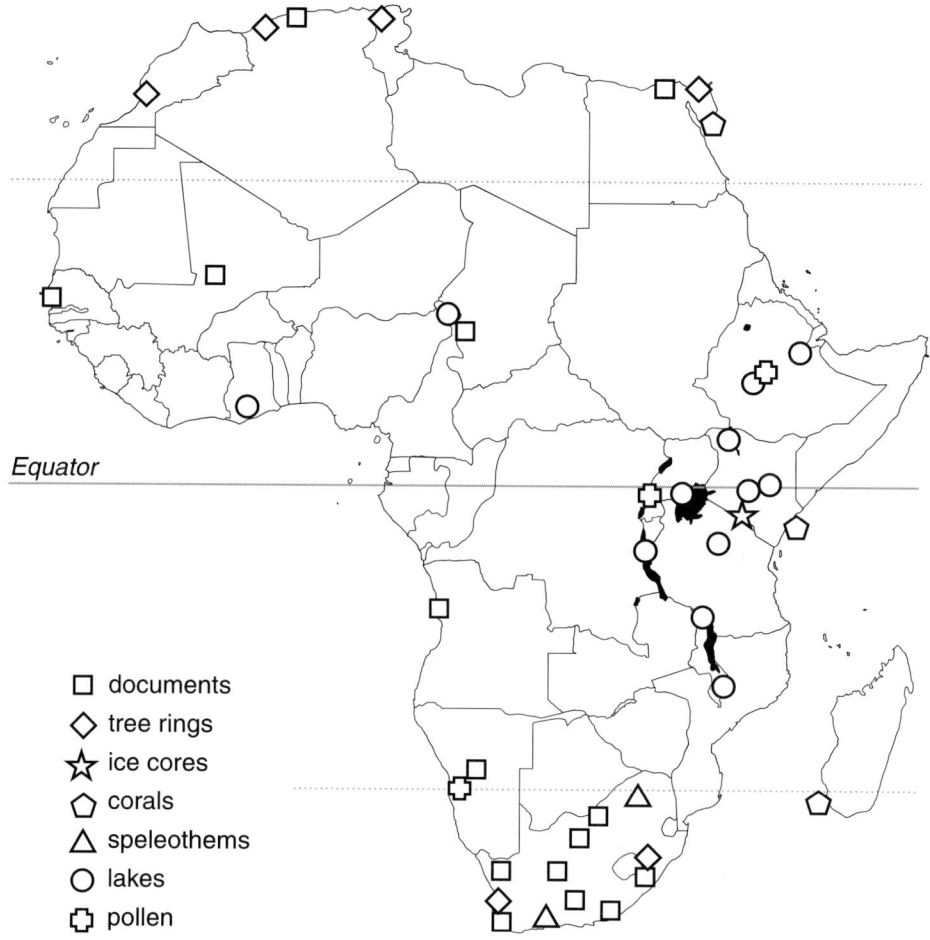

Figure 1. Distribution of high-resolution climate-proxy records on the African continent and in adjacent oceans extending beyond AD 1800. Localities of regionally integrated documentary records are approximate.

consecutive dry years punctuating a humid episode may be long remembered but rather unimportant from a long-term perspective (Nicholson 1996a). However, the scarcity in tropical Africa of natural climate-proxy archives with annual resolution (cf. below) renders each of these chronicles highly valuable, and certainly warrants their detailed evaluation. For example, a documentary record of drought and disease in coastal Angola from AD 1560 to the 1870s (Dias 1981; Miller 1982), based mainly on Portuguese colonial archives, is the only source of high-resolution climatic information from western equatorial Africa extending to before AD 1800. The documentary record with which high-resolution proxy records from tropical Africa are usually compared is the Rodah Nilometer of flood and minimum levels in the Nile River since AD 622. But this record is not without its share of problems (Nicholson 1996a), and the climatic information it is said to contain on inter-annual to century time-scales continues to be the subject of re-evaluation (Herring 1979;

Hassan 1981; Quinn 1993; Fraedrich et al. 1997; Nicholson 1996a; 1998; De Putter et al. 1998).

Dendroclimatology

Tree-ring-based climate studies in tropical Africa face significant obstacles. Limited temperature seasonality and strong inter-annual variability in the strength or duration of dry and rainy seasons causes most trees to lack the distinct annual growth rings which provide superior chronological precision to tree-ring-based climate reconstruction in temperate regions. Trees that do form well-defined growth rings in certain parts of their range (e.g., *Juniperus procera* and *Vitex keniensis* in Kenya, *Pterocarpus angolensis* in Zimbabwe) usually do not long survive the attacks of fungi and termites once dead or cut down. Consequently, there is little potential to extend calibrated tree-ring chronologies beyond the lifetime of currently living trees (Dunbar and Cole 1999). In inter-tropical Africa, possibly only Ethiopia and Zimbabwe have both the material cultural heritage and a sufficiently dry climate to yield reasonably long, cross-dated tree-ring sequences from surviving sections of old timber and furniture. So far, however, long African tree-ring records all come from South Africa, the Maghreb region of Northwest Africa, and the Sinai desert (Nicholson 1996a), except for one 145-year record of deuterium excess (a water-balance proxy) in a *Juniperus procera* tree from the Mau Escarpment in Central Kenya (Krishnamurthy and Epstein 1985).

Ice cores

Compared to South America and Asia, Africa is not well endowed with snow-capped mountains. Ice cores from Lewis Glacier on Mt. Kenya (at 4870 m asl) were among the first ones recovered from a tropical mountain, but yielded little information because intermittent melt-water percolation had evidently corrupted any previously incorporated climate signatures (Hastenrath 1981). Similar conditions may also occur in the glaciers of the Ruwenzori Mountains (4500–5100 m asl). Of Africa's three snow-capped mountains, possibly only the ice fields on Mt. Kilimanjaro (above 5500 m asl) have preserved a high-resolution proxy record of past climate change (Thompson et al. 2002), and even this record is troubled by lack of an independent chronology (Gasse 2002). While the precise timing of climate events inferred from the Mt. Kilimanjaro ice archive should be regarded with caution, it will likely remain a truly unique record of Holocene variability in atmospheric dust loading and aerosol composition in equatorial Africa.

Corals

The oxygen-isotope and trace-element composition of growth rings in massive corals provide better-than-annual resolution records of sea-surface temperature and salinity, and thus rank among the best natural archives of inter-annual to decade-scale climate variability in the tropics. In suitable coastal locations their trace-element composition can also produce annual records of river discharge and soil erosion (Cole 2003). Intensive exploration of coral reefs along Africa's coast (Gagan et al. 2000) is now yielding its first results with

~200-year climate-proxy records from the Indian Ocean coast (Cole et al. 2000) and the Red Sea (Felis et al. 2000), and one 320-year record from Madagascar (Tyson et al. 2000). About 400 years may be the practical limit of coral-based climate reconstruction, unless unique fossil coral formations are found that would permit construction of cross-dated sequences.

Speleothems

Cave speleothems have yielded several important late-Holocene climate-proxy records from South Africa (Holmgren et al. 1999; 2001; Lee-Thorp et al. 2001), the Arabian Peninsula (Burns et al. 2002) and Israel (Bar-Matthews and Ayalon, this volume), and their unexplored potential remains large. However, the known distribution of high-quality speleothems is still largely limited to Africa's extra-tropical north and south. And, like with the scarce natural lakes in those regions, the recording of climatic signals in speleothems is sometimes interrupted during phases of severe drought.

Lake sediments

Lake sediments are the most ubiquitous climate-proxy archives in tropical Africa (Fig. 1), both in space and in time. But given the long history of palaeolimnological research in Africa, still few lake-based climate records possess the time resolution and dating control required to say anything meaningful about decadal and century-scale climatic anomalies during the past 2000 years. Dating uncertainty due to poorly constrained reservoir effects or the presence of reworked terrestrial organic carbon, aggravated by missing core tops, have long compromised attempts to correlate sub-millennial climate events between study sites. With the introduction of appropriate methods to recover and date soft surface sediments, comprehensive exploration of Africa's many climate-sensitive lakes has now become feasible. The main challenge in this effort is to find those lakes (undoubtedly only a fraction of the total) which combine adequate hydrological sensitivity to decade-scale climate variability with persistence of favourable sedimentation conditions throughout the past 2000 years (Verschuren 1999).

Decadal and century-scale climate change reflected in African lake levels

Similarity of 20th-century lake-level trends across eastern equatorial Africa (Nicholson 1996a; 1998) suggests that spatial complexity in the seasonality of rainfall is underlain by more uniform patterns of long-term temporal variability controlled by large-scale atmospheric dynamics (Nicholson 2000; 2001b). Amplifier lakes (Street 1980) in the Eastern Rift Valley of Kenya (Fig. 2: Turkana, Baringo, Bogoria, and Naivasha, among others) show very similar historical trends including: (i) a major 1890s transgression leading to peak levels in the 1900s and 1910s; (ii) a long decline to a marked lowstand in the 1940s–1950s; (iii) recovery to intermediate levels, starting in the late 1950s and accelerating after heavy rainfall in 1961–1962; and (iv) low to intermediate levels in recent decades, displaying ENSO-type cyclicity. In contrast, lakes in south-eastern tropical Africa (Fig. 2: Malawi and Chilwa) show broadly opposite historical trends: (i) mid-19th century intermediate levels

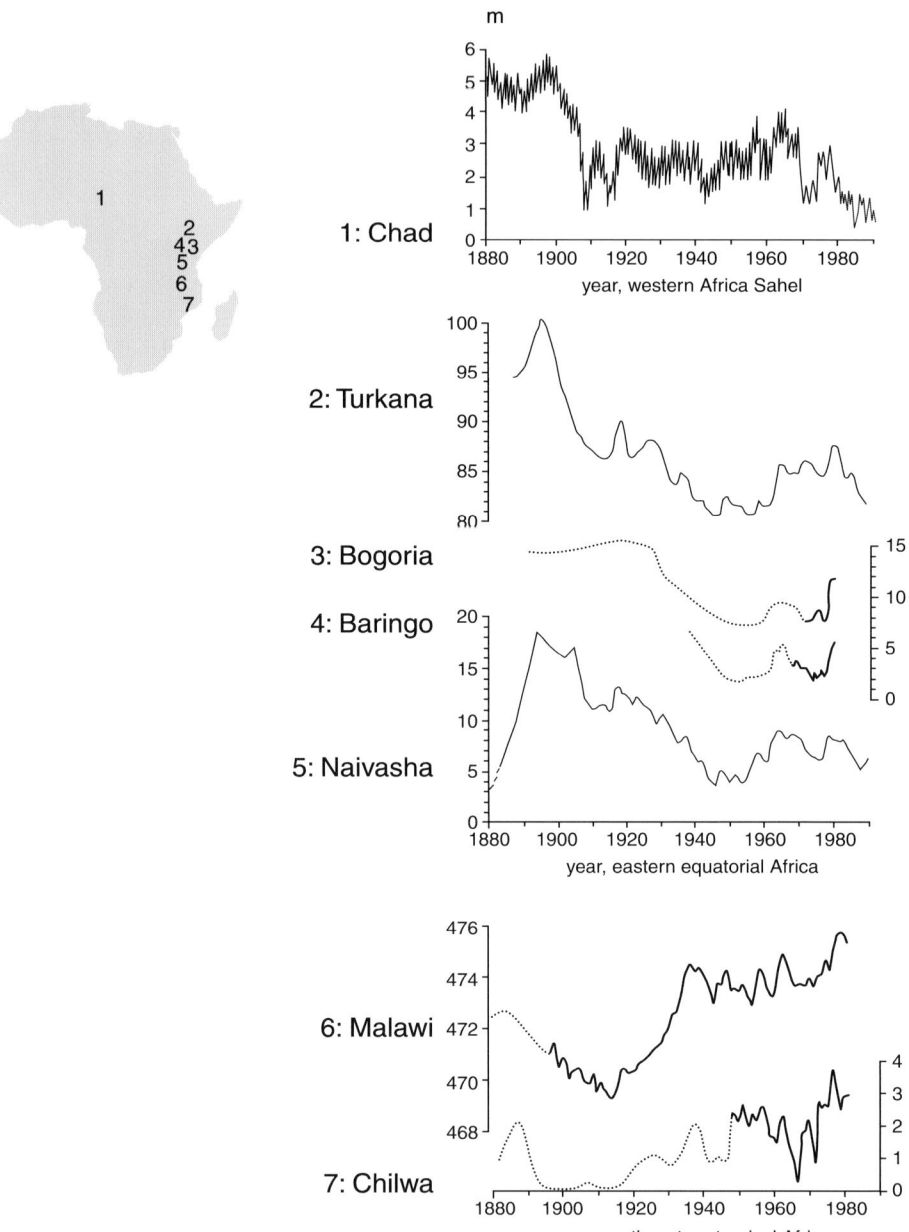

Figure 2. Historical lake-level trends in eastern and south-eastern Africa, and the Sahel region of North Africa: Chad (1; Thambiyahpillay (1983)), Turkana (2; Kolding (1992)), Baringo and Bogoria (3–4; Tiercelin et al. (1987)), Naivasha (5; Verschuren (1996)), Malawi (6; Owen et al. (1990)), and Chilwa (7; Crossley et al. (1984)). Vertical axis is metre asl in Lake Malawi, water depth in the other lakes.

continuing until the 1880s; (ii) lake-level decline in the 1890s leading to a pronounced lowstand in the 1900s and 1910s; (iii) recovery during the 1920s–1930s to reach peak level in the 1950s–1960s; and (iv) continuously high levels since then. The historical lake-level curve of Lake Chad in the Sahel region of North-Central Africa (Thambiyahpillay 1983) is also distinct (Fig. 2). Some decade-scale fluctuations are shared among all or some regions (e.g., regression during the period ~1908–1916 shared by eastern equatorial lakes and Lake Chad, and the late-1970s rise shared by eastern and south-eastern lakes), in agreement with historical patterns in the geographical distribution of rainfall anomalies (Nicholson 1993; 2001b). Importantly, this evidence for regional coherence in lake-level trends gives credence to the use of lake-level records to evaluate the regional coherence and synchroneity of decade-scale climate anomalies further back in time. Even at this fairly short time scale, past moisture-balance fluctuations appear to have been large enough to over-ride site specificity in the lakes' response, at least for amplifier lakes with large catchments, where river inflow makes a significant contribution to the water budget.

One such lake, Lake Naivasha in the Eastern Rift Valley of Central Kenya, has so far yielded the best-resolved record of moisture-balance variations in equatorial East Africa during the past two millennia with suitable time control. A combination of diatom- and chironomid-inferred salinity reconstructions with a lake-level reconstruction based on sedimentological characteristics (Verschuren et al. 2000; Verschuren 2001; D. Verschuren, K.R. Laird, H. Eggermont, B.F. Cumming, unpublished data) shows that the climate of Central Kenya over the past 1800 years was characterised by a succession of (at least) seven decade-scale episodes of aridity more severe than any drought recorded during the 20th century (Fig. 3). To facilitate correlation with climate events elsewhere, these droughts can be identified as Naivasha Drought (ND) 1 to 7, in recognition of the apparent magnitude of the hydrological anomalies involved while acknowledging uncertainty in the determination of their exact age (Verschuren 2001).

The timing of ND1, ND2 and ND3 is supported by the pre-colonial history of drought-induced famine, migration, and clan conflict recorded in the oral traditions of agriculturalist peoples from the so-called Interlacustrine Region in present-day Uganda, Rwanda, north-eastern D.R. Congo, north-western Tanzania, and southern-most Sudan (Webster (1979), see Schoenbrun (1998) for an evaluation of these data). ND1 and ND2 are also coeval with the two most important episodes of drought recorded since AD 1560 in Angola (Miller (1982), Fig. 3), suggesting that the rainfall history of Central Kenya may be representative for equatorial Africa over at least the last 500 years. Interestingly, the three episodes of positive water balance bracketed by ND1, ND2, ND3 and ND4 are broadly coeval with the Wolf (AD 1290–1350), Spörer (1450–1540), and Maunder (1645–1715) minima in solar activity. Judging from the Naivasha record, equatorial East Africa appears to have been fairly wet during the Little Ice Age (LIA) period between AD 1270 and 1750, compared to mean 20th-century climate. During the Maunder Minimum, Lake Naivasha stood higher than at any other time in the past 1800 years. The period bracketed by droughts ND4, ND5 and ND6 (~AD 870–1270) is broadly coeval with the Medieval Warm Period (MWP) in north-temperate regions, and ND7 reflects a pre-MWP drought tentatively dated to the 6th or 7th century AD. Irregularities in Naivasha's radiocarbon chronology (Verschuren 2001) suggest that accumulation in Crescent Island Crater may not have been continuous during ND5, consequently that the local record of Medieval climate variability is incomplete.

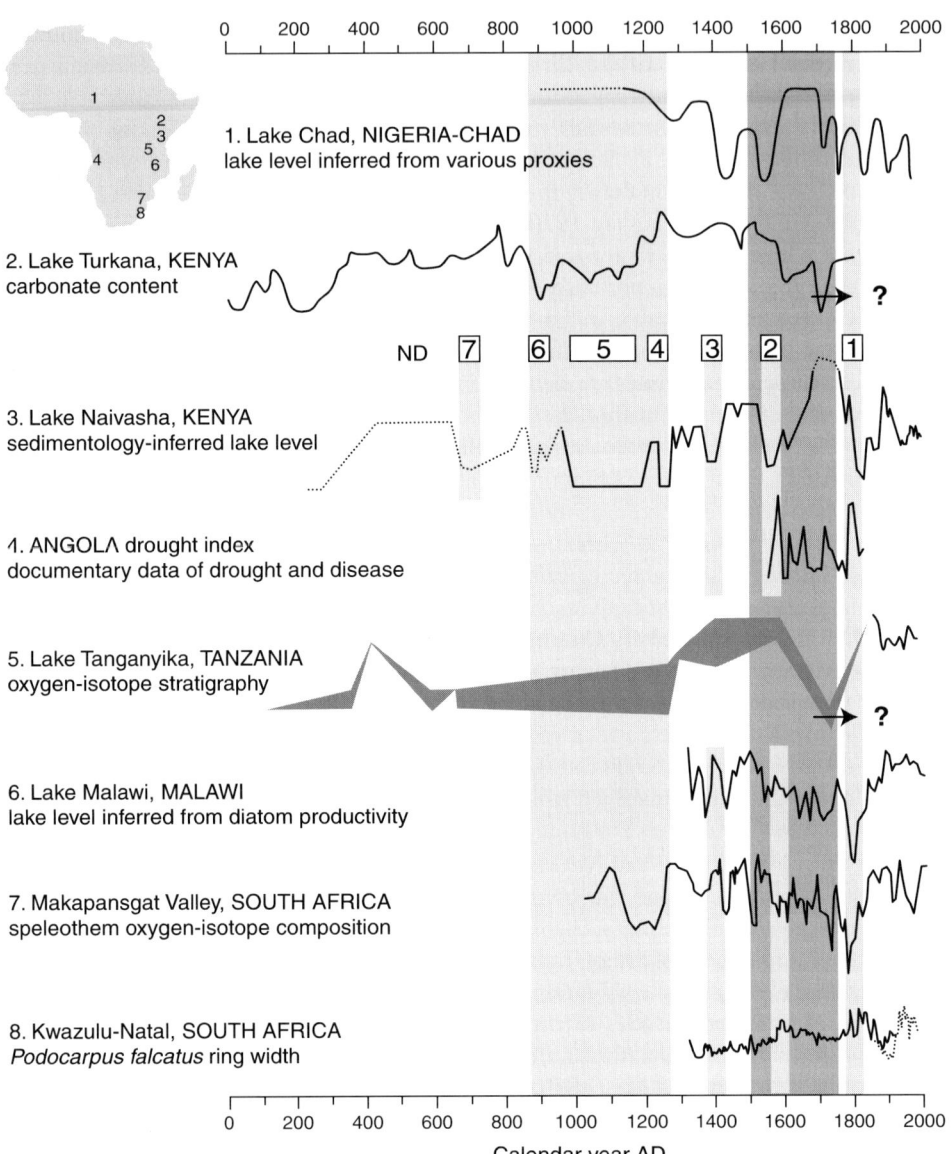

Figure 3. Lake-level reconstruction for Lake Naivasha (3; Verschuren (2001)) over the past 1800 years compared with other African proxy records presumed to reflect moisture-balance variation. From North to South: Lake Chad (1; Maley (1993)), Lake Turkana (2; Halfman et al. (1994)), Angola drought index, summed per decade (4; Miller (1982)), Lake Tanganyika (5; Cohen et al. (1997)), Lake Malawi (6; Johnson et al. (2000)), Mapakansgat speleothem oxygen isotopes (7; Holmgren et al. (1999)), and Kwazulu-Natal *Podocarpus* ring width (8; Hall (1976)).

Lake-based climate proxy data from elsewhere in East Africa, though less well resolved, generally agree with the main features of the Naivasha record. To the north, stacked records of carbonate content in Lake Turkana sediments (Halfman et al. 1994), a proxy indicator of Omo River discharge, indicate relatively dry conditions over the Ethiopian Plateau before 1700 ^{14}C yr BP (~AD 370) and during Medieval time (~AD 900–1200), while relatively humid conditions prevailed ~AD 700–900, and from ~AD 1200 to 1550 (Fig. 3). In the Ethiopian Rift, Lake Abiyata stood high until shortly before AD 1800 but then receded with the shift to drier conditions which characterised the late 18th and early 19th centuries (Legesse et al. 2002). In the Afar region bordering the Red Sea, Lake Abhé stood low during the first half of the last millennium, and then high from the 16th to 18th century (Gasse and Street 1978). As at Abiyata, this was followed by regression during much of the 19th century, and eventually halted by a late 19th-century transgression. To the west of Kenya, high-resolution fossil-diatom data from Damba Channel between the mainland and a string of islands in Lake Victoria (Stager and Mayewski 1997; Stager et al. 1997; Stager 1998) record a marked increase in benthic diatom taxa, reflecting a ~500-yr lowstand which was initially interpreted to represent LIA drought. New data from shallow, protected Pilkington Bay along the north shore of Lake Victoria, and revision of the Damba Channel chronology (Stager et al. 2003), re-assigned this inferred lowstand to between ~AD 800 and 1300, i.e., matching the timing of MWP drought at Naivasha. In the sub-humid western branch of the African Rift, Lake Tanganyika fluctuated ~25–30 m between marginally open and marginally closed states for much of the past 2800 years (Cohen et al. 1997). Patterns through time in the oxygen-isotope composition of stromatolites and molluscs suggest relative aridity from about 800 BC to AD 400 and from AD 600 to 1250, separated by a brief period of positive water balance in the 4th or 5th century (Fig. 3). After AD 1250, Lake Tanganyika rose to its overflow at about the time that a switch to positive water balance ended the MWP drought at Naivasha, and it maintained this level from AD 1350 to at least 1550. In broad agreement with the Lake Tanganyika record, pollen data from Mubwindi Swamp in Southwest Uganda show trees of the humid montane forest to reach their strongest representation between ~AD 1350 and 1500 (Marchant and Taylor 1998). Then at some time between 1550 and the early 1800s, Tanganyika fell at least 15 m (maybe up to 40 m) to perhaps its lowest level in the past 2800 years. It rebounded during a strong mid-19th century transgression, reached overflow in 1877, and has remained open since then.

Lake-based climate-proxy data from further afield display some similarity but also distinct differences with the Naivasha record. Fluctuations of Lake Chad in Central North Africa reflect precipitation over large parts of Niger, Chad, Cameroon and the Central African Republic, but most river inflow derives from humid southern portions of the drainage basin around 7–10 °N. The history of Lake Chad over the past 900 years (Maley 1973) has been revised and updated several times. Compilation of the sedimentological and palynological data (Maley 1981), anchored in three radiocarbon dates (Maley 1993) and tuned with available documentary proxy data (Maley 1989), indicate that Lake Chad stood high from before AD 1100 to 1400 and again in the 17th century, separated by low to intermediate levels during the 15th and 16th centuries when the East African lakes stood high (Fig. 3). The recent half of the Lake Chad record agrees with many historical records from West Africa, and possibly reflects a climatic opposition between West Africa and equatorial East Africa which has also occurred several times during the 20th century (Nicholson 1986; 1994; 2001b). Yet other parts of the Sahel were clearly moister than today during the 17th and 18th centuries (Nicholson 1996a), i.e., more similar to the East African pattern.

The best-resolved climate-proxy data for south-eastern tropical Africa come from the North basin of Lake Malawi, where lake history over the last 700 years is preserved in an annually-laminated sequence of diatom-silica accumulation, a proxy for nutrient up-welling that would be inversely related to lake level (Johnson et al. 2001) or may indicate intensification of northerly winds (Johnson et al. 2002; this volume). When interpreted in terms of lake-level change, the diatom-silica record suggests that Lake Malawi stood lower than today between AD 1570 and 1850, most of this period being one prolonged regression towards a pronounced lowstand shortly before AD 1800. Sediment records from the shallower South basin contain an erosional hiatus likewise testifying to an extreme lowstand (possibly >120 m below present level) from which the lake recovered during the mid-19th century (Owen et al. 1990). Consistent with these core data from both basins, archaeological and geomorphological evidence from the Lake Malawi shoreline (Crossley and Davison-Hirschmann 1981; 1982) suggest that the lake reached its highest level of the past 1700 years during the 14th and 15th centuries in a possible double-peaked event. Other highstands occurred in the 10th century and during the first half of the first millennium AD (Crossley et al. 1984). Johnson et al. (2002, this volume) suggest that the strong upwelling which stimulated diatom production in Lake Malawi's North basin from about AD 1570 to 1850 was generated by more frequent or stronger northerly winds than today, a possible indication that the Intertropical Convergence Zone migrated further southward during austral Summer coincident with Little Ice Age cooling of the Northern Hemisphere.

When comparing the high-resolution records of lakes Naivasha and Malawi, it would seem that the frequent opposition between eastern equatorial and south-eastern tropical Africa in modern patterns of both inter-annual rainfall variability (Nicholson 1986; 1996b; Ropelewski and Halpert 1989) and decade-scale rainfall and lake-level trends (Nicholson (1993, 2001b), Fig. 2) may have persisted through much of the last millennium. But this is only partly the case. First, both lakes Naivasha and Malawi experienced their most pronounced inferred lowstand of the last 700 years between AD 1780 and 1830. This timing matches records of severe 1790s drought from tropical and subtropical regions world-wide, including South Africa, Australia, and Mexico (Grove 1998). In ice-cores from Tibet (Thompson et al. 2000), evidence for failure of the South Asian Monsoon during the 1790s is recorded as the largest dust peak of the last millennium, dust which must have originated from semi-arid source regions in India and North Africa. Second, both lakes Naivasha and Malawi as well as lakes Turkana, Victoria and Tanganyika discussed above, and also the Mt. Kilimanjaro dust and oxygen-isotope records (Thompson et al. 2002), all show relatively dry conditions during the 11th to early 13th centuries, switching to a distinctly wetter climatic regime in the late 13th or early 14th century. Evidence for this transition from the Lake Malawi region is found in the temporal distribution of Iron Age ceramic traditions on the present-day Lake Malawi shoreline, which reveals major settlement gaps not only from AD 1700 to the mid-1800s but also from shortly after AD 1000 to 1300 (Owen et al. 1990). The marked shift to a positive water balance around AD 1250–1300 is not limited to tropical Africa but has also been reported from the North American Great Plains (Laird et al. 1996). Allowing for dating uncertainty in the available records, it appears to coincide with a pronounced shift in trade-wind regime over the tropical Atlantic Ocean (Black et al. 1999), cooling of the North Atlantic (Keigwin 1996), and the earliest evidence for LIA glacier advance in the European Alps (Holzhauser 1997).

The principal contrast between the Naivasha and Malawi records concerns their history of the 16th to 18th centuries. Whereas the sedimentology-based reconstruction from Lake Naivasha (Verschuren 2001) infers a prolonged trend of *increasing* moisture from the 14th to the late 17th century with a temporary reversal in the late 16th century and a dramatic ending in late 18th-century aridity, the Lake Malawi diatom-productivity record (Johnson et al. 2001) infers a prolonged trend of *decreasing* moisture starting in the 16th century and culminating in the same late 18th-century aridity. Presently, few other published studies have both the time resolution and age control to contribute meaningfully to the discussion of this apparent climatic contrast between eastern and south-eastern tropical Africa during the main phase of the Little Ice Age. The Lake Tanganyika reconstruction (Cohen et al. 1997) in its present form is compatible with both the Naivasha and Malawi records, because the period from AD 1500 to ~1850 is without data except for the probable occurrence of an extreme lowstand (Fig. 3). While we can assume this lowstand to have been coeval with the 1790s-centered lowstands of lakes Naivasha and Malawi, it remains unclear whether Tanganyika stood at its overflow level until the mid-1700s, implying a climate history similar to that of Central Kenya, or if regression started soon after AD 1500, similar to what happened at Lake Malawi. Some on-shore data from Lake Chilwa in southern Malawi are difficult to reconcile with the inferred lake-level record of Lake Malawi nearby. Differences in the age distribution of *Adansonia digitata* (baobab) trees (calculated by calibrating girth measurements by ring-width data) above and below 631 m elevation seem to indicate that a pronounced Little Ice Age-equivalent highstand of Lake Chilwa persisted at least until AD 1750 (Crossley et al. 1984). The implied humid conditions during the 17th and early 18th centuries also seem to have prevailed in the Northern Kalahari desert another 5° latitude further South, where Lake Ngami was flooded by increased discharge from the Okavango River delta (Nicholson 1996a). The Chilwa and Ngami records better agree with the alternative interpretation of the Malawi diatom-silica record (Johnson et al., this volume), which does not require a pronounced LIA lowstand.

Evidence for temperature change in Africa during the past 2000 years

A well-known obstacle to direct climatic interpretation of lake-level records is the dependence of net moisture availability on both precipitation and temperature. If tropical Africa experienced significant temperature variation over the past 2000 years, then similar temporal patterns of rainfall variability among drainage basins could still result in quite dissimilar lake-level records depending on the exact effect of a particular temperature anomaly on evaporation rates from land and water surfaces within each drainage basin. The history of Lake Naivasha, where a mostly positive water balance since ~AD 1270 appears to have culminated in a pronounced highstand coeval with the Maunder Minimum, mirrors reconstructed temperature anomalies over the past 1000 years for the Northern Hemisphere continents (Jones et al. 2001), begging the question to what extent a temperature-driven decrease of evaporation during the main phase of the Little Ice Age (~AD 1550–1750) may have contributed to the increase in moisture availability implied by the Naivasha record.

Unambiguous evidence that tropical Africa experienced significant temperature change during the last two millennia is scarce, since excursions in most climate-proxy indicators can be interpreted as evidence of both rainfall or temperature variation. A pollen record from a swamp above treeline (3600 m asl) on Mt. Arsi in south-eastern Ethiopia reveals

expansion of ericaceous shrubland at the expense of forest trees during the period ~AD
1250–1700. Bonnefille and Umer (1994) interpret it to reflect a 400 m shift of vegetation
belts down the mountain, equivalent to a maximum cooling of 2 °C (Fig. 4). At Kuiseb
River in western Namibia just south of the Tropic of Capricorn, pollen preserved in a
chronosequence of ^{14}C-dated hyrax middens (Scott 1996) show expansion of grasses and
herbs at the expense of trees during the period ~AD 1300–1800. Here this vegetation
change can be explained both by the region having been colder than today, since the frost-
sensitive trees *Acacia albida* and *Salvadora persica* approach the southern limit of their
modern distribution near Kuiseb River, and/or it having been wetter than today, since
greater rainfall favours grassland expansion there (Scott 1996). On the other hand, high-
resolution oxygen-isotope and colour variations in a cave speleothem from Mapakansgat
Valley indicate that at about the same latitude (24 °S) in north-eastern South Africa, colder
periods (up to 1 °C in the 18th century) were drier than today while warmer periods (up
to 3 °C in the 13th century) were wetter than today (Holmgren et al. (1999, 2001), Tyson
et al. (2000), Scott and Lee-Thorp (this volume); revised chronology from Lee-Thorp
et al. (2001)). Moisture-balance changes inferred from the oxygen-isotope variations in
this speleothem record broadly agree with the reconstructed fluctuations of Lake Malawi
some 15° latitude farther North (Fig. 3), supporting the reality of a long-standing, though
not necessarily continuous, inverse correlation between century-scale rainfall patterns in
eastern and south-eastern tropical Africa (Tyson et al. 2002). This inverse relationship is
also hinted at in a uniquely long ring-width record of a *Podocarpus falcatus* tree from
~28 °S in Kwazulu-Natal, Eastern South Africa (Hall 1976; Tyson 1986), which shows
strongest growth in the past 700 years to have occurred during AD 1580–1620 and AD
1770–1840, i.e., coincident with the ND1 and ND2 droughts at Lake Naivasha (Fig. 3).
Assuming that tree growth at this particular sub-tropical locality is drought-limited rather
than temperature-limited (which has yet to be confirmed), it would again indicate positive
rainfall anomalies in eastern south Africa to have coincided with drought in equatorial East
Africa.

Another potential proxy indicator for past temperature variation can be found in the
history of glacier advance and retreat on Africa's three snow-capped mountains, which
all occur within 3° latitude of the Equator. Ice cores from Lewis Glacier on Mt. Kenya
revealed that the 'eternal snow' on Mt. Kenya is only ~500–600 years old (Hastenrath
1981), suggesting that this 5200 m high mountain may have been completely ice-free under
the relatively warm, or dry, climatic conditions which prevailed in Medieval time. The
small Furtwängler Glacier on Mt. Kilimanjaro appears to be even less then 300 years
old (Thompson et al. 2002); oxygen-isotope data suggest it to be a remnant of unusually
cold, or wet, conditions prevailing during the Maunder Minimum. Moraines reflecting LIA
glacier advance are prominent both on Mt. Kenya (Karlén et al. 1999) and the Ruwenzori
Mts. (de Heinzelin 1962), now standing about 100–250 m below the present ice margin. A
continuous record of late-Holocene glacier activity on Mt. Kenya, based on the organic-
carbon content of pro-glacial lake muds in Hausberg Tarn (Karlén et al. 1999), indicates
that the glaciers stood in an advanced position from ~200 BC to AD 300 and from ~AD
650 to 850 (Fig. 4). This was followed by retreat from AD 850 to ~1250, i.e., coeval with
the ND4 to ND6 lowstands of Lake Naivasha. Although radiocarbon age control on the
last 1000 years of Hausberg Tarn's history is rather poor, it appears that the most recent,
LIA-equivalent glacier advance peaked between 400 and 250 years ago.

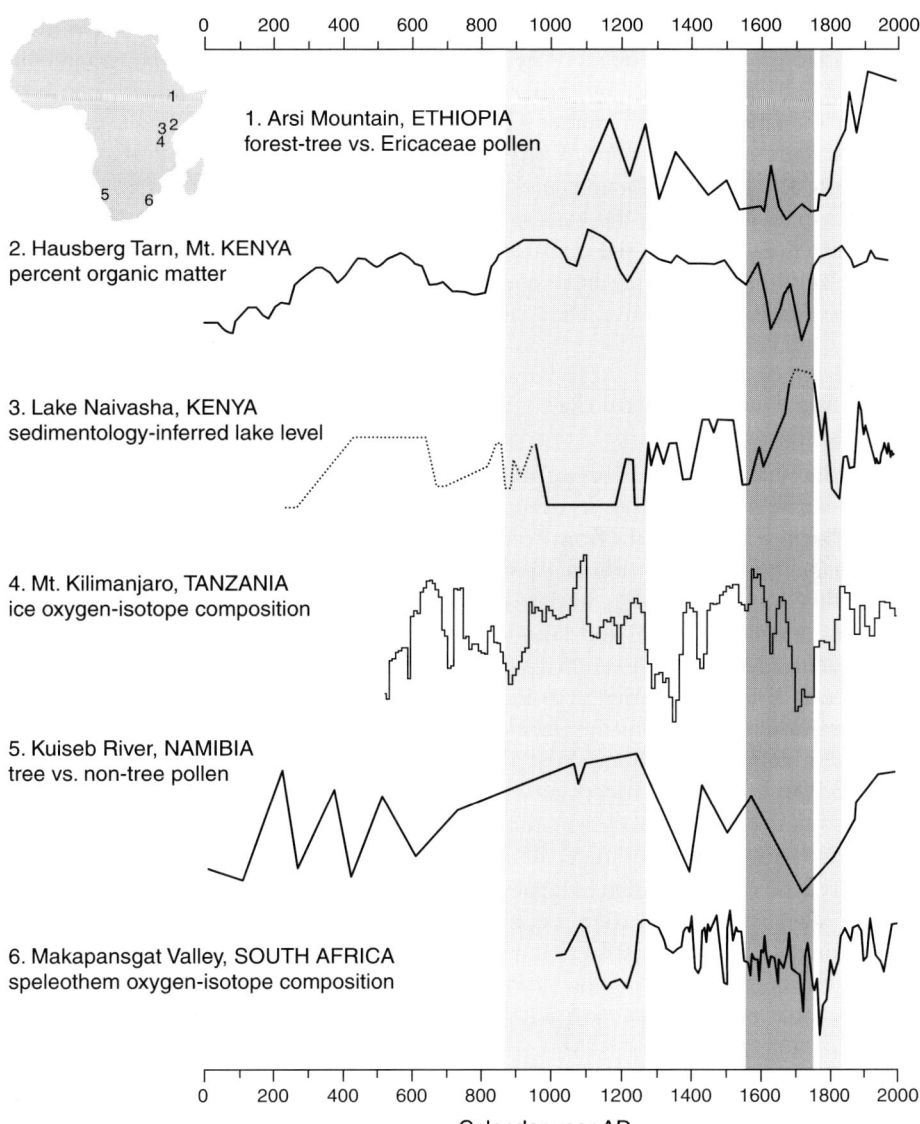

Figure 4. Lake-level reconstruction for Lake Naivasha (3; Verschuren (2001)) over the past 1800 years compared with African proxy records presumed to at least partly reflect past temperature variation. From North to South: Arsi Mountain tree pollen (1; Bonnefille and Umer (1994)), Mt. Kenya glacier activity (2; Karlén et al. (1999)), Mt. Kilimanjaro ice-core record (4; Thompson et al. (2002)), Kuiseb River tree pollen (4; Scott (1996)), and the Makapansgat speleothem record (5; Holmgren et al. (1999)).

Thompson et al. (2002) interpret pronounced $\delta^{18}O$ minima in the Mt. Kilimanjaro ice-core record around AD 1250 and during the Maunder Minimum (Fig. 4) to reflect episodes of substantial cooling. However, African rainfall $\delta^{18}O$ correlates far stronger with rainfall amount than with air temperature (Rozanski et al. 1993). Thus, Barker et al. (2001) argued that oxygen-isotope signals in fossil-diatom silica from alpine lakes on Mt. Kenya primarily reflect variations in moisture and cloud height driven by sea-surface temperature anomalies over the tropical Indian Ocean, and hence that Holocene ice advances in equatorial Africa were caused by enhanced rainfall rather than cold. This would explain obvious resemblance between the Hausberg Tarn record of Mt. Kenya glacier activity (Fig. 4; reversed) and the Naivasha lake-level record. Indeed, most of Lake Naivasha's river inflows originate in the moist highlands flanking the eastern Rift Valley, less than 80 km from Mt. Kenya. Unfortunately, it leaves open the question to what extent regional temperature variation during the last millennium may have contributed to the substantial changes in effective moisture recorded by Lake Naivasha.

Summary and concluding remarks

Reconstructing the climatic history of tropical Africa over the last 2000 years, including the environmental background of its eventful political and cultural history, is hampered by scarcity of documentary records from before the colonial period, and the limited potential of traditional high-resolution climate-proxy archives such as tree rings and ice cores. To correct this situation, various initiatives are now underway to systematically explore all suitable speleothem, coral, and lake-sediment archives throughout the continent and adjacent oceans. Lake-sediment archives in particular have great potential to document the climate history of tropical Africa in both space and time. Continuously varved sediment records are rare, however, so that the chronology of most lake-based climate reconstructions will continue to rely on a combination of lead-210 and radiocarbon dating. Still, careful site selection and application of modern paleolimnological methods may eventually yield the spatial array of high-resolution climate reconstructions needed to elucidate the mechanisms of decadal and century-scale climate variability over the African continent.

Currently available proxy data indicate that most areas of inter-tropical Africa were drier than today ~AD 900–1270, broadly coeval with Medieval warming in north-temperate regions. Drought was also widespread in the late 1700s and early 1800s, with maximum aridity possibly occurring during the 1790s. For two to three centuries before this drought, however, i.e., the period equivalent with the main phase of the Little Ice Age, rainfall anomalies in eastern and southeastern tropical Africa appear to have been inversely related, suggesting prominent influence of a forcing mechanism which generates regional rather than continent-wide rainfall patterns. In equatorial East Africa, relatively moist conditions from ~AD 1270 to the mid-18th century were interrupted at least twice by decade-scale episodes of severe aridity dated to around AD 1400 and the late 1500s, matching historical records from Angola, Uganda, Tanzania, and Rwanda describing drought-induced famine, migration, and conflict between pastoralist and agriculturalist peoples. This succession of wet and dry episodes correlates with the tree-ring reconstructed record of atmospheric radiocarbon production, indicating a discernible influence of solar forcing on Little Ice Age climate in equatorial Africa. High-resolution lake-level records can be an excellent indicator of past variations in net moisture available for natural ecosystems and human societies, but whether the high lake levels in East Africa during the Little Ice Age were sustained only by an intensified hydrological cycle, or also by reduced evaporation associated with lower temperatures, remains unclear.

New lake-based climate-proxy records from lakes Naivasha and Malawi possess appropriate time resolution and age control to serve as templates for regional analysis of decadal and century-scale climate anomalies in tropical Africa during the last 2000 years. Although age control on most other proxy records covering this period is not as good, many of them display a characteristic succession of century-scale climatic anomalies that are consistent with the patterns observed in the two high-resolution records. The combined evidence suggests that one of today's continent-scale spatial patterns of inter-annual rainfall variability, characterised by opposite anomalies in south-eastern Africa and eastern equatorial Africa (Nicholson 1986; 1993; 2001a), has also been expressed at longer time scales in the past, for instance during the period equivalent with the Little Ice Age of temperate regions. At other times, for example in Medieval times, the 1790s and early 1800s, and the early 20th century, regional patterning of rainfall anomalies seems to have been overridden by widespread drought. This apparent alternation of episodes characterised by inter-regional contrasts and episodes characterised by continent-wide spatial coherence is not unexpected, as it may simply reflect the varying influence through time of several distinct climate-forcing mechanisms (Nicholson 2000).

The reality of the reconstructed climate patterns discussed here ultimately rests on our understanding of the relationships between climate-proxy indicators and climate, which are archive-dependent and always complex. Climate inferences from each of Africa's many climate-sensitive lakes depends on: (i) justified extrapolation of a presumed or established relationship between the sedimentary proxy indicator and lake level, from the historical period through the entire period of climate inference; and (ii) existence of a simple, constant relationship between lake level and rainfall at the relevant time scale. Using these criteria, neither the Naivasha, Malawi, nor any other lake-based climate record from tropical Africa is at present beyond discussion, and many more high-resolution climate histories from other sites, using different archives and different climate-proxy indicators, will be required to elucidate the mechanisms which control African rainfall at the time-scales relevant to society. Limits on the geographical distribution and temporal coverage of high-resolution climate archives will continue to challenge efforts to reconstruct Africa's climate history with better than decade-scale resolution and age control. But steady improvements in the methods, rigour, and scale of palaeoclimate research in Africa create great promise that its treasure trove of natural climate archives will eventually realise its full potential.

Acknowledgments

The author's own research described in this paper was supported by the US National Science Foundation, The US National Oceanic and Atmospheric Administration, the Fund for Scientific Research - Flanders (Belgium), and the universities of Minnesota (USA) and Ghent (Belgium). He enjoys a Postdoctoral Fellowship of the Fund for Scientific Research - Flanders.

References

Bar-Matthews M. and Ayalon A., this volume. Speleothems as palaeoclimate indicators, a case study from Soreq Cave located in the eastern Mediterranean region, Israel. In: Battarbee R.W., Gasse F.

and Stickley C.E. (eds), Past Climate Variability through Europe and Africa. Kluwer Academic Publishers, Dordrecht, the Netherlands, pp. 363–391.

Barker P.A., Street-Perrott F.A., Leng M.J., Greenwood P.B., Swain D.L., Perrott R.A., Telford R.J. and Ficken K.J. 2001. A 14,000-year oxygen isotope record from diatom silica in two alpine lakes on Mt. Kenya. Science 292: 2307–2310.

Black D.E., Peterson L.C., Overpeck J.T., Kaplan A., Evans M.N. and Kashgarian M. 1999. Eight centuries of North Atlantic Ocean atmosphere variability. Science 286: 1709–1713.

Bond G., Kromer B., Beer J., Muscheler R., Evans M.N., Showers W., Hoffmann S., Lotti-Bond R., Hadjas I. and Bonani G. 2001. Persistent solar influence on North Atlantic climate during the Holocene. Science 294: 2130–2136.

Bonnefille R. and Umer M. 1994. Pollen-inferred climatic fluctuations in Ethiopia during the last 3000 years. Palaeogeogr. Palaeoclimatol. Palaeoecol. 109: 331–343.

Burns S.J., Fleitmann D., Mudelsee M., Neff U., Matter A. and Mangini A. 2002. A 780-year annually resolved record of Indian Ocean monsoon precipitation from a speleothem from South Oman. J. Geophys. Res.-Atm. 107 (D20): art. no. 4434.

Butzer K.W., Isaac G.L., Richardson J.L. and Washbourn-Kamau C. 1972. Radiocarbon dating of East African lake levels. Science 175: 1069–1076.

Cohen A.S., Talbot M.R., Awramik S.M., Dettman D.L. and Abell P. 1997. Lake level and paleoenvironmental history of Lake Tanganyika, Africa, as inferred from late Holocene and modern stromatolites. Bull. Geol. Soc. Am. 109: 444–460.

Cole J.E. 2003. Dishing the dirt on coral reefs. Nature 421: 705–706.

Cole J.E., Dunbar R.B., McClanahan T.R. and Muthiga N.A. 2000. Tropical Pacific forcing of decadal SST variability in the Western Indian Ocean over the past two centuries. Science 287: 617–619.

Crossley R. and Davison-Hirschmann S. 1981. Hydrology and archaeology of Lake Malawi and its outlet during the Iron Age. Palaeoecol. Afr. 13: 123–126.

Crossley R. and Davison-Hirschmann S. 1982. High levels of Lake Malawi during the late Quaternary. Palaeoecol. Afr. 15: 109–115.

Crossley R., Davison-Hirschmann S., Owen R.B. and Shaw P. 1984. Lake-level fluctuations during the last 2000 years in Malawi. In: Vogel J.C. (ed.), Late Caenozoic Paleoclimates of the Southern Hemisphere. A.A. Balkema, Rotterdam, pp. 305–316.

de Heinzelin J. 1962. Carte des extensions glaciaires du Ruwenzori (versant Congolais). Biuletyn Peryglacjalny 11: 133–139.

De Putter T., Loutre M.-F. and Wansard G. 1998. Decadal periodicities of Nile River historical discharge (622–1470) and climatic implications. Geophys. Res. Lett. 25: 3193–3196.

Dias J. 1981. Famine and disease in the history of Angola. J. Afr. Hist. 22: 349–378.

Dunbar R.B. and Cole J.E. 1999. Annual records of tropical systems (ARTS): recommendations for research. PAGES Workshop Report 99–1, 72 pp.

Eriksson M.G. 1998. Landscape and soil erosion history in Central Tanzania. Dept of Physical Geography Dissertation Series No. 12, Stockholm University.

Felis T., Pätzold J., Loya Y., Fine M., Nawar A.H. and Wefer G. 2000. A coral oxygen isotope record from the northern Red Sea documenting NAO, ENSO, and North Pacific teleconnections on Middle East climate variability since the year 1750. Paleoceanography 15: 679–694.

Fraedrich K., Jiang J., Gerstengarbe F.-W. and Werner P.C. 1997. Multiscale detection of abrupt climate change: application to River Nile flood levels. Int. J. Climatol. 17: 1301–1315.

Gagan M.K., Ayliffe L.K., Beck J.W., Cole J.E., Druffel E.R.M., Dunbar R.B. and Schrag D.P. 2000. New views of tropical paleoclimates from corals. Quat. Sci. Rev. 19: 45–64.

Gasse F. 1977. Evolution of Lake Abhé (Ethiopia and TFAI) from 70,000 BP. Nature 260: 42–45.

Gasse F. 2000. Hydrological changes in the African tropics since the Last Glacial Maximum. Quat. Sci. Rev. 19: 189–211.

Gasse F. 2002. Kilimanjaro's secrets revealed. Science 298: 548–549.

Gasse F. and Street F.A. 1978. Late Quaternary lake-level fluctuations and environments of the north-
    ern Rift Valley and Afar region (Ethiopia and Djibouti). Palaeogeogr. Palaeoclimatol. Palaeoecol.
    24: 279–325.

Gasse F. and Van Campo E. 1994. Abrupt post-glacial climate events in West Asia and North Africa
    monsoon domains. Ear. Planet. Sci. Lett. 126: 435–456.

Grove R.H. 1998. Global impact of the 1789–1793 El Niño. Nature 393: 318–319.

Gupta A.K., Anderson D.M. and Overpeck J.T. 2003. Abrupt changes in the Asian southwest monsoon
    during the Holocene and their links to the North Atlantic Ocean. Nature 421: 354–357.

Halfman J.D., Johnson T.C. and Finney B.P. 1994. New AMS dates, stratigraphic correlations and
    decadal climatic cycles for the past 4 ka at Lake Turkana, Kenya. Palaeogeogr. Palaeoclimatol.
    Palaeoecol. 111: 83–98.

Hall M. 1976. Dendroclimatology, rainfall and human adaptation in the later Iron Age of Natal and
    Zululand. Ann. Natal Mus. 22: 693–703.

Hamilton A.C. 1982. Environmental History of East Africa: A Study of the Quaternary. Academic
    Press, London. 328 pp.

Hassan F.A. 1981. Historical Nile floods and their implications for climate change. Science 112:
    1142–1145.

Hastenrath S. 1981. The Glaciers of Equatorial East Africa. Reidel, Dordrecht, the Netherlands.

Herring R.S. 1979. Hydrology and chronology: the Rodah Nilometer as an aid in dating Interlacustrine
    history. In: Webster J.B. (ed.), Chronology, Migration and Drought in Inter-lacustrine Africa.
    Longman and Dalhousie University Press, London, pp. 39–86.

Hoelzmann P., Gasse F., Dupont L.M., Salzmann U., Staubwasser M., Leuschner D.C. and Sirocko
    F., this volume. Palaeoenvironmental changes in the arid and subarid belt (Sahara-Sahel-Arabian
    Peninsula) from 150 kyr to present. In: Battarbee R.W., Gasse F. and Stickley C.E. (eds), Past
    Climate Variability through Europe and Africa. Kluwer, Dordrecht, the Netherlands, pp. 219–256.

Holmgren K., Karlén W., Lauritzen S.E., Lee-Thorp J.A., Partridge T.C., Piketh S., Repinski P.,
    Stevenson C., Svanered O. and Tyson P.D. 1999. A 3000-year high-resolution stalagmite record
    of palaeoclimate for North-Eastern South Africa. The Holocene 9: 271–278.

Holmgren K., Tyson P.D., Moberg A. and Svanered O. 2001. A preliminary 3000-year regional
    temperature reconstruction for South Africa. S. Afr. J. Sci. 97: 49–51.

Holzhauser H. 1997. Fluctuations of the Grosser Aletsch Glacier and the Gorner Glacier during the
    last 3200 years: new results. Palaoklimaforschung/Palaeoclimate Research 24: 35–58.

Johnson T.C., Barry S., Chan Y. and Wilkinson P. 2001. Decadal record of climate variability spanning
    the last 700 years in the southern tropics of East Africa. Geology 29: 83–86.

Johnson T.C., Brown E.T, Mcmanus J., Barry S., Barker P. and Gasse F. 2002. A high-resolution
    paleoclimate record spanning the past 25,000 years in Southern East Africa. Science 296: 113–
    132.

Johnson T.C., Brown E.T. and McManus J., this volume. Diatom productivity in Northern Lake
    Malawi during the past 25,000 years: implications for the position of the Intertropical Convergence
    Zone at millennial and shorter time scales. In: Battarbee R.W., Gasse F. and Stickley C.E. (eds),
    Past Climate Variability through Europe and Africa. Kluwer Academic Publishers, Dordrecht, the
    Netherlands, pp. 93–116.

Jolly D., Taylor D., Marchant R., Hamilton A., Bonnefille R., Buchet G. and Riollet G. 1997.
    Vegetation dynamics in Central Africa since 18,000 yr BP: pollen records from the interlacustrine
    highlands of Burundi, Rwanda and Western Uganda. J. Biogeogr. 24: 495–512.

Jones P.D., Osborn T.J. and Briffa K.R. 2001. The evolution of climate over the last millennium.
    Science 292: 662–667.

Karlén W., Fastook J.L., Holmgren K., Malmström M., Matthews J.A., Odada E., Risberg J., Rosqvist
    G., Sandgren P., Shemesh A. and Westerberg L.-O. 1999. Glacier fluctuations on Mount Kenya

since ~6000 cal. years BP: implications for Holocene climatic change in Africa. Ambio 28: 409–418.

Keigwin L.D. 1996. The Little Ice Age and Medieval Warm Period in the Sargasso Sea. Science 274: 1504–1508.

Kolding J. 1992. A summary of Lake Turkana: an ever-changing mixed environment. Mitt. Internat. Verein. Limnol. 23: 25–35.

Krishnamurthy R.V. and Epstein S. 1985. Tree ring D/H ratio from Kenya, East Africa and its palaeoclimatic significance. Nature 317: 160–162.

Laird K.R., Fritz S.C., Maasch K.A. and Cumming B.F. 1996. Greater drought intensity and frequency before AD 1200 in the Northern Great Plains, USA. Nature 384: 552–554.

Lamb H.F., Darbyshire I. and Verschuren D. 2003. Vegetation response to rainfall variation and human impact in Central Kenya during the past 1100 years. The Holocene 13: 282–295.

Lamb H.F., Gasse F., Benkaddour A., el Hamouti N., Van der Kaars S., Perkins W.T., Pearce N.J. and Roberts C.N. 1995. Relation between century-scale Holocene arid intervals in tropical and temperate zones. Nature 373: 134–137.

Lee-Thorp J.A., Holmgren K., Lauritzen S.-E., Linge H., Moberg A., Partridge T.C., Stevenson C. and Tyson P.D. 2001. Rapid climate shifts in the southern African interior throughout the mid to late Holocene. Geophys. Res. Lett. 28: 4507–4510.

Legesse D., Gasse F., Radakovitch O., Vallet-Coulomb C., Bonnefille R., Verschuren D., Gibert E. and Barker P. 2002. Environmental changes in a tropical lake (Lake Abiyata, Ethiopia) during recent centuries. Palaeogeogr. Palaeoclimatol. Palaeoecol. 187: 233–258.

Livingstone D.A. 1975. Late Quaternary climatic change in Africa. Ann. Rev. Ecol. Syst. 6: 249–280.

Livingstone D.A. 1980. Environmental changes in the Nile headwaters. In: Williams M.A.J. and Faure H. (eds), The Sahara and the Nile. A.A. Balkema, Rotterdam, pp. 339–359.

Maley J. 1973. Mécanisme des changements climatiques aux basses latitudes. Palaeogeogr. Palaeoclimatol. Palaeoecol. 3: 193–227.

Maley J. 1981. Etudes palynologiques dans le bassin du Tchad et paléoclimatologie de l'Afrique nord-tropicale de 30.000 ans a l'époque actuelle. Trav. Doc. ORSTOM 129: 1–586.

Maley J. 1989. L'importance de la tradition orale et des données historiques pour la reconstitution paléoclimatique du dernier millénaire sur l'Afrique nord-tropicale. In: Sud Sahara, Sahel Nord. Centre Culturel Francais d'Abidjan, pp. 53–57.

Maley J. 1993. Chronologie calendaire des principales fluctuations du lac Tchad au cours du dernier millénaire. In: Barreteau D. and Von Graffenried C. (eds), Datation et Chronologie dans le Bassin du Lac Tchad. ORSTOM, Paris, pp. 161–163.

Marchant R. and Taylor D. 1998. Dynamics of montane forest in Central Africa during the late Holocene: a pollen-based record from Western Uganda. The Holocene 8: 375–381.

Miller J. 1982. The significance of drought, disease and famine in the agriculturally marginal zones of West-Central Africa. J. Afr. Hist. 23: 17–61.

Nicholson S.E. 1986. The spatial coherence of African rainfall anomalies: inter-hemispheric teleconnections. J. Climat. Appl. Meteorol. 25: 1365–1381.

Nicholson S.E. 1993. An overview of African rainfall fluctuations of the last decade. J. Clim. 6: 1463–1466.

Nicholson S.E. 1994. Recent rainfall fluctuations in Africa and their relationships to past conditions over the continent. The Holocene 4: 121–131.

Nicholson S.E. 1996a. Environmental change within the historical period. In: Adams W.A., Goudie A.S. and Orme A.R. (eds), The Physical Geography of Africa. Oxford University Press, pp. 60–87.

Nicholson S.E. 1996b. A review of climate dynamics and climate variability in Eastern Africa. In: Johnson T.C. and Odada E. (eds), The Limnology, Climatology and Paleoclimatology of the East African Lakes. Gordon and Breach, Newark, pp. 25–56.

Nicholson S.E. 1998. Historical fluctuations of Lake Victoria and other lakes in the northern Rift Valley of East Africa. In: Lehman J.T. (ed.), Environmental Change and Response in East African Lakes. Kluwer, Dordrecht, pp. 7–29.

Nicholson S.E. 2000. The nature of rainfall variability over Africa on time scales of decades to millennia. Glob. Planet. Chan. 26: 137–158.

Nicholson S.E. 2001a. A semi-quantitative, regional precipitation data set for studying African climates of the 19th century, Part 1. Overview of the data set. Clim. Change 50: 317–353.

Nicholson S.E. 2001b. Climatic and environmental change in Africa during the last two centuries. Clim. Res. 17: 123–144.

Owen R., Crossley R., Johnson T., Tweddle D., Kornfield I., Davison S., Eccles D. and Engstrom D. 1990. Major low levels of Lake Malawi and their implications for speciation rates in cichlid fishes. Proc. Roy. Soc. Lond. B 240: 519–553.

Quinn W.H. 1993. A study of Southern Oscillation-related climatic activity for AD 622–1990 incorporating Nile River flood data. In: Diaz H.F. and Markgraf V. (eds), El Niño: Historical and Paleoclimatic Aspects of the Southern Oscillation. Cambridge University Press, pp. 119–149.

Ropelewski C.F. and Halpert M.S. 1989. Global and regional-scale precipitation patterns associated with the El Niño Southern Oscillation. Month. Weather Rev. 115: 1606–1626.

Rozanski K., Araguas-Araguas L. and Gonfiantini R. 1993. Isotopic patterns in modern global precipitation. In: Swart P.K. (ed.), Climate Changes in Continental Isotopic Records. Geophysical Monograph No. 78, American Geophysical Union, Washington D.C., pp. 1–36.

Schoenbrun D. L. 1998. A Green Place, a Good Place: Agrarian Change, Gender, and Social Identity in the Great Lakes Region to the 15th Century. Oxford University Press, Oxford, 301 pp.

Scott L. 1996. Palynology of hyrax middens: 2000 years of palaeoenvironmental history in Namibia. Quat. Internat. 33: 73–79.

Scott L. and Lee-Thorp J.A., this volume. Holocene climate trends and rhythms in Southern Africa. In: Battarbee R.W., Gasse F. and Stickley C.E. (eds), Past Climate Variability through Europe and Africa. Kluwer Academic Publishers, Dordrecht, the Netherlands, pp. 69–91.

Stager J.C. 1998. Ancient analogues for recent environmental changes at Lake Victoria, East Africa. In: Lehman J.T. (ed.), Environmental Change and Response in East African Lakes. Kluwer, Dordrecht, pp. 37–46.

Stager J.C., Cumming B.F. and Meeker L. 1997. A high-resolution 11,400-year diatom record from Lake Victoria, East Africa. Quat. Res. 47: 81–89.

Stager J.C., Cumming B.F. and Meeker L. 2003. A 10,200-year high-resolution diatom record from Pilkington Bay, Lake Victoria, East Africa. Quat. Res. 59: 172–181.

Stager J.C. and Mayewski P.A. 1997. Abrupt early to mid-Holocene climatic transition registered at the equator and the poles. Science 276: 1834–1836.

Street F.A. 1980. The relative importance of climate and local hydrogeological factors in influencing lake-level fluctuations. Palaeoecol. Afr. 12: 137–158.

Talbot M.R. and Delibrias G. 1980. A new late Pleistocene-Holocene water-level curve for Lake Bosumtwi, Ghana. Ear. Planet. Sci. Lett. 47: 336–344.

Taylor D.M. 1990. Late Quaternary pollen records from two Ugandan mires: evidence for environmental change in the Rukiga Highlands of Southwest Uganda. Palaeogeogr. Palaeoclimatol. Palaeoecol. 80: 283–300.

Taylor D., Robertshaw P. and Marchant R.A. 2000. Environmental change and political-economic upheaval in precolonial Western Uganda. The Holocene 10: 527–536.

Thambiyahpillay G.C.R. 1983. Hydrogeography of Lake Chad and environs: contemporary, historical and palaeoclimatic. Ann. Borno 1: 105–145.

Thompson L.G., Mosley-Thompson E., Davis M.E., Henderson K.A., Brecher H.H., Zagorodnov V.S., Mashiotta T.A., Lin P.-N., Mikhalenko V.N., Hardy D.R. and Beer J. 2002. Kilimanjaro ice core records: evidence of Holocene climate change in tropical Africa. Science 298: 589–593.

Thompson L.G., Yao T., Mosley-Thompson E., Davis M.E., Henderson K.A. and Lin P.-N. 2000. A high-resolution millennial record of the South Asian Monsoon from Himalayan ice cores. Science 289: 1916–1919.

Tiercelin J.J. and 27 others 1987. Le demi-graben de Baringo-Bogoria, Rift Gregory, Kenya: 30.000 ans d'histoire hydrologique et sédimentaire. Bull. Centr. Rech. Expl.-Prod. Elf-Aquitaine 11: 249–540.

Tyson P.D. 1986. Climatic Change and Variability in Southern Africa. Oxford University Press.

Tyson P.D., Karlén W., Holmgren K. and Heiss G.A. 2000. The Little Ice Age and medieval warming in South Africa. S. Afr. J. Sci. 96: 121–126.

Tyson P.D., Lee-Thorp J., Holmgren K. and Thackeray J.F. 2002. Changing gradients of climate change in Southern Africa during the past millennium: implications for population movements. Clim. Change 52: 129–135.

Verschuren D. 1996. Comparative paleolimnology in a system of four shallow, climate-sensitive tropical lake basins. In: Johnson T.C. and Odada E. (eds), The Limnology, Climatology and Paleoclimatology of the East African Lakes. Gordon and Breach, Newark, pp. 559–572.

Verschuren D. 1999. Sedimentation controls on the preservation and time resolution of sedimentary climate-proxy records from shallow fluctuating lakes. Quat. Sci. Rev. 18: 821–837.

Verschuren D. 2001. Reconstructing fluctuations of a shallow East African lake during the past 1800 years from sediment stratigraphy in a submerged crater basin. J. Paleolimnol. 25: 297–311.

Verschuren D., Laird K.R. and Cumming B.F. 2000. Rainfall and drought in equatorial East Africa during the past 1100 years. Nature 403: 410–414.

Webster J.B. 1979. Noi! Noi! Famines as an aid to Interlacustrine chronology. In: Webster J.B. (ed.), Chronology, Migration and Drought in Interlacustrine Africa. Longman and Dalhousie University Press, London, pp. 1–37.

9. LATE QUATERNARY CLIMATE CHANGES IN THE HORN OF AFRICA

MOHAMMED UMER (mohammed_umer@hotmail.com)
Department of Geology and Geophysics
Addis Ababa University
P.O. Box 3434
Addis Ababa
Ethiopia

DAGNACHEW LEGESSE (dagnu@geobs.aau.edu.et)
Department of Geology and Geophysics
Addis Ababa University
P.O. Box 3434
Addis Ababa
Ethiopia

FRANÇOISE GASSE (gasse@cerege.fr)
Centre Européen de Recherche et d'Enseignement
de Géosciences de l'Environnement (CEREGE)
Europole de l'Arbois
B.P. 80, F-13454 Aix-en-Provence cedex 4
France

RAYMONDE BONNEFILLE (bonnefille@cerege.fr)
Centre Européen de Recherche et d'Enseignement
de Géosciences de l'Environnement (CEREGE)
Europole de l'Arbois
B.P. 80, 13545 Aix-en-Provence cedex 04
France

HENRY F. LAMB (hfl@aber.ac.uk)
Institute of Geography and Earth Sciences
University of Wales
Aberystwyth, SY23 3DB
UK

MELANIE J. LENG (m.leng@nigl.nerc.ac.uk)
NERC Isotope Geosciences Laboratory
Keyworth
Nottingham, NG12 5GG
UK

R. W. Battarbee et al. (eds) 2004. *Past Climate Variability through Europe and Africa.*
Springer, Dordrecht, The Netherlands.

ANGELA L. LAMB (a.lamb@livjm.ac.uk)
School of Biological and Earth Sciences
Liverpool John Moores University
Liverpool, L3 3AF
UK

Keywords: Late Quaternary, Climate change, Horn of Africa, Ethiopia, Rift lakes, Glaciation, Holocene

Introduction

The Horn of Africa extends from the Sahara desert eastwards to the Red Sea, the Gulf of Aden and the Indian Ocean (Fig. 1). Due to its complex volcano-tectonic evolution over the past 15 million years (Mohr 1971), the region is characterised by considerable changes in elevation within short distances. The Ethiopian highlands, which rise to altitudes exceeding 4000 m above sea level, form an extensive uplifted plateau, delimited by pronounced escarpments on both east and west. In Southern Ethiopia the 1000 km-wide uplifted volcanic province is divided asymmetrically into northwestern and southeastern plateaux by the Main Ethiopian Rift which runs SSW-NNE and represents the northern end of the continental East African Rift. Eastward, the continental Afar Rift (Ethiopia and Republic of Djibouti) sits astride the Gulf of Aden — Red Sea sea-floor spreading axis, with several areas below sea level (e.g., -155 m at Lake Asal). These marked altitudinal gradients result in a wide variety of climates and environments, making the region particularly suitable for investigating past environmental change. The region is subject to the seasonal migration of the Inter-Tropical Convergence Zone (ITCZ) and is very sensitive to monsoon variability. Several studies have revealed climate variability from millennial to inter-annual timescales during the Late Quaternary (Gasse 2000).

This paper provides a detailed overview of this region complementary to those of Hoelzmann et al. (this volume) and Barker et al. (this volume). It reviews Late Quaternary palaeoenvironmental and palaeoclimatic changes, as recorded from various archives (lakes, high altitude peat-bogs, glacial landforms, palaeosols), and proxies (geomorphology, sedimentology, geochemistry, pollen, diatoms). The chronological framework of most records is based on radiocarbon ages (^{14}C yr BP). Radiocarbon ages are converted into calendar estimates (cal. yr) using the CALIB 4.0 program (Stuiver et al. 1998) back to 18,000 ^{14}C yr BP, or the polynomial equations established by Bard (1998) which allow calendar estimates back to 36,000 ^{14}C yr BP (41,516 cal. yr). Ages are first given in the time-scale provided in the original literature (^{14}C kyr BP or cal. kyr), then the calibrated scale is indicated in brackets.

Major climatic, vegetation and hydrological features

Climate and vegetation

The Horn of Africa's climate is governed by the general patterns of winds and pressure over the African continent, strongly modulated by topography and the proximity of the Indian

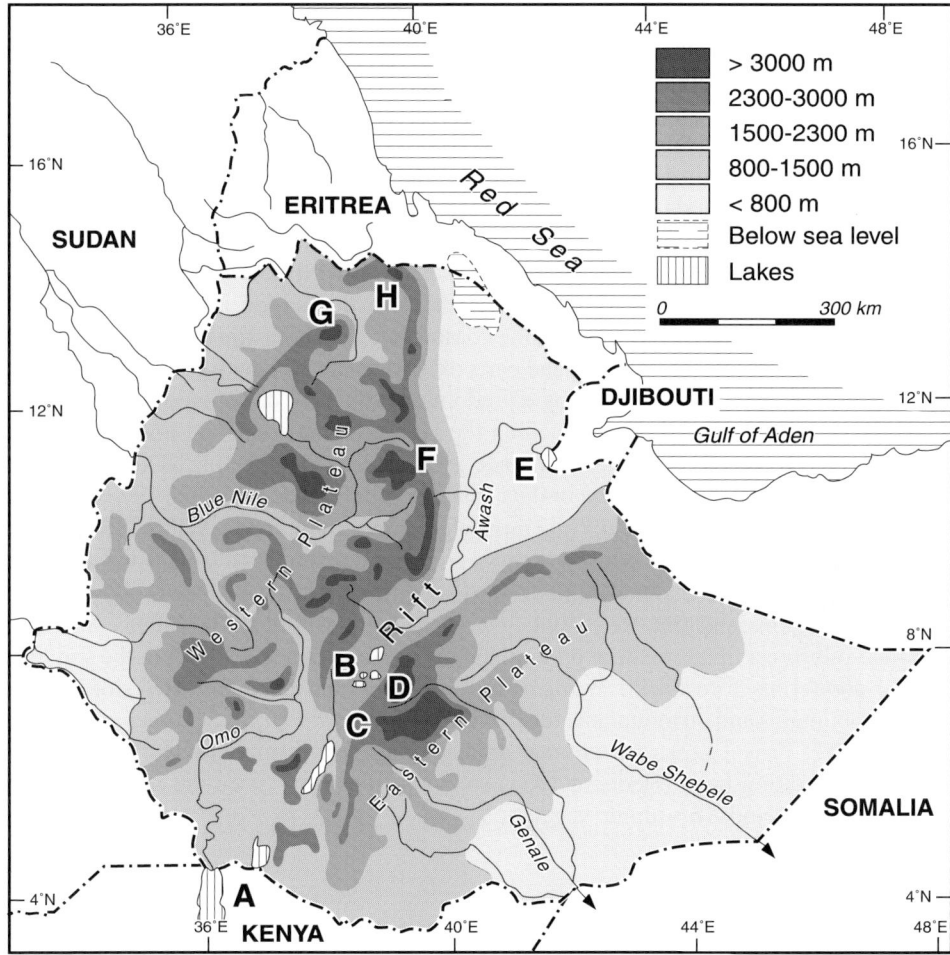

Figure 1. Location map of the studied sites in the Horn of Africa: A: Lake Turkana, B: Ziway-Shala Lakes and Lakes Tilo and Awassa, C: The Bale Mountains, D: The Arsi Mountains, E: Lake Abhé, F: Lakes Hayk and Hardibo, G: Semien mountains, H: Area of buried soils in Northern Ethiopia (Tigray).

Ocean. Almost all rainfall is ultimately derived from the Indian and Atlantic Oceans. The climate is characterised by three main seasons. The long rainy season in the summer (July-September; locally known as '*kiremt*' with 50–70% of the mean annual total) is primarily controlled by the northward movement of the ITCZ. Due to intense heating of the high plateau, convergence of the wet monsoonal air masses from the Indian and Atlantic Oceans brings much rain to the region during the summer rainy season (Griffiths 1972) when 85–95% of the food crop of the country is produced (Degefu 1987). The dry season extends between October and February ('*bega*') during which time the NE Trade Winds traversing Arabia dominate the region. The 'small rain' season ('*belg*'; 20–30% of the annual amount) occurs from March to May. It coincides with a diminution of the Arabian high, causing warm, moist air with a southerly component to flow over most of the country (Griffiths

1972). The '*belg*' rain is extremely important both from agricultural and hydrological points of view, although it accounts for only 5–15% of crop production (Degefu 1987). Generally, the time of onset and duration of rains in Ethiopia shows a latitudinal pattern associated with the seasonal shifts in the convergence zones, with maximum rainfall in the southwest. South of 5 °N, rain falls again in September and October as the ITCZ moves rapidly equatorwards. Conversely, the small rains are sporadic north of 6.5 °N. Inter-annual rainfall variability is extreme in the highlands (Seleshi and Demarée 1995). Droughts greatly affect water resources and the production of rain-fed agriculture in the region.

In the Horn of Africa rainfall increases markedly with altitude, while air temperature and evaporation decrease. At its lowest point, the Afar depression is one of the world's hottest deserts. Arid zones also comprise the coastal belt of Eritrea, the lowlands of Ogaden (Southeastern Ethiopia) and Somalia, and parts of Northeastern Kenya. The extreme aridity in the eastern coastal areas is caused by a predominant NE/SW direction of the prevailing winds preventing moist air masses reaching the land. Summer rains are low or absent in this region due to the directional divergence produced by overheating at the Ethiopian highlands, and frictional divergence in the coastal parallel winds (Griffiths 1972).

The vegetation of the region falls into three phytogeographic zones (White 1983). These are the Somalia-Maasai type with *Acacia* steppe and deciduous savannah woodland in low-altitude regions, the Sudano-Zambezian type with *Acacia-Commiphora* woodland in mid-altitude regions and the Afro-montane type at high altitudes. The latter consists of broad-leaf rainforest in the western part, mixed humid forest on wet slopes in the southern part and coniferous forest with *Podocarpus, Juniperus* and *Olea* on drier sides of the southern mountains and on mountains further north where *Podocarpus* becomes rare (Pichi-Sermolli 1957; Friis 1992). The climate gradient results in a predominantly altitudinal pattern of vegetation belts (Hedberg 1951) showing transitions from the montane forest belt to Ericaceous and Afro-alpine vegetation.

Hydrology

Major hydrological features are governed by topography, climatic gradients, and geology. Many lakes in the highlands are of volcanic origin, e.g., Lake Tana, the largest lake in Ethiopia, which is dammed by a volcanic flow, and the crater lakes at Debre Zeit which are supplied by direct precipitation and groundwater. The highlands act as water towers for the surrounding semi-arid and arid lowlands, through surface runoff and groundwater flows. The northwest plateau supplies the Blue Nile (Abbay; the outflow of Lake Tana), Tekeze (Atbara) and the Baro-Akobo (Sobat) rivers, which contribute about 85% of the main Nile flood runoff (Waterbury 1979; Conway 2000). The Wabi Shebeli and Genale rivers drain the southeastern plateau toward the Somali lowland and reach the Indian Ocean.

The rift systems are characterised by internal drainage basins. The volume and salinity of several lakes in the Main Ethiopian Rift Valley are very sensitive to changes in rainfall in the surrounding highlands. The northernmost large lake of the African Rift, Lake Turkana, depends on rainfall on the SW Ethiopian plateau through the Omo River. In the Afar, the few small, hypersaline lakes (e.g., L. Abhé, L. Asal) are relicts of large, deep lakes supplied by the Awash River flowing from Central Ethiopia, and by groundwaters percolating through the plateaux and escarpment zones. Long-term variations in precipitation and evaporation have been amplified in water-level and salinity responses of terminal lakes lying in the deepest tectonic basins.

Millennial-scale climate changes during the last glaciation

Lake records

The longest and most continuous sedimentary sequence so far obtained is from Lake Abhé, the terminal lake of the Awash river in central Afar. A 50 m-long sediment core reveals four major lacustrine episodes over at least 70,000 years (Gasse 1977; Gasse et al. 1980) but only the upper section is supported by ^{14}C dating. Reconstruction of lake-level and salinity fluctuations over the past 40,000 years was based on sedimentary and diatom studies of that core and of numerous outcrops, and palaeo-beaches (Gasse (1977), Gasse et al. (1980, 1998), Fig. 2). Following shallow, saline-alkaline conditions around 31 ^{14}C kyr BP (ca. 36 k cal. yr), Lake Abhé became a large (6,000 km^2), deep (170 m), freshwater lake between 27 and 23 ^{14}C kyr BP (ca. 31.6–27 cal. kyr). A stepwise fall to the modern lake-level occurred between ca. 23 and 17.1 ^{14}C kyr BP (27–20.2 cal. kyr). A palaeosol, with *in situ* grass remains, dated to between 17.1 and 16.1 ^{14}C kyr BP (20.2–19 cal. kyr), then developed at the core site. The lake was dry, or nearly so, during the Last Glacial Maximum (LGM). Traces of this wet Late Pleistocene episode and arid LGM are found in all the main grabens in central Afar. Lake Abhé and the neighbouring basins filled up again rapidly during the Late-Glacial and at the onset of the Holocene period.

The Main Ethiopian Rift experienced a comparable climatic evolution, as shown by geomorphic and sedimentological studies of outcropping sequences and shorelines in the Ziway-Shalla basin (Street 1979a; Gasse and Street 1978; Gasse et al. 1980; Benvenuti et al. 2002; Le Turdu et al. 1999). Today, this internal drainage basin contains four lakes of decreasing altitude and increasing salinity: L. Ziway, L. Langano, L. Abiyata, and L. Shalla. A highstand, at least 83 m above the modern Lake Shalla, occurred between ca. 26.5 and 22 ^{14}C kyr BP (31–26 cal. kyr; Street (1979a), Fig. 2). The lake then fell dramatically to levels at or below present, and remained low until about 12.5 ^{14}C kyr BP (ca. 14.5 cal. kyr). Assuming temperatures 3 to 6 °C lower than today, water balance calculations suggest a decrease in annual precipitation of 9 to 32% compared to modern during the LGM (Street 1979a; 1979b). At Lake Turkana no high lake stand deposits have been found in the interval between >40–11 cal. kyr BP suggesting that the lake level was low (Butzer et al. 1972). Johnson et al. (1987) reported evidence for a lake level as much as 60 m below its present position during this period, prior to the large Early Holocene transgression (Johnson 1996). This consistent regional LGM aridity is also reflected in significant reductions in the seasonal discharge of the Blue Nile River (Said 1993).

All lacustrine and geological evidence in the region converges to show dry conditions during the LGM. Therefore it is clear that the compilations based on the same data and showing high LGM-lake levels in Ethiopia and Djibouti (Jolly et al. 1998; Farrera et al. 1999; Kohfeld and Harrison 2000) are erroneous as a result of distortion of the original data during compilations for the purpose of model-data comparison. The major trends in Late Pleistocene climate changes as inferred from lake records can be accounted for by orbital forcing, which predicts enhanced monsoon rainfall, and thus high lake levels, during

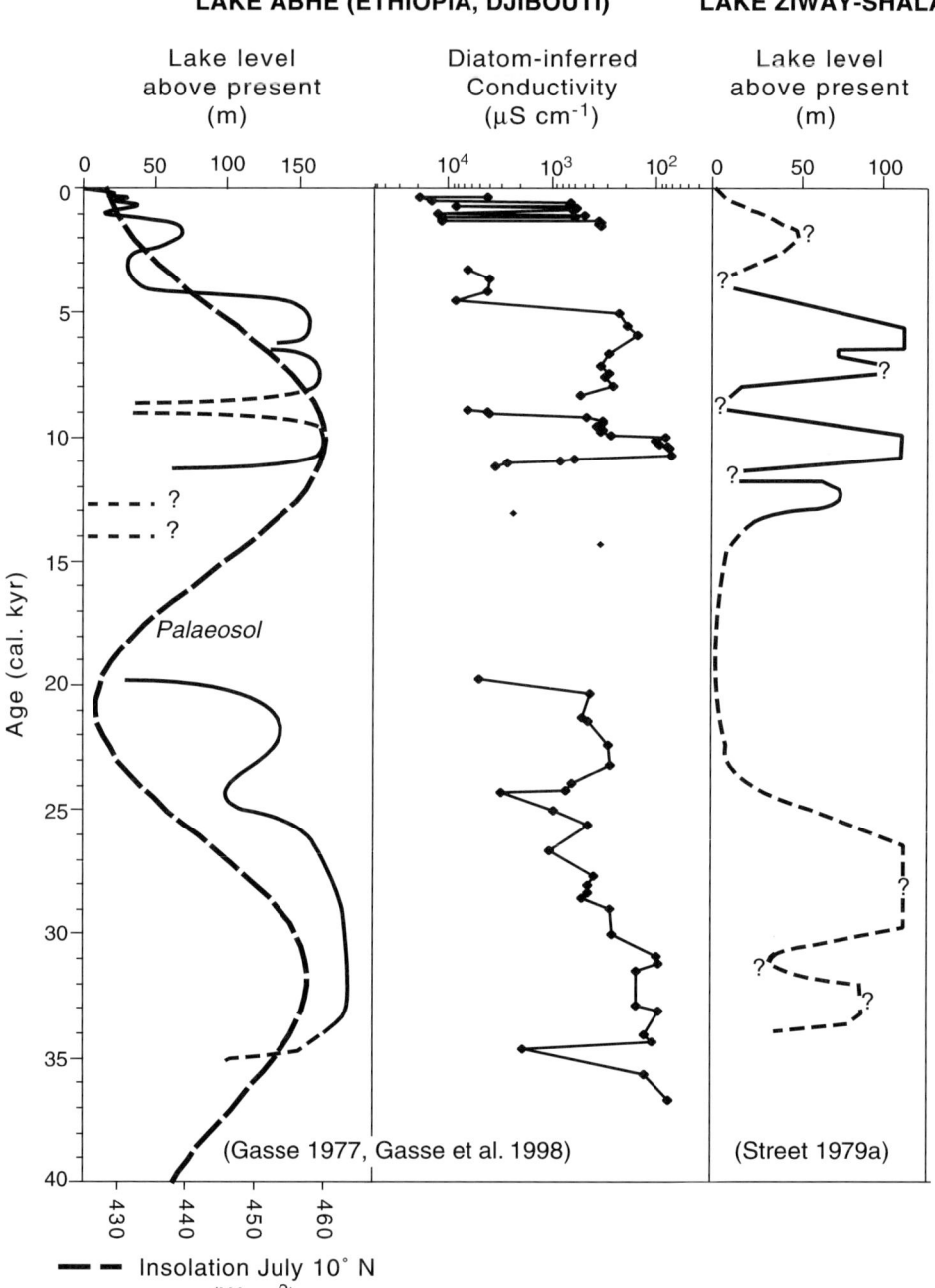

Figure 2. Climate change during the last 40 cal. kyr in the Horn of Africa.

periods of increased summer insolation around 35–30 cal. kyr BP, and dry climates during the LGM in the northern tropics (e.g., Kutzbach et al. (1993)).

Glacial geomorphology

None of the Ethiopian mountains are glaciated today, but traces of Pleistocene glaciations on the Bale, Arussi and Semien mountains have long been identified (Nillson 1940). Small ice caps and cirque glaciers were present above 3,700 m during the Late Pleistocene (Hastenrath 1977; 1989; Messerli et al. 1977; 1980; Street 1979a; Hurni 1982; 1989). Street (1979a) estimated that the firn line elevation in the Arussi mountains was around 3845 m. This is close to the estimates for Semien and Bale mountains (Messerli et al. 1977; Hurni 1989) and corresponds to a lowering of mean annual temperature by about 6 °C, assuming no change in precipitation (Street 1979a). These studies also suggest a lowering of the present-day solifluction limits (4300–4400 m) by about 800 m, to an elevation of 3600 m, due to decreased temperature (by about 7 °C) and rainfall (Hastenrath 1977; Messerli et al. 1977; Hurni 1982).

Although the Bale mountains were supposed to have contained the most extensive glaciation covering an area of about 600 km^2 (Messerli et al. 1977) however recent observation by Osmaston et al. (2003) revealed that there was a total glaciated area of only 150 km^2 (Osmaston et al. In press).

The above information is supported by a 2.6 m core taken from a swamp at ca. 4000 m near the eastern edge of the plateau just north of Badegesa Hill, outside the area of the BBM. It has been dated as 17,763 BP near the base and is currently being analysed (Zech, Universität Bayreuth, personal comm. 2003). This probably shows the limits of the LGM ice on the eastern side and should yield important information.

The age of the most recent glacial advance is unknown. As glacier development requires both cold and wet climatic conditions, this last advance may have taken place during the 35–30 cal. kyr BP wet phase, rather than during the dry LGM (Gasse et al. 1980). The age of the last deglaciation has been indirectly inferred from the study of high altitude peat-bogs. In the Arussi mountains, a peat core from a small cirque at 4070 m suggests that the Mt Badda ice-cap had totally disappeared by at least 11.5 ^{14}C kyr (13.4 cal. kyr) (Street 1979a; Hamilton 1982). An age for deglaciation of about 13–14 ^{14}C kyr BP (16.5–15 cal. kyr) was proposed from pollen analysis in the Bale mountains (Mohammed and Bonnefille (1998), Fig. 3).

A 16 m core in a small cirque lake, dammed by a glacial moraine, at 4000 m on the Bale (lake Garba Guracha or Gouratch) was recently recovered. It was taken at 6 m water depth and was dated at the base as ∼14000 yr. BP (Mohammed et al. 2003). The results of pollen, organic matter, sedimentological analysis are in press. This study also reveals that deglaciation occurred prior to 14000 yr BP.

Climate changes over the past 15,000 cal. years

Much wetter conditions than today prevailed during the early-mid Holocene in response to an orbitally-induced increase in monsoon strength. It has long been apparent that the modern vegetation altitudinal belts began to establish during the Holocene in response to

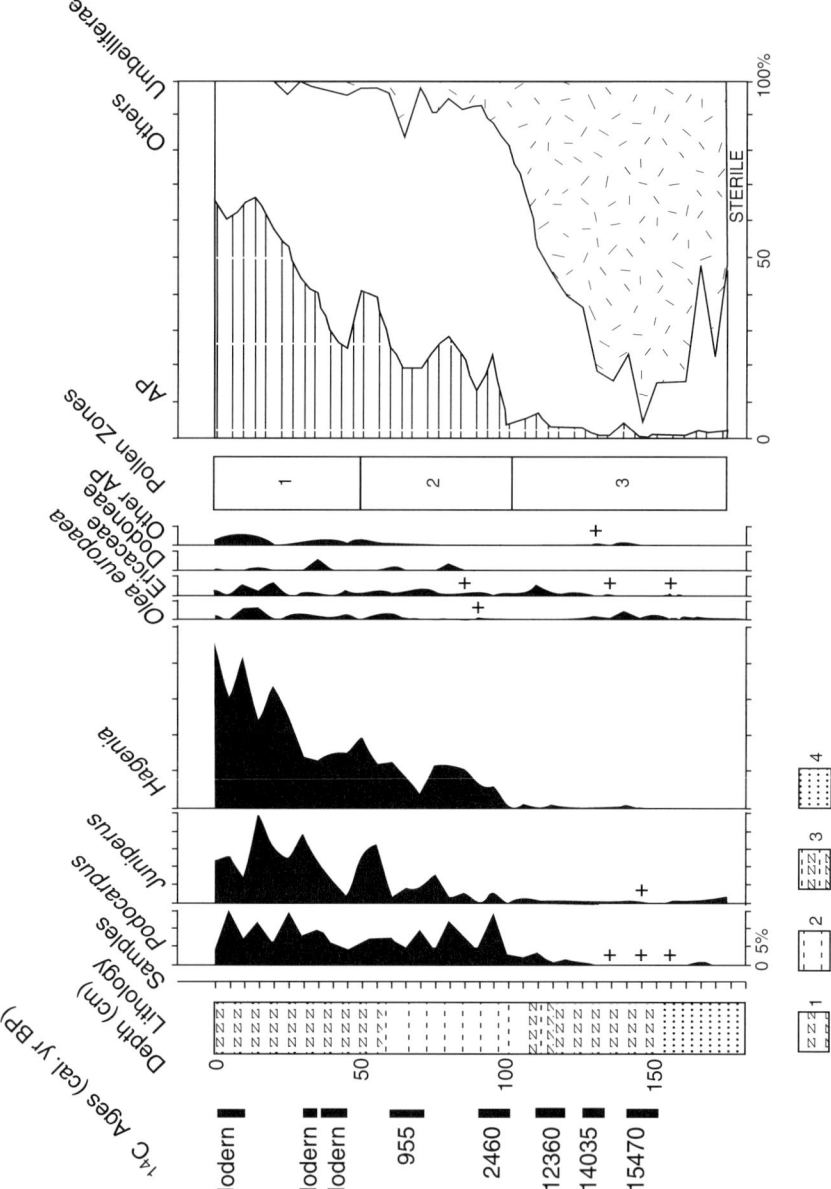

Figure 3. Pollen diagram from the Tamsaa core (Bale mountains, 3000 m; Mohammed and Bonnefille (1998)). Principal arboreal pollen taxa shown, Cyperaceae excluded. The Late Glacial landscape was occupied by herbaceous taxa after the end of glacial activity. Key to lithology: 1- fibrous peat, 2- fine peat, 3- transition zone, 4- silty to sandy organic mud.

both wetter and warmer climatic conditions (Hamilton 1982; Lézine and Bonnefille 1982; Mohammed and Bonnefille 1998). The lakes experienced early-mid Holocene high stands followed by generally low water levels during the past 5000 cal. yr (e.g., Butzer et al. (1972), Fontes and Pouchan (1975), Street and Grove (1979), Street (1979a), Williams et al. (1977), Gasse (1977), Gasse and Street (1978), Gasse and Descourtieux (1979), Owen et al. (1982)). However, most records show that vegetation and hydrological changes did not respond to the smooth sinusoidal waves of orbital forcing alone.

Lake records

Reconstructions of lake-level fluctuations from ancient shore lines
The Ziway-Shalla lake-level record (Street (1979a), Gillespie et al. (1983), Figs. 2, 4) shows that the re-establishment of wet conditions after the LGM occurred in two phases, at 14.5 and 11.5 cal. kyr BP. These two steps were thought to match the two major de-glacial warming events separated by the Younger Dryas cold event observed in high northern latitudes (Street-Perrott and Perrott 1990). The second step was by far the most significant: the lakes of the Ziway-Shalla basin merged to form a single lake that reached an outlet level and overflowed northward toward the Awash River, probably between ca. 9.5 and 8.5 ^{14}C kyr BP (10.7–9.5 cal. kyr) and from ca. 5.5 to 4.5 ^{14}C kyr BP (6.3 to 5.1 cal. kyr) when its surface was 112 m above the modern Lake Shalla level (Fig. 4). During these intervals, Lake Abhé in the Afar depression received waters from both the modern drainage area of the Awash River and the Ziway-Shalla overflow. Abhé was a large, closed, freshwater lake, 160 m deep (Gasse 1977).

The lake-level reconstruction of both Lakes Ziway-Shalla and Abhé (Street 1979a; Gasse and Street 1978; Gasse et al. 1980; Gillespie et al. 1983; Gasse 2000) show a large but short-term water-level lowering culminating at 7.8–7.2 ^{14}C kyr BP (8.7–8.1 cal. kyr), and a minor one around 5.9 ^{14}C kyr BP (6.7 cal. kyr). The first of these dry events, recorded in the lake-level curves, has been correlated with a stratigraphical level devoid of pollen in a core from lake Abiyata (Lézine and Bonnefille 1982). Lake Ziway-Shalla fell drastically after ca. 5–4.5 ^{14}C kyr BP (5.7–5.1 cal. kyr) and Lake Abhé after 4.5–4.1 ^{14}C kyr BP (5.1–4.6 cal. kyr). Fluctuations of moderate amplitude have been recorded during the Late Holocene. A rise of 42 m above the present-day Lake Shalla level was placed around 2–1.5 ^{14}C kyr BP (ca. 1.9–1.4 cal. kyr) but with large dating uncertainties (Gillespie et al. 1983). A rise of 70 m was dated between 3.5 and 1.6 ^{14}C kyr BP (3.8–1.6 cal. kyr BP) at Lake Abhé (Gasse 1977; Fontes et al. 1985). Street (1979b) estimated that mean annual precipitation in the early Holocene was 28–47% higher than modern rainfall in the Ziway-Shalla basin. For the entire catchment area of the early Holocene Lake Abhé, water and salt balance calculations suggest an increase in Precipitation-Evaporation (P-E) of ca. 25% compared to the present day (Gasse et al. 1980).

Quantitative reconstruction of water-level fluctuations of the Afar and Main Ethiopian Rift Valley lakes were among the first evidence that arid intervals interrupted the generally humid early-mid Holocene climate. Although the rapid regressive events around 8.7–8.1, 6.7, and 5.5–4.5 cal. kyr. BP appear roughly coincident with dry spells in other regions of the northern monsoon domain (Gasse and van Campo 1994), no satisfactory explanation has been proposed for these events, and for a brief return to wetter conditions during the Late

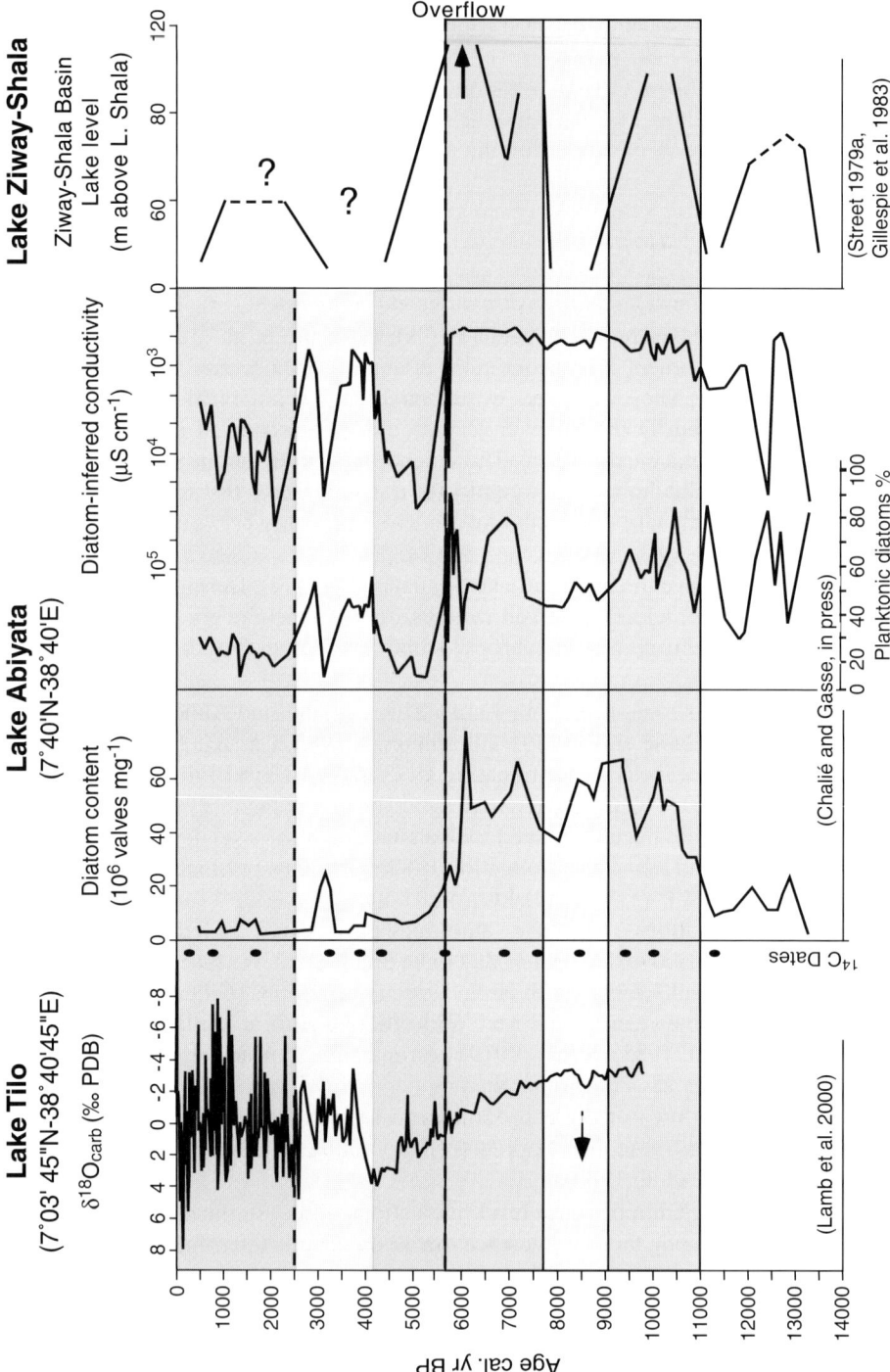

Figure 4. Climate change during the last 14 cal. kyr in the Ethiopian Rift Valley.

Holocene, which are too short to be accounted for by orbital forcing. The understanding of these abrupt changes requires more precise data on their timing, duration and intensity. Detailed lake sediment core studies have recently complemented and refined these older records.

Lake core records in the Main Ethiopian Rift

Three lakes without surface outlet have been recently investigated in the Main Ethiopian Rift Valley: (i) Lake Abiyata in the tectonic, internal drainage Ziway-Shalla basin; (ii) Lake Tilo, a small crater lake lying southwest of the Ziway-Shala basin and (iii) Lake Awassa, which occupies a caldera in the southern part of the Main Ethiopian Rift. These three lakes partly depend on groundwater inflows. The same diatom-based transfer functions (Gasse et al. 1995) were used to infer salinity changes, which in closed lakes can reflect changes in P-E, provided that the signal is not biased by local hydrological factors (Gasse et al. 1997). While diatom records from Lakes Tilo and Abiyata can be interpreted in terms of climate change, observed changes from diatom records in Lake Awassa are believed to be primarily due to pulsed input of saline groundwater, independent of climate (Telford et al. 1999).

The 6.5 ^{14}C kyr BP (7.3 cal. kyr) δ^{18}O record from carbonates in Lake Awassa does, however, show broad regional climate changes (Lamb A.L. et al. 2002). Co-varying and increasing δ^{18}O and δ^{13}O values from ~4.8 ^{14}C kyr BP (5.5 cal. kyr) suggest an aridification of climate after the early Holocene insolation maximum. After 4.0 ^{14}C kyr BP (4.5 cal. kyr), humid conditions return until after ~2.8 ^{14}C kyr BP (2.9 cal. kyr) when δ^{18}O increases again, reflecting drier conditions.

Lake Abiyata is currently a closed, saline-alkaline lake, mainly supplied by the outflow of the upstream Lakes Ziway and Langano. Although sub-surface water contributes significantly to the water supply (Tenalem Ayenew 1998; Gibert et al. 1999), lake-level and salinity are very sensitive to P-E balance in the surrounding plateaux (Vallet-Coulomb et al. 2001; Dagnachew Legesse et al. 2003). A 12.6 m-long sediment core taken in the deepest part of the lake represents the past 13,500 years. Diatom-inferred changes in water level and chemistry (Chalié and Gasse 2002) confirm that conditions generally much wetter than today prevailed from ca. 11 to 5.7 cal. kyr. There was a rapid shift toward an overall Late Holocene water deficit between 5.7 and 5 cal. kyr (ca. 5–4.4 ^{14}C kyr BP). The short-term regressions reconstructed by Gillespie et al. (1983) at 8.8–8.5 and 6.7 cal. yr BP are not apparent in the diatom record, which instead indicates a smooth regression between 9.3 and 7.3 cal. kyr BP. After 5 cal. kyr BP, temporary reversals to relatively wet conditions are observed between 4.3 and 3.6 cal. kyr (ca. 3.6–3.2 ^{14}C kyr BP) and at 2.9–2.6 cal. kyr (ca. 2.8–2.5 ^{14}C kyr BP). The diatom record suggests that events of maximum water deficit for the whole Holocene period occurred at 5.3–4.9, 3.2–3, and 2–1.8 cal. kyr BP (ca. 4.7–4.3, 3–2.8, 2.1–1.9 ^{14}C kyr; Fig. 4).

Lake Tilo is supplied by direct precipitation and by groundwater inflow, and is currently saline-alkaline. A 23 m-sediment core taken from this lake covers the past 10,000 cal. years. The core was analysed for diatoms (Telford and Lamb 1999) and stable isotopes of carbonates (Lamb et al. 2000). The diatom record shows that deep, freshwater conditions prevailed until about 4.5 ^{14}C kyr BP (ca. 5.1 cal. kyr). A marked change in the diatom assemblages and in calcite and silica deposition rates at 5.5 ^{14}C kyr BP (ca. 6.3 cal. kyr) was attributed to a decline in geothermal groundwater inflow (Telford et al. 1999). Lake salinity began to increase about 4.5 ^{14}C kyr BP in response to decreasing P-E, and reached approximately its present state ca. 2.5 ^{14}C kyr BP (2.6 cal. kyr). There is, however, a temporary reversal to lower salinity at 3.8–3.5 ^{14}C kyr BP (4.2–3.8 cal. kyr). The isotope record (Lamb et al. 2000) shows low δ^{18}O values until 5.5 ^{14}C kyr BP (ca. 6.3 cal. kyr) reflecting a combination of higher precipitation, greater spring inflow and less evaporation from the lake surface than today. This wet phase was interrupted by short periods of aridity from 7.9 to 7.6 ^{14}C kyr BP (8.8–8.5 cal. kyr) and at 5.9 ^{14}C kyr BP (6.7 cal. kyr; Fig. 4). These events are not recorded by the diatoms but are in good agreement with the Lake Ziway-Shalla and Lake Abhé lake-level reconstructions. An abrupt increase in δ^{18}O values at 4.2 ^{14}C kyr BP (4.7 cal. kyr) reflects a sharp fall in P-E which was maintained up to 3.7 ^{14}C kyr BP (4.1 cal. kyr). It was followed by a wetter episode that lasted for at least 300 years, in agreement with the diatom record. Over the last 2.5 ^{14}C kyr (2.6 cal. kyr), large swings in isotope values and the occurrence of varied laminations are interpreted in terms of anoxic bottom-water conditions, and intermittent carbonate diagenesis possibly linked to irregular overturning of the lake (Fig. 4).

The Tilo, Awassa and Abiyata core records agree with previous data in confirming the major shift from wet to dry conditions around 5.5–5 cal. kyr BP, and the occurrence (at Tilo) of short-term drier periods centred around 8.8–8.5 and at ca. 6.7 cal. kyr BP. They suggest that the return to wetter conditions after aridification at 5 cal. kyr took place at ca. 4.4–4 cal. kyr BP, earlier than previously suggested, and that the Late Holocene period has a more complex climatic history than shown by previous records.

Lake core records in the surrounding lowlands: Lake Turkana

Early investigations by Butzer et al. (1972) and Owen et al. (1982) showed that the level of Lake Turkana rose rapidly at 12–11 cal. kyr BP. Many dates from terraces, which are 60–80 m above the 1971 level, fall within the range of 10–4 ^{14}C kyr BP (ca. 11.5–4.5 cal. kyr). The lake overflowed toward the Nile River during most of the early-mid Holocene period, although a poorly-dated regressive phase occurred between 9–7 cal. kyr BP. Recent studies of sediment characteristics, carbonate stable-isotope composition and fossil diatom assemblages in piston cores have provided more detailed information on the last 6000 years of lake history. Lake Turkana began to retreat from its Early-Holocene overflow level at around 4.5 ^{14}C kyr BP (ca. 5.1 cal. kyr), and appears to have become a closed basin by 3.9 ^{14}C kyr BP (4.3 cal. kyr). Between 2.5 to 1.6 ^{14}C kyr BP (2.6–1.5 cal. kyr) the lake fell to below its modern level. Pollen and isotopic records then indicate a modest transgression to above today's level at 1.5 cal. kyr BP, followed by a gradual decline after 1 ^{14}C kyr BP (0.87 cal. kyr; Mohammed et al. (1996), Johnson (1996), Fig. 5).

Travertine and buried soil records

Early-Holocene peat and buried soil sections with wood fragments, deposited in travertine dammed basins in the Tigray region (Northern Ethiopia), indicate relatively wet climate conditions at the time. The intervening phases of soil degradation, coincident with regional lake low stands, dated at 8.3–7.6 ^{14}C kyr BP (9.2–8.5 cal. kyr), 6.6–5.8 ^{14}C kyr BP (7.5–

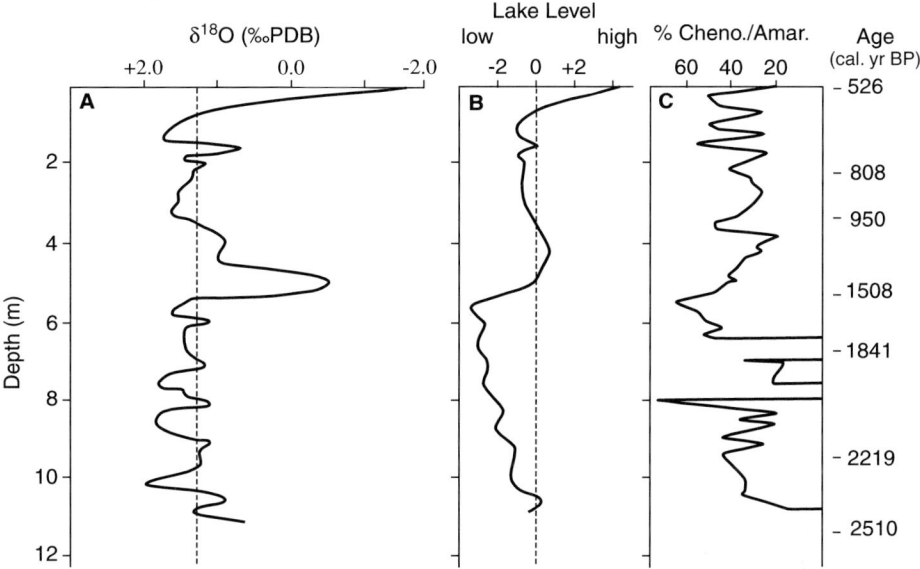

Figure 5. Proxy-climate records from Lake Turkana. a: the $\delta^{18}O$ composition of authigenic calcite in core LT-84 8P (Johnson et al. 1991). b: interpretation of lake-level history based on a model of isotopic response of lake-level to change in hydrological balance, assuming that the equilibrium value for the calcite is +1.3‰ (vertical dashed line in a). Each deviation of $\delta^{18}O$ from the equilibrium value is assumed to be proportional to the magnitude of lake-level change. c: changes in the relative abundance of Chenopodiaceae / Amaranthaceae pollen is assumed to be indicative of lake-level change.

6.6 cal. kyr) may be associated with episodes of increased aridity (Dramis and Mohammed 2003). A pause in the formation of travertine and peats in Northern Ethiopia occurred after 5 ^{14}C kyr (5.7 cal. kyr; Barakhi et al. (1998), Belay (1997)) although a phase of soil formation was noted between 4–3.5 ^{14}C kyr BP (4.5–3.8 cal. kyr BP; Machado et al. (1998)) in rather good agreement with the age of the Late-Holocene period of wetter conditions inferred from the lake core records in the Main Ethiopian Rift Valley.

Century, decadal, and interannual climatic variations

High-resolution palaeoclimate proxies and historical records of climate events covering the last centuries are still scarce from the Horn of Africa. Historical records come mainly from Ethiopian Royal chronicles (Pankhurst 1966; 1985; Degefu 1987), travellers' accounts, and indirectly from the Nile flow records at Aswan, Egypt (Tousson 1925; Hassan 1981; Ortlieb, this volume). The occurrence and impact of climate change during the Medieval Warm Period and the Little Ice Age has become the subject of research. The Nile record from AD 640 to 1921 (Hassan 1981) shows several decade-scale Nile flood variations linked to rainfall on the Ethiopian highlands. When the Nile reaches at its maximum flood stage, the White Nile contributes 10%, Atbera 20% and the Blue Nile 85% of total discharge. Otherwise the White Nile's water supply remains steady throughout the year and supplies

80% of its water to the main Nile at its lowest level (April-May). The Blue Nile from the Ethiopian highlands supplies 85% of the total annual flow. These differential flow amounts are the basis for separating signatures of temporal variation in rainfall over Ethiopia and over equatorial Africa in the Nile record. Hassan (1981) argued for a possible link between low Nile discharge and cold climate in Europe. According to this, the high frequency of famine events from the 16th to the 19th century (Pankhurst 1985; Machado et al. 1998) appears to have happened during the high latitude cold period of the Little Ice Age (Lamb 1977). Pollen data on the Ethiopian highlands also revealed a much colder climate than today (Bonnefille and Mohammed (1994), Fig. 6). Snowfall was reported above 3000 m on the Ethiopian mountains during the 17th century, but is unknown today (Hövermann 1954). This may correspond to neo-glacial activity on Mt Kenya from 1500 to 1800 AD (Rosqvist 1990). Johnson et al. (2001) for example, report that drier conditions were observed between 1570 and 1820 in tropical Africa. Huffman (1996) reports it was also colder in Southern Africa between 1300 and 1800 AD and according to Holmgren et al. (2001), Tyson et al. (2000) and Scott and Lee-Thorp (this volume) the magnitude of this Little Ice Age cooling reached about 1 °C.

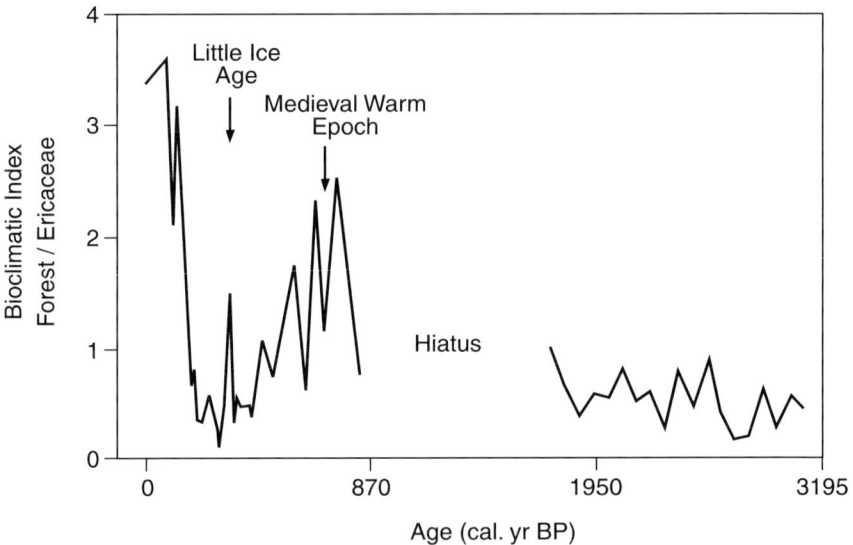

Figure 6. Bioclimatic index (ratio between arboreal and Ericaceae pollen) expressed as arbitrary units vs. time (Bonnefille and Mohammed 1994).

Recent information from Kenya and Ethiopia have shown that the general climate in Equatorial East Africa was wetter during the Little Ice Age, from AD 1270 to 1850, than today, while it was relatively dry during the Medieval Warm Epoch, from AD 1000 to 1270, (Verschuren et al. 2000; Verschuren, this volume; Darbyshire et al. 2003). However the Kenyan record also indicates that the Little Ice Age was interrupted by three periods of prolonged and severe drought (AD 1390–1420, 1560–1625 and 1760–1840) which had profound societal impact (Verschuren et al. 2000). The latter two droughts overlap with droughts in Ethiopia in 1543–1562, 1618, 1828–1829, and 1864–1866 which may have coincided with El-Niño Southern Oscillation (ENSO) events (Tsegaye-Wolde Giorgis 1997). Different mechanisms might have operated to produce the general wet climate of the Little Ice Age, interrupted by severe droughts in equatorial Africa. Further research is needed for a more precise determination of the nature, extent, timing, causes and impact of these events.

High resolution analysis of a ^{210}Pb-dated sediment-core from Lake Abiyata (Main Ethiopian Rift) shows large environmental changes over recent centuries (Dagnachew Legesse et al. 2002). A marked drought revealed by a maximum in diatom-inferred salinity is suggested at around AD 1890, and appears to be coincident with a major famine which occurred in the region at AD 1888–1892 (Pankhurst 1966; Degefu 1987). A relatively low salinity record which ended at ca. 1800 may correspond to a higher P/E ratio at the end of the Little Ice Age, as was observed in Kenya (Verschuren et al. 2000).

High inter-annual variability of rainfall during this century and particularly the occurrence of droughts in the early 1970s and 1980s, have been attributed to the ENSO phenomenon (Haile 1988) although the relative role played by ENSO and other factors influencing Ethiopian rainfall have not yet been clearly defined (Conway 2000). A recent increase in temperature has also been apparent from meteorological data (Billi 1998), from a rise in the forest limit shown in pollen diagrams on high altitude sites of Southern Ethiopia (Bonnefille and Mohammed 1994), and from observational records on the Semien mountains that indicate a rise in *Erica arborea* by about 120 m in altitude during the last three decades (Kaeppeli 1998). These observations, and especially the rapid retreat of ice from Kilimanjaro (Thompson et al. 2002), suggest that the region is being affected by warming.

During recent decades many lakes have experienced significant changes in water depth and salinity. Some lakes in the Ethiopian Rift Valley were affected by human impact, e.g., extraction of soda ash from Lake Abiyata since the mid 1980s, and irrigation activities on various rivers and lakes (Chernet 1998; Tenalem Ayenew 1998). Others have shown a tendency to rise (e.g., Lake Beseka, Lake Awassa and Lake Hora) in spite of the rainfall pattern (data from the Ethiopian Ministry of Water Resources). Further research is required to clarify the relative importance of climate, hydrology and human impact on lake-levels and chemistry in the region. The stable-isotopic composition of recent laminated sediments in Lake Hora, Debre Zeit has been used to model lake-level variations (Lamb H.F. et al. 2002). A study of the oxygen-isotope balance of the lakes revealed lake/groundwater interactions, and showed lake sensitivity to climate change (Kebede et al. 2002). Meteorological and hydrological time-series from catchments and lakes in the Ziway-Shala basin were used to model river discharge, lake-level and chloride budgets (Vallet-Coulomb et al. 2001; Dagnachew Legesse et al. 2003) in order to understand the hydrological functioning of the basin at local as well as regional scales. Coupling of catchment and lake water-balance models should clarify the response of lake-level and chemistry to changes in land use as well as to climate variability in the basin. This will be achieved by sensitivity analyses of the different parameters of the hydrological models at different temporal and spatial scales. Once the models are well calibrated for the present, they can then be applied to reconstruct past climatic and environmental conditions or to predict future scenarios.

Summary

Our knowledge of climatic variability in the Horn of Africa has considerably increased over the past few years, especially through detailed analysis of lake sediment cores, pollen and buried soils. Many questions remain to be answered, however. The main points arising from this review are as follows:

1. At the millennial time-scale, orbitally-induced changes in summer solar radiation account for a large part of observed environmental changes in the region. Orbital precession appears to be a satisfactory explanation for enhanced monsoonal rainfall and thus high lake levels around 35–30 cal. kyr BP, during the early-mid Holocene, and for dry LGM conditions. One of the most spectacular responses to the precessional cycle was the wetting of the present arid lowlands, as they accumulated precipitation from huge watersheds extending from high to low altitudes. For instance, in the central Afar alone (ca. 11–12 °N; 40°40'–42°40' E; ca. 24,000 sq. km), surface and groundwater-fed lakes extended over ca. 8,000 km² and stored ca. 600 km³ of freshwater at ∼30 and 10 cal. kyr. However, no long records are available to estimate the role of orbital forcing in climate change over a full glacial-interglacial cycle, but the region offers potential sites (e.g., Lake Abhé, central Afar) for long-term palaeoclimatic studies. A recent seismic and sedimentary study within the Lake Abiyata basin (Le Turdu et al. 1999) suggests the presence of a 600 m-thick sedimentary sequence with an age estimated at 750–330 kyr.

2. The development of glaciers in the Ethiopian highlands during the Late Pleistocene was an important event, but the age during which glacial advance took place is problematic. New dating techniques, e.g., cosmogenic isotopes, could be applied to date moraines and answer this question.

3. The generally wet early-mid Holocene episode ended abruptly at ∼5.5–5 cal. kyr, despite a gradual decrease in summer insolation. An abrupt termination of the north African "humid period" is recorded at the same time in the Sahara-Sahel (Hoelzmann et al., this volume) and in marine sediments off West Africa (de Menocal 2000). According to current coupled vegetation-atmosphere models, the aridification of the Sahara was an abrupt response to gradual change in insolation forcing which requires strongly non-linear ocean temperature and vegetation feedbacks (Claussen et al. 1999). Such mechanisms may account for the major mid-Holocene shift to dry conditions in the Horn of Africa.

4. Other mechanisms such as changes in solar activity (Beer et al. 2000) should be invoked to explain the short-term dry events that punctuated the Early Holocene wet period, and the complexity of Late Holocene climate. In particular, we need to constrain the timing of the dry spell at ∼8.7–8 cal. kyr, observed in several lake records, to clarify its possible relationship to the brief cooling dated at 8.2 cal. kyr evident in the Greenland ice core records. The sparse data on the regional manifestation of the Little Ice Age and the Medieval Warm Period are controversial. Future research should aim at obtaining annual to decadal resolution records from tree rings, stalagmites, and laminated lake sediments

5. Meteorological time-series in the region are too scarce and too short to reveal any significant cyclicity or major climate trends. Nevertheless, instrumental records provide the required information for validating ecological or hydrological system models which should be developed to analyse the response of individual systems and proxies to climate changes. Together with the development of climate-system models with finer spatial and

temporal scales, this will allow better predictions of how the region will respond to global climate variability and to changes in land-use and water consumption.

Acknowledgments

Recent research on palaeoclimate in Ethiopia has been achieved through collaborative research between the Department of Geology and Geophysics of Addis Ababa University and institutions in France, the UK and Italy. We would like to thank the Centre National de la Recherche Scientifique (CNRS) and the French Ministry of Foreign Affairs as well as the European Union, the UK Natural Environment Research Council (NERC) and the Italian co-operation for financial support. Addis Ababa University provided institutional support. The authors would like to thank all who have contributed to these projects. The PAGES-PEP III and African Pollen Database (APD) projects have laid down the scientific issues upon which the research has been based. Comments from the reviewers, D. Verschuren and A.C. Stevenson, helped to improve the manuscript.

References

Barakhi O., Brancaccio L., Calderoni G., Coltorti M., Dramis F. and Mohammed M.U. 1998. The Mai Mekdan sedimentary sequence — A reference point for the environmental evolution of the highlands of Northern Ethiopia. Geomorphology 23: 127–128.

Bard E. 1998. Geochemical and geophysical implications of the radiocarbon calibration. Geoch. Cosmoch. Acta. 62/12: 2025–2038.

Barker P., Talbot M.R., Street-Perrott F.A., Marret J., Scourse J. and Odada E., this volume. Late Quaternary climatic variability in intertropical Africa. In: Battarbee R.W., Gasse F. and Stickley C.E. (eds), Past Climate Variability through Europe and Africa. Kluwer Academic Publishers, Dordrecht, the Netherlands, pp. 117–138.

Beer J., Mende W. and Stellmacher R. 2000. The role of the sun in climate forcing. Quat. Sci. Rev. 19: 403–415.

Benvenuti M., Carnicellli S., Belluomini G., Dainelli N., Grazia Di S., Ferrari G.A., Iasio C., Sagri M., Ventra D., Balemual A. and Seifu K. 2002. The Ziway-Shalla lake basin (main Ethiopian Rift, Ethiopia): a revision of basin evolution with special reference to the Late Quaternary. Journal of African Earth Science 35: 247–269.

Belay T. 1997. Variabilities of catena on degraded hill slopes of Watiya catchment, Wollo, Ethiopia. SINET: Ethiopian Journal of Science 20: 151–175.

Billi P. 1998. Climatic change. In Land Resources Inventory and Environmental Changes analysis and their application to Agriculture in the Lake region (Ethiopia). In: Sagri M. (ed.), Co-ordinator. Report submitted to EC. Project STD 3 contract no Ts 3-CT 92-76 (1993–1998). pp. 108–122.

Bonnefille R. and Mohammed U. 1994. Pollen inferred climatic fluctuations in Ethiopia during the last 3000 yrs. Palaeogeogr. Palaeoclim. Palaeoecol. 109: 331–343.

Butzer K.W., Isaac G.L., Richardson J.L. and Washbourn-Kamau C. 1972. Radiocarbon dating of East African lake levels. Science 175: 1069–1076.

Chalié F. and Gasse F. 2002. A 13,500 years diatom record from the Tropical East African Rift Lake Abiyata (Ethiopia). Palaeogeogr. Palaeoclim. Palaeoecol. 187: 259–283.

Chernet T. 1998. Etude des mécanismes de minéralisation en fluorure et éléments associés de la région du Rift Ethiopien. Ph.D. thesis, Université d'Avignon, Avignon, France. 205 pp.

Claussen M., Kubatzki C., Brovkin V., Ganopolski A., Hoelzmann P. and Pachur H.J. 1999. Simulation of an abrupt change in Saharan vegetation in the mid-Holocene. Geophys. Res. Lett. 26: 2037–2040.

Conway D. 2000. Some aspects of climate variability in the northeast Ethiopian Highlands — Wollo and Tigray. SINET: Ethiopian Journal of Science. 23: 139–161.

Dagnachew Legesse, Vallet-Coulomb C. and Gasse F. 2003. Hydrological response of a catchment to climate and land use changes in tropical Africa: Case study South Central Ethiopia. J. Hydr. 275: 67–85.

Dagnachew Legesse, Gasse F., Radakovitch O., Vallet-Coulomb C., Bonnefille R., Verschuren D., Gibert E. and Barker P. 2002. Environmental changes in a tropical lake system (Ziway-Shalla basin; Main Ethiopian Rift) over the last centuries — Preliminary results from short cores from Lake Abiyata. Palaeogeogr. Palaeoclim. Palaeoecol. 187: 233–258.

Darbyshire I., Lamb H.F. and Umer M. 2003. Forest clearance and regrowth in Northern Ethiopia during the last 3000 years. The Holocene 13: 553–562.

Degefu W. 1987. Some aspects of meteorological drought in Ethiopia. In: Glantz M.H. (ed.), Drought and Hunger in Africa. Cambridge University Press, Cambridge. pp. 23–36.

DeMenocal P., Ortiz J., Guilderson T., Adkins J., Sarntheim M., Baker L. and Yarusinsky M. 2000. Abrupt onset and termination of the African Humid period: rapid climate responses to gradual insolation forcing. Quat. Sci. Rev. 19: 347–361.

Dramis F., Mohammed U.M., Calderoni G. and Mitiku H. 2003. Holocene climate phases from buried soils in Tigray (northern Ethiopia). Quaternary Research 60: 274–283

Farrera I., Harrison S.P., Prentice I.C., Ramstein G., Guiot J., Bartlein P.J., Bonnefille R., Bush M., Cramer W., Von Grafenstein U., Holmgren K., Hoogmhiemstra H., Hope G., Jolly D., Lauritzen S.E., Ono Y., Pinot S., Stute M. and Yu G. 1999. Tropical climates at the Last Glacial Maximum: a new synthesis of terrestrial palaeoclimate data. I. Vegetation, lake-levels and geochemistry. Clim. Dyn. 15: 823–856.

Fontes J.C. and Pouchan P. 1975. Les cheminees du lac Abhe (TFAI): station hydroclimatique de l'Holocène. C.R. Acad. Sci. Paris. 280D: 383–386.

Fontes J.C., Gasse F., Camara E., Millet B., Saliege J.F. and Steinberg M. 1985. Late Holocene changes in Lake Abhé hydrology (Ethiopia - Djibouti). Zeitschrift für Gletscherkunde und Glazialgeologie. 21: 89–96.

Friis I. 1992. Forests and forest trees of Northeast tropical Africa. Their natural habitats and distribution patterns in Ethiopia, Djibouti and Somalia. Kew Bull. Add. Ser. 15 I-v: 1–396.

Gasse F. 1977. Evolution of Lake Abhé (Ethiopia and T.F.A.I.) from 70,000 BP. Nature 265: 42–45.

Gasse F. 2000. Hydrological changes in the African tropics since the Last Glacial Maximum. Quat. Sci. Rev. 19: 189–211.

Gasse F. and Descourtieux C. 1979. Diatomées et évolution de trois milieux éthiopiens d'altitude différente, au cours du Quaternaire supérieur. Palaeoecology of Africa. 11: 117–134.

Gasse F. and Street F.A. 1978. Late Quaternary lake level fluctuations and environments of the northern rift valley and Afar region (Ethiopia and Djibouti). Paleogeogr. Paleoclim. Paleoecol. 25: 145–150.

Gasse F. and van Campo E. 1994. Abrupt Post-Glacial climate events in West Asia and North African monsoon domains. Ear. Planet. Sci. Lett. 126: 435–456.

Gasse F., Barker P.A., Gell P.A., Fritz S.C. and Chalié F. 1997: Diatom-inferred salinity in palaeolakes: an indirect tracer of climatic change. Quat. Sci. Rev. 16: 547–563.

Gasse F., Bergonzini L., Chalié F., Gibert E., Massault M. and Mélières F. 1998. Paléolacs et paléoclimats aux pourtours de l'océan Indien occidental depuis 25 ka BP. In: Causse C. and Gasse F. (eds), Hydrologie et Géochimie Isotopique. Proceedings of the Int. Symp. In Memory of Jean-Charles Fontes, Paris, June 1st and 2nd 1995. Paris. pp. 147–176.

Gasse F., Juggins S. and Ben Khelifa L. 1995. Diatom-based transfer functions for inferring past hydrochemical characteristics of African lakes. Palaeogeogr. Palaeoclim. Palaeoecol. 117: 31–54.

Gasse F., Rognon P. and Street F.A. 1980. Quaternary history of the Afar and Ethiopian Rift lakes. In: Williams M.A.J. and Faure H. (eds), The Sahara and the Nile. Balkema, Rotterdam, pp. 161–400.

Gibert E., Travi Y., Massault M., Chernet T., Barbecot F. and Laggoun Defarge F. 1999. Comparing carbonate and organic AMS-C-14 ages in Lake Abiyata sediments (Ethiopia): Hydrochemistry and paleoenvironmental implications. Radiocarbon 41: 271–286.

Gillespie R. Street-Perrott F.A. and Switzur R. 1983. Post glacial arid episodes in Ethiopia have implications for climate prediction. Nature. 306: 680–683.

Griffiths J.F. 1972. Ethiopian highlands. In: Landsberg H.E. (ed.), World Survey of Climatology 10. Elsevier, Amsterdam, pp. 369–381.

Haile T. 1988. Causes and characters of drought in Ethiopia. Ethiopian Journal of Agricultural Science 10: 1–2, pp. 85–97.

Hamilton A.C. 1982. Environmental history of East Africa. A study of the Quaternary. Academic Press, London, pp. 328.

Hassan F. 1981. Historical Nile Floods and their implication for climatic change. Science. 212: 1142–1145.

Hastenrath S. 1977. Pleistocene mountain glaciation in Ethiopia. Journal of Glaciology. 28: 309–313.

Hedberg O. 1951. Vegetation belts of East African mountains. Svenssk Botanisk Tidiskrift 45: 140–204.

Hoelzmann P., Gasse F., Dupont L.M., Salzmann U., Staubwasser M., Leuschner D.C. and Sirocko F., this volume. Palaeoenvironmental changes in the arid and subarid belt (Sahara-Sahel-Arabian Peninsula) from 150 ka to present. In: Battarbee R.W., Gasse F. and Stickley C.E. (eds), Past Climate Variability through Europe and Africa. Kluwer Academic Publishers, Dordrecht, the Netherlands, pp. 219–256.

Holmgren K., Tyson P.D., Moberg A. and Svanered O. 2001. A preliminary 3000-year regional temperature reconstruction for South Africa. South African Journal of Science 97: 49–51.

Hövermann J. 1954. Ue ber die Hohenlage der Schneegrenze in Aethiopien und ihre Schwankungen in historischer Zeit. Nachr. Akad. Wiss Göttingen 11a (6): 11–137.

Huffman T.N. 1996. Archaeological evidence for climatic change during the last 2000 years in Southern Africa. Quat. Int. 33: 55–60.

Hurni H. 1982. Hochgebriirge von Semien- Äthiopien. Vol II. Klima und Dynamic der Höhenstufung von der letzten Kaltzeit bis zur Gegenwart (Teil Ii gemeinsam mit Peter stähli). Bern.

Hurni H. 1989. Late Quaternary of Simien and other mountains in Ethiopia. In: Mahaney W.C. (ed.), Quaternary and Environmental Research on East African Mountains. Balkema, Rotterdam, pp. 105–120.

Johnson T.C. 1996. Sedimentary processes and signals of past climate change in the large lakes of East African Rift Valley. In: Johnson T.C. and Odada E.O. (eds), The Limnology, Climatology and Paleolimnology of the East African Lakes. Gordon and Breach, Amsterdam, pp. 367–412.

Johnson T.C., Halfman J.D., Rosendahel B.P. and Lister G.S. 1987. Climatic and tectonic effects on sedimentation in a rift valley lake: evidence from high resolution seismic profiles, Lake Turkana, Kenya. Geol. Soc. Am. Bul. 98: 439–447.

Johnson T.C., Barry S., Chan Y. and Wilkinson P. 2001. Decadal record of climate variability spanning the past 700 yr in the Southern Tropics of East Africa. Geology 29: 83–86.

Jolly D., Harrison S.P., Damnati B. and Bonnefille R. 1998. Simulated climate and biomes of Africa during the Late Quaternary: comparison with pollen and lake status data. Quat. Sci. Rev. 17: 629–657.

Kaeppeli M. 1998. Regeneration and age structure of relict ericaceous forests: a dendrochronolog-ical study near the timberline in the Simien Mountains, Ethiopia. Diploma Thesis, Institute of Geography, University of Bern, Switzerland. 127 pp. + appendices.

Kebede S., Lamb H.F., Telford R.J., Leng M.J. and Umer M.U. 2002. Lake ground water relationships, oxygen isotope balance and climate sensitivity of the Bishoftu crater lakes, Ethiopia. In: Odada E. and Olago D. (eds), The East African Great Lakes Region: Limnology, Paleoclimatology and Biodiversity. Advances in Global Research Series, Kluwer.

Kohfeld I.E. and Harrison S.P. 2000. How well can we simulate past and present climates? Evaluating the models using global paleoenvironmental data sets. Quat. Sci. Rev. 19: 321–346.

Kutzbach J.E., Guetter P.J., Behling P.J. and Delin R. 1993. Simulated climatic changes: results of COHMAP climate-model experiments. In: Global Climates since the Last Glacial Maximum. University of Minnesota Press, Minneapolis, pp. 5–11.

Lamb A.L., Leng M.J., Lamb H.F. and Mohammed M.U. 2000. A 9000-yr oxygen and carbon isotope record of hydrological change in a small crater lake. The Holocene 10: 167–177.

Lamb A.L., Leng M.J, Lamb H.F., Telford R.J. and Mohammed M.U. 2002. Climatic and non-climatic effects on the $\delta^{18}O$ and $\delta^{13}C$ composition of Lake Awassa, Ethiopia, during the last 6.5 ka. Quat. Sci. Rev. 21: 2199–2211.

Lamb H.F., Kebede S., Leng M.J., Ricketts D., Telford R.J. and Umer M.U. 2002. Origin and isotopic composition of aragonite laminae in an Ethiopian crater lake. In: Odada E. and Olago D. (eds), The East African Great Lakes Region: Limnology, Paleoclimatology and Biodiversity. Advances in Global Research Series, Kluwer.

Le Turdu C., Tiercelin J.J., Gibert E., Travi Y., Lezzar K.E., Richert J.P., Massault M., Gasse F., Bonnefille R., Decobert M., Gensous B., Jeudy V., Tamrat E., Mohammed M.U., Martens K., Atnafu B., Chernet T., Williamson D. and Taieb M. 1999. The Ziway Shalla basin system, Main Ethiopian Rift: Influence of volcanism, tectonics and climatic forcing on basin formation and sedimentation. Palaeogeogr. Palaeoclim. Palaeoecol. 150: 135–177.

Lezine A.M. and Bonnefille R. 1982. Diagramme pollinique Holocene d'un sondage du lac Abiyata (Ethiopie, 7°42' Nord). Pollen et Spore. 3–4: 463–480.

Machado J.M., Perez-Gonzalez A. and Benito G. 1998. Paleo-environmental changes during the last 4000 yrs in Tigray, Northern Ethiopia. Quat. Res. 49: 312–321.

Messerli B., Hurni H., Kienholz H. and Winiger M. 1977. Bale mountains: largest Pleistocene Mountain glacier system of Ethiopia. X INQUA congress. Birmingham. Abstracts: 1.

Messerli B., Rognon P.H. and Winiger M. 1980. The Saharan and East African uplands during the Quaternary. In: Williams M.A.J. (ed.), The Sahara and the Nile. Balkema, Rotterdam, pp. 87–148.

Mohammed M.U., Bonnefille R. and Johnson T.C. 1996. Pollen and isotopic records of Late Holocene sediments from lake Turkana, N. Kenya. Palaeogeogr. Palaeoclimat. Palaeoecol. 119: 371–383.

Mohammed M.U. and Bonnefille R. 1998. A Late Glacial to Late Holocene pollen record from a high land peat at Tamsaa, Bale mountains, South Ethiopia. Glob. Planet. Chan. 16–17: 121–129.

Mohammed M.U., Tiercelin J.J., Gibert E., Hureau D., Lézine A.M., Lamb H.F and Bonnefille R. 2003. Paleoenvironmental Evolution of the High Altitude South-Eastern Ethiopian Highlands during the Last 13,000 years. Poster presented at the XVIth INQUA Congress held in Reno USA in July 2003.

Mohr P.A. 1971. The Geology of Ethiopia. University Press, Addis Ababa.

Nillson E. 1940. Ancient changes of climate in British East Africa and Abyssinia — Geografisker Annaler 22: 1–79.

Ortlieb L., this volume. Historical chronology of ENSO and the Nile flood record. In: Battarbee R.W., Gasse F. and Stickley C.E. (eds), Past Climate Variability through Europe and Africa. Kluwer Academic Publishers, Dordrecht, the Netherlands, pp. 257–278.

Osmaston H., Mitchell W. and Osmaston N., in press. Quaternary glaciation of the Bale mountains, Ethiopia. Geomorphology.

Owen R.B., Barthelme J.W., Renaut R.W. and Vincens A. 1982. Paleolimnology and archaeology of Holocene deposits north-east of Lake Turkana. Nature 298: 523–528.

Pankhurst R. 1966. The great Ethiopian famine of 1888–1892: A new assessment. Journal of the History of Medicine and Applied Sciences 21: 39–92.

Pankhurst R. 1985. The history of famine and Epidemics in Ethiopia prior to the twentieth century. Addis Ababa University. Report to Relief and Rehabilitation Commission. Addis Ababa, Ethiopia.

Pichi-Sermolli R.E.G. 1957. Una carta geobotanica dell'Africa Orientale (Eritrea, Ethiopia, Somalia).- Webbia. 13: 15–132.

Rosqvist G. 1990. Quaternary Glaciations in Africa. Quat. Sci. Rev. 9: 281–297.

Said R. 1993. The River Nile. Pergamon, Oxford, 320 pp.

Scott L. and Lee-Thorp J.A., this volume. Holocene climatic trends and rhythms in Southern Africa. In: Battarbee R.W., Gasse F. and Stickley C.E. (eds), Past Climate Variability through Europe and Africa. Kluwer Academic Publishers, Dordrecht, the Netherlands, pp. 69–91.

Seleshi Y. and Demarée G.R. 1995. Rainfall variability in the Ethiopian and Eritrean highlands and its links with the southern oscillation. J. Biogeogr. 22: 945–952.

Street F.A. 1979a. Late Quaternary Lakes in the Ziway-Shalla Basin, Southern Ethiopia. Cambridge University. 493 pp.

Street F.A. 1979b. Late Quaternary precipitation estimates for the Ziway-Shalla basin, Southern Ethiopia. Palaeoecology of Africa 11: 135–143.

Street F.A. and Grove A.T. 1979. Global maps of lake-level fluctuations since 30,000 yr BP. Quat. Res. 12: 83–118.

Street-Perrott F.A. and Perrott R.A. 1990. Abrupt climate fluctuations in the tropics, the influence of Atlantic Ocean circulation. Nature 343: 607–611.

Stuiver M., Reimer P.J., Bard E., Beck J.W., Burr G.S., Hughen K.A., Kromer B., McCormac G., Van der Plicht J. and Spurk M. 1998. INTCAL98 radiocarbon age calibration, 24,000-0 cal BP. Radiocarbon. 40 (3): 1041–1083.

Telford R.J. and Lamb H.F. 1999. Ground water mediated response to Holocene climate change recorded by diatom stratigraphy of an Ethiopian crater lake. Quat. Res. 52: 63–75.

Telford R.J., Lamb H.F. and Mohammed M.U. 1999. Diatom derived paleoconductivity estimates for Lake Awassa, Ethiopia: evidence for pulsed inflow of saline ground water. J. Paleolim. 21: 409–421.

Tenalem Ayenew 1998. The hydrological system of the Lake District basin, Central main Ethiopian Rift. Ph.D. thesis, ITC, University of Amsterdam, The Netherlands. no. 54. ISBN 90-6164-158-6, 259 pp.

Thompson L.G., Mosley-Thompson E., Davis M.E., Henderson K.A., Brecher H.H., Zagorodnov V.S., Mashiotta T.A., Lin P.N., Mikhalenko V.N., Hardy D.R. and Beer J. 2002. Kilimanjaro ice core records: evidence of Holocene climate change in tropical Africa. Science 298: 589–593.

Tsegaye Wolde-Georgis 1997. El-Niño and drought early warning in Ethiopia. IJAS, issue no. 2: using science against famine: food security, famine early warning and El-Niño.

Tousson O. 1925. L'Histoire du Nile. Mem. Inst. Egypte. 9 Part 2.

Tyson P.D., Karlen W., Holmgren K. and Heiss G.A. 2000. The Little Ice Age and medieval warming in South Africa. South African Journal of Science 96: 121–126.

Vallet-Coulomb C., Dagnachew Legesse, Gasse F., Travi Y. and Chernet T. 2001. Lake evaporation estimates in tropical Africa from limited meteorological data. J. Hydrol. 245: 1–18.

Verschuren D., Laird K.R. and Cumming B.F. 2000. Rainfall and drought in equatorial East Africa during the past 1,100 years. Nature 403: 410–414.

Verschuren D., this volume. Decadal and century-scale climate variability in tropical Africa during the past 2000 years. In: Battarbee R.W., Gasse F. and Stickley C.E. (eds), Past Climate Variability through Europe and Africa. Kluwer Academic Publishers, Dordrecht, the Netherlands, pp. 139–158.

Waterbury J. 1979. Hydropolitics of the Nile Valley. Syracuse University Press, New York.

White F. 1983. The vegetation of Africa. A descriptive memoir to accompany the UNESCO /AETFAT
    vegetation map of Africa. Paris.
Williams M.A.J., Bishop P.M., Dakin F.M., Gillespie R. 1977. Late Quaternary lake levels in Southern
    Afar and the adjacent Ethiopian Rift. Nature 267: 690–693.

10. PALAEOENVIRONMENTS, PALAEOCLIMATES AND LANDSCAPE DEVELOPMENT IN ATLANTIC EQUATORIAL AFRICA: A REVIEW OF KEY SITES COVERING THE LAST 25 KYRS

HILAIRE ELENGA (hilaire_elenga@yahoo.fr)
Faculté des Sciences
Université Marien Ngouabi
BP 69 Brazzaville
Congo

JEAN MALEY (jmaley@isem.univ.montp2.fr)
Paléoenvironnements et Palynologie
Université Montpellier-2
34095, Montpellier
France

ANNIE VINCENS (vincens@cerege.fr)
Centre Européen de Recherche et d'Enseignement
de Géosciences de l'Environnement (CEREGE)
Europole de l'Arbois
B.P. 80, F-13545 Aix-en-Provence cedex 4
France

ISABELLE FARRERA (farrera@isem.univ.montp2.fr)
Paléoenvironnements et Palynologie
Université Montpellier-2
34095 Montpellier
France

Keywords: Pollen, Late Quaternary, Equatorial Africa, Cameroon, Congo, Forest, Vegetation dynamics, Climate changes, Human impact.

Introduction

The pollen data available for the Atlantic equatorial African region indicate that during the Late Quaternary, major climate changes occurred and caused important modifications in tropical lowland rain-forest in terms of composition and distribution (Maley 1991; Elenga et al. 1991; 1994; Farrera et al. 1999; Vincens et al. 1999). Nevertheless, despite a wealth of proxy evidence from various disciplines, there are very few terrestrial sites dated

R. W. Battarbee et al. (eds) 2004. *Past Climate Variability through Europe and Africa.*
Springer, Dordrecht, The Netherlands.

continuously since the Last Glacial Maximum (LGM) to the Late Holocene in this area (e.g., Lanfranchi and Schwartz (1990), Servant and Servant-Vildary (2000)). The oldest radiocarbon-dated sites are from Lake Barombi Mbo in Cameroon (Brenac 1988; Maley 1991; Maley and Brenac 1998a) and the Ngamakala Pond in the Congo (Elenga et al. 1994), although most Holocene sequences from this area start during the mid-Holocene (Vincens et al. 1999). This paper reviews the results available for some key sites, allowing regional correlations for major Late Quaternary periods. We have focused on the vegetation changes and the climate related to these changes. Numerous radiocarbon dates allow precise palaeoenvironmental reconstructions, especially since the mid-Holocene to be made. In this paper all ^{14}C dates are uncalibrated.

Data

Within the large equatorial lowland rain-forest ecosystem, the pollen data-set only contains sequences from Southern Cameroon and Southern Congo, no data being available from Gabon and PDR Congo, ex Zaïre (Elenga et al. 2000). Five well radiocarbon-dated pollen records in which the LGM and/or the Holocene periods are well-defined, are used in this paper. In addition, some important results from a poorly dated sequence (Bilanko) are included. The geographical location of the sites is given on Figure 1 and Table 1. The detailed description of the pollen diagrams and the lithology of the cores are published elsewhere (Elenga et al. 1991; 1994; 1996; Giresse et al. 1994; Vincens et al. 1994; 1998; Reynaud-Farrera 1995; Reynaud-Farrera et al. 1996; Maley 1996; Maley and Brenac 1998a). The oldest sequences are those from Barombi Mbo and Ngamakala. The respective ages of these cores are 28 kyr and 25 kyr BP. Only three sites (Kitina, site 1; Ossa, site 5; and Barombi Mbo, site 6; Fig. 1), are today located in the true lowland tropical rain-forest, which is characterised by an annual rainy season of 8–9 months or more. The other sites are presently located in secondary grasslands and wooded grasslands (Sinnda, sites 2; Ngamakala, site 3; and Bilanko, site 4; Fig. 1). As a consequence of this low geographical coverage of records, the palaeoclimatic interpretations proposed in this paper must be considered as a first approximation.

Table 1. Selected pollen sites from Atlantic Equatorial Africa. Geographical location and main references.

Site	Latitude	Longitude	Altitude	References
1. Kitina	4°15'S	11°59'E	150 m	Elenga et al. 1996
2. Sinnda	3°50'S	12°48'E	128 m	Vincens et al. 1994; 1998
3. Ngamakala	4°04'S	15°23'E	400 m	Elenga et al. 1994
4. Bilanko	3°31'S	15°21'E	600 m	Elenga et al. 1991
5. Ossa	3°50'N	10°01'E	8 m	Reynaud-Farrera et al. 1996; Van Geel et al. 1998
6. Barombi Mbo	4°40'N	9°24'E	300 m	Maley and Brenac 1998a

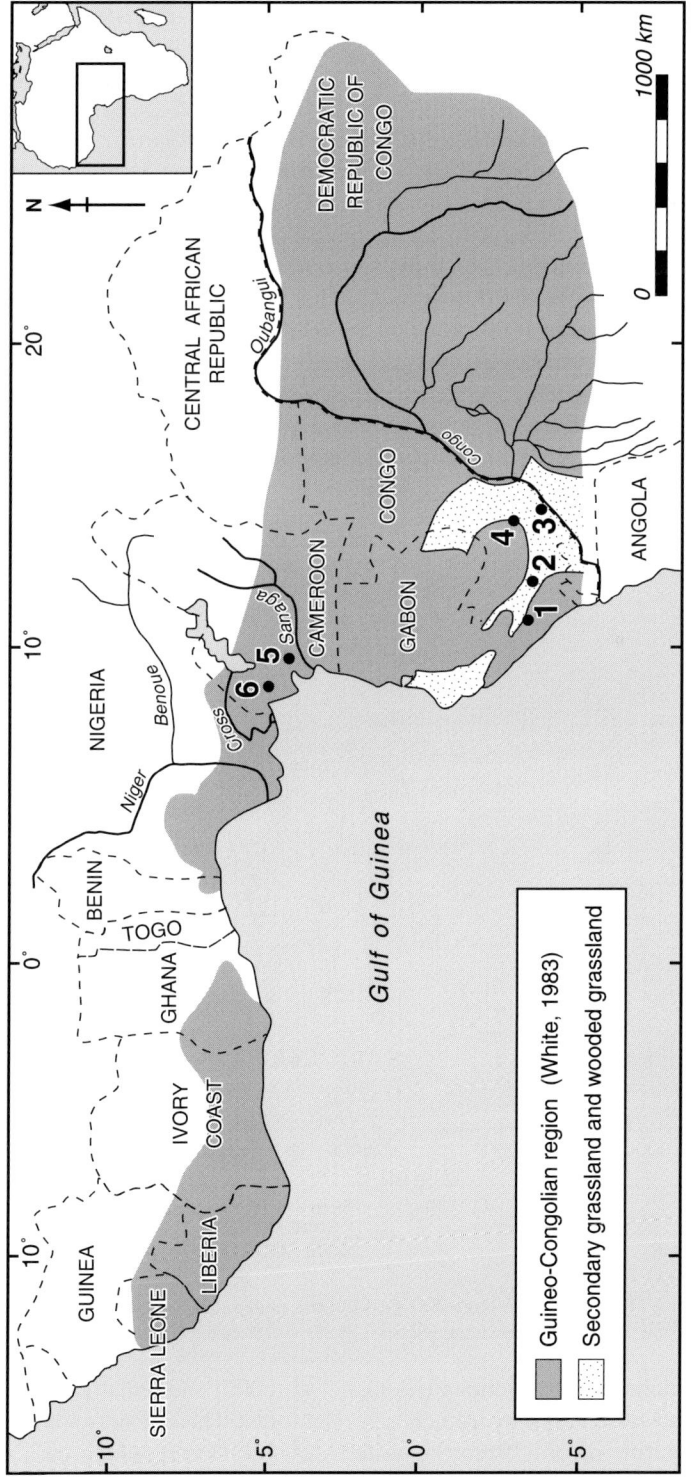

Figure 1. Location map of the selected pollen sites in Atlantic Equatorial Africa: 1, Kitina; 2, Sinnda; 3, Ngamakala; 4, Bilanko; 5, Ossa; 6, Barombi Mbo.

Results and discussion

Late Glacial Maximum

The two pollen sequences representing this period, Barombi Mbo (Figs. 2, 3 and 4) and Ngamakala (Fig. 5), indicate that, at these sites, the prevailing vegetation types were semi-deciduous forests and grasslands during the LGM. The data do not show that forest formations were completely replaced by an open landscape (Elenga et al. 1994; Maley and Brenac 1998a). At Barombi Mbo, the co-existence of typical African mountain elements such as *Olea* and *Podocarpus* until 10 kyr BP (Fig. 4), found today on the Cameroon mountains at altitudes between 1600 and 3000 m (White 1983) and of lowland forest trees, could be explained by the persistence of humid conditions favourable to forest ecosystems, probably as refugia (Maley 1987; 1996). Today isolated stands of *Podocarpus latifolius* occur at lower elevations in Southern Cameroon, Equatorial Guinea and on the Chaillu Massif in Congo (Maley et al. 1990; Parmentier and Maley 2001).

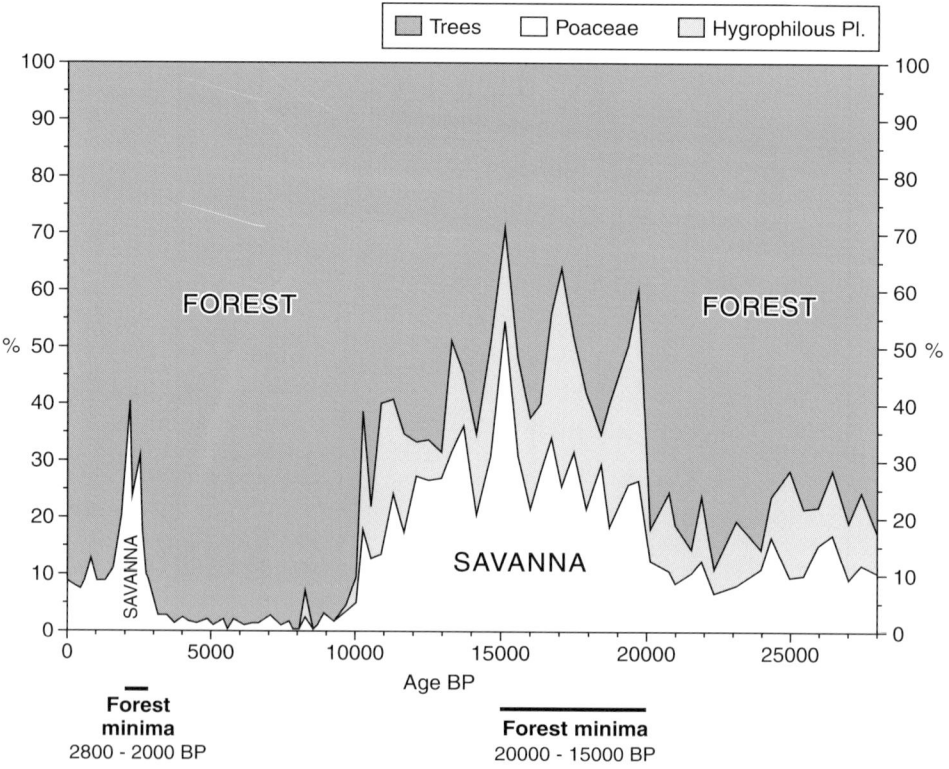

Figure 2. Lake Barombi Mbo, Cameroon (core BM-6): synthetic pollen diagram (the pollen sum includes all taxa, excluding spores and damaged grains) (adapted from Maley and Brenac (1998a)).

Pollen-based biome reconstructions (Elenga et al. 2000) show that during the LGM, tropical rain-forest was replaced by tropical seasonal forest. These results are in agreement with those from marine records from the eastern Atlantic Ocean, which indicate a large

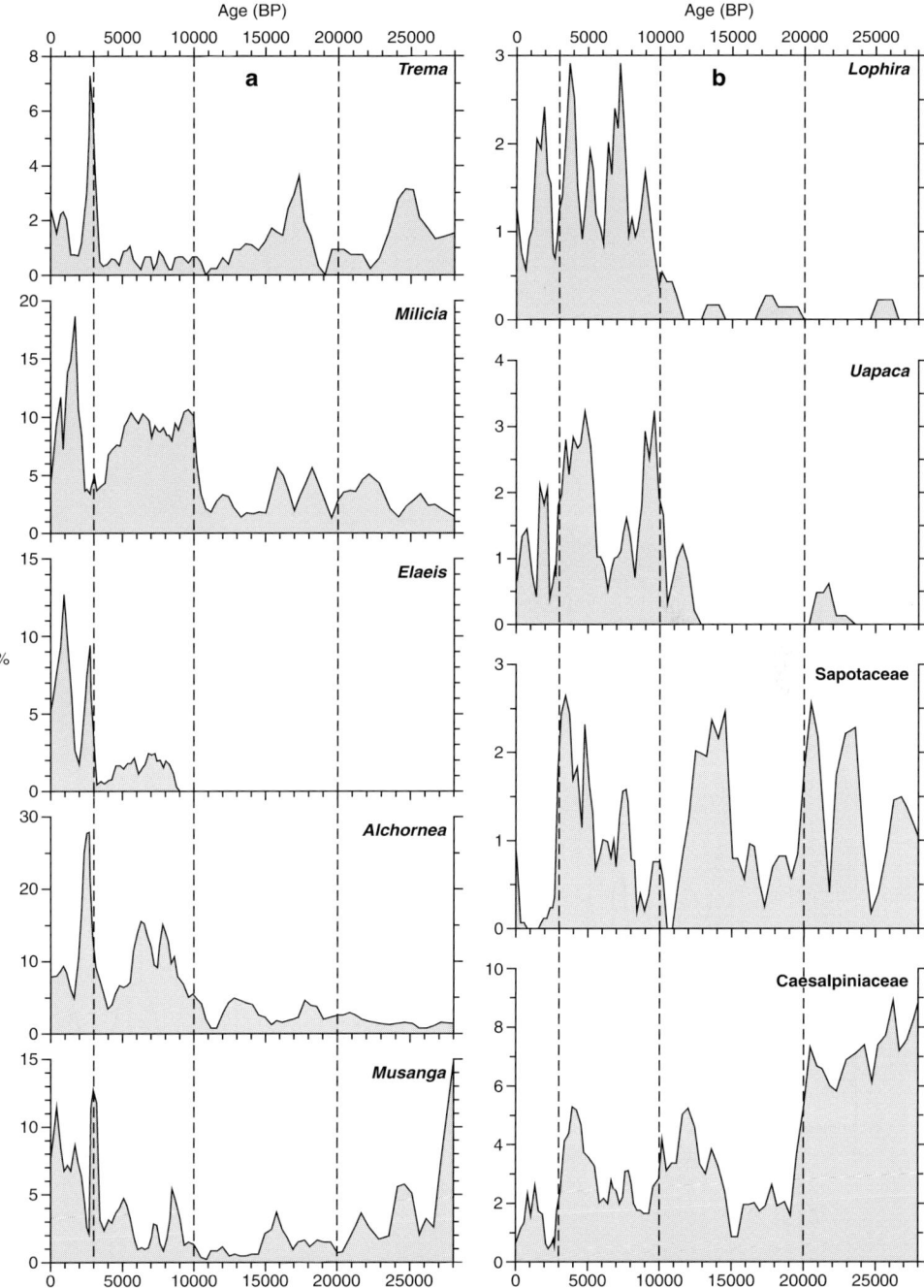

Figure 3. Lake Barombi Mbo, Cameroon (core BM-6): Pollen variations (%) of a) typical tree taxa from pioneers forests; b) typical tree taxa from mature forests (the pollen sum includes all taxa, excluding spores and damaged grains) (adapted from Maley and Brenac (1998a)).

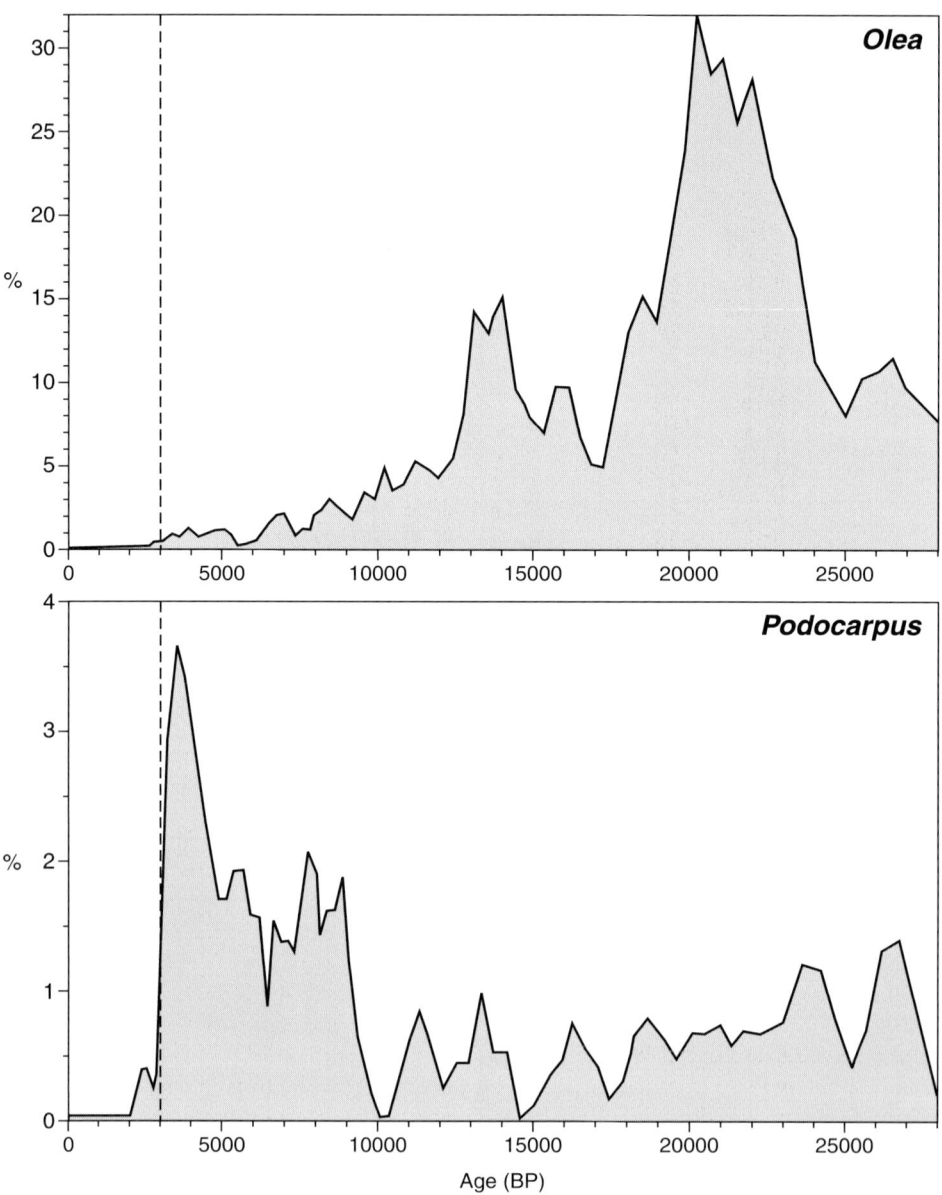

Figure 4. Lake Barombi Mbo, Cameroon (core BM-6): Pollen variations (%) of 2 typical afromontane forest trees, *Podocarpus cf. latifolius* and *Olea capensis* (the pollen sum includes all taxa, excluding spores and damaged grains) (adapted from Maley and Brenac (1998a)).

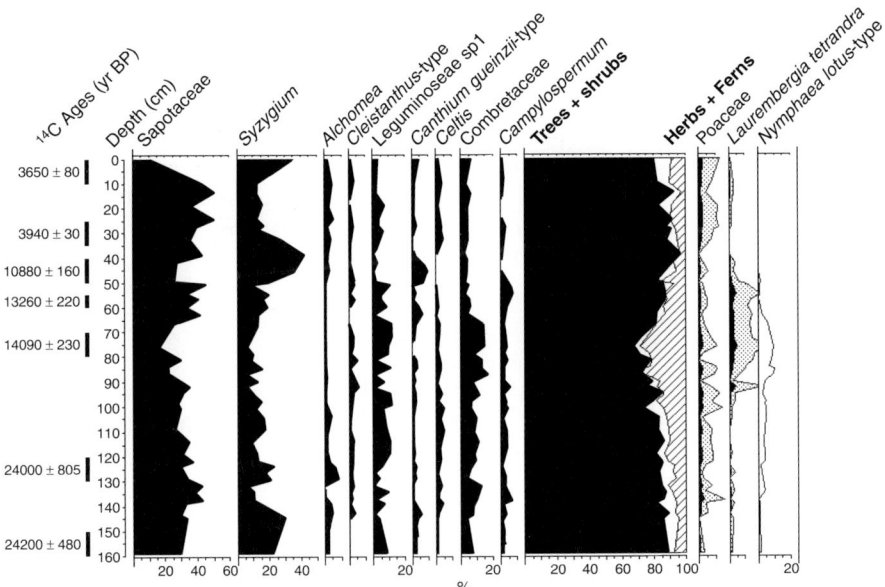

Figure 5. Ngamakala Pond, Congo (core Gama 4): synthetic pollen diagram (the pollen sum includes all taxa, excluding damaged grains) (adapted from Elenga et al. (1994)).

decline of rain-forest in West and Atlantic Equatorial Africa during the cold stages, but without complete disappearance (Bengo and Maley 1991; Lézine and Vergnaud-Grazzini 1993; Jahns 1996; Dupont et al. 1998; Jahns et al 1998; Marret et al. 1999).

The shift of Afro-montane taxa to lower altitudes during the LGM has been related to a pronounced decrease in sea-surface temperature (SST) in the Gulf of Guinea (Maley 1989; Maley and Elenga 1993). Marine investigations have shown that during this period large increases in the trade-winds induced first, strong coastal and equatorial upwellings and productivity (Jansen et al. 1984), and second, a large SST fall (Mix et al. 1986). This cooling slowed down the monsoon but greatly increased the low stratiform clouds which currently produce fogs, humidity and cool conditions on lowland hills and cliffs where Afro-montane taxa can survive (Maley and Elenga 1993). The humidity brought by these clouds is intercepted by the vegetation and then, by flowing down the hills, is made available for lower altitude mesophilous forests (e.g., as is observed at present on Mount Nimba, Ivory Coast (Schnell 1952)). These processes may have relevance for the functioning of forest refuges during the LGM (Maley 1987; 1989; 1996).

In inter-tropical African regions, the decrease in temperature required to produce the observed vegetation shift has been estimated to between 3 to 8 °C (Maley and Livingstone 1983; Maley 1987; 1989; 1991; 1996; Brenac 1988; Elenga et al. 1991; Vincens 1991; Bonnefille et al. 1992; Vincens et al. 1993). Nevertheless, low atmospheric CO_2 concentration during this period could also potentially have contributed to such a lowering of the montane tree-line and explain the occurrence of relatively cold-tolerant trees such as *Ilex* and *Olea* (Jolly and Haxeltine 1997).

However, a comparison of two rather close sites, at similar altitudes in Southern Congo, Ngamakala (site 3) and Bilanko (site 4, Fig. 1), shows a large development of mountain elements (*Ilex, Olea* and *Podocarpus*) in the sediments from Bilanko before ca. 11 kyr BP (Elenga et al. 1991), but their complete absence in those from Ngamakala (Elenga et al. 1994) dated from 24 kyr BP to the Middle Holocene (Fig. 5). Such a result would indicate that, at low latitudes near the Equator, the shift of these elements was not a widespread phenomenon but probably localised related to stratiform cloud development (Maley and Elenga 1993) (see above). This local character is reflected at the present day by the numerous isolated sites with *Olea* and *Podocarpus* across the regions close to the Gulf of Guinea (White 1978; 1983; Maley et al. 1990).

The Holocene

Except for the sequence from Lake Barombi Mbo (Figs. 2, 3 and 4), no well radiocarbon-dated evidence is available for the Glacial/Interglacial transition in the Atlantic equatorial lowland dense forest. Pollen data from this site indicate a first successional phase of pioneer trees at around 14 kyr BP, suggesting a trend towards humid conditions at this time (Maley and Brenac 1998a) (Fig. 3a). The greatest expansion of dense forest occurred during the interval 9.5–3 kyr BP. Around this site the extension of the forest was characterised by variations in the main tree taxa of mature (or "primary") forest with a mean periodicity of about 2000 to 2500 years (Maley and Brenac 1998a) (Fig. 3b). In the other sites, with a basal date of ca. 6–5 kyr BP, pollen data also show the occurrence of rain-forest taxa belonging to Caesalpiniaceae, Euphorbiaceae and Sapotaceae, even in areas presently covered by open grassland (Vincens et al. 1994; 1998; Elenga et al. 1996; Reynaud-Farrera et al. 1996) (Figs. 5 to 6). These data, supplemented by geomorphological and palaeobotanical studies carried out in the Congo and Cameroon (Kadomura et al. 1986; Deschamps et al. 1988), and by marine pollen records off the Gulf of Guinea (Caratini and Giresse 1979; Bengo and Maley 1991; Jahns 1996; Marret et al. 1999; Dupont et al. 2000), indicate a widespread expansion of dense forest during early and mid-Holocene in West and Atlantic equatorial Africa. In East Africa, high altitude pollen sequences show that the maximum forest cover is registered from ca. 9 kyr BP until ca. 4 kyr BP (Bonnefille and Riollet 1988; Taylor 1990; 1992; 1993; Bonnefille et al. 1995; Jolly et al. 1997; Marchant et al. 1997) when pollen-inferred precipitation was high (Bonnefille and Chalié 2000), and this compares well with the results from the Atlantic equatorial lowland area described above. All these data indicate a strong increase in precipitation over Intertropical Africa during the early Holocene, which is corroborated by a synchronous rise of African lake levels at this time (Street and Grove 1979; Johnson 1996).

From ca. 4 kyr BP, an increase in typical semi-deciduous and heliophilous forest taxa such as *Celtis, Alchornea* and *Macaranga* is registered in the southern Congolese sector of the Central African lowland (Sinnda: Vincens et al. (1994, 1998), Kitina: Elenga et al. (1996)) (Figs. 7 and 8). At the same time, mineralogical studies carried out at these two sites indicate a decrease in quartz and kaolinite flux and an increase in calcite and talc, interpreted as a progressive reduction of rainfall (Elenga et al. 1996; Bertaux et al. 2000). In the northern part of the forest domain, Barombi Mbo and Ossa in Southern Cameroon, the phenomenon was different and characterised by an increase in the evergreen rain-forest taxa, particularly those belonging to the Caesalpiniaceae, with a maximum between 4 and

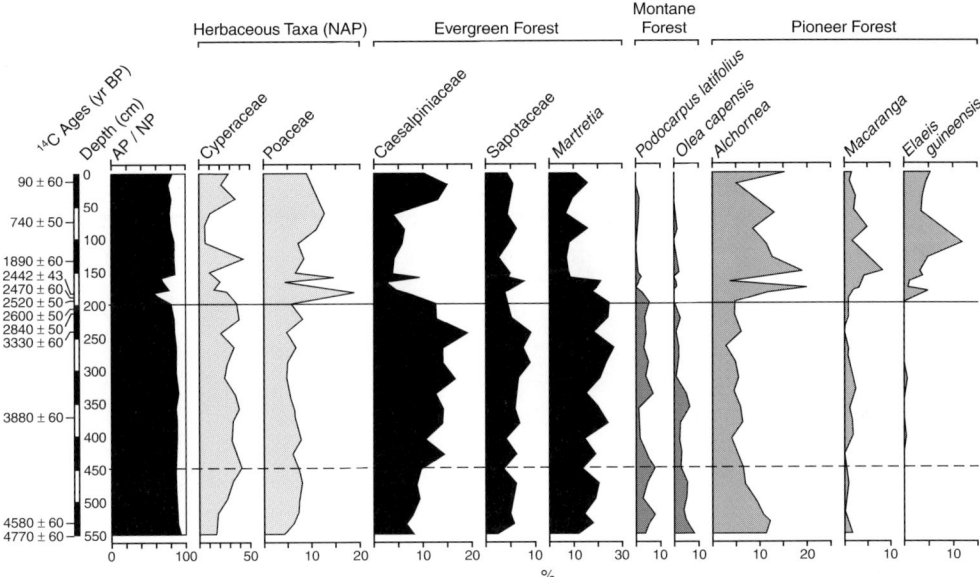

Figure 6. Lake Ossa, Cameroon (core OW-4): synthetic pollen diagram (the pollen sum includes all taxa, excluding spores and damaged grains) (adapted from Reynaud-Farrera et al. (1996), van Geel et al. (1998)).

3 kyr BP (Reynaud-Farrera et al. 1996; Maley 1997; Maley and Brenac 1998a) (Figs. 3b and 6). Moreover, at Ossa, diatom analysis indicates that the lake remained relatively high until ca. 3 kyr BP (Nguetsop 1997; Nguetsop et al. 1998), whereas, in the southern part of the forest domain, lake Sinnda completely dried up after ca. 3.8 kyr BP (Vincens et al. 1994; 1998) (Fig. 7), in phase with low lake levels registered in many parts of tropical Africa (e.g., Street and Grove (1979), Servant and Servant-Vildary (1980), Talbot et al. (1984)).

This pattern of change is linked to long-term and seasonal monsoon variations which were induced by SST variations (Fontaine and Bigot 1993; Fontaine et al. 1998). The earliest phase, between about 4 and 2.8 kyr BP, was associated with relatively low SSTs, sharply reduced from the early and mid Holocene (Morley and Dworetzky 1993). The low SST produced mainly stratiform clouds. In Western Congo, these clouds probably evolved towards non-rain clouds. Because the present day four-month dry season here is characterised by the quasi-permanence of these cloud types (Saint Vil 1984), one can infer that this dry season began to dominate during this period. However, during the same period, the development of the evergreen forest over West and South Cameroon can be explained by the evolution of the stratiform clouds towards rainy nimbostratus types (Maley 1997; Maley et al. 2000; Maley 2001).

The major Holocene vegetation change registered in Atlantic equatorial Africa, is evidenced between ca. 3 and 2 kyr BP. In the humid forested sites at present (Barombi Mbo, Kitina and Ossa), this change occurred between 2.7 and 2.5 kyr BP, when Poaceae (ex. Gramineae) and heliophilous pioneer taxa abruptly increase, indicating a rapid phase of rain-forest regression, with local extension of savannas or of other open formations (Elenga et al. 1996; Reynaud-Farrera et al. 1996; Maley and Brenac 1998a; van Geel et al. 1998) (Figs. 2, 8 and 6). At Sinnda, presently the driest site, this event was not registered as a

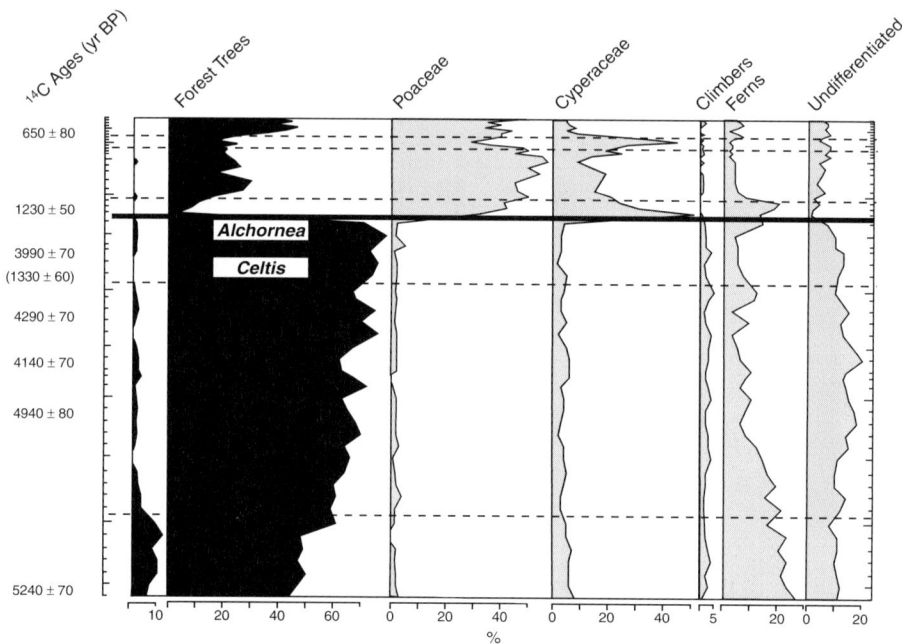

Figure 7. Lake Sinnda, Congo (core SN-2): synthetic pollen diagram (the pollen sum includes all taxa, excluding damaged grains) (adapted from Vincens et al. (1998)).

complete drying up of the lake at this time (Vincens et al. 1998; Bertaux et al. 2000). This phase of forest regression seems to be general in the Atlantic equatorial forest domain, because its repeated occurrence at every studied site. So, it has been interpreted as a deterioration of climate towards arid conditions rather than an anthropogenic impact on vegetation. This arid episode is also evidenced by the lowering of lake Ossa (Nguetsop 1997) and by sedimentological studies of fluviatile or soil deposits, showing the presence of very coarse strata dated between 3 and 2 kyr BP (Maley and Brenac 1998b; Maley and Giresse 1998; Maley 2001; 2002).

However, the period of open vegetation was relatively short because at the same time a strong increase in colonising pioneer forest taxa also occurred (Figs. 3a, 8 and 6) (Elenga et al. 1996; Reynaud-Farrera et al. 1996; Maley and Brenac 1998a; van Geel et al. 1998). Moreover, this event had a direct impact not only on the structure and distribution of forests in Atlantic equatorial Africa, but also on regional human populations. Indeed, the large openings of the rain-forest domain ca. 3–2 kyr BP probably facilitated and may have triggered the major Bantu migration from areas presently north of the forest, to those now south of it, from Cameroon, through Gabon then Congo (Schwartz 1992). This late Holocene climatic deterioration, also registered at many East African sites (e.g., Hamilton (1982), Vincens (1989, 1993), Taylor (1990, 1992), Jolly and Bonnefille (1992)), appears to be a global event, related to solar forcing evidenced by a sudden and sharp rise in the atmospheric ^{14}C content (van Geel et al. 1998; 1999; Renssen et al. 2000). Over Central Africa, this abrupt climatic deterioration appears to be linked to an increase in the dry

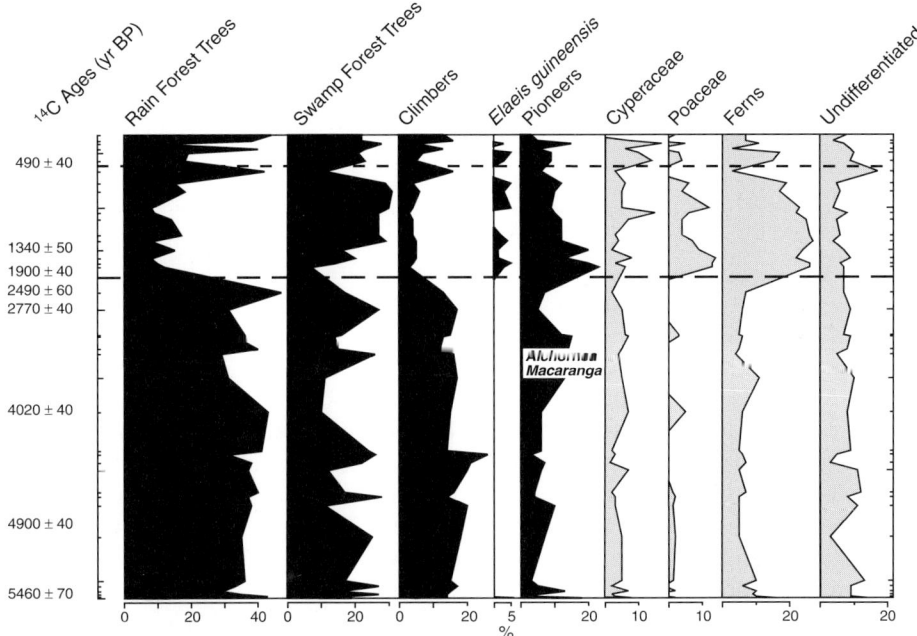

Figure 8. Lake Kitina, Congo (core KT-3): synthetic pollen diagram (the pollen sum includes all taxa, excluding damaged grains) (adapted from Elenga et al. (1996)).

season caused by the strengthening of the harmattan. Indeed at this time the sediments of lake Ossa registered maximum influx of allochthonous diatoms brought by the dust and dry haze originating from the central part of the Chad basin (Nguetsop et al. 1998).

During the last 2 kyr BP, the return of wetter conditions was favourable to a new expansion of forests in Atlantic equatorial Africa. However, the pattern of regeneration was not synchronous from one site to another. In contrast to a forest regressive phase which can happen very quickly, a forest recolonisation is a longer process characterised, first, by a pioneer phase with a "secondary" type of vegetation, and then, a mature phase with several types of "primary" vegetation (Oldeman 1983; Pascal 1995). The delay between the destruction and the restoration has been referred to as an hysteresis (Ritchie 1986). For these reasons, the regeneration of primary forest, was not synchronous at the different sites studied, and probably also reflects the distribution of the residual forest during the maximum destructive phase (Maley 2001). But, if the process of forest restoration was variable in time, it was active during the last 2 millennia at all the sites, and was clearly linked to a return of more humid conditions as confirmed by the refilling of Lake Sinnda from 1.3 kyr BP onwards (Vincens et al. 1998) (Fig. 7). However, several short drier oscillations occurred during the last millennium, particularly during the interval spanned by the Little Ice Age, around 0.6 kyr (Vincens et al. 1998). During the long phase of forest restoration, the Bantu people have exploited the oil palm (*Elaeis guineensis*), as evidenced by the large occurrence of its pollen grains (Figs. 3a, 8 and 6) and nuts in the sediments (Richards 1986; Maley 1999; Vincens et al. 1999). However, in the traditional African societies, this palm tree is never planted and it grows first as a pioneer tree; its exploitation takes place later when the tree is mature (Maley and Chepstow-Lusty 2001).

The process of forest expansion was continuous during the 20th century, as evidenced by the recent progression of forest ecosystems on savannas which occurs both in South Cameroon and Congo, despite an increase in human activity (Youta Happi 1998; Youta Happi et al. 2000; Achoundong et al. 2000; Guillet et al. 2000; 2001; Schwartz et al. 2000).

Conclusions

A synthesis of the history of forest and related climate during the Late Pleistocene and Holocene in Atlantic equatorial Africa is proposed in Table 2. Although forest formations may differ regionally, this synthesis shows that, in spite of the scarcity of the pollen data, the dynamics of the Atlantic equatorial African lowland forest ecosystem appear to respond to global climatic changes. During the LGM, cool temperatures allowed the migration of montane elements to low altitudes, but their co-existence with lowland rain-forest taxa indicates the persistence of locally humid conditions. The early and mid-Holocene was characterised by a general widespread distribution of rain-forest formations in relation to a large increase in precipitation. A relatively short phase of forest retreat and vegetation opening occurred between 3 and 2 kyr BP linked to an arid event. Despite human activities, during the last 2 millennia a general dynamism of the forest ecosystem has caused a new expansion of the forest domain. These observations show clearly that climate has been the most important factor in environmental change.

Table 2. Summary of former climates and vegetation fluctuations during the last 28 kyr BP in Atlantic Equatorial Africa.

Ages kyrs BP	Climate	Vegetation
0 (Present day)	Wet and warm	Forest extension, in progress
2/0	Wetter and warmer/cooler	Forest extension
3/2	Dry and cooler/warmer	Abrupt forest opening, savanna extension and phases of pioneer development
4.5/3	Wetter (north) and cooler	In South Cameroun, increase of Caesalpiniaceae In upland, extension of montane forest (*Podocarpus*)
9.5/4.5	Dryer (south) and cooler Humid and warm	In South Congo, increase of semi-deciduous taxa Maximum forest extension
12/10	Transition (warmer and wetter)	Forest recolonization with large oscillations
20/12	Colder and dryer (temperature ca. 3–4 °C lower)	Extension of montane grassland and savanna in lowland Treeline ca. 1000 m lower Montane taxa in lowland forest Lowland forest refugia
30/20	Relatively moist Temperature relatively lower	Lowland forest with more Caesalpiniaceae Montane taxa in lowland forest

Summary

A review of Late Pleistocene and Holocene data derived from several pollen sequences recovered in Atlantic equatorial African sediments between 5 °S and 5 °N, and 9 °E and 15 °E, show similarities in vegetation change and climatic variation between the various localities studied. The local persistence of rain-forest in the area, even at the LGM is recorded, suggesting the occurrence of glacial forested refuges. During this period, the presence of some afro-montane taxa such as *Ilex*, *Olea* and *Podocarpus* indicates the existence of localised areas with a cool, but humid, climate. More humid conditions during the early and mid-Holocene are evidenced by the expansion of dense forest. Drier conditions ca. 3–2 kyr BP produced large changes in floristic composition, structure and geographical distribution of the vegetation. A new expansion of pioneer forest began post 2 kyr BP, with locally more humid forest formations since ca. 900–600 yrs BP, in spite of an increase in human activity.

Acknowledgments

The research presented in this paper was undertaken with the support from several programmes mainly: GEOCIT (Géodynamique du Climat Intertropical; IRD [ex. ORSTOM] and NSF); ECOFIT (Ecosystèmes et Paléoécosystèmes des Forêts Intertropicales; IRD, CNRS and CEA). Some of the pollen data are presently stored and available in the African Pollen Database (APD) managed by the INCO CEE project and the UNESCO PICG 431. This paper is Institut des Sciences de l'Evolution de Montpellier (ISEM/CNRS UMR-5554) contribution 2004-025.

References

Achoundong G., Youta Happi J., Guillet B., Bonvallot J. and Kamgang Beyala V. 2000. Formations et évolutions des recrûs sur savanes. In: Servant M. and Servant-Vildary S. (eds), Dynamique à Long Terme des Écosystèmes Forestiers Intertropicaux. UNESCO, Paris, pp. 31–41.

Bengo M.D. and Maley J. 1991. Analyse des flux polliniques sur la marge sud du Golfe de Guinée depuis 135.000 ans. C.R. Acad. Sci. Paris, sér. 2, 313: 843–849.

Bertaux J., Schwartz D., Vincens A., Sifeddine A., Elenga H., Mansour M., Mariotti A., Fournier M., Martin L., Wirrmann D. and Servant M. 2000. Enregistrement de la phase sèche d'Afrique Centrale vers 3000 ans BP par la spectrométrie IR dans les lacs Sinnda et Kitina (Sud-Congo). In: Servant M. and Servant-Vildary S. (eds), Dynamique à Long Terme des Écosystèmes Forestiers Intertropicaux. UNESCO, Paris, pp. 43–49.

Bonnefille R. and Riollet G. 1988. The Kashiru pollen sequence (Burundi): palaeoclimatic implications for the last 40,000 yr BP in tropical Africa. Quat. Res. 30: 19–35.

Bonnefille R., Chalié F., Guiot J. and Vincens A. 1992. Quantitative estimates of full glacial temperatures in equatorial Africa from palynological data. Clim. Dyn. 6: 251–257.

Bonnefille R., Riollet G., Buchet G., Icole M., Lafont R., Arnold M. and Jolly D. 1995. Glacial/interglacial record from tropical Africa, high resolution pollen and carbon data at Rusaka, Burundi. Quat. Sci. Rev. 14: 917–936.

Bonnefille R. and Chalié F. 2000. Pollen-inferred precipitation time-series from equatorial mountains, Africa, the last 40 kyr BP. Glob. Planet. Chan. 26: 25–50.

Brenac P. 1988. Evolution de la végétation et du climat dans l'Ouest Cameroun entre 25.000 et 11.000 ans BP. Actes X° Symposium Ass. Palynologues Langue Française, Trav. Sect. Sci. et Techn. Inst. Français Pondichéry 25: 91–103.

Caratini C. and Giresse P. 1979. Contribution palynologique à la connaissance des environnements continentaux et marins du Congo à la fin du Quaternaire. C.R. Acad. Sci. Paris, sér. D, 288: 379–382.

Deschamps R., Guillet B. and Schwartz D. 1988. Découverte d'une flore forestière Mi-Holocène (5800–3100) conservée in situ sur le littoral Ponténégrin (R.P. du Congo). C.R. Acad. Sc. Paris, sér. 2, 306: 615–618.

Dupont L., Marret F. and Winn K. 1998. Land-sea correlation by means of terrestrial and marine palynomorphs from the equatorial East Atlantic: phasing of SE trade wind and the oceanic productivity. Palaeogeogr. Palaeoclimatol. Palaeoecol. 142: 51–84.

Dupont L., Jahns S., Marret F. and Ning S. 2000. Vegetation change in equatorial West Africa: time-slices for the last 150 ka. Palaeogeogr. Palaeoclimatol. Palaeoecol. 155: 95–122.

Elenga H., Vincens A. and Schwartz D. 1991. Présence d'éléments forestiers montagnards sur les Plateaux Batéké (Congo) au Pléistocène supérieur: nouvelles données palynologiques. Palaeocology of Africa 22: 239–252.

Elenga H., Schwartz D. and Vincens A. 1994. Pollen evidence of Late Quaternary vegetation and inferred climate changes in Congo. Palaeogeogr. Palaoeclimatol. Palaeoecol. 109: 345–356.

Elenga H., Schwartz D., Vincens A., Bertaux J., de Namur C., Martin L., Wirrmann D. and Servant M. 1996. Diagramme pollinique holocène du lac Kitina (Congo): mise en évidence de changements paléobotaniques et paléoclimatiques dans le massif forestier du Mayombe. C.R. Acad. Sci. Paris, sér. 2a, 232: 403–410.

Elenga H., Peyron O., Bonnefille R., Jolly D., Cheddadi R., Guiot J., Andrieu V., Bottema S., Buchet G., de Beaulieu J.-L., Hamilton A.C., Maley J., Marchant R., Perez-Obiol R., Reille M., Riollet G., Scott L., Straka H., Taylor D., Van Campo E., Vincens A. and Jonson H. 2000. Pollen-based biome reconstruction for Southern Europe and Africa 18,000 yrs BP. J. Biogeogr. 27: 621–634.

Farrera I., Harrison S.P., Prentice I.C., Ramstein G., Guiot J., Bartlein P.J., Bonnefille R., Bush M., Cramer W., von Grafenstein U., Holmgren K., Hooghiemstra H., Hope G., Jolly D., Lauritzen S.E., Ono Y., Pinot S., Stute M. and Yu G. 1999. Tropical climates at the Last Glacial Maximum: a new synthesis of terrestrial palaeoclimate data. I. Vegetation, lake-levels and geochemistry. Clim. Dyn. 15: 823–856.

Fontaine R. and Bigot S. 1993. West-African rainfall deficits and sea surface temperatures. Int. J. Climatol. 13: 271–286.

Fontaine B., Trzaska S. and Janicot S. 1998. Evolution of the relationship between near global and Atlantic SST modes and the rainy season in West Africa: statistical analyses and sensitivity experiments. Clim. Dyn. 14: 353–368.

Giresse P., Maley J. and Brenac P. 1994. Late Quaternary palaeoenvironments in the Lake Barombi Mbo (West Cameroon) deduced from pollen and carbon isotopes of organic matter. Palaeogeogr. Palaeoclimatol. Palaeoecol. 107: 65–78.

Guillet B., Maman O., Achoundong G., Mariotti A., Girardin T., Schwartz D. and Youta Happi J. 2000. Essai d'interprétation de la dynamique de la mosaïque forestière dans la zone de contact forêt-savane du Sud-Est Cameroun. In: Servant M. and Servant-Vildary S. (eds), Dynamique à Long Terme des Écosystèmes Forestiers Intertropicaux. UNESCO, Paris, pp. 169–174.

Guillet B., Achoundong G., Youta Happi J., Kamgang Kabeyene V., Bonvallot J., Riera B., Mariotti A. and Schwartz D. 2001. Agreement between floristic and soil organic carbon isotope ($^{13}C/^{12}C$, ^{14}C) indicators of forest invasion of savannas during the last century in Cameroon. J. Trop. Ecol. 17: 809–832.

Hamilton A.C. 1982. Environmental History of East Africa — A Study of Quaternary. Academic Press, London, 311 pp.

Jahns S. 1996. Vegetation history and climatic changes in West Equatorial Africa during the Late Pleistocene and the Holocene based on a marine pollen diagram from the Congo fan. Veget. Hist. and Archaeobot. 5: 207–213.

Jahns S., Hüzls M. and Sarnthein M. 1998. Vegetation and climate history of west equatorial Africa based on a marine pollen record off Liberia (site GIK 16776) covering the last 400,000 years. Rev. Palaeobot. Palynol. 102: 277–288.

Jansen J.H.F., Van Weering T.C.E., Gieles R. and Van Iperen J. 1984. Middle and Late Quaternary oceanography and climatology of the Zaïre-Congo fan and the adjacent Eastern Angola basin. Neth. J. Sea Res. 17: 201–249.

Johnson T.C. 1996. Sedimentary processes and signals of past climatic change in the large lakes of East African Rift Valley. In: Johnson T.C. and Odada E.O. (eds), The Limnology, Climatology and Paleoclimatology of the East African Lakes. Gordon and Breach Publishing, Amsterdam, pp. 367–412.

Jolly D. and Bonnefille R. 1992. Histoire et dynamique du marécage tropical de Ndurumu (Burundi), données polliniques. Rev. Palaeobot. Palynol. 75: 133–151.

Jolly D. and Haxeltine A. 1997. Effect of low glacial atmospheric CO_2 on tropical African montane vegetation. Science 276: 786–788.

Jolly D., Taylor D., Marchant R., Hamilton A., Bonnefille R., Buchet G. and Riollet G. 1997. Vegetation dynamics in Central Africa during the late glacial and Holocene periods: pollen records from the interlacustrine highlands of Burundi, Rwanda and Western Uganda. J. Biogeogr. 24: 495–512.

Kadomura H., Hori N., Kuete M., Tamura T., Omi G., Haruki M. and Chujo H. 1986. Late Quaternary environmental changes in Southern Cameroon: a synthesis. In: Kadomura H. (ed.), Geomorphology and Environmental Changes in Tropical Africa: Case Studies in Cameroon and Kenya. Hokkaido University, Sapporo, pp. 145–158.

Lanfranchi R. and Schwartz D. (eds) 1990. Paysages Quaternaires de l'Afrique Centrale Atlantique. Didactiques, ORSTOM, Paris, 535 pp.

Lézine A.-M. and Vergnaud-Grazzini C. 1993. Evidence of forest extension in West Africa since 22000 BP: A pollen record from eastern tropical Atlantic. Quat. Sci. Rev. 12: 203–210.

Maley J. and Livingstone D.A. 1983. Extension d'un élément montagnard dans le sud du Ghana (Afrique de l'Ouest) au Pléistocène Supérieur et à l'Holocène Inférieur: premières données polliniques. C.R. Acad. Sci. Paris, Sér. 2, 196: 1287–1292.

Maley J. 1987. Fragmentation de la forêt dense humide africaine et extension de biotopes montagnards au Quaternaire récent: nouvelles données polliniques et chronologiques. Implications paléoclimatiques et biogéographiques. Palaeocology of Africa 18: 307–334.

Maley J. 1989. Late Quaternary climatic changes in the African rain-forest: the question of forest refuges and the major role of sea surface temperature variations. In: Leinen M. and Sarnthein M. (eds), Palaeoclimatology and Palaeometeorology: Modern and Past Patterns of Global Atmospheric Transport. NATO Adv. Sci. Inst., Ser. C, Math. Phys. Sci. 282, pp. 585–616.

Maley J., Caballé G. and Sita P. 1990. Etude d'un peuplement résiduel à basse altitude de *Podocarpus latifolius* sur le flanc Congolais du Massif du Chaillu. Implications paléoclimatiques et biogéographiques. Etude de la pluie pollinique actuelle. In: Lanfranchi R. and Schwartz D. (eds), Paysages Quaternaires de l'Afrique Centrale Atlantique. Didactiques, ORSTOM, Paris, pp. 336–352.

Maley J. 1991. The African rain-forest vegetation and palaeoenvironment during the Late Quaternary. Climatic Change 19: 79–88.

Maley J. and Elenga H. 1993. Le rôle des nuages dans l'évolution des paléoenvironnements montagnards de l'Afrique Tropicale. Veille Climatique Satellitaire 46: 51–63.

Maley J. 1996. The African rain-forest: main characteristics of changes in vegetation and climate from the upper Cretaceous to the Quaternary. Proceed. R. Soc. Edinburg, Biol. Sc.104B: 31–73.

Maley J. 1997. Middle to Late Holocene changes in tropical Africa and other continents. Paleomonsoon and sea surface temperature variations. In: Dalfes H.N. et al. (eds), Third Millenium BC Climate Change and Old World Collapse. NATO ASI Series, Global Environmental Change, Springer-Verlag, Berlin, pp. 611–640.

Maley J. and Brenac P. 1998a. Vegetation dynamics, palaeoenvironments and climatic changes in the forest of Western Cameroon during the last 28,000 years BP. Rev. Palaeobot. Palynol. 99: 157–178.

Maley J. and Brenac P. 1998b. Les variations de la végétation et des paléoenvironnements du sud Cameroun au cours des derniers millénaires. Etude de l'expansion du Palmier à huile. In: Bilong P. and Vicat J.P. (eds), Géosciences au Cameroun. Publication occas. GEOCAM n°1, Presses Universitaires de Yaoundé, Cameroun, pp. 85–87.

Maley J. and Giresse P. 1998. Etude d'un niveau argileux organique du Mayombe (Congo occidental) riche en pollens d'*Elaeis guineensis* et daté d'environ 2800 ans BP. Implications pour les paléoenvironnements de l'Afrique Centrale. In: Bilong P. and Vicat J.P. (eds), Géosciences au Cameroun. Publication occas. GEOCAM n°1, Presses Universitaires de Yaoundé, Cameroun, pp. 77–84.

Maley J. 1999. L'expansion du palmier à huile (*Elaeis guineensis*) en Afrique centrale au cours des trois derniers millénaires: nouvelles données et interprétations. In: Bahuchet S. et al. (eds), L'homme et la Forêt Tropicale. Trav. Soc. Ecol. Humaine, Marseille, pp. 237–254.

Maley J., Brenac P., Bigot S. and Moron V. 2000. Variations de la végétation et des paléoenvironnements en forêt dense africaine au cours de l'Holocène. Impact de la variation des températures marines. In: Servant M. and Servant-Vildary S. (eds), Dynamique à Long Terme des Écosystèmes Forestiers Intertropicaux. UNESCO, Paris, pp. 205–220.

Maley J. 2001. La destruction catastrophique des forêts d'Afrique centrale survenue il y a environ 2500 ans exerce encore une influence majeure sur la répartition actuelle des formations végétales. In: Robbrecht E., Degreef J. and Friis I. (eds), Plant Systematics and Phytogeography for the Understanding of African Biodiversity. Syst. Geogr. Plantes, Meise, 71: 777–796.

Maley J. and Chepstow-Lusty A. 2001. *Elaeis guineensis* Jacq. (oil palm) fluctuations in Central Africa during the late Holocene: climate or human driving forces for this pioneering species? Veg. Hist. and Archaeobot 10: 117–120.

Maley J. 2002. A catastrophic destruction of African forests about 2,500 years ago still exerts a major influence on present vegetation formations. Brighton University, Bull. Inst. Develop. Studies 33: 13–30.

Marchant R., Taylor D. and Hamilton A. 1997. Late Pleistocene and Holocene history at Mubwindi Swamp, Southwest Uganda. Quat. Res. 47: 316–328.

Marret F., Scourse J., Jansen J.H.F. and Schneider R. 1999. Changements climatiques et paléocéanographiques en Afrique centrale atlantique au cours de la dernière déglaciation: contribution palynologique. C.R. Acad. Sci. Paris, sér. 2, 329: 721–726.

Mix A.C., Ruddiman W.F. and McIntyre A. 1986. Late Quaternary paleoceanography of the tropical Atlantic. 2: The seasonal cycle of sea surface temperatures, 0–20,000 years BP. Paleoceanography 1: 339–353.

Morley J.J. and Dworetzky B.A. 1993. Holocene temperature patterns in the South Atlantic, Southern, and Pacific Oceans. In: Wright H.E. et al. (eds), Global Climates since the Last Glacial Maximum. University of Minnesota Press, pp. 125–135.

Nguetsop V.F. 1997. Evolution des environnements de l'Ouest Cameroun depuis 6000 ans d'après l'étude des diatomées actuelles et fossiles dans le lac Ossa. Implications paléoclimatiques. Unpublished Thesis, Mus. Nat. Hist. Naturelle, Paris, 277 pp.

Nguetsop V.F., Servant M. and Servant-Vildary S. 1998. Paléolimnologie et paléoclimatologie de l'Ouest-Cameroun au cours des 5000 dernières années, à partir de l'étude des diatomées du lac Ossa. C.R. Acad. Sci. Paris, Sér. 2, 327: 39–45.

Oldeman R.A. 1983. Tropical rain-forest, architecture, silvigenesis and diversity. In: Sutton S.L. et al. (eds), Tropical Rain-Forest: Ecology and Management. Blackwell, Oxford, pp. 139–150.

Parmentier I. and Maley J. 2001. L'arbre ou le pigeon, ou le pigeon et l'arbre? Deux arbres afromontagnards recensés sur des inselbergs en Guinée Equatoriale. Canopée, J. Prog. Européen ECOFAC 19: 6–9.

Pascal J.P. 1995. Quelques exemples de problèmes posés à l'analyste et au modélisateur par la complexité de la forêt tropicale humide. Revue Ecologie, Terre et Vie 50: 237–249.

Renssen H., van Geel B., van der Plicht J. and Magny M. 2000. Reduced solar activity as a trigger for the start of the Younger Dryas? Quat. Int. 68–71: 373–383.

Reynaud-Farrera I. 1995. Histoire des paléoenvironnements forestiers du Sud-Cameroun à partir d'analyses palynologiques et statistiques de dépôts holocènes et actuels. Unpublished Thesis, University of Montpellier-2, 198 pp.

Reynaud-Farrera I., Maley J. and Wirrmann D. 1996. Végétation et climat dans les forêts du sud-ouest Cameroun depuis 4770 ans BP. Analyse pollinique des sédiments du lac Ossa. C.R. Acad. Sci. Paris, Sér. 2a, 322: 749–755.

Richards K. 1986. Preliminary results of pollen analysis of a 6000 year core from Mboandong, a crater lake in Cameroon. Hull Univ. Geogr. Dept., Misc. Ser. 32: 14–28.

Ritchie J.C. 1986. Climate change and vegetation response. Vegetatio 67: 65–74.

Saint-Vil J. 1984. La grande saison sèche au Gabon. Ann. Univ. Nat. Gabon 5: 107–119.

Schnell R. 1952. Végétation et flore des monts Nimba. Vegetatio 3: 350–406.

Schwartz D. 1992. Assèchement climatique vers 3000 BP et expansion Bantu en Afrique centrale atlantique: quelques réflexions. Bull. Soc. géol. France 163: 353–361.

Schwartz D., Elenga H., Vincens A., Bertaux J., Mariotti A., Achoundong G., Alexandre A., Belingard C., Girardin C., Guillet B., Maley J., de Namur C., Reynaud-Farrera I. and Youta Happi J. 2000. Origine et évolution des savanes des marges forestières en Afrique Centrale Atlantique (Cameroun, Gabon, Congo). Approches aux échelles millénaires et séculaires. In: Servant M. and Servant-Vildary S. (eds), Dynamique à Long Terme des Écosystèmes Forestiers Intertropicaux. UNESCO, Paris, pp. 325–338.

Servant M. and Servant-Vildary S. 1980. L'environnement quaternaire du bassin du Tchad. In: Williams M.A.J. and Faure H. (eds), The Sahara and the Nile. Balkema, Rotterdam, pp. 133–162.

Servant M. and Servant-Vildary S. (eds) 2000. Dynamique à Long Terme des Écosystèmes Forestiers Intertropicaux. UNESCO, Paris, 434 pp.

Street F.A. and Grove A.T. 1979. Global maps of lake fluctuations since 30000 BP. Quat. Res. 261: 385–390.

Talbot M.R., Livingstone D.A., Palmer P.G., Maley J., Melack J.M., Delibrias G. and Gulliksen S. 1984. Preliminary results from sediment cores from lake Bosumtwi, Ghana. Palaeoecology of Africa 16: 173–192.

Taylor D.A. 1990. Late Quaternary pollen records from two Uganda mires: evidence for environmental change in the Rukiga highlands of Southwest Uganda. Palaeogeogr. Palaeoclimatol. Palaeoecol. 80: 283–300.

Taylor D.A. 1992. Pollen evidence from Muchoya Swamp, Rukiga Highlands (Uganda), for abrupt changes in vegetation during the last ca. 21,000 years. Bull. Soc. géol. France 163: 77–82.

Taylor D.A. 1993. Environmental change in montane Southwest Uganda: a present record for the Holocene from Ahakagyezi Swamp. The Holocene 3/4: 324–332.

Van Geel B., Van Der Plicht J., Kilian M.R., Klaver E.R., Kouwenberg J.H.M., Renssen H., Reynaud-Farrera I. and Waterbolk H.T. 1998. The sharp rise of δ^{14}C ca. 800 cal BC: possible causes, related climatic teleconnections and the impact of human environments. Radiocarbon 40, 1: 535–550.

Van Geel B., Rapopov O.M., Renssen H., van der Plicht J., Dergachev V.A. and Meijer H.A.J. 1999. The role of solar forcing upon climate change. Quat. Sci. Rev. 18: 331–338.

Vincens A. 1989. Paléoenvironnements du basin Nord-Tanganyika (Zaïre, Burundi, Tanzanie) au cours des 13 derniers mille ans: apport de la palynologie. Rev. Palaeobot. Palynol. 61: 69–88.

Vincens A. 1991. Late Quaternary vegetation history of the South Tanganyika basin. Climatic implication in South Central Africa. Palaeogeogr. Palaeoclimatol. Palaeoecol. 86: 207–226.

Vincens A. 1993. Nouvelle séquence pollinique du Lac Tanganyika: 30,000 ans d'histoire botanique et climatique du bassin Nord. Rev. Palaeobot. Palynol. 78: 381–394.

Vincens A., Chalié F., Bonnefille R., Guiot J. and Tiercelin J.J. 1993. Pollen-derived rainfall and temperature estimates from Lake Tanganyika and their implication for Late Pleistocene water levels. Quat. Res. 40: 343–350 and Erratum 1994, 41: 253–254.

Vincens A., Buchet G., Elenga H., Fournier M., Martin L., de Namur C., Schwartz D., Servant M. and Wirrmann D. 1994. Changement majeur de la végétation du lac Sinnda (Vallée du Niari, Sud-Congo) consécutive à l'assèchement climatique Holocène supérieur: apport de la palynologie. C.R. Acad. Sci. Paris, sér. 2, 318: 1521–1526.

Vincens A., Schwartz D., Bertaux J., Elenga H. and de Namur C. 1998. Late Holocene climatic changes in Western Equatorial Africa inferred from pollen from lake Sinnda, Southern Congo. Quat. Res. 50: 34–45.

Vincens A., Schwartz D., Elenga H., Reynaud-Farrera I., Alexandre A., Bertaux J., Mariotti A., Martin L., Meunier J.-D., Nguetsop F., Servant M., Servant-Vildary S. and Wirrmann D. 1999. Forest response to climate changes in Atlantic Equatorial Africa during the last 4000 years BP and inheritance on the modern landscapes. J. Biogeogr. 26: 879–885.

White F. 1978. The afromontane region. In: Werger M.J.A. and Van Bruggen A.C. (eds), Biogeography and Ecology of Southern Africa. Junk, The Hague, pp. 463–513.

White F. 1983. The Vegetation of Africa. UNESCO/AETFAT/UNSO, Maps and Memoir, UNESCO, Paris, 356 pp.

Youta Happi J. 1998. Arbres contre graminées: la lente invasion de la savane par la forêt au Centre-Cameroun. Unpublished Thesis, Univ. Paris IV, 237 pp.

Youta Happi J., Hotyat M. and Bonvallot J. 2000. La colonisation des savanes par la forêt à l'Est du Cameroun. In: Servant M. and Servant-Vildary S. (eds), Dynamique à Long Terme des Écosystèmes Forestiers Intertropicaux. UNESCO, Paris, pp. 423–428.

11. ASPECTS OF NIGERIAN COASTAL VEGETATION IN THE HOLOCENE: SOME RECENT INSIGHTS

M ADEBISI SOWUNMI (bisisowunmi@hotmail.com)
Department of Archaeology and Anthropology
University of Ibadan
Ibadan
Nigeria

Keywords: Holocene, South-western Nigeria, Coastal, Vegetation history, Dahomey Gap, Palynological evidence

Introduction

A palynological study of an 11 m terrestrial core from the Badagry area in coastal South-Western Nigeria was undertaken as part of an international project: "The Dahomey Gap: vegetation history and archaeobotany of the forest/savanna boundary in Bénin and South-Western Nigeria." The project is the joint effort of a team comprising an archaeobotanist (German), a botanist (Béninois), and four palynologists (Béninoise, German, Moroccan, and Nigerian). The Dahomey Gap is an unusual portion of the Guinean-Congolian forest zone, comprising a mosaic of savanna and the drier type lowland rain-forest. This Gap today stretches from the western-most part of Southern Nigeria through Bénin (formerly Dahomey) and Togo to the eastern-most part of Ghana. It effectively partitions the forest zone into western and eastern blocks (Plate 2).

As reviewed by Hamilton (1976), Martin (1990) and Maley (1996), biogeographers, studying the African tropical forest biota, plants, mammals, birds, insects, butterflies and molluscs, were the first to recognise centres of endemism in West and Central Africa. These centres, which "show a considerably higher species diversity than surrounding forest areas" (Martin 1990) are believed to have been refuges during the arid phases of the Pleistocene when the rain-forest was drastically reduced and fragmented. The issue of exactly where the refuges were is as yet not completely resolved. In the most current reconstruction, based on the work of several authors, these refuges during the peak of the last major arid phase are believed to have been in Liberia, Southwest Ghana and Southeast Côte d'Ivoire in the west, and South Cameroon, Gabon and Congo-Zaïre basin in the east (Maley (1996), Fig. 5).

The region presently known as the Dahomey Gap, referred to above, would have been part of the intervening areas from which the forest disappeared. The history and cause of this Gap are as yet not known with certainty. On the basis of palynological and limnological evidence from Lake Bosumtwi, Ghana, Maley (1996) suggested that the Dahomey Gap was closed in the early and middle Holocene, but probably re-opened from 3700/3800 to 3000 years BP. From their study of a marine core, west of the Niger Delta, Dupont and

R. W. Battarbee et al. (eds) 2004. *Past Climate Variability through Europe and Africa.*
Springer, Dordrecht, The Netherlands.

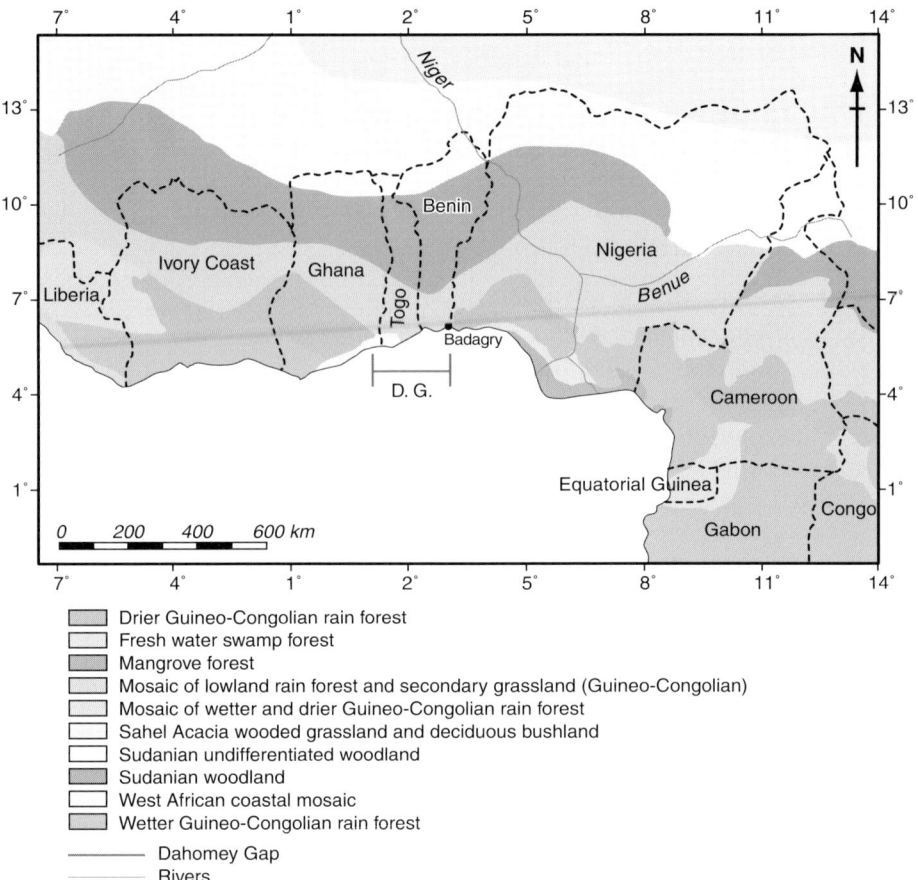

Plate 2. Vegetation map of West Africa (modified after Lawson (1986)). Colour version of this Plate can be found in Appendix, p.628

Weinelt (1996) inferred that this Gap "most probably did not exist" in the early and middle Holocene. More recent studies of marine cores from the coast of West and Central Africa led Dupont et al. (2000) also to conclude that the Dahomey Gap was closed in the early and middle Holocene. Unfortunately the last 6,000 years was not represented in the marine cores studied by Dupont et al. (2000), hence they had no direct evidence regarding the late Holocene history of this Gap. From the above review, it is clear that while there is a concurrence of indirect evidence, which suggests that the Gap was closed in the early and middle Holocene, there is as yet no direct botanical evidence from the area of the Gap, to shed light on the late Quaternary history of this Gap.

With regard to the cause of this Gap, a climatic factor, i.e., decreased rainfall, has been proffered. Dupont et al. (2000), for example, suggested that reduced rainfall along the coast in the last few millennia, exacerbated by human activities, led to the establishment of this Gap. Even though today, the mean annual rainfall in the region of the Gap (1,000 mm–1,500 mm) is less than that of the regions to its west and east (2,000 mm–3,500 mm)

(Hayward and Oguntoyinbo 1987), the rainfall is sufficient to support semi-deciduous forest. A similar enigma is the Togo Gap, "an anomalous climatic region in the south-eastern corner of Ghana and in Southern Togo" (Hayward and Oguntoyinbo 1987). The Togo Gap is within the Dahomey Gap, and its mean annual rainfall is 1,000 mm. The two hypotheses proffered to explain the lower precipitation in the Togo Gap also apply to the Dahomey Gap. According to the first hypothesis, less precipitation reaches the region because that section is aligned parallel to the direction of the moisture-laden south-west monsoon winds. The second hypothesis ascribes the anomaly to lower sea-surface temperature, which is due to the upwelling of cold waters in July-October (see Hayward and Oguntoyinbo (1987)). But as Hayward and Oguntoyinbo (1987) argued, neither of the two hypotheses provides a "convincing explanation."

Consequently, one of the main objectives of the Dahomey Gap project referred to earlier here, is to ascertain the late Quaternary history of this unique stretch of vegetation using terrestrial cores from the region. Hitherto there has been no study of suitable terrestrial cores from this region. Studies on the history of the rain-forest based on terrestrial cores from West Africa and North Central Atlantic Africa, such as those from Lake Bosumtwi, Ghana (Maley and Livingstone 1983; Talbot et al. 1984; Maley 1987; Maley 1996), and Lakes Mboandong (Richards 1986), Barombi (Maley 1996; Maley and Brenac 1998), and Ossa (Reynaud-Farrera et al. 1996), Cameroon, cannot, *ipso facto*, provide the required direct evidence.

This paper considers palynological evidence on the vegetation history of the Badagry area from the latter part of the Early Holocene through to the Late Holocene periods. The evidence is discussed in the light of previous ecological studies of the savannas in this area (Pugh (1953), in Adejuwon (1970) and Adejuwon (1977)) and of the very recent palynological study of sediments from adjacent Southern Bénin (Tossou 2002).

Materials and methods

Forty three samples were analysed from an 11 m core drilled at Ahanve, Badagry, South-Western Nigeria, (6° 25' 58" N, 2° 46' 29" E; 0.2 masl), at a distance of about 20 km from the Nigerian-Bénin border (Plate 2, Fig. 1), and below a water depth of about 50 cm. The site is a *Typha*-dominated fresh-water swamp, annually flooded by the Badagry Creek. The probable catchment area comprises the vegetation complexes around the site, including the Atlantic coast, the Badagry area, and, to a less extent, the region traversed by the River Yewa and Badagry Creek (see Fig. 1). Sub-samples for analysis were taken at intervals that ranged from 2 cm through 10 to 70 cm, depending on the degree of homogeneity of the colour of sediments as determined with the Munsell soil colour chart. The visual observation was corroborated later during the microscopic examination of the treated samples. One gram each of the sub-samples was given the standard treatment with KOH, HCL, HF, and acetolysis. A tablet of *Lycopodium* spores was added to each sample after the KOH treatment. The mountant used was silicone oil. An average number of 300 grains was counted per sub-sample, except for those from sections where the pollen content was extremely low, presumably due to post-depositional destruction by oxidation (see below).

The pollen of *Rhizophora* was always excluded from the pollen sum because it was over-represented, while those of *Alchornea*, *Elaeis guineensis* and Poaceae were excluded for sub-samples between 40 cm and 10 cm where they too were over-represented. The pollen

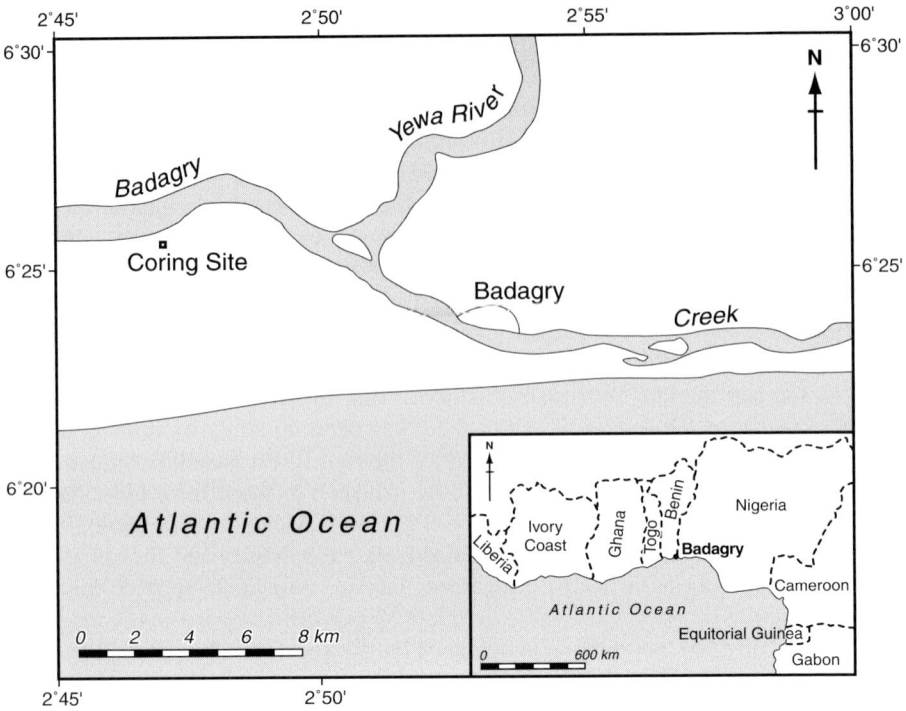

Figure 1. Map of Badagry area showing coring site with mosaic of vegetation types and scattered cultivation.

diagrams were prepared using the TILIA and TILIA graph computer software (Grimm 1991).

Present-day environment of the site

The site is a fresh-water swamp dominated by *Typha australis* and is annually flooded by the Badagry Creek. The River Yewa flows into this creek (Fig. 1). The vegetation of this *Typha* swamp is burnt annually in the dry season. At the time of coring (February, 2001), the swamp had been burnt, and the surface was a thick cushion of charred plant material. Around the periphery of the swamp are isolated clusters of the mangrove swamp fern, *Acrostichum aureum*, along with *Alchornea* and *Mimosa pigra*, while the fern, *Cyclosorus striatus* is abundant. *Rhizophora* is completely absent. The *Typha* swamp is surrounded by *Cocos nucifera* and *Elaeis guineensis* (coconut and oil palms, respectively).

Next is a mosaic of: (i) semi-deciduous and secondary forests, the components of which include *Newbouldia laevis*, *Pycnanthus angolensis*, *Alchornea cordifolia* and *Elaeis guineensis*, the latter being the most predominant; (ii) fresh-water swamp forest with *Phoenix reclinata*, *Raphia vinifera* and abundant *Anthocleista*, *inter alia*; and (iii) herbaceous fresh-water swamp, dominated by *Thalia welwitschii* and the sedge *Cyperus articulatus*, while *Nymphaea lotus* occurs in pockets of open water.

On the basis of an interpretation of aerial photographs and "intensive ground observations", Adejuwon (1970) recognized three physiognomic types of coastal savannas, which, according to him, cover about 50% of the Badagry area. According to Adejuwon (1977), the savanna is a vegetation unit, regardless of floral composition, in which grasses form the dominant ground layer. First is the treeless savanna, or open grassland, in which the tree-layer is insignificant. Second are thickets above a ground layer of grasses. *Alchornea cordifolia* tangles are the commonest dominants of such thickets. Third is park savanna in which the tree-layer comprises mainly old oil palm trees (*Elaeis guineensis*). Adejuwon (1970) did not report the presence of typical savanna trees, such as characterise the main savanna zones of Nigeria. However, during a reconnaissance of the area by the present author in September, 2002, the occurrence of some typical derived savanna trees, such as *Vitellaria paradoxa*, *Parkia biglobosa*, *Vitex doniana*, *Bridelia ferruginea*, *Annona senegalensis* and *Borassus aethiopum*, was noted in the sandy areas of the park savanna, in particular *Borassus aethiopum* often occurs in clusters.

On the raised, sandy, beach ridges (altitude 0.4 masl), extending from the sea inwards and including areas near coastal villages, there is also savanna vegetation, made up of extensive thickets of *Xanthozyllum zanthoxyloides* as well as typical derived savanna trees, predominantly *Annona senegalensis*, *Bridelia ferruginea* and *Vitex doniana*, with the grass, *Imperata cylindrica*. Derived savanna, as first defined by Jones (1945) in Hopkins (1962) and further elaborated upon by Keay (1959a), is the drier type of semi-deciduous forest degraded to savanna and maintained "through over-cultivation and over-burning." (Hopkins 1962). It is savanna woodland, with relicts of lowland rain-forest, and occurs as a zone between the rain-forest and the climatic climax savannas (see Plate 2). It has several species in common with the southern-most part of the latter, the so-called Southern Guinea savanna zone of Keay (1959b). It is zoned along with Southern and Northern Guinea Savanna zones of Keay (1959b) as a mosaic of Guineo-Congolian lowland rain-forest and secondary grassland (Plate 2). This derived savanna is rightly regarded as anthropogenic, and can be formed within a few decades or centuries. Experiments on savanna burning in South-Western Nigeria vividly demonstrate the anthropogenic nature of this vegetation type. When completely protected from farming and fire, it reverted back to forest within a few decades (Hopkins 1965).

Cultivated in the nearby villages are common crops such as *Manihot utilissima*, *Mangifera indica*, *Artocarpus communis*, and *Cocos nucifera*. With the crops are found weeds of farmland such as *Commelina* sp., *Emilia sonchifolia*, and *Chromolaena odorata*. Wayside weeds include *Ipomoea involucrata*, *Urena lobata*, and *Triumfetta rhomboidea*.

Along the coast and extending to varying distances inland are coconut plantations. As Adejuwon (1970) pointed out, the ground below the coconut trees may be bare or with a thin cover of plants, including the grass *Imperata cylindrica*. In large openings, common food crops such as cassava are cultivated.

Between the main lagoon and the ocean is a coastal strand of shrubs such as *Drepanocarpus lunatus*, *Hibiscus tiliaceus* and *Conocarpus erectus*, (Keay 1959b), and creepers such as *Ipomoea pes-caprae* and *Paspalum vaginatum* (Adejuwon 1970).

The climate of the Badagry area is equatorial. The mean monthly temperature is about 27 °C, while the relative humidity "is seldom below 70%" (Adejuwon 1970). The mean annual rainfall is 1,500 mm.

Results and discussion

Lithology

The entire column sampled is silt. Sand was struck between 11 m and 12 m, at which point coring became impossible. Nine informal lithological units are recognised on the basis of colour, black or very dark grey alternating with reddish-brown, dark reddish-black or dark reddish-grey. Colour determination was done using the Munsell soil colour chart (Table 1).

Table 1. Lithological units defined in the Ahanve core, Badagry, South Western Nigeria.

Depth (cm.)	Units	Colour
0–48	I	Black, dark grey/grey
48–68	II	Reddish black - dark reddish grey
68–138	III	Black
138–148	IV	Black - dark reddish brown
148–500	V	Black
500–558	VI	Dark reddish brown
558–678	VII	Very dark grey - black
678–688	VIII	Dark reddish brown
688–1100	IX	Very dark grey alternating with black

Radiocarbon dates

Six AMS-generated radiocarbon dates were obtained from the Leibniz Laboratory for age determination and isotope research, Christian-Albrechts University, Kiel, Germany (Table 2). The six dates are internally consistent. The estimated, average sediment accumulation rate for depths 156–1088 cm is 32.2 cm/100 years, the range being 29.2 cm to 42.6 cm/100 years. These rates indicate a sufficiently continuous sediment accumulation to permit interpolation. In contrast, the estimated, average accumulation rate between 32 and 156 cm (4.7 cm/100 years) is much lower, suggestive of a probable hiatus in accumulation. Consequently, no interpolation was done for this section of the core, nor extrapolation for the portion above it. In the text that follows, interpolated dates are not accompanied by standard errors.

Table 2. AMS ^{14}C dates for Ahanve core, Badagry, South Western Nigeria.

Depth (cm.)	Material dated	Lab. No.	^{14}C date (BP)	δ^{13}C
32–34	Silty peat + charred plant fragments	KIA 17574	3109 ± 26	−24.57 ± 0.37
154–156	Plant fragments	KIA 17575	5682 ± 32	−26.45 ± 0.24
402–404	Silty peat + plant fragments	KIA 17576	6409 ± 34	−27.65 ± 0.22
518–520	Silty peat + wood	KIA 13968	6681 ± 42	−26.56 ± 0.35
676–678	Wood + silty peat + shell fragments?	KIA 17577	7172 ± 33	−24.86 ± 0.17
1086–1088	Wood	KIA 13966	8576 ± 48	−26.06 ± 0.09

Palynomorph types

Altogether 141 pollen and spore types were recognised, of which 104 were identified. Identification was done through detailed comparisons with the reference pollen slides collection at the Palynology Unit of the department of Archaeology and Anthropology, University of Ibadan, pollen and spore slides of floral materials and sporophylls collected by the author in the environment of the site, as well as the following publications: Salard-Cheboldaeff (1980, 1981, 1982, 1983, 1984, 1985, 1986, 1987), and Sowunmi (1973, 1995).

Palynomorphs were identified to family, genus or species level. The forms that have been identified to species level are those which: (i) have distinctive pollen types, e.g., *Berlinia craibiana, Ceiba pentandra, Elaeis guineensis, Leea guineensis, Rhizophora* spp., and *Triplochiton scleroxylon*; (ii) belong to monospecific genera, e.g., *Crateranthus talbotii*; or (iii) are the only species of their respective genera known to occur naturally in Nigeria, e.g., *Bosquiea angolensis, Pandanus candelabrum, Pentadesma butyracea, Phoenix reclinata, Tieghemella heckelii* and *Trema guineensis*.

Phyto-ecological groups

Nine phyto-ecological groups were recognised on the basis of the present-day natural distribution of the respective taxa (Hutchinson and Dalziel 1954–1968; Keay et al. 1960; Keay et al. 1964).

Pollen Zones

Three pollen zones are distinguishable visually in the pollen diagram (Fig. 2).

Pollen Zone I (1100 cm–185 cm; ca. 8,576 ± 48 BP - ca. 5,780 BP)
This Zone covers about 83.2% of the 11 m core, and comprises 34 of the 43 sub-samples analysed. It is characterised by the prevalence of three types of forest: (i) *Rhizophora* mangrove; (ii) lowland, semi-deciduous and secondary rain-forest, with a few woody, savanna elements; and (iii) fresh-water swamp forest. *Rhizophora* was the most predominant component (71.3–96.0%). Rain-forest elements, including *Entada gigas* type, *Alstonia, Pentaclethra macrophylla, Celtis, Bosquiea angolensis, Irvingia, Trema guineensis* and *Pycnanthus angolensis*, mostly constituted 10% to 35.8% of respective pollen sums. The fresh-water swamp forest species, which constitute for the most part 9.6%–49.9% of the pollen sums, include *Uapaca* spp., *Dalbergia, Macaranga, Leea guineensis, Nauclea, Calamus deëratus* and *Pentadesma butyracea*. The few savanna species represented are *Bridelia ferruginea/scleroneura* type, which was the commonest, constituting 1.2–11.8% of the pollen sums, *Keetia venosa* type (syn. *Canthium venosum*), *Pericopsis* (syn. *Afrormosia*), *Pseudocedrella* and *Vitex* cf. *doniana* type. Grasses were absent at certain intervals and where they occurred, the representation was comparatively low (an average of 5.1%). The occurrence of the woody savanna species did not always coincide with that of grasses. These savanna species most probably occurred interspersed among the forest elements. Thus, it seems there was no distinct savanna vegetation.

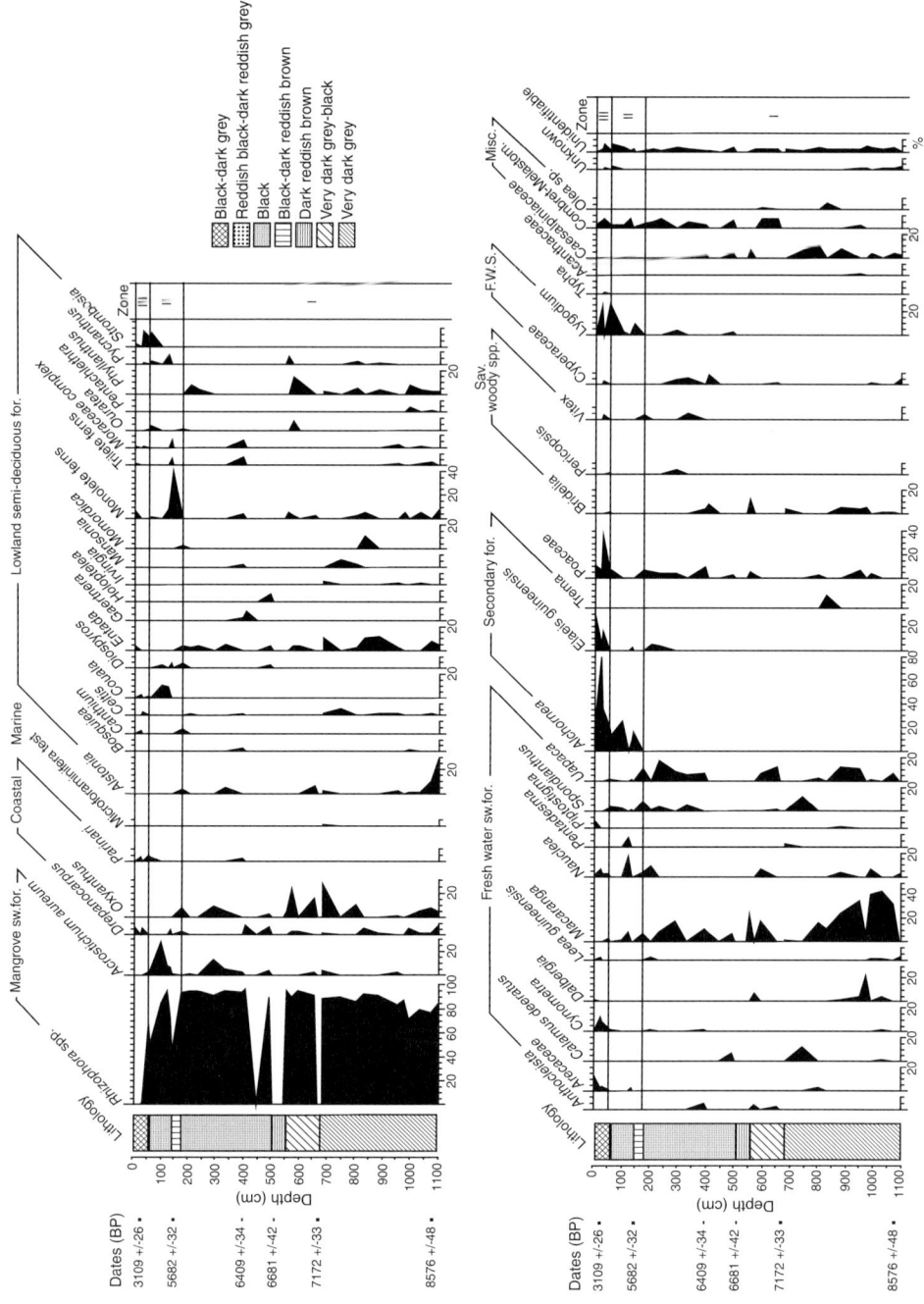

Figure 2. Ahanve core, Badagry, South-West Nigeria. Pollen diagram of selected taxa. (a) sw. for. = swamp forest; (b) for. = forest; (c) Sav. woody spp. = savanna woody species; (d) F.W.S. = fresh-water swamp.

The sediments deposited during the periods ca. 7,200 BP - ca. 7,140 BP and ca. 6,775 BP–6660 BP are devoid of palynomorphs, except for traces of *Rhizophora* and of a variety of other forms, such as *Nauclea* type, Asteraceae and *Brachystegia*. The pollen grains of the latter two are corroded. The colour of these sediments is dark reddish-brown, suggestive of the presence of oxidized mineral compounds. In contrast, the sediments of the respective layers immediately above and below each of those two sections are black or very dark grey, indicative of accumulation under reducing environments. They have very high values of *Rhizophora* and of other taxa. Consequently, the paucity of palynomorphs in the reddish-brown sediments, and the corrosion of some of the palynomorphs are strongly suggestive of a post-depositional destruction of most of the pollen and spores in an aerobic environment. Probably the dry seasons were more prolonged during those periods.

Consistently high and peak values of *Rhizophora* (91.2–96.0%) occurred between ca. 6,640 BP and ca. 5,780 BP. This is indicative of an extensive occurrence of *Rhizophora* mangrove in tidally-flooded swamps. The occurrence of micro-foraminiferal tests, although in comparatively low proportions (0.1–1.0%), is evidence of marine influence in the area. This period seems to have been the wettest period throughout the length of time covered by the core (ca. 8,576 ± 48 BP to some time after ca. 3,109 ± 26 BP).

Pollen Zone II (185–60 cm; ca. 5,780 BP - prior to ca. 3,109 ± 26 BP)

Zone II comprises six sub-samples. The events that occurred in this zone are transitional between those in Zones I and III. Three sets of marked vegetation changes can be deciphered. The first changes, which occurred between ca. 5,780 BP and ca. 5,682 ± 32 BP are, notably, drastic reductions in rain-forest (from 18.5% to 1.2%), fresh-water swamp forest (from 29.6% to 1.2%) and *Rhizophora* (from 94.2% to 43.2%), accompanied by sharp rises in both smooth, monolete spores, most probably *Cyclosorus striatus*, (from 0.0% to 39.8%), and the fresh-water swamp fern, *Lygodium* (from 0.0% to 9.5%). There was also a notable rise in *Alchornea*, a notable pioneer in the re-colonisation of land following the spatial reduction of rain-forest, (from 0.0% to 15.8%). These developments suggest that the semi-deciduous and fresh-water swamp forests became drier and more open, most probably due to a short-lived, drier climatic interval. It is of interest to note that Maloney (1992) pointed out that an increase in *Lygodium* spores is an indication of the disturbance of vegetation. Admittedly Maloney's (1992) observation is with reference to faraway South-East Asia, but the similarity with developments at Ahanve is sufficiently striking for one to infer that the high increase in *Lygodium* recorded here also reflects a disturbance of forests. This inference is supported by the concomitant increase in *Alchornea*. The reduction in *Rhizophora* is indicative of a slight marine regression. It is very significant to note that even though the forests became drier and more open, there is no evidence of the establishment of savanna vegetation.

The second set of changes took place some time after ca. 5,682 ± 32 BP and prior to ca. 3,109 ± 26 BP (140–110 cm). After the drastic decline of forests referred to above, there was a period of regeneration, marked by increases, reaching peak values, in semi-deciduous forest (25.6%), fresh-water swamp forest (33%) and *Rhizophora* (95.8%), along with a marked drop in fern spores (from 36.2% to 8.3%). These changes are indicative of forest expansion as wetter conditions were restored. This period probably is coincident, at least in part, with the Nouakchottian marine transgression of ca. 5,500–4,000 BP.

The third set of changes occurred subsequently (110–60 cm). There was another phase of reduction in semi-deciduous and fresh-water swamp forests, concomitantly with increases in *Lygodium* and *Alchornea* (up to 26.3% and 25%, respectively). The increases in the latter two taxa are higher than those recorded for the earlier, drier phase, and most probably reflect a greater disturbance of the forest vegetation. There were fluctuations in the occurrence of *Rhizophora*, all to values less than the peak in the section immediately preceding, suggestive of an overall reduction in tidal floods. The vegetation changes indicate the prevalence of a late Holocene dry phase.

Pollen Zone III (60–10 cm; just before ca. 3,109 ± 26 BP to after ca. 3,109 ± 26 BP)

This zone, the uppermost analysed part of the core, is the shortest. Only three sub-samples were studied. The sediment in the topmost 10 cm of the core contains mostly charred plant materials. At least four notable series of vegetation changes occurred in this zone, indicative of the following environmental conditions: the persistence of the late Holocene dry phase, a further reduction in tidal flooding owing to increased marine regression, and, an influx of fresh-water.

Firstly, just before ca. 3,109 ± 26 BP, *Rhizophora* declined very sharply from 67.0% to a mere trace (1.7%), which is equivalent to its absence from the site (see below). Indeed, *Rhizophora* had disappeared completely by ca. 3,109 ± 26 BP. The complete disappearance of *Rhizophora* is indicative of an accentuation of the marine regression that was evident in Zone II. The complete disappearance of *Rhizophora* was accompanied by both a phenomenal rise in the fresh-water swamp fern, *Lygodium* (0.0% to 23.5%) and an increase in fresh-water swamp forest (18.3% to 23.4%), notably with *Nauclea*, *Cynometra*, *Dalbergia*, *Piptostigma*, *Phoenix* and other Arecaceae. Furthermore, an initial, slight increase in rain-forest (11.1% to 19.1%) was followed by a decrease, down to 11.6%. The forest species include *Ouratea*, *Canthium*, *Celtis*, Moraceae complex, *Ceiba pentandra*, *Pycnanthus*, and *Paullinia pinnata*. The last three taxa are found especially in secondary forests today. *Alchornea* and *Elaeis* were probably among these secondary forest species, in view of the phenomenal increase in their pollen, as mentioned below.

Secondly, there was the establishment of the three physiognomic types of coastal savannas recognised by Adejuwon (1970). The development of these savannas is evidenced by unprecedented and sharp increases in Poaceae (from 7.1% to 35.7%), *Alchornea* (21.2% to 80.0%), and *Elaeis guineensis* (from 2.4–5.4% at lower levels in the core to 15.1%). Among the grass pollen was a new form, seen only once before, at depth 500 cm below the surface, with an estimated date of ca. 6,645 BP. This new form of grass pollen is larger-sized, with a much thicker annulus. It is at the moment not possible to say what this new grass species is; however, its occurrence is suggestive of an additional new source of grasses. Thus, *Rhizophora* was replaced by the coastal savannas. This is the first evidence of the presence of distinct savanna vegetation at the site and in its vicinity since ca. 8,576 ± 48 BP. Derived savanna was probably also in place, as evidenced by the occurrence of the pollen of *Pericopsis laxiflora* type (syn. *Afrormosia laxiflora*) and *Vitex*, along with increased grass pollen.

Thirdly, *Typha* and Cyperaceae fresh-water swamps were established, although the pollen of *Typha*, the dominant plant at the site today, is grossly under-represented in the pollen spectra, a mere 0.8%.

Fourthly, some time after ca. 3,109 ± 26 BP, there was an upsurge in *Elaeis* (29.7%, the highest value recorded throughout the core), while *Alchornea* registered a drastic decline (from 80% to 19.8%). This upsurge in *Elaeis* will be discussed later. The semi-deciduous and fresh-water swamp forests stabilised, with a marked decline in *Lygodium*.

As a result of the changes outlined above, the landscape became more open by ca. 3,109± 26 BP and from thence onwards.

The environmental and regional significance of the disappearance of Rhizophora *at the site and in its vicinity*

According to Ng and Sivasothi (2001), mangroves are tropical and sub-tropical, coastal, salt-water, inter-tidal forest communities. Of all the variants of tropical rain-forests, only the mangrove is under the direct influence of sea-water. It thus has been referred to as "rainforest by the sea." *Rhizophora* is one of the four commonest "true mangrove" genera, which, ecologically, are restricted to tidal swamps (Lauri and Gibson 2000; Feller and Sitnik 2001). Most mangrove swamps are dominated by the cosmopolitan red mangrove, *Rhizophora* (Schuster 1999). This is the case in Nigeria. Of the coastal wetlands, mangroves are found closest to the sea, the red mangrove being the most seaward, in the so-called seaward tidal zone, followed by *Avicennia* (Lauri and Gibson 2000). Consequently, *Rhizophora* is an excellent indicator of sea level or of flooding by tidal water. It is a particularly good palaeoecological indicator because it is a prolific pollen producer and the pollen is well dispersed in the air. Studies in the present author's laboratory have shown that *Rhizophora harrisonii* produces an average number of 66,432 and 531,460 grains per anther and flower, respectively. As can be inferred from Sowunmi's (1981) study of fossil sediments from on-shore Eastern Niger delta, Nigeria, strata where *Rhizophora* pollen constitutes over 40% of the pollen spectrum can be regarded as reflecting an appreciable occurrence of this mangrove, and hence tidal flooding, i.e., nearness to the coast. In contrast, values less than 40% indicate a very limited occurrence and a greatly reduced incursion of marine water, or increased distance from the coast. It is very striking to note that Edorh (pers. comm.), based on the study of modern surface samples in some coastal areas in West Africa, indicated that where *Rhizophora* is present, its pollen is not less than 40% of the pollen sum.

Rhizophora was predominant in the vegetation of the site and its vicinity between ca. 8,576 ± 48 BP and just before ca. 3,109 ± 26 BP. It disappeared by ca. 3,109 ± 26 BP. This total absence of *Rhizophora* in the late Holocene is not likely to have been due to a post-depositional destruction of its pollen, along with other pollen, as was the case during the middle Holocene (between ca. 7,200 BP and ca. 7,140 BP, and ca. 6,775 BP to ca. 6,660 BP) when there were only minor traces of some pollen types and the absence of most others. In the late Holocene, in sharp contrast to the situation during the middle Holocene, other palynomorphs were very well represented, with a phenomenal increase in the fresh-water fern, *Lygodium*. Furthermore, the sediment in the late Holocene was black to dark grey and not reddish-brown. Thus, the non-occurrence of *Rhizophora* pollen most probably was caused by a total destruction of the red mangrove, due to a much reduced inundation by sea water, coupled with a greater flooding by fresh-water. This fresh-water effectively replaced or suppressed the hyper-saline tidal water, resulting in a marked hydrological change in the soil. This change was not conducive to the survival of *Rhizophora*. Hence, *Rhizophora* was replaced by coastal savannas, nearer the coast, and by fresh-water swamp and swamp forest farther inland. The palynological evidence here lends strong support to the suggestion by Pugh (1953) in Adejuwon (1970), that the greyish-black humus or pseudo-humus mixed with the silvery, sandy soils in the lower parts of the Porto-Novo (Bénin)-Badagry beach-sand ridges are remnants of the mud swamp in which a more extensive mangrove vegetation once flourished.

The submission of Blasco et al. (1996) provides corroborative evidence for the suggestion made by the present author regarding the disappearance of *Rhizophora*. Based on studies of the mortality of coastal mangrove at several locations in the Pacific, West and East Africa, and Asia, Blasco et al. (1996) argued: "these coastal ecosystems are so specialized that any minor variation in their hydrological or tidal regimes causes noticeable mortality." Today, the Ahanve site is a *Typha australis*-dominated fresh-water swamp, which is seasonally flooded by the Badagry creek. The discontinuous clumps of *Acrostichum aureum* around the periphery of the site are relicts of the abundant mangrove swamp, which extended to the area of the site in the early through to the late Holocene. A comparable, natural destruction of mangrove swamp has been reported from Kosi Bay, South Africa, where two of the five mangrove species are the red mangrove (*Rhizophora mucronata*), and the white mangrove (*Avicennia marina*): "At Kosi Bay in 1965 the natural closing of the estuary mouth for five months resulted in" (upstream fresh-water) "flooding of the mangrove swamp (aided by Cyclone Claude) and caused a mass mortality of the mangroves from which they are at present slowly recovering." (Greig J.C.). Aspects of the palynological study of the Holocene vegetation history of the Innisfail coastal plain, Northeast Queensland, Australia, by Crowley and Gagan (1995), are even more comparable in this context. Crowley and Gagan (1995) found that extensive, *Rhizophora*-dominated mangroves became established in their area of study ca. 7,000 BP, in response to marine transgression. Subsequently, however, due to inundation by fresh-water, the salt intrusion in the upper reaches of the Mulgrave estuary was depressed. The development of hyper-salinity in the upper tidal zone was thus prevented. The mangroves were then succeeded by either fresh-water swamp or swamp-forest, depending on drainage conditions.

It is significant to note that from her study of four cores from Southern Bénin, Tossou (2002) recorded a very similar, abrupt decline and disappearance of *Rhizophora* between ca. $6,093 \pm 59$ BP and $2,254 \pm 49$ BP. In the Niger delta, Sowunmi (1981) recorded fluctuations in the occurrence of *Rhizophora* after ca. 2,900 BP all to levels below the maximum attained ca. 5,300 BP. Furthermore, Caratini and Giresse (1979) reported that in the Congo, *Rhizophora* was less represented after ca. 3,000 BP than at ca. 5,000 BP. Tossou (2002) reviewed additional evidence from Senegal, Nigeria and Côte d'Ivoire, which places the Holocene history of *Rhizophora* in a regional context. For much of the West African coast, *Rhizophora* was very abundant in the early and middle Holocene, becoming very extensive during the maximum phase of the Nouakchottian transgression, ca. 5,500 BP. In the late Holocene, *Rhizophora* either disappeared or became drastically reduced. Results from the present study are in consonance with this regional, Holocene history of *Rhizophora*.

This disappearance or drastic decline of *Rhizophora* can be attributed to a marine regression and the accompanying building up of beach-sand ridges. Einsele et al. (1977), for example, indicated that there was a lowering of the sea level to -3.5 ± 0.5 m along the Mauritanian coast ca. 4,100 BP, and possibly one or two more oscillations between 4,000 and 1,500 BP, i.e., the Taffolian regression. Furthermore, a thick sandbar was built on the shoreline along the estuary of the Senegal River area between ca. 4,000 and 1,800 BP

(Faure et al. 1980). Between ca. 4,180 ± 140 BP and 2,700 ± 100 BP, aeolian beach sands were deposited in South-East Ghana (Talbot 1981). The marine regression and building of beach-sand ridges are likely to have been regional phenomena. Hence, one can surmise that the Porto-Novo (Bénin)-Badagry beach-sand ridges most probably developed during this period. As outlined earlier, the marked decline and subsequent complete disappearance of *Rhizophora* occurred during a dry phase in the late Holocene. This dry phase has been recorded for other parts of the West African region, based on geomorphological and palynological evidence (see Sowunmi (2002), and Tossou (2002), for recent reviews).

Pollen of woody, savanna species

Among the derived or Guinea savanna elements easily observable in today's vegetation of the Badagry area are *Borassus aethiopum*, *Parkia biglobosa*, *Annona senegalensis*, *Vitellaria paradoxa*, *Bridelia ferruginea*, and *Vitex doniana*. The distinctive pollen grains of the first three were not found in the Ahanve core. It is likely that their pollen grains are dispersed by means other than the wind. In addition to the pollen of *Bridelia* and *Vitex*, the only other pollen of savanna, woody species identified are those of *Keetia venosa* type (syn. *Canthium venosum*), *Pericopsis* (syn. *Afrormosia*) and *Pseudocedrella*. Only *Bridelia* featured fairly regularly throughout the core. It thus seems that woody, derived or Guinea savanna elements are grossly under-represented in the core. Consequently, their presence in the vegetation is likely to have been inadequately reflected, particularly from ca. 3,109 ± 26 BP onwards when savannas became established in the study area.

Fungal spores

A variety of fungal spores, unicellular, bicellular and multicellular, were found in the core. A detailed study of these spores is in progress. Suffice it to say that many of them bear great resemblance to littoral fungal spores recovered by Salard-Cheboldaeff and Locquin (1980), from Tertiary deposits along the Gulf of Guinea in coastal Cameroon and which were found to be similar to extant species of Adelomycetes and Ascomycetes, among others.

Origin and status of the coastal savannas in South-Western Nigeria

As mentioned earlier, in the coastal areas west of the Niger Delta there are, among the rain-forest vegetation, stretches of three types of savanna, as defined by Adejuwon (1970), i.e., the treeless savanna or open grassland in which the tree layer is insignificant, thickets above a ground layer of grasses, *Alchornea cordifolia* tangles being the commonest dominants, and, park savanna in which the tree-layer comprises mainly old oil palm trees (*Elaeis guineensis*). It is noteworthy that these coastal savanna patches are limited to the west of the Niger Delta, i.e., they are not known to occur in Central and Eastern Niger Delta.

There are two noteworthy attempts to explain the development and status of these coastal savannas. Both Pugh (1953) in Adejuwon (1970), and Adejuwon (1977) regard the grassy, treeless savanna as a seral stage in the transition of mangroves to oil palm bushes and forests. According to these two authors, mangroves die off and are replaced by grasses when silting raises the terrain above salt-water. The grasses are in turn succeeded by the oil palm forest. Adejuwon (1970) suggests that the park savanna, which usually surrounds the treeless grasslands and is adjacent to oil palm forests, is a consequence of two processes. In the first process, the oil palm forest is prevented from succeeding the treeless savanna. The second process is the deterioration of oil palm forest to savanna. Both processes, according to Adejuwon (1970) are due to annual burning by natural and anthropogenic fires.

Palynological data from the present study have provided evidence on the simultaneous origin of the coastal and adjacent derived savannas during the late Holocene drier phase. The coastal savannas were established just prior to ca. 3,109 ± 26 BP, when grasses, *Alchornea* and *Elaeis* became abundant, replacing *Rhizophora*, which disappeared completely as a result of a marine regression. At about the same time secondary forest and derived savanna developed where the semi-deciduous forest declined.

As evidenced here, the establishment of these savannas was a natural development. However, the subsequent maintenance and expansion of at least the park and derived savannas can be attributed to human action. The phenomenal rise in *Elaeis guineensis* to a peak of 29.7% some time after ca. 3,109±26 BP, is probably due to an enhanced proliferation of the oil palm in openings created in the forest through human action. The oil palm tree occurs naturally in open spaces within the secondary forest. It is significant to note that the other pioneer taxon, *Alchornea*, decreased sharply (from 80% to 19.8%), at the time when *Elaeis* increased tremendously. *Elaeis guineensis* is a valuable tree, the different parts of which are used for various purposes in Southern Nigeria (Sowunmi 2002). Sowunmi (1999) has argued that the rapid spread of *Elaeis guineensis* in Ghana and Nigeria, from about 3,800 BP–3,200 BP onwards, is due to human action, which brought about an expansion of this pioneer species. Sowunmi (1999), contrary to a very curious misinterpretation of her argument by Maley (2001), did not state that *Elaeis guineensis* was cultivated in the past, but that conditions which favour its expansion were created through the opening up of forests for the cultivation of other crops. In other words, human action has contributed (and still contributes) to an expansion of the naturally-occurring oil palm. Alabi (1999) in a brief review of some recent works, outlined both palynological and anthracological (i.e., microscopic study of charcoal fragments) evidence from the Sahel region of West Africa and from Ghana, which shows that human impact on the natural vegetation became noticeable from ca. 3,000 BP.

There is archaeological evidence of the use of fire by the Late Stone Age occupants of an excavated Site in the Badagry area, east of Ahanve (Alabi 1999). Charcoal and charred palm kernels occur throughout the 190 cm thick occupation layer. A ground stone axe was found along with the charred palm kernels and charcoal at a level above the one dated to 2,670±90 BP. Alabi (1999) opined that these finds indicate "the clearing and burning of the forest probably preparatory to planting." It is therefore possible that humans in the Badagry area also had an influence on the natural vegetation from some time after ca. 3,109 ± 26 BP. Some farming is practised today in the area, as observed by the present author. It would be illuminating to ascertain more precisely the antiquity of farming there, on the basis of another independent line of evidence.

With regard to the derived savanna, it seems there were limited patches of it within the semi-deciduous forest from ca. 8,576 ± 48 BP up until just before ca. 3,109 ± 26 BP, judging by the presence of woody savanna species and the low occurrence of grasses along with forest elements. However, starting from some time in the late Holocene, owing to both natural and anthropogenic factors, this derived savanna expanded as evidenced by the marked increase in grasses, including one form from a new source.

The vegetation history of this coastal part of Nigeria thus has provided some direct evidence in support of Adejuwon's (1970) attempt to explain the widespread occurrence of derived savanna in the tropics. In supporting the thesis that derived savanna is essentially an anthropogenic community, Adejuwon (1970) had suggested that the extensive occurrence of this type of savanna in tropical lands worldwide is enhanced by the presence of natural savannas. He then went on to infer that naturally-occurring patches of savanna probably were initially present in the forests of tropical lands, much earlier than the arrival of farming communities. The burning of those patches could have resulted in the wide expansion of the savanna.

The Dahomey Gap

The middle and late Holocene vegetation history of South Western Nigeria as revealed in the core studied here (6° 25′ 58″ N, 2° 46′ 29″ E) is very comparable with that recently obtained from Southern Bénin by Tossou (2002). In three of the cores studied by Tossou (2002), from locations between 6° 29′ 42″ N, 2° 22′ 42″ E and 6° 36′ 25″ N, 2° 35′ 43″ E, *Rhizophora* disappeared around 2,500 BP and was replaced by a grassy savanna, the "prairie". Furthermore, the semi-deciduous forest, which was in place between ca. 7,500 BP and 2,500 BP declined at the end of that period. In its place was established an open vegetation of herbs, grasses, *Alchornea* and *Elaeis guineensis*, with forest relics. This open vegetation has rightly been recognized by Tossou (2002) as part of the Dahomey Gap. Although the forest in coastal, South-Western Nigeria did not undergo as marked a decline as that in Southern Bénin, the mosaic of vegetation established just prior to ca. 3,109 ± 26 BP, comprising (i) secondary forest, (ii) coastal savannas, consisting of abundant *Alchornea*, *Elaeis guineensis* and grasses, *inter alia*, and (iii) derived savanna, can be regarded as the eastern-most extension of the Dahomey Gap. As pointed out earlier, these savannas do not occur in Central and Eastern Niger delta.

The results obtained here indicate that though the part of the Dahomey Gap in South-Western Nigeria initially was established as a result of drier climatic conditions, which set in during the late Holocene, human action from some time after ca. 3,109 ± 26 BP has contributed to its maintenance and expansion.

This study therefore provides direct evidence, which supports the suggestion by Dupont et al. (2000), that the Gap was probably established in the last few millennia due to reduced rainfall along the coast, but that human activities exacerbated it.

Conclusion

The study of a terrestrial core, near the Nigerian-Bénin border has shown that from ca. 8,576 ± 48 BP to some time just prior to ca. 3,109 ± 26 BP, the predominant vegetation in coastal South-Western Nigeria comprised abundant forests: *Rhizophora*-dominated mangrove forest in the tidal zone, fresh-water swamp forest in swampy areas under the influence of fresh-water, and, the drier, semi-deciduous forest on well-drained terrain. Interspersed within the semi-deciduous forest were some woody, savanna species with a sparse grass cover at certain intervals. Prominent among the fresh-water swamp forest species are *Calamus deëratus, Dalbergia, Macaranga, Spondianthus* and *Uapaca.* The most notable semi-deciduous forest species are *Alstonia, Celtis, Coula, Entada, Phyllanthus, Pycnan-thus,* and *Strombosia,* while *Bridelia* is the commonest savanna element, along with *Keetia venosa* type (syn. *Canthium venosum*), *Pericopsis* (syn. *Afrormosia*), *Pseudocedrella, Vitex,* and grasses.

At some time just prior to ca. 3,109 ± 26 BP, significant vegetation changes occurred: *Rhizophora* disappeared completely and was replaced by fresh-water swamp as well as three types of coastal savanna comprising, *inter alia,* abundant grasses, *Alchornea,* and *Elaeis guineensis.* Secondary forest, dominated by *Alchornea* and *Elaeis guineensis,* and derived savanna developed as the semi-deciduous forest declined slightly. These vegetation changes are, primarily, attributable to the onset of a slightly drier climate as well as marine regression, which led to the cessation of marine influence at the site. The site then became seasonally inundated by fresh-water, as is the condition today.

The mosaic of a more open semi-deciduous forest, and the coastal and derived savannas is regarded as the eastern-most extension of the Dahomey Gap. This is the first, direct palynological indication that a part of the Dahomey Gap became established in the western-most part of South-Western Nigeria just prior to ca. 3,109 ± 26 BP.

There is an indication, from palynological evidence, that from some time after ca. 3,109 ± 26 BP human action probably contributed to the phenomenal rise in *Elaeis guineensis,* which resulted in the extension of the coastal park savanna, and perhaps the expansion of the derived savanna as well. Archaeological finds of a ground stone axe, charred palm kernels, and charcoal, from a site in the Badagry area would seem to support this indication.

Summary

The palynological study of an 11 m, terrestrial core from Ahanve, a coastal site in South-Western Nigeria, adjacent to the eastern-most portion of Bénin Republic (formerly Da-homey), was undertaken with a view to shedding more light on aspects of the vegetation history of the Dahomey Gap and the Guinean rain-forest near the coast.

The prevalent vegetation communities in the area from ca. 8,576 ± 48 BP to some time just prior to ca. 3,109 ± 26 BP comprised lowland, semi-deciduous forest along with some woody, savanna species, fresh-water swamp and abundant mangrove forests. Low and discontinuous occurrence of grasses indicates the absence of savanna vegetation. Between ca. 5,780 BP and ca. 5,682 ± 32 BP marked decreases in the three types of forest, and phenomenal increases in monolete fern spores, *Lygodium* and *Alchornea* suggest a short-lived, drier climatic interval.

Just prior to ca. 3,109 ± 26 BP, *Rhizophora* spp. declined very sharply and, soon after, disappeared completely; the mangrove was replaced by fresh-water swamp, an expanded fresh-water swamp forest and coastal savannas comprising, *inter alia,* abundant grasses, *Alchornea,* and *Elaeis guineensis.* The drier forest became more open and derived savanna developed. The mosaic of more open forest and savannas, established just prior to ca. 3,109± 26 BP, constitutes the eastern-most extension of the Dahomey Gap.

Though climatic and geomorphological factors seem to be the prime causes of the late Holocene vegetation changes, palynological and archaeological evidence suggests that from some time after ca. 3,109 ± 26 BP human action probably became an additional causative factor. The palynological results obtained here are, in essence, comparable with recent ones from adjacent Southern Bénin.

Acknowledgments

The generous financial grant from the Volkswagen-Stiftung of Germany, which made the realisation of this project possible, is most gratefully acknowledged. Thanks are also due to Dr Katharina Neumann of J.W. Goethe-Universität, Frankfurt-am-Main, Germany, for her initiative and able co-ordination of the project. The author's participation at the PAGES/PEP III conference at Aix-en-Provence, France, where an initial version of this paper was presented, was fully sponsored by The European Science Foundation. The author deeply appreciates this assistance. Thanks are due to Dr. Ulrich Salzmann for useful discussions held with him. I am grateful to Frau Gulla Schenk for numerous forms of assistance with the practicalities of the project. Many thanks are due to Mr. Jean-Pierre Cazet, who very kindly and painstakingly prepared an initial set of pollen diagrams. I thank Dr Anne-Marie Lézine, through whose kind auspices those pollen diagrams were produced. The assistance of Dr. Eric Grimm who provided the TILIA and TILIA graph software used by the present author in preparing the pollen diagrams featured in this publication, and his numerous prompt responses to electronic communication are gratefully acknowledged. I am deeply grateful to Mr. 'Layo Oyelakin and Dr Fabiyi of the University of Ibadan's GIS laboratory, for the production of the two maps. Finally, I am particularly grateful to Messrs P. C. Opara and E. Nwagbara who prepared the pollen slides, sometimes under very difficult conditions and to Dr. R. A. Alabi and Mr. P. C. Opara for their invaluable assistance with the coring.

References

Adejuwon J.O. 1970. The ecological status of coastal savannas in Nigeria. Journ. Trop. Geog. 30: 1–10.

Adejuwon J.O. 1977. A biogeographical survey of the dynamics of savanna vegetation in Nigeria. Nigerian Geographical Journal 14: 31–48.

Alabi R.A. 1999. Human-environment relationship: A synthesis of ethnoarchaeological evidence of human impact on the environment of the Badagry coastal area of Southwestern Nigeria. J. Sci. Res. 5 (1): 25–31.

Blasco F., Saenger P. et al. 1996. Mangroves as indicators of coastal change. Catena 27 (3-4): 167–178. (Source: Bibliography of sea-level change and mangroves, compiled by Andrea Schwarzbach. http://posssun.murdoch.edu.au/~mangrove/sea_level.html).

Caratini C. and Giresse P. 1979. Contribution palynologique à la connaissance des environnements continentaux et marins du Congo à la fin du Quaternaire. C.r. Acad. Sc., Paris 288, série D: 379–382.

Crowley G.M. and Gagan M.K. 1995. Holocene evolution of coastal wetlands in wet-tropical Northeastern Australia. Holocene 5 (4): 385–399.

Dupont L.M., Jahns S., Marret F. and Ning S. 2000. Vegetation change in equatorial West Africa: time slices for the last 150 ka. Palaeogeogr. Palaeoclim. Palaeoecol. 155: 95–122.

Dupont L.M. and Weinelt M. 1996. Vegetation history of the savanna corridor between the Guinean and the Congolian rain forest during the last 150,000 years. Veget. Hist. Archaeobot. 5: 273–292.

Einsele G., Herms D. and Schwartz U. 1977. Variation du niveau de la mer sur la plate-forme continental et la côte Mauritaniene vers la fin de la glaciation de Würm et à l'Holocène. Bull. Association Sénégal et Quaternaire Ouest Africa, Dakar 51: 35–48.

Faure H., Fontes J.C., Hebrard L., Monteillet J. and Pirazzoli P.A. 1980. Geoidal change and shore-level tilt along Holocene estuaries: Sénégal River area, West Africa. Science 210: 421–423.

Feller I.C. and Sitnik M. (eds) 2002. *Mangrove Ecology: A Manual for a Field Course. A Field Manual Focused on the Biocomplexity of Mangrove Ecosystems.* Smithsonian Institution. (Source: http://www.mangroves.si.edu/manual/manual.htm).

Greig J.C. *Mangroves of Kosi.* www.wildnetafrica.com: Courtesy of African Wildlife, Volume 36, No 4/5.

Grimm E.C. 1991. TILIA and TILIA GRAPH. Illinois State Museum, Research and Collection Centre, Springfield, Illinois, USA.

Hamilton A. 1976. The significance of patterns of distribution shown by forest plants and animal in tropical Africa for the reconstruction of Upper Pleistocene palaeoenvironments: a review. Palaeoecololology of Africa 9: 63–97.

Hayward D. and Oguntoyinbo J. 1987. The Climatology of West Africa. Hutchinson, London, 271 pp.

Hopkins B. 1962. Vegetation of the Olokemeji Forest Reserve, Nigeria. I. General features of the Reserve and the research sites. J. Ecol. 50: 559–598.

Hopkins B. 1965. Observations of savanna burning in the Olokemeji forest reserve, Nigeria. J. Applied Ecol. 2 (2): 367–382.

Hutchinson J. and Dalziel J.M. 1954–1968. Flora of West Tropical Africa Vol. I, Parts 1 and 2 (Second edition, revised by R.W.J. Keay); Vols. II and III, Parts 1 and 2 (Second edition, edited by F.N. Hepper), Crown Agents for Oversea Governments and Administrations, London, 295 pp., 828 pp.; 544 pp., 276 pp., 574 pp.

Keay R.W.J. 1959a. Derived savanna: derived from what? Bull. I.F.A.N. 21 Ser. A: 427–438.

Keay R.W.J. 1959b. An Outline of Nigerian Vegetation. Fed. Govt. of Nigeria, Lagos, Nigeria. 46 pp.

Keay R.W.J., Onochie C.F.A. and Stanfield D.P. 1960. Nigerian Trees. Vol. I Federal Department of Forest Research, Ibadan, Nigeria. 529 pp.

Keay R.W.J. and Onochie C.F.A. 1964. Nigerian Trees. Vol. II Federal Department of Forest Research, Ibadan, Nigeria. 495 pp.

Lauri B. and Gibson J. 2000. Mangroves. San Diego Natural History Museum. (Source: http://www.oceanoasis.org/fieldguide/mangrove.html).

Lawson G.W. (ed.) 1986. Plant Ecology in West Africa: Systems and Processes. John Wiley and Sons, Ltd., Chichester, 357 pp.

Maley J. 1987. Fragmentation de la fôret dense humide africaine et extension des biotopes montagnards au Quaternaire récent: nouvelles données polliniques et chronologiques, implications paléoclimatiques et biogéographiques. Palaeoecology of Africa 18: 307–336.

Maley J. 1996. The African rain forest — main characteristics of changes in vegetation and climate from the Upper Cretaceous to the Quaternary. Proc. r. Soc. Edinburgh 104b: 31–73.

Maley J. 2001. *Elaeis guineensis* Jacq. (oil palm) fluctuations in Central Africa during the late Holocene: climate or human driving forces for this pioneering species? Veget. Hist. Archaeobot. 10: 117–120.

Maley. J. and Brenac P. 1998. Vegetation dynamics, palaeoenvironments and climatic changes in the forests of West Cameroon during the last 28,000 years. Rev. Palaeobot. Palynol. 99: 157–188.

Maley J. and Livingstone D.A. 1983. Extension d'un element montargnard dans le sud du Ghana (Afrique de l'Ouest) au Pléistocene supérieur et à l'Holocène inférieur. Premières données pollinique. C.r. Acad. Sci. Paris 296, Ser. II: 1287–1292.

Maloney B.K. 1992. Late Holocene climatic change in SouthEast Asia, the palynological evidence and its implications for archaeology. World Archaeology 24 (1): 25–34.

Martin C. 1990. The Rainforests of West Africa: Ecology — Threats — Conservation. Birkhaüser Verlag.

Ng P.K.L. and Sivasothi N. 2001. Introduction: What is a mangrove? In: A Guide to mangroves of Singapore, Peter K.L. Ng and Sivasothi N. (eds), Singapore Science Centre. (Source: http://mangrove.nus.edu.sg/guidebooks/text/1002html).

Reynaud-Farrera I., Maley J. and Wirrmann D. 1996. Végétation et climat dans les fôrets du sud-ouest Cameroun depuis 4770 ans BP: analyse pollinique des sediments du lac Ossa. C.r. Acad. Sci. Paris Sér. 2 (322): 749–755.

Richards K. 1986. Preliminary results of pollen analysis of a 6000 year core from Mboandong, a crater lake in Cameroun. In: Baker R.G.E., Richards K. and Rimes C.A. (eds), The Hull University Cameroun Expedition: 1981–1982. University of Hull, Department of Geography, Misc. Ser. 30: 14–28.

Salard-Cheboldaeff M. 1980. Palynologie Camerounaise I. Pollen de la mangrove et des fourrés arbustifs côtiers. 105e Congrès nationale des Sociétés savantes, Caen, Sciences 1: 233–247.

Salard-Cheboldaeff M. 1981. Palynologie Camerounaise II. Grains de pollen de la forêt litorale de basse altitude. 106e Congrès nationale des Sociétés savantes, Perpignan, Sciences 1: 125–136.

Salard-Cheboldaeff M. 1982. Palynologie Camerounaise III. Grains de pollen de la forêt dense humide de basse et moyenne altitude. 107e Congrès nationale des Sociétés savantes, Brest, Sciences 1: 127–141.

Salard-Cheboldaeff M. 1983. Palynologie Camerounaise IV. Grains de pollen de la forêt dense humide de moyenne altitude. 108e Congrès nationale des Sociétés savantes, Grenoble, Sciences 1 (2): 117–129.

Salard-Cheboldaeff M. 1984. Palynologie Camerounaise V. Grains de pollen de la forêt dense humide semi- caducifoliée de moyenne altitude. 109e Congrès nationale des Sociétés savantes, Djon, Sciences 2: 19–35.

Salard-Cheboldaeff M. 1985. Palynologie Camerounaise VI. Grains de pollen des savanes péri-forestières. 110e Congrès nationale des Sociétés savantes, Montpellier Sciences 2: 231–248.

Salard-Cheboldaeff M. 1986. Palynologie Camerounaise VII. Grains de pollen des savanes arbustives et arbores, voire boisées, de l'Adamaoua. 111e Congrès nationale des Sociétés savantes, Poitiers, Sciences 2: 59–80.

Salard-Cheboldaeff M. 1987. Palynologie Camerounaise VIII. Grains de pollen des savanes boisées et forêts claires sèches. 111e Congrès nationale des Sociétés savantes, Lyon, Sciences 3: 47–63.

Salard-Cheboldaeff M. and Locquin M.V. 1980. Champiognons presents au Tertiaire le long du littoral de l'Afrique équatoriale. 105e Congrès nationale des Sociétés savantes, Caen, Sciences 1: 183–195.

Schuster G.A. 1999. Lecture Topics in Coral Reef — Mangrove swamps. (Source: www.biology.eku.edu/Schuster/bio/mangrove.htm).

Sowunmi M.A. 1973. Pollen grains of Nigerian Plants. I. Woody species. Grana 13: 145–186.

Sowunmi M.A. 1981. Late Quaternary environmental changes in Nigeria. Pollen Spores 23 (1): 125–148.

Sowunmi M.A. 1995. Pollen of Nigerian Plants. II. Woody species. Grana 34: 120–141.

Sowunmi M.A. 1999. The significance of the oil palm (*Elaeis guineensis*) in the late Holocene environments of West and West Central Africa: a further consideration. Veget. Hist. Archaeobot. 8: 199–210.

Sowunmi M.A. 2002. Environmental and human responses to climatic events in West and Central Africa during the late Holocene. In: Hassan F.A. (ed.), Droughts, Food and Culture: Ecological Change and Food Security in Africa's Later Prehistory. Kluwer Academic/Plenum Publishers, Dordrecht, pp. 95–104.

Talbot M.R. 1981. Holocene changes in tropical wind intensity and rainfall: evidence from Southeast Ghana. Quat. Res. 16: 201–220.

Talbot M.R., Livingstone D.A., Palmer P.G., Maley J., Melack J.M., Delibrias G. and Gulliksen S. 1984. Preliminary results from sediment cores from Lake Bosumtwi, Ghana. Palaeoecology of Africa 16: 173–192.

Tossou M. 2002. Recherche palynologique sur la vegetation Holocène du Sud-Bénin (Afrique de l'Ouest). Unpublished doctoral thesis, Université de Lome, Faculté des Sciences, 133 pp + Annexures.

12. PALAEOENVIRONMENTAL CHANGES IN THE ARID AND SUBARID BELT (SAHARA-SAHEL-ARABIAN PENINSULA) FROM 150 KYR TO PRESENT

PHILIPP HOELZMANN (philipp.hoelzmann@leco.de)
Max-Planck-Institut für Biogeochemie
Postfach 100164, 07701 Jena
Germany
Currently at
LECO Instrumente GmbH
Marie Bernays-Ring 31
41199 Moenchengladbach
Germany

FRANÇOISE GASSE (gasse@cerege.fr)
Centre Européen de Recherche et d'Enseignement
de Géosciences de l'Environnement (CEREGE)
Europole de l'Arbois
B.P. 80, F-13454 Aix-en-Provence cedex 4
France

LYDIE M. DUPONT (dupont@allgeo.uni-bremen.de)
Universität Bremen
Fachbereich Geowissenschaften
P.O. Box 30440, D-28334 Bremen
Germany

ULRICH SALZMANN
(ulrich.salzmann@zmt.uni-bremen.de)
J.W. Goethe-Universität
Seminar für Vor- und Frühgeschichte
Archäologie und Archäobotanik Afrikas
Grüneburgplatz 1, 60323 Frankfurt /M.
Germany
Currently at
Zentrum für Marine Tropenoekologie (ZMT)
(Center For Tropical Marine Ecology)
Fahrenheitstrasse 6
28359 Bremen
Germany

R. W. Battarbee et al. (eds) 2004. *Past Climate Variability through Europe and Africa.*
Springer, Dordrecht, The Netherlands.

MICHAEL STAUBWASSER
(michael.staubwasser@earth.ox.ax.uk)
Department of Earth Sciences
University of Oxford
Parks Road, OX1 3PR
Oxford
UK

DIRK C. LEUSCHNER (dcleu@mail.uni-mainz.de)
Institut für Geophysik und Geologie
Universität Leipzig
Talstraße 35
04103 Leipzig
Germany

FRANK SIROCKO (sirocko@mail.uni-mainz.de)
Institut für Geowissenschaften
Universität Mainz
Becherweg 21
55099 Mainz
Germany

Keywords: Palaeoclimatology, Palaeohydrology, Palynology, Palaeoceanography, Quaternary, Sahara, Sahel, Arabian Peninsula, Arabian Sea

Introduction

The PEP III Arid to Subarid Belt includes the largest hot desert in the world, the Sahara-Arabian desert and the Sahel zone. The region of interest extends south of the Atlas Mountains and south and east of the Mediterranean Sea to approximately $10°N$ and shows a broadly zonal pattern with a varying seasonal distribution of precipitation. In the north (ca. 20–23 °N), rainfall results from the southward displacement of the mid-latitude westerlies during winter whereas the south is governed by seasonal northward migration of the Intertropical Convergence Zone (ITCZ). Contraction and expansion phases of these presently semi-arid to hyper-arid desert areas result from significant changes in local precipitation. Palaeoenvironmental records from Northern Africa (north of 10 °N) and the surrounding seas document long-term changes in the magnitude and extent of the African monsoon in response to orbitally-forced changes in insolation. However, marine records as well as terrestrial palaeohydrological indicators (e.g., lakes, speleothems, rivers, pollen and charcoal) show that there have been changes in the hydrological cycle superimposed on the long-term waxing and waning of the monsoon which cannot be explained exclusively by changes in insolation. These fluctuations in space, time and magnitude were on a regional to continental scale.

Here, we review available data on near-surface palaeohydrological indicators and vegetational changes in arid North Africa and the Arabian Peninsula as well as changes in the intensity of the South Asian Monsoon identified from marine sediments of the Arabian Sea. A comparison of regional environmental changes can clarify relations between the environment and changes in the Earth's climate system. Each data-set is initially presented independently because they represent heteregeneous records from different regions and time periods and thereby emphasise their potential to provide evidence of continental chronostratigraphic palaeoenvironmental changes. Data-sets of lake status and vegetational change are complementary as they strongly reflect hydrological variation. Deep-sea sediments from the Arabian Sea were used to generate continuous records of oceanic upwelling, continental humidity, and dust and river discharge, that are closely related to palaeoenvironmental changes on the surrounding continents. After presenting the individual data-sets we compare the palaeoclimatic reconstructions derived from the different types of evidence.

Early work established that periods in which the hyper-arid desert of the Sahara was most extensive was correlated with full-glacial periods, while humid periods with lush vegetation were related to a smaller desert (Van Campo 1975). However, terrestrial palaeobotanical evidence from the desert and semi-desert regions of North Africa and Arabia for periods older than 10.0 ^{14}C kyr (11.4 cal. kyr) is extremely rare. Reconstructions of the Saharan-Sahelian vegetation and that of Arabia (Lioubimtseva et al. 1998; Farrera et al. 1999; Petit-Maire and Bouysse 2000; Elenga et al. 2000) have to rely on geomorphological and lake-status data (e.g., Gasse (2000), Williams et al. (2000), Swezey (2001)), or on data from marine sediments (e.g., Dupont (1993)). For the Holocene period, terrestrial palaeobotanical sites provide a better coverage. Surface palaeohydrological data from the late Pleistocene are particularly sparse but groundwater records provide interesting complementary information. In general, proxy data from terrestrial sites for Northern Africa and the Arabian Peninsula are often limited by too few continuous records, a spatially-biased scatter of sites, and dating difficulties. Nevertheless, groundwater-fed lakes, such as those which prevailed in the Sahara and Sahel can be sensitive indicators, provided that the recharge area is restricted, the flow patterns are known, and the response time to precipitation change is short (Gasse et al. 1990; Street-Perrott and Perrott 1993; Gasse 2000).

The chronologies of the majority of the palaeo-records (Fig. 1) are based on radiocarbon dating. For consistent comparison and correlation between different palaeo-records, the age referred to in the original paper is cited and the corresponding midpoint of the 1σ range from the probability method on the calibrated timescale is given in brackets. For the conversion between "^{14}C kyr" and "cal kyr" the radiocarbon calibration program CALIB (Stuiver and Reimer 1993) was applied which uses, in version, 4.3 the calibration curve IntCal98 (Stuiver et al. 1998).

Lacustrine records and other near-surface palaeohydrological indicators

From the penultimate interglacial to the last glacial maximum

Northern Africa
In the Northern Sahara, evidence from the "Continental Intercalaire" aquifer points to a major recharge period between ca. 45 to 23.5 ^{14}C kyr when noble gas-inferred recharge temperature was 2–3 °C lower than the modern mean annual temperature (Guendouz et al. 1998). In general, Saharan fossil waters show progressive depletion in ^{18}O and D from West to East (Sonntag et al. 1978; Joseph et al. 1992; Sultan et al. 1997; Edmunds et al., this

volume), strongly suggesting that North Africa must have received precipitation from wet Atlantic air masses crossing the Northern Sahara from West to East during wet Pleistocene phases. In the Sahel, recharge of confined aquifers, such as the Middle aquifer of the Chad Formation (Northern Nigeria) occurred between 28–23.2 ^{14}C kyr, and then ceased (Edmunds et al. 1999). Noble gas studies suggest recharge temperatures generally lower than at present within the range of 2–3 °C for the N Sahara (Guendouz et al. 1998), 4 °C for SW Egypt (Sonntag et al. 1982) and at least 5–6 °C in the Sahel (Edmunds et al. 1999). In contrast to that of the Sahara, the source of precipitation in the Sahel must be of tropical convective origin as fossil groundwaters (south of 20 °N) show meridional variations (Thorweihe 1990). The slightly higher D-excess, in comparison to northern Sahelian palaeowaters, is typical for tropical summer rains in the Sahel zone, which has received tropical rain in all time periods but with highly variable amounts during specific time periods (Sonntag et al. 1978). The end of the last cool and wet period of aquifer recharge in the Sahara and Sahel was attributed to the onset of Last Glacial Maximum (LGM) arid conditions (Edmunds et al. 1999; Gasse 2000).

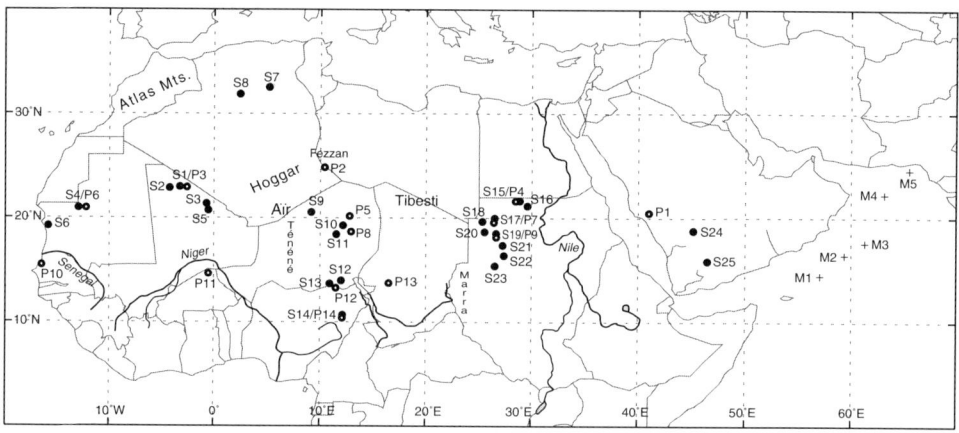

Figure 1. Schematic map of the PEP III Arid to Sub-arid Belt showing selected sites with palaeolake (S1-S23 = •; cf. Table 1, Plates 3a and 3b), palaeovegetation (P1-P14 = o; cf. Table 1 and Plate 4) and marine records (M1-M5; +; cf. Figs. 3 and 5).

The chronology of the few palaeolake records of the Sahara-Sahel (Servant 1983; Rognon 1987; Street-Perrott et al. 1989; Street-Perrott and Perrott 1993; Pachur 1993; Gibert et al. 1990) related to these recharge periods lack continuous sedimentation and consistent chronologies, as they are often based on very few (≤3) associated radiocarbon dates. Additionally, for radiocarbon dating the relative influence of artefacts or contamination becomes more important with increasing age and may result in erroneous radiocarbon chronologies (Fontes and Gasse 1989). Late Pleistocene authigenic carbonate precipitation within palaeolakes was identified for the N. Sahara (N. Egypt) between 32–19 ^{14}C kyr (organic and carbonate fractions dated; Pachur (1993), Pachur et al. (1987), Pachur (unpubl.)) and may point to a southward displacement of the westerlies during glacial times bringing more precipitation to Northern Africa. This accords with the time of late Pleistocene aquifer recharge (Guendouz et al. 1998) and may correspond to the youngest Pleistocene palaeolake-forming episode dated by U/Th. U/Th dating of molluscs from the

Table 1. Sites and references with palaeolake (S1-S23) and palaeovegetation (P1-P14) records used in Figure 1, Plates 3a, 3b and 4

	Site name	Reference(s)
	West	
S1	Taoudenni-Agorgott	Aucour 1988; Fabre and Petit-Maire 1988; Petit-Maire et al. 1987; Petit-Maire 1991
S2	Wadi Haijad	Petit-Maire et al. 1987; Gasse 2002
S3	Tagnout-Chaggaret	Petit-Maire and Riser 1983; 1981; Hillaire-Marcel et al. 1983
S4	Chemchane	Lézine et al. 1990; Lézine 1993
S5	Ine Kousamene	Petit-Maire and Riser 1983; Hillaire-Marcel et al. 1983
S6	Erg Akchar	Deynoux et al. 1993
S7	Sebkha Mellala	Gasse et al. 1990; Gibert et al. 1990; Gasse 2002
S8	Hassi el Mejnah	Fontes et al. 1985; Gasse et al. 1987; Gasse 2002
	Central	
S9	Tin Ouffadene	Dubar 1988; Fontes and Gasse 1991; Gasse 2002
S10	Kawar-Bilma	Servant 1983; Baumhauer 1991
S11	Fachi-Dogonboulo	Servant 1983; Baumhauer 1991
S12	Bougdouma	Gasse et al. 1990; Téhet et al. 1990; Fontes and Gasse 1991; Gasse 2002
S13	Bal Lake	Holmes et al. 1999a
S14	Lake Tilla	Salzmann et al. 2002
	East	
S15	Selima Oasis	Haynes et al. 1989; Pachur et al. 1987
S16	Dry Selima	Pachur and Wünnemann 1996; Abell and Hoelzmann 2000
S17	Oyo	Ritchie et al. 1985
S18	Wadi Fesh-Fesh	Pachur and Hoelzmann 1991; Hoelzmann et al. 2001
S19	El Atrun	Ritchie 1987; Pachur et al. 1987
S20	Lower Wadi Howar	Pachur and Kröpelin 1987; Kröpelin 1993
S21	Lake Gureinat	Pachur and Hoelzmann 1991; Hoelzmann 1993
S22	Ridge Lake T175	Pachur and Wünnemann 1996
S23	Meidob Hills	Pachur and Wünnemann 1996
	Arabian Peninsula	
S24	Mundafan	McClure 1976
S25	al-Hawa	Lézine et al. 1998
	Arabian Peninsula	
P1	An Nafud	Schulz and Whitney 1986
	Sahara	
P2	Djebel Acacus	Schulz 1994; Trevisan Grandi et al. 1998
P3	Taoudenni	Schulz 1991; Cour and Duzer 1976
P4	Selima Oasis	Haynes et al. 1989
P5	Seguedine, Kawar	Schulz 1994
P6	Chemchane	Lézine 1987; 1989
P7	Oyo	Ritchie et al. 1985
P8	Bilma	Schulz et al. 1994
P9	El Atrun	Ritchie 1987; Jahns 1995
	Sahel	
P10	Niayes	Lézine 1987; 1989
P11	Mare d'Oursi	Ballouche and Neumann 1995
P12	Manga Grasslands	Salzmann and Waller 1998
P13	Tjeri	Maley 1981; 1983
	Sudan	
P14	Lake Tilla	Salzmann 2000; Salzmann et al. 2002

N.W. Sahara (Causse et al. 1988; 1989; 1991; Vita-Finzi and Richards 1991) and authigenic carbonates from Southern Tunisia (Zouari et al. 1998) and the E. Sahara (Szabo et al. 1989; 1995) place the most recent Pleistocene humid phases in the N Sahara at ca. 30, 45, 65–90 and 120–150 kyr.

The Arabian Peninsula

A strong southwest monsoon during the Last Interglacial (Marine Isotope Stage (MIS) substage 5e) is inferred from a marine pollen record (Van Campo et al. 1982; Prell and Van Campo 1986) and is also documented by a speleothem record from Northern Oman from ca. 125–117 kyr (Burns et al. 1998). Considerably wetter conditions are attributed to a shift of the convergence of northwesterlies and southwesterlies during the summer monsoon north of its present position. The postglacial transition from wet to dry occurred at ca. 117 kyr and thereafter more arid climates prevailed in southern Arabia until some time before 9.7 kyr (Burns et al. 1998).

However, Late Pleistocene humid conditions during the period between 30–19.0 ^{14}C kyr on the Arabian Peninsula are inferred from well-established hydrological records obtained from palaeolakes (McClure 1976; Schulz and Whitney 1986) or calcites from groundwater (Clark and Fontes 1990). This period accords with late-Pleistocene aquifer recharge (Guendouz et al. 1998) and authigenic carbonate precipitation within N. Saharan palaeolakes, for which similar radiocarbon chronologies exist (Pachur 1993; Pachur et al. 1987).

The last glacial maximum

The LGM is an arid period marked by aeolian deflation and sand mobilization in the whole Sahara-Sahel (Swezey 2001). Consistent terrestrial palaeohydrological records that extend into the LGM are extremely scarce, as most of the small lakes of the late Pleistocene period were desiccated and the aquifer recharge stopped.

In Northwest Africa, a sedimentary record of a crater lake at 10.5 °N/12.25 °E in the sudanian zone of Nigeria shows low lake levels in association with clastic sedimentation during that period (S14, Plates 3a and 3b, Salzmann et al. (2002)). In the Chad Basin (8–22 °N/9–21 °E), Lake Chad was considerably reduced. Model simulations computed the extension of Lake Chad at the LGM as 7% of its modern area (Adams and Tetzlaff 1985).

In Northeast Africa, the Jebel Marra crater lake (13 °N/24.5 °E) experienced lower lake levels at 17.2 ^{14}C kyr (ca. 20.5 cal. kyr) after having reached an overflow high level at some stage before 40 ^{14}C kyr (Williams et al. 1980). During the LGM and probably as late as 12.0 ^{14}C kyr (ca. 14.0 cal. kyr) the lower White Nile valley also revealed evidence of aridity, and as a result the White Nile was a strongly seasonal river while the present Sudd swamps did not exist (Williams et al. 2000). Eastward, lakes depending on the Ethiopian highlands present the best documented palaeolake records from North-Eastern Africa (Lake Abhé at 11.5 °N and Ziway-Shala at 6.5–7.6 °N; Gasse (1977), Gillespie et al. (1983), Umer et al. (this volume)). These lakes were low during the LGM after having experienced a step-wise shrinking between 27.0 to 20.0 ^{14}C kyr (Lake Abhé) and subsequent desiccation lasting until 12.0 ^{14}C kyr (ca. 14.0 cal. kyr; Gasse (1977), Gasse and Street (1978), Gillespie et al. (1983), Le Turdu et al. (1999)).

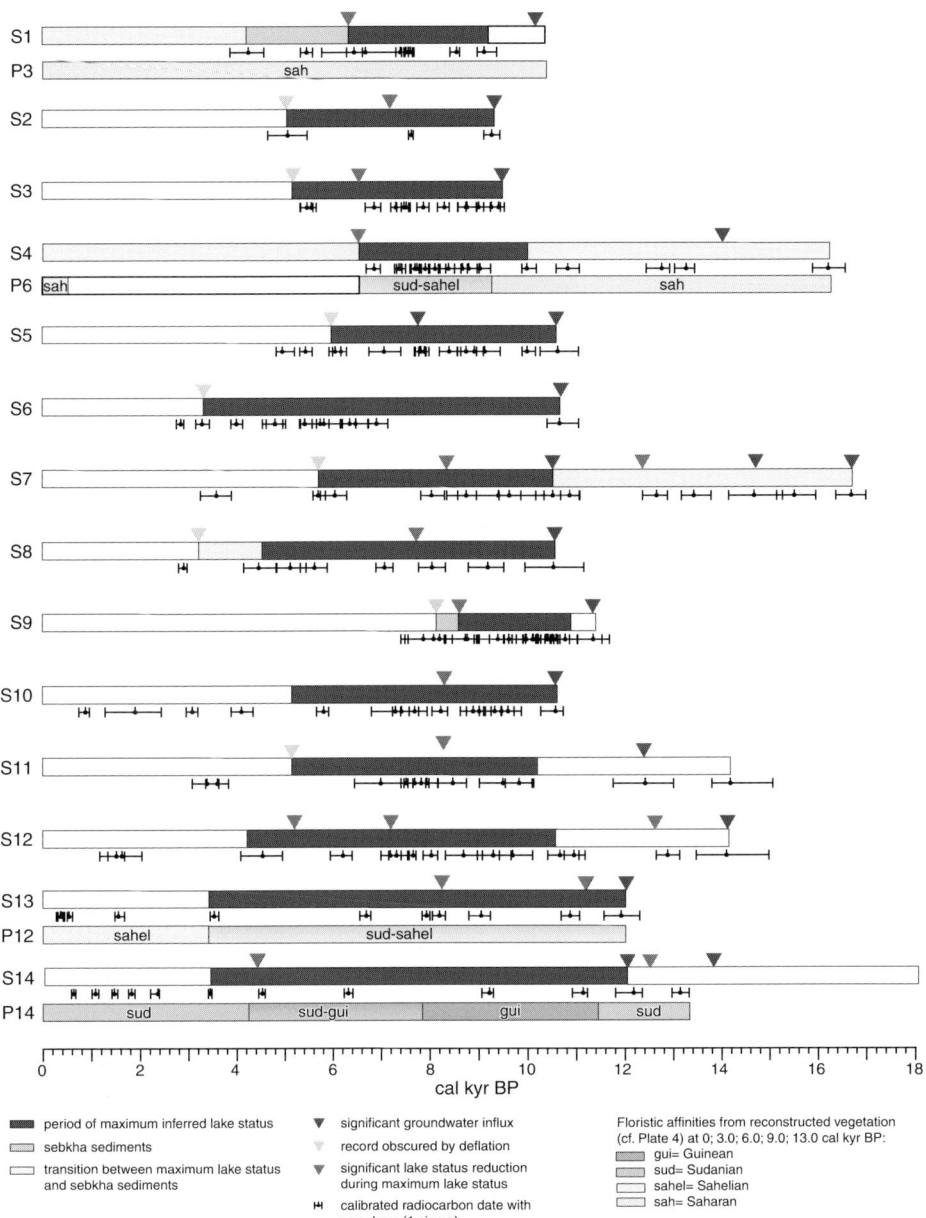

Plate 3a. Observed surface-near palaeohydrological changes over the last 18.0 cal. kyr for selected sites of the Western and Central Sahara/Sahel. Shaded areas represent periods of maximum inferred moisture (dark blue), transition between maximum lake status and sebkha environments (-pale grey,) and sebkha environments (mid blue). Triangles show centre-points of major events: significant groundwater influx (solid dark blue), lake-level reductions (red), and the age of truncation of the record by deflation (solid yellow). Radiocarbon dates from palaeolake records were calibrated according to Stuiver et al. (1998) using CALIB 4.3 (Stuiver and Reimer 1993). The midpoint of the 1 sigma range taken from the probability method is shown and error bars define the age range. Floristic affinities (cf. Plate 4) are also shown for selected palaeolake sites. Colour version of this Plate can be found in Appendix, p.629

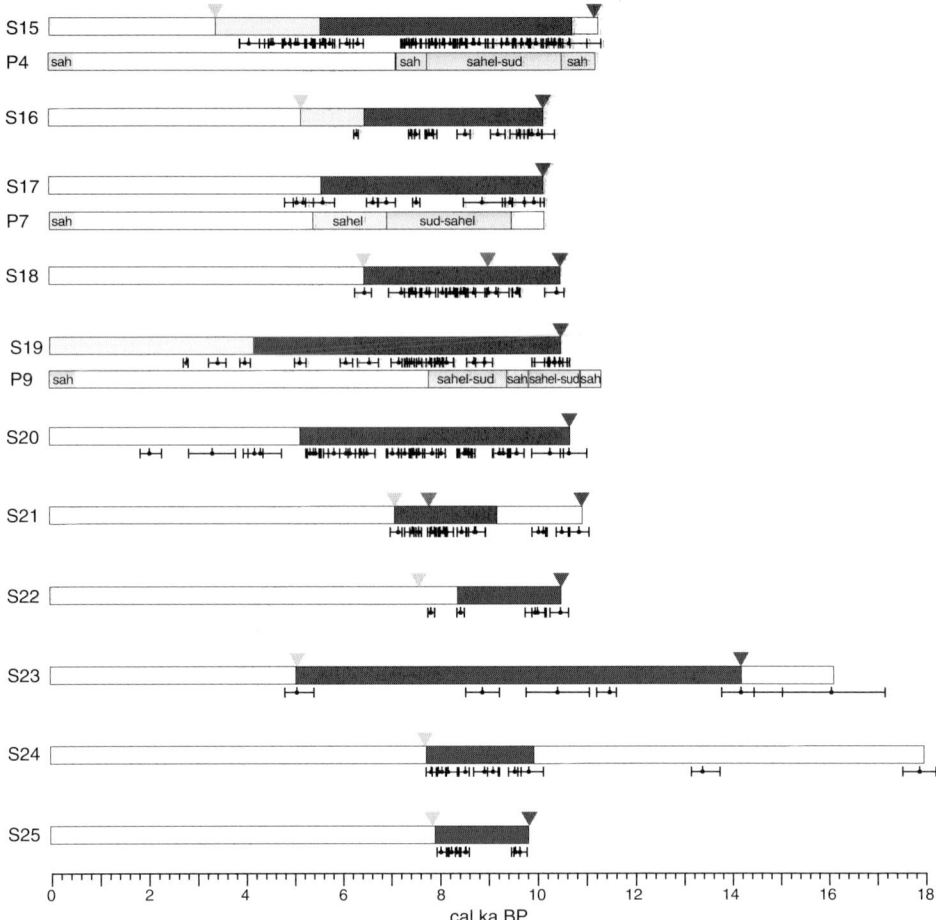

Plate 3b. Same as a) but for Eastern Sahara/Sahel and Arabian Peninsula. Colour version of this Plate can be found in Appendix, p.630

Late Pleistocene - Holocene Transition

The post-glacial hydrological change towards wetter conditions in Northern Africa is documented by step-wise lake-level rises in response to both insolation forcing and feedback processes with changes in oceanic circulation and sea-surface conditions (Gasse et al. 1990; Street-Perrott and Perrott 1990; Gasse and Van Campo 1994; Durand 1995; Gasse 2000). Deviating from gradual insolation forcing, lake-level rises were interrupted at several sites by a short-term return to drier and cooler conditions around 11.0–10.0 ^{14}C kyr (ca. 13.0 to 11.4 cal. kyr). This period coincides with the Younger Dryas (YD) interval, which has been reported from several sites of Northern Africa (Street-Perrott and Roberts 1983; Roberts et al. 1994; Gasse and van Campo 1994; Gasse 2000).

Sites showing regressional events centred around 10.2 ^{14}C kyr (ca. 12.0 cal. kyr; Plates 3a and 3b) are widespread over Northern Africa and exist in both monsoonal and Mediterranean sectors suggesting a common cause that involves a temporary decrease in ocean-to-continent moisture flux (Street-Perrott and Perrott 1990; Roberts et al. 1994; Gasse 2000).

In Northwest Africa, the record of Sebkha Mellala in the Northern Sahara (S7; Plates 3a and 3b) shows an increase in precipitation in the Saharan Atlas beginning around 14.0 ^{14}C kyr (ca. 17.0 cal. kyr) with short-term wet pulses (at 14.0 and 12.8 ^{14}C kyr; ca. 16.8 and 15.1 cal. kyr) followed by an arid period at 10.6–10.3 ^{14}C kyr (ca. 12.6–12.1 cal. kyr) with low lake levels (Fontes et al. 1985; Gasse et al. 1990). Along the southern and eastern margins of the Ténéré desert (15°–20 °N/9 °E) pre-Holocene lacustrine sediments date to ca. 11.0 ^{14}C kyr (ca. 13.0 cal. kyr; Servant and Servant-Vildary (1980), Gasse et al. (1990)). At Adrar Bous at the boundary between the Ténéré and Aïr mountains (20 °N/9 °E), a first flooding is registered as early as 13.0 ^{14}C kyr (ca. 15.6 cal. kyr; Williams et al. (1987), Dubar (1988), Gasse and Fontes (1992), Gasse (2002)). A similar timing for the onset of post-glacial wetter conditions was also recorded at 12.0 ^{14}C kyr (ca. 14.0 cal. kyr; Lézine (1987)) near the west coast of Africa, e.g., at Chemchane (S4; Plates 3a and 3b), but there, early-mid Holocene lacustrine deposits are absent (until ca. 7.0 ^{14}C kyr; ca. 7.8 cal. kyr) due to lower sea levels and a lower water table (Chateauneuf et al. 1986; Lézine and Casanova 1989). In the lowlands of the Chad basin, wetter conditions were registered within a short period around ca. 12.0 ^{14}C kyr (ca. 14.0 cal. kyr) at Bahr-el-Ghazal (16 °N/18 °E) and at Termit (16.1 °N/11.3 °E; Servant and Servant-Vildary (1980), Gasse et al. (1990)). The best records are those from the Manga grasslands (13.3 °N/11 °E; P12; see below 'Vegetational Changes') of Bal Lake (12.0 cal. kyr; S13; Plates 3a and 3b), Kajemarum Oasis (12.0–10.8 cal. kyr), and Bougdouma (S12; Plates 3a and 3b; ca. 12.0 ^{14}C kyr; ca. 14.1 cal. kyr), that point to a similar timing of the onset of wetter conditions (Gasse et al. 1990; Holmes et al. 1997; 1999a). Marked regressive events around 10.7 ^{14}C kyr (ca. 12.8 cal. kyr) were identified at the interdunal depressions of Bougdouma (S12; Plates 3a and 3b; Gasse and Van Campo (1994)) and Termit (Servant and Servant-Vildary 1980). Regressive events were also found at Bal Lake for 11.2 cal. kyr (S13; Plates 3a and 3b; Holmes et al. (1999b)). Further south, in the Sudanian zone of Nigeria, a slow rise in lake levels of Lake Tilla (S14; Plates 3a and 3b) is recorded for ca. 12.0 ^{14}C kyr (ca. 14.0 cal. kyr), which was interrupted by a distinct dry event between 11.0–10.2 ^{14}C kyr (ca. 13.0 to 11.9 cal. kyr) attributed to the YD (Salzmann et al. 2002).

In central-saharan Africa, wetter than present conditions at 15.0 ^{14}C kyr (ca. 18.0 cal.kyr) were registered in sediments of the Trou au Natron crater in the Tibesti mountains (21 °N; Maley et al. (1970), Maley (2000)). Fluvio-lacustrine activity between 16.0 to 7.0 ^{14}C kyr (ca. 19.0 to 7.8 cal. kyr; Jäkel (1979)) is documented in the so-called "middle terrace" of the Enneri-Bardague wadi system in the western part of the Tibesti mountains. During this period the wadi reached more than 60 km further downstream compared to the modern position of its playa. Speleothem formation occurred solely between ca. 17.5–12.3 kyr in South-Western Fezzan (ca. 25 °N/11 °E Libya; Cremaschi (1998)) and may document the largest water supply in the region during the last 15,000 years including the Holocene (Carrara et al. 1998).

Eastward, higher lake levels are registered in craters in the mountainous regions of Jebel Marra as early as 16.3 ^{14}C kyr (ca. 19.4 cal. kyr; 13 °N; lake level +8.0 m; Williams et al. (1980)) and in the Meidob Hills at ca. 12.0 ^{14}C kyr (ca. 14.0 cal. kyr; S23; Plates 3a and 3b; Pachur and Wünnemann (1996)). Further south-east, Lake Abhé (11.5 °N/42 °E) recorded low lake-levels as late as 11.0 ^{14}C (ca. 13.0 cal. kyr) followed by a rise to its highest level shortly after ca. 10.0 ^{14}C kyr (ca. 11.4 cal. kyr; Gasse (1977)). This is consistent with the nearby Ziway-Shala basin (Gasse and Street 1978) where a first wet pulse was centred at ca. 11.0 ^{14}C (ca. 13.0 cal. kyr) followed by a major lake level rise at ca. 10.0 ^{14}C kyr (ca. 11.4 cal. kyr).

Early- to Mid-Holocene

During the early-mid Holocene, substantially wetter conditions prevailed over much of Northern and Eastern Africa and the Arabian Peninsula, as documented by numerous palaeohydrolohical records (Plates 3a and 3b). This was primarily due to the intensification of the African and Indian monsoons resulting from earth orbital changes which increased the latitudinal and seasonal distribution of insolation (Kutzbach and Street-Perrott 1985; COHMAP Members 1988; Street-Perrott and Perrott 1993). This large shift towards warmer and wetter conditions with a maximum in the P-E balance is evidenced by numerous palaeohydrological records throughout the entire Sahel and Sahara (Plates 3a and 3b; Hillaire-Marcel et al. (1983), Petit-Maire and Riser (1983), Pachur and Kröpelin (1987), Baumhauer (1991), Fontes and Gasse (1991), Pachur and Hoelzmann (1991), Thiemeyer (1992), Petit-Maire et al. (1993), Jelinowska et al. (1997), Gasse (1998), Gasse (2002)).

However, in contrast to the gradual changes in insolation, the onset of this period, often referred to as the 'African Humid Period' (deMenocal et al. 2000a), was abrupt and occurred within decades to centuries, as documented by palaeolake records (e.g., Haynes et al. (1989), Pachur and Hoelzmann (2000), Gasse (2002)). In addition, several records show that this period was punctuated by dry spells of varying amplitude, e.g., substantial decreases in P-E around 8.4–8.0 cal. kyr (7.6–7.4 ^{14}C kyr), 7.5–7.0 cal. kyr (6.6–6.1 ^{14}C kyr), and 4.2–4.0 cal. kyr (3.9–3.7 ^{14}C kyr) in the northern monsoon region (Gasse and Van Campo 1994; Gasse 2000).

Northwest Africa.
Sites studied in the North-Western Sahara (S7 and S8; Plates 3a and 3b) show similar timing with major lacustrine episodes at 9.3–7.3 ^{14}C kyr (10.5–8.0 cal. kyr) and 6.6–4.0 ^{14}C kyr (7.5–4.5 cal. kyr), separated by an arid period centred around 7.2 ^{14}C kyr (8.0 cal. kyr; Fontes et al. (1985), Gasse et al. (1987), Gasse (2002)). The lakes were supplied by water from the Atlas mountains and local rainfall. The isotopic composition of the groundwater of these sites suggests that the isotopic composition of Holocene rainfall had a composition close to that of present-day rainfall, which falls as winter precipitation (Gasse 2002). In SW-Fezzan (24–25 °N) lacustrine deposits date between 8.5–5.3 ^{14}C kyr (ca. 9.5–6.1 cal. kyr) with an assumed interruption of lacustrine sedimentation from 8.0–7.0 ^{14}C kyr (ca. 8.9–7.8 cal. kyr; Cremaschi (1998)) similar to sites of the Northwestern Sahara. The Central Sahara north of 20 °N is characterized by periods of lacustrine deposits from 9.5–6.5 ^{14}C kyr (ca. 10.9–7.5 cal. kyr) and between 5.5–4.5 ^{14}C kyr (ca. 6.3–5.1 cal. kyr; Petit-Maire and Riser (1981), Hillaire-Marcel et al. (1983)). According to Lézine et al. (1990) the 'humid' maximum occurred around 8.5 ^{14}C kyr (ca. 9.5 cal. kyr). Southward, several sites of the Sahel of Western Africa show two early- to mid-Holocene episodes of high lake levels at 9.3–7.7 ^{14}C kyr (ca. 10.5–8.4 cal. kyr) and 6.4–4.5 ^{14}C kyr (ca. 7.3–5.2 cal. kyr), which are separated by deposition of swamp/wetland and/or aeolian sediments (Plates 3a

and 3b; Servant (1983), Servant and Servant-Vildary (1980)), although this event is not reflected in the vegetation records of that site (Salzmann and Waller 1998). Further south, perennial lacustrine conditions without substantial lake-level reductions throughout the early- to mid-Holocene are inferred from sediment and vegetation records from Lake Tilla (S14/P14; Plates 3a, 3b and 4; Salzmann and Waller (1998), Salzmann et al. (2002)).

Northeast Africa
In the Eastern Sahara (East of 16 °E) a consistent early-Holocene lacustrine phase occurred between 9.5 to ca. 6.0 ^{14}C kyr (ca. 10.9–6.8 cal. kyr; S15, S17, S19, S20; Plates 3a and 3b; Ritchie et al. (1985), Haynes et al. (1989), Kröpelin (1993)), which is also well-documented by palaeolakes situated outside hydrologically favourable locations such as oases (S16, S18, S21, S22; Plates 3a and 3b; Pachur and Hoelzmann (1991), Hoelzmann (1993)). The onset of this wet phase occurred within a very short time period of a few decades at ca. 9.5–9.3 ^{14}C kyr (ca. 10.9–10.5 cal. kyr) and is documented in most sedimentary records by a sharp transition from sandy to peaty and/or lacustrine deposits. In contrast with the Western Sahara-Sahel, humid conditions persisted, without systematic short-term lake-level fluctuations. This is probably due to the region's strong continentality and the stability of its ecosystem (Pachur and Hoelzmann 2000). In the Eastern Sahara, gradients of decreasing rainfall existed from West to East and from South to North as demonstrated by the co-existence of playas north of 23 °N/east of 24 °E and palaeolakes south of 21 °N/west of 24 °E (Pachur et al. 1990; Pachur and Hoelzmann 2000). Comparison of stable isotope analyses of sahelian to south-saharan groundwater (12–15 °N; Thorweihe (1990)) with lacustrine sediments (17–19 °N; Hoelzmann (1992), Abell et al. (1996)) show extremely depleted δ^{18}O-values that can be explained by intense tropical summer rainfall. However, north Saharan groundwater originates from wet air masses which were transported by the western drift across the continent (Sonntag et al. 1978). The Holocene boundary between winter and summer (monsoonal) rainfall is difficult to determine. Isotopic studies of lacustrine sediments along a N-S transect indicate that this boundary may be placed between 21–23 °N (Abell et al. 1996; Abell and Hoelzmann 2000).

Further south-east, records from Lake Abhé and Ziway-Shala depending on the Ethiopian highlands show that the wet early-mid Holocene period was interrupted by a large regressive event around 8.5–8.0 cal. kyr and a minor dry spell around 7.0–6.5 cal. kyr, before the major drop of lake-levels around 5.5–5.0 cal. kyr (Gasse and Street 1978; Gillespie et al. 1983; Gasse and van Campo 1994; Umer et al., this volume).

The Arabian Peninsula
An early-Holocene wet period on the Arabian Peninsula lasting from ca. 8.8 to 5.3 ^{14}C kyr (ca. 9.9 to 6.1 cal. kyr) has been inferred from a number of palaeolake sites (S24 and S25; McClure (1976), Zarins et al. (1979), Schulz and Whitney (1986), Lézine et al. (1998)) and other palaeohydrological evidence such as travertines, fluvial terraces, palaeosols and speleothems (Al Sayari and Zötl 1978; Jado and Zötl 1984; Clark and Fontes 1990; Roberts and Wright 1993; Burns et al. 1998). However, regional differences in the timing of on- and off-set exist. Lacustrine episodes recorded from the northern Arabian Peninsula (23–25.5 °N) lasted until ca. 5.3 ^{14}C kyr (ca. 6.1 cal. kyr; Schulz and Whitney (1986)), considerably later than those identified in other regions of the Arabian Peninsula (ca. 8.8–7.2 ^{14}C kyr; ca. 9.9–8.0 cal. kyr; McClure (1976), Lézine et al. (1998)). The Mediterranean

Sea probably influenced the climatic development of the northern part of the Arabian Peninsula, in such a way that more frequent cyclonic activities occurred and interacted with increased summer rain of monsoonal origin (Schulz and Whitney 1986; Lézine et al. 1998). For other regions of the Arabian Peninsula the cyclonic (Mediterranean) activity was of minor importance since they were mainly influenced by increased southwest monsoon activity and a northward shift of the ITCZ during the early Holocene (Burns et al. 1998; Lézine et al. 1998).

The best Holocene record currently available in the Arabian Peninsula is the high resolution $\delta^{18}O$ record from a U-Th-dated stalagmite in Northern Oman (23 °N), which documents the variations in the tropical circulation and monsoon rainfall between 9.6 and 6.1 kyr (Burns et al. 1998; Neff et al. 2001). This record shows several shifts in the monsoon strength, the major shifts occurs around 8.1 kyr. The excellent correlation between the $\delta^{18}O$ record and the $\delta^{14}C$ record suggests that one of the primary controls on centennial- to decadal changes in monsoon intensity during this time are variations in solar radiation.

Mid- to Late-Holocene aridification

The mid- to late-Holocene aridification is well documented at selected sites albeit a large number of records is truncated by intensive deflation during the present arid to hyper-arid period. In contrast to its rapid onset, a time-transgressive fading is observed that can be related to the ecosystem's buffer capacity, which was strongly influenced by an exponential recession curve of near-surface aquifers (Sonntag et al. 1982). Despite its time-transgressive character the transition to present-day's arid to hyper-arid climate may have occurred in two steps with two arid episodes at 5.9–4.7 ^{14}C kyr (6.7–5.5 cal. kyr) and 3.7–3.3 ^{14}C kyr (4.0–3.6 cal. kyr). The two steps are shown by comparing different sites (Gasse 2002) and radiocarbon dating of Saharan surface freshwater indicators (Guo et al. 2000). The latter of the two aridification steps was more severe and is represented in the majority of palaeoenvironmental records.

Important lacustrine regressions are observed throughout the Sahara between 4.5–4.0 ^{14}C kyr (ca. 5.2–4.5 cal. kyr; Plates 3a and 3b; Servant (1983), Petit-Maire and Riser (1983), Pachur and Kröpelin (1987), Street-Perrott et al. (1989), Pachur and Hoelzmann (1991), Gasse and Van Campo (1994)). However, the climatic development was not uniform throughout Northern Africa and regional differences are strong. According to the South to North gradient of decreasing rainfall identified in the Eastern Sahara (Pachur and Hoelzmann 2000), the climatic deterioration started earlier in the north at 22 °N ca. 5.7–4.5 ^{14}C kyr (ca. 6.5–5.2 cal. kyr; S15 and S16; Plates 3a and 3b) and later in the south at 17 °N ca. 4.5–3.8 ^{14}C (ca. 4.5–4.2 cal. kyr; S19, El Atrun; Pachur et al. (1987)).

In the southern savanna zones the mid-Holocene aridification is registered at Lake Tilla (S14; Plates 3a and 3b) by a gradual decrease in lake-levels after 7.0 ^{14}C kyr (ca. 7.8 cal. kyr) which accelerated at ca. 4.0–3.8 ^{14}C kyr (ca. 4.5–4.2 cal. kyr; Salzmann et al. (2002)). Similar results are reported from high resolution records of the West African Sahel (Bal Lake (S13; Plates 3a and 3b) and Kajemarum Oases) where a reduction of the P/E-ratio occurred between ca. 3.3–2.1 ^{14}C kyr (3.6–2.1 cal. kyr) after a mid-Holocene perennial freshwater phase (Holmes et al. 1997; 1999b).

For the Arabian Peninsula, the mid-Holocene aridification is often not documented in the lacustrine indicators itself, but by truncation of the records and the absence of evidence

(Schulz and Whitney 1986; Lézine et al. 1998). However, speleothem and travertine records from Oman both show an abrupt change to more arid conditions at 6.2 kyr (U/Th-dated) by a sharp increase in $\delta^{18}O$ (Burns et al. 1998; Neff ct al. 2001) or a change in travertine-type formation (Clark and Fontes 1990). After 6.2 kyr continental records related to surface hydrology are absent for the Arabian Peninsula.

A return to wetter conditions in the Sahara-Sahel has been registered in several Saharan palaeolakes (e.g., Servant (1983), Gasse and Van Campo (1994), Maley (1997), Gasse (2000)), palaeo-wadis (S20; Plates 3a and 3b; Kröpelin (1993)), and rivers, including the Nile River (Hassan 1997). However, the episode shows variations of minor amplitude and its timing varies at different sites between ca. 3.8–3.0 ^{14}C (ca. 4.2–3.2 cal. kyr). Obviously, local and site-specific characteristics become more important. It is generally accepted that after 3.0 ^{14}C kyr (ca. 3.2 cal. kyr) arid or even hyper-arid conditions similar to present prevailed. However, high resolution, continuous records combining several palaeohydrological indicators show that this assumption might be too simplistic. For example, Holmes et al. (1997, 1999a, 1999b) showed marked environmental changes took place over the last 5.5 kyr in the Manga Grasslands (see below 'Vegetational Changes') and inferred a more variable climate since ca. 1.5 ^{14}C kyr (1.5 cal. kyr) with episodic recurrence of Sahel droughts.

Similar to the changes in the vegetation (see below 'Vegetation Changes'), the temporal variation between different sites indicates a trend towards aridification with site-specific amplitudes due to individual basin characteristics. This trend, rather than a single abrupt event, produced regression and finally desiccation of the palaeolakes. Deflation set in, often truncating the palaeorecord.

Vegetational changes

From the penultimate interglacial to the last glacial maximum

Northwest Africa
Pollen as part of the dust input from N.W. Africa to sediments of the East-Atlantic is predominantly aeolian and depends on transport by the northeasterly trade winds and the Saharan Air Layer (Sarnthein et al. 1981; Hooghiemstra 1988a; 1988b). The latter wind system carries pollen from the Southern Sahara and the Sahel. Consequently, the marine pollen record enables inferences to be made about the latitudinal position of the southern desert fringe. A series of marine pollen records from 35 °N to 9 °S register the fluctuations and latitudinal position of vegetation zones in the western part of NW Africa (Fig. 2; Dupont (1993)).

Northward extensions of Sudanian savannas follow the more humid phases of MIS 5 (130–74 kyr). The larger northward excursion occurred between 130 and 120 kyr, and the smaller ones between 110–100 kyr and 90–80 kyr (Lézine and Casanova 1991; Hooghiemstra and Agwu 1992). From ca. 125 to ca. 115 kyr, the Saharan-Sahelian boundary shifted from a northern position at ca. 23 °N to as far south as 15 °N, but a position farther south of 15° is not confirmed by a marine core at 14 °N (Lézine 1991). During the rest of Stage 5, the Saharan-Sahelian boundary migrated back and forth between ca. 21°–19 °N (Dupont and Hooghiemstra 1989). The Mediterranean-Saharan transition zone probably formed a narrow belt between 28–32 °N along the southern fringe of the Atlas Mountains. The north boundary of the Western Sahara shifted from ca. 29 °N along the coast to ca. 30 °N more inland. (Hooghiemstra et al. 1992).

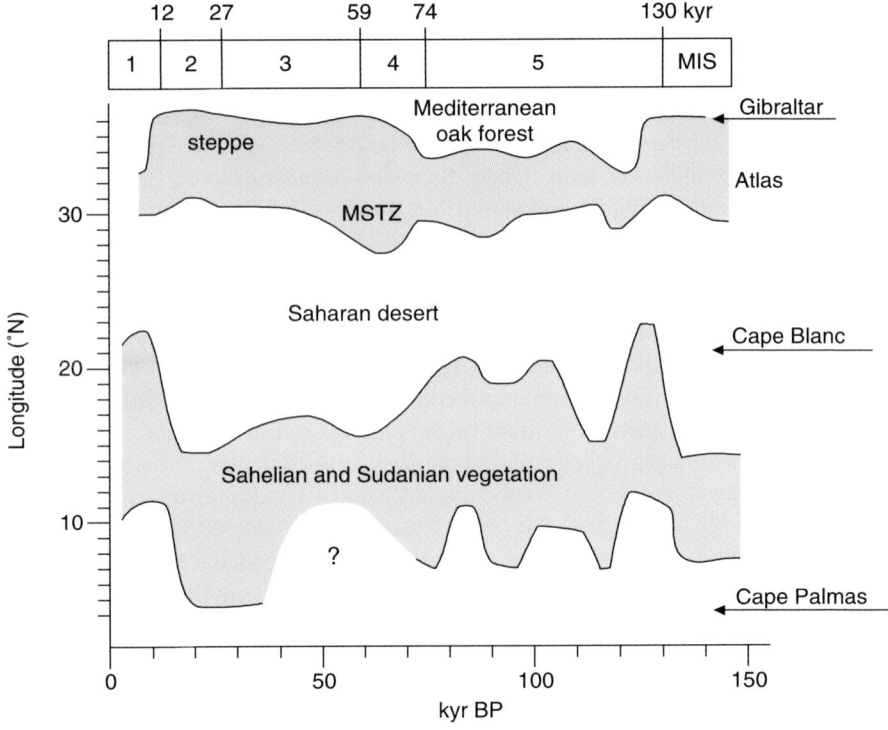

Figure 2. Schematic representation of latitudinal shifts of vegetation zones in NW Africa during the last Glacial-Interglacial climate cycle (adapted after Dupont (1993)). MIS: Marine Isotope Stage; MSTZ: Mediterranean Saharan Transitional vegetation Zone.

A large Mediterranean-Saharan transitional steppe with *Artemisia* and *Ephedra* existed between 74–59 kyr (MIS 4) reaching probably as far south as 25° to 30 °N in the Western Maghreb and south of the Atlas mountains. The Mediterranean forest was either non-existent in North Africa or scattered in a few refugia (Hooghiemstra et al. 1992). A marine core at 14 °N registered pollen from the desert vegetation, but hardly any arboreal pollen (Lézine 1991). At ca. 60 kyr, the southern boundary of the desert may have been around 15–16 °N. After 60 kyr, marine pollen diagrams indicate a slight increase of vegetation in the Western Maghreb in the form of scattered oak and pine forest. However, steppes with *Artemisia*, Chenopodiaceae, liguliflorous composites, and *Plantago* prevailed.

After 30 kyr (late MIS 3), the area of the Mediterrananean forest reduced and the northern boundary of the Saharan desert is estimated at ca. 28 °N by the coast to ca. 32 °N inland (Hooghiemstra et al. 1992). Marine records from the Gulf of Gabès indicate that a steppe with Poaceae, *Artemisia*, Chenopodiaceae, and *Tamarix* occurred in Southern Tunisia. At ca. 27 kyr *Artemisia* decreased. In contrast, the north of Tunisia probably was forested with *Quercus faginea*, *Pinus*, some *Cedrus*, *Ulmus*, *Alnus*, and *Salix* (Brun 1987; 1989; Ben Tiba and Reille 1982). The southern limit of the desert shifted northwards to 16–17 °N. A slight increase in Sudanian and Guinean taxa between 52 and 44 kyr is recorded in a marine diagram at 14 °N (Lézine 1991).

During the LGM the vegetation in the Sahara must have been extremely sparse. One pollen sample from the Tibesti mountains is dated around 17.0 ^{14}C kyr (ca. 20.3 cal. kyr) and mainly contains fern spores (Maley 1983). Based on geomorphological evidence, it is concluded that lake levels of the Lake Chad area were extremely low, and a semi-desert grassland persisted (Maley 1977; Servant and Servant-Vildary 1980). For the LGM, the Saharan-Sahelian boundary is estimated to have reached as far south as 14 °N in Senegal and 12 °N in East Nigeria (Rossignol-Strick and Duzer 1979; Dupont and Hooghiemstra 1989; Völkel 1989; Lézine 1991). North of the Sahara (in the Middle Atlas), herb-rich grasslands and steppe-like vegetation with *Artemisia* and Chenopodiaceae might have prevailed during the LGM (Lamb et al. 1989; Hooghiemstra et al. 1992). In North Tunisia, an open steppe vegetation with Chenopodiaceae, *Artemisia*, and *Ephedra* dominated, but locally stands of oak occurred (Ben Tiba and Reille 1982; Brun 1989).

Eastern Africa and Arabian Peninsula
Palynological evidence of the vegetation of N.E. Africa and the Arabian Peninsula comes from a marine pollen record from the Arabian Sea (Van Campo et al. 1982; Prell and Van Campo 1986). The record shows marked high relative amounts of pollen from the Sudanian savanna and humid tropical taxa signalling a strong southwest monsoon during the Last Interglacial (MIS substage 5e). During the glacial periods (MIS 6 and 4-2), relative amounts of pollen from the Mediterranean steppe increase and the humid tropical taxa decrease indicating a shift of the main pollen source areas from East Africa to the Arabian Peninsula.

The change in the relative importance of the wind systems is corroborated by a multi-tracer study of sediments from the Arabian Sea. During the LGM, strong northern winds, both from the Northwest (Arabia) and the Northeast (Persian Gulf) prevailed over weak southwest monsoons. After the glacial, the southwest monsoon became stronger and the northeast wind decline (Sirocko 1996; Sirocko et al. 2000). The pollen signal of Arabian Sea sediments, having its source area during the LGM mainly in Arabia, indicates strong northwesterly winds and increased aridity (Van Campo et al. 1982; Prell and Van Campo 1986).

A Pleistocene record of phytoliths from the Middle Awash, Ethiopia, dated to the Middle Stone Age, shows a grassland formation with scattered woody elements (Barboni et al. 1999). Also in Ethiopia, the pollen record of Lake Abiyata (Lézine 1982) indicates lower temperatures than today in a semi-arid environment with *Artemisia* and Ericaceae. The sequence shows a discontinuity in the pollen sedimentation between 30 and 10 kyr, suggesting a desert environment during that period. At Nafud, Saudi Arabia, pollen records of Pleistocene lakes indicate a semi-desert environment between 34 and 24 kyr. The lakes desiccated after 24 kyr (Schulz and Whitney 1986).

Late Pleistocene-Holocene Transition

Marine pollen records from the Atlantic coast (Hooghiemstra et al. 1992; Dupont 1993) and the Gulf of Gabes (Brun 1991) show that the hyper-arid climatic conditions of the LGM terminated between ca. 13.0 and 12.0 ^{14}C kyr (ca. 15.6–14.0 cal. kyr) with the spread southward of Mediterranean vegetation. Around the same period lakes with pollen bearing sediments began to form at the southern margin of the Sahara and the adjacent savanna zone. Nevertheless, climate still remained rather dry until the onset of the Holocene. While a semi-desert rich in Chenopodiaceae and *Artemisia* may have prevailed in Southern Tunisia (Brun 1991), the southern margin of the Sahara in Mauritania (Sebkha Chemchane) at ca. 21 °N was covered by desert vegetation with *Tamarix*, *Ephedra* and *Artemisia* (Lézine 1989). Further to the south, an open grass-savanna with few Sahelian and Sudanian arboreal taxa (e.g., Combretaceae, *Celtis*, *Alchornea*) prevailed in Western Senegal at ca. 15–16 °N (Lézine 1989). Open grass-savanna also occurred in Northeastern Nigeria at latitudes of the present-day Sahelian (13 °N) and Sudanian vegetation zones (10 °N; Salzmann and Waller (1998), Salzmann (2000)). An abrupt return to drier climatic conditions, which can be attributed to the Younger Dryas period, occurred at the Nigerian sites between ca. 11.0–10.2 ^{14}C kyr (ca. 13.0–11.9 cal. kyr).

Early to Mid-Holocene humid period

The onset of the Holocene is characterised by a shift towards wetter conditions. At numerous sites rising lake levels are accompanied by an increasing representation of Sudanian and Guinean taxa which rapidly spread northwards (Table 1; Plate 4). In the Sudanian zone of Northeastern Nigeria at 10 °N a dense Guinean savanna became established between ca. 8.5–6.8 ^{14}C kyr (ca. 9.5–7.6 cal. kyr) (Salzmann 2000). At the coastal sites of the Niayes and the Senegal river in Western Senegal a mesophilous Guinean forest may have reached up to 16 °N during the early Holocene (Lézine 1988; Michel and Assémien 1969–1970; Medus 1984; P10; Plate 4). At the more continental sites in the Sahel of Nigeria, dense Guinean swamp forests became established in interdune depressions but the surrounding savanna with Sudanian affinities remained relatively open (Salzmann and Waller 1998). At Tjeri (ca. 14 °N, Chad; P13; Plate 4), the maximum northward extension of the Sahelian zone can be detected between 7.0–5.0 ^{14}C kyr (ca. 7.8–5.7 cal. kyr) whereas the increasing presence of Sudano-Guinean taxa was interpreted in terms of rising fluvial transport from the Chari-Logone river system into Lake Tchad (Maley 1981). In contrast the pollen diagram from Mare d'Oursi in Burkina Faso (ca. 14 °N; P11; Plate 4) shows almost no vegetation change during the early- to mid-Holocene humid period (Ballouche and Neumann 1995). The vegetation remained strictly Sahelian with no Sudanian or Guinean elements present. Similar findings have also been reported, from mid-Holocene charcoal records, in Central Sudan situated at ca. 15 °N (Barakat 1995a). These results demonstrate that the model of shifting vegetation zones is too simplistic. For many of the palaeoecological sites the reconstructed vegetation has no modern analogue. Thus the zonation and composition of vegetation during the early Holocene appears to have been different from that which exists today (e.g., Schulz (1994), Ballouche and Neumann (1995), Salzmann and Waller (1998)).

Data for reconstructing the vegetation history at the northern limit of the Sahara are extremely scarce. Marine cores suggest that during the early Holocene this limit was situated only slightly south of its present-day position, whereas the transitional zone with *Artemisia*-semi-desert was reduced due to the southward migration of Mediterranean forests (Hooghiemstra et al. 1992). However, the semi-deserts of Southern Tunisia may have persisted throughout the Holocene (Brun 1991). On the other hand, studies of a mid-Holocene peat layer from Djerba (34 °N) identified Mediterranean woodland with *Juniperus* on the island around 4300 ^{14}C kyr (ca. 4.9 cal. kyr) (Damblon and Vanden Berghen 1993).

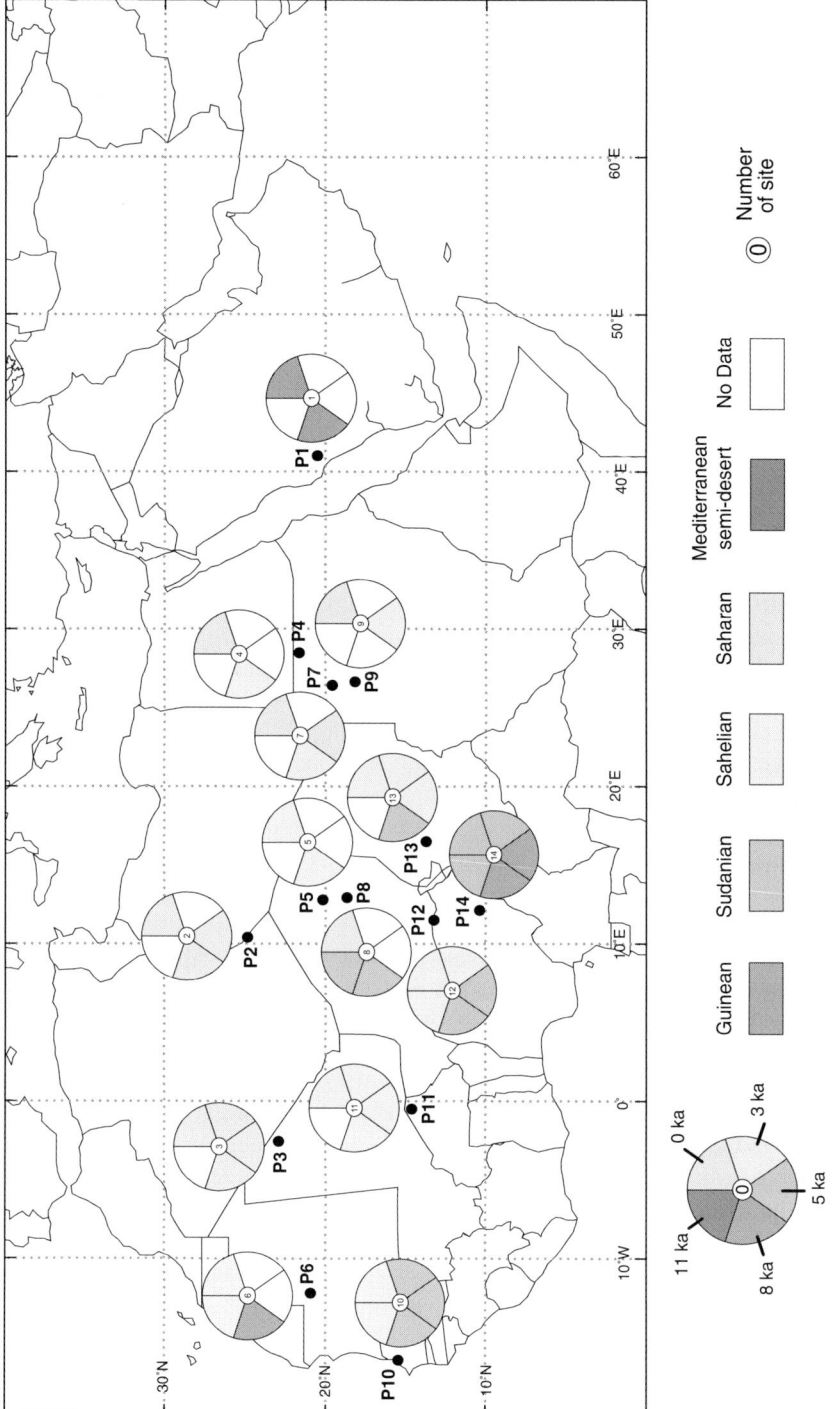

Plate 4. Floristic affinities of reconstructed vegetation from selected terrestrial pollen diagrams for the time slices 0 kyr, 3.0 ^{14}C kyr (ca. 3.2 cal. kyr); 5.0 ^{14}C kyr (ca. 5.7 cal. kyr), 8.0 ^{14}C kyr (ca. 9.0 cal. kyr), and 11.0 ^{14}C kyr (ca. 13.0 cal. kyr). Colour version of this Plate can be found in Appendix, p.631

In early palynological studies the presence of northern elements in pollen spectra from the Central Sahara had been interpreted as reflecting a southward migration of Mediterranean and temperate taxa into the Saharan mountains (Pons and Quézel 1958; Quézel and Martinez 1958–59). Later, investigations of the modern pollen rain revealed that these studies probably underestimated the significance of long distance transport (Cour and Duzer 1976; Schulz 1990). Investigations from the Hoggar massif demonstrate that, with a dominance of *Pistacia atlantica* and *Olea laperrini* during the mid-Holocene, the vegetation of the Saharan mountains closely resembled that of the present-day (Thinon et al. 1996; Barakat 1995c).

Pollen records from the West (Taoudenni, Niger, 20 °N; P3; Plate 4; Segedim, Mali, 22 °N) of the Northern Sahara indicate that between 8.0–6.0 ^{14}C kyr (ca. 8.9–6.8 cal. kyr) the desert-savanna transition zone with *Maerua-Acacia* and *Acacia-Panicum* savannas was located between 20 °N and 22 °N (Schulz 1991; Cour and Duzer 1976). In the East, desert persisted at Djebel Acacus (ca. 25 °N; P2; Plate 4) in Southern Libya (Schulz 1994; Mercuri 1999; Trevisan Grandi et al. 1998) and at Nabta Playa (ca. 22,5 °N) in Southern Egypt (Wasylikowa 1997; Barakat 1995b), while north of 25 °N a semi-desert shrub vegetation with winter rains prevailed in Egypt (Neumann 1989). However, at all sites the wetter climate throughout the early- to mid-Holocene supported a denser and more diverse vegetation. The comparatively late occurrence of tropical elements in the charcoal samples from Djebel Acacus (Castelletti et al. 1999) and Abu Ballas (24 °N, Egypt) was attributed by Neumann (1989) and Neumann and Uebel (2001) to low temperatures during the early Holocene which hampered the northward spread of southern species. Evidence for low temperatures during the early Holocene is also provided by high percentages of the mountain olive tree *Olea capensis* in pollen diagrams from Lake Tilla (Nigeria, 10 °N; Salzmann (2000); P14; Plate 4) and Lake Bosumtwi (Ghana, 6 °N; Maley and Livingstone (1983)) in Western Africa.

In the Sahara-Sahel sites south of 22 °N, charcoal and pollen analysis indicate that a Sudanian savanna with *Terminalia*, *Crataeva*, *Ximenia*, *Hymenocardia* and *Alchornea* covered the Ténéré desert at ca. 19 °N during the early Holocene. This suggests an increase in annual precipitation of at least 350 mm (Neumann 1992; Schulz 1994). In the Mauritanian desert at ca. 21 °N a Sudano-Sahelian savanna became established between 8.3–6.8 ^{14}C kyr (ca. 9.3–6.8 cal. kyr; Lézine (1989); P6; Plate 4). Similar vegetation changes were also recorded in the Eastern Sahara. For the early Holocene, a Sahelian or Sahelo-Sudanian savanna between ca. 18–21 °N is documented in the pollen and charcoal sequences from Northern Sudan (Selima Oasis; P4; Plate 4: Haynes et al. (1989); Bir El Atrun; P9; Plate 4: Ritchie (1987), Jahns (1995); Oyo depression; P7; Plate 4: Ritchie et al. (1985), Selima Sand Sheet and Wadi Shaw: Neumann (1989)).

For the Arabian Peninsula at Nafud (20.5 °N; P1; Plate 4) a semi-desert shrub vegetation comparable to the modern but with a denser herb cover indicates a stronger influence of winter rains (Mediterranean cyclones) or an interaction of these with monsoonal rains between 8.4–5.4 ^{14}C kyr (9.4–6.2 cal. kyr; Schulz and Whitney (1986)).

Mid- to Late-Holocene aridification

At the Saharan sites vegetation change in response to drier conditions occurred after ca. 7.0–6.0 ^{14}C kyr (ca. 7.8–6.8 cal. kyr). Desert became established ca. 5.0–4.0 ^{14}C kyr (ca. 5.7–4.5 cal. kyr) when most of the lakes dried out (e.g., Ritchie et al. (1985), Haynes et al. (1989), Lézine (1989), Schulz (1991)). In Libya and Southern Tunisia semi-desert shrub communities similar to the modern vegetation predominated at ca. 2.0 ^{14}C kyr (ca. 2.0 cal. kyr; Gale et al. (1993)).

In the savanna zones south of the Sahara, mid-Holocene aridification is marked by a decline in Sudano-Guinean taxa. In Western Senegal (Lézine 1989) and in Northeastern Nigeria (Salzmann 2000), Sudanian and Sudano-Guinean savanna became established after ca. 7.0 ^{14}C kyr (ca. 7.8 cal. kyr). At the Sahelian sites of Tjeri (Maley 1981) and the Manga Grasslands (Salzmann and Waller 1998; P12; Plate 4) the representation of 'southern pollen taxa' decreases after ca. 5.0 ^{14}C kyr (ca. 5.7 cal. kyr). Between ca. 4.5–2.4 ^{14}C kyr (ca. 5.2–2.4 cal. kyr) a renewed expansion of Sudano-Guinean forest is recorded from Western Senegal (Lézine 1989). However, this return to wetter conditions does not seem to have affected the more continental savanna zone of Nigeria and Burkina Faso. Here a shift towards a drier climate is recorded for ca. 3.8–3.0 ^{14}C kyr (ca. 3.2 cal. kyr), resulting in the establishment of a savanna which, in terms of floristic composition, closely resembles the present vegetation (Salzmann and Waller 1998; Salzmann 2000; Ballouche and Neumann 1995). A similar change occurred in Western Senegal at ca. 2.0 ^{14}C kyr (ca. 2.0 cal. kyr) (Lézine 1989). The temporal variation between these sites suggests that vegetation change during the Late Holocene was produced by an increasing trend towards aridification with different site specific amplitudes, rather than by a single abrupt event.

The role of human activity during the Holocene is the subject of controversy. Human impact has been emphasised as being responsible for the progressive degradation of savannas and the formation of the modern cultural landscape by clearing, pastoralism, agriculture and iron smelting. Those processes might have started during the mid- and late-Holocene (e.g., Ballouche and Neumann (1995), Schulz (1994), Barakat (1995a)). However, the detection of anthropogenic activities in pollen diagrams is strongly hampered by the absence of unambiguous anthropogenic indicators (Waller and Salzmann 1999). The major vegetational change at the onset of the Late Holocene, which eventually produced the modern vegetation zones, can be confidently attributed to climatic change.

Evolution of the South Asian monsoon since the penultimate glaciation — a synthesis from Arabian Sea records

The present day South Asian Monsoon and its influence on the Arabian Sea

Arabian Sea oceanography at the present day is dominated by the South Asian Monsoon. The South Asian Monsoon is characterised by a seasonal reversal of winds associated with distinct wet and dry periods on the Indian subcontinent (Webster et al. (1998), Fig. 3). These winds affect surface and thermocline circulation in the Arabian Sea (Flagg and Kim 1998; You 1998). They affect or even control important hydrographic parameters such as thermocline ventilation, sea-surface temperature (SST), sea-surface salinity, as well as plankton production and dust deposition from the Arabian deserts (Fig. 4). Ocean upwelling in the western Arabian Sea (Swallow 1984) is perhaps the most notable feature of the summer monsoon season (SW monsoon), and is associated with high surface production (Brock and McClain 1992). The resulting high organic flux to the deep (Rixen et al. 1996) and the naturally low oxygen content of Central Indian Water, the largest source of water in

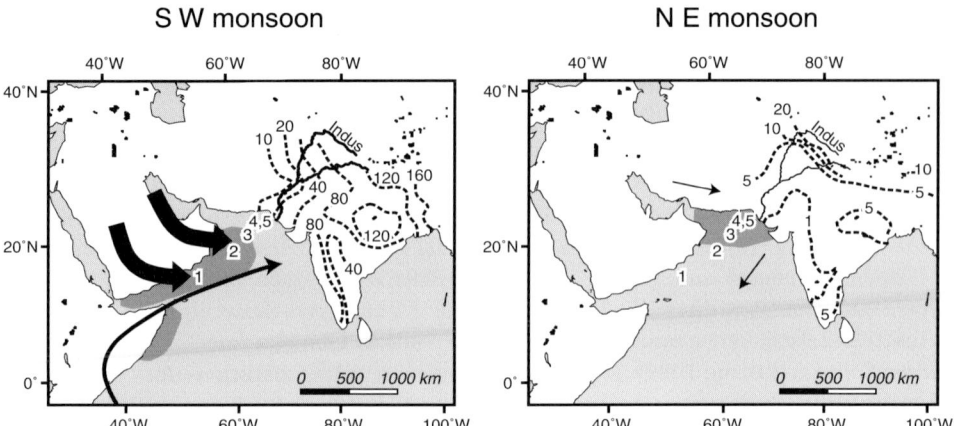

Figure 3. General features of the South Asian Monsoon and its impact on the Arabian Sea. Arrows: dominating wind direction. Shaded areas: upwelling and sea surface cooling (SW monsoon), sea surface cooling and convective mixing (NE monsoon). Precipitation contours after Rao (1981). Cores referred to in the text: 1 = 74 KL, 2 = NIOP 492, 3 = 70 KL, 4 = NIOP 464, 5 = 63 KA.

the Arabian Sea thermocline, maintain a pronounced oxygen minimum zone in the Arabian Sea (You 1998; Morrison et al. 1999). SSTs in the Arabian Sea are highest during the spring inter-monsoon season (Levitus and Boyer 1994). Upwelling in the western Arabian Sea during the SW monsoon cools the sea surface by up to 3 °C compared to the preceding spring inter-monsoon. The winter monsoon (N.E. monsoon) at present sees up to 4 °C cooler SSTs in the northern Arabian Sea than the following spring inter-monsoon (Levitus and Boyer 1994), and is associated with convective mixing and moderate bio-production (Banse 1994).

The Arabian Sea is surrounded by arid and semi-arid lands, which provide a huge aeolian sediment supply from the Arabian Peninsula to the western Arabian Sea (Sirocko and Sarnthein 1989), fluvial detritus with locally considerable amounts of tectonically expelled mud along the northern margin (Staubwasser and Sirocko 2001), and both fluvial and aeolian material in the eastern Arabian Sea / Indus fan region (Sirocko and Sarnthein 1989; Prins et al. 2000). Dust raised from Mesopotamia and central Arabia is blown to the Arabian Sea primarily during the spring inter-monsoon and S.W. monsoon season by northwesterly winds overriding the SW monsoon jet (Fig. 3). Deposition onto the sea floor occurs mostly during the SW monsoon, when organic matter scavenges the suspended detritus from the water column (Clemens 1998).

From the penultimate interglacial to the last glacial maximum

A number of studies from the western Arabian Sea have sought to reconstruct palaeo-production, upwelling, and dust flux from proxies measured in deep-sea sediment cores in order to infer past changes in the intensity of the South Asian Monsoon, in particular the moisture-bringing SW monsoon. On a glacial to interglacial time-scale, the general picture is a dominating influence of orbital parameters on the evolution of the South Asian Monsoon and Arabian Sea oceanography (SW monsoon upwelling and palaeoproduction,

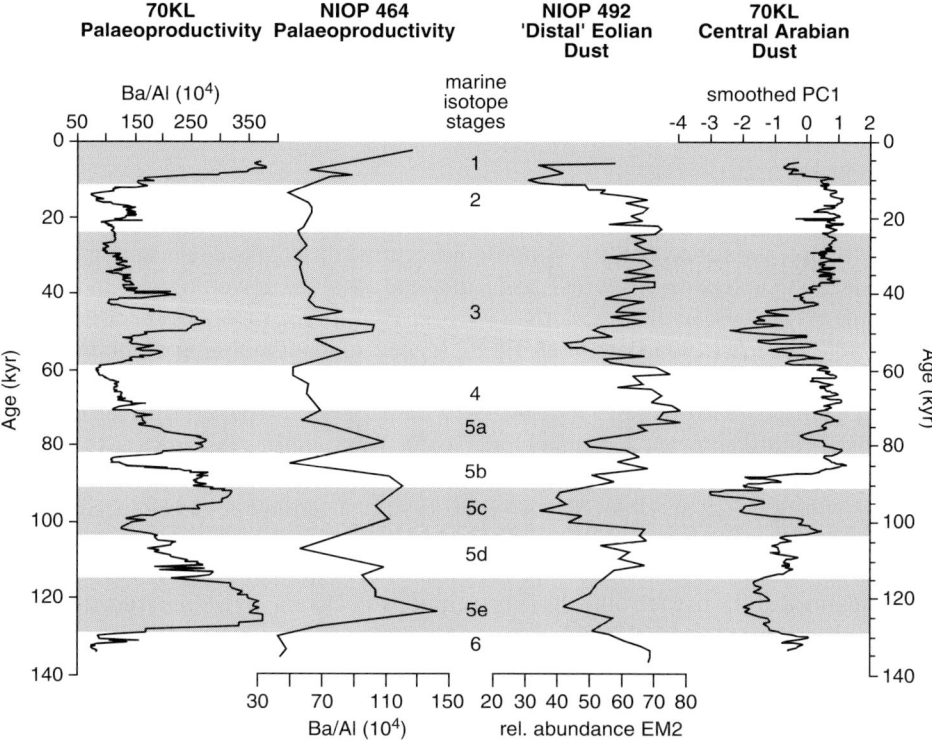

Figure 4. Records of Arabian Sea palaeoproduction and dust supply from the Arabian Peninsula during the last glacial cycle in relation to oxygen isotope stages and substages. Ba/Al records are from the western Arabian Sea (core 70 KL, Leuschner et al. (in press)) and the northern Arabian Sea (core NIOP 464, Reichart et al. (1997)). The first principal component (PC1) from the geochemical record of core 70 KL represents the relative amount of lithogenic dust derived from the Central Arabian Region (Leuschner et al., in press), and the 'distal' aeolian dust end member (EM2) from NIOP 492 is estimated by numerical modelling of grain-size distributions (Prins and Weltje 1999).

NE monsoon convective mixing), and on dust deposition, with most of the variability observed in the precession (23 kyr) frequency band (Clemens and Prell 1991; Altabet et al. 1995; Reichart et al. 1998). High sea-surface production prevailed throughout the Arabian Sea during the warm stages MIS 5e, 5c, 5a, early 3, and the Holocene (Fig. 4). This suggests that upwelling, and by inference the South Asian Monsoon, at these times was similar to its present day state. Dust transport into the Arabian Sea reconstructed from two different sites yield consistent results (Fig. 4). Warm stages see a much reduced deposition of dust from the Arabian Peninsula, which implies more humid conditions there. Large quantities of dust were deposited during the cold stages (MIS 5d, 5b, 4, and the LGM). These findings are in good agreement with enhanced dune activity within the Arabian Peninsula, especially for MIS 5d and 4 (Preusser, pers. comm.). High resolution sedimentary records of total organic carbon and bulk $CaCO_3$ from the Arabian Sea show a millennial-scale variability during MIS 3 that very much resembles the Dansgaard/Oeschger climate oscillation observed in Greenland ice cores (Schulz et al. 1998; Leuschner and Sirocko 2000). Although the relation of these records to a specific feature of the South Asian Monsoon remains ambiguous, such a similarity indicates a close coupling between northern high-latitude climate and sediment supply to the Arabian Sea during cold stages. In particular, dust flux to the Arabian Sea since the LGM appears to have followed a 1450-yr cycle, comparable to observed climate variability in other northern hemispheric climate records (Sirocko et al. 1996).

Deglaciation and Holocene

A step-wise intensification of the SW monsoon occurred during the deglaciation, as seen in planktonic $\delta^{18}O$, alkenone-based SST, and faunal and chemical upwelling/palaeoproductivity records everywhere in the Arabian Sea (Sirocko et al. 1993; 1996; Overpeck et al. 1996; Zonneveld et al. 1997; Sonzogni et al. 1998). It appears that the timing of the two largest and most commonly observed steps is comparable to the deglaciation sequence of the northern hemisphere (Sirocko 1996; Cayre and Bard 1999), if local effects (see below) on ^{14}C-derived stratigraphy are accounted for (Fig. 5). This confirms the previously proposed link between climate in the high-latitude of the northern hemisphere and the South Asian Monsoon (Overpeck et al. 1996; Sirocko et al. 1996). Nevertheless, the Holocene level of SW monsoon intensity was apparently not reached before roughly 9.4 cal. kyr (Fig. 5). The pattern of several successive climate steps in the monsoon region since the LGM, on average a millennium apart from each other, is in principle also inherent in records from the ENSO-dominated eastern subtropical Pacific, which suggests a close coupling of these two climate systems over that time interval (Heusser and Sirocko 1997). But detailed comparison of palaeo-monsoon records within the Arabian Sea or with palaeoclimate records elsewhere must incorporate local proxy response (Anderson and Prell 1992; Delaygue et al. 2001) and the effects of seasonal upwelling and convective mixing in the Arabian Sea on any ^{14}C-derived stratigraphy. A number of ^{14}C studies suggest that the reservoir ^{14}C age of the Arabian Sea surface varied from 640 ^{14}C years at the beginning of the last century to 850 ^{14}C years during the mid- Holocene, to as much as 1140 ^{14}C yrs during the deglaciation and the early Holocene (Uerpmann 1991; von Rad et al. 1999; Staubwasser et al. 2002).

So far, only a few records exist for the Holocene evolution of the South Asian Monsoon. During the early and mid Holocene, upwelling and palaeoproduction in the western Arabian Sea was enhanced compared to the present (Anderson and Prell 1992; Overpeck et al. 1996). Stable oxygen isotope records on planktonic foraminifera from the eastern Arabian Sea indicate significant changes in river discharge from the Indian subcontinent on a centennial time-scale (Sarkar et al. 2000; Staubwasser et al., unpub.). Across the 8.2 cal. kyr event, where continental records from Africa and Asia indicate drier conditions (Gasse and van Campo 1994; Stager and Mayewski 1997), river discharge decreased (heavier $\delta^{18}O$), and the amount of dust deposited from Arabia sharply increased (Fig. 5). Another brief pulse of dust deposition into the Gulf of Oman occurred at 4.1 cal. kyr, and has been linked to a major drought event in Mesopotamia, coincident with the end of the Akkadian civilisation (Cullen et al. 2000).

Discussion and conclusions

Palaeoenvironmental changes within the arid to sub-arid belt of N Africa and the Arabian Peninsula, as summarised in this chapter, occurred on millenial to decadal time-scales. In

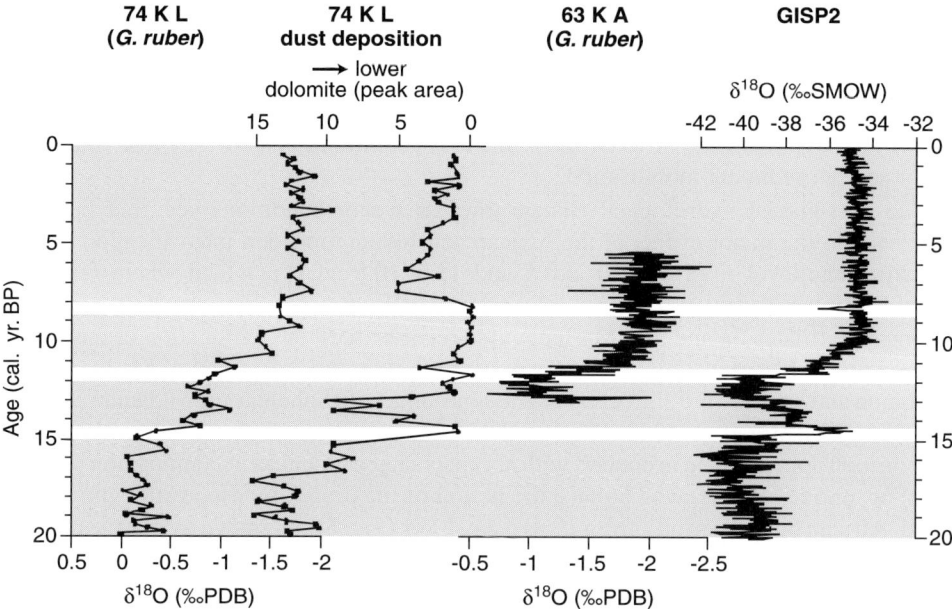

Figure 5. Stable oxygen isotopes (*G. ruber*) and dust deposition in the western Arabian Sea (core 74KL), $\delta^{18}O$ (*G. ruber*) 150 km NW of the Indus delta (core 63 KA), along with the GISP2 $\delta^{18}O$ (ice) record (Grootes and Stuiver 1997). A reservoir correction of 850 yrs (Uerpmann 1991) was applied to 74 KL ^{14}C ages (previous correction was 400 yrs, Sirocko et al. (1993)), and a variable reservoir correction between 740 and 1140 yrs was applied to 63 KA ^{14}C ages (Staubwasser et al., unpub.) before conversion to calibrated ages (Stuiver et al. 1998). The two largest events during the deglaciation and the 8.2 cal. kyr event are marked.

contrast to marine records which provide continuous, well-dated and high-resolution data, palaeohydrological proxy data from terrestrial sites are often limited by too few continuous records and a spatially-biased scatter of sites. Additionally, it must be emphasized that, at present state, for most terrestrial palaeo-data sites the chronologies show low time resolution and relatively large dating uncertainties, which limit the time resolution between >1000 yr (Late Pleistocene and LGM) to ca. 250 yr (Holocene). However, for the Late Pleistocene/Holocene transition and mainly the Holocene, well-dated terrestrial records are available and a consistent scheme of palaeoenvironmental changes can be put together, which then can be correlated to well-dated, high-resolution data from marine records.

Periods in which the hyper-arid areas of the Sahara were most extensive must be correlated to full glacial periods as shown by marine pollen and sedimentary records (Figs. 2, 4 and 5). Correlation of humid periods and restricted desert areas is more differentiated and relies on reconstructions of Saharan-Sahelian vegetation and that of Arabia (Fig. 2 and Plate 4) in close relation to geomorphological and lake status data (Plates 3a, 3b).

Marine pollen records indicate most extensive hyper-arid areas of the Sahara during glacial periods (MIS 6, 4, 2; Fig. 2) when stronger northerly trade winds on the North African continent and the Arabian Peninsula led to a decrease in humid tropical taxa and an increase in relative pollen amounts from the Mediterranean steppe. During more humid periods (MIS early 5, 3) restricted desert areas and northward extensions of Sudanian savannas occurred and were related to monsoon intensification. Prior to the LGM palaeolake-forming episodes occurred in the N Sahara around 150–120, 90–65, 45, and 30 kyr. The youngest two of these episodes indicate that the recharge occurred under cooler than at present conditions when the source of precipitation in the Sahel was of tropical convective origin, whereas the Sahara received precipitation from wet Atlantic air masses. The end of the cool and wet periods of aquifer recharge mark the onset of the LGM arid conditions as shown by low lake levels and/or aeolian sediment mobilisation.

The post-glacial hydrological change towards wetter conditions occurred in correspondence to the major events of the African and Indian monsoon intensification by two step-wise lake-level rises at 15.0 ± 0.5 and 11.5–10.8 cal. kyr, which where separated by an arid interval centered around 12.4 cal. kyr and corresponds to the Younger Dryas Chronozone (Plates 3a and 3b). Marine pollen records mark the termination of the hyperarid conditions between 15.6 and 14.0 cal. kyr with a southward spread of Mediterranean vegetation and a northward shift of open grass-savanna with Sahelian and Sudanian arboreal taxa (Hooghiemstra et al. 1992; Dupont 1993). This two-step deglacial intensification of the monsoon appears to be in concert with abrupt changes in sediment composition (aeolian input) of marine records, which place the period of maximum monsoon strength, the so-called 'African humid period', from 10.4–5.5 cal. kyr (Fig. 5). As shown by marine records, the maximum monsoon intensification was reached ca. 9.4 cal. kyr (Fig. 5) and corresponds in timing to the highest Holocene lake-levels and the maximum Holocene northward shift of vegetation (Plates 3a, 3b and 4). However, this northward shift of vegetation was patchy as the distribution of vegetation followed hydrologically-favoured migration paths such as mountains, rivers, wadis and oases. Additionnally, temporal variations in vegetation changes between sites suggest that these occurred with site-specific timing and amplitudes and for many sites the reconstructed vegetation lacks modern analogues so that the zonation and composition of vegetation may have been different from what exists today. Therefore, the model of shifting vegetation zones is too simplistic.

The early- to mid-Holocene humid period was punctuated by dry spells, e.g., at ca. 8.4–8.0 cal. kyr as registered in terrestrial and marine records (Plates 3a and 3b, Fig. 5). However, this event is not clearly traced in the E. Sahara probably due to its strongly continental position being remote from the Atlantic and Indian monsoon source regions so that local, site-specific characteristics become more important. In contrast to its abrupt onset, the weakening of the monsoon during the mid- to late-Holocene was more gradual and started earlier in the North (Plate 3b). However, despite this time-transgressive fading (ca. 5.7–4.2 cal. kyr) of the period of maximum monsoon strength, the Sahara-Sahel region experienced sharp reductions in effective moisture at 6.7–5.5 cal. kyr and 4.0–3.6 cal. kyr. These drier events bracket the end of the 'African humid period' which has been identified in marine records at ca. 5.5 cal. kyr. Vegetation changes in response to drier conditions were produced by an increasing trend towards aridification with different site specific amplitudes and temporal variations after ca. 7.8–6.8 cal. kyr and accelerated between ca. 4.2–2.4 cal. kyr. Human impact might have started during the mid- and late-Holocene but its role as being responsible for the degradation of savannas and formation of the modern cultural landscape is the subject of controversy. However, the major vegetation change at the onset of the late Holocene, which led eventually to the modern vegetation zones, occurred in concert with anthropogenically-independent palaeohydrological changes and can therefore be confidently attributed to climatic change. A large number of sites is truncated by the intensive deflation during the present arid to hyper-arid period and therefore the documentation of the time-transgressive fading of the Holocene wet phase is limited to few sites. A return to a period of wetter conditions has been recorded between 4.2–3.2 cal. kyr at several palaeolakes, wadis and rivers of the arid - semi-arid belt but the timing and intensity are increasingly influenced by local and site-specific characteristics as the amplitude of palaeohydrological changes decrease. After ca. 3.2 cal. kyr arid or even hyper-arid conditions prevailed, but high-resolution, continuous records from the Sahel show that marked environmental changes occurred and climate was more variable since 1.5 cal. kyr.

The early to mid-Holocene palaeoenvironmental changes within the arid to sub-arid belt of N. Africa represent the largest anomaly of the atmosphere-biosphere system during the last 12,000 years. Although this anomaly has been subject to intensive modeling studies (Joussaume et al. 1999; Broström et al. 1998; Kutzbach and Liu 1997) none of the models so far has been able to simulate the full extent of land surface changes to account for the observations. Model-data comparisons show that the General Circulation Models do not generate as arid conditions over desert areas as the palaeodata suggest (Pinot et al. 1999; Braconnot et al. 2000). Also, modelling of the position and extent of biomes such as 'tropical xerophytic bush/savanna' and 'warm grass/shrub' (Jolly et al. 1998a; 1998b) tend to underestimate the latitudinal shifts of the West-African vegetation zones. As strong amplifications of aridity and humidity by atmosphere-vegetation feedbacks have been modelled for the mid-Holocene (Texier et al. 1997; Claussen et al. 1999), this bias between observations (Pachur and Altmann 1997; Hoelzmann et al. 1998) and models has been attributed to insufficient incorporation of synergistic interactions between feedback-mechanisms in the biosphere and the resulting changes in land-surface conditions into atmosphere-ocean models (e.g., Braconnot et al. (1999, this volume), de Noblet-Ducoudré et al. (2000)).

A synchronously coupled atmosphere-ocean-vegetation climate model, which was configured for a transient simulation of the last 9.0 kyr, simulated an abrupt and robust humid-arid transition at 5.5 kyr that occurred within 500 years (Claussen et al. 1999). This abrupt transition was triggered by subtle and gradual variations in the Earth's orbit which were strongly amplified by atmosphere-vegetation feedbacks in the subtropics. The timing of the transition was mainly governed by a global interplay between atmosphere, ocean, sea-ice, and vegetation (Claussen et al. 1999). However, this transient simulation does not consider the palaeoenvironmental changes being observed in terrestrial and marine palaeo-data, such as (i) the reduced effective moisture during a short-term period around 8.2 cal. kyr; (ii) the two-stepped character of an aridification trend occurring at 6.7–5.5 cal. kyr and 4.0–3.6 cal.; and (iii) the return to wetter conditions of low amplitude before and after 3.2 cal. kyr.

Large changes in effective moisture have strong consequences on the physics of the land surface notably in the tropics and subtropics. Thus changes in precipitation patterns may be dominating climate change in the arid to subarid belt of N. Africa and the Arabian Peninsula. Further palaeoenvironmental research should focus on, firstly, obtaining solid chronologies at a better time resolution, and secondly, quantitative reconstructions of hydrological cycles and moisture transport patterns.

The complex interactions and feedback mechanisms between the atmosphere, the oceans, the cryosphere, and the terrestrial systems can be explored best by models which couple these compartments synchronously or by Earth system Models of Intermediate Complexity (EMICs) which integrate the interacting components of the climate system (Claussen et al. 2001). In combination, results from palaeoclimate studies and integrative modeling approaches should lead to a better understanding of the response of environmental changes in the tropics and subtropics and their influence on the global climate system.

Summary

Periods in which the hyper-arid areas of the Sahara and the Arabian Peninsula were most extensive must be correlated to full glacial periods as shown by marine sedimentary and pollen records. Large excursions of the Saharan-Sahelian boundary (between 15° and 21 °N) occurred during glacial-interglacial transitions. Prior to the LGM (ca. 23–28 kyr) palaeolake-forming and aquifer recharge episodes occurred in the N. Sahara around 150–120, 90–65, 45, and 30 kyr. The youngest of these episodes indicate that the recharge occurred under cooler than at present conditions. The end of the last cool and wet period of aquifer recharge marks the onset of the LGM arid conditions as shown by low lake levels and/or aeolian sediment mobilisation. Marine sedimentary records and terrestrial data (palaeolake and pollen records) suggest that the postglacial intensification of the monsoon and wetter conditions on the continent occurred in concert and in two abrupt events at about 15.0 ± 0.4 and 11.5–10.8 cal. kyr, which were separated by an arid interval centered around ca. 12.4 cal. kyr (Younger Dryas Chronozone).

During the Holocene period of strong monsoons (10.4–5.5 cal. kyr) the maximum monsoon intensification was reached ca. 9.4 cal. kyr and corresponds in timing to the highest Holocene lake-levels and the maximum Holocene northward distribution of vegetation, which was patchy as it followed hydrologically favoured migration paths. For many sites the reconstructed vegetation lacks modern analogues and therefore the zonation and composition of vegetation during the Holocene were different from that which exist today, showing that a model of shifting vegetation zones is too simplistic. The early-mid Holocene humid period was punctuated by dry spells, e.g., at ca. 8.4–8.0 cal. kyr as registered in marine and some terrestrial records. Vegetation changes in response to drier conditions were produced by an increasing trend towards aridification after ca. 7.8–6.8 cal. kyr and accelerated between ca. 4.2–2.4 cal. kyr. Sharp lake-level reductions at 6.7–5.5 and 4.0–3.6 cal. kyr document a time-transgressive fading of the Holocene wet phase, while the end of this so-called African Humid Period is indicated in marine records at ca. 5.5 cal. kyr. The major vegetational change at the onset of the late Holocene, which led eventually to the modern vegetation zones, occurred in concert with anthropogenically-independent palaeohydrological changes and is therefore attributed mainly to climatic change. After ca. 3.2 cal. kyr arid to hyper-arid conditions prevailed, but a return to wetter conditions of minor amplitude is registered in Sahelian sites and show that environmental changes occurred and climate was variable.

Acknowledgments

The responsible co-authors for the different sections of this review are: Philipp Hoelzmann and Françoise Gasse (Near-surface palaeohydrological indicators and lacustrine records); Ulrich Salzmann and Lydie Dupont (Vegetational changes); and Michael Staubwasser, Dirk Leuschner, and Frank Sirocko (Evolution of the south Asian monsoon since the Penultimate glaciation — a synthesis from Arabian Sea records). Thorough reviews by M.A.J. Williams and an anonymous reviewer greatly improved this paper.

References

Abell P., Hoelzmann P. and Pachur H.-J. 1996. Stable isotope ratios of gastropod shells and carbonate sediments of NW Sudan as palaeoclimatic indicators. Palaeoecol. Afr. 24: 33–52.

Abell P. and Hoelzmann P. 2000. Holocene paleoclimates in Northwestern Sudan: stable isotope studies on molluscs. Glob. Planet. Chan 26: 1–12.

Adams L.J. and Tetzlaff G. 1985. The extension of Lake Chad at about 18,000 yr BP. Zeitschr. Gletscherk. Glaz. 21: 115–123.

Al-Sayari S.S. and Zötl J.G. 1978. Quaternary Period in Saudi Arabia. Springer-Verlag, New York, 179 pp.

Altabet M.A., François F., Murray D.W. and Prell W.L. 1995. Climate-related variations in denitrification in the Arabian Sea from sediment $^{15}N/^{14}N$ ratios. Nature 373: 506–509.

Anderson D.M. and Prell W.L. 1992. The structure of the southwest monsoon winds over the Arabian Sea during the late Quaternary: Observations, simulations, and marine geological evidence. J. Geophys. Res. 97: 15481–15487.

Aucour A.M. 1988. Sédimentologie des dépôts lacustres holocènes de la région de Taoudenni (Mali). Implications paléoclimatiques. Ph.D. thesis, Université Aix-Marseille, France, 165 pp.

Ballouche A. and Neumann K. 1995. A new contribution to the Holocene vegetation history of the West African Sahel: pollen from Oursi, Burkina Faso and charcoal from three sites in Northeast Nigeria. Veget. Hist. Archaeobot. 4: 31–39.

Banse K. 1994. On the coupling of hydrography, phytoplankton, zooplankton, and settling organic particles offshore in the Arabian Sea. In: Lal B.U. (ed.), Biogeochemistry of the Arabian Sea. Indian Academy of Sciences, Delhi, pp. 27–63.

Barakat H.N. 1995a. Middle Holocene vegetation and human impact in Central Sudan: charcoal from the Neolithic site at Kadero. Veget. Hist. Archaeobot. 4: 101–108.

Barakat H.N. 1995b. Charcoals from Neolithic site at Nabta Playa (E-75-6), Egypt. Acta Palaeobot. 35: 163–166.

Barakat H.N. 1995c. Plant macroremains from Z'Bib N Elias. A subfossil midden from a prehistoric cave in the Hoggar, Central Sahara. Acta Palaeobot. 35: 99–103.

Barboni D., Bonnefille R., Alexandre A. and Meunier J.D. 1999. Phytoliths as paleoenvironmental indicators, West Side Middle Awash Valley, Ethiopia. Palaeogeogr. Palaeoclim. Palaeoecol. 152: 87–100.

Baumhauer R. 1991. Palaeolakes of the South Central Sahara: problems of palaeoclimatological interpretation. Hydrobiologia 214: 347–357.

Ben Tiba B. and Reille M. 1982. Recherches pollen analytiques dans les montagnes de Kroumirie (Tunisie septentrionale): premiers résultats. Ecologica Medit. VIII 4: 75–80.

Braconnot P., Joussaume S., Marti O. and de Noblet N. 1999. Synergistic feedbacks from ocean and vegetation on the African monsoon response to mid-Holocene insolation. Geophys. Res. Lett. 26: 2481–2484.

Braconnot P., Joussaume S., de Noblet N. and Ramstein G. 2000. Mid-Holocene and Last Glacial Maximum African monsoon changes as simulated within the Paleoclimate Modelling Intercomparison Project. Glob. Planet. Chan. 26: 51–66.

Braconnot P., Harrison S.P., Joussaume S., Hewitt C.D., Kitoch A., Kutzback J.E., Liu Z., Otto-Bliesner B., Syktus J. and Weber N., this volume. Evaluation of PMIP coupled ocean-atmosphere simulations of the mid-Holocene. In: Battarbee R.W., Gasse F. and Stickley C.E. (eds), Past

Climate Variability through Europe and Africa. Kluwer Academic Publishers, Dordrecht, the Netherlands, pp. 515–533.

Brock J.C. and McClain C.R. 1992. Interannual variability in phytoplankton blooms observed in the northwestern Arabian Sea during the southwest monsoon. J. Geophys. Res. 96: 733–750.

Broström A., Coe M., Harrison S.P., Gallimore R., Kutzbach J.E., Foley J., Prentice I.C. and Behling P. 1998. Land surface feedbacks and paleomonsoons in Northern Africa. Geophys. Res. Lett. 25: 3615–3618.

Brun A. 1987. Etude palynologique des limons organiques du site de l'Oued el Akarit (sud Tunisien). Bull. assoc. franc. étud. Quat. 1987-1: 19–25.

Brun A. 1989. Microflores et paléovégétations en Afrique du Nord depuis 30 000 ans. Bull. Soc. geol. France 8V, 25–33.

Brun A. 1991. Reflexions sur les Pluviaux et Arides au Pléistocène supérieur et à l'Holocène en Tunisie. Palaeoecol. Afr. 22: 157–170.

Burns S.J., Matter A., Frank N. and Mangini A. 1998. Speleothem-based paleoclimate record from Northern Oman. Geology 26: 499–502.

Carrara C., Cremaschi M. and Quinif Y. 1998. The travertine deposits in the Tadrart Acacus (Libyan Sahara) nature and age. In: Cremaschi M. and Di Lernia S. (eds), Wadi Teshuinat Palaeoenvironment and Prehistory in South-Western Fezzan (Libyan Sahara). All'Insegna del Giglio, Firenze, CNR, Quaderni di Geodinamica Alpina e Quaternaria, Milano, pp. 59–66.

Castelletti L., Castiglioni E., Cottini M. and Rottoli M. 1999. Archaeobotanical analysis of charcoal, wood and seeds. In: Di Lernia S. (ed.), The Uan Afuda Cave. Arid Zone Monographs 1. Edizioni All'Insegna del Giglio, Firenze, pp. 131–148.

Causse C., Conrad G., Fontes J.C., Gasse F., Gibert E. and Kassir A. 1988. Le dernier "Humide" pléistocène du Sahara nord-occidental daterait de 80–100 000 ans. C.R. Acad. Sc. Paris 306: 1459–1464.

Causse C., Coque R., Fontes J.C., Gasse F., Gibert E., Ben Ouezdou H. and Zouari K. 1989. Two high levels of continental waters in the southern Tunisian chotts at about 90 and 150 ka. Geology 17: 922–925.

Causse C., Coque R., Fontes J.C., Gasse F., Gibert E., Ben Ouezdou H. and Zouari K. 1991. A reply to Drs Richards and Vita-Finzi. Two high levels of continental waters in the Southern Tunisian chotts at about 90 and 150 ka. Geology 29: 95–96.

Cayre O. and Bard E. 1999. Planktonic foraminiferal and alkenone records of the last deglaciation from the eastern Arabian Sea. Quat. Res. 52: 337–342.

Chateauneuf J.J., Faure H. and Lézine A.M. 1986. Facteurs contrôlant la genèse et la destruction des tourbes tropicales du littoral ouest-africain. Doc. Bur. Rech. geol. minières 110: 77–91.

Clark I.D. and Fontes J.C. 1990. Paleoclimatic reconstruction in Northern Oman based on carbonates from hyperalkaline groundwaters. Quat. Res. 33: 320–336.

Claussen M., Kubatzki C., Brovkin V., Ganopolski A., Hoelzmann P. and Pachur H.-J. 1999. Simulation of an abrupt change in Saharan vegetation in the mid-Holocene. Geophys. Res. Lett. 26: 2037–2040.

Claussen M., Mysak L.A., Weaver A.J., Crucifix M., Fichefet T., Loutre M.F., Weber S.L., Alcamo J., Alexeev V.A., Berger A., Calov R., Ganopolski A., Goosse H., Lohmann G., Lunkeit F., Mokhov I.I., Petoukhov V. and Wang Z. 2001. Earth system models of intermediate complexity: closing the gap in the spectrum of climate systems models. Clim. Dyn. 18: 579–586.

Clemens S.C. 1998. Dust response to seasonal atmospheric forcing: Proxy evaluation and calibration. Palaeoceanography 13: 471–490.

Clemens S.C. and Prell W.L. 1991. Late Pleistocene variability of Arabian Sea summer monsoon winds and continental aridity: Eolian records from the lithogenic compounds of deep-sea sediments. Palaeoceanography 5: 109–145.

COHMAP Members 1988. Climatic changes of the last 18,000 years: observations and model simulation. Science 241: 1043–1052.

Cour P. and Duzer D. 1976. Persistance d' un climat hyperaride au Sahara central et méridional au cours de l'Holocène. Rev. Geogr. Phys. Geol. Dyn. 18: 175–198.

Cremaschi M. 1998. Late Quaternary geological evidence for environmental changes in South-Western Fezzan (Libyan Sahara). In: Cremaschi M. and di Lernia S. (eds), Wadi Teshuinat Palaeoenvironment and Prehistory in South-Western Fezzan (Libyan Sahara), All'Insegna del Giglio, Firenze, CNR, Quaderni di Geodinamica Alpina e Quaternaria, Milano, pp. 13–48.

Cullen H.M., deMenocal P.M., Hemming S., Hemming G., Brown F.H., Guilderson T. and Sirocko F. 2000. Climate change and the collapse of the Akkadian empire: Evidence from the deep sea. Geology 28: 379–382.

Damblon F. and Vanden Berghen C. 1993. Etude paléo-écologique (Pollens et Macrorestes) d'un depôt tourbeux dans l'île de Djerba, Tunisie méridionale. Palynosciences 2: 157–172.

Delaygue G., Bard E., Rollion C., Jouzel J., Stievenard M., Duplessy J.-C. and Ganssen G. 2001. Oxygen isotope/salinity relationship in the northern Indian Ocean. J. Geophys. Res. 106: 4565–4574.

deMenocal P., Ortiz J., Guilderson T. and Sarnthein M. 2000a. Coherent high- and low-latitude climate variability during the Holocene Warm Period. Science 288: 2198–2202.

de Noblet-Ducoudré N., Claussen M. and Prentice I.C. 2000. Mid-Holocene greening of the Sahara: first results of the GAIM 6000 year BP experiment with two asynchronously coupled atmosphere/biome models. Clim. Dyn 16: 643–659.

Deynoux M., Kocurek G., Benan Ahmed C.A., Crabaugh M., Havholm K. and Pion J.C. 1993. Stratigraphie séquentielle en milieu désertique. Exemple de l'erg Akchar en Mauritanie occidentale (Afrique de l'Ouest). C.R. Acad. Sc. Paris 317: 1199–1205.

Dubar C. 1988. Eléments de paléohydrologie de L'Afrique Saharienne: les dépôts quaternaires d'origine aquatique du Nord-est de L'Aïr (Niger, PALHYDAF, site 3). Thesis, Université de Paris-Sud, Orsay, France, 188 pp.

Dupont L.M. 1993. Vegetation zones in NW Africa during the Brunhes Chron reconstructed from marine palynological data. Quat. Sci. Rev. 12: 189–202.

Dupont L.M. and Hooghiemstra H. 1989. The Saharan-Sahelian boundary during the Brunhes chron. Acta Bot. Neerl. 38: 405–415.

Durand A. 1995. Conséquences géomorphologiques de phénomènes néotectoniques dans le bassin du lac Tchad: Modifications du réseau hydrographique et origine du pseudo-rivage du Mégatchad dans la région du Kadzell (République du Niger). C.R. Acad. Sc. Paris 321: 223–229.

Edmunds W.M., Fellman E. and Baba Goni I. 1999 Environmental change, lakes and groundwater in the Sahel of Northern Nigeria. J. Geol. Soc. Lond. 156: 345–356.

Edmunds W.M., Dodo A., Djoret D., Gasse F., Gaye C.B., Goni I.B., Travi Y., Zouari K. and Zuppi G.M., this volume. Groundwater as an archive of climatic and environmental change: Europe to Africa. In: Battarbee R.W., Gasse F. and Stickley C.E. (eds), Past Climate Variability through Europe and Africa. Kluwer Academic Publishers, Dordrecht, the Netherlands, pp. 279–306.

Elenga H., Peyron O., Bonnefille R., Jolly D., Cheddadi R., Guiot J., Andrieu V., Bottema S., Buchet G., De Beaulieu J.-L., Hamilton A.C., Maley J., Marchant R., Perez-Obiol R., Reille M., Riollet G., Scott L., Straka H., Taylor D., Van Campo E., Vincens A., Laarif F. and Jonson H. 2000. Pollen-based biome reconstruction for Southern Europe and Africa 18,000 yr BP. J. Biogeogr. 27: 621–634.

Fabre J. and Petit-Maire N. 1988. Holocene climatic evolution at 22–23 °N from two paleolakes in the Taoudenni area (Northern Mali). Palaeogeogr. Palaeoclim. Palaeoecol. 65: 133–148.

Farrera I., Harrison S.P., Prentice I.C., Ramstein G., Guiot J., Bartlein P.J., Bonnefille R., Bush M., Cramer W., Von Grafenstein U., Holmgren K., Hooghiemstra H., Hope G., Jolly D., Lauritzen S.-E., Ono Y., Pinot S., Stute M. and Yu G. 1999. Tropical climates at the Last Glacial Maximum: a new synthesis of terrestrial palaeoclimate data. I. Vegetation, lake-levels and geochemistry. Clim. Dy15: 823–856.

Flagg C.N. and Kim H.S. 1998. Upper ocean currents in the northern Arabian Sea from shipboard ADCP measurements collected during the 1994–1996 U.S. JGOFS and ONR programs. Deep-Sea Res. II 45: 1917–1959.

Fontes J.C. and Gasse F. 1989. On the ages of humid Holocene and late Pleistocene phase of North Africa - Remarks on "Late Quaternary climatic reconstruction for the Maghreb (North Africa)" by P. Rognon. Palaeogeogr. Palaeoclim. Palaeoecol. 70: 393–398.

Fontes J.C. and Gasse F. 1991. PALHYDAF Programme. Objectives, methods, major results. Palaeogeogr. Palaeoclim. Palaeoecol. 84: 191–215.

Fontes J.C., Gasse F., Callot Y., Plaziat J.C., Carbonel P., Dupeuble P.A. and Kacmarska I. 1985. Freshwater to marine-like environments from Holocene lakes in Northern Sahara. Nature 317: 608–610.

Gale S.J., Gilbertson D.D., Hoare P.G., Hunt C.O., Jenkinson R.D.S., Lamble A.P., O' Toole C.O., van der Veen M. and Yates G. 1993. Late Holocene environmental change in the Libyan pre-desert. J. Arid Env. 24: 1–19.

Gasse F. 1977. Evolution of Lake Abhé (Ethiopia and TFAI), from 70,000 b.p. Nature 265: 42–45.

Gasse F. 1998. Water resources variability in tropical and subtropical Africa in the past. In: Servat E. et al. (eds), Water Resources Variability in Africa during the 20th Century, IAHS 252: pp. 97–106.

Gasse F. 2000. Hydrological changes in the African tropics since the Last Glacial Maximum. Quat. Sci. Rev. 19: 189–211.

Gasse F. 2002. Diatom-inferred salinity and carbonate oxygen isotopes in Holocene waterbodies of the Western Sahara and Sahel (Africa). Quat. Sci. Rev. 21: 737–767.

Gasse F. and Fontes J.C. 1992. Climatic changes in Northwest Africa during the last deglaciation. In: Bard E. and Wallace W.S. (eds), The Last Deglaciation: Absolute and Radiocarbon Chronologies. Nato ASI Series 12, Springer-Verlag, Berlin, pp. 295–325.

Gasse F. and Street F.A. 1978. Late Quaternary lake-level fluctuations and environments of the Northern Rift Valley and Afar region (Ethiopia and Djibouti). Palaeogeogr. Palaeoclim. Palaeoecol. 24: 279–325.

Gasse F. and van Campo E. 1994. Abrupt post-glacial climate events in West Asia and North Africa monsoon domains. Earth Planet. Sci. Lett. 126: 435–456.

Gasse F., Fontes J.C., Plaziat J.C., Carbonnel P., Kaczmarska P., De Deckker P., Soulié-Märsche I. and Callot Y. 1987. Biological remains, geochemistry and stable isotopes for the reconstruction of environmental and hydrological changes in the Holocene lakes from North Sahara. Palaeogeogr. Palaeoclim. Palaeoecol. 60: 1–46.

Gasse F., Téhet R., Durand A., Gilbert E. and Fontes J.C. 1990. The arid-humid transition in the Sahara and the Sahel during the last deglaciation. Nature 346: 141–146.

Gibert E., Arnold M., Conrad G., De Deckker P., Fontes J.C., Gasse F. and Kassir A. 1990. Retour des conditions humides au Tardiglaciaire au Sahara septentrional (Sebkha Mellala, Algérie). Bull. Soc. Géol. France 6: 497–504.

Gillespie R., Street-Perrott F.A. and Switsur R. 1983. Post-glacial arid episodes in Ethiopia have implications for climate prediction. Nature 306: 680–682.

Grootes P.M. and Stuiver M. 1997. Oxygen 18/16 variability in Greenland snow and ice with 10^3 to 10^5 year time resolution. J. Geophys. Res. 102: 26455–26470.

Guendouz A., Moulla A., Edmunds W.M., Shand P., Poole J., Zouari K. and Mamou A. 1998. Palaeoclimatic information contained in groundwaters of the Grand Erg Oriental, North Africa. Proceedings Vienna Symposium 1997, IAEA SM 349/43: 555–571.

Guo Z., Petit-Maire N. and Kröpelin S. 2000. Holocene non-orbital climatic events in present-day arid areas of Northern Africa and China. Glob. Planet. Chan. 26 (1-3): 97–103.

Hassan F.A. 1997. Nile floods and political disorder in early Egypt. In: Dalfes H.N., Kukla G. and Weiss H. (eds), Third Millenium BC Climate Change and Old World Collapse. Nato ASI Series 149, Springer-Verlag, Berlin, pp. 1-23.

Haynes C.V., Eyles C.H., Pavlish L.A., Ritchie J.C. and Rybak M. 1989. Holocene paleoecology of the Eastern Sahara: Selima Oasis. Quat. Sci. Rev. 8: 109–136.

Heusser L.E. and Sirocko F. 1997. Millennial pulsing of environmental change in Southern California from the past 24 k.y: A record of Indo-Pacific ENSO events? Geology 25: 243–246.

Hillaire-Marcel C., Riser C., Rognon P., Petit-Maire N., Rosso J.C. and Soulié-Märsche I. 1983. Radiocarbon chronology of Holocene hydrologic changes in Northeastern Mali. Quat. Res. 20: 145–164.

Hoelzmann P. 1992. Palaeoecology of Holocene lacustrine sediments within the West Nubian Basin, SE-Sahara. Würzb. Geogr. Abh. 84: 59–71.

Hoelzmann P. 1993. Holozäne Limnite im NW-Sudan. Ph.D. thesis, Freie Universität Berlin, 1–191.

Hoelzmann P., Jolly D., Harrison S., Laarif F., Bonnefille R. and Pachur H.-J. 1998. Mid-Holocene land-surface conditions in Northern Africa and the Arabian peninsula: a data set for the analysis of biogeophysical feedbacks in the climate system. Glob. Biogeoch. Cycles 12: 35–51.

Hoelzmann P., Keding B., Berke H., Kröpelin S. and Kruse H.J. 2001. Environmental change and archaeology: lake evolution and human occupation in the Eastern Sahara during the Holocene. Palaeogeogr. Palaeoclim. Palaeoecol. 169: 193–217.

Holmes J.A., Allen M.J., Street-Perrott F.A., Ivonovich M., Perrott R.A. and Waller M.P. 1999b. Late Holocene palaeolimnology of Bal Lake, Northern Nigeria, a multidisciplinary study. Palaeogeogr. Palaeoclim. Palaeoecol. 148: 169–185.

Holmes J.A., Street-Perrott F.A., Allen M.J., Fothergill P.A., Harkness D.D., Kroon D. and Perrott R.A. 1997. Holocene palaeolimnology of Kajemarum Oasis, Northern Nigeria: an isotopic study of ostracodes, bulk carbonate and organic carbon. J. Geol. Soc. Lond. 154: 311–319.

Holmes J.A., Street-Perrott F.A., Perrott R.A., Stokes S., Waller M.P., Huang Y., Eglington G. and Ivanovich M. 1999a. Holocene landscape evolution of the Manga Grasslands, NE Nigeria: evidence from palaeolimnology and dune chronology. J. r. Geol. Soc. 156: 357–368.

Hooghiemstra H. 1988a. Changes of major wind belts and vegetation zones in NW Africa 20,000–5000 yr B.P., as deduced from a marine pollen record near Cap Blanc. Rev. Palaeobot. Palynol. 55: 101–140.

Hooghiemstra H. 1988b. Palynological records from northwest African marine sediments: a general outline of the interpretation of the pollen signal. Phil. Trans. r. Soc. Lond. B318: 431–449.

Hooghiemstra H. and Agwu C.O.C. 1988. Changes in the vegetation and trade winds in equatorial Northwest Africa 140,000–70,000 yr BP as deduced from two marine pollen records. Palaeogeogr. Palaeoclim. Palaeoecol. 66: 173–213.

Hooghiemstra H., Stalling H., Agwu C.O.C. and Dupont L.M. 1992. Vegetational and climatic changes at the northern fringe of the Sahara 240,000–5000 years BP: Evidence from 4 marine pollen records located between Portugal and the Canary Islands. Rev. Palaeobot. Palynol. 74: 1–53.

Jado A.R. and Zötl J.G. 1984. Quaternary Period in Saudi Arabia. Springer-Verlag, New York, USA, 278 pp.

Jäkel D. 1979. Run-off and fluvial formation processes in the Tibesti mountains as indicators of climatic history in the Central Sahara during the late Pleistocene and Holocene. Palaeoecol. Afr. 11: 13–43.

Jahns S. 1995. A Holocene pollen diagram from El Atrun, Northern Sudan. Rev. Veget. Hist. Archaeobot. 4: 23–30.

Jelinowska A., Tucholka P., Massault M., Gasse F. and Mélières F. 1997. Environmental changes deduced from magnetic properties of Late Pleistocene and Holocene lake sediments, Lake Faguibine, Mali. Eur. J. Env. Engin. Geophys. 2: 191–203.

Jolly D., Harrison S.P., Damnati B. and Bonnefille R. 1998a. Simulated climate and biomes of Africa during the late Quaternary: comparison with pollen and lake status data. Quat. Sci. Rev. 17: 629–657.

Jolly D., Prentice I.C., Bonnefille R., Ballouche A., Bengo M., Brenac P., Buchet G., Burney D., Cazet J.P., Cheddadi R., Edorh T., Elenga H., Elmoutaki S., Guiot J., Laarif F., Lamb H., Lézine A.M., Maley J., Mbenza M., Peyron O., Reille M., Reynaud-Farrera I., Riollet G., Ritchie J.C., Roche E., Scott L., Ssemmanda I., Straka H., Umer M., Van Campo E., Vilimumbalo S., Vincens A. and Waller M. 1998b. Biome reconstruction from pollen and plant macrofossil data for Africa and the Arabian peninsula at 0 and 6 ka. J. Biogeogr. 25: 1007–1027.

Joseph A., Frangi J.F. and Aranyossy J.F. 1992. Isotope characteristics of meteoric water and groundwater in the Sahelo-Sudanes zone. J. Geophys. Res. 97: 7543–7551.

Joussaume S., Taylor K.E., Braconnot P., Mitchell J.F.B., Kutzbach J.E., Harrison S.P., Prentice I.C., Broccoli A.J., Abe-Ouchi A., Bartlein P.J., Bonfils C., Dong B., Guiot J., Herterich K., Hewitt C.D., Jolly D., Kim J.W., Kislov A., Kitoh A., Loutre M.F., Masson V., McAvaney B., McFarlane N., de Noblet N., Peltier W.R., Peterschmitt J.Y., Pollard D., Rind D., Royer J.F., Schlesinger M.E., Syktus J., Thompson S., Valdes P., Vettoretti G., Webb R.S. and Wyputta U. 1999. Monsoon changes for 6000 years ago: results of 18 simulations from the Paleoclimate Modeling Intercomparison Project (PMIP). Geophys. Res. Lett. 26: 859–862.

Kröpelin S. 1993. Zur Rekonstruktion der spätquartären Umwelt am Unteren Wadi Howar (Südöstliche Sahara/NW Sudan). Berl. Geogr. Abh. 54: 1–193.

Kutzbach J.E. and Street-Perrott F.A. 1985. Milankovitch forcing of fluctuations in the level of tropical lakes from 18 to 0 kyr BP. Nature 317: 130–134.

Kutzbach J.E. and Liu Z. 1997. Response of the African monsoon to orbital forcing and ocean feedbacks in the middle Holocene. Science 278: 440–443.

Lamb H.F., Eicher U. and Switsur V.R. 1989. An 18,000-year record of vegetational, lake-level and climatic change from the Middle Atlas, Morocco. J. Biogeogr. 16: 65–74.

Le Turdu C., Tiercelin J.J., Gibert E., Travi Y., Lezzar K.-E., Richert J.-P., Massault M., Gasse F., Bonnefille R., Decobert M., Gensous B., Jeudy V., Tamrat E., Umer M., Martens K., Atnafu B., Chernet T., Williamson D. and Taieb M. 1999. The Ziway-Shala lake basin system, Main Ethiopian Rift: influence of volcanism, tectonics, and climatic forcing on basin formation and sedimentation. Palaeogeogr. Palaeoecol. Palaeoclim. 150: 135–177.

Leuschner D.C. and Sirocko F. 2000. The low-latitude monsoon climate during Dansgaard/Oeschger cycles and Heinrich events. Quat. Sci. Rev. 19: 243–254.

Leuschner D.C., Sirocko F., Schettler G. and Garbe-Schönberg D., in press. Geochemical implications for changing the dust supply by the Indian Monsoon system to the Arabian Sea during the last glacial cycle. In: Smykatz-Kloss S., Henningsen F. and Zöller L. (eds), Paleoecology of Quaternary Drylands, Lecture Notes in Earth Sciences. Springer-Verlag, Stuttgart.

Levitus S. and Boyer T.P. 1994. World Ocean Atlas 1994, vol. 4, Temperature NOAA Atlas NESDIUS4. US Department of Commerce, Washington D.C., 117 pp.

Lézine A.-M. 1982. Etude palynologique des sédiments quaternaires du lac Abiyata (Ethiopie). Palaeoecol. Afr. 14: 93–98.

Lézine A.M. 1987. Paléoenvironnemnts végétaux d'Afrique nord-tropicale depuis 12000 BP Ph.D. thesis, University of Aix-Marseille 2: 1–180.

Lézine A.M. 1988. Les variations de la couverture forestière mésophile d'Afrique occidentale au cours de l'Holocène. C.R. Acad. Sci. Paris Sér. II 307: 439–445.

Lézine A.M. 1989. Late Quaternary vegetation and climate of the Sahel. Quat. Res. 2: 317–334.

Lézine A.M. 1991. West African paleoclimates during the last climatic cycle inferred from an Atlantic Deep-Sea Pollen record. Quat. Res. 35: 456–463.

Lézine A.M. 1993. Chemchane, histoire d' une Sebkha. Sécheresse 4: 25–30.

Lézine A.-M. and Casanova J. 1989. Pollen and hydrological evidence for the interpretation of past climates in tropical West Africa during the Holocene. Quat. Sci. Rev. 8: 45–55.

Lézine A.-M. and Casanova J. 1991. Correlated oceanic and continental records demonstrate past climate and hydrology of North Africa (0–140 ka). Geology 19: 307–310.

Lézine A.M., Casanova J. and Hillaire-Marcel C. 1990. Across an early Holocene humid phase in Western Sahara: Pollen and isotope stratigraphy. Geology 18: 264–265.

Lézine A.M., Saliège J.-F., Robert C. and Wertz F. 1998. Holocene lakes from Ramlat as-Sab'atayn (Yemen) illustrate the impact of the monsoon activity in Southern Arabia. Quat. Res. 50: 290–299.

Lioubimtseva E., Simon B., Faure H., Faure-Denard L. and Adams J.M. 1998. Impacts of climatic change on carbon storage in the Sahara-Gobi desert belt since the Last Glacial Maximum. Glob. Planet. Chan. 16-17: 95–105.

Maley J. 1977. Palaeoclimates of Central Sahara during the early Holocene. Nature 269: 573–577.

Maley J. 1981. Etudes palynologiques dans le bassin du Tchad et paléoclimatologie de l'Afrique nord tropicale de 30,000 ans a l'époque actuelle. Travaux et Documents de l'ORSTOM 129: Paris, 586 pp.

Maley J. 1983. Histoire de la végétation et du climat l'Afrique nord-tropicale au Quaternaire récent. Bothalia 14: 377–389.

Maley J. 1997. Middle to late Holocene changes in tropical Africa and other continents: Paleomonsoon and sea surface temperature variations. In: Dalfes H.N., Kukla G. and Weiss H. (eds), Third Millenium BC Climate Change and Old World Collapse. Nato ASI Series 149, Springer-Verlag, Berlin, pp. 611–640.

Maley J. 2000. Last Glacial Maximum lacustrine and fluviatile Formations in the Tibesti and other Saharan mountains, and large-scale climatic teleconnections linked to the activity of the Subtropical Jet Stream. Glob. Planet. Chan. 26 (1-3): 121–136.

Maley J., Cohen J., Faure H., Rognon P. and Vincent P.M. 1970. Quelques formations lacustres et fluviatiles associeés à différentes phases du volcanisme au Tibesti (nord du Chad). Cah. ORSTOM Sér. Geol. 2: 127–152.

Maley J. and Livingstone D.A. 1983. Extension d'un élément montagnard dans le sud du Ghana (Afrique de l'Ouest) au Pléistocène supérieur et à l'Holocène inférieur: premières données polliniques. Comptes Rendus Académie des Sciences, Série II 296: 1287–1292.

McClure H.A. 1976. Radiocarbon chronology of late Quaternary lakes in the Arabian Desert. Science 263: 755.

Medus J. 1984. Essai de reconstitution de la végétation et du climat Holocènes de la cote septentrionale du Sénégal. Rev. Palaeobot. Palynol. 41: 31–38.

Mercuri A.M. 1999. Palynological analysis of the Early Holocene sequence. Arid Zone Archaeol. 1: 149–181.

Michel P. and Assemien P. 1969–1970. Etudes sédimentologique et palynologique des sondages de Bogué (basse vallée du Sénégal) et leur interprétation morphoclimatiques. Rev. Géom. dyn. 19: 97–113.

Morrison J.M., Codispoti L.A., Smith S.L., Wishner K., Flagg C., Gardner W.D., Gaurin S., Naqvi S.W.A., Maghnani V., Prosperie L. and Gundersen J.S. 1999. The oxygen minimum zone in the Arabian Sea during 1995. Deep-Sea Res. 46: 1903–1931.

Neff U., Burns S.J., Mangini A., Mudelsee M., Fleitmann D. and Matter A. 2001. Strong coherence between solar variability and the monsoon in Oman between 9 and 6 kyr ago. Nature 411: 290–293.

Neumann K. 1989. Holocene vegetation of the Eastern Sahara: charcoal from prehistoric sites. Afr. Archaeol. Rev. 7: 97–116.

Neumann K. 1992. Une flore soudanienne au Sahara central vers 7000 BP: les charbons de bois de Fachi, Niger. Bulletin Société Botanique France 139, Actual. bot. 2-4: 565–569.

Neumann K. and Uebel D. 2001. The cold early Holocene in the Acacus: evidence from charred wood. In: Garcea E. (ed.), Uan Tabu in the Settlement History of the Libyan Sahara. Arid Zone Monographs 2. Edizione all'Insegna del Giglio, Rome, pp. 211–213.

Overpeck J., Anderson D., Trumbore S. and Prell W. 1996. The southwest Indian Monsoon over the last 18,000 yrs. Clim. Dynam. 12: 213–225.

Pachur H.J. 1993. Paläodrainagesysteme im Sirte-Becken und seiner Umrahmung. Würzb. Geogr. Arb. 87: 17–34.

Pachur H.J. and Hoelzmann P. 1991. Paleoclimatic implications of late Quaternary lacustrine sediments in Western Nubia. Sudan. Quat. Res. 36: 257–276.

Pachur H.J. and Hoelzmann P. 2000. Late Quaternary palaeoecology and palaeoclimates of the Eastern Sahara. J. Afr. Earth Sci. 30: 929–939.

Pachur H.J. and Kröpelin S. 1987. Wadi Howar: paleoclimatic evidence from an extinct river system in the South-Eastern Sahara. Science 237: 298–300.

Pachur H.J. and Wünnemann B. 1996. Reconstruction of the palaeoclimate along 30 °E in the Eastern Sahara during the Pleistocene/Holocene transition. Palaeoecol. Afr. 24: 1–32.

Pachur H.J. and Altmann N. 1997. The Quaternary (Holocene, ca. 8000 a B.P.). In: Schandelmeier H. and Reynolds P.O. (eds), Palaeogeographic-Palaeotectonic Atlas of North-Eastern Africa, Arabia, and Adjacent Areas. Balkema, Rotterdam, pp. 111–125.

Pachur H.-J., Kröpelin S., Hoelzmann P., Goschin M. and Altmann N. 1990. Late Quaternary fluvio-lacustrine environments of Western Nubia. Berl. geowiss. Abh. (A) 120.1: 203–260.

Pachur H.J., Röper H.-P., Kröpelin S. and Goschin M. 1987. Late Quaternary Hydrography of the Eastern Sahara. Berl. geowiss. Abh. 75.2: 331–384.

Petit-Maire N. and Bouysse P. 2000. Geological records of the recent past, a key to the near future world environments. Episodes 23: 230–246.

Petit-Maire N. and Riser J. 1981. Holocene lake deposits and palaeoenvironments in Central Sahara, Northeastern Mali. Palaeogeogr. Palaeoclim. Palaeoecol. 35: 45–61.

Petit-Maire N. and Riser J. 1983. Sahara ou Sahel? Quaternaire récent du bassin de Taoudenni (Mali). Luminy, Marseille, France, 179 pp.

Petit-Maire N., Page N. and Marchand J. 1993. The Sahara in the Holocene: Map 1/5,000,000. Laboratoire de Géologie du Quaternaire CNRS, Marseille, France.

Petit-Maire N. 1991. Paléoenvironnements du Sahara. Lacs Holocènes à Taoudenni (Mali). Editions CNRS, Paris, France, 212 pp.

Petit-Maire N., Fabre J., Carbonel P., Schulz E. and Aucour A.M. 1987. La dépression de Taoudenni (Sahara malien) à l'Holocène. Géodynamique 2: 127–160.

Pinot S., Ramstein G., Harrison S.P., Prentice I.C., Guiot J., Stute M. and Joussaume S. 1999. Tropical paleoclimates at the Last Glacial Maximum: comparison of Paleoclimate Modeling Intercomparison Project (PMIP) simulations and paleodata. Clim. Dynam. 15: 857–874.

Pons A. and Quézel P. 1958. Premières remarques sur l'étude palynologique d' un guano fossile du Hoggar. C.R. Acad. Sc. Paris 244: 2290–2292.

Prell W.L. and Van Campo E. 1986. Coherent response of Arabian Sea upwelling and pollen transport to late Quaternary monsoonal winds. Nature 323: 526–528.

Prins M.A., Postma G., Cleveringa J., Cramp A. and Kenyon N.H. 2000. Controls on sediment supply to the Arabian Sea during the late Quaternary. Marine Geol. 169: 327–349.

Prins M.A. and Weltje G.J. 1999. End-member modeling of siliciclastic grain-size distributions; the late Quaternary record of eolian and fluvial sediment supply to the Arabian Sea and its paleoclimatic significance. In: Harbaugh J.W., Watney W.L., Rankey E.C., Slingerland R., Goldstein R.H. and Franseen E.K. (eds), Numerical Experiments in Stratigraphy; Recent Advances in Stratigraphic and Sedimentologic Computer Simulation, SEPM Spec. Publ., Tulsa, OK, pp. 91–111.

Quézel P. and Martinez C. 1958–59. Le dernier interpluvial au Sahara central. Essai de chronologie palynologique et paléoclimatique. Lybica 6-7; 211–227.

Rao Y.B.1981. The climate of the Indian subcontinent. In: Takahashi K.A. (ed.), World Survey of Climatology, Vol. 9: Climates of Southern and Western Asia. Elsevier, Amsterdam, pp. 67–118.

Reichart G.J., den Dulk M., Visser H.J., van der Weijden C.H. and Zachariasse W.J. 1997. A 225 kyr record of dust supply, paleoproductivity and the oxygen minimum zone from the Murray Ridge (northern Arabian Sea). Palaeogeogr. Palaeoclim. Palaeoecol. 134: 149–169.

Reichart G.J., Lourens L.J. and Zachariasse W.J. 1998. Temporal variability in the northern Arabian Sea Oxygen Minimum Zone (OMZ) during the last 225,000 years. Paleoceanography 13: 607–621.

Ritchie J.C. 1987. A Holocene pollen record from Bir Atrun, Northwest Sudan. Pollen et Spores 29: 391–410.

Ritchie J.C., Eyles C.H. and Haynes C.V. 1985. Sediment and pollen evidence for an early to mid-Holocene humid period in the Eastern Sahara. Nature 314: 352–354.

Rixen T., Haake B., Ittekkot V., Guptha M.V.S., Nair R.R. and Schluessel P. 1996. Coupling between SW monsoon-related surface and deep ocean processes as discerned from continuous particle flux measurements and correlated satellite data. J. Geophys. Res. 101: 28569–28585.

Roberts N. and Wright H.E. 1993. Vegetational, lake-level, and climate history of the Near East and Southwest Asia. In: Wright H.E., Kutzbach J.E., Webb III T., Ruddiman W.F., Street-Perrott F.A. and Bartlein P. (eds), Global Climates since the Last Glacial Maximum. University of Minnesota Press, Minneapolis, pp. 194–220.

Roberts N., Lamb H.F., El Hamouti N. and Barker P. 1994. Abrupt Holocene hydro-climatic events: palaeolimnological evidence from North-West Africa. In: Millington A.C. and Pye K. (eds), Environmental Change in Drylands: Biogeographical and Geomorphological Perspectives. John Wiley and Sons, New York, pp. 163–175.

Rognon P. 1987. Aridification and abrupt climatic events on the Saharan northern and southern margins, 20,000 Y BP to present. In: Berger W.H. and Labeyrie L.D. (eds), Abrupt Climatic Change. Reidel, Dordrecht, pp. 209–220.

Rossignol-Strick M. and Duzer D. 1979. Late Quaternary pollen and dinoflagellate cystes in marine cores off West Africa. "Meteor" Forschungs-Ergebnisse C30: 1–14.

Salzmann U. 2000. Are savannas degraded forests? — A Holocene pollen record from the Sudanian zone of NE-Nigeria. Veget. Hist. Archaeobot. 9: 1–15.

Salzmann U. and Waller M. 1998. The Holocene vegetational history of the Nigerian Sahel based on multiple pollen profiles. Rev. Palaeobot. Palynol. 100: 39–72.

Salzmann U., Hoelzmann P. and Morcinek I. 2002. Late Quaternary climate and vegetation of the sudanian zone of NE-Nigeria. Quat. Res. 58: 73–83.

Sarkar A., Ramesh R., Somayajulu B.L.K., Agnihotri R., Jull A.J.T. and Burr G.S. 2000. High resolution Holocene monsoon record from the eastern Arabian Sea. Earth Planet. Sci. Lett. 177: 209–218.

Sarnthein M., Tetzlaff G., Koopmann B., Wolter K. and Pflaumann U. 1981. Glacial and interglacial wind regimes over the eastern subtropical Atlantic and North-West Africa. Nature 293: 193–196.

Schulz E. 1990. Zwischen Syrte und Tschad. Der aktuelle Pollenniederschlag in der Sahara. Berl. Geogr. Stud. 30: 193–220.

Schulz E. 1991. Holocene environments in the Central Sahara. Hydrobiolgia 214: 359–365.

Schulz E. 1994. The southern limit of the Mediterranean vegetation in the Sahara during the Holocene. Hist. Biol. 9: 134–156.

Schulz E. and Whitney W. 1986. Upper Pleistocene and Holocene lakes in the An Nafud, Saudi Arabia. Hydrobiolia 143: 175–190.

Schulz H., von Rad U. and Erlenkeuser H. 1998. Correlation between Arabian Sea and Greenland climate oscillations of the past 110,000 years. Nature 393: 54–57.

Servant M. 1983. Séquences continentales et variations climatiques: évolution du bassin du Tchad au Cénozoique supérieur. O.R.S.T.O.M. 159: 1–573.

Servant M. and Servant-Vildary S. 1980. L' environnement quaternaire du bassin du Tchad. In: Williams M.A. and Faure H. (eds), The Sahara and the Nile. Balkema, Rotterdam, pp. 132–162.

Sirocko F. 1996. The evolution of the monsoon climate over the Arabian Sea during the last 24,000 years. Palaeoecol. Afr. 24: 53–69.

Sirocko F. and Sarnthein M. 1989. Wind-borne deposits in the Northwestern Indian Ocean: records of Holocene sediment versus satellite data. In: Leinen M. and Sarnthein M. (eds), Palaeoclimatology and Palaeometeorology: Modern and Past Patterns of Global Atmospheric Transport. Kluwer Academic Publishers, Dordrecht, pp. 401–433.

Sirocko F., Garbe-Schönberg C.-D., McIntyre A. and Molfino B. 1996. Teleconnections between the subtropical monsoons and high-latitude climates during the last deglaciation. Science 272: 526–529.

Sirocko F., Garbe-Schönberg D. and Devey C. 2000. Processes controlling trace element geochemistry of Arabian Sea sediments during the last 25,000 years. Glob. Planet. Chan. 26: 217–303.

Sirocko F., Sarnthein M., Erlenkeuser H., Lange H., Arnold M. and Duplessy J.C. 1993. Century-scale events in monsoonal climate over the past 24,000 years. Nature 364: 322–324.

Sonntag C., Thorweihe U. and Rudolph J. 1982. Isotopenuntersuchungen zur Bildungsgeschichte saharaischer Paläowässer. Geomethodica 7: 55–78.

Sonntag C., Thorweihe U., Rudolph J., Löhnert E.P., Junghans C., Münnich K.O., Klitzsch E. and El Shazly E.M. 1978. Isotopic identification of saharian groundwater, groundwater formation in the past. Palaeoecol. Afr. 12: 159–171.

Sonzogni C., Bard E. and Rostek F. 1998. Tropical sea-surface temperatures during the last glacial period: a view based on alkenones in Indian Ocean sediments. Quat. Sci. Rev. 17: 1185–1201.

Stager J.C. and Mayewski P.A. 1997. Abrupt early to mid-Holocene climatic transition registered at the equator and the poles. Science 276: 1834–1836.

Staubwasser M. and Sirocko F. 2001. On the formation of laminated sediments on the continental margin off Pakistan: the effects of sediment provenance and sediment redistribution. MarGeol. 172: 43–56.

Staubwasser M., Sirocko F., Grootes P.M. and Erlenkeuser H. 2002. Asian monsoon climate change and radiocarbon in the Arabian Sea during early and mid Holocene. Paleoceanography 17: 1063–1073.

Street-Perrott F.A. and Perrott R.A. 1990. Abrupt climate fluctuations in the tropics: the influence of Atlantic Ocean circulation. Nature 343: 607–610.

Street-Perrott F.A. and Perrott R.A. 1993. Holocene vegetation, lake levels, and climate of Africa. In: Wright H.E., Kutzbach J.E., Webb III T., Ruddiman W.F., Street-Perrott F.A. and Bartlein P. (eds), Global Climates since the Last Glacial Maximum. University of Minnesota Press, Minneapolis, pp. 318–356.

Street-Perrott F.A. and Roberts N. 1983. Fluctuations in closed-basin lakes as an indicator of past atmospheric circulation patterns. In: Street-Perrott F.A., Beran M. and Ratcliffe R.D. (eds), Variations in the Global Water Budget. Reidel, Dordrecht, pp. 331–345.

Street-Perrott F.A., Marchand D.S., Roberts N. and Harrison S.P. 1989. Global Lake-Level Variations from 18,000 to 0 Years Ago: A Palaeoclimatic Analysis. Report DOE/ER/60304-H1, Washington, 213 pp.

Stuiver M. and Reimer P.J. 1993. Extended ^{14}C data b and revised Calib 3.0 ^{14}C age calibration program. Radiocarbon 35: 215–230.

Stuiver M., Reimer P.J., Bard E., Beck J.W., Burr G.S., Hughen K.A., Kromer B., McCormac G., van der Plicht J. and Spurk M. 1998. INTCAL 98 radiocarbon age calibration. Radiocarbon 40: 1041–1084.

Sultan M., Stucchio N., Hassan F.A., Hamdan M.A., Mahmoud A.N., Alfy Z. and Stein T. 1997. Precipitation source inferred from stable isotopic composition of Pleistocene groundwater and carbonate deposition in the Western desert. Qual. Res. 48: 29–31.

Swallow J.C. 1984. Some aspects of the physical oceanography of the Indian Ocean. Deep-Sea Res. 31: 639–650.

Swezey C. 2001. Eolian sediment responses to late Quaternary climate changes: temporal and spatial patterns in the Sahara. Palaeogeogr. Palaeoclim. Palaeoecol. 167: 119–155.

Szabo B.J., Haynes C.V. and Maxwell T.A. 1995. Ages of Quaternary pluvial episodes determined by uranium-series and radiocarbon dating of lacustrine deposits of Eastern Sahara. Palaeogeogr. Palaeoclim. Palaeoecol. 113: 227–242.

Szabo B.J., McHugh W.P., Schaber C.G., Haynes C.V. and Breed C.S. 1989. Uranium-Series dated authigenic carbonates and Acheulian sites in Southern Egypt. Science 243: 1053–1056.

Téhet R., Gasse F., Durand A., Schroeter P. and Fontes J.-C., 1990. Fluctuations climatiques du Tardiglaciaire à l'Actuel au Sahel (Bougdouma, Niger méridional). C.R. Acad. Sc. Paris 311: 253–258.

Texier D., De Noblet N., Harrison S., Haxeltine A., Jolly D., Laarif F., Prentice I. and Tarasov P. 1997. Quantifying the role of biosphere-atmosphere feedbacks in climate change: coupled model simulations for 6000 yrs BP and comparison with paleodata for Northern Eurasia and Northern Africa. Clim. Dy 13: 865–882.

Thiemeyer H. 1992. On the age of the Bilma ridge. A new ^{14}C record from Konduga area, Borno State, NE-Nigeria. Zeitschr. Geom. N.F. 36: 113–118.

Thinon M., Ballouche A. and Reille M. 1996. Holocene vegetation of the Central Saharan Mountains: the end of a myth. Holocene 6-4: 457–462.

Thorweihe U. 1990. The Nubian aquifer. In: Said R. (ed.), The Geology of Egypt. Balkema, Rotterdam, 601–722.

Trevisan Grandi G., Mariotti Lippi M. and Mercuri A.M. 1998. Pollen in dung layers from rockshelters and caves of Wadi Teshuinat (Libyan Sahara). In: Cremaschi M. and Di Lernia S. (eds), Wadi Teshuinat — Palaeoenvironment and Prehistory in South-Western Fezzan (Libyan Sahara), pp. 95–106.

Uerpmann H.-P. 1991. Radiocarbon dating of shell middens in the Sultanate of Oman. PACT, 29 (IV.5), pp. 335–347.

Umer M., Legesse D., Gasse F., Bonnefille R., Lamb H., Leng M. and Lamb A., this volume. Late Quaternary climate changes in the Horn of Africa. In: Battarbee R.W., Gasse F. and Stickley C.E. (eds), Past Climate Variability through Europe and Africa. Kluwer Academic Publishers, Dordrecht, the Netherlands, pp. 159–180.

Van Campo E., Duplessy J.C. and Rossignol-Strick M. 1982. Climatic conditions deduced from a 150-kyr oxygen isotope-pollen record from the Arabian Sea. Nature 296: 56–59.

Van Campo M. 1975. Pollen analyses in the Sahara. In: Wendorf F. and Marks A.E. (eds), Problems in Prehistory: North Africa and the Levant. Southern Methodist University Press, Dallas, pp. 45–64.

Vita-Finzi C. and Richards G.W. 1991. Comment on "Two high levels of continental waters in the southern Tunisian chotts at about 90 and 150 ka". Geology 29: 94–95.

Völkel J. 1989. Formation of dunes and pedogenesis as palaeoclimatic indicators in the eastern part of the Republic of Niger (Sahara and Sahel). Palaeoecol. Afr. 20: 37–54.

von Rad U., Schaaf M., Michaels K., Schulz H., Berger W.H. and Sirocko F. 1999. A 5000-yr record of climate change in varved sediments from the oxygen minimum zone off Pakistan, northeastern Arabian Sea. Quat. Res. 51: 39–53.

Waller M. and Salzmann U. 1999. Holocene vegetation changes in the Sahelian zone of NE-Nigeria: The detection of anthropogenic activity. Palaeoecol. Afr. 26: 85–102.

Wasylikowa K. 1997. Flora of the 8000 years old archaeological site E-75-6 at Nabta Playa, Western Desert, Southern Egypt. Acta Palaeobot. 37–2: 99–205.

Webster P.J., Magana V.O., Palmer T.N., Shukla J., Tomas R.A., Yanai M. and Yasunari T. 1998. Monsoons: Processes, predictability, and the prospects of prediction. J. Geophys. Res. 103: 14451–14510.

Williams M.A.J., Adamson D.A., Williams F.M., Morton W.H. and Parry D.E. 1980. Jebel Marra volcano: a link between the Nile Valley, the Sahara and Central Africa. In: Williams M.A.J. and Faure H. (eds), The Sahara and the Nile. Quaternary Environments and Prehistoric Occupation in Northern Africa. Balkema, Rotterdam, pp. 305–337.

Williams M.A.J., Abell P.I. and Sparks B.W. 1987. Quaternary landforms, sediments, depositional environments and gastropod isotope ratios at Adrar Bous, Tenere Desert of Niger, South-Central Sahara. Geol. Soc. Spec. Publ. 35: 105–125.

Williams M.A.J., Adamson D., Cock B. and McEvedy R. 2000. Late Quaternary environments in the White Nile region, Sudan. Glob. Planet. Chan. 26: 305–316.

You Y. 1998. Seasonal variations of thermocline circulation and ventilation in the Indian Ocean. J. Geophys. Res. 102: 10391–10422.

Zarins J., Ibrahim M., Potts D. and Edens C. 1979. Preliminary report on the survey of the central province. 1978. Atlal 3: 14–19.

Zonneveld K.A.F., Ganssen G., Troelstra S., Gerard J.M., Versteegh J.M. and Visscher H. 1997. Mechanisms forcing abrupt fluctuations of the Indian Ocean summer monsoon during the last deglaciation. Quat. Sci. Rev. 16: 187–201.

Zouari K., Chkir N. and Causse C. 1998. Pleistocene humid episodes in southern Tunisian chotts. In: IAEA (ed.), Isotope Techniques in the Study of Environmental Change. Proceedings Series, Vienna, 543–554.

13. HISTORICAL CHRONOLOGY OF ENSO AND THE NILE FLOOD RECORD

LUC ORTLIEB (luc.ortlieb@bondy.ird.fr)
UR PALEOTROPIQUE
(Paléoenvironnements tropicaux et variabilité climatique)
Institut de Recherche pour le Développement
Centre IRD-Ile de France
32 avenue Henri-Varagnat
F-93143 Bondy cedex
France

Keywords: ENSO, El Niño, Documentary climatology, Little Ice Age, Nile, South America, Teleconnections

Introduction

At present the El Niño system is one of the primary causes of inter-annual climatic variability, not only in tropical regions but also on a global level (Ropelewsky and Halpert 1987; Philander 1989; Kiladis and Diaz 1989; Glynn 1990). A considerable effort is dedicated to the understanding of the El Niño-Southern Oscillation (ENSO) system and its interactions with other kinds of climatic variations, at the decadal/interdecadal (e.g., the Pacific Decadal Oscillation) and centennial (Little Ice Age, Medieval Warm Period) scales (Anderson et al. 1992; Bradley and Jones 1992; Allan et al. 1997; Diaz and Markgraf 2000). Under societal and political pressure, several international research programs aim to decipher the relationships that link ENSO and the present global warming of the planet (Ropelewesky 1992; Cane et al. 1997; Timmermann et al. 1999). These scientific challenges and questions require that long series of climate variations, much longer than those available from instrumental data alone, be compiled and studied. A renewed interest for the climate evolution, at local, regional and global scales, during the last few centuries, is thus justified by the need to understand better the climatic particularities of the Little Ice Age (LIA) and the transition period to the modern situation (19th and 20th centuries) (e.g., Grove (1988), Bradley and Jones (1992, 1993), Allan and d'Arrigo (1999), Allan (2000), Verschuren (this volume), Scott and Lee-Thorp (this volume)). These studies should help to determine how much of the recent global warming results from anthropogenic activities as against a natural post-LIA evolution. Deciphering the relationships between the global state of the planet and the El Nino system (frequency and intensity of individual events, teleconnection links) requires us to document as precisely as possible former climate variations and ENSO

R. W. Battarbee et al. (eds) 2004. *Past Climate Variability through Europe and Africa.*
Springer, Dordrecht, The Netherlands.

manifestations, during warm and cold periods, such as the Medieval Warm Period and the LIA respectively, and to compare them to the present (instrumentally controlled) climate variability (Bradley and Jones 1992; 1993; Diaz and Pulwarty 1994; Jones et al. 2001).

Palaeoclimatologists are thus strongly invited to contribute reconstructions of inter-annual climate variability during the last two thousand years (see PAGES program and related sub-programs like the PEP transects), and more specifically during the last few centuries (CLIVAR program). Regarding ENSO chronology and inter-annual climate vari-ability, research is particularly active through studies of coral sequences (e.g., Tudhope et al. (1995), Dunbar et al. (1996), T. Quinn et al. (1996), Dunbar and Cole (1999), Corrège et al. (2000, 2001)), tree-rings (e.g., Lough and Fritts (1985), Michaelsen (1989), d'Arrigo and Jacoby (1991), Cook (1992)) and tropical ice cores (e.g., Thompson et al. (1984, 1992, 2000), Thompson and Mosley-Thompson (1989), Thompson (1992)) (Annual Records of Tropical Systems program). At present, the longest published sequences, that encompass more than the last few centuries, were produced by dendroclimatologists from subtropical regions or mid-latitude areas, like Northern Mexico and Southern USA (e.g., Michaelsen (1989), d'Arrigo and Jacoby (1991), Stahle and Cleaveland (1993), Stahle et al. (1998)). For various reasons, the ice-cores and coral records have not yet yielded fully reliable sequences of ENSO variability during the past few centuries.

Documentary archive studies on former El Niño manifestations were initiated more than 20 years ago by the late William (Bill) Quinn († 1993) and are now being developed in several tropical and subtropical regions. Most researchers working on ENSO variability (prior to the 20th Century) heavily relied upon the chronological sequences of El Niño events produced by Quinn and collaborators, and many of them calibrated their palaeo-climatic proxies on Quinn's sequences (e.g., Peterson (1990), Enfield and Cid (1991), Anderson et al. (1992), Thompson et al. (1992), Dunbar et al. (1994, 1996)). Up to now, the documentary approach for the reconstruction of El Niño occurrences during the last few centuries remains unavoidable, in spite of intrinsic methodological limitations. Historical documentary research deals with a great variety of records including reports of droughts, heavy rainfalls, storms or climate anomalies, streamflows variations, crop yields, marine resource availability and travel time on known maritime routes (Foley 1957; Anonymous 1981; Hamilton and Garcia 1987; Quinn et al. 1987; Grove 1988; Zhang and Crowley 1989).

One of the major problems with any of the above-mentioned approaches based on climate proxies or written documents, deals with the spatio-temporal representativity of the records: where (in the world) should we look for the most appropriate palaeo-ENSO record through time? Which region registered recordable ENSO manifestations that were not affected by modifications of the teleconnection patterns observed nowadays? One way to address this problem is to compare records from different areas, which may or may not involve the same categories of proxies (Enfield 1992). In an attempt to reconstruct a so-called "global" ENSO record, Quinn (1992, 1993) compared chronological sequences of climate anomalies in China, India, Australia, Egypt and South-America. Similarly, Whetton and colleagues (Whetton and Rutherfurd 1994; Whetton et al. 1996) intended to assess an ENSO chronology of the last three centuries by comparing data from Java, China, India, Egypt and South America. In these studies, a particular interest was dedicated to the intercomparison of sequences encompassing the last few centuries, focusing on the two extremes of the ENSO system, i.e., the Nile flow record and the South-American El Niño occurrences. Diaz and Pulwarty (1992) also analysed, from a statistical point of view, the historical ENSO signals from the Nile and South American regions.

This paper aims to provide a closer look at this intercomparison between the records from the western and eastern ends of the ENSO "see-saw". Recently published data (Ortlieb 1999; 2000) and on-going studies on South American El Niño records show that the "classical" series of historical El Niño events produced by Quinn (Quinn et al. 1987; Quinn and Neal 1992; Quinn 1993) are less well-founded than previously assumed. These studies, and recent historical reconstructions of climate anomalies in Argentina (Prieto 1994; Prieto et al. 1999), suggest that variations in the teleconnection patterns within South America did occur during the last few centuries. As a consequence, the establishment of an over-all "South-American" El Niño record still requires much more work before a synthetic regional sequence be validated. These new developments obviously affect previous studies bearing upon E-W intercomparisons, across the entire ENSO system. The paper focuses on several methodological problems and raises a few questions regarding the validation and the interpretation of raw documentary data (both from the Nile and South America regions). Clearly, it does not pretend to solve the problem of the validity of the Nile flood record as a reliable ENSO indicator. In the framework of the inter-PEP activities, this is a contribution of a participant of the PEPI transect to the PEP III programme.

The historical El Niño chronologies in South America

Quinn's work

Quinn and collaborators (Quinn et al. 1978; 1987) produced the first chronological sequence of El Niño events for the last 450 years based on written archives of climatic and oceanographic anomalies from the Pacific coast of South America. Their study was based upon a compilation of published reports of climate anomalies along the arid coast of Peru such as strong rainfall episodes, anomalously short (or long) travel times of vessels along known routes, crop failures, inundations and flood destructions, or epidemic diseases. In their earliest published paper (1987) most of the information concerned Peru and Southern Ecuador, and only a few data came from nearby regions (Bolivian Altiplano or Chile). The result was a historical chronology of El Niño events since 1525, with an evaluation of the intensity of the major events of the 16th and 17th centuries, and of all events (including those of moderate and weak strength) of the 1800–1987 period. Subsequently, Quinn and Neal (1992) amended this chronological sequence, incorporating additional documentary data from other regions of South America and extended the evaluation of strength to all ENSO events since 1525. In 1993, in his last published work, Quinn slightly modified the previous chronological sequence of El Niño events based on South American documentary evidence and proposed a sequence of "global ENSO events". This last sequence resulted from a compilation of historical data on Indian and Eastern Asia droughts, Australian precipitation records, Nile low floods and the above mentioned South-American record. The last published sequence of Quinn's "regional El Niño events" based on South American evidence, with an indication of the inferred intensity of every event, is reproduced here as the fourth column of Table 1.

Table 1. Comparison of historical documentary data for the reconstruction of ENSO events in South America and low Nile flows between 1525 and 1900, according two different sources (Quinn 1992; 1993; Whetton and Rutherfurd 1994). Boxes indicate years of positive correlation between Nile flood data and at least two of the three South American records.

Central Chile	Peru+ Ecuad.	Argentina	S. America El Niño		Low flows of the Nile	
Ortlieb, in prep.	O. & H. In prep.	Prieto et al.	*Quinn, 1993*		W & R 1994	Quinn 1993
			1525-	*m*		
			-E1526			
			1531-	*m*		
			-E1532			
1535 ?			*1535*	*m+*		
	1539 ?		*1539-*			
			-1540-	*m/S*		1540
			-1541			
1544			*1544*	*m+*		1544
	1546 ?		**1546-**	**S**		
			-1547			
			1552	**S**		
						<u>1553</u>
			1558-			
			-1559-	*m/S*		
			-1560-			
			-E1561			
			1565	*m+*		
			1567-	**S+**		<u>1567</u>
	1568 ?		**-1568**			
1574	1574 ?		**1574**	**S**		
	1578		<u>**1578-**</u>	**VS**		<u>1578</u>
			-E1579			
1581			*1581-*	*m+*		
			-1582			
			1585	*m+*		
1588		**1588**				
			1589-		NO	1589
			-1590-	*m/S*		
			-1591			
	1596		*1596*	*m+*		
1597 ?					DATA	
			1600	**S**		**1600**
1604 ?			*1604*	*m+*		**1604**
		1606				
		1607	**1607-**	**S**		**1607**
			-1608			
1609						

Table 1. (continued) Comparison of historical documentary data for the reconstruction of ENSO events in South America and low Nile flows between 1525 and 1900, according two different sources (Quinn 1992; 1993; Whetton and Rutherfurd 1994). Boxes indicate years of positive correlation between Nile flood data and at least two of the three South American records.

Central Chile	Peru+ Ecuad.	Argentina	S. America El Niño		Low flows of the Nile
		1613			
			1614	*S*	
	1617 ?				
			1618-	*S*	1618
1619		1619	*-1619*		
1621 ?			*1621*	*m+*	**1621**
	1622 ?				
	1624 ?		*1624*	*S+*	
		1625			
		1628			
		1629			
			1630	*m*	**1630**
					1631
			1635	*S*	1635
		1636			
			1640-	*m*	**1640**
		1641	*-1641*		**1641**
1647			*1647*	*m+*	
			1650	*m*	**1650**
			1652	*S+*	
1655 ?			*1655*	*m*	
			1661	*S*	1661
		1665			
		1669			
		1671	*1671*	*S*	
		1676			
		1677			
	1678 ?	1678			
		1679			
		1680			
		1681	*1681*	*S*	
		1684	*1684*	*m+*	
1686	1686				
1687 ?		1687	*1687*	*S+*	
1688 ?					
			1692	*S*	
					1694
			1695	*m*	1695
1697 ?			*1697*	*m+*	

Table 1. (continued) Comparison of historical documentary data for the reconstruction of ENSO events in South America and low Nile flows between 1525 and 1900, according two different sources (Quinn 1992; 1993; Whetton and Rutherfurd 1994). Boxes indicate years of positive correlation between Nile flood data and at least two of the three South American records.

Central Chile	Peru+ Ecuad.	Argentina	S. America El Niño		Low flows of the Nile	
	1701		*1701*	*S+*		
					1703	
			1704	*m*		
			1707-			
			-1708-	*m/S*		
		1709	*-1709*		1709	
1712 ?						
			1713	*m*		**1713**
			1715-	*S*	1715	**1715**
			-1716		1716	**1716**
1718 ?	1718 ?		*1718*	*m+*		
	1720		*1720*	*VS*	1720	
1723 ?			*1723*	*m+*	1723	1723
					1725	**1725**
	1728		*1728*	*VS*		
1730 ?						
					1731	1731
			1734	*m*	1734	
1737	1737 ?		*1737*	*S*	1737	1737
					1743	
1744 ?			*1744*	*m+*		
1746		1746				
	1747	1747	*1747*	*S+*		
1748	1748 ?					
1751 ?			*1751*	*m+*		
					1752	
					1753	
			1754-	*m*		
			-1755			
			1758	*m*		**1758**
					1759	
		1760				
	1761		*1761*	*S*		
					1762	1762
		1763				
1764						
			1765	*m*	1765	**1765**
					1766	**1766**
1768 ?		1768	*1768*	*m*		

Table 1. (continued) Comparison of historical documentary data for the reconstruction of ENSO events in South America and low Nile flows between 1525 and 1900, according two different sources (Quinn 1992; 1993; Whetton and Rutherfurd 1994). Boxes indicate years of positive correlation between Nile flood data and at least two of the three South American records.

Central Chile	Peru+ Ecuad.	Argentina	S. America El Niño		Low flows of the Nile	
		1771				
			1772	*m*	1772	**1772**
		1775				
		1776	*1776-*		1776	1776
		1777	*-1777-*	*S*		
			-E1778			
		1782	*1782-*	*S*	1782	**1782**
1783		*1783*	*-1783*		1783	**1783**
	1784	**1784**			1784	**1784**
	1785	1785	*1785-*	*m+*		**1785**
			-1786			
		1788				
1791	**1791**	**1791**	*1791*	*VS*	1791	**1791**
		1792				**1792**
						1793
		1794				**1794**
						1795
						1796
						1797
		1798				
					1799	1799
		1801				
		1803	*1803-*	*S+*		**1803**
	1804		*-1804*			
			1806-	*m*		1806
			-1807			1807
		1810	*1810*	*m*		
			1812	*m+*		
	1814	1814	*1814*	*S*		
	1815 ?	**1815**				
1817	1817 ?	**1817**	*1817*	*m+*		
1819	1819	1819	*1819*	*m+*		
1820		**1820**				
1821	1821 ?	**1821**	*1821*	*m*		
		1823				
	1824 ?	**1824**	*1824*	*m+*	1824	**1824**
1825 ?					1825	**1825**
1827		1827				
1828	**1828**		*1828*	*VS*	1828	1828

Table 1. (continued) Comparison of historical documentary data for the reconstruction of ENSO events in South America and low Nile flows between 1525 and 1900, according two different sources (Quinn 1992; 1993; Whetton and Rutherfurd 1994). Boxes indicate years of positive correlation between Nile flood data and at least two of the three South American records.

Central Chile	Peru+ Ecuad.	Argentina	S. America El Niño		Low flows of the Nile	
1829						
			1830	*m*	1830	1830
	1832 ?		*1832*	*m+*		
1833					1833	**1833**
			1835		1835	**1835**
						1836
1837	1837		*1837*	*m+*	1837	**1837**
						1838
					1839	**1839**
1841		**1841**				
1843						
1844 ?	1844 ?		*1844-*			
1845	**1845**	1845	*-1845-*	*S*	1845	**1845**
		1846	*-E1846*			
1850	1850 ?		*1850*	*m*	1850	1850
1851						
1852 ?			*1852*	*m*	1852	1852
			1854	*m*		
1855		1855			1855	**1855**
1856						
	1857 ?		*1857-*	*m*		
1858			*-1858*		1858	1858
					1859	1859
1860		1860	*1860*	*m*		
	1862 ?		*1862*	*m-*		
1864	**1864**	**1864**	*1864*	*S*	1864	**1864**
	1866 ?		*E1866*	*m+*		
			L1867-	*m+*	1867	
1868			*-1868*		1868	**1868**
	1871	**1871**	*1871*	*S+*		
					1873	1873
			1874	*m*		
		1875				
		1876				
1877	1877-	**1877**	*1877-*	*VS*	1877	**1877**
1878 ?	-1878	1878	*-1878*			
		1879				
1880		1880	*1880*	*m*	1880	
						1882

Table 1. (continued) Comparison of historical documentary data for the reconstruction of ENSO events in South America and low Nile flows between 1525 and 1900, according to two different sources (Quinn 1992; 1993; Whetton and Rutherfurd 1994). Boxes indicate years of positive correlation between Nile flood data and at least two of the three South American records.

Central Chile	Peru+ Ecuad.	Argentina	S. America El Niño		Low flows of the Nile	
1884 ?	**1884**		*1884*	*S+*		1884
		1885			1886	
1887 ?	1887?		*L1887-*			
1888	1888 ?		*-1888-*	*m*	1888	**1888**
			-E1889			
1891	**1891**		*1891*	*VS*	1891	
					1893	
	1897 ?		*1897*	*m+*	1897	
1899-	1899		*1899-*	*S*	1899	**1899**
-1900	1900 ?		*-E1900*			

Col. 1, 2 & 3: Years with El Niño manifestations in Central Chile (Ortlieb 1994, unpubl.), Northern Peru and Southern Ecuador (Ortlieb 2000; Ortlieb and Hocquenghem, in prep.), and NW Argentina (Prieto 1994; Prieto et al. 1998; 1999). In bold, years for which archival data suggest **strongest El Niño events**.

Col. 4 & 5: Latest «regional» (South America and eastern Pacific) El Niño chronology published by Quinn (1993). EN occurrence within year: E = early (Jan-March) or L = late (Sep.-Dec.). EN event intensity: m: medium ; **S: Strong; VS: Very strong**.

Col. 6 : Low Nile flows according to Hassan, in Whetton and Rutherfurd (1994): years with flood height at least 0.5 std dev. below average.

Col. 7: Nile flood deficits according Quinn (1992, 1993), re-interpreted here as: weak (grade 2), **strong** (grades 3 & 4), **very strong** (grade 5). Shaded areas pinpoint concordance between col. 6 & 7 data.

Re-examination of the documentary evidence in Peru

In 1992, Hocquenghem and Ortlieb, and more recently Ortlieb (2000), re-examined the whole set of documents used by Quinn and co-authors (Quinn et al. 1987; Quinn and Neal 1992), and added some newly found archival data concerning Peru and Southern Ecuador. It was thus shown that many so-called El Niño events (as reconstructed by Quinn) were poorly assessed and may not have occurred, and that the intensity of a number of events was probably lower than previously interpreted. In the revision of the documentary data base, a major emphasis was put on the evidence of rainfall anomalies in the coastal area of Northern Peru and Southern Ecuador. Those years during which drought was reported in that area (Piura and Guayaquil) are not considered as El Niño years. As instrumental observation indicates that Rimac river floods are not related to El Niño, it is interpreted that former floods of this river in Lima should not be considered as manifestations of past El Niño events. Similarly, statistical studies on rainfall anomalies along the coast of Southern Peru do not show clear-cut relationships with ENSO occurrences. Hence, all the documented former climate anomalies available from Central or Southern Peru which were not coeval with documented precipitation excess in Northwestern Peru were not interpreted as reliable manifestations of El Niño conditions. It was also shown that cold conditions, drought or heavy rainfall events in Bolivia (particularly in the Potosi area, where a long documentary record is available) cannot either be used for the reconstruction of former El Niño events. Finally, years for which only epidemic diseases were reported, without any other indications of climatic conditions, were not considered as El Niño years. Along these restrictive lines, Ortlieb (1999, 2000) thus produced a consolidated, updated, chronological sequence of El Niño manifestations in Peru, in which was specifically mentioned the kind and location of the proxy information used, as well as the precise position of the quotations referred to by Quinn and co-authors or by himself. He also discussed how and why a number of years should not be viewed as El Niño years, and why the intensity of many events was reduced with respect to Quinn's interpretation. All the quotes from published and unpublished sources, with detailed comments (regarding their intrinsic reliability, relevance with respect to ENSO reconstruction, or their originality), is in preparation (Ortlieb and Hocquenghem, unpublished). The sequence of El Niño events identified in Peru, as indicated in the second column of Table 1, is an updated version of the Ortlieb (2000) study.

The number of reconstructed El Niño events during the 1525–1900 period in Peru (Table 1, column 2) appears to be significantly lower than those proposed by Quinn (1993) for the whole South American record (column 4).

The documentary evidence in Chile

Instrumental data available since 1876 indicate a significant correlation between the amount of winter precipitation in Central Chile and the Southern Oscillation Index (Quinn and Neal 1983; Deser and Wallace 1987; Aceituno 1988; Rutllant and Fuenzalida 1991). For more than a century, El Niño years have been characterised in Santiago and Valparaiso by wetter winters, while La Niña years are almost systematically marked by rainfall deficits. This region of South America, in which both El Niño and La Niña conditions can be registered, thus offers a more complete potential record of the ENSO fluctuations than that of Peru in which only warm ENSO events are liable to be registered in written archives. Therefore, a similar study to the one mentioned for Peru is in process in Chile with the aim to document the yearly positive and negative rainfall anomalies from the 16th to 19th centuries (Ortlieb, unpublished). A preliminary chronological sequence of El Niño years recorded in Chile, compiled from different sources (Ortlieb 1994) had shown a co-occurrence of positive rainfall anomalies between Northern Peru and Central Chile from 1817 onward, and a general lack of co-occurrence during the 16th to early 19th centuries. This situation suggests that the teleconnection patterns which drive the relationship between rainfall excess and ENSO was modified at the beginning of the 19th Century (Ortlieb 1999; 2000). It was assumed that this change probably reflected a different configuration of the atmospheric circulation pattern.

The first column of Table 1 lists the years which experienced rainfall excess in Central Chile, and which are thus considered possible El Niño years. The sequence presented is based *pro-parte* on previous work (Ortlieb 1994) and on an on-going study of documentary data (Ortlieb, unpublished). As mentioned above, there are very few cases of co-occurrence of rainfall excess in Chile and Peru before the beginning of the early 19th Century, and it can be added that no systematic lead or lag, by one year, is observed between the Peruvian and

Chilean sequences. How should these discrepancies be interpreted? Is the Peruvian record more reliable than the Chilean record, as suggested by Ortlieb (2000)? Or should all the years with rainfall excess, either in Chile or in Peru, be considered as FNSO manifestations (as implicitly implied by Quinn and Neal (1992), Quinn (1993))? Until a thorough compilation of the documentary evidence from Central Chile is completed, these questions will have to remain open. Another approach may be to compare these two sequences to records from other South-American regions which are under the influence of ENSO.

Historical records of ENSO impacts in Argentina

Documentary records on the variability of snowfall in the Andes of Northwestern Argentina and of rainfall anomalies in the Mendoza area were recently studied (Prieto 1994; Prieto et al. 1998; 1999), and provide additional information on regional ENSO impacts. The area concerned, which is located at the latitude of Central Chile, across the Andean Cordillera, is under the influence of the southern westerlies and of the Pacific circulation pattern (including ENSO). Prieto and collaborators analysed historical series on former streamflow in the Mendoza area. High streamflows are interpreted as an indication of heavy snowfall on the cordilleran summits and as rainfall excess on the eastern flank of the Andes, in accordance with present day observations during warm ENSO events.

The third column of Table 1 indicates the years of high precipitation (snow and rainfall) during the 16th–19th centuries in the Mendoza area, on the eastern flank of the Andes. In some cases, documentary evidence suggests co-occurrences of precipitation anomalies on both sides of the Andes, which are thought to represent ENSO events.

Towards a development of a South-American historical ENSO record

Until now, in South-America, the reconstruction of a series of ENSO occurrences during the last few centuries has depended heavily on documentary data. On-going work on ice cores and coral sequences has not yet yielded reliable data. A Galapagos coral sequence that covers the span from 1586 to 1953 AD was intended to provide a sequence of El Niño manifestations (Dunbar et al. 1994), but this sequence was not well constrained in time, being offset by several years, and the yearly sampling for stable isotope analyses was of too low resolution to identify the occurrence and intensity of ENSO-related anomalies which can only be detected through monthly or bi-monthly sampling (Ortlieb 1999; 2000). Similarly, the reconstruction of ENSO events from Andean ice-core data still poses many problems; one of them is the fact that the snow accumulation on the Andes is more related to Atlantic than Pacific circulation (Ortlieb and Macharé 1993). Another interesting approach, developed by the tree-ring laboratory of Mendoza (Argentina), concerns the dendroclimatology of trees from Southern and Central Chile and Argentina (e.g., Boninsegna (1990), Villalba (1994), Villalba et al. (1998, 2001)). Reconstruction of former El Niño manifestations of the past centuries can be tracked through indications of humidity excess in the Chilean tree rings and high air temperature in the Patagonian tree rings.

In spite of Quinn's efforts, a rigorous assessment of the historical record of ENSO events, based on documentary data and encompassing the last few centuries in South-America, is not yet available. The approach which consisted in reconstructing former El

Niño events from documentary evidence gathered in different, widely separated, regions of Northeast Brazil, Chile and Peru, without paying attention to indications of non-El Niño manifestations in one or more of the other regions is questionable (Ortlieb 2000). The large number of El Niño events proposed by Quinn (e.g., 1993) is related to the fact that manifestations of the warm phase of ENSO, in one or the other of several South-American regions, were used without consideration of coeval climatic conditions in other ENSO-sensitive areas. Table 1 clearly shows that the documentary records in three different areas (Chile, Peru and N.W. Argentina) are poorly correlated, particularly before the early 19th Century. There may be biases due to the documentary approach (fragmentary and discontinuous information, reliability of sources, climatic interpretation of the reported features) but these cannot explain all the observed discrepancies. As previously suggested, part of the misfit between the chronological sequences seems to be related to a different pattern of atmospheric circulation during and after the Little Ice Age. In the present state of knowledge (before completion of a thorough analysis of the Chilean record), the Peruvian sequence is viewed as the best documented record. The problem of correlation between the Northern Peru record and the subtropical record (Central Chile and N.W. Argentina) will benefit from an integration of dendroclimatic data and a detailed (year-to-year) comparison with the documentary record from Chile and Argentina.

The Nile flood historical record

Nile floods relationship with the Southern Oscillation

The Nile floods, which occur between July and October, are largely dependent on the Blue Nile hydrology which is controlled by rainfall on the Ethiopian plateaus, and to a lesser extent on the Blue Nile input (Popper 1951; Shahin 1985; Williams et al. 1986; Janowiak 1988, and others). The Ethiopian contribution, including that of the Atbara River, has been estimated to about 68% (Fairbridge 1984). The White Nile supplies water mainly from March to June, and is at that time the major contributor of the downstream Nile (Brooks 1926; Hassan 1981; Fairbridge 1984). Since the work of Walker (1910, 1924) and Bliss (1925), it is commonly accepted that, through their dependence on the Ethiopian rainfall and the macro-scale pressure changes that drive the monsoon, the Nile floods are linked to the Southern Oscillation (Walker and Bliss 1932). This relationship was further studied and documented by numerous authors (Popper 1951; Griffiths 1972; Hassan 1981; Whetton et al. 1990; Quinn 1992; Diaz and Pulwarty 1992; Wang and Eltahir 1999). Ethiopia may be viewed as the western end of the large Southern Oscillation see-saw system of which the Pacific coast of South America forms the eastern extremity (Quinn 1992). When SOI (Walker and Bliss 1932) is strongly positive, a large low pressure system is centered on the Arabian Sea, and heavy rains fall on the Ethiopian highlands. Inversely, when SOI is strongly negative (during El Niño events), the low pressure system is weakened and/or displaced to the east, and the summer monsoon rains are below normal. Quinn (1992, Table 6.3) compiled documentary and instrumental evidence, for the period 1824–1972, on the satisfactory correlation between weak Nile floods, Indonesian monsoon and Australian drought, deficient India summer monsoon, on one hand, and the South American El Niño on the other hand.

During the twentieth Century, the record of the inter-annual variations in the Nile stream-flow was established on the basis of maximum annual height readings at El Cairo (Roda gauge), monthly and annual river discharge at Aswan and in other locations along the White Nile, main Nile, Blue Nile and Atbara river. In spite of the successive disturbances caused by the construction of the first Aswan Dam (1902) and subsequent development (1912 and 1934), and of the High Aswan Dam (1964) (Shahin 1985), it has been possible to convert discharge values measured in different locations into Nile River flood height equivalents. These calculations allowed a calibration of the Nile low flows with instrumented climatological data and correlations with manifestations of the Southern Oscillation impacts in other region of the world.

Data bases and interpretation of the Nile flood record by Quinn

The first numerical data available on the annual Nile floods are dated AD 622. Nile flood data were gathered and tabulated by Toussoun (1925) and Hurst and Phillips (1933). Later on Popper (1951) used documentary information (notably from Ibn Taghri Birdi, Ibn al-Hijazi and Ibn Aibak) to correct and modify Toussoun's data. Popper's tabulated data-set is almost complete from AD 622 to 1469. Very few data are available between AD 1470 and 1700. There is another major gap during the Napoleonic occupation, between 1800 and 1825. Then at the beginning of the twentieth Century, several coincident factors, like the construction of the first major dams and a well-documented alteration of the precipitation regime, complicated the picture.

In his 1992 paper, Quinn produced a chronology of the Nile flood records which extends from AD 622 to 1522 and another one (already mentioned) from 1824 to 1972. In these two sets of low Nile floods, which were established using Toussoun's and Popper's tables and other documentary material, Quinn evaluated a "degree of deficiency (1 to 5)" of the Nile floods and indicated a confidence rating (1 to 5) to his own reconstruction. In his (latest) 1993 paper, Quinn incorporated a Nile flood chronology covering the intermediate period between 1522 and 1824 and added data for the period encompassing 1973–1990.

In the presentation of his reconstruction of the Nile flood chronology, Quinn (1992, 1993) did not enter into much detail (as he had done for the El Niño historical reconstruction, giving references of published papers). He mentioned that some of the breaks in the Nilometer record were compensated by interpretation of anecdotal reports from varied sources (beside Toussoun (1925), Popper (1951)) including Walford 1879, Lyons 1906, Jarvis 1935, Hurst 1957, Bell 1971, Shahin 1985. However, Quinn did discuss the difficulties met in the interpretation of the available raw data. Examples of the major problems and caveats dealt with by Quinn include:

- Where (on which scale) and how (with which units) were the original measurements made?

- How can data based on threshold regimes for agricultural purposes be converted to scale measurements?

- Did the sediment accumulation rates (within the Nile valley or in the delta) evolve in time, at the inter-annual, decadal or centennial scales, and how did this affect the flood height measurements?

- What was, for every set of data, the reference level, with respect to the present mean sea level (inter-calibration of the vertical position of the different Nilometers)?

- What was the date of the observations after conversion from the Muhammadan lunar calendar to the AD calendar years (knowing that for every 32 solar years, the 33rd lunar year should be skipped, a procedure which has not always been followed by the different chroniclers)?

Quinn's (1992, 1993) sequence of Nile floods was thus based on a compilation of data which were expressed in grades of reduction in maximum flood height at Cairo, with respect to long term averages. He determined five categories (grades) in the departure of annual maximum height of the Nile above sea level: 1 (0.27–0.53 m), 2 (0.54–0.80 m), 3 (0.81–1.07 m), 4 (1.08–1.34 m) and 5 (over 1.35 m). These categories were also expressed as departures from annual cumulative discharge amount, for July to October, below a specified annual average amount for the river discharge sites considered. On this scale the fifth grade thus corresponds to the most severe rainfall deficit in the Ethiopian plateaus. In Table 1 (7th column), Quinn's sequence of low Nile flows is reproduced, with the following transformations:

- grade 1 (slight deficit of flood height) was obviated (35 occurrences in the whole record);

- grade 2 is viewed as a weak flood deficit;

- grade 3 and 4 are grouped and viewed as a strong flood deficit (years in bold);

- grade 5 is considered as a very strong deficit (years in bold and underlined).

Interpretation of the Nile flood record by Whetton et al. (1990)

Beside Quinn's work, another Nile flood record has been produced and interpreted in terms of its ENSO relationship by Whetton and collaborators (Whetton et al. 1990; Whetton and Rutherfurd 1994). They used a Nile flood height data-set from F. Hassan (Washington State University). This record only covers the periods: 1587–1625, 1720–1800 and 1824–1921 and is expressed in meters above sea level at Roda Island (Cairo). For inter-comparison and calibration purposes, these authors extended the record to 1984 by computing data on the river discharge at Aswan (prior to the dam construction, in 1964) and at Dongola (upstream, for the most recent period). They verified the good correlation between the Cairo record and the Aswan and Dongola records using an overlapping period (1871–1921). Their approach was different from that of Quinn (1992, 1993).

Although part of the original data overlaps with data used by Quinn, Whetton and Rutherfurd (1994) applied quite different statistical treatments, and expressed the results in different ways. These authors did not seek to quantify yearly anomalies, but rather distinguished years for which the Nile floods were above, or below, the average annual flood height (by at least a 0.5 standard deviation). Besides, they considered data prior to AD 1700 as unreliable, despite the fact that they included the period 1587–1625 (see Whetton and Rutherfurd (1994), Fig. 2).

Whetton and Rutherfurd's (1994) reconstructed Nile flood data are shown in the 6th column of Table 1. The years indicated are those for which the annual flood height is at least 0.5 std. dev. below the mean. It can be noted that, unlike Quinn, these authors did not attempt to extrapolate or reconstruct Nilometer data for missing years or for the period 1800–1823.

Comparison of the Nile low-flow sequences

Since Whetton and Rutherfurd were reluctant to include data prior to 1700 AD, the inter-comparison with the sequence reconstructed by Quinn (Table 1) is limited to the 18th and 19th centuries. As a large part of the original data is common to both reconstructions, one should expect a good match. In fact there is a majority (37) of coincidences, which are indicated by shaded boxes in Table 1. It must be added that parts of some years identified as low Nile-flood events by Whetton and Rutherfurd were also recognised as such by Quinn (1992, 1993) but with a lower grade "1" (not included in Table 1). Several matching cases correspond to pairs (or triplets) of years (1715–16, 1765–66, 1782–83–84, 1824–25 and 1858–59) and may correspond to manifestations of multi-year ENSO events.

The comparison of the two sequences that were analysed using different methods should produce a consolidated record of low flows of the Nile during the 18th and 19th centuries. The result, shown in Table 1, is rather encouraging. A large number of years recognised as of very low flows by Quinn (1993) (in bold) were also viewed as low-flow years by Whetton and Rutherfurd (1994). In 12 other cases, there is a coincidence of weaker low-flows (grade "2" of Quinn) in both sequences. All this gives some indirect confidence in the record proposed by Quinn (alone) for the 16th and 17th centuries.

Discussion

The identification of climatic anomalies in the last few centuries based on documentary records is hampered by several methodological limitations, but at the same time it is a powerful instrument for the study of inter-annual events like ENSO. In this respect, it is significant that the sequences of ENSO occurrences based on documentary data (Quinn et al. 1987; Quinn and Neal 1992; Quinn 1993) remain the main basis for the calibration of other proxies (Ortlieb 1999). The reduction of some of the intrinsic limitations of historical climate records relies upon comparisons of chronological sequences from different areas which are expected to respond similarly to a common climatic forcing (ENSO in the present case). This inter-regional approach has been followed by previous authors (Quinn 1993; Whetton and Rutherfurd 1994; Whetton et al. 1996). In this paper, we do not consider their respective conclusions, but focus on their respective reconstructions of the chronological sequence in the western extremity of the large-scale Southern Oscillation see-saw. One of the aims was to evaluate the reliability of the reconstruction of the Nile flood sequence. Another objective was to check whether recent findings relative to South-American El Niño manifestations, at the other extremity of the see-saw, correlate with the Nile flood record.

As the Nile flood record (Nilometer) has been viewed as one of the longest records of inter-annual hydrologic and climatic variability (Bliss 1925; Popper 1951; Fairbridge 1984), it is relevant to determine whether a clear and unequivocal relationship can be established between the low-flood events with El Niño occurrences during the last few centuries. If it were the case, it would confirm that the Nile record is particularly useful for the long-term reconstruction of the warm phase of ENSO, potentially since AD 622.

For the common period covered by the two reconstructed sequences of low flows of the Nile (by Whetton and Rutherfurd (1994), Quinn (1993)), i.e., AD 1700–1900, there is a fair correlation for the regular and very low floods (shaded areas in Table 1). However, the Whetton and Rutherfurd sequence includes a number of occurrences of low flows which do not coincide with any other ENSO manifestations from South America (in columns 1, 2 or 3, of Table 1), namely: 1703, 1743, 1752, 1753, 1759, 1886 and 1893. From this point of view, Quinn's Nile flood sequence seems better linked to the South American record (only in three cases are there are matches: 1713, 1758, and 1838). The differences of statistical treatment and of interpretation of the original sources by the two groups of workers may explain the lack of match for a series of years, but it cannot be clearly determined which sequence is more reliable. We therefore consider only the years for which both groups demonstrate a deficit in the annual flood (shaded boxes) as being reliable.

Regarding the intercomparison between the low Nile flows (shaded boxes in 6th and 7th columns) and the South American records (1st, 2nd and 3rd columns), several observations can be made. First, there is a good number of cases in which low flows of the Nile do not coincide with any ENSO-like manifestation in Peru, Chile or Argentina (1715–16, 1725, 1731, 1762, 1765–66, 1772, 1799, 1830, 1839, 1859, 1873 and 1882). As this represents almost half of the cases of the consolidated occurrences of low flows of the Nile (shaded boxes), there is reason to question the robustness of the co-occurrence between signals on both sides of the see-saw.

Second, the co-occurrences of low Nile flows and of El Niño-like manifestations in at least two of the three countries considered (Peru, Chile and Argentina) are indicated in Table 1 with closed boxes: 1737, 1783–84, 1791, 1824, 1828, 1837, 1845, 1850, 1855, 1864, 1877, 1888 and 1899. In several of theses cases, the reconstructed intensity of the climatic anomalies was strong or very strong (1783, 1791, 1845, 1864, 1877, 1888 and 1899). These co-occurrences of equivalent strength give some support to the large-scale correlations. However, it is noted that a few strong events in South America do not coincide with extreme low flows of the Nile (1814, 1871 and 1891).

Third, considering the whole period covered in Table 1 (16th to 19th centuries), it appears that it is during the 19th Century, and more precisely since 1824 that a fair correlation is observed between evidence of precipitation excess in one or several countries of South America and low Nile flows. During the eighteenth Century only three such coincidences were observed (1737, 1783 and 1791). The better correlation in the 19th Century may be due to higher quality and reliability of the more recent data. However, this might not be the major cause. We recall that it is also since about 1817 that there is a satisfactory correlation between the Peruvian and Chilean records of El Niño-related climatic anomalies (Ortlieb 1999; 2000). A possible explanation for this situation within South-America is that some changes occurred in the teleconnection pattern in the early nineteenth Century. The better correlation also observed between the Nile flood data and the South-American record since 1824 supports the interpretation that the alteration of the teleconnection pattern may have operated at a larger scale. It is hypothesised that this possible rearrangement of the atmospheric circulation is linked to the end of the Little Ice Age. Further, studies on the changes that accompanied the transition from the Little Ice Age to present-day conditions, and on the interaction between secular climatic changes and the ENSO system should help to clarify these teleconnection changes. Several recent studies dedicated to the evolution through time of the strength of the relationship between the Indian monsoon and ENSO

(Hoerling et al. 1997; Kumar et al. 1999a; b) may enlighten the processes involved in these modifications.

Summary

The reconstruction of ENSO (El Niño-Southern Oscillation) event occurrences in the course of the last few centuries is a prerequisite for a better understanding of the major source of inter-annual climate variability and of the interactions between ENSO and modes of longer-range climate variations. As the annual Nile floods are linked with the western end of the Southern Oscillation see-saw, and as their records extend (with some gaps) back to the 6th Century AD, this kind of documentary register presents a high potential for palaeo-ENSO studies. Several previous studies aimed to use at least parts of this record to verify its ability to reflect former El Niño occurrences, as identified in South America and eventually to assess the longest chronological sequence of historical ENSO events. Recent re-evaluations and updates of the South American record of ENSO events justify a close re-examination of previous attempts at correlation with the low Nile flood record. The reconstructed sequences of ENSO events proposed by Quinn (1992, 1993) and Whetton and Rutherfurd (1994), for the period 1525–1900, were thus compared to the new updated data from Peru, Chile, and Argentina. Occurrences of large-scale ENSO conditions are thus confirmed for the years: 1737, 1783–84, 1791, 1824, 1828, 1837, 1845, 1850, 1855, 1864, 1877, 1888, and 1899.

The present study shows that there is a fair correlation between low Nile flow events and excess of precipitation in Northern Peru, Central Chile and Northwestern Argentina in the major part of the 19th Century as well as during the 20th Century. On both ends of the Southern Oscillation system, at least since ca. 1824 synchronous signals of the warm phase of ENSO were recorded. Prior to the 19th Century, no clear correlation is observed. This poses a problem of reliability of the data, on either side of the see-saw. On grounds of previous (and on-going) work in South-America, where a similar situation is found (better correlation since the early nineteenth Century) it is envisaged that the lack of correlation of ENSO impacts between the Ethiopian plateaus and Southwestern South-America can be related to large scale modifications of the teleconnection system of ENSO.

Acknowledgements

This is a contribution from the UR "PALEOTROPIQUE", IRD, France. The author thanks the editors of the present volume for their invitation to contribute this chapter, and two reviewers who provided useful comments and suggestions.

References

Aceituno P. 1988. On the functioning of the Southern Oscillation in the South American sector, Part 1: Surface climate. Monthly Weather Review 116: 505–524.

Allan R.J. 2000. ENSO and climatic variability in the past 150 years. In: Diaz H. and Markgraf V. (eds), El Niño and the Southern Oscillation: Multiscale Variability and Global and Regional Impacts, Cambridge University Press, Cambridge, pp. 3–55.

Allan R., Lindesay J. and Parker D. 1997. El Niño-Southern Oscillation and climatic variability. CSIRO Publishing, Collingwood (Victoria, Australia).

Allan R. and d'Arrigo R. 1999. 'Persistent' ENSO sequences: how unusual was the 1990–1995 El Niño? The Holocene 9: 101–118.

Anderson R.Y., Soutar A. and Johnson T.C. 1992. Long-term changes in El Niño/Southern Oscillation: Evidence from marine and lacustrine sediments. In: Diaz H.F. and Markgraf V. (eds), El Niño, Historical and Paleoclimatic Aspects of the Southern Oscillation, Cambridge University Press, Cambridge, pp. 419–433.

Anonymous 1981. Five hundred years of wetness/dryness (rainfall) in China. Meteorological and Scientific Research Institute, Beijing.

Bell B. 1971. The first Dark Age in Egypt. American Journal of Archaeology 75: 1–26.

Bliss E.W. 1925. The Nile flood and world weather. Royal Meteorological Society Memoir 1, pp. 79–84.

Boninsegna J.A. 1990. Santiago de Chile winter rainfall since 1220 as being reconstructed by tree rings. In: Rabassa J. (ed.), Quaternary of South America and Antarctic Peninsula, 6: 67–87.

Bradley R.E. and Jones P.D. 1992. Climate since AD 1500. Routledge, New York.

Bradley R.E. and Jones P.D. 1993. "Little Ice Age" summary temperature variations: their nature and relevance to recent global warming trends. The Holocene 3: 367–376.

Brooks C.E.P. 1926. Climate through the Ages. Ernest Benn, London.

Cane M.A., Clement A.C., Kaplan A., Kushnir Y., Murtugudde R., Pozdnyakov D., Seager R. and Zebiak S. 1997. Twentieth-Century sea surface temperature trends. Science 275: 957–960.

Cook E.R. 1992. Using tree rings to study past El Niño/Southern Oscillation influences on climate. In: Diaz H.F. and Markgraf V. (eds), El Niño, Historical and Paleoclimatic Aspects of the Southern Oscillation. Cambridge University Press, Cambridge, pp. 203–214.

Corrège T., Delcroix T., Récy J., Beck J.W., Cabioch G. and Le Cornec F. 2000. Evidence for stronger El Niño-Southern Oscillation (ENSO) events in a mid-Holocene massive coral. Paleoceanography 15: 465–470.

Corrège T., Quinn T., Delcroix T., Le Cornec F., Récy J. and Cabioch G. 2001. Little Ice Age sea surface temperature variability in the southwest tropical Pacific. Geophys. Res. Lett. 28: 3477–3480.

d'Arrigo R.D. and Jacoby G.C. 1991. A 1 000-year record of winter precipitation from Northwestern New Mexico, USA: A reconstruction from tree-rings and its relationship to El Niño and the Southern Oscillation. The Holocene 1: 95–101.

Deser C. and Wallace J.M. 1987. El Niño events and their relation to the Southern Oscillation: 1925–86. J. Geophys. Res. 92: 14189–14196.

Diaz H. and Pulwarty R.S. 1992. A comparison of Southern Oscillation and El Niño signals in the tropics. In: Diaz H.F. and Markgraf V. (eds), El Niño, Historical and Paleoclimatic Aspects of the Southern Oscillation. Cambridge University Press, Cambridge, pp. 175–192.

Diaz H. and Pulwarty R.S. 1994. An analysis of the time scales of variability in centuries-long ENSO sensitive records in the last 1 000 years. Climatic Change 26: 317–342.

Diaz H. and Markgraf V. (eds) 2000. El Niño and the Southern Oscillation: Multiscale Variability and Global and Regional Impacts. Cambridge University Press, Cambridge, 496 pp.

Dunbar R.B., Linsley R.K. and Wellington G.M. 1996. Eastern Pacific corals monitor El Niño/Southern Oscillation, precipitation, and sea surface temperature variability over the past 3 centuries. In: Jones P.D., Bradley R.S. and Jouzel J. (eds), Climate Variations and Forcing Mechanisms of the Last 2000 Years. Springer-Verlag, Berlin, NATO ASI Series 1 (41), pp. 373–405.

Dunbar R. and Cole J.E. 1999. Annual Records of Tropical Systems (ARTS), Recommendation for research. PAGES-CLIVAR workshop report, Series 99-1, 72 pp.

Dunbar R.B., Wellington G.M., Colgan M.W. and Glynn P.W. 1994. Eastern Pacific sea surface temperature since 1600 AD: The $\delta^{18}O$ record of climate variability in Galápagos corals. Paleoceanography 9 (2): 291–315.

Enfield D.B. 1992. Historical and prehistorical overview of El Niño/Southern Oscillation. In: Diaz H.F. and Markgraf V. (eds), El Niño, Historical and Paleoclimatic Aspects of the Southern Oscillation. Cambridge University Press, Cambridge, pp. 95–117.

Enfield D.B. and Cid L. 1991. Low frequency changes in El Niño-Southern Oscillation. Journal of Climate 4: 1137–1146.

Fairbridge R.W. 1984. The Nile floods as a global climate/solar proxy. In: Mörner N.A. and Karlén W. (eds), Climatic Changes on a Yearly to Millennial Basis. Reidel, Dordrecht, pp. 181–190.

Foley J.C. 1957. Droughts in Australia: Review of Records from Earliest Years of Settlement to 1955. Australian Bureau of Meteorology Bulletin 34, 281 pp.

Glynn P.W. (ed.) 1990. Global Ecological Consequences of the 1982–83 El Niño-Southern Oscillation. Elsevier, Amsterdam.

Griffiths J.F. 1972. Ethiopian highlands. In: Griffiths J.F. (ed.), Climates of Africa, World Survey of Climatology, 10. Elsevier, New York, pp. 369–381.

Grove J.M. 1988. The Little Ice Age. Methuen, London. 498 pp.

Hamilton K. and Garcia R.R. 1987. El Niño-Southern Oscillation events and their associated midlatitude teleconnections, 1531–1841. Bulletin of the American Meteorological Society 67: 1354–1362.

Hassan F.A. 1981. Historical Nile floods and their implications for climatic change. Science 212: 1142–1145.

Hocquenghem A.-M. and Ortlieb L. 1992. Eventos El Niño y lluvias anormales en la costa del Perú: siglos XVI-XIX. Bulletin de l'Institut Français d'Etudes Andines 21: 197–278.

Hoerling M.M., Kumar A. and Zhong M. 1997. El Niño, La Niña and the nonlinearity of their teleconnections. Journal of Climate 119: 1769–1786.

Hurst H.E. 1957. The Nile. Constable, London.

Hurst H.E. and Phillips P. 1933. The Nile Basin, volume III, Ten-day Mean and Monthly Mean Gauge Readings of the Nile and its Tributaries. Physical Department, Ministry of Public Works, Government Press, Cairo.

Janowiak J.E. 1988. An investigation of interannual rainfall variability in Africa. Journal of Climatology 1: 240–255.

Jarvis C.W. 1935. Flood-stage records of the River Nile. Transcriptions of the American Society of Civil Engineers 101: 1012–1071.

Jones P.D., Osborn T.J. and Briffa K.R. 2001. The evolution of climate over the last Millennium. Science 292: 662–667.

Kiladis C.N. and Diaz H.Z. 1989. Global climatic anomalies associated with extremes in the Southern Oscillation. Monthly Weather Review 2: 1069–1090.

Kumar K.K., Rajagopalan B. and Cane M. 1999a. On the weakening relationship between the Indian Monsoon and ENSO. Science 284: 2156–2159.

Kumar K.K., Kleeman R., Cane M. and Rajagopalan B. 1999b. Epochal changes in Indian monsoon-ENSO precursors. Geophys. Res. Lett. 26: 75–78.

Lough J.M. and Fritts H.C. 1985. The Southern Oscillation and tree rings, 1600–1961. Journal of Climatology and Applied Meteorology 24: 952–966.

Lyons H.G. 1906. The physiography of the River Nile and its basin. Survey Dept, Cairo.

Michaelsen J. 1989. Long period fluctuations in El Niño amplitude and frequency reconstructed from tree rings. In: Peterson D.H. (ed.), Aspects of Climate Variability in the Pacific and the Western Americas. Geophysical Monograph 55, pp. 69–74.

Ortlieb L. 1994. Las mayores precipitaciones históricas en Chile central y la cronología de eventos "ENSO" en los siglos XVI-XIX. Revista Chilena de Historia Natural 67 (3): 117–139.

Ortlieb L. 1999. "Calibration" studies for ENSO events of the last few centuries. Lettre PIGB-PMRC, April 1999, pp. 29–37 + pp. VI.

Ortlieb L. 2000. The documentary historical record of El Niño events in Peru: An update of the Quinn record (sixteenth through nineteenth centuries). In: Diaz H. and Markgraf V. (eds), El Niño and the Southern Oscillation: Multiscale Variability and Global and Regional Impacts. Cambridge University Press, Cambridge, pp. 207–295.

Ortlieb L. and Macharé J. 1993. Former El Niño events: Records from Western South America. Glob. Planet. Chan. 7: 181–202.

Peterson D.H. (ed.) 1990. Aspects of Climate Variability in the Pacific and the Western Americas. Geophysical Monograph 55, Washington.

Philander S.G.H. 1989. El Niño, La Niña, and the Southern Oscillation. Academic Press, San Diego.

Popper W. 1951. The Cairo Nilometer. University of California Press, Berkeley.

Prieto M.R. 1994. Reconstrucción del clima de América del Sur mediante fuentes históricas; Estado de la cuestión. Revista del Museo de Historia Natural de San Rafael (Mendoza) XII (4): 323–342.

Prieto M.R., Herrera R. and Dussel P. 1998. Clima y disponibilidad hídrica en el sur de Bolivia y noroeste de Argentina entre 1560 y 1710; Los documentos españoles como fuente de datos ambientales. Bamberger Geographische Schriften 15: 35–56.

Prieto M.R., Herrera R. and Dussel P. 1999. Historical evidences of streamflow fluctuations in the Mendoza River, Argentina, and their relationships with ENSO. The Holocene 9: 473–481.

Quinn T.M., Crowley T.J. and Taylor F.W. 1996. New stable isotope results from a 173-year coral from Espiritu Santo Island, Vanuatu. Geophys. Res. Lett. 23: 3413–3416.

Quinn W.H. 1992. A study of Southern Oscillation-related climatic activity for AD 622–1900 incorporating Nile River flood data. In: Diaz H.F. and Markgraf V. (eds), El Niño, Historical and Paleoclimatic Aspects of the Southern Oscillation. Cambridge University Press, Cambridge, pp. 119–149.

Quinn W.H. 1993. The large-scale ENSO event, the El Niño, and other important features. Bulletin de l'Institut Français d'Etudes Andines 22: 13–34.

Quinn W.H., Zopf D.O., Short K.S. and Kuo Yang R.T. 1978. Historical trends and statistics of the Southern Oscillation, El Niño and Indonesian droughts. Fish. Bull. 76: 119–149.

Quinn W.H. and Neal V.T. 1983. Long-term variations in the Southern Oscillation, El Niño, and Chilean subtropical rainfall. Fish. Bull. 81: 363–374.

Quinn W.H., Neal V.T. and Antunez de Mayolo S. 1987. El Niño occurrences over the past four and a half centuries. J. Geophys. Res. 93: 14449–14461.

Quinn W.H. and Neal V.T. 1992. The historical record of El Niño events. In: Bradley R.S. and Jones P.D. (eds), Climate since AD 1500. Routledge, London, pp. 623–648.

Ropelewsky C.F. 1992. Predicting El Niño events. Nature 356: 476–477.

Ropelewsky C.F. and Halpert M.S. 1987. Global and regional scale precipitation patterns associated with El Niño/Southern Oscillation. Monthly Weather Review 115: 1606–1626.

Ruttlant J. and Fuenzalida H. 1991. Synoptic aspects of the Central Chile rainfall variability associated with the Southern Oscillation. Int. J. Climat. 111: 63–76.

Scott L. and Lee-Thorp J.A. 2003. Holocene climatic trends and rhythms in Southern Africa. In: Battarbee R.W., Gasse F. and Stickley C.E. (eds), Past Climate Variability through Europe and Africa. Kluwer Academic Publishers, Dordrecht, the Netherlands, pp. 69–91.

Shahin M. 1985. Hydrology of the Nile Basin. Elsevier, Amsterdam.

Stahle D.W. and Cleaveland M.K. 1993. Southern Oscillation extremes reconstructed from tree-rings of the Sierra Madre Occidental and Southern Great Plains. Journal of Climate 6: 1247–1252.

Stahle D.W., d'Arrigo R.D., Cleaveland M.K., Krusic P.J., Allan R.J., Cook E.R., Cole J.E., Dunbar R.B., Therell M.D., Gay D.A., Moore M., Stokes M.A., Burns B.T. and Thompson L.G. 1998. Experimental multiproxy reconstructions of the Southern Oscillation. American Meteorological Society Bulletin 79: 2137–2152.

Thompson L.G., Mosley-Thompson E. and Arnao B.M. 1984. El Niño-Southern Oscillation events, recorded in the stratigraphy of the tropical Quelccaya Ice Cap, Peru. Science 226: 50–53.

Thompson L.G. and Mosley-Thompson E. 1989. One-half millennia of tropical climate variability as recorded in the stratigraphy of the Quelccaya Ice Cap, Peru. In: Peterson D.H. (ed.), Aspects of Climate Variability in the Pacific and the Western Americas. Geophysical Monograph 55, pp. 15–31.

Thompson L.G. 1992. Ice-core evidence from Peru and China. In: Bradley R.S. and Jones P.D. (eds), Climate since AD 1500. Routledge, London, pp. 517–548.

Thompson L.G., Mosley-Thompson E. and Thompson P.A. 1992. Reconstructing interannual climate variability from tropical and subtropical ice-core records. In: Diaz H.F. and Markgraf V. (eds), El Niño, Historical and Paleoclimatic Aspects of the Southern Oscillation, Cambridge University Press, Cambridge, pp. 295–322.

Thompson L.G., Henderson K.A., Mosley-Thompson E. and Lin P.N. 2000. The tropical ice core record of ENSO. In: Diaz H. and Markgraf V. (eds), El Niño and the Southern Oscillation: Multiscale Variability and Global and Regional Impacts. Cambridge University Press, Cambridge, pp. 325–556.

Timmermann A., Oberhuber J., Bacher A., Esch M., Latif M. and Roeckner E. 1999. Increased El Niño frequency in a climate model forced by future greenhouse warming. Nature 398: 694–697.

Toussoun O. 1925. Mémoire sur l'Histoire du Nil. Mémoires de l'Institut d'Egypte, Le Caire.

Tudhope A.W., Shimmield G.B., Chilcott C.P., Jebb M., Fallick A.E. and Dalgleish A.N. 1995. Recent changes in the far western equatorial Pacific and their relationship to the Southern Oscillation; oxygen isotope records from massive corals, Papua, New Guinea. Earth and Planetary Earth Sciences 136: 575–590.

Verschuren D. 2003. Decadal and century-scale climatic variability in tropical Africa during the past 2000 years. In: Battarbee R.W., Gasse F. and Stickley C.E. (eds), Past Climate Variability through Europe and Africa. Kluwer Academic Publishers, Dordrecht, the Netherlands, pp. 139–158.

Villalba R. 1994. Fluctuaciones climáticas en latitudes medias de América del Sur durante los últimos 1000 años: sus relaciones con la Oscilación del Sur. Revista Chilena de Historia Natural 67: 453–461.

Villalba R., Cook E.R., Jacoby G.C., d'Arrigo R., Veblen T.T. and Jones P.D. 1998. Tree-ring based reconstructions of Northern Patagonia precipitation since AD 1600. The Holocene 8: 659–674.

Villalba R., d'Arrigo R., Cook E.R., Jacoby G.C. and Wiles G. 2001. Decadal-scale climatic variability along the extratropical western coast of the Americas: Evidence from tree-ring records. In: Markgraf V. (ed.), Interhemispheric Climate Linkages. Academic Press, New York, pp. 155–172.

Walford C. 1879. Famines of the World, Past and Present. Burt Franklin, New York.

Walker G.T. 1910. Correlation in seasonal variations of weather, II. Sunspots and temperature. Memoir of the India Meteorological Department 21, part II, pp. 22–45.

Walker G.T. 1924. Correlation in seasonal variations of weather, IX. A further study of world weather. Memoir of the India Meteorological Department 24, part IX, pp. 275–332.

Walker G.T. and Bliss E.W. 1932. World weather, V. Memoir of the Royal Meteorological. Society 4, pp. 53–84.

Wang G. and Eltahir E.A.B. 1999. Use of ENSO information in medium- and long-range forecasting of the Nile floods. Journal of Climate 12: 1726–1737.

Whetton P.H., Adamson D.A. and Williams M.A.J. 1990. Rainfall and river flow variability in Africa, Australia and East Africa linked to El Niño-El Southern Oscillation events. Geological Society of Australia Symposium, Proceedings 1, pp. 1134–1137.

Whetton P.H. and Rutherfurd I. 1994. Historical ENSO teleconnections in the Eastern Hemisphere. Climatic Change 28: 221–253.

Whetton P.H., Allan R.J. and Rutherfurd I. 1996. Historical ENSO teleconnections in the Eastern Hemisphere: comparison with latest El Niño series of Quinn. Climatic Change 32: 103–109.

Williams M.A.J., Adamson D. and Baxter J.T. 1986. Late Quaternary environments in the Nile and
    Darling Basins. Australian Geographical Studies 24: 128–144.
Zhang J. and Crowley T.J. 1989. Historical climate records in China and reconstruction of past
    climates. Journal of Climate 2: 833–849.

14. GROUNDWATER AS AN ARCHIVE OF CLIMATIC AND ENVIRONMENTAL CHANGE: EUROPE TO AFRICA

W. MIKE EDMUNDS (wme@btopenworld.com)
British Geological Survey
Crowmarsh Gifford
Wallingford
Oxon, OX10 8BB
UK

ABDELKADER DODO
Universite Abdou Moumouni
Faculte des Sciences
Departement de Geologie
BP 13316 Niamey
Niger

DJAIRA DJORET (fsciences@sdntcd.undp.org)
Faculte des Sciences Exactes et Appliques
BP 1027 N'Djamena
CHAD

FRANÇOISE GASSE (gasse@cerege.fr)
Centre Européen de Recherche et d'Enseignement
de Géosciences de l'Environnement (CEREGE)
Europole de l'Arbois
B.P. 80, F-13454 Aix-en-Provence cedex 4
France

CHEIKH B. GAYE (c.b.gaye@iaea.org)
International Atomic Energy Agency
Wagramerstasse 5
A1040 Vienna
Austria

IBRAHIM B. GONI (ibgoni@infoweb.abs.net)
University of Maiduguri
Maiduguri
Borno State
Nigeria

R. W. Battarbee et al. (eds) 2004. *Past Climate Variability through Europe and Africa.*
Springer, Dordrecht, The Netherlands.

YVES TRAVI (yves.travi@univ-avignon.fr)
Laboratoire Hydrogeologie
Faculte des Sciences
33 Rue Pasteur
84000 Avignon
France

KAMEL ZOUARI (kamel.zouari@rnu.tn)
Ecole Nationale d'Ingenieurs
3038 Sfax
Tunisia

GIAN-MARIA ZUPPI (zuppi@tin.it)
Department of Environmental Science
University of Venice
Ca' Foscari
Santa Marta
Dorsoduro 2137
30123 Venice
Italy

Keywords: Groundwater, Palaeohydrology, Africa, Isotopes, Palaeoclimate, Noble gases, Hydrogeochemistry, Holocene, Pleistocene

Introduction

Droughts and wet periods in recent earth history have a clear link with human habitation and migration notably in the Middle East and in Africa. Water availability in the form of perennial rivers, lakes and springs dictated the first settlement patterns. Reference is found in the first written records to dramatic climatic changes affecting water availability (Lambert and Millard 1969) and a fascinating challenge exists to reconstruct and to relate the hydrological records to the early historical and archaeological evidence (Issar 1990; Hassan 1997) as well as the rapidly growing body of data on palaeoclimate.

Groundwater is emerging as an archive at both low and mid-latitudes of past climatic and hydrological change, which may be used alongside other proxy data. Indirect evidence of the palaeohydrology in the late Pleistocene and Holocene, has been deduced from various sources especially lake sediments (Fontes and Gasse 1991; Gasse 2000; Hoelzmann et al. 2000; this volume) and speleothems (Bar-Matthews et al. 1997). In contrast to other archives such as ice cores or tree rings, which contain high-resolution information, data available in large groundwater bodies are of low resolution (typically ± 1000 yr). This is due to the advection or dispersion of any climatic input signal in the water body. Many groundwater data are obtained from pumped samples where sample intervals may extend over tens of metres. Nevertheless, specific indications of palaeo-temperature, air mass circulation and vegetation history may be retained in a range of different chemical and isotopic signals

(Fontes et al. 1993a; Stute and Schlosser 1993), notably in confined aquifers where sequential changes may be recorded along flow lines, or in the stratification of phreatic aquifers. Dated groundwaters are important since these contain the direct evidence of prolonged wet episodes. Even the absence of dated waters over a specific time interval may indicate periods of drought (Sonntag et al. 1978). The correlation between groundwater records and aeolian deposition in semi-arid/arid regions (Stokes et al. 1998; Swezy 2001) can also provide complimentary evidence of wet and dry intervals.

Moisture in the unsaturated zone may under favourable circumstances also contain records of past environments and climate at decadal to millennial scale resolution, contained mainly as variations in salinity and in stable isotope enrichments (Edmunds and Tyler 2002). Such records are found in porous media in areas of low moisture flux, notably beneath modern arid or semi-arid areas. The resolution of unsaturated (vadose) zone records will depend on the dispersion of the signal (Cook et al. 1992) but decadal scale records may be retained, as in West Africa, over several hundred years (Edmunds and Gaye 1997), or at the millennial scale over the late Pleistocene (Tyler et al. 1996).

Some of the classical studies of hydrogeological systems that contain palaeo-waters have been conducted in Northern Africa and in Western Europe over the past three decades. In this paper the evidence contained in unsaturated zone profiles and especially in phreatic and confined aquifers in semi-arid and arid areas of Northern Africa is used to demonstrate the current possibilities in using groundwater archives. The results from Northern Africa are then compared with the evidence found in palaeowaters in Europe and elsewhere along the PEP III transect.

Tools for groundwater archive studies

Inert tracers (stable isotopes ($\delta^{18}O$, δ^2H)), together with noble gases, NO_3, Cl, Br/Cl, are the main palaeoenvironmental indicators, and retain different information on the input conditions (Herczeg and Edmunds 1999). The importance of these tracers for reconstruction is greatly enhanced if absolute chronometers are available.

Radiocarbon is the primary tool for reconstruction of groundwater records which depend upon a reliable chronology. Over the timescales of interest (10^3- $>10^6$ yr) other specialised options for absolute dating, such as ^{39}Ar and possibly ^4He, also exist (Loosli et al. 1999) although for the focus on Holocene records and the Pleistocene/Holocene transition these other tools are inappropriate. Many caveats apply to the use of ^{14}C in groundwaters due to difficulties of knowing input conditions and the water-rock interactions involved, especially where carbonate minerals are present along flow paths (Clark and Fritz 1997). In the predominantly non-carbonate aquifers of many large basins, however, age correction may be applied with caution to provide calibration of the flow sequence. Changes in radiocarbon activities along flow lines, expressed as pmc (percent modern carbon), allow relative timescales to be established, especially using $\delta^{13}C$ and supporting geochemical evidence to model the sources of the carbon. However due to mixing and/or reaction it may not be possible to resolve ages within a few thousand years. Care is also needed to verify the closed system conditions in the aquifers since other crustal sources of CO_2 may be present (Andrews et al. 1994). The $\delta^{13}C$ values may also be used to infer near surface phenomena including reactions in the soil and with surface crusts and changes in C_3-C_4 vegetation types.

The wide variations in oxygen-18 and deuterium observed in precipitation at the present day are sensitive indicators of change and complexity in temperature, precipitation patterns and air mass circulation, especially in the Sahara/Sahel region (Dray et al. 1983). Past rainfall stored as palaeo-groundwater, together with the other hydrological archives such as ice, provide evidence of former climatic conditions (Rozanski et al. 1997). Climatic changes are expressed primarily as: (i) isotopic depletion relative to modern groundwaters with reference to the meteoric water line; (ii) change in the deuterium excess, signifying changes in humidity in the air mass as it detaches from its primary oceanic source moving over arid regions; and (iii) local condensation and evaporation effects within clouds or in falling rain. Oxygen isotopic enrichment due to near-surface evaporation is also found in many groundwaters signifying that rates of aquifer recharge are likely to be low.

Chloride, in contrast to water, is conservative and so changes in chloride concentrations with time, where these can be measured, are usually good indicators of changes in aridity/wetness. The combined use of chloride and the stable isotopes of water ($\delta^{18}O$, $\delta^{2}H$), moreover, provides a powerful technique for studying past environments in groundwaters. Over continental areas groundwater solutes are dominantly of atmospheric origin and are concentrated in proportion to evaporation during recharge. The large freshwater reserves in some basins of modern arid zones are therefore *a priori* indicators of wetter climates. With lowered sea levels during the late Pleistocene and lasting until some 18–10 kyr BP, there was also the opportunity for freshwater to advance offshore relative to the present day coastlines and to displace saline formation waters (Edmunds and Milne 2001). The present-day distribution of groundwater salinity therefore provides clues to climate variations. In shallow aquifers higher salinity is mainly a legacy of the onset of more arid conditions during the past 4000 yr.

Noble gas contents (Ne, Ar, Kr, Xe) of groundwater (corrected for excess air) under closed system conditions reflect the annual mean air temperature. They form the most reliable indicator of palaeo-temperature in groundwaters and may be used to help interpret the significance of changes in the stable isotope ratios, which may not always be related to temperature (Andrews 1993; Andrews et al. 1994; Stute and Schlosser 1993; Stute and Talma 1998).

Nitrate remains inert in the presence of dissolved oxygen and may retain the signature of the environmental conditions at the time of recharge (Edmunds and Gaye 1997; Hartsough et al. 2001). In Africa high nitrate concentrations, quite frequently exceeding World Health Organisation (WHO) limits for drinking waters, are considered to result naturally from fixing by leguminous vegetation. In reducing waters, the former existence of higher nitrate concentrations may be determined by enhanced N_2/Ar ratios (Andrews et al. 1994). The ratio of Br/Cl may also be used to fingerprint the atmospheric (as well as geological) origins of the Cl (Edmunds 1996; Davis et al. 1998) and Br enrichment usually indicates biomass decay, including forest fires (Goni et al. 2001).

Unsaturated zone archives in Northern Africa

Unsaturated or vadose zones offer an opportunity over aquifers to preserve climate signals with a resolution of decades to centuries (Cook et al. 1992). Salinity changes, proportional to rainfall properties and recharge rates may be preserved under piston flow conditions to depths of ca. 10 to 300 m. Below the soil, mixing processes in the vadose zone are often small, velocities of water are low and combined with the great depth of many arid vadose zones, long-past infiltration events (as well as recent oscillations) may be discernible. Recent progress towards the use of the unsaturated zone as an archive has been reviewed by Edmunds and Tyler (2002) and several examples are to be found in Africa (Senegal, Nigeria, Niger, Cyprus and Tunisia), two of which are used here as illustration.

Several profiles obtained from Northern Senegal have been interpreted as archives of recharge, climatic and environmental change for periods up to 500 years (Edmunds et al. 1992; Cook et al. 1992). One profile (L3) has been calibrated using good instrumental records for rainfall and river flow. The profile record (Fig. 1a) is 108 years and, assuming that the piston flow model applies, the peaks in Cl at 4–6 m and 6–13 m correspond respectively to periods of drought in the 1970s and 1980s and the lower one in the 1940s. Another peak in the 1900s also reflects a recorded drought period. The unsaturated zone profile is compared with the rainfall record at St Louis (some 80 km from the research site) dating back to the 1890s (Gac 1990) and with the Senegal River with records over a similar period (Olivry 1983). Whereas the correlation with the rainfall records is moderately good, the correlation with the river flow, representing the regional influence is better. It is also possible that rainfall intensity is a critical factor in transmission of the rainfall signal to the profile. The correspondence with the main wet phase from 1920–1940 is well shown in all sets of data. During the dry episodes the recharge rate reduced to around 4 mm/yr but during the wet phases this rose to as high as 20 mm.

A further profile at Louga, Northern Senegal (L10), which contains higher chloride concentrations than in the L3 profile, is extrapolated beyond the instrumental record (Fig. 1b). This profile is compared with a reconstruction of water level variations of Lake Chad based on sedimentological and palynological data (Maley 2000). The profiles have been aligned using a chloride fallout of $1 \, \mathrm{g \, m^{-2} yr^{-1}}$ since 1950 and of $0.7 \, \mathrm{g \, m^{-2} yr^{-1}}$ before 1950; the lower value has been arbitrarily chosen to provide a best fit with the Lake Chad record. The data are here expressed as cumulative chloride against time where the conversion to years is possible using the values of fallout given above. There is a reasonable correlation between both archives for events at the century scale, although there are clear limits to using the Lake Chad record, which relies on drainage history from the south. Numerical modelling of the profile data show that there is no unique recharge history that can be determined from the profile (Cook et al. 1992). However the total recharge that caused the lowering of Cl concentrations above 30, 15 and 5 mg/cm^2 cumulative chloride (300, 150 and 20 yr BP) must have been 150, 50 and 50 mm/yr respectively.

An unsaturated zone record of 70 m has also recently been described from Southern Niger that represents a recharge history of over 700 yr (Bromley et al. 1997). This is beyond the instrumental record and over this timescale presents the possibility that such records could provide the most reliable chronology of hydrological events in such areas. So far, no records in excess of 1000 yr are recorded in the semi-arid areas (>250 mm rainfall) of Northern Africa, although archives, possibly in excess of this may be found in the more arid areas.

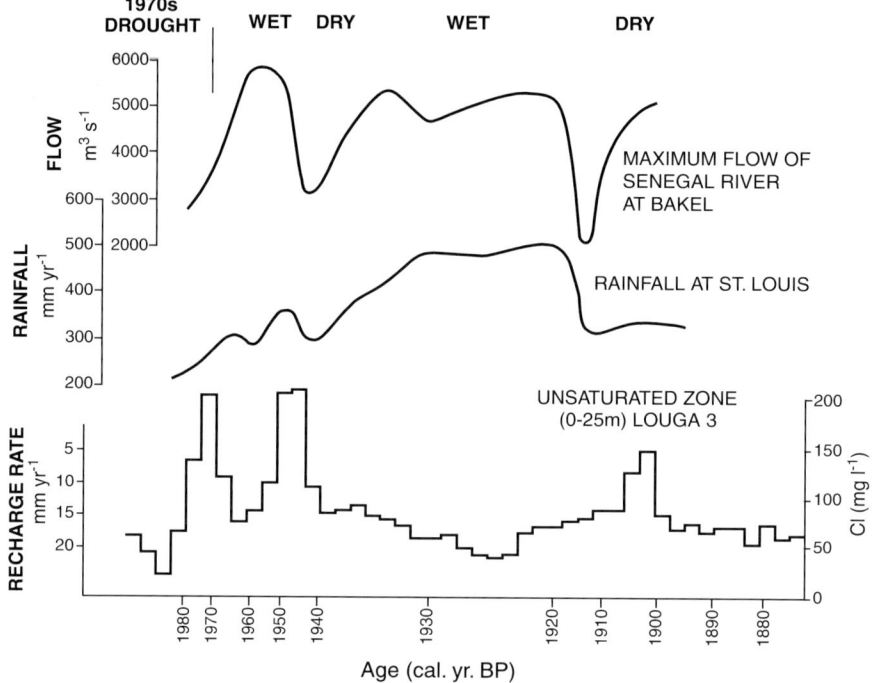

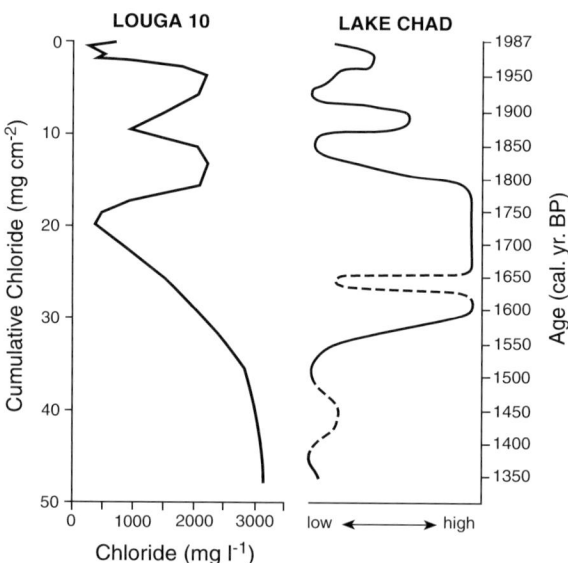

Figure 1. Unsaturated zone chloride profiles L3 and L10 from Louga region, Senegal. a) L3 - an unsaturated zone profile (25 m) showing timescale of the past 120 yr with corresponding recharge rates. The curves are matched against records of rainfall at St Louis and flow of the Senegal River. b) L10 profile expressed as cumulative Cl against chloride where the linear timescale (using a Cl flux of 1 g Cl m^{-2} yr^{-1}) is compared with the lake Chad record over the past 600 yr.

Main aquifers

Approach adopted

There is a wide literature on the isotopic and chemical composition of groundwaters in Northern Africa obtained over the past 30 years. For this paper a selection of data containing both radiocarbon as well as oxygen and hydrogen stable isotope compositions has been made, providing a good geographical coverage (Plate 5); age corrections are not used for the main analysis in view of the uncertainties that may be introduced. A restriction has been placed by selecting, as far as possible, groundwaters from non-carbonate sedimentary aquifers. In these aquifers the measured radiocarbon activities are likely to be proportional to age. Stable carbon isotope ratios have been used to screen the data further and generally those results where the values are more enriched than –6.00‰, indicating dilution with non-active (marine carbonate) carbon by water-rock interaction, have been rejected. Authors' data-sets are used at face value and for more detailed explanations the original papers should be referred to. Plots of radiocarbon activities expressed as percent modern carbon (pmc) are used to provide a relative timescale to evaluate the significance of the stable isotope ($\delta^{18}O$) compositions (Fig. 2a-f). In these plots three timescales, Modern, Holocene and late Pleistocene are recognised. Further interpretation is then provided using selected plots of $\delta^{18}O$ vs. $\delta^{2}H$ and a consideration of noble gas recharge temperatures (NGRT) where these exist.

Sources of data

Western and Northwestern Africa

The Saïs Plain, Northern Morocco contains records of dated groundwaters from a Quaternary lacustrine carbonate sequence (Kabbaj et al. 1978) with $\delta^{13}C$ values (-13 to $-17‰$) indicating little or no reaction with the carbonate matrix. In Senegal, data are from the Cretaceous Maastrichtian aquifer with a few results also from the overlying Oligo-Miocene (Faye 1994; Faye et al. 1993). For Mali, results from Fontes et al. (1991) contain a transect from the Niger river near Toumbouctou northwards. Only those data from north of the Azaouad Ridge are included here since those to the south are considered to have recharged from northwards (Holocene) flooding of the Niger. Also included are data from Western Mali on the line from Koulikoro north to Nara (Dinçer et al. 1984) where the aquifers are mainly found in sandstones and schists of Cambrian age; the $\delta^{13}C$ values are all more negative than $-10‰$.

Central North Africa, Sahara and Sahel

The large sedimentary aquifers of the Continental Intercalaire (CI) and Complexe Terminal (CT) in Algeria and Tunisia contain confined groundwaters which retain sequential evidence of past recharge conditions along flow lines of some 800 km with a record probably extending back over 100,000 years. Radiocarbon dated waters from the CI and the CT aquifer (Guendouz et al. 1998; Edmunds et al. 2003) are used here (Fig. 2d), together with results from the CI from Southern Tunisia. Groundwaters from the Saharan region of Niger (Fig. 2b) are represented by those in the south from the Illumeden Basin (Le Gal La Salle 1992), and from the north from Arlit (Dodo and Zuppi 1999) and the wider Irhazer region

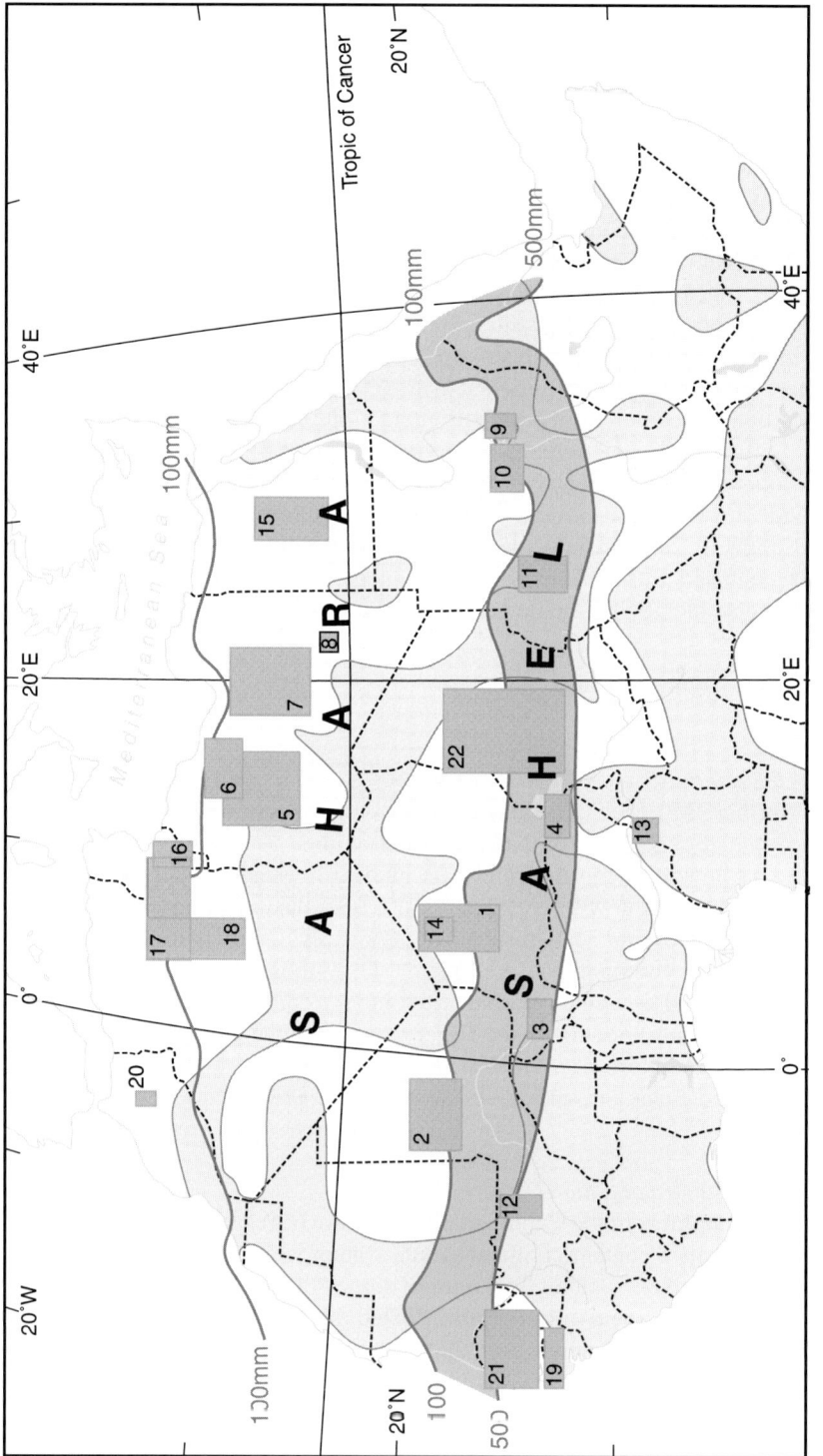

Plate 5. The distribution of sedimentary basins in Africa (pale yellow shading) and the location of studies used as data sources for the present paper: 1) and 2) Azaouad (Fontes et al. 1991); 3) Illumeden (Le Gal La Salle 1992); 4) Chad Basin; 5) Western Libya (Salem et al. 1980); 6) Northern Libya (Srdoč et al. 1980); 7) Sirt Basin (Edmunds and Wright 1979); 8) Kufra Basin (Edmunds and Wright 1979); 9) Butana region (Darling et al. 1991); 10) Kordofan (Groening et al. 1993); 11) Darfur (Groening et al. 1993); 12) Kolokoni-Nara (Dincer et al. 1983); 13) Garoua (Njitchoua et al. 1993); 14) Irhazer (Andrews et al. 1993); 15) Western Desert (Thorweihe 1982); 16) and 17) Continental Intercalaire (Guendouz et al. 1998); 18) Complexe Terminal (Edmunds et al., in press); 19) Casamance (Faye et al. 1993); 20) Saïs (Kabbaj et al. 1978); 21) N Senegal (Faye 1994); 22) Chad basin (UNESCO 1972). Colour version of this Plate can be found in Appendix, p.632

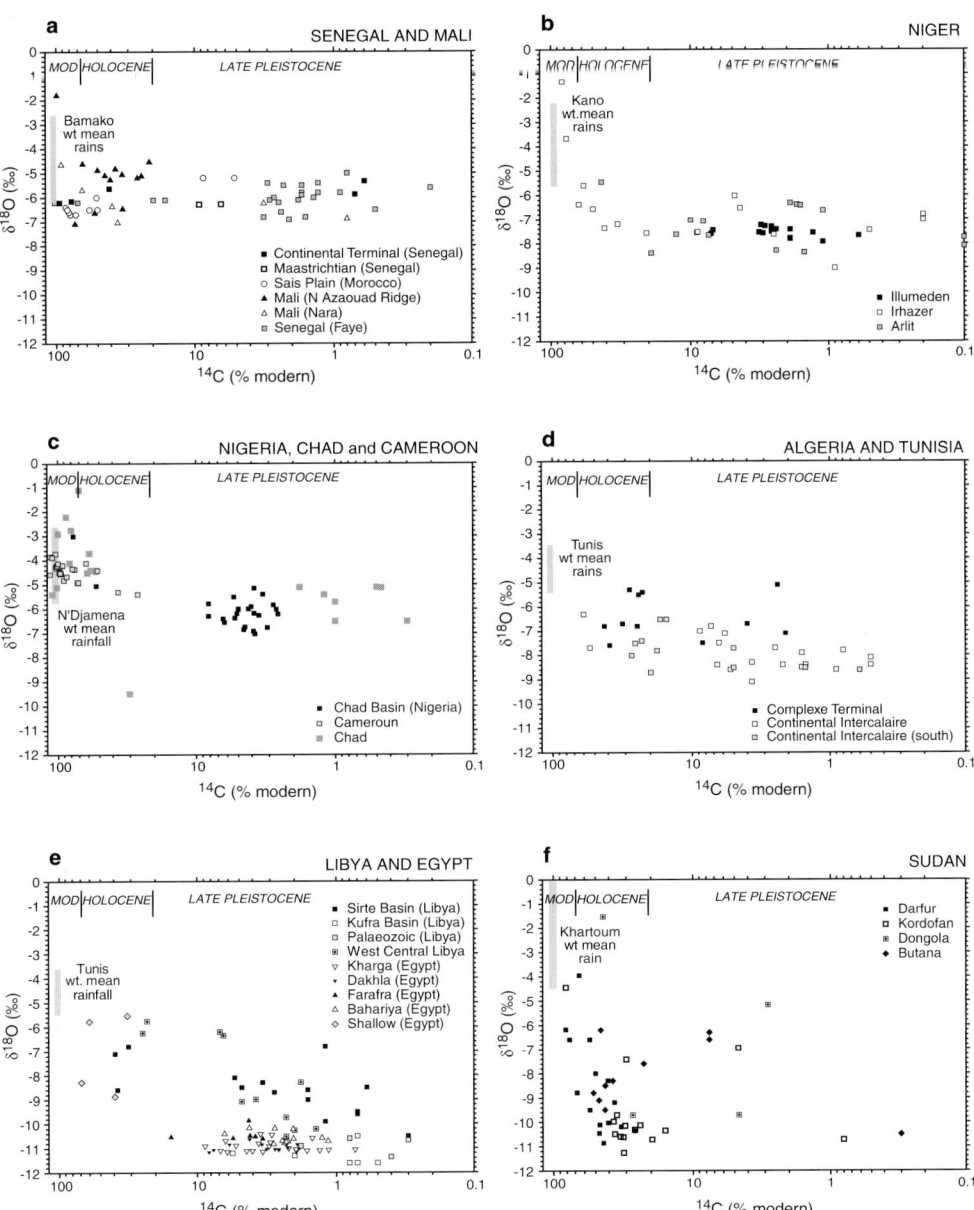

Figure 2. The change in $\delta^{18}O$ during the late Pleistocene and Holocene as shown by dated groundwaters (^{14}C as percent modern carbon): a) Senegal and Mali; b) Niger; c) Nigeria, Chad and Cameroon; d) Algeria and Tunisia; e) Libya and Egypt; f) Sudan.

(Andrews et al. 1994). In the latter area it was proposed that some disturbance of the ages may have been caused by dilution with CO_2 from deep sources. Pleistocene groundwaters (Fig. 2c) are found in the Quaternary Middle and Lower Zone aquifers of Nigeria, near Lake Chad which were emplaced during lowered levels of Lake Chad, but groundwaters of early Holocene age are not found (or preserved) here, recharge having been prevented by the Holocene rise in Lake Chad lake levels (Edmunds et al. 1997; 1999) or subsequently displaced by modern rain. Dated groundwaters in Chad are available from the original studies on the basin by UNESCO (1972). In Cameroon only groundwaters of Holocene age have been recognised (Nijtchoua et al. 1993) in sandstones of the Garoua region.

North-East Africa, Sahara and Sahel
This region is dominated by the vast Nubian Cretaceous Sandstone aquifer (Plate 5) in the Kufra Basin, Northern Sudan and the Western Desert of Egypt. The Nubian system is then overstepped in Libya by Tertiary sedimentary basins, which like the Nubian system are mainly continental in origin but with marine facies (containing brackish water) nearer to the coast. Data used here (Fig. 2e) are from freshwaters from the continental aquifers, which preserve mainly initial carbon isotope inputs with little or no modification by water-rock interaction, although some Holocene waters (Edmunds and Wright 1979) may have reacted with active calcretes, affecting age correction. Data from Northern Libya are from the Sirt Basin (Edmunds and Wright 1979), and from the Murzuq and other basins in Western and Central Libya (Srdoč et al. 1980; Salem et al. 1980). Data from Egypt (Thorweihe 1982) are from the phreatic or semi-confined aquifers feeding the major oases of the Western Desert in the Nubian sandstone, which may be compared with data from Southern Libya (Kufra) in Edmunds and Wright (1979). Data from the same aquifer in Northern Sudan (Fig. 2f) are from Darling et al. 1987 (Butana), as well as from Darfur and Kordofan (Groening et al. 1993).

Discussion of isotope/age relationships

Some general observations may be made from the radiocarbon/stable isotope diagrams (Fig. 2a-f):
    i) There is an overall trend in late Pleistocene palaeowater $\delta^{18}O$ composition from west to east of some 4.5‰, from around -6‰ to -10.5‰ (cf. Fig. 2a-3f). This confirms a trend first observed by Sonntag et al. (1978, 1980) and is evident both in the aquifers of North Africa and also in parts of Africa south of the Sahara in the Illumeden basin of Niger for example. It describes an overall continental effect with an Atlantic moisture source moving from west to east (as at the present day where this is mainly limited to north of 22 °N) where the air mass evolved according to Rayleigh (closed system) fractionation with residual isotopically lighter rains further to the east. The former widespread extent of these rains recorded in groundwaters beneath the present Sahara points to the shift of the Atlantic jet stream well to the south during most of the late Pleistocene, as well as a weakening of the south-west monsoon. The more enriched isotopic signatures of groundwaters in Nigeria/Cameroon, compared with waters to the north suggests that, during part of the late Pleistocene, monsoon rains derived from the Gulf of Guinea still prevailed south of 16 °N. In Senegal, the similarity of the isotopic compositions reflects the continuity of

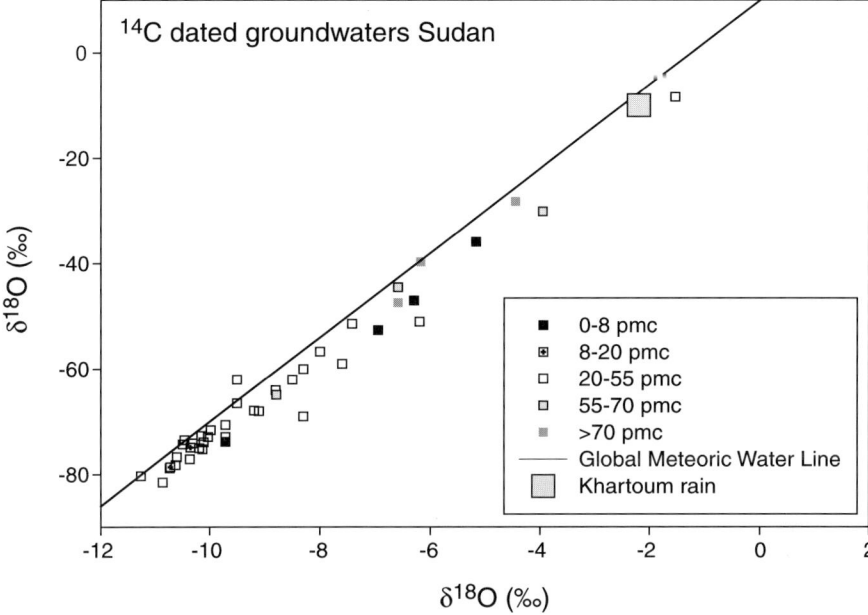

Figure 3. Plot of δ^{18}O vs. δ^2H for dated groundwaters from Sudan as well as modern rainfall from Khartoum.

the maritime influence but with recharge derived from the south-west monsoon activity of variable strength.

ii) Within the late Pleistocene palaeo-waters of North Africa there are also distinct trends on a north to south basis, seen most clearly in the groundwaters in Libya and Egypt (Fig. 2e). All these waters are isotopically light as compared with the groundwater nearer the Atlantic source or further south. Groundwater in the Kufra Basin (Nubian Sandstone) is isotopically the lightest in North Africa (−11.5‰) and this compares with the Sirte Basin to the north where the palaeo-waters are some 3‰ more enriched (Edmunds and Wright 1979). Thus each sedimentary basin seems to have a distinctive composition, which supports the likelihood of local evolution of groundwater (as well as the lack of hydraulic continuity between the Kufra and Sirte basins). Groundwaters from the Egyptian oases (Thorweihe 1982) lie within the range −10 to −11‰, distinct from the Sirte Basin at the same latitude and more akin to Kufra Basin compositions. Although there is the possibility that the continental effect could have led to the easternmost enrichments as seen in Egypt, the vast reserves of freshwater are also anomalous in that they are found at the extreme of the evolution of the Atlantic air mass source where lower rainfall amounts would be expected. An additional possibility is that some recharge from the south-west within the hydraulically continuous Nubian sandstone took place, with a superimposed altitude effect the result of heavier rains and surface runoff from the Tibesti mountains. The effects of the mountain areas elsewhere in Northern Africa may also have contributed to isotopically light runoff.

iii) A gap in the record exists in most of the data-sets for groundwaters with ^{14}C of approximately 5–15 pmc. This was first noticed for north African ^{14}C data-sets by Geyh and Jäkel (1974) and demonstrated for groundwaters by Sonntag et al. (1978, 1980). This is interpreted as an arid interlude, discussed further below, coinciding with the LGM in Europe. This effect is obscured slightly due to mixing of stratified waters, but is observable over most of Northern Africa (except possibly in the data from Niger which may be masked by the effect of higher rainfall in the Saharan mountains).

iv) Holocene groundwaters show markedly different properties across the continent both in relation to those from the late Pleistocene, as well as trends during the Holocene itself. In Morocco, the Holocene groundwaters are isotopically lighter (around 1–2‰) relative to those from humid periods of the late Pleistocene. In Senegal there is a similar if subdued tendency less marked than to the north. This may be explained by the maritime situation of both countries where the influence of the Atlantic moisture (and sea-surface temperature effects) is felt directly. The relative isotopic enrichment during the late Pleistocene is the result of the change in the ocean composition due to the lighter isotopes being enriched in the ice caps. This is the only noticeable effect over the past 30 kyr in the coastal areas and the otherwise similar isotope compositions demonstrate the constancy of the westerly Atlantic air masses, and the south westerly monsoons (which varied in intensity), over the whole period.

Elsewhere in Northern Africa, modern and most Holocene groundwaters have isotopic compositions considerably more enriched relative to the late Pleistocene pluvial periods. The Holocene, on the basis of extensive evidence from other proxy data, was a period with short (millennial scale) duration, but intense wet phases, which were not synchronous across the continent, from 11.4 to 5.4 kyr BP (Gasse 2000), prior to the dessication that led to the present day conditions commencing around 4.5 kyr BP. The weighted mean (3σ) compositions of modern rainfall from stations across Africa, based on the IAEA GNIP data base (*http://www.iaea.or.at:80/programs/ri/gnip/gnipmain.htm*), are plotted in Figure 2a-f; the weighted means are likely to represent the heaviest rains contributing to groundwater recharge at the present day. A very large range in modern compositions across Africa is evident and it is also clear that the compositions of the Holocene groundwaters are depleted when compared to the modern rains. This is a general expression of the amount and intensity of the rainfall in the Holocene as compared with the present-day arid climates. There are, however, wide variations in the isotopic compositions with time during the Holocene, especially in Eastern Africa (Libya, Egypt, Sudan); some of these cases are discussed in more detail below.

In Figure 3 a plot of $\delta^{18}O$ vs. $\delta^{2}H$ for the dated groundwaters from Sudan shows a wide variation in composition of recharge throughout the late Pleistocene and Holocene. It is significant that all waters lie just below the meteoric water line with similar enrichment to the modern rainfall compositions represented by the Khartoum rainwater. This is good evidence that the composition of the monsoon rainfall source has remained constant but with different intensity from late Pleistocene to modern times.

Noble gas recharge temperatures

Reconnaissance measurements of noble gas recharge temperatures (NGRT) were carried out in Northern Africa by Rudolph et al. (1984) and subsequently more extensive palaeotemperature measurements have been carried out in Niger (Le Gal La Salle 1992), Senegal (Faye

Table 1. Summary of noble gas recharge temperature (NGRT) results for Northern Africa

Location	Mean NGRT and range	Modern annual mean air temperature	ΔT (LGM) °C
Illumeden (Niger)	19.6 (16–22)		>7
Algeria (Guendouz et al. 1998)	16.9 (12–24)	21.0	2–3
Nigeria (Chad Formation)	22.0 (20.3–27.5)	28.5	6–7
Egypt	23.7 (21.8–25.6)	21.8–23.5	2–3

1992), Algeria (Guendouz et al. 1998) and Nigeria (Edmunds et al. 1999). The feature of most of the results is the wide scatter in the measured data. This may be due to measurement problems with gas loss between the over-pressured aquifers and the surface (sampling at high pressure is not often possible in many wells in the region), as well as problems relating to the actual conditions of recharge, especially the presence of excess air. A summary of results is given in Table 1. These have been calculated relative to the mean annual air temperature in most cases, although present day shallow groundwater temperatures may be a better reference. Possibly the best set of data so far is from Nigeria (Edmunds et al. 1999) where the palaeo-temperatures were found to have been some 6–7 °C lower than present day shallow groundwater temperatures; this is supported by data from the Illumeden basin in Niger (Le Gal La Salle 1992). This compares with 5.3 °C for the near coastal Stampriet aquifer, Namibia (Stute and Talma 1998). It is becoming apparent that groundwaters point to a clear record of global cooling at the LGM of around 5 °C (Loosli et al. 1998) but that the cooling towards the centre of the continents may have been 1 to 2 °C greater.

Nitrate - index of past vegetation cover

The presence of high nitrate concentrations in unconfined aerobic groundwaters is now well established across Northern Africa, both in palaeo-waters up to 35,000 years old, as well as modern waters. High nitrate is also a feature of some palaeo-groundwaters in Southern Africa (Heaton 1984). The widespread distribution in space and time supports the idea that N-fixing vegetation such as *Acacia* species is the primary source and has persisted across much of Northern Africa in the Holocene and Pleistocene, involving primarily the northwards expansion of the Sudano-Guinean vegetation cover during the wet periods and its subsequent decline (Edmunds 1999). At the present day the generation of nitrate within the soil zone, associated with leguminous vegetation, can be detected as high NO_3/Cl ratios in unsaturated zone moisture profiles (Edmunds and Gaye 1997). In Senegal the average modern nitrate in the shallow wells is 11.1, whereas in Nigeria it is 7.7 mg l^{-1} NO_3-N.

A summary of nitrate concentrations in ^{14}C-dated waters from Northern Africa is given in Figure 4. High nitrate groundwaters were recorded during the first exploration for groundwater in the Sirte and Kufra basins in Eastern Libya (Edmunds and Wright 1969), occurring widely in aerobic waters contained mainly in the phreatic aquifers. The nitrate concentrations in many groundwaters exceed recognised potable limits with maxima above 40 mg l^{-1} NO_3-N. In the Butana region of Sudan, the phreatic Nubian sandstone (Darling et al. 1987) aquifer contains fresh water of Holocene age (10,300 to 5,600 yr BP) containing nitrate concentrations up to 14 mg l^{-1}. In Mali groundwaters of Holocene age, mainly found

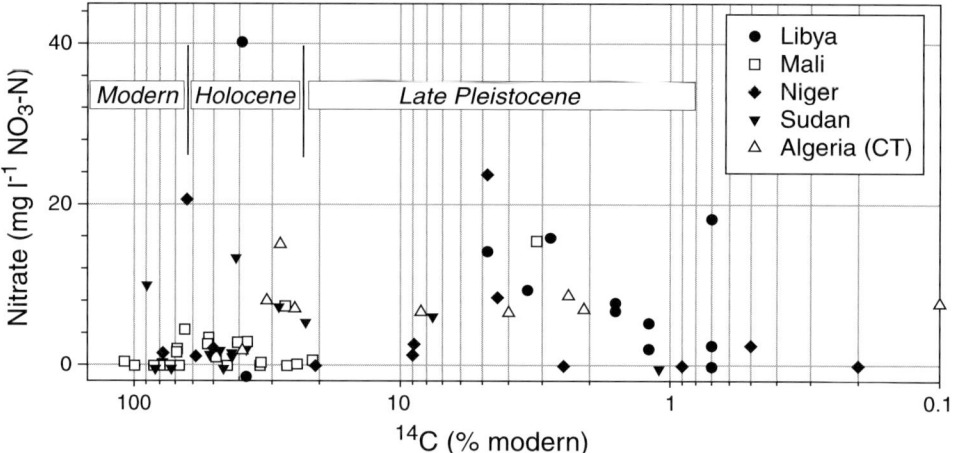

Figure 4. High nitrate concentrations preserved in the dated groundwaters of Northern Africa.

in the Continental Intercalaire of the Azaouad depression, contain nitrate concentrations up to 7.3 mg l^{-1} NO$_3$-N, although some of these may be derived partly from contemporaneous Niger River floods. The confined groundwaters contain a significant excess of dissolved nitrogen with respect to air saturation that is considered the product of denitrification:

$$4NO_3 + 5CH_2O + 4H^+ = 2N_2 + 5CO_2 + 7H_2O.$$

The N$_2$/Ar ratios have been corrected for excess air using the noble gas ratios and then used to calculate the amount of NO$_3$ that was converted to N$_2$ gas. An equivalent of up to 10.2 mg l^{-1} NO$_3$-N in the Mali aquifer has been reduced in this way (Fontes et al. 1991).

Aerobic groundwaters of Holocene age contained in the multilayer aquifer of the Agadez and Dabla sandstones in N.W. Niger contain nitrate up to 23 mg l^{-1} NO$_3$-N (Andrews et al. 1994). However most groundwaters are confined and anaerobic with nitrate absent. It is calculated, as in Mali, using the corrected N$_2$/Ar ratios, that these anaerobic groundwaters contain equivalent NO$_3$-N concentrations up to 6.8 mg l^{-1}. In Algeria aerobic groundwaters (Late Pleistocene to Holocene) along a south to north groundwater profile of some 500 km in Mio-Pliocene sands (Complexe Terminal), contain nitrate concentrations typically of between 5 and 8 mg l^{-1} NO$_3$-N (Guendouz et al. 2003). High nitrate concentrations (up to 17.8 mg l^{-1} NO$_3$-N) are also recorded in aquifers in the Arlit region of N. Niger (Dodo and Zuppi 1999).

A picture is emerging, therefore, of widespread leguminous (nitrate-fixing) vegetation such as *Acacia* species having persisted across much of Northern Africa in the Holocene and Pleistocene, being a northwards expansion of the Sudano-Guinean vegetation cover. This was the dominant vegetation until its decline some 4000 years BP. Traces of this vegetation are still to be found in the desert regions and the Central Sahara mountains. The groundwater evidence thus supports other evidence for a northward shift of vegetation belts of some 500 km during the Holocene wet phases (Lezine 1989) and possibly much of the late Pleistocene.

Examples of groundwater age relationships relating to climate

Although significant trends have been described at the continental scale, local variations are important since the regional or local topography, geology and hydrography may induce further isotopic and chemical changes and store additional climatic and environmental information. Two examples are given.

Groundwater stratification (Libya)

Confined aquifers in the Sahara and Sahel region contain clear evidence of late Pleistocene recharge, but Holocene groundwaters are rare in confined aquifers. Diffuse Holocene recharge may be preserved near surface in phreatic aquifers, being undetected by drilling programmes that produce from well below the water table. Distinct episodes of recharge from perennial rivers may however be found. An example is provided in the Sirt Basin in East-Central Libya (Edmunds and Wright 1979). During exploration studies in the mainly unconfined post-Middle Miocene (PMM) aquifer a distinct body of very fresh groundwater (<50 mg l^{-1} Cl) around 100 m deep was found cross-cutting the general NW-SE trend of salinity increase (Fig. 5). This feature, around 10 km in width, may be traced in a roughly NE-SW direction for around 130 km where the depth to the water table is currently around 30–50 m. Because of the good coverage of water supply wells associated with hydrocarbon exploration in this region a three-dimensional impression can be gained of the water quality. It is clear that this feature is fresh-water that must have been formed by recharge beneath a Holocene river. No obvious traces of the course of this river were found in this area, which had undergone significant erosion, although Neolithic artefacts and other remains testify to settlement during the Holocene.

The regional, relatively mineralised, groundwaters from the Sirt and Kufra basins gave radiocarbon values of 0.7–5.4% modern carbon (pmc) and show evidence of age stratification (Fig. 5). The waters marking the river recharge gave values from 37.6–51.2 pmc and are also distinctive in their isotopic and chemical composition. These younger waters with enriched $\delta^{13}C$ values gave ages ranging from 5000–7800 years (uncorrected ages since it was argued that reaction with carbonates in the soil zone or with calcretes would have influenced the ^{14}C acquisition).

Evidence from shallow wells in the vicinity (Fig. 4) proved that a thin layer of fresh and probably younger water was also present at the water table indicating that, simultaneously with the river flows, some direct recharge was also occurring at a regional scale, although this is unobserved in many deeper drilled wells. Coinciding with the hydrogeological work, palaeo-environmental studies were being carried out, which traced the line of a large, now-inactive river system (Wadi Behar Belama) from the Tibesti Mountains (Pachur et al. 1987). Thus, converging evidence from both above and below ground demonstrates that in the middle Holocene a significant wet phase occurred. Some regional diffuse recharge to the aquifer took place but the main recharge was from perennial river flow until around 5000 yr BP. Palaeontological evidence including that of large mammals shows that the wadi system was active until at least 3500 BP (Pachur 1975) after which time the present arid conditions (<20 mm/yr) commenced with no recorded evidence of modern recharge having reached the water table. At the present day in Sudan a similar recharge zone from the Nile River

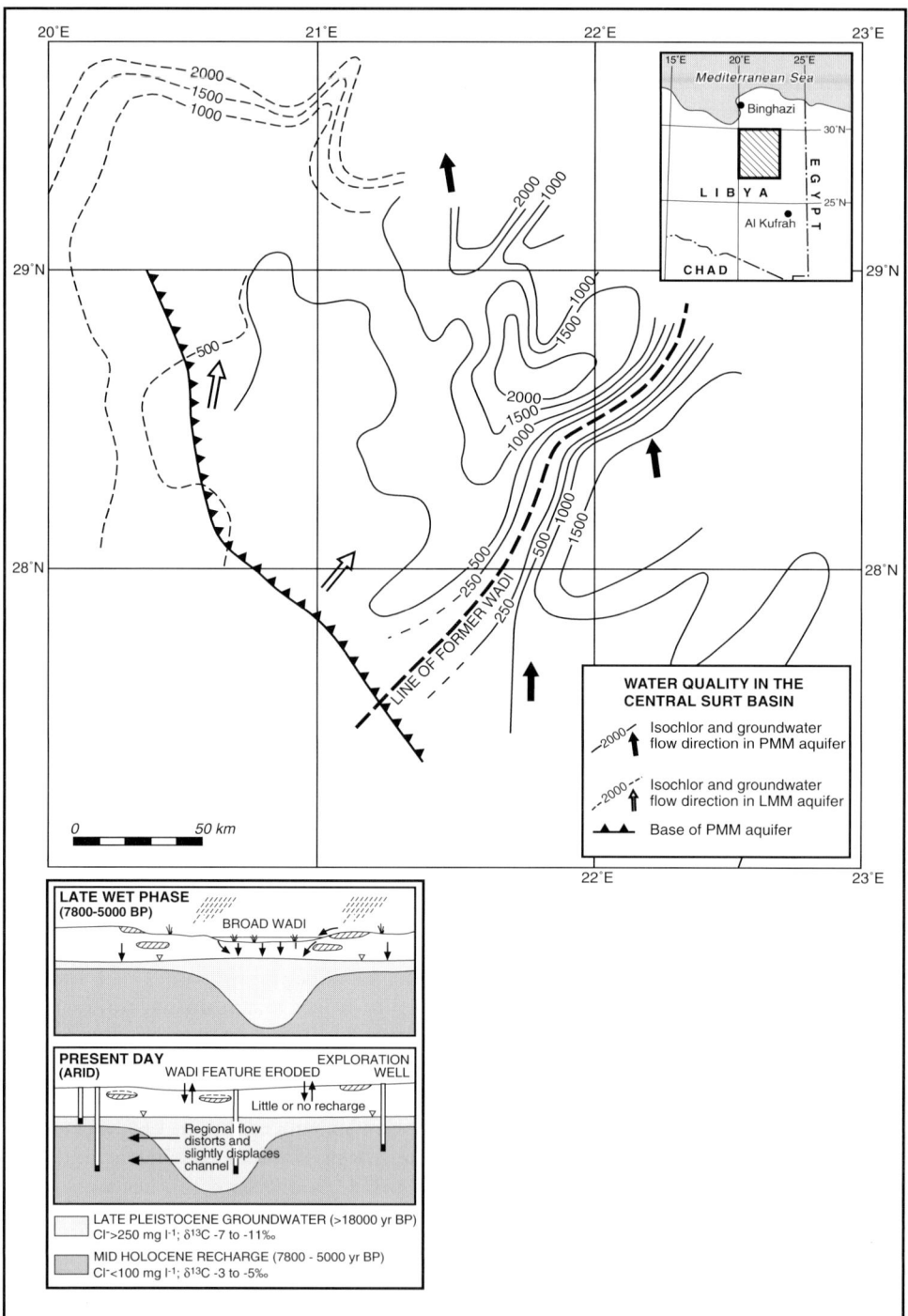

Figure 5. Fresh-water body derived from Holocene recharge by a palaeo-river is recognisable from isotopic and chemical evidence, superimposed on the regional groundwaters of late-Pleistocene age in Central Libya.

may be observed, detected both in the isotopic and chemical compositions (Darling et al. 1987).

The Lake Chad Basin (Nigeria) - past hydrology and climate

Palaeohydrological evidence for the Chad Basin has been deduced from sedimentological (Durand 1982; 1995) and palaeolimnological evidence (Maley 1980; Servant and Servant-Vildary 1980; Gasse et al. 1990). On the basis of diatoms and fluvio-lacustrine sediments it has been established that the Holocene contained two intervals (9000–8000 and around 6000 yr BP) significantly wetter than the present day (Servant and Servant-Vildary 1980). The interval corresponding to the LGM in Europe (20 000–13 000) was a distinctly drier than normal period in the Sahel although the period prior to this ca. 20 000–40 000 yr BP was also wetter than at the present day. It is commonly agreed that the present era since about 3500 yr is one of the most arid periods in the dateable record. In the early Holocene the levels of Lake Chad were significantly higher and the lake had a far greater extent than today (Fig. 6). There is general agreement that its maximum extent occurred around 9000 yr BP in the mid-Holocene. A maximum level of ca. 320 m has been proposed by several authors (cf. Servant and Servant-Vildary (1980)) although this is challenged by Durand (1982, 1995) who suggests a maximum of ca. 290 m, around 10 m higher than its modern level, on the basis of neotectonic analysis.

Artesian groundwaters are found in the Plio-Pleistocene lacustrine sedimentary aquifers beneath and adjacent to Lake Chad and these have been studied extensively in the Nigeria-Niger border region (Carter et al. 1963; Edmunds et al. 1999). There are three major aquifer units (Fig. 7). The Lower Zone aquifer (inter-bedded sands and clays) has a maximum thickness of around 90 m. The main, Middle Zone aquifer has a maximum-recorded thickness of about 31 m, comprising fine to very coarse-grained, poorly-graded and mostly uncemented sands. The initial hydraulic gradients established during the extensive studies of the early 1960s showed a general flow direction towards Lake Chad. The Upper Zone consists of inter-bedded sands and clays and forms an aquifer of variable importance, which underlies the whole region.

The isotopic and geochemical characteristics of the Chad Formation aquifers have been investigated (Edmunds et al. 1998; 1999) along a representative 180 km line of section along the flow line from Maidiguri to Baga on the shores of Lake Chad (Fig. 8). The groundwater in the Middle and Lower Zone aquifers is very fresh with concentrations of Cl below 100 mg l^{-1} indicative of an active hydrological system.

The δ^{13}C values (around $-14‰$) show very little variation across the aquifer implying a single controlling process for the sources of dissolved carbon and allowing ages to be derived for the groundwaters, using δ^{13}C values of 1‰ and $-26‰$ for carbonate and soil CO_2 respectively. The activities of ^{14}C (Fig. 8) are rather uniform within the Middle Zone aquifer at between 2.6 and 6.3% modern carbon (pmc). The corrected ages for the Middle Zone aquifer lie in a relatively narrow range, from 18,600 to 24,000 yr BP, and in the Lower Zone the aquifer has ages of 23,500 yr BP at depth (Edmunds et al. 1998).

The groundwaters from both the Middle and Lower Zone aquifers (Fig. 8) have stable isotope compositions, which are depleted by about 1–2‰ compared with modern rainfall and Holocene groundwaters (Fig. 2c) and there is a consistent trend within the aquifer for lighter isotopic values in the deeper groundwaters.

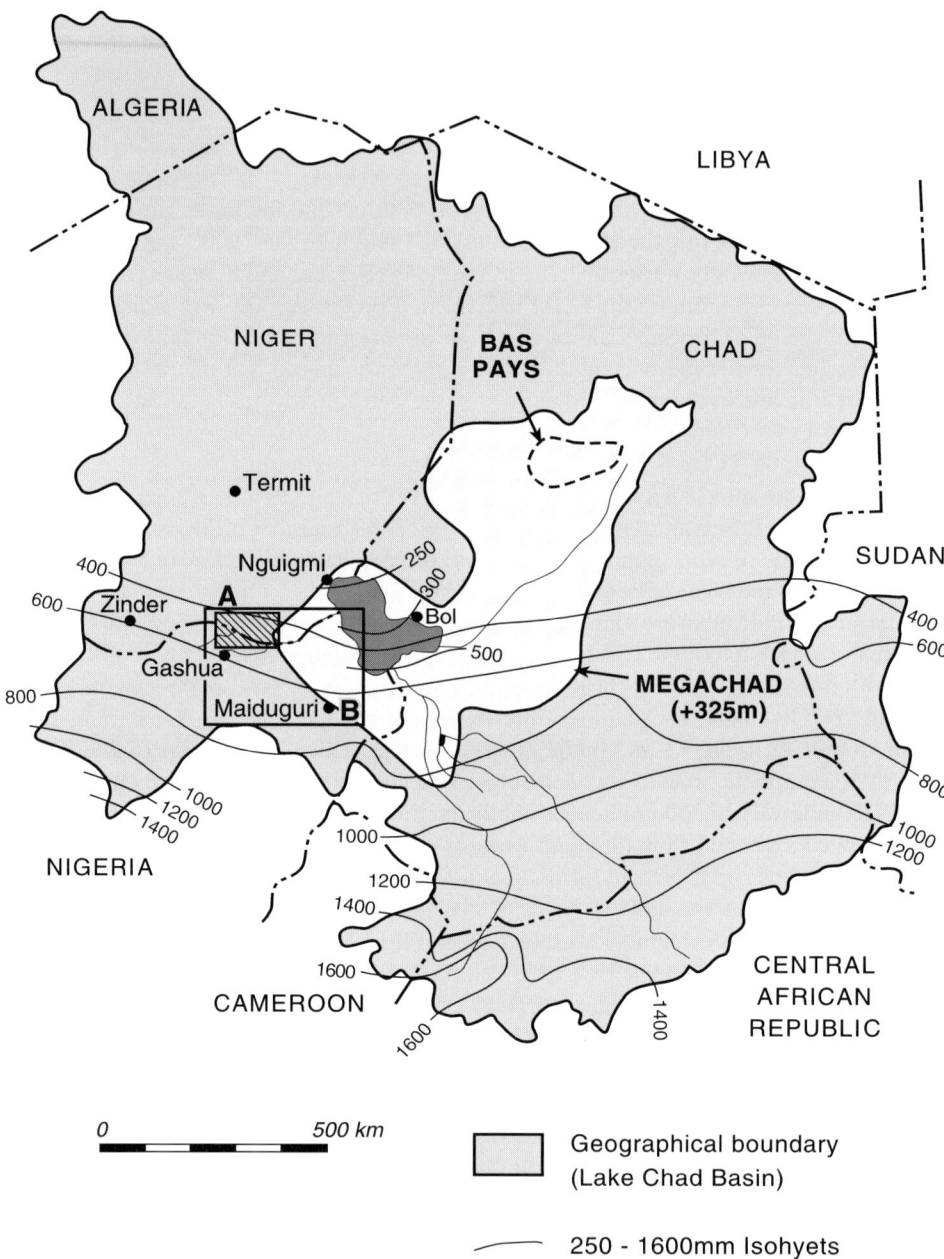

Figure 6. Map of the Chad basin showing the present day and former (Holocene) maximum proposed extent of the Lake. The likely former discharge area **Pays Bas (also called Bodele)** in the Republic of Chad.

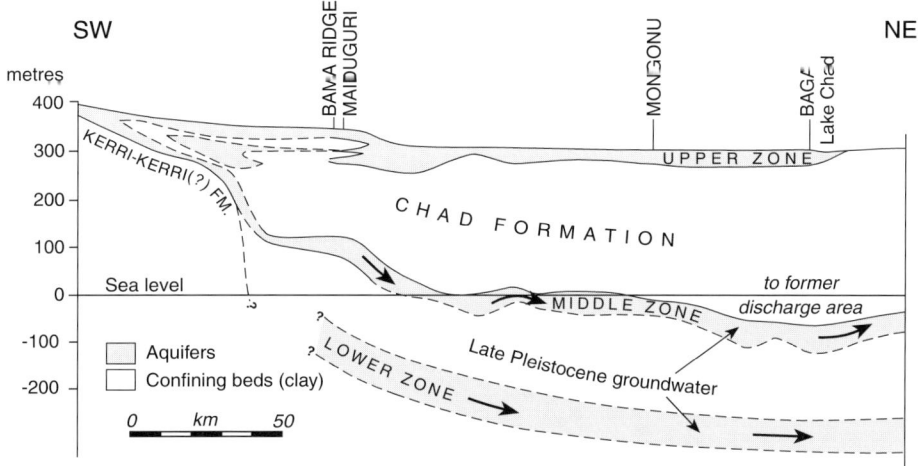

Figure 7. Geological cross-section through the Plio-Quaternary sediments of the Nigerian sector of the Chad basin running SW-NE from Maiduguri to Lake Chad.

The noble gas (Ne, Ar, Kr and Xe) concentrations, after correction for excess air (Andrews 1991), have been used to derive former recharge temperatures (NGRT). In the Middle and Lower Zone aquifers rather consistent palaeo-temperatures of $22.0 \pm 1.5\,°C$ are recorded. This compares with the temperatures of groundwater measured in shallow wells from the same area which have a mean value of $28.2 \pm 1.5\,°C$.

The isotope and noble gas results thus provide evidence for climatic change and changes of the recharge source with time, adding to the wealth of indirect evidence on palaeohydrology from the sedimentary record. The noble gas results in the deep aquifers provide strong evidence for cooler climatic conditions during the period 18–24,000 yr BP, recording temperatures at least $6\,°C$ lower than at the present day; the amount of lowering of temperature is consistent with that measured elsewhere in Africa (Table 1). The 2‰ lighter values for $\delta^{18}O$ in the Middle and Lower Zone aquifers are also consistent with cooler Late Pleistocene recharge conditions.

The geochemical and isotopic evidence from the Middle Zone aquifer points therefore to a significant period of recharge having occurred between 24 and 20,000 yr BP. Rapid flow velocities $(40\,m\,yr^{-1})$ are inferred from the ^{14}C age gradient over the 180 km section and it is considered that an outlet, in the Pays Bas depressions of Chad, controlled this flow during the Late Pleistocene. The absence of groundwaters with younger ages implies that all effective recharge ceased quite suddenly at the end of the LGM probably due to the rise of lake levels forming the Lake Mega Chad covering the Pays Bas outlet during the early Holocene.

Discussion

The debate over the significance of the isotopic signatures in groundwaters beneath the Sahara and adjacent areas originated with Degens (1962) who recognised these as pluvial

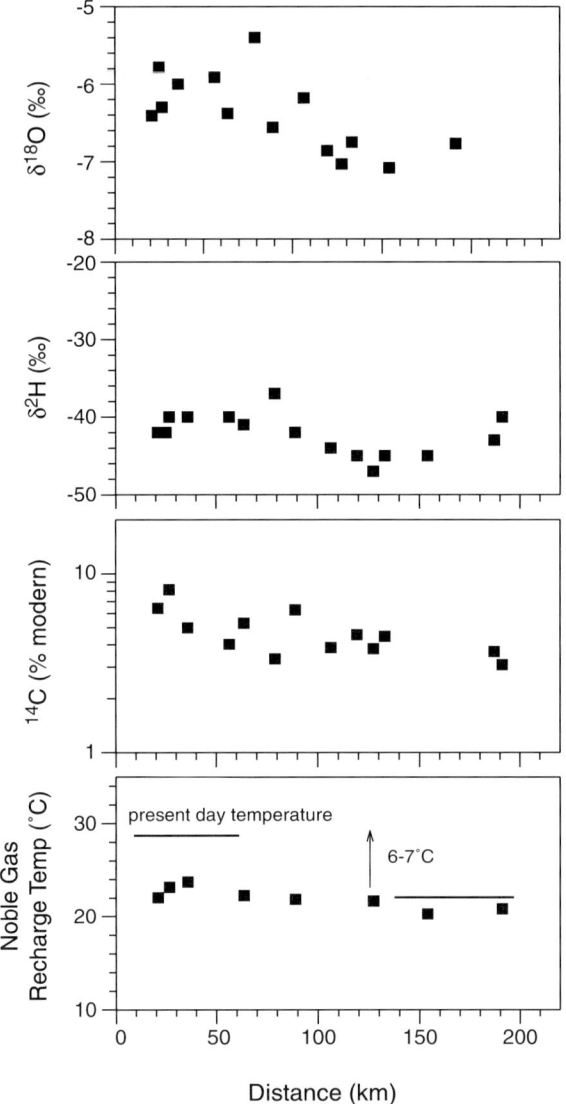

Figure 8. Profiles along the line of section (Fig. 7) showing stable isotopes, radiocarbon (as pmc) and noble gas recharge temperatures.

waters from past climates, supported by the first radiocarbon evidence. Since then contro-versy has centred on the evidence for the dominance of westerly Atlantic air masses as the source of rainfall recharge (Sonntag et al. 1978; 1980; Sultan et al. 1997), as opposed to the monsoon source(s) derived from the Gulf of Guinea (or possibly the Indian Ocean). It is now clear that both systems were operative, but the question remains on the persistence and intensity of the westerly flow and the intensity and progression north of the monsoon. To derive information from the groundwater systems it is vital to consider the chronology of the recorded events: how late Pleistocene, early Holocene and Modern waters differ from each other both temporally and between locations.

It is also clear from the isotopic record (most groundwaters are aligned to slopes close to those of the global meteoric water line (GMWL) and biased towards the lightest iso-topic compositions and the heaviest events) and the chemistry (low initial salinities) that groundwaters record the heaviest rains. Exceptions occur, however, when recharge from river systems takes place, commonly in the Holocene. The formation of large lake systems in the Holocene may have led to recharge from evaporated sources although the formation of large lakes may also have inhibited recharge as in the vicinity of Lake Chad (Edmunds et al. 1999). The feedback of moisture from lakes may locally have modified the humidity and hence the isotopic composition of the atmosphere and be recorded in groundwater although any evidence for this is not yet clear. The variation of air saturation induces a small reduction of temperature, allowing increasing atmospheric humidity. Such effects have been measured in modern irrigated systems in West Africa (Taupin et al. 2000).

In N.W. Sudan (Hoelzman et al. 2000) the groundwater contributions to the Holocene lake systems were relatively small. On the basis of lake dimensions the rainfall in the early Holocene is calculated to between 500–900 mm yr^{-1}. Direct recharge was likely to have been restricted to sand-covered areas as indicated by the unsaturated zone rates of the present day. Lake basins were restricted mainly to low permeability areas, although a general rise in water table would have led to lake systems developing within low lying areas with thin unsaturated zones as in the Manga area around Lake Chad at the present day. Discharge zones (equivalent to the depressions found as present day oases) from the large confined aquifers would have remained active throughout the whole period.

The lowering of land and sea-surface temperatures led globally to a displacement of climatic zones southwards by about 500–800 km relative to the present day during the late Pleistocene, and culminating in the LGM when aridity was at a maximum over the Sahel and Sudan-Guinean zones (Kutzbach and Street-Perrott 1985). At the Holocene climatic optimum, these zones were estimated to have penetrated north of their present positions some 500–800 km, to about 23 °N. This orbital forcing of climate produced a time of increased summer insolation in the northern Hemisphere, corresponding to the Holocene transition, leading to an increase in intensity in monsoon activity. This was a time of abrupt climatic change, although not all of the observed changes in the available records are considered to be due to orbital forcing (Gasse 2000).

The isotopic evidence from groundwaters near to the West African coastline during the late Pleistocene reflects primarily the changing compositions of the oceans, but not their surface temperature. Continuous recharge occurred with similar wind directions. Only the noble gas ratios provide evidence for cooler glacial climates in these groundwaters. Across the continent the isotopic evidence indicates that changes took place both related to altitude as well as longitude. Considering the isotopic content of groundwater from Ténéré and Tassili aquifers (Saighi et al. 2001) a latitude gradient of 0.7 per mil per 100 km and an altitude gradient of 0.4‰ per 100 m is indicated. These gradients were evaluated for the same recharge period, taking into account the geographical position of the recharge areas in the Hoggar massif.

The contrast between recharge conditions across Africa before and after the LGM is quite striking. This contrast increases from west to east and is expressed as the strong variation in isotope composition within the Holocene rains (over 9‰ in $\delta^{18}O$ in Sudan) as well as their overall compositions, more depleted than modern rainfall but more enriched than the more extensive pluvial compositions, which characterised the late Pleistocene. The effect is most strikingly displayed in groundwaters from the Sudan especially in Kordofan and Darfur but also in Butana, where Holocene groundwaters are well represented (in contrast with Libya and Egypt to the north). The bulk of these groundwaters have isotopic compositions ($\delta^{18}O$ of -9.5 to $-11.5‰$), significantly lighter than most late Pleistocene waters, being most abundant in the early Holocene age group. The youngest waters are still generally lighter than the weighted-mean isotope compositions of the present day. This feature is not observed elsewhere, although admittedly the database for Holocene waters from Chad to the west, even Nigeria, is meagre.

The explanations for this anomaly are not clear, although it has been argued that this would be a feature of colder condensation from high altitude clouds during more intense monsoon activity (Fontes et al. 1993b). Such an effect is seen, although with extreme rarity, at the present day in some individual heavy rainfall events, as recorded in the IAEA precipitation database (http://www.iaea.or.at:80/programs/ri/gnip/gnipmain.htm). If this were the case it would point to a regional shift in the heavier monsoon rain belt further to the east. A shift to the east of the south-western monsoon rather than moisture derived from the Indian Ocean is favoured on the isotopic evidence, since the latter would be strongly enriched (Travi and Chernet 1998).

The groundwater evidence for Holocene wet periods is strongly paralleled by the excellent and well documented sedimentary and archaeological records across the Sahara (Gasse 2000; Petit-Maire and Riser 1982; Pachur et al. 1987) as well as in the mountain regions (Maley 2000) where humid, lower temperature conditions maintained lake formation even through the LGM. The increased intensity of early Holocene monsoonal activity is well recorded for example in the gastropods and associated carbonates in the former lakes of N.W. Sudan (Abell and Hoelzmann 2000).

The depleted isotope compositions associated with the Holocene monsoon rains are not to be confused with the very (isotopically) light rains derived by Rayleigh fractionation of the westerly moisture across Northern Africa and typified by the groundwater compositions found at Kufra and the oases of the western desert of Egypt. These rains were restricted to the main pluvial prior to about 22 kyr BP. Following the LGM the westerly rainfall belt shifted north along the Mediterranean; the evidence from the phreatic aquifers in North Africa points to the Holocene rains being volumetrically less important than in the Sahara/Sahel belt to the south.

Evidence from comparable non-carbonate aquifers in Europe also records strongly contrasting palaeoclimatic conditions controlling recharge to aquifers (Edmunds and Milne 2001; Loosli et al. 2001). In Estonia, strongly depleted isotopic signatures record recharge under pressure directly beneath the ice cap. In UK and in Switzerland there is evidence of a recharge gap interpreted as the presence of frozen ground at the time of the LGM. However, in the UK and elsewhere recharge in the late Pleistocene contains a depleted isotope signature of variable magnitude compared to the present day indicating cooler climate, confirmed by the noble gas recharge temperatures (NGRT) with a mean cooling of 5–6 °C. In Portugal there is also evidence from NGRTs of cooling of 5–7 °C, as well as continuous recharge, although in coastal Portugal a slight enrichment in $\delta^{18}O$ reflects the proximity to the ocean and the constancy of the S.W. Atlantic air circulation over the whole of the late Pleistocene and Holocene, as observed in the case of Morocco. Thus, the groundwater evidence in the northern hemisphere records a gradation at the LGM from glacial recharge, through the absence of recharge due to permafrost, to continuous rainfall recharge at Mediterranean latitudes and then desiccation in the modern Sahara region, coincident with greater aridity.

Conclusions

The late Pleistocene groundwaters in Northern Africa record evidence of cooler climates and significant recharge prior to the LGM. Air mass circulation over Africa during the late Pleistocene was significantly different from the present day with evidence, shown clearly in the groundwater archive, of a reinforcement and southward shift of the Atlantic westerly flow across the present Sahara during the period. A corresponding decline of monsoon rains occurred at this time. Evidence then is found for a northward extension of the African monsoon, with increased intensity notably during the early to mid-Holocene coinciding with a retreat of the Atlantic system to the north. The extent of cooling at the LGM recorded in the noble gas ratios was up to 7 °C. There is clear evidence in the isotopic signature of expansion in the early Holocene of rivers such as the Niger as well as an extensive groundwater-fed fluvial system north of the Tibesti, remnants of which today are recorded in the phreatic aquifers. The groundwater isotopic evidence in different places records strong variations in humidity of the air masses supplying moisture across the continent at different times over the past 30,000 yrs.

The past climate and hydrology in Africa is also mirrored by the palaeo-groundwater archives in Europe, where significant replenishment of aquifers took place during the pluvial periods and where general cooling of around 5–7 °C is also recorded in the noble gas recharge temperatures. In coastal Portugal, as in Morocco, however the lack of any stable isotope depletion indicates the constancy of the S.W. Atlantic air circulation at mid-latitudes over the whole of the late Pleistocene and Holocene, as well as proximity to the oceans.

Modern recharge conditions, comparable with the instrumental records are recorded at a decadal scale in the moisture of the unsaturated zone and in certain areas of Africa it may be possible to use these to reconstruct a continuous record of groundwater recharge and hence climate variations during the Holocene.

Summary

Groundwaters of known age contained in major aquifer systems in sedimentary basins are of specific value in determining low resolution (± 1000 yr) characteristics of past climates, specifically palaeo-temperature, air mass origins, humid/arid transitions, rainfall intensity, as well as some aspects of past soil and vegetation cover. Additionally, the unsaturated zone may, under favourable circumstances, contain higher resolution records (10–100 yr) of droughts and more humid intervals, contained as concentrations of chloride, supported by isotope measurements, in moisture profiles. Results from Northern Africa indicate the predominance of westerly air flow during the late Pleistocene, characterised also by a continental effect of about 4.5‰ $\delta^{18}O$, possibly modified by altitude effects of the mountains. Greater aridity is recorded by the absence of dated groundwaters over most of N. Africa during the Last Glacial Maximum (LGM). An intensification of the African monsoon during the early Holocene is apparent from isotopically-light groundwaters found especially over Sudan; however, over the past 6000 years there was a shift to the more enriched composition of modern times.

Local conditions have been important in recording palaeoclimatic and palaeohydrological events. Surface runoff from major river systems in Libya is recorded for the Holocene. In Nigeria the absence of Holocene recharge in the Chad formation is attributed to the extension of lake Mega-Chad over former discharge areas. Although the LGM is marked in N. Africa by an arid phase, this contrasts with the Atlantic regions of Morocco and Portugal where continuous recharge occurred, with isotopic enrichment prior to the LGM linked to the change in oceanic composition. Maximum cooling around the LGM of 5–7 °C is recorded in the noble gas recharge temperatures. Evidence in African groundwaters is matched by complementary data from Northern Europe where a recharge gap is found corresponding to permafrost conditions.

Acknowledgments

This work is based in part on the EC GASPAL project (ENV4-CT97-0591) for which financial support is acknowledged (http://www.bgs.ac.uk/hydrogeology/gaspal). The constructive reviews by Jonathan Holmes and Bernd Wünnemann are gratefully acknowledged. This paper is published with the permission of the Executive Director, British Geological Survey, Natural Environment Research Council.

References

Abell P.I. and Hoelzmann P. 2000. Holocene palaeoclimates in Northwestern Sudan: stable isotope studies on molluscs. Glob. Planet. Chan. 26: 1–12.

Andrews J.N. 1993. Isotopic composition of groundwaters and palaeoclimate at aquifer recharge. In: IAEA (eds), Isotope Techniques in the Study of Past and Current Environmental Changes in the Hydrosphere and Atmosphere. IAEA. Vienna, pp. 271–292.

Andrews J.N., Fontes J.-Ch., Aranyossy J.-F., Dodo A., Edmunds W.M., Joseph A. and Travi Y. 1994. The evolution of alkaline groundwaters in the Continental Intercalaire aquifer of the Irhazer Plain, Niger. Water Resour. Res. 30: 45–61.

Bar-Matthews M., Ayalon A. and Kaufman A. 1997. Late Quaternary paleoclimate in the eastern Mediterranean Region from stable isotope analysis of speleothems at Soreq Cave, Israel. Quat. Res. 47: 155–168.

Bromley J., Edmunds W.M., Fellman E., Brouwer J., Gaze S.R., Sudlow J. and Taupin J.-D. 1997. Rainfall inputs and direct recharge to the deep unsaturated zone of Southern Niger. J. Hydrol. 188: 139–154.

Carter J.D., Barber W. and Tait E.A. 1963. The geology of parts of Adamawa, Bauchi and Bornu provinces in North-Eastern Nigeria. Bulletin 30, Geol. Surv. Nigeria.

Clark I.D. and Fritz P. 1997. Environmental Isotopes in Hydrogeology. Lewis, Baton-Rouge, 328 pp.

Cook P.G., Edmunds W.M. and Gaye C.B. 1992. Estimating palaeorecharge and palaeoclimate from unsaturated zone profiles. Wat. Resour. Res. 28: 2721–2731.

Darling W.G., Edmunds W.M., Kinniburgh D.G. and Kotoub S. 1987. Sources of recharge to the Basal Nubian Sandstone Aquifer, Butana Region, Sudan. In: IAEA (eds), Isotope Techniques in Water Resources Development. IAEA, Vienna, pp. 205–224.

Davis S.N., Whittemore D.O. and Fabryka-Martin J. 1998. Uses of chloride/bromide ration in studies of potable water. Ground Water 36: 338–350.

Degens E.T. 1962. Geochemische Untersuchungen von Wässern aus der ägyptischen Sahara, Geolog. Rundschau. 52: 625–639.

Dinçer T., Dray M., Zuppi G.M., Guerre A., Tazioli G.S. and Traore S. 1984. L'alimentation des eaux souterraines de la zone Kolokani-Nara au Mali. In: IAEA (eds), Isotope Hydrology, 1983: Proc. Internat. Symp. on Isotope Hydrology in Water Resources Development. IAEA. Vienna, pp. 341–365.

Dodo A. and Zuppi G.M. 1999. Variabilité climatique durant le Quaternaire dans la nappe du Tarat (Arlit, Niger), C.r. Acad. Sci., Paris, 328: 371–379.

Dray M., Gonfiantini R. and Zuppi G.M. 1993. Isotopic composition of groundwater in the Southern Sahara. In: IAEA (eds), Palaeoclimates and Palaeowaters: A Collection of Environmental Isotope Studies. IAEA, Vienna, pp. 187–199.

Durand A. 1982. Oscillations of Lake Chad over the past 50 000 years: new data and new hypothesis. Palaeogeogr. Palaeoclimatol. Palaeoecol. 39: 37–53.

Durand A. 1995. Sédiments quaternaires et changements climatiques au Sahel central (Niger et Tchad). Africa Geosci. Rev. 2: 323–614.

Edmunds W.M. 1996. Bromide geochemistry in British groundwaters. Mineral. Mag. 60: 275–284.

Edmunds W.M. 1999. Groundwater nitrate as a palaeo-environmental indicator. In: Armannsson H. (ed.), Geochemistry of the Earth's Surface. Proc. 5th International Symposium on the Geochemistry of the Earth's Surface, Reykjavik, Iceland. Balkema, Rotterdam, pp. 35–38.

Edmunds W.M. and Wright E.P. 1979. Groundwater recharge and palaeoclimate in the Sirte and Kufra basins, Libya. J. Hydrol. 40: 215–241.

Edmunds W.M., Gaye C.B. and Fontes J.-Ch. 1992. A record of climatic and environmental change contained in interstitial waters from the unsaturated zone of Northern Senegal. In: IAEA (eds), Isotope Techniques in Water Resources Development, 1991. IAEA, Vienna, pp. 533–549.

Edmunds W.M. and Gaye C.B. 1997. High nitrate baseline concentrations in groundwaters from the Sahel. J. Envir. Qual. 26: 1231–1239.

Edmunds W.M. and Milne C.J. (eds). 2001. Palaeowaters in Coastal Europe: Evolution of Groundwater since the Late Pleistocene. Geol. Soc. Special Publ. 189. Geol. Soc., London, 332 pp.

Edmunds W.M., Fellman E. and Goni I.B. 1999. Lakes, groundwater and palaeohydrology in the Sahel of NE Nigeria: evidence from hydrogeochemistry. J. Geol. Soc., London, 156: 345–355.

Edmunds W.M. and Tyler S.W. 2002. Unsaturated zones as archives of past climates: towards a new proxy for continental regions. Hydrogeol. J., 10: 216–228.

Edmunds W.M., Guendouz A.H., Mamou A., Moulla A.S., Shand P. and Zouari K. (2003). Ground-water evolution in the Continental Intercalaire aquifer of Southern Algeria and Tunisia: trace element and isotopic indicators. Appl. Geochem. 18: 805–822.

Faye A. 1994. Recharge et palaeorecharge des aquifères profonds du bassin du Senegal. Apport des isotopes stables et radioactifs de l'environnement et implications palaeohydrologiqes et palaeoclimatiques. Thesis Doc. ès Sciences. Université Cheikh Anta Diop. Dakar.

Faye A., Tandia A.A., Travi Y., Le Priol J. and Fontes J.-Ch. 1993. Apport des isotopes de l'environnement à la connaissance des aquifères de Casamance (extreme sud du Sénégal). In: IAEA (eds), Les Ressources en Eau au Sahel. Tech. Doc. 721, IAEA, Vienna, pp. 123–132.

Fontes J.-Ch. and Gasse F. 1991. PALHYDAF (Palaeohydrology in Africa) program: objectives, methods, major results. Palaeogeogr. Palaeoclimatol. Palaeoecol. 84: 191–215.

Fontes J.-Ch., Andrews J.N., Edmunds W.M., Guerre A. and Travi Y. 1991. Palaeorecharge by the Niger River (Mali) deduced from groundwater chemistry. Water Resour. Res. 27: 199–214.

Fontes J.-Ch., Stute M., Schlosser P. and Broecker W.S. 1993a. Aquifers as archives of palaeoclimate. Eos, 74: 21–22.

Fontes J.-Ch., Gasse F. and Andrews J.N. 1993b. Climatic conditions of Holocene groundwater recharge in the Sahel zone of Africa. In: IAEA (eds), Isotopic Techniques in the Study of Past and Current Environmental Changes in the Hydrosphere and Atmosphere. IAEA, Vienna, pp. 231–248.

Gac J.-Y. 1990. Le haut bassin versant du fleuve Sénégal. Unpubl. Rep. CCE Project (EQUESEN).

Gasse F. 2000. Hydrological changes in the African tropics since the Last Glacial Maximum. Quat. Sci. Rev. 19: 189–211.

Gasse F., Tehet R., Durand R., Gibert E. and Fontes J.-Ch. 1990. The arid-humid transition in the Sahel during the last deglaciation. Nature 346: 141–146.

Geyh M.A. and Jäkel D. 1974. Spätpleistozäne und Holozäne Klimageschichte der Sahara aufgrund zugänglicher ^{14}C Daten. Z. Geomorphol. 18: 82–98.

Goni I.B., Fellmann E. and Edmunds W.M. 2001. Rainfall geochemistry in the Sahel region of Northern Nigeria. Atmos. Envir. 35: 4331–4339.

Groening M., Sonntag C. and Suckow A. 1993. Isotopic evidence for extremely low groundwater recharge in the Sahel zone of Africa. In: Thorweihe U. (ed.), Geoscientific Research in North-East Africa. Balkema, Rotterdam, pp. 671–676.

Guendouz A., Moulla A.S., Edmunds W.M., Shand P., Poole J., Zouari K. and Mamou A. 1998. Palaeoclimatic information contained in groundwaters of the Grand Erg Oriental, N. Africa. In: IAEA (eds), Isotope Techniques in the Study of Past and Current Environmental Changes in the Hydrosphere and Atmosphere. IAEA, Vienna, pp. 555–571.

Guendouz A., Moulla A.S., Edmunds W.M., Shand P., Zouari K. and Mamou A. (2003). Hydrogeo-chemical and isotopic evolution of the Complexe Terminal groundwaters in the Algerian Sahara. Hydrogeol. J. (in press).

Hartsough P., Tyler S.W., Sterling J. and Walvoord M. 2001. A 14.6 kyr record of nitrogen flux from desert soil profiles as inferred from vadose zone pore waters. Geophys. Res. Lett. 28: 2955–2958.

Hassan F.A. 1997. Nile floods and political disorder in Early Egypt. In: NATO (eds), Third Millennium BC Climate Change and Old World Collapse. NATO ASI Series 1, 49: 1–24.

Heaton T.H.E. 1984. Sources of the nitrate in phreatic groundwater in the Western Kalahari. J. Hydrol. 67: 249–259.

Herczeg A.L. and Edmunds W.M. 1999. Inorganic ions as tracers. In: Cook P.G. and Herczeg A.L. (eds), Environmental Tracers in Subsurface Hydrology. Kluwer, Boston, pp. 31–77.

Hoelzmann P., Kruse H.-J. and Rottinger F. 2000. Precipitation estimates for the eastern Saharan palaeomonsoon based on a water balance model of the West Nubian palaeolake basin. Glob. Planet. Chan. 26: 105–120.

Hoelzmann P., Gasse F., Dupont L.M., Salzmann U., Staubwasser M., Leuschner D.C. and Sirocko F., this volume. Palaeoenvironmental changes in the arid and subarid belt (Sahara-Sahel-Arabian Peninsula) from 150 kyr to present. In: Battarbee R.W., Gasse F. and Stickley C.E. (eds), Past Climate Variability through Europe and Africa. Kluwer Academic Publishers, Dordrecht, the Netherlands, pp. 219–256.

Issar A.S. 1990. Water Shall Flow from the Rock: Hydrology and Climate in the Lands of the Bible. Springer-Verlag, Berlin, 213 pp.

Kabbaj A., Zeryouhi I., Carlier C. and Marce A. 1978. Contribution des isotopes du milieu a l'étude de grands aquifères du Maroc. In: International Symposium on Isotope Hydrology, Vienna. Vol. II, pp. 491–524.

Kutzbach J.E. and Street-Perrott F.A. 1985. Milankovitch forcing of fluctuations in the level of tropical lakes from 18 to 0 kyr BP. Nature 317: 130–134.

Lambert W.G. and Millard A.R. 1969. Atra-Hasis, the Babylonian Story of the Flood. Clarendon Press, Oxford, 198 pp.

Le Gal La Salle C. 1992. Circulation des eaux souterraines dans l'aquifère captif du Continental Terminal — Bassin des Illumeden, Niger. Thesis D. ès Sci., Univ. Paris Sud (Orsay).

Lezine A.-M. 1989. Late Quaternary vegetation and climate of the Sahel. Quat. Res. 32: 317–334.

Loosli H.H., Lehmann B., Aeschbach-Hertig W., Kipfer R., Edmunds W.M., Eichinger L., Rozanski K., Stute M. and Vaikmae R. 1998. Tools used to study palaeoclimate help in water managenent. Eos 79: 581–582.

Loosli H.H., Lehmann B. and Smethie W.M. 1999. Noble gas radioisotopes (^{37}Ar, ^{85}Kr, ^{39}Ar, ^{81}Kr). In: Cook P.G. and Herczeg A.L. (eds), Environmental Tracers in Subsurface Hydrology. Kluwer, Boston, pp. 379–396.

Loosli H.H., Aeschbach-Hertig W., Barbecot F., Blaser P., Darling W.G., Dever L., Edmunds W.M., Kipfer R., Purtschert R. and Walraevens K. 2001. Isotopic methods and their hydrogeochemical context in the investigation of palaeowaters. In: Edmunds W.M. and Milne C.J. (eds), Palaeowaters of Coastal Europe: Evolution of Groundwater since the Late Pleistocene. Geol. Soc. Special Publ. 189. Geol. Soc., London, pp. 193–212.

Maley J. 1980. Études Palynologiques dans le Bassin du Tchad et Paléoclimatologie de l'Afrique Nord Tropicale de 30 000 ans à l'Époque Actuelle. Trav. Doc. ORSTOM, 129, 586 pp.

Maley J. 2000. Last Glacial maximum lacustrine and fluviatile formations in the Tibesti and other Saharan mountains, and large scale teleconnections linked to the activity of the Subtropical Jet Stream. Glob. Plan. Chan. 26: 105–120.

Njitchoua R., Fontes J.-Ch., Dever L. Naah E. and Aranyossy J. 1993. Recharge naturelle des eaux souterraines du bassin des gres de Garoua. (Nord Cameroun). In: IAEA (eds), Les Ressources en Eau au Sahel. Tech. Doc. 721, IAEA, Vienna, pp. 133–146.

Olivry J.C. 1983. Le point en 1982 sur l'évolution de la sécheresse en Sénégambie et aux Iles du Cap-Vert: Examen de quelques de séries de longue durée (débits et précipitations). Cah. ORSTOM Sér. Hydrol. 20: 47–69.

Pachur H.-J. 1975. Zur spätpleistozänen und holozänen Formung auf der Nordabdachung des Tibesti Gebirges. Die Erde 106: 21–46.

Pachur H.-J., Röper H.-P., Kröpelin S. and Goschin M. 1987. Late Quaternary hydrography of the Eastern Sahara. Berliner geowiss. Abh. A 75.2: 331–384.

Pachur H.-J. and Wünnemann B. 1996. Reconstruction of the palaeoclimate along 30 °N in the Eastern Sahara during the Pleistocene/Holocene transition. Palaeoecol. Africa 24: 1–32.

Petit-Maire N. and Riser J. 1982. Sahara ou Sahel: Quaternaire Récent du Bassin de Taoudenni, Mali. CNRS. Lamy. Marseille, 473 pp.

Rozanski K., Johnsen S.J., Schotterer U. and Thompson L.G. 1997. Reconstruction of past climates from stable isotope records of palaeo-precipitation preserved in continental archives. Hydrol. Sci. J. 42: 725–745.

Rudolph J., Rath K. and Sonntag C. 1984. Noble gases and stable isotopes in ^{14}C dated waters from Central Europe and the Sahara. In: IAEA (eds), Isotope Hydrology, 1983: Proc. Internat. Symp. on Isotope Hydrology in Water Resources Development. IAEA, Vienna, pp. 467–477.

Saighi O., Michelot J.L. and Filly A. 2001. Isotopic characteristics of meteoric water and groundwater in the Ahaggar Massif (Central Sahara). In: IAEA (eds), Isotope Techniques in Water Resource Investigations in Arid and Semi-arid Regions. IAEA, Vienna, pp. 7–25.

Salem O., Visser J.-H., Dray M. and Gonfiantini R. 1980. Groundwater flow patterns in the Western Libyan Arab Jamahirya evaluated from isotopic data. In: IAEA (eds), Arid-zone Hydrology: Investigations with Isotopic Techniques. IAEA, Vienna, pp. 165–179.

Servant M. and Servant-Vildary S. 1980. L'environnement quaternaire du bassin du Tchad. In: Williams M.A.J and Faure H. (eds), The Sahara and the Nile. A.A. Balkema. Rotterdam, pp. 133–162.

Sonntag C., Klitsch E., Lohnert E.P., Munnich K.O., Junghans C., Thorweihe U., Weistroffer K. and Swailem F.M. 1978. Palaeoclimatic information from D and ^{18}O in ^{14}C-dated North Saharian groundwaters, groundwater formation from the past. In: IAEA (eds), Isotope Hydrology. IAEA, Vienna, pp. 569–580.

Sonntag C., Thorweihe U., Rudolph J., Lohnert E.P., Junghans C., Munnich K.O., Klitsch E., El Shazly E.M. and Swailem F.M. 1980. Isotopic identification of Saharan groundwaters, groundwater formation in the past. Palaeoecol. Africa 12: 159–171.

Srdoč D., Sliepčević A., Obelić B., Horvatinčić N., Moser H. and Stichler W. 1980. Isotope investigations as a tool for regional hydrogeological studies in the Libyan Arab Jamahiriya. In: Arid-zone Hydrology: Investigations with Isotopic Techniques. IAEA, Vienna, pp. 569–580.

Stokes S., Maxwell T.A., Haynes C.V. and Horrocks J. 1998. Latest Pleistocene and Holocene sand-sheet construction in the Selima Sand Sea, Eastern Sahara. In: Alsharan A.S., Glennie K.W., Whittle G.L. and Kendall C.G.St.C. (eds), Quaternary Deserts and Climatic Change. Balkema, Rotterdam, pp. 175–183.

Stute M. and Schlosser P. 1993. Principles and applications of the noble gas paleothermometer, In: Swart P.K., Lohmann K.C., McKenzie J. and Savin S. (eds), Climate Change in Continental Isotopic Records. Am. Geophys. Union. Geophys. Monograph 78: 89–100.

Stute M. and Talma S. 1998. Glacial temperatures and moisture transport regimes reconstructed from noble gases and $\delta^{18}O$, Stampriet aquifer, Namibia. In: IAEA (eds), Isotope Techniques in the Study of Past and Current Environmental Changes in the Hydrosphere and Atmosphere. IAEA, Vienna, pp. 307–318.

Sultan M., Sturchio N., Hassan F., Hamdan M.A.R., Mahmood A.M., El Alfy Z. and Stein T. 1997. Precipitation source inferred from stable isotope composition of Pleistocene groundwater and carbonate deposits in the western desert of Egypt. Quat. Res. 48: 29–37.

Swezey C. 2001. Eolian sediment responses to late Quaternary climatic changes: temporal and spatial patterns in the Sahara. Palaeogeogr. Palaeoclimatol. Palaeoecol. 167: 119–155.

Taupin J.-D., Coudrain-Ribstein A., Gallaire R., Zuppi G.M. and Filly A. 2000. Rainfall characteristics ($\delta^{18}O$, $\delta^{2}H$, ΔT and ΔH_r) in Western Africa, regional scale and influence of irrigated areas. J. Geophys. Res. 105: 11911–11924.

Thorweihe U. 1982. Hydrogeologie des Dakhla Beckens (Ägypten). Berliner geowiss. Abh. (A), 38: 1–58.

Travi Y. and Chernet T. 1998. Fluoride contamination in the lakes region of the Ethiopian Rift: origin, mechanism and evolution. In: IAEA (eds), TECDOC-1046. IAEA, Vienna, pp. 95–105.

Tyler S.W., Chapman J.B., Conrad S.H., Hammermeister D.P., Blout D.O., Miller J.J., Sully M.J. and Ginani J.N. 1996. Soil-water flux in the southern Great Basin, United States: Temporal and spatial variations over the last 120 000 years. Water Resour. Res. 32: 1481–1499.

UNESCO. 1972. Investigations of groundwater in the Lake Chad basin. Unpublished Report.

Wright E.P. and Edmunds W.M. 1969. Distribution and origin of nitrate in the groundwaters. In: Hydrogeological Studies in Central Cyrenaica, Kingdom of Libya. Report to Libyan Govt. British Geological Survey. Keyworth.

15. MEDITERRANEAN SEA PALAEOHYDROLOGY AND PLUVIAL PERIODS DURING THE LATE QUATERNARY

NEJIB KALLEL (nejib.kallel@fss.rnu.tn)
Faculté des Sciences de Sfax
B.P. 802, 3018 Sfax
Tunisia

JEAN-CLAUDE DUPLESSY
(jean-claude.duplessy@lsce.cnrs-gif.fr)
Laboratoire des Sciences du Climat et de l'Environnement
Laboratoire Mixte CNRS-CEA
Parc du CNRS 91 198
Gif-sur-Yvette cedex
France

LAURENT LABEYRIE
(laurent.labeyrie@lsce.cnrs-gif.fr)
Laboratoire des Sciences du Climat et de l'Environnement
Laboratoire Mixte CNRS-CEA
Parc du CNRS 91 198
Gif-sur-Yvette cedex
France

MICHEL FONTUGNE (michel.fontugne@lsce.cnrs-gif.fr)
Laboratoire des Sciences du Climat et de l'Environnement
Laboratoire Mixte CNRS-CEA
Parc du CNRS 91 198
Gif-sur-Yvette cedex
France

MARTINE PATERNE (martine.paterne@lsce.cnrs-gif.fr)
Laboratoire des Sciences du Climat et de l'Environnement
Laboratoire Mixte CNRS-CEA
Parc du CNRS 91 198
Gif-sur-Yvette cedex
France

Keywords: Mediterranean Sea, Deep-sea cores, Isotopes, Modern analogue technique, Palaeohydrology, Sapropels, Pluvial periods, Holocene, Climatic cycle, Quaternary

307

R. W. Battarbee et al. (eds) 2004. *Past Climate Variability through Europe and Africa.*
Springer, Dordrecht, The Netherlands.

Introduction

Quaternary climatic changes deeply modified the environments of middle and high latitudes: continental ice sheets grew and collapsed, large temperature fluctuations occurred in temperate latitudes and major precipitation changes occurred in subtropical latitudes. In particular wet episodes developed in the Mediterranean region during the late Quaternary. These periods of enhanced wetness are indicated by the presence of palaeo-lacustrine deposits which can be found today in the Sahara and in the North of Africa (Bar-Matthews et al. 1997; Petit-Maire et al. 1991; Callot and Fontugne 1992; Street and Grove 1979; Adamson et al. 1980; Nicholson and Flohn 1980; Ritchie et al. 1985; Ritchie and Haynes 1987) and are reflected by former high lake levels and high river discharge over the continent bordering the Northern Mediterranean Sea (Harrison et al. 1991; Starkel 1991). It has been suggested that these wet time intervals coincide with times of higher seasonal contrast in the Northern Hemisphere with warmer summers and colder winters (Kutzbach and Guetter 1986; de Noblet et al. 1996; Masson et al. 1999). Numerical experiments have been conducted using atmospheric general circulation models to investigate the sensitivity of the African monsoon to changes in incoming solar radiation at the top of the atmosphere. For the Mediterranean region, none of the models was able to reproduce the observed enhanced precipitation during the middle Holocene (de Noblet et al. 1996; Masson et al. 1999; Braconnot et al. 2000). The origin of these changes remains, therefore, largely unknown. In the Mediterranean Sea, organic rich horizons, called sapropels, intercalated in the normal carbonate oozes of the basin are associated with periods of high summer insolation of the Holocene and the Pliocene-Pleistocene periods (Rossignol-Strick 1985; Hilgen 1987; 1991; Hilgen et al. 1993). This increase in the preservation of deposited organic matter is probably related to the establishment of bottom water anoxia resulting from the lowering of the sea-surface salinity and the stratification of the Mediterranean water column (Bradley 1938; Cita et al. 1977; Williams et al. 1978; Thunell and Williams 1989).

Past changes in the local fresh-water budget of the Mediterranean region would also affect the surface salinity of the Mediterranean Sea. The study of past Mediterranean hydrological variations and the comparison with North Atlantic records would permit a better quantification of the fluctuations of its fresh-water balance and of climatic changes of the surrounding continents. Such data could also help to validate models coupling ocean and atmosphere and better evaluate the role of the Mediterranean Sea in sub-tropical climate change.

In this paper, we present a comprehensive synthesis of our recent studies on the past hydrological changes of the Mediterranean Sea (Kallel et al. 1997a; 1997b; 2000) and their relation with climatic change, which occurred in the Mediterranean area during the last two climatic cycles. In addition, we have generated a new detailed climatic record of the Holocene in a Western Mediterranean Sea core to determine the timing of the Holocene wet period. We therefore estimated variations of sea-surface temperature and salinity in the Mediterranean Sea during the last 200,000 years. Four cores recovered in the major basins were used. We first show that Mediterranean sediments record both changes observed in the North Atlantic temperature and salinity and in the local Mediterranean fresh-water budget by using a new data of a well-dated core recovered in the Tyrrhenian Sea. We then reconstruct a map of Mediterranean Sea surface salinity (SSS) at the time of the last sapropel (middle Holocene). These data, combined with the sea-surface temperature (SST) estimates, are used to propose a mode of formation for the sapropel. Finally, we have tested the validity of this model for five other sapropels, which occurred during the last 200,000 years.

Methods

Modern analogue technique

To derive the Mediterranean SST, we developed a comprehensive reference modern data-base containing planktonic foraminiferal counts from 128 core tops from the Mediterranean Sea and 123 core tops from the North Atlantic Ocean (Kallel et al. 1997a). We used the modern analogue technique (Hutson 1980; Prentice 1980; Prell 1985; Overpeck et al. 1985) to compare fossil planktonic foraminiferal samples with this data-base and to identify, for each fossil association, the ten best modern analogues. We have carried out an extensive analysis of the Mediterranean fossil samples using this technique and we obtained modern analogues for Mediterranean fossil samples for both cold and warm periods. Continuous surface palaeo-temperature estimates are attainable by these methods in this basin. Our objective therefore was to compare those estimates with the isotopic composition of con-temporary planktonic foraminiferal assemblages to reconstruct the surface water oxygen isotopic composition in this sea following the methods of open ocean studies (Duplessy et al. 1992; 1993). As the $\delta^{18}O$ of sea-water is linearly linked to salinity (Craig and Gordon 1965), it is hence also possible to reconstruct past variations of the surface salinities in this basin (Kallel et al. 1997a; b).

Calibration of the foraminiferal oxygen isotopes and reconstruction of sea-water oxygen isotope composition

Isotopic fractionation in the "carbonate-water" system depends on temperature (Epstein et al. 1953; Shackleton 1974). In contrast to the deep ocean, the annual variation of the sea-surface temperature can be very important. In the Mediterranean Sea, this amplitude is near to 14 °C (Levitus 1982). To derive a reliable reconstruction of the past sea-water oxygen isotope composition and salinity we must know the exact season of planktonic foraminiferal development in this sea by using their isotopic composition.

Two planktonic foraminiferal species are generally abundant in Mediterranean Sea cores: *Globigerina bulloides* dominates in the western basin, whereas *Globigerinoides ruber* dominates in the eastern basin. Ecological studies of these two species conducted in the Mediterranean Sea mainly by Pujol and Grazzini (Pujol and Vergnaud-Grazzini 1989; 1995; Vergnaud-Grazzini et al. 1989) show that the *G. bulloides* bloom is restricted to spring, whereas *G. ruber* grows mainly during autumn. We measured the $\delta^{18}O$ value of both *G. bulloides* from 34 core tops and *G. ruber* from 40 core tops. Isotopic temperatures have been calculated using the palaeo-temperature equation of Shackleton (1974). The resulting isotopic temperatures were compared with those prevailing during the annual cycle (Levitus 1982). A statistical analysis shows that the best linear fit is obtained between the isotopic temperature of *G. bulloides* and April-May Levitus SST and between the isotopic temperature of *G. ruber* and October-November (Kallel et al. 1997a).

Tyrrhenian Sea SST and SSS variation during the last 18 000 yrs

In the North Atlantic Ocean, north-south fluctuations in the position of the polar front, which separates the warm saline sub-tropical waters in the south from cold less saline polar waters in the north, were associated during the last deglaciation with spectacular changes in temperature and salinity of the North Atlantic Surface Water (Bard et al. 1987; Duplessy et al. 1992) which constitutes the source of the Mediterranean Surface Water. To understand better the effect of these changes on the climate and the hydrology of the Mediterranean Sea, we have reconstructed the temperature and salinity variation of the western Mediterranean Sea during the last 18,000 ^{14}C yrs using the Tyrrhenian Sea core KET80-19. This core, which has a high sediment accumulation rate (Kallel et al. 1997b), has been selected to generate a new high-resolution climate record (Fig. 1).

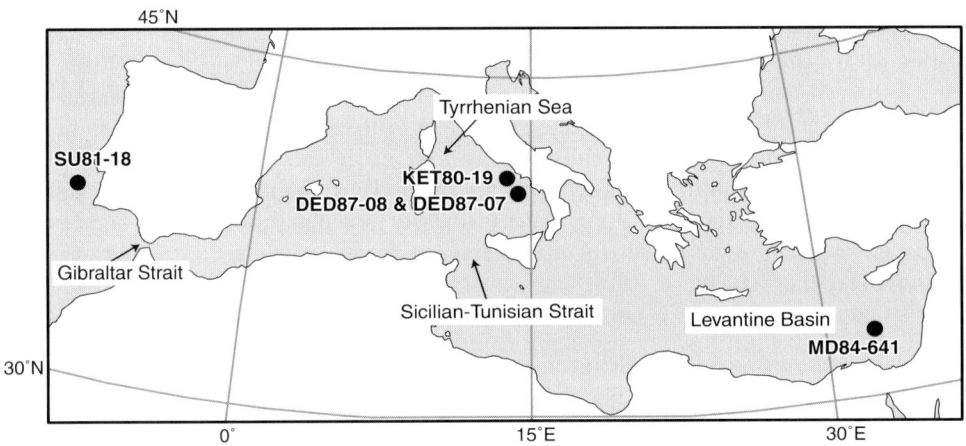

Figure 1. Core location: KET80-19 (40°33′N; 13°21′E; 1920 m water depth), SU81-18 (37°46′N; 10°11′W; 3135 m), DED87-07 (39°41′N; 13°35′E; 2970 m), DED87-08 (39°42′N; 13°34′E; 2965 m) and MD84-641 (33°02′N; 33°38′E; 1375 m).

Figure 2 displays the April-May SST estimates and the dissimilarity coefficient (which measures the mean degree of similarity between each fossil assemblage and the corresponding ten best modern analogues) in the Tyrrhenian Sea core. Modern analogues are available in the data-base during most of the last deglaciation and the Holocene interglacial period when the dissimilarity coefficient was usually lower than 0.2. Reconstruction is of a lesser quality during the glaciation when the dissimilarity coefficient is higher than 0.2 but rarely exceeds 0.25. The chronology of core KET80-19 was derived from five marine tephra layers, geochemically correlated to ^{14}C-dated continental volcanic deposits (Paterne et al. 1986; 1988; Fontugne et al. 1989; Kallel et al. 1997b) and from six new ^{14}C ages obtained at Gif-sur-Yvette from accelerator mass spectrometry (AMS) dating on peaks of planktonic foraminiferal abundances (Fig. 2). A 400 yr correction was made to conventional

foraminiferal ^{14}C ages to take into account the difference in radiocarbon content between the total dissolved CO_2 in the modern Mediterranean surface waters and the CO_2 of the atmosphere (Bard 1988).

The SST variation shows a pattern similar to that of the North Atlantic Ocean (Bard et al. 1987; Duplessy et al. 1991a). From 18 kyr BP (Last Glacial Maximum (LGM)) to the late Holocene, April-May SST in the Tyrrhenian Sea and mean annual SSTs in the Atlantic Ocean increased by about 4–5 °C (Fig. 3a). Mediterranean SST estimates are lower than those of the LGM by about 3 °C between about 15,000 and 13,000 ^{14}C yr BP. This cold event, associated in the North Atlantic Ocean with the Heinrich event H_1 (Bond et al. 1993; Cortijo et al. 1995), is more pronounced off Portugal (the cooling is of about 7 °C). In a same manner than in the North Atlantic Ocean, the warming following this period was interrupted in the Tyrrhenian Sea by a spectacular climatic deterioration and an abrupt return to glacial temperatures between about 11,000 and 10,000 ^{14}C yrs at the same epoch as the Younger Dryas (Mangerud 1970; Duplessy et al. 1981; Bard et al. 1987). During the middle Holocene, from about 8,500 to 7,500 ^{14}C yrs BP and between about 7,000 and 4,800 ^{14}C yrs BP, the Tyrrhenian SST experienced another inferred cooling of about 2 to 3 °C, which is not observed in the North Atlantic Ocean.

During the LGM and the last deglaciation, the best analogues of Tyrrhenian Sea fossil fauna come from the North Atlantic Ocean. Those of Holocene age were generally found in the Mediterranean Sea itself, with the exception of the two mid-Holocene cold periods (Kallel et al. 1997b). During the latter periods, the Tyrrhenian Sea fauna was dominated by *Neogloboquadrina pachyderma*-right coiling ($\approx$60%). This species is considered an indicator of cold sub-polar conditions (Bé and Tolderlund 1971). Nevertheless, in the Mediterranean Sea, previous studies interpreted the strong increase in the relative abundances of *N. pachyderma*-right coiling during the middle Holocene as an indication of change in the position of the pycnocline in the water column, rather than in the temperature of surface waters (Rohling et al. 1995; Targarona 1997). In agreement with the latter hypothesis, the application of the $U^{k'}_{37}$ index to derive past Holocene SST values in the Tyrrhenian Sea shows a warming trend when the modern analogue technique indicates a cooling of the surface water (Sbaffi et al. 2001). However, the statistical analysis that we made on the Tyrrhenian Sea Holocene fossil fauna dominated by *N. pachyderma*-right coiling indicates that this assemblage is very similar to that observed in the modern reference data-base of the Bay of Biscay. In the modern Mediterranean Sea, this species is dominant in the Gulf of Lion, where winter SSTs are close to 13 °C (Pujol and Vergnaud-Grazzini 1995). Although the same annual SST variation pattern is observed at both locations (Levitus 1982), the planktonic foraminiferal assemblages show a significant difference. *Globorotalia truncatulinoides*-left coiling (l.c.), which reaches up to 20% in surface sediment samples from the Gulf of Lion, is scarce in the modern Bay of Biscay. This species is missing in middle Holocene fossil samples from the Tyrrhenian Sea. A similar mid-Holocene cooling can therefore be reconstructed in the Tyrrhenian Sea using either Bay of Biscay or Gulf of Lion modern analogues if *G. truncatulinoides*-l.c. is excluded.

Taking into account the SST estimates and the foraminiferal $\delta^{18}O$ values, we reconstructed the surface water $\delta^{18}O$ variations in the Tyrrhenian Sea since the LGM by solving the palaeo-temperature equation (Epstein et al. 1953; Shackleton 1974). In the Mediterranean Sea, surface water $\delta^{18}O$ (δw) changes reflect both changes in the $\delta^{18}O$ value of the incoming Atlantic water and variations of the local fresh-water budget (Precipitation plus

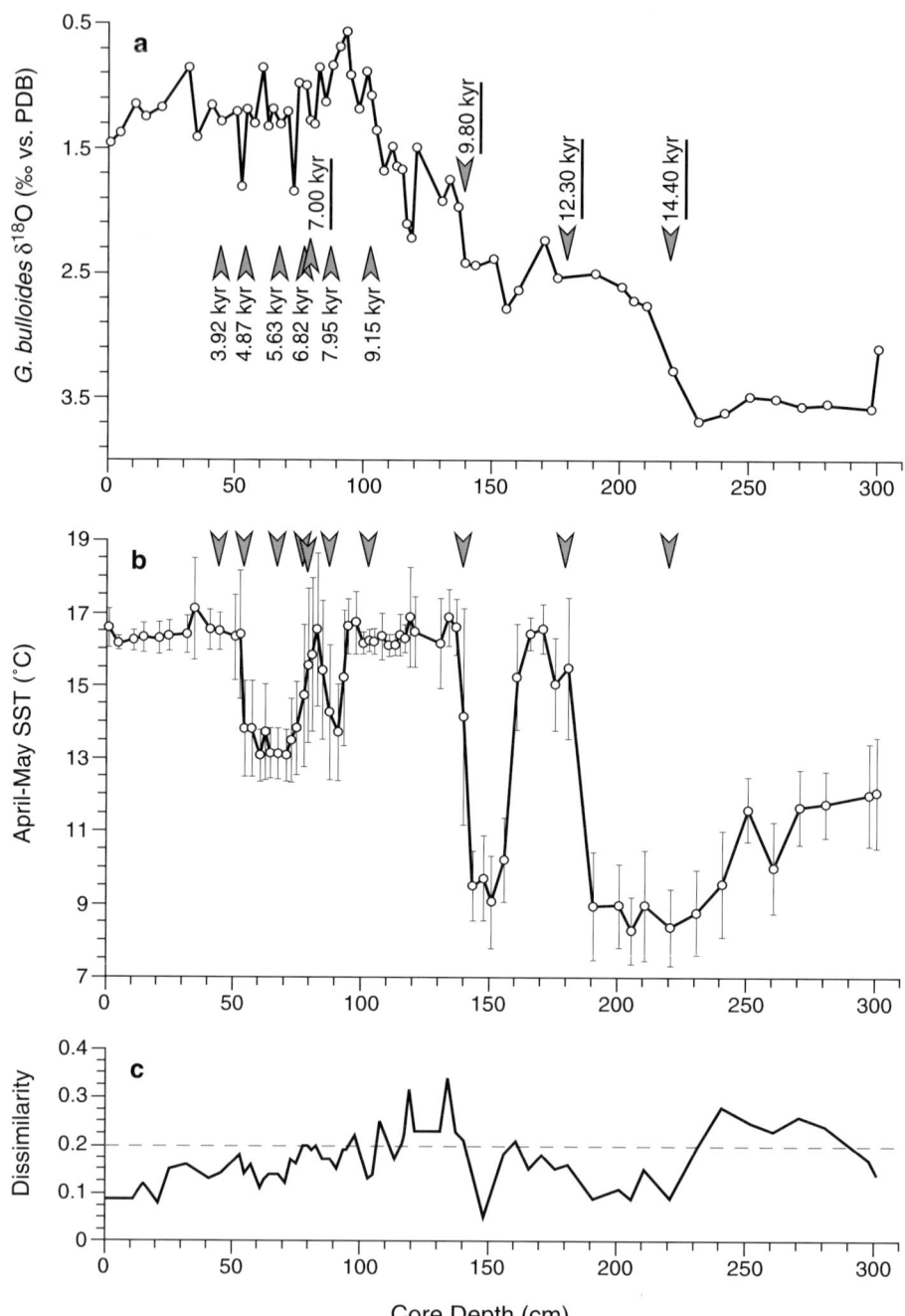

Figure 2. Climatic records versus depth of core KET80-19: (a) oxygen isotope of *G. bulloides*; (b) April-May Sea Surface Temperature and their error bars; (c) dissimilarity coefficient. Good modern analogues are indicated by a dissimilarity coefficient lower than 0.2 (dashed line). Planktonic foraminifera AMS ^{14}C and tephra-layer (underlined) ages obtained on core KET80-19 are also shown.

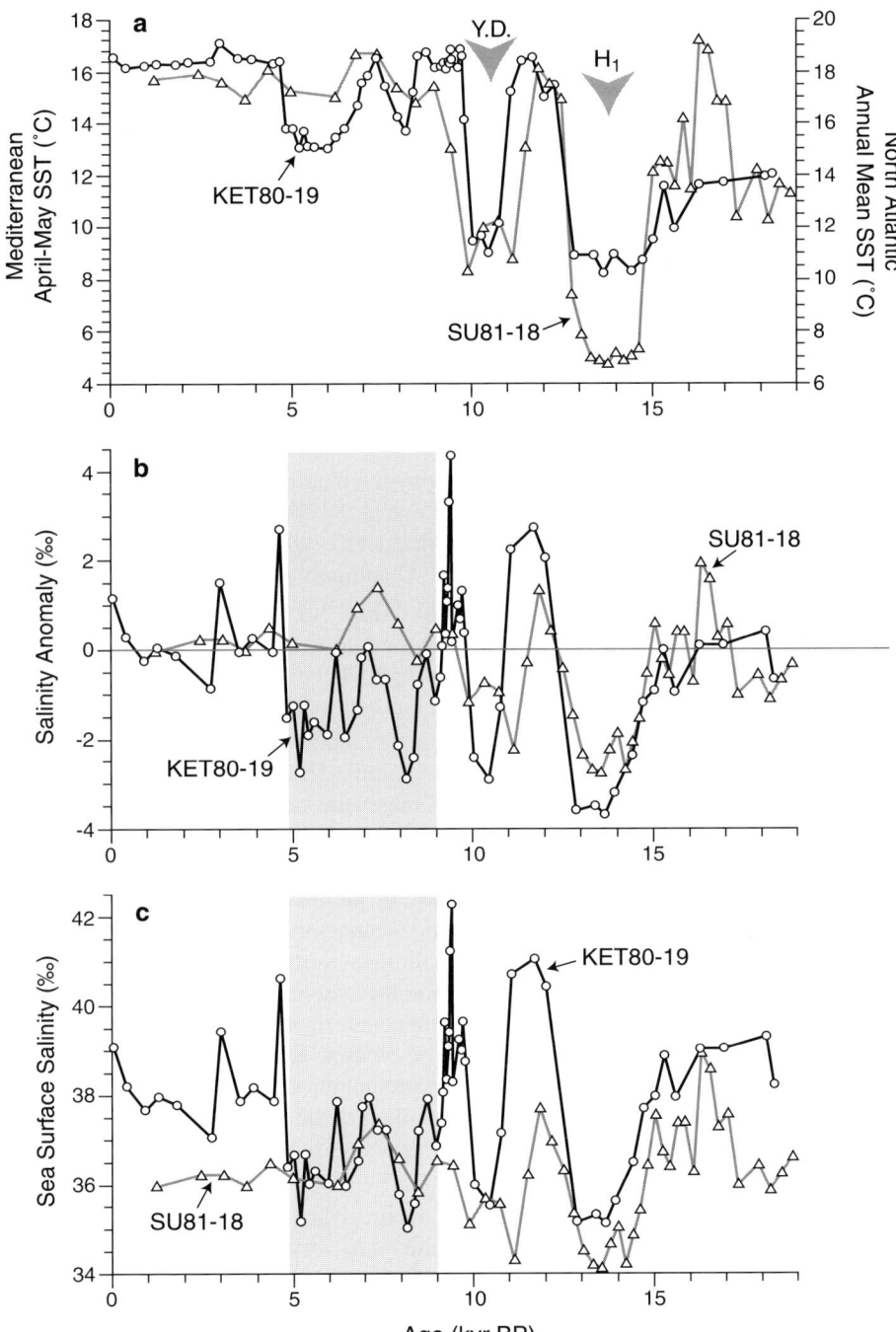

Figure 3. Climatic records: (a) April May and Annual Mean Sea Surface temperature in Tyrrhenian Sea and in core SU81-18 (North Atlantic Ocean); Comparison of (b) salinity anomalies and (c) surface salinity records of the Tyrrhenian Sea (core KET80-19) and the North Atlantic Ocean (core SU81-18). Grey intervals correspond to the period of change in the fresh-water budget of the western Mediterranean Sea.

continental Runoff minus Evaporation, $P + R - E$) of the Mediterranean basin (Kallel et al. 1997b). The $\delta^{18}O$ of the Atlantic surface water varies mainly with continental ice volume and patterns of sea surface circulation.

The $\delta^{18}O$ variation of the North Atlantic surface water, associated with the continental ice-sheet melting during the last deglaciation has been estimated by Duplessy et al. (1991b) using the Barbados sea-level curve (Fairbanks 1989). As the Mediterranean Sea is linked to the North Atlantic Ocean through the Gibraltar Strait, we suppose that the meltwater effect was recorded in a similar manner in the two regions. We therefore estimated the local Mediterranean Sea surface $\delta^{18}O$ variations with respect to the modern situation as the difference between calculated and modern sea-water $\delta^{18}O$, corrected for the isotopic effect of meltwater injection on the North Atlantic surface water.

$$\delta^{18}O\text{anomaly} = \text{calculated } \delta w - (\delta^{18}O\text{modern} + \text{meltwater } \delta^{18}O\text{signal}).$$

We have then used the modern sea-water oxygen-isotope/salinity relationships for the western Mediterranean Sea (changes in local δw of 0.41‰ when local δw is lower than 1.5‰ and of 0.2‰ when local δw is higher than 1.5‰ will result in 1‰ change in local salinity, Kallel et al. (1997a)) to convert the $\delta^{18}O$ anomaly to a salinity anomaly (Fig. 3b).

To generate the Tyrrhenian Sea salinity record (Fig. 3c) we take into account the global and local effects on salinity. The salinity change due to ice-volume variation has been estimated assuming that salinity increased by 1 unit when the sea-water $\delta^{18}O$ increased by 1.2‰ (Labeyrie et al. 1987; Shackleton 1987; Fairbanks 1989).

Core KET80-19 exhibits three strong negative salinity anomalies during the cooling synchronous to H_1, the Younger Dryas and the middle Holocene cooling events (3.4, 2.5 and 2‰ respectively, Fig. 3b). To understand better the origin of these salinity changes, we compared them with those obtained from the core SU81-18 recovered in the North Atlantic Ocean off Portugal (Duplessy et al. 1992; 1993). The first two events were found to be associated with a similar salinity decrease of the incoming Atlantic water, whereas the Holocene event, absent in the North Atlantic Ocean, must be of regional origin. In fact, during the latter epoch, Tyrrhenian surface salinities were similar to those of the North Atlantic Ocean off Gibraltar suggesting a strong increase in the fresh-water input into the western Mediterranean Sea that was sufficient to counterbalance the evaporation effect and to prevent any surface salinity increase within the basin (Kallel et al. 1997b). Thus, these data can only be explained by an increase of precipitation in the western basin and the surrounding continents. It seems also that the cooling of the SST in the Northern part of the Western basin at that time can be responsible for a significant decrease in the evaporation by about 0.4 m/year (Masson et al. 1999).

Over the southern and eastern Mediterranean borderlands, various environmental markers show that pluvial conditions, related to the "African Humid Period" in the Sahara, prevailed from about 10,000 to 5,000 years BP. Simultaneously, enhanced wetness over the continent bordering the northern Mediterranean Sea, was reflected by high lake levels and higher river discharge. The optimum of these wet conditions coincides also with the appearance of the last sapropel in the Eastern Mediterranean at about 8,000 yrs BP. It is an organic-rich horizon intercalated in the normal marl oozes of the basin associated with a negative peak in the planktonic foraminiferal $\delta^{18}O$ records. To better understand the hydrological conditions of the whole Mediterranean during the last sapropel, we will reconstruct the surface salinity map of this basin at that time.

Hydrology of the Mediterranean Sea during the last sapropel

At the time of the last sapropel event (S_1) which is indicated by the lowest Holocene foraminiferal $\delta^{18}O$ values, SSTs were lower than today by about 1.5 °C in the Alboran Sea and 2.5 °C in the Tyrrhenian Sea. In the Sicilian-Tunisian Strait and the Eastern Sea, SSTs are not significantly different from the modern ones (Kallel et al. 1997a).

Taking into account the SST pattern, we reconstructed Mediterranean surface water $\delta^{18}O$ at the time of the last sapropel event using the lowest Holocene foraminiferal $\delta^{18}O$ values from 47 deep sea cores. We have then converted these values to salinity using the δw/salinity relationship of the Mediterranean and mapped the resultant averages by basin (Fig. 4b). The Mediterranean salinity map during S_1 shows that values were of the same order over the whole Mediterranean Sea. This pattern contrasts with the modern situation (Fig. 4a) in which evaporation dominates and surface water oxygen isotopic composition and salinity increase from west to east, and suggests that the fresh-water budget (precipitation plus runoff minus evaporation, P+R−E) was nearly equilibrated over the whole Mediterranean. Therefore, the Mediterranean Sea was not a concentration basin at that time. Increasing river runoff from all the continents surrounding the Mediterranean Sea, together with a change in the fresh-water budget over the whole Mediterranean area provide the best explanation for the surface water salinity decrease reconstructed during S_1.

Changes in the fresh-water budget of the Mediterranean Sea have had a great impact on its intermediate and deep circulation. Under modern conditions, the fresh-water budget of the Mediterranean Sea is negative and the resulting strong salinity increase within the basin is responsible for the establishment of a strong density gradient between the Western Intermediate water and the North Atlantic subsurface water at the level of the Gibraltar sill (Béthoux 1984). This value is now equivalent to about 1.65‰. The intensity of the Mediterranean outflow depends on this density gradient. During the middle Holocene pluvial period, a strong surface salinity decrease was recorded in the Mediterranean Sea (Fig. 4c). This induced a decrease in the density gradient between both sides of the Gibraltar sill. Mediterranean outflow was therefore drastically reduced, resulting in an increase of the residence time and reduction of the ventilation of the Mediterranean intermediate and deep water.

Because of the 2.5 °C surface water cooling in the Northern part of the Western Sea during S_1, the intermediate water temperature would be close to 10.5 °C (present temperature: 13 °C). By contrast, in the Eastern basin, no significant temperature change was observed during the Holocene and salinities were equivalent to those in the west. The density gradient was thus inverted over the Sicilian-Tunisian Strait at the time of the last sapropel inducing a circulation reverse over this sill. Intermediate and deep water of the Eastern basin originating from the low ventilated Western Sea, presented densities that cannot be reached at the surface in the Eastern basin. This lay at the origin of the water column stratification and the establishment of a permanent pycnocline between surface and deep waters. This structure hence inhibited the dissolved oxygen supply at depth and promoted the establishment of anoxic conditions at the bottom of the Eastern Sea.

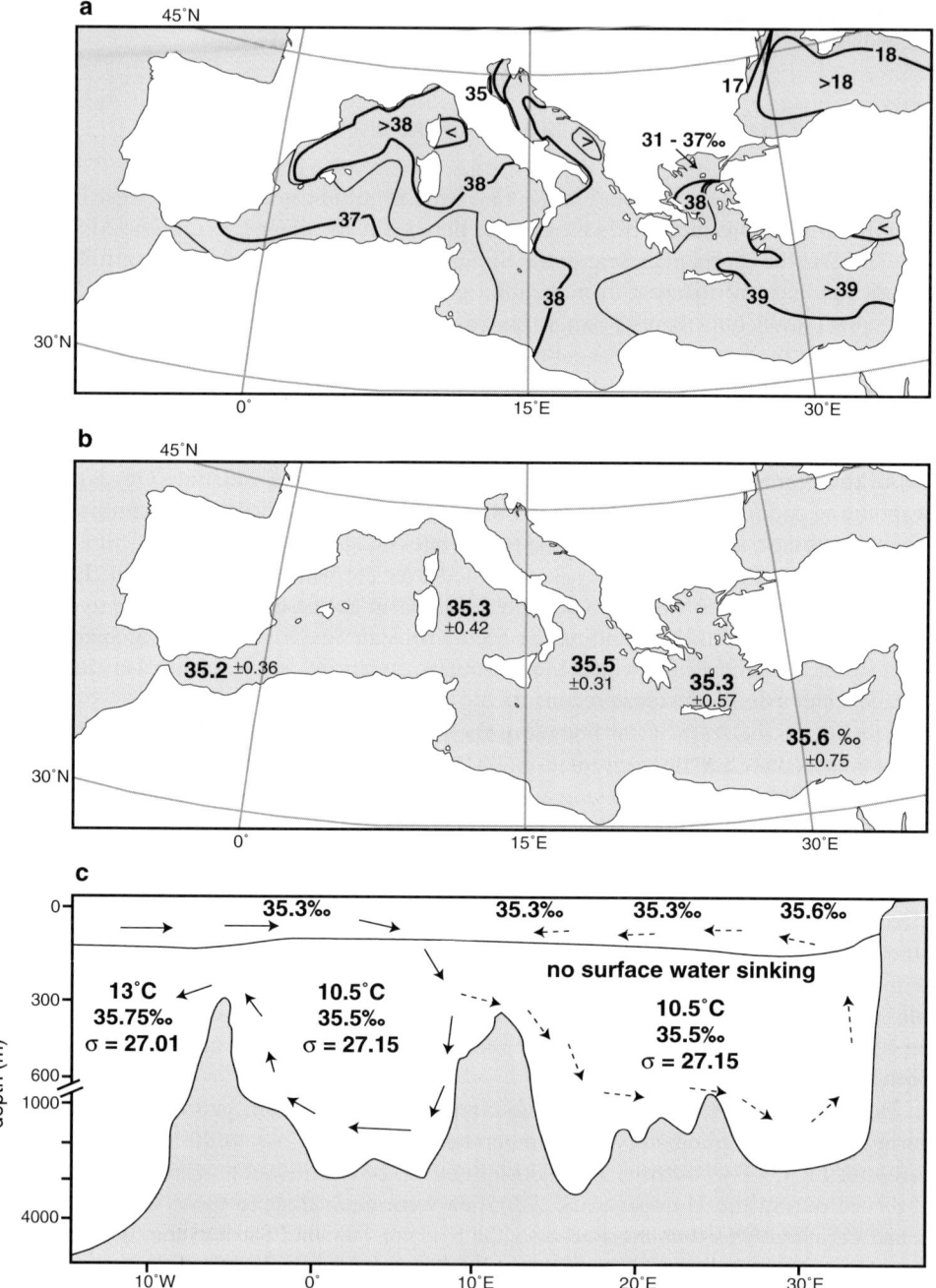

Figure 4. (a) Modern Mediterranean Sea surface salinities; (b) mean surface salinities in the Mediterranean Sea at the time of the last sapropel event (S_1); (c) West-East cross section displaying the circulation pattern of the Mediterranean Sea at the time of the last sapropel.

Hydrological conditions associated with sapropel occurrences in the Mediterranean Sea during the last 200,000 years

Three Mediterranean deep-sea cores have been studied to reconstruct the hydrological conditions at the time of deposition of sapropels during the time interval 200–60 kyr BP. Core MD84-641 was raised from the Levantine basin. Cores DED87-07 and DED87-08 come from the Tyrrhenian Sea (Fig. 1). Cores MD84-641 and DED87-08 were previously described by Fontugne and Calvert (1992), Paterne et al. (1986, 1988) and Tric et al. (1992). Core DED87-07 was recovered at the same location than core KET80-04 (Kallel et al. 2000). It has been chosen for this study because of its higher resolution during MIS-5. The chronology of the three cores was obtained by correlation of their planktonic foraminiferal $\delta^{18}O$ records with the SPECMAP stack (Martinson et al. 1987). Core MD84-641 covers the whole studied time interval. Core DED87-07 extends to isotope stage 6 (Paterne et al. 1986; Tric et al. 1992). Core DED87-08, from a nearby location is disturbed above stage 5e, but extends to stage 7. We pieced both records together to generate a complete $\delta^{18}O$ record from the Tyrrhenian Sea (Plate 6a).

Organic carbon content was also measured in core MD84-641, as high values ($\geq 2\%$) are closely associated with sapropel layers (Plate 6d). The record exhibits five periods of high values, at about 80, 96, 122, 170, and 195 kyr BP, corresponding to sapropels S_3 to S_7.

SSTs were estimated using the Mediterranean transfer function (Kallel et al. 2000). Results show that from 200 to 60 kyr BP, April-May SST varied, sometimes rapidly, between 7.5 °C and 18 °C in the Tyrrhenian Sea (Plate 6b). The lowest temperatures were associated with stage 6 and the end of Stage 7 (7 to 11 °C). SST estimates for isotope stage 5e (~ 17.5 °C) were about 1 °C warmer than those of the upper Holocene. The rest of isotope stage 5 is relatively warm, with a mean temperature close to 14 °C although some cold events appeared within this stage. In the Levantine basin, except for two large events within MIS 6 which occurred between 180 and 170 kyr BP and at ~ 140 kyr BP., the amplitude of October-November SST variations is smaller than that observed in the Tyrrhenian Sea. Sapropel S_6 is associated with the first cooling. Marine isotope stage 5 has a mean October-November temperature of 22.5 °C, which is 1 °C lower than the modern one. However, this period was divided by two coolings of 6 °C and 2 °C associated with sapropels S_4 and S_5 respectively.

Foraminiferal $\delta^{18}O$ and SST records for each core were used to estimate past variations of surface water $\delta^{18}O$, using the palaeo-temperature equation (Plate 6c). Mediterranean surface water $\delta^{18}O$ variations reflect both changes in the $\delta^{18}O$ value of the incoming Atlantic water and variations of the fresh-water budget of the Mediterranean basin (Kallel et al. 1997b). As no long record of the $\delta^{18}O$ variations of the water penetrating at Gibraltar is presently available, we used the Mediterranean δw records as a proxy for surface salinity variations, without attempting to quantify them in terms of salinity units.

Our results show a strong δw/salinity decrease in both basins associated with each sapropel. These δw decreases reflect a drastic change in the local fresh-water budget and were comparable to that observed during the last sapropel (S_1) centred at about 8 kyr BP. Cooling events observed during some sapropels, specially sapropel S_6, can be associated with a significant decrease in the evaporation amount. In any case, our data suggest that sapropels S_3 to S_7, like S_1, occurred when precipitation plus runoff were equivalent to or exceeded evaporation, so that the Mediterranean Sea was no longer a concentration basin. The low surface salinity was responsible for the water column stratification and resulted

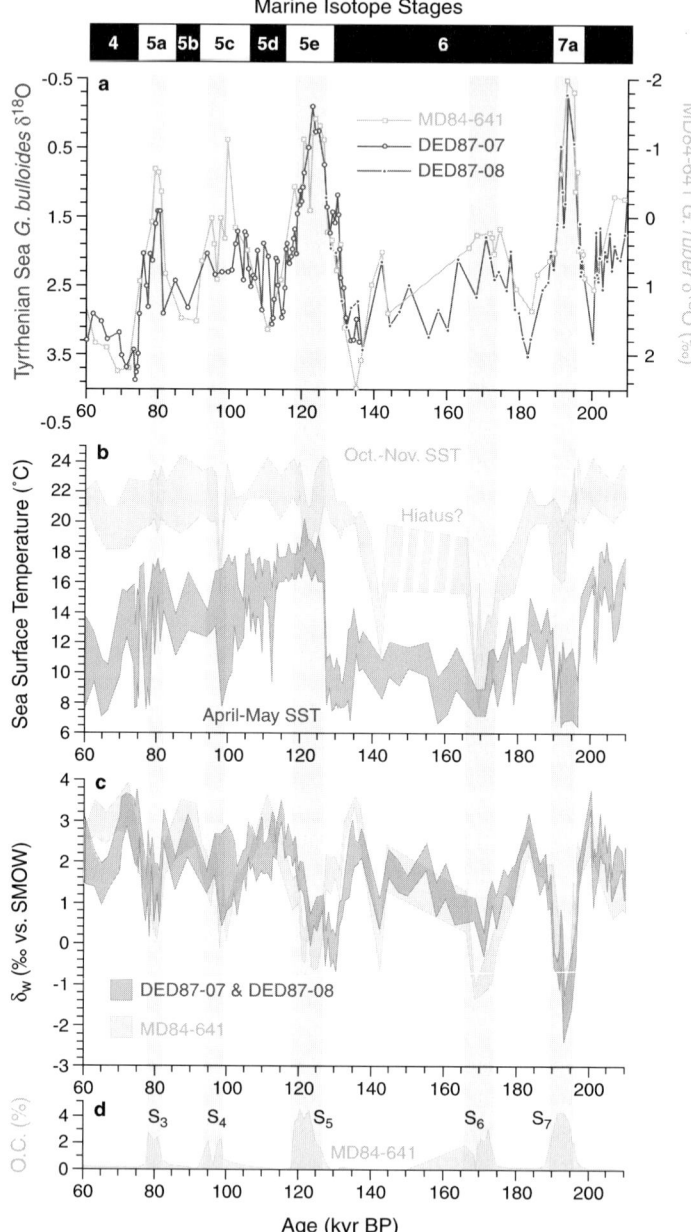

Plate 6. Tyrrhenian and Levantine Seas climatic records plotted against time: (a) oxygen isotope records of *G. bulloides* (Tyrrhenian Sea cores DED87-07 & DED87-08) and *G. ruber* (Levantine basin core MD84-641); (b) upper and lower confidence margins of April-May (Tyrrhenian Sea cores DED87-07 and DED-87-08) and October-November (core MD84-641) sea surface temperature (SST) estimates; (c) upper and lower confidence margins (depending on the error on SST estimates and on the foraminiferal isotope analytical error; Kallel et al. (1997b)) of the sea-surface water oxygen isotopic composition (δw) estimates; (d) organic carbon record in the Levantine Sea core MD84-641. Light blue intervals delimit the periods of significant increase in the percentage of the organic carbon (sapropels) in the Eastern Mediterranean Sea. Colour version of this Plate can be found in Appendix, p.633

in a significant reduction of the rate of deep water oxygenation and sapropel formation in the eastern basin. Other low δw values were also observed during a brief event at $\sim$140 kyr BP and the 6/5e transition, before the establishment of sapropel S_5. As these events are not associated with an increase of the organic carbon content in the sediment (sapropel formation), they would be induced by a salinity decrease of the incoming Atlantic surface water at the time as the first Heinrich event (H_1) and the Younger Dryas (Kallel et al. 1997b).

Correlation with continental records and conclusions

Our data suggest that all the sapropel events of the last 200 kyr were associated with enhanced precipitation over the whole Mediterranean basin. Such changes should be reflected by humid conditions over the neighbouring continents (Rohling and Hilgen 1991; Calvert et al. 1992; Rohling and De Rijk 1999). In Southern Europe, the last two climatic cycles are covered by several long pollen sequences: including Valle di Castiglione in Central Italy (Follieri et al. 1990), Ioannina (Tzedakis 1993) and Tenaghi Philippon (Wijmstra 1969) in Greece. Pollen spectra show a strong increase in arboreal pollen during the periods corresponding to the isotope stages 1, 5a, 5c, 5e and 7, which coincide with the occurrence of sapropels S_1, S_3, S_4, S_5 and S_7 respectively. These periods are considered as warm and humid. In the southern borderlands of the Mediterranean Sea, no vegetation record is available and well-dated environmental markers are very scarce. However, several indicators suggest the occurrence of a pluvial period during the last interglaciation: fossil shorelines attributed to this period are present along the coasts of Northern Africa (Pirazzoli 1987; Jedoui et al. 2001) and in Sardinia (Kindler et al. 1997). They are predominantly made of more or less consolidated fine quartz sands, which are associated with an increase in continental runoff. In addition, the development of an extensive palaeo-lake in Libya at the same time (Gaven et al. 1981) is in agreement with the hypothesis of a strong pluvial event.

Oxygen and carbon isotope records from speleothems in Israel (Bar-Matthews et al. 2000); dated by a large number of U/Th Thermal Ionisation Mass Spectrometry (TIMS) age determinations are used to reconstruct the rainfall and vegetation history in the eastern Mediterranean region during the last climatic cycle. Results show that the ages of the lowest $\delta^{18}O$ values (enhanced precipitation) and the highest $\delta^{13}C$ values (enhanced rock weathering due to large water flux) correspond to the estimated ages of the eastern Mediterranean sapropel layers S_1 to S_5.

During glacial isotopic stage 6, the Valle di Castiglione, Ioannina and Tenaghi Philippon pollen sequences exhibit a decrease in the arboreal pollen/non-arboreal pollen percentages. These data suggest particularly cold conditions. However, at Tenaghi Philippon, Wijmstra et al. (1990) and Mommersteeg et al. (1995), recognised during the time of sapropel S_6 the occurrence of a pollen assemblage representing open forest vegetation associated with a cooler, relatively wet climate with rain throughout the year despite the glacial conditions over Europe. On the other hand, the oxygen isotope record obtained on Argentarola cave stalagmite (Tyrrhenian coast of Italy) dated by U/Th TIMS between 180 and 170 kyr indicates an increase of precipitation in the Western Mediterranean Sea during the deposition of the sapropel event 6 (Bard et al. 2002).

Our data suggest therefore that sapropels S_1 to S_7 occurred when precipitation plus runoff were equivalent to or exceeded evaporation, so that the Mediterranean Sea was no longer a concentration basin. In most cases, this hydrological budget should rest on enhanced rainfall. The weakening of the Mediterranean intermediate and deep circulation and the interruption of winter overturning in the eastern basin were responsible for the sapropel formation in the eastern Mediterranean Sea. Sapropel deposits depend closely on the Mediterranean Sea hydrology and cannot be used to delimit wet periods that were geographically limited to a singular part of the Mediterranean region. Moreover, beyond these exceptional wet events, precipitation might also have increased during some periods in the Mediterranean Sea without being sufficient to balance the evaporation budget (Paterne et al. 1999). In such a case, the Mediterranean Sea would have acted as a concentration basin and form waters with densities higher than the subsurface water off Gibraltar. The active renewal of its deep and intermediate waters would prevent the establishment of anoxic conditions on the bottom and the preservation of organic matter. Only a high resolution reconstruction of the temporal and spatial evolution of the surface palaeo-salinities of the Mediterranean Sea and the comparison with the North Atlantic records would allow these hypotheses to be verified.

Summary

We have used the modern analogue technique to derive a sea surface temperature (SST) record of Mediterranean deep sea cores. The isotopic and sea surface temperature records of planktonic foraminifera were used to estimate the oxygen isotopic composition of surface water (δw) in the Tyrrhenian Sea and Levantine basins. Our results show a strong δw/salinity decrease in both basins associated with each sapropel of the last two climatic cycles. These δw decreases reflect a drastic change in the local fresh-water budget. A strong precipitation increase transformed the whole Mediterranean Sea into a non-concentration basin. The low surface salinity was responsible for the water column stratification. This resulted in a significant reduction of the rate of deep water oxygenation and the promotion of sapropel formation in the eastern basin. Other low δw values were also observed during a brief event at $\sim$140 kyr BP, the 6/5e transition, before the establishment of sapropel S_5, and during the last deglaciation. These events are not associated with an increase of the organic carbon content in the sediment (sapropel formation) and are induced by a salinity decrease of the incoming Atlantic surface water.

Acknowledgements

We thank B. Le Coat, J. Tessier for their help in the isotopic analyses and two anonymous reviewers for the detailed and constructive review of the manuscript. This work was supported by Laboratoire 3E (Tunisia) and by CNRS, CEA, PNEDC and the European program CLIVAMP. N.K. gratefully acknowledges the support of Ministère Tunisien de l'Enseignement Supérieur et de la Recherche Scientifique (DGRST), l'Ambassade de France à Tunis, l'Institut Français de Coopération, Tunis (Projet CMCU n° 00/F 1003) and PEP III program.

References

Adamson D.A., Gasse F., Street F.A. and Williams M.A.J. 1980. Late Quaternary history of the Nile. Nature 288: 50–55.

Bard E. 1988. Correction of accelerator mass spectrometry ^{14}C ages measured in planktonic foraminifera: paleoceanographic implications. Paleoceanography 3: 635–645.

Bard E., Arnold M., Maurice P., Duprat J., Moyes J. and Duplessy J.C. 1987. Retreat velocity of the North Atlantic polar front during the last deglaciation determined by ^{14}C accelerator mass spectrometry. Nature 328: 791–794.

Bard E., Delaygue G., Rostek F., Antonioli F., Silenzi S. and Schrag D.P. 2002. Hydrological conditions over the western Mediterranean basin during the deposition of the cold sapropel 6 (ca. 175 kyr BP). Earth Planet. Sci. Lett. 202: 481–494.

Bar-Matthews M., Ayalon A. and Kaufman A. 1997. Late Quaternary paleoclimate in the Eastern Mediterranean region from stable isotope analysis of speleothems at Soreq Cave, Israel. Quat. Res. 47: 155–168.

Bar-Matthews M., Ayalon A. and Kaufman A. 2000. Timing and hydrological conditions of Sapropel events in the Eastern Mediterranean, as evident from speleothems, Soreq cave, Israel. Chem. Geol. 169: 145–156.

Bé A.W.H. and Tolderlund D.S. 1971. Distribution and ecology of living foraminifera in surface waters of the Atlantic and Indian Oceans. In: Funnel B.M. and Riedel W.R. (eds), The Micropaleontology of Oceans, pp. 105–149.

Béthoux J.P. 1984. Paléo-hydrologie de la Méditerranée au cours des derniers 20 000 ans. Oceanol. Acta 7: 43–48.

Bond G., Broecker W., Johnsen S., McManus J., Labeyrie L., Jouzel J. and Bonani G. 1993. Correlations between climate records from North Atlantic sediments and Greenland ice. Nature 365: 143–147.

Braconnot P., Joussaume S., de Noblet N. and Ramstein G. 2000. Mid-Holocene and Last Glacial Maximum African monsoon changes as simulated within the Paleoclimate Modelling Intercomparison Project. Glob. Planet. Chan. 26: 51–66.

Bradley W.H. 1938. Mediterranean sediments and Pleistocene sea levels. Science 88: 376–379.

Callot Y. and Fontugne M. 1992. Les étagements de nappes dans les paléolacs holocènes du nord-est du Grand Erg Occidental (Algérie). C.R. Acad. Sci., Paris 315: 471–477.

Calvert S.E., Nielsen B. and Fontugne M.R. 1992. Evidence from nitrogen isotope ratios for enhanced productivity during formation of Eastern Mediterranean sapropels. Nature 359: 223–225.

Cita M.B., Vergnaud-Grazzini C., Robert C., Chamley H., Ciaranfi N. and d'Onofrio S. 1977. Paleoclimatic record of a long deep sea core from the eastern Mediterranean. Quat. Res. 8: 205–235.

Cortijo E., Yiou P., Labeyrie L. and Cremer M. 1995. Sedimentary record of rapid climatic variability in the North Atlantic Ocean during the last glacial cycle. Paleoceanography 10: 911–926.

Craig H. and Gordon A. 1965. Deuterium and oxygen 18 variations in the ocean and the marine atmosphere. In: Tongiorgi E. (ed.), Stable Isotopes in Oceanic Studies and Paleotemperatures. Spoleto, CNR, Pisa, pp. 9–130.

de Noblet N., Braconnot P., Joussaume S. and Masson V. 1996. Sensitivity of simulated Asian and African summer monsoons to orbitally induced variations in insolation 126, 115 and 6 k BP. Clim. Dyn. 12: 589–603.

Duplessy J.C., Delibrias G., Turon J.L., Pujol C. and Duprat J. 1981. Deglacial warming of the Northeastern Atlantic Ocean: correlation with the paleoclimatic evolution of the European continent. Palaeogeogr. Palaeoclimat. Palaeoecol. 35: 121–144.

Duplessy J.C., Labeyrie L.D., Juillet-Leclerc A., Maitre F., Duprat J. and Sarnthein M. 1991a. Surface salinity reconstruction of the North Atlantic Ocean during the last glacial maximum. Oceanol. Acta 14: 311–324.

Duplessy J.C., Bard E., Arnold M., Shackleton N.J., Duprat J. and Labeyrie L.D. 1991b. How fast did the ocean-atmosphere system run during the last deglaciation? Earth Planet. Sci. Lett. 103: 41–54.

Duplessy J.C., Labeyrie L.D., Arnold M., Paterne M., Duprat J. and van Weering T.C.E. 1992. Changes in surface salinity of the North Atlantic Ocean during the last deglaciation. Nature 358: 485–487.

Duplessy J.C., Bard E., Labeyrie L.D., Duprat J. and Moyes J. 1993. Oxygen isotope records and salinity changes in the Northeastern Atlantic Ocean during the Last 18 000 years. Paleoceanography 8: 341–350.

Epstein S., Buchsbaum R., Lowenstam H.A. and Urey H.C. 1953. Revised carbonate-water isotopic temperature scale. Geol. Soc. Am. Bull. 64: 1315–1325.

Fairbanks R.G. 1989. A 17 000-year glacio-eustatic sea level record: influence of glacial melting rates on the Younger event and deep ocean circulation. Nature 342: 637–642.

Follieri M., Magri D. and Narcisi B. 1990. A comparison between lithostratigraphy and palynology from the lacustrine sediments of Valle di Castiglione (Roma) over the last 0.25 MA. Mem. Soc. Geol. It. 45: 889–891.

Fontugne M., Paterne M., Calvert S.E., Murat A., Guichard F. and Arnold M. 1989. Adriatic deep water formation during the Holocene: implication for the reoxygenation of the deep Eastern Mediterranean Sea. Paleoceanography 4: 199–206.

Fontugne M. and Calvert S.E. 1992. Late Pleistocene variability of the carbon isotopic composition of organic matter in the Eastern Mediterranean: monitor of changes in carbon sources and atmosphere CO_2 concentrations. Paleoceanography 7: 1–20.

Gaven C., Hillaire-Marcel C. and Petit-Maire N. 1981. A Pleistocene lacustrine episode in South-Eastern Libya. Nature 290: 131–133.

Harrison S.P., Prentice I.C. and Bartlein P.J. 1991. What climate models can tell us about the Holocene palaeoclimates of Europe. In: Frenzel B. (ed.), Evaluation of Climate Proxy Data in Relation to the European Holocene. Gustav Fischer Verlag, Stuttgart and Jena, New York, pp. 285–299.

Hilgen F.J. 1987. Sedimentary rhythms and high-resolution chronostratigraphic correlations in the Mediterranean Pliocene. Newsl. Stratigraph. 17: 109–127.

Hilgen F.J. 1991. Astronomical calibration of Gauss to Matuyama sapropels in the Mediterranean and implication for the geomagnetic polarity time scale. Earth Planet. Sci. Lett. 104: 226–244.

Hilgen F.J., Lourens L.J., Berger A. and Loutre M.F. 1993. Evaluation of the astronomically calibrated time scale for the late Pliocene and earliest Pleistocene. Paleoceanography 8: 549–565.

Hutson W.H. 1980. The Aghulas current during the late Pleistocene: analysis of modern analogs. Science 207: 64–66.

Jedoui Y., Kallel N., Labeyrie L., Reyss J.L., Montacer M. and Fontugne M. 2001. Variabilité climatique rapide lors du dernier Interglaciaire (stade isotopique marin 5e) enregistrée dans les sédiments littoraux du Sud-Est tunisien. C.R. Acad. Sci., Paris 333: 733–740.

Kallel N., Paterne M., Duplessy J.C., Vergnaud-Grazzini C., Pujol C., Labeyrie L., Arnold M., Fontugne M. and Pierre C. 1997a. Enhanced rainfall on Mediterranean region during the last sapropel event. Oceanol. Acta 20: 697–712.

Kallel N., Paterne M., Labeyrie L.D., Duplessy J.C. and Arnold M. 1997b. Temperature and salinity records of the Tyrrhenian sea during the last 18 000 years. Palaeogeogr. Palaeoclimat. Palaeoecol. 135: 97–108.

Kallel N., Paterne M., Fontugne M., Labeyrie L.D., Duplessy J.C. and Montacer M. 2000. Mediterranean pluvial periods and sapropel formation during the last 200 000 years. Palaeogeogr. Palaeoclimat. Palaeoecol. 157: 45–58.

Kindler P., Davaud E. and Strasser A. 1997. Tyrrhenian coastal deposits from Sardinia (Italy): a petrographic record of high sea levels and shifting climate belts during the last interglacial (isotopic substage 5e), Palaeogeogr. Palaeoclimatol. Palaeoecol. 133: 1–25.

Kutzbach J.E. and Guetter P.J. 1986. The influence of changing orbital parameters and surface boundary conditions on climate simulations for the past 18 000 years. J. Atmos. Sci. 43: 1726–1759.

Labeyrie L.D., Duplessy J.C. and Blanc P.L. 1987. Variations in mode of formation and temperature of oceanic deep waters over the past 125 000 years. Nature 327: 477–482.

Levitus S. 1982. Climatological atlas of the world ocean. NOAA Professional Paper No. 13, Rockville, Maryland, USA.

Mangerud J. 1970. Late Weichselian vegetation and ice-front oscillations in the Bergen District, Western Norway. Nor. Geogr. Tidsskr. 24: 121–148.

Martinson D.G., Pisias N.G., Hays J.D., Imbrie J., Moore T.C. Jr. and Shackleton N.J. 1987. Age dating and the orbital theory of ice ages: development of a high-resolution 0 to 300 000 year chronostratigraphy. Quat. Res. 27: 1–29.

Masson V., Cheddadi R., Braconnot P., Joussaume S., Texier D. and PMIP Participants. 1999. Mid-Holocene climate in Europe: what can we infer from PMIP model data comparison? Clim. Dyn. 15: 163–182.

Mommersteeg H.J.P.M., Loutre M.F., Young R., Wijmstra T.A. and Hooghiemstra H. 1995. Orbital forced frequencies in the 975 000 year pollen record from Tenagi Philippon (Greece). Clim. Dyna. 11: 4–24.

Nicholson S.E. and Flohn H. 1980. African environmental and climatic changes and the general atmospheric circulation in late Pleistocene and Holocene. Climatic Change 2: 313–348.

Overpeck J.T., Webb III T. and Prentice I. 1985. Quantitative interpretation of fossil pollen spectra: dissimilarity coefficients and the method of modern analogs. Quat. Res. 23: 87–108.

Paterne M., Guichard F., Labeyrie J., Gillot P.Y. and Duplessy J.C. 1986. Tyrrhenian Sea tephrachronology of the oxygen isotope record for the past 60 000 years. Mar. Geol. 72: 259–285.

Paterne M., Guichard F. and Labeyrie J. 1988. Explosive Activity of the South Italian volcanoes during the past 80 000 years as determined by marine tephrochronology. Jour. Volcan. and Geother. Res. 34: 153–172.

Paterne M., Kallel N., Labeyrie L., Vautravers M., Duplessy J.C., Rossignol-Strick M., Cortijo E., Arnold M. and Fontugne M. 1999. Hydrological relationships between the North Atlantic Ocean and the Mediterranean Sea during the past 15–75 kyr. Paleoceanography 14: 626–638.

Petit-Maire N., Burollet P.F., Ballais J.L., Fontugne M., Rosso J.C. and Lazaar A. 1991. Paléoclimats holocènes du Sahara septentrional. Dépôts lacustres et terrasses alluviales en bordure du Grand Erg Oriental à l'extrême-sud de la Tunisie. C.R. Acad. Sci. Paris 312: 1661–1666.

Pirazzoli P.A. 1987. Sea-level changes in the Mediterranean. In: Tooley M.J. and Shennan I. (eds), Sea-Level Changes. Basil Blackwell, Oxford, pp. 152–181.

Prell W. 1985. The stability of low-latitudes sea surface temperatures: an evaluation of the CLIMAP reconstruction with emphasis on the positive SST anomalies. Technical Report. TR025, United States Department of Energy, Washington, D.C., 60 pp.

Prentice I.C. 1980. Multidimensional scaling as a research tool in Quaternary palynology: a review of theory and methods. Rev. Palaeobot. Palynol. 31: 71–104.

Pujol C. and Vergnaud-Grazzini C. 1989. Palaeoceanography of the last deglaciation in the Alboran Sea (Western Mediterranean). Stable isotopes and planktonic foraminiferal records. Marine Micropal. 15: 153–179.

Pujol C. and Vergnaud-Grazzini C. 1995. Distribution patterns of live planktic foraminifers as related to regional hydrography and productive systems of the Mediterranean Sea. Marine Micropal. 25: 187–217.

Ritchie J.C. and Haynes C.V. 1987. Holocene vegetation zonation in the Eastern Sahara. Nature 330: 645–647.

Ritchie J.C., Eyles C.H. and Haynes C.V. 1985. Sediment and pollen evidence for an early to mid-Holocene humid period in the Eastern Sahara. Nature 314: 352–354.

Rohling E.J. and Hilgen F.J. 1991. The eastern Mediterranean climate at times of sapropel formation: a review. Geol. in Minjbouw 70: 253–264.

Rohling E.J. and De Rijk S. 1999. Holocene Climate Optimum and Last Glacial Maximum in the Mediterranean: the marine oxygen isotope record. Mar. Geol. 153: 57–75.

Rohling E.J., Dulk M.D., Pujol C. and Vergnaud-Grazzini C. 1995. Abrupt hydrographic change in the Alboran Sea (western Mediterranean) around 8 000 yrs BP Deep Sea Res. 42: 1609–1619.

Rossignol-Strick M. 1985. Mediterranean Quaternary sapropels, an immediate response of the African Monsoon to variation of insolation. Palaeogeogr. Palaeoclim. Palaeoecol. 49: 237–263.

Sbaffi L., Wezel F.C., Kallel N., Paterne M., Cacho I., Ziveri P. and Shackleton N.J. 2001. Response of the pelagic environment to palaeoclimatic changes in the central Mediterranean Sea during the Late Quaternary. Mar. Geol. 178: 39–62.

Shackleton N.J. 1974. Attainment of isotopic equilibrium between ocean water and the benthonic foraminifera genus Uvigerina: isotopic changes in the ocean during the last glacial. Colloque CNRS n°219, Centre National de la Recherche Scientifique, Paris, pp. 203–210.

Shackleton N.J. 1987. Oxygen isotopes, ice volume and sea-level. Quat. Sci. Rev. 6: 183–190.

Starkel L. 1991. Fluvial environments as a source of information on climatic changes and human impact in Europe. In: Frenzel B. (ed.), Evaluation of Climate Proxy Data in Relation to the European Holocene. Gustav Fischer Verlag, Stuttgart, Jena and New York, pp. 241–254.

Street F.A. and Grove A.T. 1979. Global maps of lake-level fluctuations since 30 000 yr. BP Quat. Res. 12: 83–118.

Targarona J. 1997. Climatic and Oceanographic Evolution of the Mediterranean Region over the Last Glacial-Interglacial Transition. LLP Contributions series No. 7, Utrecht, 123 pp.

Thunell R.C. and Williams D.F. 1989. Glacial-Holocene salinity changes in the Mediterranean Sea: hydrographic and depositional effects. Nature 338: 493–496.

Tric E., Valet J.P., Tucholka P., Paterne M., Labeyrie L., Guichard F., Tauxe L. and Fontugne M. 1992. Paleointensity of the geomagnetic field during the last 80 000 years. J. Geophys. Res. 97: 9337–9351.

Tzedakis P.C. 1993. Long-term tree populations in Northwest Greece through multiple Quaternary climatic cycles. Nature 364: 437–440.

Vergnaud-Grazzini C., Caralp M., Faugères J.C., Gonthier E., Grousset F., Pujol C. and Saliège J.F. 1989. Mediterranean outflow through the Strait of Gibraltar since 18 000 years BP Oceanol. Acta 12: 305–324.

Wijmstra T.A. 1969. Palynology of the first 30 meters of a 120 m deep section in Northern Greece. Acta Bot. Neerl. 18: 511–527.

Wijmstra T.A., Young R. and Witte H.J.L. 1990. An evaluation of the climatic conditions during the Late Quaternary in Northern Greece by means of multivariate analysis of palynological data and comparison with recent phytosociological and climate data. Geol. Mijnbouw 69: 243–252.

Williams D.F., Thunell R.C. and Kennett J.P. 1978. Periodic fresh-water flooding and stagnation of the Eastern Mediterranean Sea during the Late Quaternary. Science 201: 252–254.

16. PALAEOENVIRONMENTAL CHANGES IN THE MEDITERRANEAN REGION 250–10 KYR BP

DONATELLA MAGRI (donatella.magri@uniroma1.it)
Dipartimento di Biologia Vegetale
Università "La Sapienza"
P.le Aldo Moro, 5, 00185 Roma
Italy

NEJIB KALLEL (nejib.kallel@fss.rnu.tn)
Faculté des Sciences de Sfax
B.P. 802, 3018 Sfax
Tunisia

BIANCAMARIA NARCISI (narcisi@casaccia.enea.it)
ENEA - C.R. Casaccia
PO Box 2400, 00100 Roma AD
Italy

Keywords: Mediterranean region, Climate variability, Upper Pleistocene, Sedimentary records, Palaeoenvironmental research

Introduction

The Mediterranean region is characterised by two features that make it especially interesting for palaeoclimatic studies over the last 250,000 years. On the one hand it has yielded a number of long terrestrial records, in some cases spanning with continuity more than one interglacial-glacial cycle. These complement the Mediterranean Sea records, providing a useful measure for comparison with global climatic signals. On the other hand, the palaeoclimatic evidence shows considerable complexity and sometimes conflicting features. These are likely to be a reflection of the large degree of heterogeneity that also at present characterises the geography and climate of the region.

The Mediterranean region: modern climatic variability

In the Afro-European transect PEP III, the general term "Mediterranean region" indicates a wide geographical area including Europe south of the Alps and Africa north of the Sahara desert, and extending from the eastern margin of the Atlantic Ocean to Iran (Fig. 1). The

325

R. W. Battarbee et al. (eds) 2004. *Past Climate Variability through Europe and Africa.*
Springer, Dordrecht, The Netherlands.

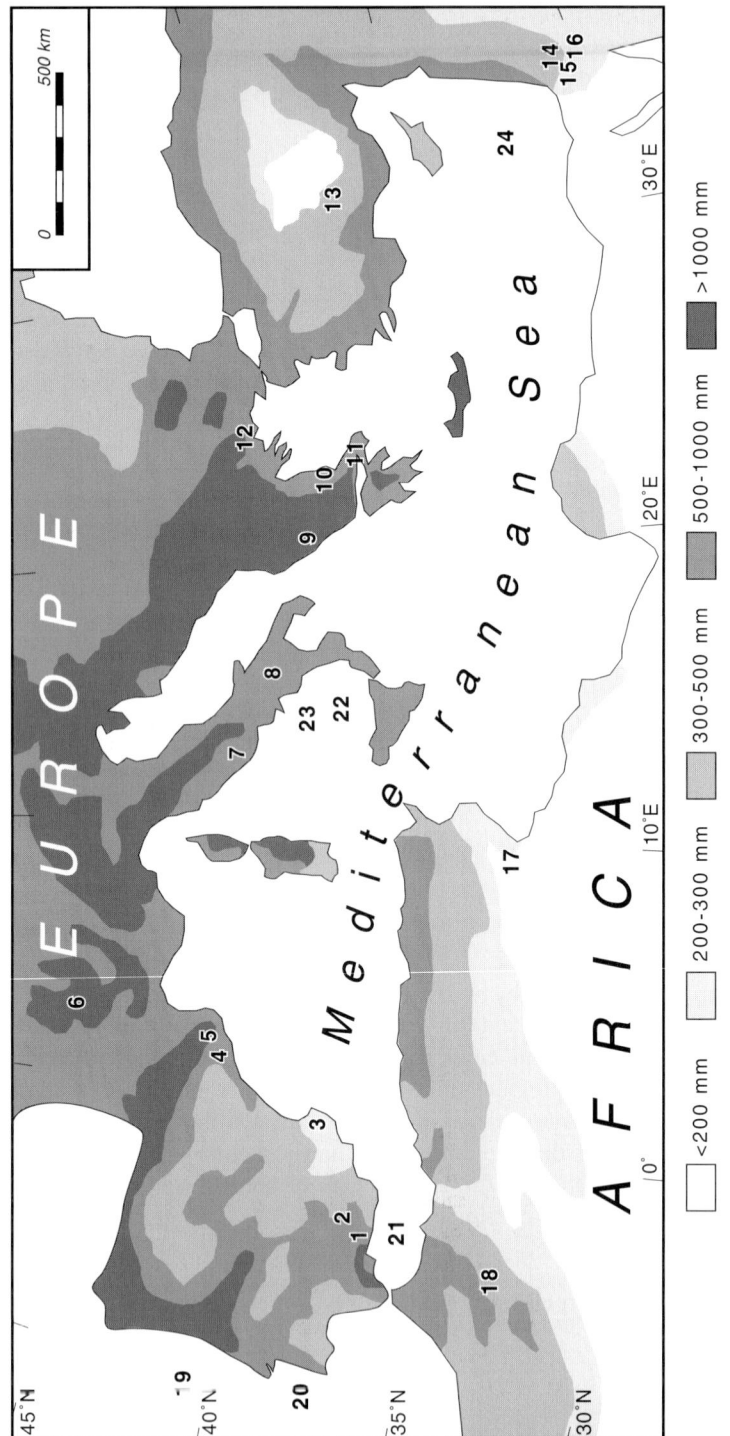

Figure 1 Mean annual precipitation around the Mediterranean basin (redrawn from Milliman et al. (1992)) and location of sites considered in the text: 1. Padul (Pons and Reille 1588); 2. Carihuela Cave (Carrión et al. 1999; 2000); 3. Beneito Cave (Carrión et al. 1999); 4. Abric Romaní (Burjachs and Julià 1994); 5. Lake Banyoles (Pérez-Obiol and Julià 1994); 6. Velay maars (Reille et al. 2000); 7. Lazio lakes (Follieri et al. 1988; Calderoni et al. 1994; Lowe et al. 1996; Ramrath et al. 1999a; Magri and Sadori 1999); 8. Lago Grande di Monticchio (Watts et al. 1996; Allen et al. 2000); 9. Ioannina (Tzedakis 1994); 10. Lake Xinias (Digerfeldt et al. 2000); 11. Lake Kopais (Tzedakis 1999; Okuda et al. 2001); 12. Tenaghi Philippon (Wijmstra 1969; Wijmstra and Smit 1976); 13. Konya Basin (Kuzucuoğlu et al. 1999; Roberts et al. 1999); 14. Jerusalern (Frumkin et al. 1999); 15. Soreq cave (Bar-Matthews et al. 1997; 1999; 2000); 16. Lake Lisan (Stein et al. 1997; Schramm et al. 2000; Bartov et al. 2002); 17. Matmata Plateau (Dearing et al. 2001); 18. Tigalmamine (Lamb et al. 1989); 19. Core MD952039 (Roucoux et al. 2001); 20. Core MD952042 (Sánchez Goñi et al. 1999; 2000a; b ; 21. Core MD952043 (Cacho et al. 1999; Sánchez Goñi et al. 2002); 22. Core KET 8003 (Rossignol-Strick and Planchais 1989; Paterne et al. 1999); 23. Cores KET 8004 and DED 8704 (Kallel et al. 2000); 24. Core MD 84-641 (Kallel et al. 2000).

Mediterranean climate is characterised by a prolonged and intense summer drought of at least two to three months, while rainfall occurs mainly in autumn and winter. This feature, considered at a global scale, represents a singularity. Temperature varies widely with latitude, altitude and continentality.

Although the whole area lies within the temperate climatic belt, and the term "Mediterranean region" immediately evokes a gentle climate with mild winters and sunny summers, the circum-Mediterranean areas are characterised by a marked climatic variability due to the contemporary presence of: (i) atmospheric perturbations originating outside the Mediterranean (Siberian, Saharan, North Atlantic and tropical Atlantic); (ii) cyclonic and anti-cyclonic thermal centres originating within the Mediterranean, a water body sufficiently large (about 2.5 million km^2) and deep (down to over 5100 m) to moderate the climate of the surrounding land; and (iii) atmospheric disturbances due to local orographic features, as considerable mountain chains are found in many Mediterranean countries. In addition, the Middle East region is influenced by the monsoonal system originating in the Indian Ocean. The effects of these different climatic factors interfere in different ways in space and time, by adding to or contrasting each other, generally with seasonal patterns.

Reconstructions of palaeoenvironmental variability cannot omit consideration of the balance between extra-regional or even global climatic factors influencing the Mediterranean and the regional and local phenomena. Awareness of such important climatic factors as distance from the Atlantic Ocean, the bathymetry of the Mediterranean Sea and relief, which have not significantly changed over the last hundreds of thousands of years, may appreciably help our understanding of the past climate variability due to global changes. A good example is provided by the pattern of modern precipitation values, ranging from a hundred millimetres per year at the margin of desert areas to over two thousand millimetres per year in mountainous areas of the Italian and Balkan peninsulas (Fig. 1). In the past, these values have certainly varied in relation to global climatic changes; however it may be reasonably supposed that significant climatic differences from one site to the other, due to the geographic characters of the region, have been maintained.

Palaeoenvironmental proxy-data: archives and approaches

Along the PEP III transect, the Mediterranean Basin is probably the geographical region where the collection of long continuous records has been the most conspicuous, particularly in the last ten years. This is certainly due to the interest of the scientific community in a region that is very sensitive to climate change, and therefore with a very high potential for significant advances in the interpretation of past climatic processes. Moreover, the Mediterranean region has proved particularly suitable for the study of long records, as it was never covered by extensive glaciations. It offers the chance to study uninterrupted sedimentary records during whole glacial periods, and in some cases during more than one interglacial-glacial cycle. In addition to the marine cores from the Mediterranean Sea, in continental areas long continuous records have been investigated not only from lakes, but also from speleothems, as their deposition was not inhibited by freezing during the glacials.

In the Mediterranean region a variety of sedimentary records have proved suitable for late-Quaternary palaeoenvironmental studies. These include investigations on sediments from different types of lacustrine basins (e.g., Pons and Reille (1988), Lamb et al. (1989),

Pérez-Obiol and Julià (1994), Guilizzoni and Oldfield (1996), Stein et al. (1997), Allen et al. (1999, 2000), Kuzucuoğlu et al. (1999), Roberts et al. (1999), Allen and Huntley (2000)), cave infillings (e.g., Burjachs and Julià (1994), Carrión et al. (1999), Woodward and Goldberg (2001)), speleothems (e.g., Bar-Matthews et al. (1997, 1999), Bar-Matthews and Ayalon (this volume), Frumkin et al. (1999)), travertines (e.g., Pentecost (1995)) and marine cores (e.g., Cita et al. (1977), Vergnaud-Grazzini et al. (1977, 1986), Williams et al. (1978), Rossignol-Strick (1985), Fontugne and Calvert (1992), Rohling and De Rijk (1999), Combourieu Nebout et al. (1999)) (Fig. 1). In many cases these are high-resolution, accurately-dated, multi-proxy records, often replicated in the same site or region, as in the case of the long lacustrine records from the Massif Central in France (Reille et al. 2000) and from the Lazio region in Italy (Calderoni et al. 1994; Follieri et al. 1998; Lowe et al. 1996; Ramrath et al. 1999a), and of the long oxygen isotopic record in speleothems from Israel (Bar-Matthews et al. 1997; Frumkin et al. 1999). In addition, a great number of investigations on coastal deposits (e.g., Sivan et al. (1999)), glacial deposits (e.g., Giraudi and Frezzotti (1997)), loess-palaeosol successions (e.g., Dearing et al. (2001), Günster and Skowronek (2001), Gvirtzman and Wieder (2001)) and fluvial sediments (e.g., Amorosi et al. (1999), Macklin et al. (2002)), all around the Mediterranean have substantially contributed in describing the Quaternary environments as very variable from site to site and highly sensitive to climatic changes.

Major difficulties arise when attempting to collect and collate all these data, which are often based on a very wide range of proxy data (geological, sedimentological, geophysical, geochemical and palaeobiological) and apply different methodological approaches. Moreover, they are sometimes dispersed in inaccessible publications and written in a considerable variety of languages.

A further element of uncertainty is the chronology of the sedimentary sequences. As the age of the last two interglacial-glacial cycles extends well beyond the reach of radiocarbon dating, other methods are often used that are not always easily compared with each other, including U/Th (e.g., Frumkin et al. (1999), Bar-Matthews et al. (2000), Schramm et al. (2000)); K/Ar and Ar/Ar dating (e.g., Roger et al. (1999), Ton-That et al. (2001)), annual lamination dating (e.g., Zolitschka and Negendank (1996), Ramrath et al. (1999a)), and synchronisation with orbital parameters e.g. (Follieri et al. 1988; Rossignol-Strick and Paterne 1999). Tephra layers, which are very frequent in the Mediterranean records (Narcisi and Vezzoli 1999) are a particularly useful tool for dating and making direct correlations between sites. In the case of lack of chronological information obtained directly from the study material, long-distance correlations with curves generally considered of global significance (e.g., SPECMAP stacked $\delta^{18}O$ record (Imbrie et al. 1984), GRIP ice core (Dansgaard et al. 1993)) are often suggested, with the aim of solving local problems of interpretation of results. However, this solution does not take into account the phase lags and response times between the local and the global records. In addition there is a risk of circular argument when a record chronologically bound to other records on the basis of their respective climate signals is used for further interpretations on how the local environment has responded to global climate variability.

The difficulty of setting the sequences in well-defined chronostratigraphical frameworks strongly reduces the number of records that can be advantageously used to infer past environmental variability in the Mediterranean. So, even if the number of studied sites is considerable, only a few of them are sufficiently long, continuous and well-dated to be considered fundamental for palaeoenvironmental interpretation and to support the setting of fragmentary data from other sites.

Another main limitation is that not all proxy data can be directly compared from site to site even within the same region. An example from the Italian peninsula may illustrate the question. Enhanced deposition of clastic material occurred in the presence of steppic vegetation during the pleniglacial in the crater lake of Lago Grande di Monticchio (Zolitschka and Negendank 1996), but not in the crater lake of Valle di Castiglione (Lazio region), where accumulation of detrital matter is typical of interglacial and interstadial forested periods (Follieri et al. 1990), suggesting that climatic information obtained from different sites should be interpreted taking into account local lithological and geomorphological conditions. Similarly, it is not always easy to define unequivocal responses of lake levels to climatic changes at a Mediterranean geographical scale. Thus, during the last glacial maximum there is evidence of high lake stands at sites in Central Italy (Giraudi 1989) and in the Near East (Bartov et al. 2002), while in Central Greece low lake levels are found (Digerfeldt et al. 2000). It clearly appears that multiple lines of evidence and high-quality data-sets with improved dating are needed. Moreover, generalisations of the lake-level status over the whole Mediterranean region are hazardous, because of local or regional climatic variability, geomorphological factors and tectonic history.

Among the climate proxies, pollen records appear generally suitable for comparisons within the circum-Mediterranean regions. In fact, although the floristic composition may be very different from site to site depending on the history of individual taxa, the variations of the vegetation structure, indicated by the general trends of deforestation and afforestation, are generally related to climatic changes, at least until human activity significantly modified the landscape. Although temperature variations between glacials and interglacials have certainly had some influence on vegetation, the main driving factor for forest expansion all around the Mediterranean was water availability, which is at present the main limiting factor for Mediterranean vegetation. Palaeohydrological evidence from other proxy data, such as isotopic records from speleothems (Frumkin et al. 1999; Bar-Matthews et al. 1999) and marine cores from the Mediterranean Sea (Kallel et al. 1997a; 2000; this volume) are of extreme importance to avoid the circular reasoning of inferring past humidity from vegetation changes and then using the reconstructed climate to interpret the observed vegetation variability. Similarly, it is particularly important to base the comparisons of long Mediterranean pollen records on independent chronologies. In this sense the Holocene pollen records from the circum-Mediterranean regions very clearly highlight that marked differences in the timing and extent of the early postglacial forest expansions and late-Holocene natural deforestation patterns are observed from site to site (Follieri et al. 2000), suggesting that considerable timing differences are to be expected also during forested periods of previous interglacials.

The response of the Mediterranean Sea to the climatic changes which affected the surrounding continental areas has been studied by the analysis of deep sea cores recovered from the basin. Many climatic signals have been used; among them, the oxygen isotope record of planktonic foraminifera ($\delta^{18}O$), sea surface temperature (SST) estimates, the organic carbon content of the sediment and the micropalaeontological analyses of benthic foraminifera (Cita et al. 1977; Williams et al. 1978; Rossignol-Strick 1985; Fontugne and Calvert 1992; Calvert et al. 1992; Kallel et al. 1997a; Cacho et al. 1999; Jorissen 1999). In the Mediterranean Sea, oxygen isotopic records show a pattern similar to that observed in the open ocean and display the classical glacial-interglacial alternation (Emiliani 1955). Correlation between the Mediterranean oxygen isotope records and the open ocean records is therefore reliable. During glacial periods, continental ice-sheet expansion (with a negative $\delta^{18}O$) is responsible for the increase of the mean oxygen isotope composition of the ocean water. As the Mediterranean Sea is linked to the open ocean through the Strait of Gibraltar, this signal is also recorded in the basin. However, the amplitude of glacial-interglacial $\delta^{18}O$ change in planktonic foraminifera is stronger in the Mediterranean basin than in the Ocean, and within the Mediterranean Sea this amplitude is higher in the eastern than in the western basin (Cita et al. 1977; Vergnaud-Grazzini et al. 1977; 1986; Williams et al. 1978; Rossignol-Strick 1985; Fontugne and Calvert 1992; Kallel et al. 2000).

On the other hand, Quaternary sediments of the Mediterranean Sea show organic-rich layers (sapropels), which are black sediments containing abundant planktonic foraminifera, but are devoid of benthic fossils (McCoy 1974; Cita et al. 1977; Williams et al. 1978; Parisi and Cita 1982; Nolet and Corliss 1990; Fontugne and Calvert 1992). These sapropels appeared in both eastern and western basins, but they have been mainly studied in the eastern basin. During these events, the shallowest dwelling species exhibit a significant $\delta^{18}O$ decrease (Cita et al. 1977; Vergnaud-Grazzini et al. 1977; 1986; Williams et al. 1978; Rossignol-Strick 1985; Fontugne and Calvert 1992; Rohling and De Rijk 1999), which may be explained by either increased sea surface temperature (SST) or higher precipitation (with a low $^{18}O/^{16}O$ ratio).

By estimating independently past SST variations (Kallel et al. 1997a; Cacho et al. 1999), foraminiferal $\delta^{18}O$ can be used to estimate past variations of surface water $\delta^{18}O$ (δ_w) by solving the palaeotemperature equation (Epstein et al. 1953; Shackleton 1974; Duplessy et al. 1991). Mediterranean surface water $\delta^{18}O$ variations reflect both changes in the $\delta^{18}O$ value of the incoming Atlantic water and variations of the freshwater budget of the Mediterranean basin (Kallel et al. 1997b) and Mediterranean δ_w records can be used as a proxy for surface salinity variations.

Finally, within the PEP III transect the Mediterranean region has the invaluable potential of direct land-sea correlations when pollen analysis is made of marine cores. Vegetation changes can be directly framed in the marine isotopic chronology and used to indicate phase relationships between global ice volume and palaeoenvironmental changes on land. Future pollen studies on long marine records will certainly resolve some of the problems presently encountered, including lower resolution than for terrestrial records, incompleteness during the glacial periods in the eastern Mediterranean, where pollen grains are well preserved only in the sapropel layers (Cheddadi and Rossignol-Strick 1995a), and the lack of reference continental pollen sites in close proximity to the marine core, making it difficult to assess the reliability of representation of pollen changes in marine cores with respect to the neighbouring terrestrial situation.

Timescales of palaeoenvironmental variability

The long time interval considered in Time Stream 2, from 250 kyr to the Holocene, inevitably concerns a variety of palaeoenvironmental problems, which are tackled in different degrees of detail. In fact, while only sparse continuous records are available for the whole period, an increasing number and variety of proxy data have been collected for the most recent time periods from all over the Mediterranean. While the reduced number of continuous terrestrial records for the glacial period corresponding to Marine Isotope Stage (MIS) 6 allows only a rough evaluation of its duration and importance, the wealth of data available for the last glacial period allows reconstruction of millennial-scale variations and correlation with regional and extra-regional climatic events.

Although it is clear that climate variability is the effect of a multitude of mechanisms interacting at various spatial and temporal scales, we are forced to schematise our knowledge and identify a number of major climate themes at specific geographical and temporal scales. We will therefore discuss a few examples of climate variability in the Mediterranean region at three different timescales (10^5, 10^4 and 10^3 years), which involve different palaeoclimatic topics, quality and quantity of archives, methodological approaches, and interpretations in terms of dynamics of climate variability (trends, cycles, events).

The 100,000-year time-scale

In addition to a few marine cores from the Mediterranean Sea (Cheddadi and Rossignol-Strick 1995b; Rossignol-Strick and Paterne 1999), five long lacustrine sequences between 38 °N and 45 °N (Bouchet/Praclaux (Reille et al. 2000), Valle di Castiglione: (Follieri et al. 1988), Ioannina 249 (Tzedakis 1994), Kopais (Okuda et al. 2001), Tenaghi Philippon (Wijmstra 1969, Wijmstra and Smit 1976)) provide an opportunity to develop and correlate with each other complete, high-resolution records of vegetation changes over multiple interglacial-glacial cycles. The time-scale of the sequences, although supported by various independent sources of evidence, including radiocarbon, U-series and Ar/Ar dating, counts of annual laminations, correlation with orbital parameters and palaeomagnetic analyses, is eventually tuned to the well-established marine chronostratigraphy (Tzedakis et al. 1997). In this way the opportunity to develop a terrestrial chronostratigraphical scheme for Southern Europe is missed in favour of a solution that is inadequate from the stratigraphical point of view (Gibbard and West 2000), but more easily used by the palaeoclimatic scientific community, as it is based on the well-known marine notation when referring to specific terrestrial temperate stages. On the other hand, a sufficiently precise terrestrial time-scale for the two last interglacial-glacial cycles is still far from being developed despite recent important advances in geochronological tools.

Although not strictly time parallel, the stratigraphical scheme in which the long south European pollen records are now framed (Tzedakis et al. 1997) raises a number of issues deserving further discussion:

i) divergence in the relative amplitude of oscillations recorded by pollen and oxygen isotope curves, suggesting, for example, that major vegetation changes occurred in the Mediterranean region even when ice volume changes appear to have been moderate (Tzedakis et al. 1997);

ii) different behaviour of various tree taxa during successive stages, importance of their diagnostic value for chronostratigraphical purposes and limitations due to the inherent variability of vegetation patterns as determined by local factors and historical aspects (Tzedakis et al. 2001);

iii) identification of the ecological processes underlying the vegetation-orbital relationships, for an improved understanding of the interaction between the biosphere and the climatic system (Magri and Tzedakis 2000);

iv) significance of the long Mediterranean terrestrial records for interpreting fragmentary records from a wider geographic area, including Central-Northern Europe and North Africa.

On the marine side the following issues may be highlighted:

i) difference in the timing of the hydrological changes reconstructed from marine cores in the two major Mediterranean basins. This difference is now well established for the Holocene wet period (Kallel et al., unpublished).

ii) significance of the Mediterranean hydrological records during the last two climatic cycles. A long isotopic record off Gibraltar is necessary to separate changes due to the North Atlantic hydrology variations from the local Mediterranean signal which is linked to the freshwater budget changes in the sea.

The 10,000-year time-scale

Special attention has been paid in the last few years to the study of records of the last interglacial in the Mediterranean, reflecting the interest that this time-period has received by the European scientific community, as an analogue for the present interglacial and a potential tool for predicting future climatic trends. The effort made to recognize events of climatic instability in Central Europe during the Eemian, and to evaluate their length and extent (e.g., Field et al. (1994), Thouveny et al. (1994), Cheddadi et al. (1998), Kukla et al. (1997)) has been paralleled by new high-resolution studies in the Mediterranean regions, where most of the last interglacial records have the advantage of being part of longer sequences, offering a more secure chronostratigraphical setting and allowing preservation of the immediately preceding and following climatic events.

Outstanding results from the Mediterranean region include:

i) Isotopic and palynological data from lacustrine sediments in Greece (Frogley et al. 1999; Tzedakis 1999; 2000) confirm patterns of climatic instability both during the late-glacial interval and in the second part of the interglacial, and reveal considerable regional variability, which appears to reflect closely differences in climatic regimes found in Greece today and suggesting the presence of similar climatic patterns during the last interglacial.

ii) Pollen analyses of a marine core off Portugal (Sánchez Goñi et al. 1999; 2000b) provide direct land-sea correlations and show that the Eemian interglacial, as recognized by the pollen assemblages of the southwestern Iberian margin, spans from the lightest isotopic values of stage-5e (ca. 126 kyr BP) to the heavier isotopic values towards the 5e-5d transition.

iii) Many Mediterranean deep sea cores have been studied to reconstruct the hydrological conditions during the last 200 kyr (Fontugne and Calvert 1992; Paterne et al. 1986; 1988; Tric et al. 1992; Kallel et al. 2000; this volume). The records reveal six periods of high organic carbon values, at about 8, 80, 96, 122, 170, and 195 kyr BP, corresponding to sapropels S_1 to S_7. Strong δ_w/salinity decreases in both western and eastern basins are found to be associated with each sapropel (Kallel et al. 2000; this volume). These δ_w decreases reflect a drastic change in the local freshwater budget and suggest that sapropels occurred when precipitation plus runoff was high and nearly equalled or superseded evaporation, so that the Mediterranean Sea was no longer a concentration basin. The low surface salinity was responsible for the water column stratification. It resulted in a significant reduction of the rate of deep water oxygenation and sapropel formation in the eastern basin.

iv) Well-dated oxygen isotope records from speleothems in Israel (Frumkin et al. 1999; Bar-Matthews et al. 1997) show that the lowest $\delta^{18}O$ values of the last 170,000 years are found during the last interglacial, parallelling ice and marine records; the combination of low $\delta^{18}O$ values and high $\delta^{13}C$ values is interpreted as indicative of high rainfall, which would also agree with the causes for the coeval formation of sapropels in the eastern Mediterranean Sea (Bar-Matthews et al. 2000; Bar-Matthews and Ayalon, this volume).

v) Radiochemically dated coastal marine deposits in Southeastern Tunisia (Jedoui et al. 2001) are composed of two distinct lithostratigraphic units separated by an erosion surface. Humid conditions at the beginning of the last interglacial were responsible for a supply of terrigenous material and a siliciclastic sedimentation. The regression of these wet conditions during the second half of the last interglacial favoured carbonate sedimentation.

The 1000-year time-scale

A major question is the evaluation of the possible impact that North Atlantic millennial scale climate variability, including Dansgaard-Oeschger cycles and Heinrich events, may have had on the Mediterranean climate and environment. For the interval corresponding to MIS 3 a considerable number of continuous records are available for the Mediterranean region, enhancing also the quality of information (variety of proxy data, detail of analysis and chronological control).

These have led to different views and methodological approaches. Here we discuss the main results concerning the study of the Heinrich events.

Paterne et al. (1999) working on the ^{14}C AMS well-dated core KET80-03 recovered in the Tyrrhenian Sea, and Cacho et al. (1999) working on the Alboran Sea core MD952043, identify various types of imprints of Heinrich events in the Mediterranean sediments. In the Tyrrhenian Sea, these events are marked by cold surface water temperatures (micropalaeontological transfer function) and low salinities. The hydrological modifications observed in the North Atlantic Ocean were directly transferred to the Mediterranean Sea without any amplification as indicated by the salinity gradient between the two marine areas which was equivalent to the modern one (Paterne et al. 1999). Following each of these events, a rapid rise in temperature of Mediterranean surface waters occurred as in the North Atlantic Ocean. However, surface salinities increased more strongly in the Mediterranean Sea than in the Atlantic Ocean. This modification of the local Mediterranean hydrology was probably linked to a decrease in the precipitation and/or to an increase of the evaporation. After that, the Mediterranean climate becomes again gradually cooler and wetter. In the Alboran Sea (Cacho et al. 1999), SST evolution inferred from the study of C_{37} alkenones shows extremely rapid warming and cooling events through the last glacial period. Five prominent cooling episodes are found to be associated with Heinrich events H1 to H5.

Recent palynological analyses of marine cores (Paterne et al. 1999; Roucoux et al. 2001; Sánchez Goñi et al. 2000a; 2002) suggest that the effects of Heinrich events are not only registered in the sea, but also in the terrestrial environments, on the basis of direct land/sea correlations. Other authors correlate millennial scale climatic events on land to the Heinrich events of the Atlantic Ocean on the basis of independent chronologies, taking into account radiocarbon ages, varve counting, calculation of sediment accumulation rates and tephrochronology (Ramrath et al. 1999b) or ^{230}Th-^{234}U (TIMS) ages (Bar-Matthews et al. 1999). Still other authors, working on Mediterranean terrestrial sequences showing clear millennial-scale-variability during the last glacial period, prefer to be cautious in suggesting long-distance correlations with extra-regional marine or ice-core records, so as not to favour the dissemination of unproven knowledge (Magri 1999; Magri and Sadori 1999).

On the whole, the results of these investigations are not always in good agreement. For example, the calculated time-intervals of the events recognised as H3, H4 and H5 in the marine cores MD95-2042 (Sánchez Goñi et al. 2000a) and KET 80-03 (Paterne et al. 1999), and in the crater lake of Lago Grande di Monticchio (Watts et al. 1996) do not even overlap. This observation confirms the difficulty of correlating these climatic events from different records and so extending the "event stratigraphy" (Walker et al. 1999; Lowe et al. 2001) to time intervals at the limit of the radiocarbon method.

In addition to chronological problems, the studied sites in the Mediterranean region point to various climatic patterns in relation to the Heinrich events. For example off the Portuguese coast (Sánchez Goñi et al. 2000a) the peaks of Heinrich events are associated with vegetation features interpreted as occurring in a cold, dry climate, while in the Tyrrhenian core KET 80-03 cold sea-surface temperatures and low salinity are inferred, with cool and relatively humid summers and low seasonal thermal contrast (Paterne et al. 1999).

Present knowledge of the possible effects of the Heinrich events in the Mediterranean suggests some general considerations, which may be applied also to other climatic patterns: the data hitherto collected are still too sparse to be conclusive, leading to contradictory interpretations of the climatic variability. However, conflicting evidence might also be amplified by the complex geographical setting of the Mediterranean region.

Discussion and highlights of prospective research development

Compared to the other regions of the PEP III transect, the Mediterranean displays characters that raise specific questions to be considered in the present and future discussion of the past climatic variability over the last interglacial-glacial cycles. In particular, the modern geographic and climatic situation draws attention to three major topics: (i) a mosaic of climatic and environmental situations on land; (ii) a significant climatic influence of the Mediterranean Sea; and (iii) seasonal interference of extra-regional, regional and local climatic phenomena. To some extent, the data hitherto collected show that the effects of these different sources of climatic variability can be traced also in the palaeoenvironmental records of the last two interglacial-glacial cycles.

The geographical variability of the records is a common topic for all fields of palaeoen-vironmental research in the Mediterranean, including geological, sedimentological, geo-physical, geochemical and palaeobiological investigations; it is clear that an attentive consideration of modern local conditions of each site in comparison with other studied Mediterranean areas is indispensable and preliminary to any palaeoclimatic interpretation.

The Mediterranean Sea appears to have played an important climatic role in the past as at present, for example affecting the precipitation cycles of the eastern Mediterranean regions (Frumkin et al. 1999; Bar-Matthews and Ayalon, this volume).

Climatic seasonality certainly underwent important changes in the past. For example in the marine cores, SST reconstructions show that seasonality was weaker during cold periods (Paterne et al. 1999; Kallel et al., this volume). This aspect of palaeoclimatic research, which is too often neglected, has great potential for the reconstruction of past environmental variability, especially in the Mediterranean. The pollen ratio of deciduous versus evergreen trees and shrubs, successfully applied to Holocene sequences in the Iberian Peninsula (Jalut et al. 2000), may be a possible interpretative approach of past seasonality changes also for long pollen records.

Extra-regional climatic influences in the Mediterranean region are testified by many studies documenting the effects of the North Atlantic marine and atmospheric circulation, above all in the western Mediterranean, and to some extent also by the monsoonal regime in the eastern Mediterranean. However, other climatic influences need to be explored in more detail, for example the importance of atmospheric perturbations from the Sahara, which may have played a role in dust loading (Narcisi 2000) and in vegetation changes (Magri and Parra 2002) in the western Mediterranean through glacial-interglacial cycles. In general, comparisons of the Mediterranean region with the neighbouring regions of the PEP III transect and particularly the sub-tropical and northern regions of Africa are still largely unexplored and deserve special attention in future research topics.

From the methodological point of view, the most recent palaeoclimatic research in the Mediterranean region indicates that although many published records provide high quality data, it is clear that none of the record on its own may be considered entirely complete and reliable, and that new progress may come from a comparison of different records and different proxies. It appears of primary importance not only to assess the similarities among sites within the Mediterranean regions and with extra-regional long records of global change, but also to discuss differences and inconsistencies. This will help to highlight not only the frequency and duration of the climatic fluctuations, but also the underlying processes and trends.

Summary

This chapter presents an overview of the current state of palaeoenvironmental research in the Mediterranean region over the last 250 kyr. It includes information from different geographical contexts and different fields of research, and critically reviews methodological approaches and chronostratigraphical correlations. The mosaic of climatic and environmental situations on land, the significant influence of the Mediterranean Sea and the seasonal impacts of extra-regional, regional and local climatic phenomena are shown to fashion not only the modern environment, but also its past climatic variability. Some examples of climate variability at three different time-scales (10^5, 10^4 and 10^3 years), which involve different palaeoclimatic themes, the quality and quantity of different archives, different methodological approaches and their interpretation in terms of climate dynamics are discussed. Possible future directions of research, specific to this region, are highlighted.

Acknowledgements

DM acknowledges the financial support from the Commission of the European Communities (project FOSSILVA, EVK2-CT-1999-00036). NK thanks the Tunisian DGRST, the Institut Français de Coopération at Tunis (CMCU project n° 00/F1003) and the LSCE at

Gif-sur-Yvette. Comments from M. Follieri, I. Parra, P.C. Tzedakis and an anonymous referee are gratefully acknowledged.

References

Allen J.R.M. and Huntley B. 2000. Weichselian palynological records from Southern Europe: correlation and chronology. Quat. Int. 73/74: 111–125.

Allen J.R.M., Brandt U., Brauer A., Hubberten H.-W., Huntley B., Keller J., Kraml M., Mackensen A., Mingram J., Negendank J.F.W., Nowaczyk N.R., Oberhänsli H., Watts W.A., Wulf S. and Zolitschka B. 1999. Rapid environmental changes in Southern Europe during the last glacial period. Nature 400: 740–743.

Allen J.R.M., Watts W.A. and Huntley B. 2000. Weichselian palynostratigraphy, palaeovegetation and palaeoenvironment; the record from Lago Grande di Monticchio, Southern Italy. Quat. Int. 73/74: 91–110.

Amorosi A., Colalongo M.L., Fusco F., Pasini G. and Fiorini F. 1999. Glacio-eustatic control of continental-shallow marine cyclicity from Late Quaternary deposits of the southeastern Po Plain, Northern Italy. Quat. Res. 52: 1–13.

Bar-Matthews M. and Ayalon A., this volume. Speleothems as palaeoclimate indicators, a case study from Soreq cave located in the Eastern Mediterranean Region, Israel. In: Battarbee R.W., Gasse F. and Stickley C.E. (eds), Past Climate Variability through Europe and Africa. Kluwer Academic Publishers, Dordrecht, the Netherlands, pp. 363–391.

Bar-Matthews M., Ayalon A. and Kaufman A. 1997. Late Quaternary paleoclimate in the Eastern Mediterranean region from stable isotope analysis of speleothems at Soreq cave, Israel. Quat. Res. 47: 155–168.

Bar-Matthews M., Ayalon A. and Kaufman A. 2000. Timing and hydrological conditions of sapropel events in the Eastern Mediterranean, as evident from speleothems, Soreq cave, Israel. Chem. Geol. 169: 145–156.

Bar-Matthews M., Ayalon A., Kaufman A. and Wasserburg G.J. 1999. The Eastern Mediterranean paleoclimate as a reflection of regional events: Soreq cave, Israel. Ear. Planet. Sci. Lett. 166: 85–95.

Bartov Y., Stein M., Enzel Y., Agnon A. and Reches Z. 2002. Lake levels and sequence stratigraphy of Lake Lisan, the late Pleistocene precursor of the Dead Sea. Quat. Res. 57: 9–21.

Burjachs F. and Julià R. 1994. Abrupt climatic changes during the last glaciation based on pollen analysis of the Abric Romani, Catalonia, Spain. Quat. Res. 42: 308–315.

Cacho I., Grimalt J.O., Pelejero C., Canals M., Sierro F.J., Flores J.A. and Shackleton N. 1999. Dansgaard-Oeschger and Heinrich event imprints in Alboran Sea paleotemperatures. Paleoceanography 14: 698–705.

Calderoni G., Carrara C., Ferreli L., Follieri M., Gliozzi E., Magri D., Narcisi B., Parotto M., Sadori L. and Serva L. 1994. Palaeoenvironmental, palaeoclimatic and chronological interpretations of a late-Quaternary sediment core from Piana di Rieti (Central Apennines, Italy). Giorn. Geologia 56: 43–72.

Calvert S.E., Nielsen B. and Fontugne M.R. 1992. Evidence from nitrogen isotope ratios for enhanced productivity during formation of Eastern Mediterranean sapropels. Nature 359: 223–225.

Carrión J.S., Munuera M., Navarro C., Burjachs F., Dupré M. and Walker M.J. 1999. The palaeoecological potential of pollen records in caves: the case of Mediterranean Spain. Quat. Sci. Rev. 18: 1061–1073.

Carrión J.S., Munuera M., Navarro C. and Sáez F. 2000. Paleoclimas e historia de la vegetación cuaternaria en España a través del análisis polínico. Complutum 11: 115–142.

Cheddadi R. and Rossignol-Strick M. 1995a. Improved preservation of organic matter and pollen in eastern Mediterranean sapropels. Paleoceanography 10: 301–309.

Cheddadi R. and Rossignol-Strick M. 1995b. Eastern Mediterranean Quaternary paleoclimates from pollen and isotope records of marine cores in the Nile cone area. Paleoceanography 10: 291–300.

Cheddadi R., Mamakowa K., Guiot J., de Beaulieu J.-L., Reille M., Andrieu V., Granoszewski W. and Peyron O. 1998. Was the climate of the Eemian stable? A quantitative climate reconstruction from seven European pollen records. Palaeogeogr. Palaeoclim. Palaeoecol. 143: 73–85.

Cita M.B., Vergnaud-Grazzini C., Robert C., Chamley H., Ciaranfi N. and d'Onofrio S. 1977. Paleoclimatic record of a long deep sea core from the eastern Mediterranean. Quat. Res. 8: 205–235.

Combourieu Nebout N., Londeix L., Baudin F., Turon J.-L., von Grafenstein R. and Zahn R. 1999. Quaternary marine and continental paleoenvironments in the western Mediterranean (Site 976, Alboran Sea): palynological evidence. Proceedings of the Ocean Drilling Program, Scientific Results 161: 457–468.

Dansgaard W., Johnsen S.J., Clausen H.B., Dahl-Jensen D., Gundestrup N.S., Hammer C.U., Hvidberg C.S., Steffensen J.P., Sveinbjörnsdottir A.E., Jouzel J. and Bond G. 1993. Evidence for general instability of past climate from a 250-kyr ice-core record. Nature 364: 218–220.

Dearing J.A., Livingstone I.P., Bateman M.D. and White K. 2001. Palaeoclimate records from OIS 8.0-5.4 recorded in loess-palaeosol sequences on the Matmata Plateau, Southern Tunisia, based on mineral magnetism and new luminescence dating. Quat. Int. 76/77: 43–56.

Digerfeldt G., Olsson S. and Sandgren P. 2000. Reconstruction of lake-level changes in lake Xinias, Central Greece, during the last 40000 years. Palaeogeogr. Palaeoclim. Palaeoecol. 158: 65–82.

Duplessy J.C., Labeyrie L.D., Juillet-Leclerc A., Maître F., Duprat J. and Sarnthein M. 1991. Surface salinity reconstruction of the North Atlantic ocean during the last glacial maximum. Oceanologica Acta 14: 311–324.

Emiliani C. 1955. Pleistocene temperatures. J. Geol. 63: 538–578.

Epstein S., Buchsbaum R., Lowenstam H.A. and Urey H.C. 1953. Revised carbonate-water isotopic temperature scale. Geol. Soc. Am. Bull. 64: 1315–1325.

Field M., Huntley B. and Müller H. 1994. Eemian climate fluctuations observed in a European pollen record. Nature 371: 779–783.

Follieri M., Magri D. and Sadori L. 1988. 250,000-year pollen record from Valle di Castiglione (Roma). Pollen Spores 30: 329–356.

Follieri M., Magri D. and Narcisi B. 1990. A comparison between lithostratigraphy and palynology from the lacustrine sediments of Valle di Castiglione (Roma) over the last 0.25 MA. Mem. Soc. geol. it. 45: 889–891.

Follieri M., Giardini M., Magri D. and Sadori L. 1998. Palynostratigraphy of the last glacial period in the volcanic region of Central Italy. Quat. Int. 47–48: 3–20.

Follieri M., Roure J.M., Giardini M., Magri D., Narcisi B., Pantaleon-Cano J., Pérez-Obiol R., Sadori L. and Yll E.I. 2000. Desertification trends in Spain and Italy based on pollen analysis. In: Balabanis P., Peter D., Ghazi A. and Tsogas M. (eds), Mediterranean Desertification — Research Results and Policy Implications. Vol. 2. European Commission, Bruxelles, pp. 33–44.

Fontugne M. and Calvert S.E. 1992. Late Pleistocene variability of the carbon isotopic composition of organic matter in the Eastern Mediterranean: monitor of changes in carbon sources and atmosphere CO_2 concentrations. Paleoceanography 7: 1–20.

Frogley M.R., Tzedakis P.C. and Heaton T.H.E. 1999. Climate variability in Northwest Greece during the last interglacial. Science 285: 1886–1889.

Frumkin A., Ford D.C. and Schwarcz H.P. 1999. Continental oxygen isotopic record of the last 170,000 years in Jerusalem. Quat. Res. 51: 317–327.

Gibbard P.L. and West R.G. 2000. Quaternary chronostratigraphy: the nomenclature of terrestrial sequences. Boreas 29: 329–336.

Giraudi C. 1989. Lake levels and climate for the last 30,000 years in the Fucino area (Abruzzo-Central Italy) — A review. Palaeogeogr. Palaeoclim. Palaeoecol. 70: 249–260.

Giraudi C. and Frezzotti M. 1997. Late Quaternary glacial events in the Central Apennines, Italy. Quat. Res. 48: 280–290.

Guilizzoni P. and Oldfield F. (eds) 1996. Palaeoenvironmental analysis of Italian crater lake and Adriatic sediments (PALICLAS). Mem. Ist. it. Idrob. 55: 1–357.

Günster N. and Skowronek A. 2001. Sediment-soil sequences in the Granada Basin as evidence for long- and short-term climatic changes during the Pliocene and Quaternary in the Western Mediterranean. Quat. Int. 78:17–32.

Gvirtzman G. and Wieder M. 2001. Climate of the last 53,000 years in the eastern Mediterranean, based on soil-sequence stratigraphy in the coastal plain of Israel. Quat. Sci. Rev. 20: 1827–1849.

Imbrie J., Hays J.D., Martinson D.G., McIntyre A., Mix A.C., Morley J.J., Pisias N.G., Prell W.L. and Shackleton N.J. 1984. The orbital theory of Pleistocene climate: support from a revised chronology of the marine [18]O record. In: Berger A.L., Imbrie J., Hays J., Kukla G. and Saltzman B. (eds), Milankovitch and Climate. Reidel, Dordrecht, pp. 269–305.

Jalut G., Esteban Amat A., Bonnet L., Gauquelin T. and Fontugne M. 2000. Holocene climatic changes in the Western Mediterranean, from South-East France to South-East Spain. Palaeogeogr. Palaeoclim. Palaeoecol. 160: 255–290.

Jedoui Y., Kallel N., Labeyrie L., Reyss J.-L., Montacer M. and Fontugne M. 2001. Variabilité climatique rapide lors du dernier Interglaciaire (stade isotopique marin 5e), enregistrée dans les sédiments littoraux du Sud-Est tunisien. C.R. Acad. Sci. Paris / Earth Planet. Sci. 333: 733–740.

Jorissen F.J. 1999. Benthic foraminiferal successions across late Quaternary Mediterranean sapropels. Mar. Geol. 153: 91–101.

Kallel N., Paterne M., Duplessy J.C., Vergnaud-Grazzini C., Pujol C., Labeyrie L., Arnold M., Fontugne M. and Pierre C. 1997a. Enhanced rainfall on Mediterranean region during the last sapropel event. Oceanologica Acta 20: 697–712.

Kallel N., Paterne M., Labeyrie L.D., Duplessy J.C. and Arnold M. 1997b. Temperature and salinity records of the Tyrrhenian sea during the last 18000 years. Palaeogeogr. Palaeoclim. Palaeoecol. 135: 97–108.

Kallel N., Paterne M., Fontugne M., Labeyrie L.D., Duplessy J.C. and Montacer M. 2000. Mediterranean pluvial periods and sapropel formation during the last 200,000 years. Palaeogeogr. Palaeoclim. Palaeoecol. 157: 45–58.

Kallel N., Duplessy J., Labeyrie L., Fontugne M. and Paterne M., this volume. Mediterranean Sea palaeohydrology and pluvial periods during the Late Quaternary. In: Battarbee R.W., Gasse F. and Stickley C.E. (eds), Past Climate Variability through Europe and Africa. Kluwer Academic Publishers, Dordrecht, the Netherlands, pp. 307–324.

Kukla G., McManus J.F., Rousseau D.-D. and Chuine I. 1997. How long and stable was the last interglacial? Quat. Sci. Rev. 16: 605–612.

Kuzucuoğlu C., Bertaux J., Black S., Denèfle M., Fontugne M., Karabiyikoğlu M., Kashima K., Limondin-Lozouet N., Mouralis D. and Orth P. 1999. Reconstruction of climatic changes during the Late Pleistocene, based on sediment records from the Konya Basin (Central Anatolia, Turkey). Geological J. 34: 175–198.

Lamb H.F., Eicher U. and Switsur V.R. 1989. An 18,000-year record of vegetation, lake-level and climatic change from Tigalmamine, Middle Atlas, Morocco. J. Biogeogr. 16: 65–74.

Lowe J.J., Hoek W.Z. and INTIMATE Group 2001. Inter-regional correlation of palaeoclimatic records for the last glacial-interglacial transition: a protocol for improved precision recommended by the INTIMATE project group. Quat. Sci. Rev. 20: 1175–1187.

Lowe J.J., Accorsi C.A., Bandini Mazzanti M., Bishop A., Van der Kaars S., Forlani L., Mercuri A.M., Rivalenti C., Torri P. and Watson C. 1996. Pollen stratigraphy of sediment sequences from lakes Albano and Nemi (near Rome) and from the Central Adriatic, spanning the interval from oxygen isotope stage 2 to the present day. Mem. Ist. it. Idrobiol. 55: 71–98.

Macklin M.G., Fuller I.C., Lewin J., Maas G.S., Passmore D.G., Rose J., Woodward J.C., Black S., Hamlin R.H.B. and Rowan J.S. 2002. Correlation of fluvial sequences in the Mediterranean basin over the last 200 ka and their relationship to climate change. Quat. Sci. Rev. 21: 1633–1641.

Magri D. 1999. Late-Quaternary vegetation history at Lagaccione near Lago di Bolsena (Central Italy). Rev. Paleobot. Palynol. 106: 171–208.

Magri D. and Parra I. 2002. Late Quaternary western Mediterranean pollen records and African winds. Earth Planet. Sci. Lett. 200: 401–408.

Magri D. and Sadori L. 1999. Late Pleistocene and Holocene pollen stratigraphy at Lago di Vico (Central Italy). Veg. Hist. Archaeob. 8: 247–260.

Magri D. and Tzedakis P.C. 2000. Orbital signatures and long-term vegetation patterns in the Mediterranean. Quat. Int. 73/74: 69–78.

McCoy F.W. 1974. Late Quaternary Sedimentation in the Eastern Mediterranean Sea. Doctorate, Harvard University, Cambridge, Mass., 132 pp.

Milliman J.D., Jeftić L. and Sestini G. 1992. The Mediterranean Sea and climate change — an overview. In: Jeftić L., Milliman J.D. and Sestini G. (eds), Climatic Change and the Mediterranean. Edward Arnold, London, pp. 1–14.

Narcisi B. 2000. Late Quaternary Eolian Deposition in Central Italy. Quat. Res. 54: 246–252.

Narcisi B. and Vezzoli L. 1999. Quaternary stratigraphy of distal tephra layers in the Mediterranean — an overview. Glob. Plan. Chan. 21: 31–50.

Nolet G.J. and Corliss B.H. 1990. Benthic foraminiferal evidence for reduced deep water circulation during sapropel deposition in the eastern Mediterranean. Mar. Geol. 94: 109–130.

Okuda M., Yasuda Y. and Setoguchi T. 2001. Middle to Late Pleistocene vegetation history and climatic changes at Lake Kopais, Southeast Greece. Boreas 30: 73–82.

Parisi E. and Cita M.B. 1982. Late Quaternary paleoceanographic changes recorded by deep-sea benthos in the western Mediterranean ridge. Geogr. Fis. Dinam. Quat. 5: 102–114.

Paterne M., Guichard F., Labeyrie J., Gillot P.Y. and Duplessy J.C. 1986. Tyrrhenian Sea tephrochronology of the oxygen isotope record for the past 60,000 years. Mar. Geol. 72: 259–285.

Paterne M., Guichard F. and Labeyrie J. 1988. Explosive activity of the south Italian volcanoes during the past 80,000 years as determined by marine tephrochronology. J. Volcanol. geotherm. Res. 34: 153–172.

Paterne M., Kallel N., Labeyrie L., Vautravers M., Duplessy J.-C., Rossignol-Strick M., Cortijo E., Arnold M. and Fontugne M. 1999. Hydrological relationships between the North Atlantic Ocean and the Mediterranean Sea during the past 15–75 kyr. Paleoceanography 14: 626–638.

Pentecost A. 1995. The Quaternary travertine deposits of Europe and Asia Minor. Quat. Sci. Rev. 14: 1005–1028.

Pérez-Obiol R. and Julià R. 1994. Climatic change on the Iberian Peninsula recorded in a 30,000-yr pollen record from Lake Banyoles. Quat. Res. 41: 91–98.

Pons A. and Reille M. 1988. The Holocene- and Upper Pleistocene pollen record from Padul (Granada, Spain): a new study. Palaeogeogr. Palaeoclim. Palaeoecol. 66: 243–263.

Ramrath A., Nowaczyk N.R. and Negendank J.F.W. 1999a. Sedimentological evidence for environmental changes since 34,000 years BP from Lago di Mezzano, Central Italy. J. Paleolim. 21: 423–435.

Ramrath A., Zolitschka B., Wulf S. and Negendank J.F.W. 1999b. Late Pleistocene climatic variations as recorded in two Italian maar lakes (Lago di Mezzano, Lago Grande di Monticchio). Quat. Sci. Rev. 18: 977–992.

Reille M., de Beaulieu J.-L., Svobodova H., Andrieu-Ponel V. and Goeury C. 2000. Pollen analytical biostratigraphy of the last five climatic cycles from a long continental sequence from the Velay region (Massif Central, France). J. Quat. Sci. 15: 665–685.

Roberts N., Black S., Boyer P., Eastwood W.J., Griffiths H.I., Lamb H.F., Leng M.J., Parish R., Reed J.M., Twigg D. and Yiğitbaşioğlu H. 1999. Chronology and stratigraphy of Late Quaternary sediments in the Konya Basin, Turkey: results from the KOPAL project. Quat. Sci. Rev. 18: 611–630.

Roger S., Féraud G., de Beaulieu J.-L., Thouveny N., Coulon Ch., Cochemé J.J., Andrieu V. and Williams T. 1999. ^{40}Ar/^{39}Ar dating on tephra of the Velay maars (France): implications for the Late Pleistocene proxy-climatic record. Earth Planet. Sci. Lett. 170: 287–299.

Rohling E.J. and De Rijk S. 1999. Holocene climate optimum and last glacial maximum in the Mediterranean: the marine oxygen isotope record. Mar. Geol. 153: 57–75.

Rossignol-Strick M. 1985. Mediterranean Quaternary sapropels, an immediate response of the African monsoon to variation of insolation. Palaeogeogr. Palaeoclim. Palaeoecol. 49: 237–263.

Rossignol-Strick M. and Paterne M. 1999. A synthetic pollen record of the eastern Mediterranean sapropels of the last 1 Ma: implications for the time-scale and formation of sapropels. Mar. Geol. 153: 221–237.

Rossignol-Strick M. and Planchais N. 1989. Climate patterns revealed by pollen and oxygen isotope records of a Tyrrhenian sea core. Nature 342: 413–416.

Roucoux K.H., Shackleton N.J., de Abreu L., Schönfeld J. and Tzedakis P.C. 2001. Combined marine proxy and pollen analyses reveal rapid Iberian vegetation response to North Atlantic millennial-scale climate oscillations. Quat. Res. 56: 128–132.

Sánchez Goñi M.F., Eynaud F., Turon J.L. and Shackleton N.J. 1999. High resolution palynological record off the Iberian margin: direct land-sea correlation for the last interglacial complex. Earth Planet. Sci. Lett. 171: 123–137.

Sánchez Goñi M.F., Turon J.-L., Eynaud F. and Gendreau S. 2000a. European climatic response to millennial-scale changes in the atmosphere-ocean system during the last glacial period. Quat. Res. 54: 394–403.

Sánchez Goñi M.F., Turon J.-L., Eynaud F., Shackleton N.J. and Cayre O. 2000b. Direct land/sea correlation of the Eemian, and its comparison with the Holocene: a high-resolution palynological record off the Iberian margin. Geol. Mijnbouw 79: 345–354.

Sánchez Goñi M.F., Cacho I., Turon J.-L., Guiot J., Sierro F.J., Peypouquet J.-P., Grimalt J.O. and Shackleton N.J. 2002. Synchroneity between marine and terrestrial responses to millennial scale climatic variability during the last glacial period in the Mediterranean region. Clim. Dyn. 19: 95–105.

Schramm A., Stein M. and Goldstein S.L. 2000. Calibration of the ^{14}C time scale to >40 ka by ^{234}U-^{230}Th dating of Lake Lisan sediments (last glacial Dead Sea). Earth Planet. Sci. Lett. 175: 27–40.

Shackleton N.J. 1974. Attainment of isotopic equilibrium between ocean water and the benthonic foraminifera genus *Uvigerina*: isotopic changes in the ocean during the last glacial. Colloque CNRS n. 219, Centre National de la Recherche Scientifique, Paris, pp. 203–210.

Sivan D., Gvirtzman G. and Sass E. 1999. Quaternary stratigraphy and paleogeography of the Galilee coastal plain, Israel. Quat. Res. 51: 280–294.

Stein M., Starinsky A., Katz A., Goldstein S.L., Machlus M. and Schramm A. 1997. Strontium isotopic, chemical, and sedimentological evidence for the evolution of Lake Lisan and the Dead Sea. Geochim. Cosmochim. Acta 61: 3975–3992.

Thouveny N., de Beaulieu J.-L., Bonifay E., Creer K.M., Guiot J., Icole M., Johnsen S., Jouzel J., Reille M., Williams T. and Williamson D. 1994. Climate variations in Europe over the past 140 kyr deduced from rock magnetism. Nature 371: 50–506.

Ton-That T., Singer B. and Paterne M. 2001. ^{40}Ar/^{39}Ar dating of latest Pleistocene (41 ka) marine tephra in the Mediterranean Sea: implications for global climate records. Earth Planet. Sci. Lett. 184. 645–658.

Tric E., Valet J.P., Tucholka P., Paterne M., Labeyrie L., Guichard F., Tauxe L. and Fontugne M. 1992. Paleointensity of the geomagnetic field during the last 80,000 years. J. Geophys. Res. 97: 9337–9351.

Tzedakis P.C. 1994. Vegetation change through glacial-interglacial cycles: a long pollen sequence perspective. Phil. Trans. r. Soc., Lond. B 345: 403–432.

Tzedakis P.C. 1999. The last climatic cycle at Kopais, Central Greece. J. Geol. Soc., Lond. 156: 425–434.

Tzedakis P.C. 2000. Vegetation variability in Greece during the last interglacial. Geol. Mijnbouw 79: 355–367.

Tzedakis P.C., Andrieu V., de Beaulieu J.-L., Crowhurst S., Follieri M., Hooghiemstra H., Magri D., Reille M., Sadori L., Shackleton N.J. and Wijmstra T.A. 1997. Comparison of terrestrial and marine records of changing climate of the last 500,000 years. Earth Planet. Sci. Lett. 150: 171–176.

Tzedakis P.C., Andrieu V., de Beaulieu J.-L., Birks H.J.B., Crowhurst S., Follieri M., Hooghiemstra H., Magri D., Reille M., Sadori L., Shackleton N.J. and Wijmstra T.A. 2001. Establishing a terrestrial chronological framework as a basis for biostratigraphical comparisons. Quat. Sci. Rev. 20: 1583–1592.

Vergnaud-Grazzini C., Ryan W.B.F. and Cita M.B. 1977. Stable isotope fractionation, climate change and episodic stagnation in the eastern Mediterranean during the late Quaternary. Mar. Micropaleont. 2: 353–370.

Vergnaud-Grazzini C., Devaux M. and Znaidi J. 1986. Stable isotope "anomalies" in the Mediterranean Pleistocene records. Mar. Micropaleont. 10: 35–69.

Walker M.J.C., Björck S., Lowe J.J., Cwynar L., Johnsen S., Knudsen K.-L., Wohlfarth B. and INTIMATE Group 1999. Isotopic 'events' in the GRIP ice core: a stratotype for the Late Pleistocene. Quat. Sci. Rev. 18: 1143–1150.

Watts W.A., Allen J.R.M. and Huntley B. 1996. Vegetation history and palaeoclimate of the last glacial period at Lago Grande di Monticchio, Southern Italy. Quat. Sci. Rev. 15: 133–153.

Wijmstra T.A. 1969. Palynology of the first 30 metres of a 120 m deep section in Northern Greece. Acta bot. neerl. 18: 511–527.

Wijmstra T.A. and Smit A. 1976. Palynology of the middle part (30–78 metres) of the 120 m deep section in Northern Greece (Macedonia). Acta bot. neerl. 25: 297–312.

Williams D.F., Thunell R.C. and Kennett J.P. 1978. Periodic freshwater flooding and stagnation of the Eastern Mediterranean Sea during the Late Quaternary. Science 201: 252–254.

Woodward J.C. and Goldberg P. 2001. The sedimentary records in Mediterranean rockshelters and caves: archives of environmental change. Geoarchaeology 16: 327–354.

Zolitschka B. and Negendank J.F.W. 1996. Sedimentology, dating and palaeoclimatic interpretation of a 76.3 ka record from Lago Grande di Monticchio, Southern Italy. Quat. Sci. Rev. 15: 101–112.

17. HOLOCENE CLIMATE, ENVIRONMENT AND CULTURAL CHANGE IN THE CIRCUM-MEDITERRANEAN REGION

NEIL ROBERTS (cnroberts@plymouth.ac.uk)
School of Geography
University of Plymouth
Plymouth, PL4 8AA
UK

TONY STEVENSON (a.c.stevenson@ncl.ac.uk)
Faculty of Law
Environment and Social Sciences
University of Newcastle
Newcastle upon Tyne, NE1 7RU
UK

BASIL DAVIS (b.davis@ncl.ac.uk)
Department of Geography
University of Newcastle
Newcastle upon Tyne, NE1 7RU
UK

RACHID CHEDDADI (epd@dialup.francenet.fr)
European Pollen Database
Place de la République
13200 Arles
France

SIMON BREWSTER (simon.brewer@wanadoo.fr)
Centre Européen de Recherche et d'Enseignement
de Géosciences de l'Environnement (CEREGE)
Europole de l'Arbois
B.P. 80, 13545 Aix-en-Provence cedex 04
France

ARLENE ROSEN (a.rosen@ucl.ac.uk)
Institute of Archaeology
University College London
31-34 Gordon Square
London, WC1H OPY
UK

R. W. Battarbee et al. (eds) 2004. *Past Climate Variability through Europe and Africa.*
Springer, Dordrecht, The Netherlands.

Keywords: Mediterranean, Holocene, Palaeoclimate, Vegetation, Modelling, Archaeology, Culture change, Biomisation

Introduction

The Mediterranean basin, which lies between 30° and 46 °N, is the largest area of the world to experience a climate of summer drought, winter rain of cyclonic origin and a mean annual temperature of 15 ± 5 °C (Köppen type Cs). Its flora is distinctive and adapted to both periodic desiccation and burning (Allen 2001). The region also has an exceptionally long and rich history of human use and abuse, stretching back to the advent of Neolithic farming in Southwest Asia at the start of the Holocene. The complex history of cultural-environmental relations around the Mediterranean "Lake" can create serious difficulties in distinguishing climate change from human impact in many proxy-data records (Bottema et al. 1990; Grove and Rackham 2001; Roberts 2002). In particular, once complex societies emerged during the Bronze Age between 5000 and 3000 BP, vegetation disturbance starts to become clearly visible in pollen diagrams. In compensation, the region offers a wealth of written archival and archaeological records back to before 2500 BP.

In this chapter we are concerned with PEP III time-stream 1 (cf. Gasse and Battarbee (this volume)) for the Mediterranean sector and we summarise recent progress in addressing the following key objectives:

(i)   to obtain high resolution, accurately dated, proxy records of sub-decadal climate variability linked to the North Atlantic Oscillation (NAO), El Niño Southern Oscillation (ENSO), solar and other forcings, from tree-ring, historical, crater lake and other sources;

(ii) to identify intra-regional patterns of climate variability, such as the east and west Mediterranean precipitation see-saw;

(iii) to examine the expression and timing of climate changes coeval with the northern European Little Ice Age and Medieval Warm Period;

(iv) to examine what is the nature and causes of periods of abrupt climate change that are evident from many lower latitude records (e.g., in inter-tropical African lake levels), and which may be linked to high magnitude events (e.g., major volcanic eruptions);

(v)   to assess how climate variability over recent millennia affected biota and natural ecosystems in the circum-Mediterranean lands;

(vi) to consider how climate variability has affected human activity and society over these time-scales, and to assess how far climate change has been influential in the development of prehistoric and classical civilisations in the Mediterranean Basin.

Data sources

To understand these issues, well-dated records of climatic variability over annual-to-centennial time-scales are needed, including data from both western and eastern ends of the basin in order to capture the historically-documented "Mediterranean Oscillation" in precipitation. Data sources on Holocene climatic change in the Mediterranean sector are notably wide-ranging, and include pollen and other palaeobotanical data (e.g., Pons (1981), van Zeist and Bottema (1991), Neumann (1992)), palaeolimnology (e.g., Roberts and Wright (1993), Harrison and Digerfeldt (1993), Zolitschka et al. (2000)), speleothems (e.g., McDermott et al. (1999), Frumkin et al. (1999), Bar-Matthews and Ayalon (this volume)), and written historical data (e.g., Camuffo and Enzi (1992), Grove and Contiero (1995), Barriendos Vallvé and Martín-Vide (1998), Rodrigo et al. (2000)). Palaeoceanographic records from Mediterranean deep-sea cores (see Kallel et al. (this volume)) and shallower-water marine environments (e.g., Trincardi et al. (1996)) provide an important complement to terrestrial palaeoclimatic archives, particularly at centennial to millennial scale time-scales, while geomorphological, soil-stratigraphic and archaeological evidence offer valuable supplementary data sources. Figure 1 shows the geographical coverage of two key palaeoclimate data sources around the Mediterranean basin, namely tree-rings and stable isotope analyses based on speleothems and lake sediments, along with exemplar records of each. Dendrological records have been studied throughout the region but climatic calibration of tree-ring data has, until recently, been attempted mainly for sites in the western part of the Mediterranean basin (e.g., Serre-Bachet et al. (1992)). Indeed, not only tree-ring but also pollen data are unevenly distributed across the circum-Mediterranean lands, and both are notably deficient in North Africa outside Morocco.

A key feature of the Mediterranean region is its role in linking and overlapping data sources between the tropical African and higher-latitude European zones of the PEP III transect. In the former, Holocene climate changes have been dominated by fluctuations in water balance as reflected, for example, in lake-level records (Verschuren, this volume), while the climatic history of the latter is largely dominated by pollen, tree-ring, peat and other palaeobotanical records (Barber et al., Snowball et al., this volume). The Mediterranean lands incorporate both types of climate archive. They also offer a comparison with the climatically similar Cape region of Southern Africa, and this provides a means of assessing the synchronism/diachronism of Holocene environmental change between the Northern and Southern Hemispheres (Scott and Lee-Thorp (this volume), see also Roberts et al. (2001a)).

Recent and historic climate variability

The Mediterranean climate is influenced by many of the same conditions that also influence the regions lying to the north and south. Precipitation is largely cyclonic of Atlantic origin, although local cyclogenesis occurs, for example, south of Genoa and around Cyprus. Consequently inter-annual variability in precipitation is strongly influenced by the synoptic conditions that also determine rainfall in temperate Europe. Observational records of precipitation in the western Mediterranean show a strong positive correlation with the NAO index. However, examination of the spatio-temporal pattern has indicated the existence of an historical see-saw oscillation between the east and west Mediterranean in atmospheric pressure distribution and precipitation. This is revealed, for example, by a negative correlation in the 500 hPA geopotential height between Algiers and Cairo from 1946 and 1989 (Palutikof et al. 1996). A similar see-saw teleconnection for winter precipitation also

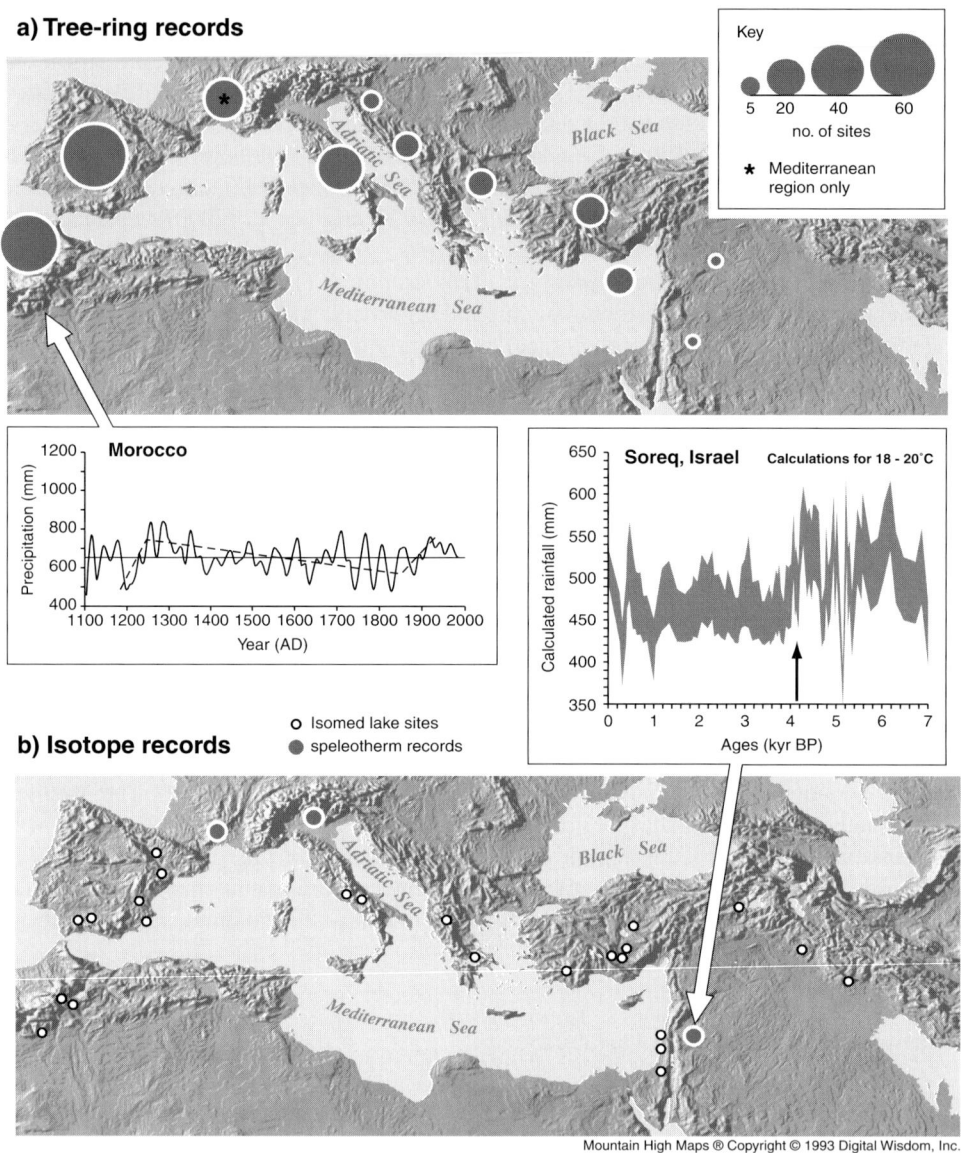

Figure 1. a) Distribution of tree-ring study sites around the Mediterranean basin (data: International tree-ring data base and other sources), along with synthetic 900-year record for precipitation changes in the Middle Atlas, Morocco (after Till and Guiot (1990)); b) Location of non-marine stable isotope records from lakes (from ISOMED working group web-site) and speleothems, along with isotope-inferred rainfall variations for the mid-and late-Holocene from Soreq Cave, Israel (after Bar-Matthews et al. (1998)). The arrow marks the proposed ca. 2200 BC drought event (see text).

emerges from an examination of longer instrumental time-series (Oldfield and Thompson, this volume). Precipitation records around the Mediterranean reach back to the mid-18th century, with spatial coverage provided by over 100 precipitation series spanning the last 150 years. These show a wetting trend in winter precipitation from the mid-19th century in the N. and W. Mediterranean (+0.1 mm/month yr^{-1}) contrasting with a drying trend in the S. and E. Mediterranean (−0.1 mm/month yr^{-1}). Instrumental precipitation records for the north-eastern sector of the Mediterranean show a weak negative correlation with the NAO index, a link that may also be evident in river-discharge and tree-ring records (Cullen and deMenocal 2000; D'Arrigo et al. 2001). Mediterranean temperature records also show a strong NAO influence (Mann 2002).

The influence of the Saharo-Arabian arid zone to the south is felt primarily during summer, when anticyclonic conditions migrate north to create seasonal drought. Summer monsoonal rains rarely reach as far north as the Mediterranean Sea so that precipitation sources in the Maghreb and the Sahel are decoupled and rainfall time-series show no correlation (Lamb and Peppler 1991).

Further back in the last millennium, historical documentary and proxy-climate records indicate greater synchroneity in the lower-frequency power spectrum for variations in precipitation and temperature across the whole Mediterranean basin. While inter-regional differences may exist in the precise timing of these shifts, the overall trend shows maximum wetness around AD 1250–1400 overlapping with the second half of the European Medieval Warm Period, followed by gradual decline towards cooler and drier conditions ca. AD 1700–1850 during the later part of the European Little Ice Age. Evidence for this includes stable isotope fluctuations in high accumulation-rate marine sediment cores from southern Levantine basin (Schilman et al. 2001) and the Gulf of Taranto (Cini Castiglione et al. 1999), the latter also showing a strong 11-year solar variability signal. A similar trend emerges from the western Mediterranean, for example, in dendro-climatic reconstructions from the Middle Atlas of Morocco (Till and Guiot 1990) (see Fig. 1a).

Pollen records of vegetation change since the mid-Holocene

Over a longer time-scale, it has often been assumed that the mid-Holocene might offer a pre-disturbance baseline for Mediterranean vegetation and landscapes. A large number of pollen studies indicate that much of the Mediterranean at that time was dominated by temperate deciduous woodland in areas today occupied by drought-adapted evergreens and shrubs (Huntley 1988; Prentice et al. 1996; Prentice et al. 2000). A major problem is that climate models continue to be unable to simulate conditions cool and wet enough to explain this mid-Holocene vegetation distribution (Prentice et al. 1998; Masson et al. 1999). These climate models are the same as those used to predict future change in the region, and their failure to reproduce mid-Holocene conditions potentially undermines the reliability of those future predictions. To investigate the evidence of the mid-Holocene vegetation, we provide here a new synthesis of pollen data from throughout the Mediterranean at 0 kyr and 6 kyr ^{14}C BP. Vegetation change has been investigated using biomisation of pollen spectra and particular efforts made to provide directly comparable maps based on data from the same site locations for the two time periods.

Pollen data

A total of 172 pollen sites from the Mediterranean area was used (Plate 7). Sites were selected on the basis of robust dating (radiocarbon, annual laminations, palaeomagnetics) to identify the 6000 BP sample. This has led to the exclusion of some sites included in the Huntley and Birks (1983) data-set and subsequent studies (Biome 6000 project (Prentice et al. 2000)) which had no absolute dating control. Cross-correlation of an adjacent dated core with an undated core from the same site on the basis of pollen stratigraphy has occasionally been used. In exceptional cases (Lake Van, Ghab), cross-correlation with more distant sites was used where this was suggested by other authors (Bottema 1995; Rossignol-Strick 1995). As well as improved chronological control, the data-set provides a much-improved spatial coverage compared to earlier studies, particularly over Turkey and Spain.

Where possible, original pollen counts from the European Pollen Database (EPD), published data, and the authors' own investigations have been used for the analysis. In common with other pollen synthetic studies, these data have been supplemented by digitised pollen data from published diagrams to improve the geographical coverage. Digitisation was undertaken where original counts were not available, and utilised the full spectra from the published diagram. 60% of sites are based on original counts, and the remainder digitised. The 6000 BP ('6 kyr') and modern ('0 kyr') time windows were defined following convention as ± 500 ^{14}C yr BP (COHMAP, Biome 6000). Only the sample closest to the target time was selected. To allow comparison between the modern and 6 kyr time periods at each site, surface samples were used where core-top samples were not within the appropriate time frame. Surface samples were obtained from the EPD (Joel Guiot (pers. comm.)) and the authors. Age-depth relationships were evaluated at all sites and the most appropriate model fitted.

Biomisation methods

The presentation of pollen data in terms of biomes provides a useful basis on which to map vegetation change, as well as compare with the results of other studies connected with the Biome 6000 project (Prentice et al. 2000). There have been two principal biomisation schemes developed for Europe (Prentice et al. 1996; Peyron et al. 1998), with additional regional biomisation schemes for Eurasia (Tarasov 1998a; b) and Africa (Jolly et al. 1998) that overlap or append the study area. Although overall the Prentice et al. (1996) scheme worked well, a number of problems emerged, especially in the Mediterranean where tundra was reconstructed in locations with large amounts of Ericales, Poaceae and Cyperaceae. The inclusion of only five non-arboreal taxa in the scheme also gave rise to problems in reconstructing the forest-steppe boundary, as well as in distinguishing warm and cold steppe biomes. These problems were addressed by Tarasov et al. (1998b) who separated steppe into warm and cold varieties, and added an additional 30 non-arboreal taxa. Further refinements were made by Peyron et al. (1998), whose method employed a further five non-arboreal taxa and a flexible PFT (Plant Functional Type) to biome allocation method. In both biomisations, evergreen oak is used to define warm-mixed from temperate-deciduous biomes. For this study we have used the Peyron et al. (1998) biomisation as more appropriate for the Mediterranean vegetation environment.

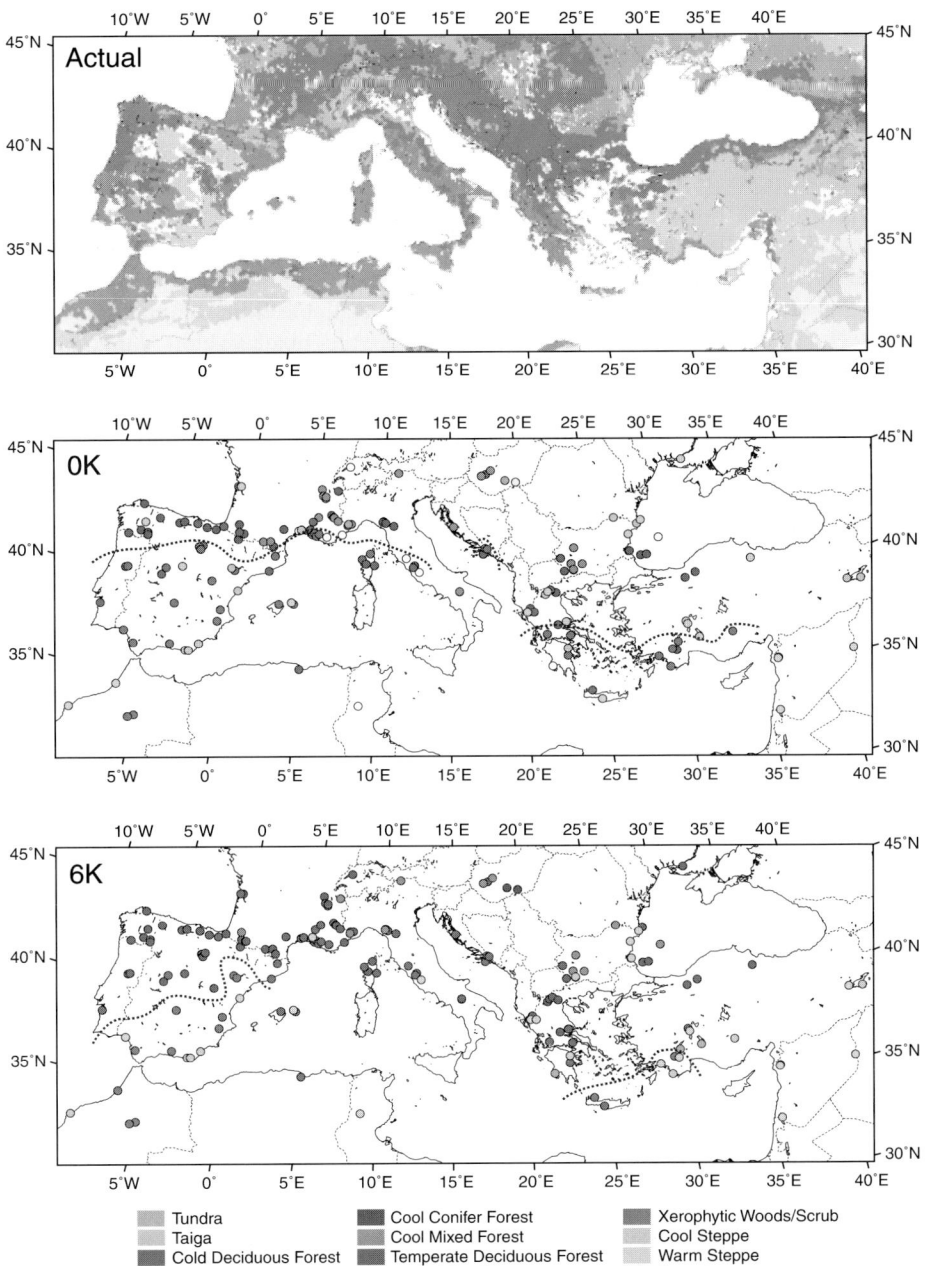

Plate 7. (a) Actual biome distribution from a high-resolution satellite-derived dataset (Olson 1994) global ecosystems legend; (b) Mediterranean biomes for 0–500 BP (0 kyr) based on pollen data biomised according to Peyron et al. (1998); (c) pollen-derived biomes for 6000 ± 500 yr BP (6 kyr). White circles at 0 kyr indicate sites where no pollen data are available for this time period. The dashed line marks the inferred northern limit of xerophytic woodland/scrub. Colour version of this Plate can be found in Appendix, p.634

To provide an independent point of reference for the pollen biome data a map of modern biome distribution for the study area was required. We have used the Olson (1994) Global Ecosystems Legend scheme, which is available as a high resolution data-set (1 km^2) from the EDC Global Land Cover Classification (GLCC), based on the IGBP-DIS AVHRR (v2.0) data-set. The GLCC data have a spatial resolution of 1 km^2 pixel size. Biomes have been assigned to each pixel on the basis of the dominant biome within a 10 km radius of the target pixel (total 276 pixels). This aggregation allows clearer mapping, a more realistic assessment of the dominant vegetation without too much distortion due to anthropogenic vegetation types, and is also probably a better approximation of the source area of the pollen samples. Where less than two diagnostic units were available from the pixels around the target pixel, no biome assignment was made and the target pixel was left blank.

The resulting map (Plate 7a) compares well with other sources that reflect actual vegetation, such as UNESCO-FAO (1970) map. As expected, it shows greater areas of steppe in Spain and Turkey than is shown on many potential vegetation maps, although the limits of xerophytic woodland/scrub correspond well to those of evergreen oak and olive associations. One of the most difficult biomes to define is the warm-mixed forest, which has no clear delineating classification within the Olson (1994) scheme. Overall, most of the pixels have been successfully classified, and blank pixels are rare. Some of the larger areas of anthropogenic land-use, which it was not possible to classify, include parts of inland Spain, the Po valley, Evros valley (Greece/Turkey) and the upper Euphrates (Syria/Turkey).

Results

Validation of the pollen biomisation technique can be shown through a visual comparison of the biome map and 0 kyr pollen biomes (Plate 7a and b). These generally show a very good level of agreement, with most of the major biome zones delineated by the distribution of sites. Previous reconstructions of biome distributions for 6 kyr that have included the Mediterranean area (Prentice et al. 1996; Jolly et al. 1998; Prentice et al. 2000) showed xerophytic woodland and steppe biomes as confined to North Africa. In contrast, the results of our new analysis show these biomes present at sites throughout the Mediterranean (Plate 7c). This change in distribution can only be partly attributed to differences between biomisation schemes. Another reason is the reclassification of *Quercus* as evergreen rather than deciduous from several key sites in Mediterranean Spain as a result of recent re-analyses (e.g., Pons and Reille (1988), Carrión and Dupré (1996)). The previous misclassification of oak as deciduous in this region may have had implications for climate reconstructions based on the Huntley and Birks (1983) data-set (e.g., Masson et al. (1999)). In contrast to the Prentice et al. (1996) biomisation scheme, xerophytic woodland biomes are now reconstructed at 6 kyr in several parts of semi-arid Spain. Relative to the west, sites with xerophytic woodland biomes in the eastern Mediterranean show a more restricted and southern distribution at 6 kyr. Changes in other cool and temperate biomes agree with previous studies indicating a more southern distribution of these biomes at 6 kyr. Temperate deciduous biomes are reconstructed south of their current position across Central Spain and Greece, as well as at increased altitude in the Pyrenees, Alps and Balkans where cool mixed and cool conifer biomes are found today.

Although forest and woodland biomes are the predominant biomes at the sample sites at 6 kyr, steppic biomes are nevertheless still important. Steppe is recorded at sites around

Southern and Eastern Spain, the Rhone delta, Italy, Bulgaria, Southern and Western Greece, Turkey and North Africa. A number of these sites represent unstable coastal or lacustrine environments dominated by weedy annuals, which by 0 kyr have stabilised sufficiently (either naturally or by human intervention) to allow woodland to predominate. Other sites show a persistence of steppe in Southern Morocco, eastern mainland Spain and the Balearics, Southern Greece and Southwest Asia.

These biomisation results need to be treated cautiously. Different schemes can produce very different results, and the method still lacks proper validation. None the less, the patterns shown in Plate 7 display good regional coherence and are in accord with more traditional ecological interpretations of pollen diagrams, for example, in showing a late-Holocene expansion of Mediterranean sclerophyll taxa. This study suggests that taxa generally spread from refugia in the south to the north, rather than east to west, in accord with the findings of Jalut et al. (1997). It is clear that some of the most significant mid-Holocene refugia were in Southern Spain, although this may be also true of Southern Italy where there is still a lack of palaeoecological information for this period.

Climatic reconstruction since the mid-Holocene

How much can the spread of xerophytic Mediterranean vegetation over the past six millennia be ascribed to climatic change, and how much to human action? Individual pollen records show that the timing and nature of human impact upon mid-Holocene vegetation varied around the Mediterranean basin. In Southwest Turkey, for example, pollen diagrams indicate that regional forest clearance and a period of cultural landscape creation took place from ca. 3000 to ca. 1250 BP (Eastwood et al. 1998). In this case the forest subsequently regenerated, but its species composition was changed to one dominated by pine. In parts of the Middle Atlas mountains of Morocco, by contrast, cedar forests have continued unbroken right through the later Holocene (Lamb and van der Kaars 1995). In other areas, forests were not clear-felled but modified and used as a managed resource for grazing, woodfuel and other products; for example, the dehesa system for using the Southern Spanish oak woodlands probably goes back to Bronze Age times (Stevenson and Harrison 1992).

This spatio-temporal complexity in human impact makes reliable statistical inferences about past climate from Mediterranean pollen-based palaeo-vegetation data more problematic than for longer time periods (e.g., Magri et al. (this volume)). One way to control for human disturbance is to compare pollen-inferred climate variables with other proxy-climate evidence from the same lake sediment cores, notably that related to changes in lake hydrology and regional water balance. Pollen versus lake-level comparisons have been attempted where the two sets of data overlap geographically both at a regional scale (e.g., Jolly et al. (1998)) and for individual site records. An example of the latter from the Mediterranean comes from Tigalmamine in Morocco, for which reconstructions have been made of both lake water-level fluctuations and pollen-based vegetation during the Holocene (Lamb et al. 1995; Lamb and van der Kaars 1995). The latter in turn have been calibrated to generate statistical estimates of winter and summer temperature, and annual precipitation (Cheddadi et al. 1998). Pollen-based estimates show higher temperatures than at present during the early Holocene in both winter and summer, with maximum Holocene precipitation occurring between ca. 5500 and 2500 ^{14}C yr BP (Fig. 2). Lake-level changes over the same time period show a general trend towards deeper water (=wetter) conditions

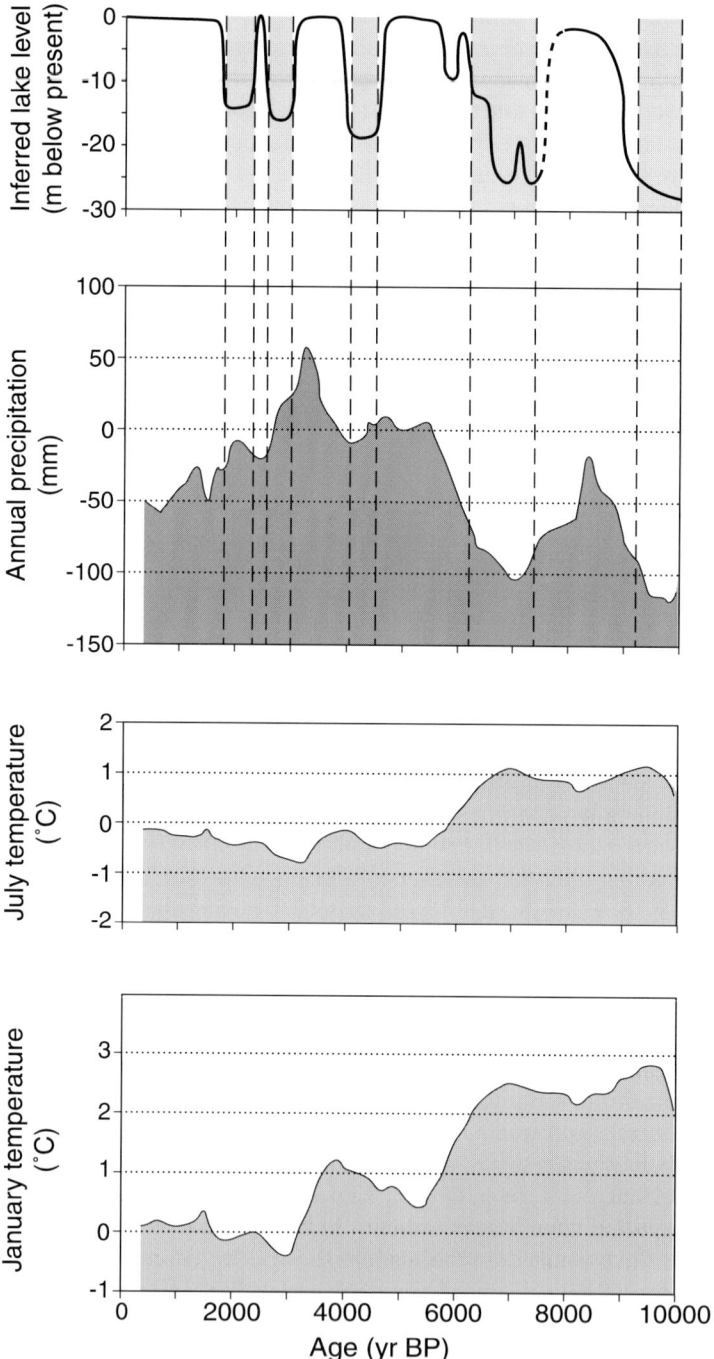

Figure 2. Pollen-inferred climate reconstruction for Tigalmamine, Morocco (after Cheddadi et al. (1998)) compared with the water-level history of the same lake (after Lamb et al. (1995)). Periods of lake-level regression, interpreted as the result of climatic aridity, are shaded.

towards the present-day, interrupted by a series of abrupt regressive (arid) events. These abrupt events of centennial duration appear to have been too short to cause lasting impact on forest composition (Lamb and van der Kaars 1995). Although there are similarities between the histories of climate inferred from pollen and lake levels, there are also notable divergences. Significantly, one of the most notable is during the last 1300 years, when lake-levels were high but precipitation values inferred from pollen data show a decline. Even with a catchment that is little impacted by agriculture, this divergence, and associated increases in clay inwash and % grass pollen entering the lake, suggest that as a proxy climate record, the upper part of the Tigalmamine pollen sequence may be inaccurate. By inference, other pollen sequences that are more obviously subject to human disturbance would provide an even less faithful reconstruction of past climatic conditions.

Elsewhere in the Mediterranean, from Southern Spain through Sicily and into Turkey, coupled pollen-palaeolimnological records suggest that more favourable moisture-balance conditions prevailed during the mid-Holocene compared to today (Reed et al. 2001; Sadori and Narcisi 2001; Roberts et al. 2001b; Wick et al. 2003). This is also given support by a number of other proxy climate records including stable isotope measurements on speleothems and land-snails from the southern Levant (Fig. 1b), (Goodfriend 1999). This would imply that in most regions, the shift to a more drought-adapted vegetation over the last 6000 years has been due both to human impact and to an overall trend towards climatic desiccation, particularly during the summer months. Regional climate modelling experiments suggest that these two trends may not have been unconnected, because summer precipitation in the Mediterranean is sensitive to land-cover changes, both through albedo effects and through recycling of soil and plant moisture back into the atmosphere. Reale and Shukla (2000) imposed a "greener" land cover on the southern shores of the Mediterranean for the Roman-Classical period based on a range of historical evidence, and found that these changed boundary conditions were sufficient to alter the pattern and amount of summer rainfall in their GCM experiment. Building on this work, Gates and Ließ (2001) undertook a sensitivity analysis involving simulated deforestation and afforestation of Mediterranean landscapes and also found that summer precipitation was altered significantly in their climate modelling experiments. This feedback suggests that large-scale reduction in forest cover around the Mediterranean during the late Holocene may have been sufficient to diminish summer rainfall levels.

Cultural impacts and abrupt climatic change

A wide range of natural "archives" from the circum-Mediterranean therefore demonstrate that climatic fluctuations were a regular occurrence throughout the Holocene, manifested at different degrees of magnitude and duration. Some of these fluctuations were quite abrupt and would have impacted upon human societies in both positive and negative ways. They range from events of centennial duration, such as those reflected in the water-level history of lakes like Tigalmamine (Fig. 2) and the Dead Sea (Frumkin et al. 1991), through to much shorter annual-to-decadal climatic excursions. An example of the latter is provided by the tree-ring series from Central Anatolia which shows a period of anomalously favourable growth conditions dated to ca. 1650 BC (Kuniholm et al. 1996; Manning et al. 2001). The atmospheric disruption responsible for this anomaly may, in turn, have been a result of the major volcanic eruption that destoyed much of the Aegean island of Thera at this time.

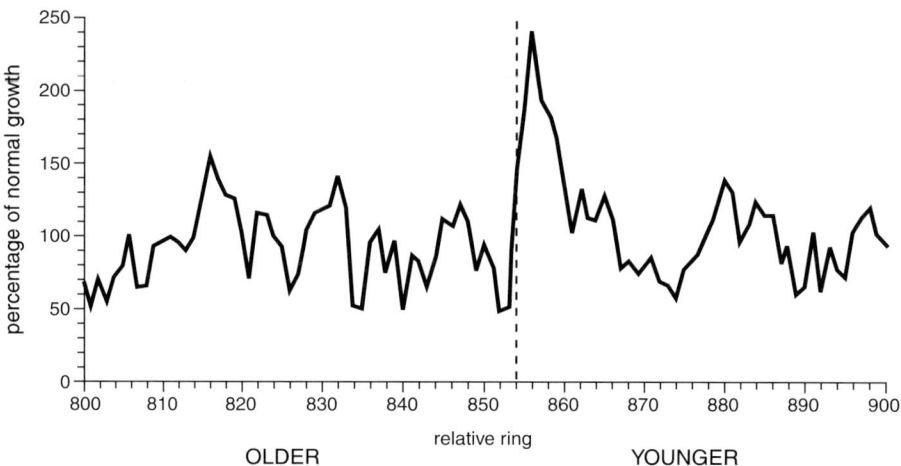

Figure 3. Central Anatolian tree-ring widths for a 100-year period during the 2nd millennium BC (after Kuniholm et al. (1996)). The abrupt growth anomaly starting with ring 854 has been linked to the "Minoan" eruption of Thera and has been dated by Manning et al. (2001) to 1650 +4/-7 BC.

However, it is noteworthy that the climatic anomaly lasted less than a decade before normal tree-growth conditions were restored (Fig. 3), so that direct societal impacts are likely to have been similarly short-lived.

The example above was associated with a shift to more favourable bio-climatic conditions (increased precipitation and cloud cover), but more attention has been given in the literature to possible drought events that would have caused stress on societies through food and water shortage. The assumption that abrupt adverse climate change would have had a direct causal effect leading to widespread land abandonment by people may not always apply in the Holocene Period, however, because the solutions for adapting to climatic change became much more complex with the development of sedentary societies (Rosen and Rosen 2001). While much has been made of cases in which proxy evidence for climatic changes correspond to collapse of civilisation and abandonment of towns and cities, much less notice has been taken of the contemporary towns and cities in the same regions which carried on with no visible effects. Such is the case, for example, with the apparent co-occurrence of a late third millennium BC arid climatic episode and collapse of Early Bronze Age societies in the Near East (Weiss et al. 1993; Dalfes et al. 1997; Cullen et al. 2000). This suggests a far more interesting phenomenon than a simple list of site abandonment. We can learn more about human responses to abrupt climate change in both the past and the present by examining and comparing cases in which solutions to climatic degradation both succeed and fail given differing levels of socio-economic and technological development. Equally, we may ask why some drought events appear to have had major socio-economic consequences, while other climatic events that were equally severe are not associated with any noticeable disruption in the archaeological or historical record. For example, in the Soreq Cave speleothem record (Fig. 1b, and Bar-Matthews and Ayalon (this volume)), the drought event ca. 2200 BC appears to be only one of several climatic fluctuations leading towards generally drier conditions, yet the other severe rainfall fluctuations which punctuated the 3rd and later 4th millennia BC are not known to be associated with any widespread cultural discontinuities.

Pre-industrial complex societies, such as the Early Bronze Age societies of the Near East, typically approached problems of climate change by using a combination of cosmological as well as technological solutions (McIntosh et al. 2000). Even during non-drought conditions, the Mediterranean climatic regime was one of strong seasonality with wet winters and generally dry summers, and, as today, would have suffered from drought years. Early Bronze Age II and III societies in the southern Levant (ca. 3100–2200 BC) were well-adapted to this regime in spite of the relatively low level of agricultural technology. They had no canal irrigation or elaborate water conservation expertise such as dams, rock-cut wells, or cisterns as was true of populations in the region in later time periods. However, they did have an environmental advantage that did not exist in subsequent periods, namely a higher base flow in the streams and active alluviating floodplains (Rosen 1991; Rosen and Halpern, unpublished). The Early Bronze Age population took advantage of this situation by buffering rainfall cereal cultivation with floodwater farming (Rosen 1995). This was the technological solution to living in an environment with periodic droughts. The social solution was a system of grain taxation and redistribution controlled by the elite managers of the society. Evidence for this comes from the massive granaries discovered at the site of Beit Yerah at the southern end of Lake Kinneret. Finally, cosmological protection was ensured by fertility cults such as that evoked by the figure of the wheat-harvest god Tamuz at Arad, and major public expenditure on temple-building. Such structures increased in size and scale from the EB II to the EB III, perhaps a function of increasing droughts and climatic insecurity throughout the third millennium BC.

At the end of the third millennium Early Bronze Age societies in the Near East were struck by severe droughts. This was accompanied by a drop in the base-flow of streams throughout the region (Rosen 1991; 1995) and the subsequent incision of the stream channels leaving the formerly wet alluvial floodplains high and dry. At the same time many sites were abandoned throughout the southern Levant as well as in Northern Syria and other parts of the Fertile Crescent. Much research has focused on these abandonments in a cautionary tale demonstrating the helplessness of pre-industrial societies in the wake of severe climatic stress (Weiss et al. 1993). However, there were also sites with evidence for continued occupation in both Northern Syria (e.g., Brak, Mozan) and in the southern Levant (e.g., Iktanu, Iskander). How then did the subsistence strategies of the populations of these sites differ from those that failed to survive the end of the third millennium BC? One could ask the same question for the Middle Bronze Age populations that moved into the void several hundred years later, building cities with large populations within a similar environmental regime to that which led to the downfall of their predecessors. The answer lies to some degree in the resilience of these early complex societies and the degree to which they were willing to accept new technologies, particularly advanced systems of water manipulation, and new social institutions. The failure of the Early Bronze Age societies of the southern Levant came about because of their inability to change their technology. This cultural and technological conservatism might have been a function of responses on the part of the managing elite who made decisions based on their perceptions of the crisis within their own unique cultural framework.

As this case-study demonstrates, cultural collapse by no means represents the only possible societal response to climatic deterioration, but is one of several alternatives mediated by prevailing socio-economic, political and ideological circumstances, that are historically and geographically specific. In Biblical texts, for example, the prophets used drought and crop failure to admonish the Jewish people for sinning against divine will (Carroll 1986), while in 15th-century Catholic Spain, similar climatic adversity was blamed on God's anger at the presence of a Moslem minority, who became the target of violent riots (Mackay 1981). Understanding the relationship between climate and culture therefore involves more than demonstrating a chronological coincidence between reconstructed climatic changes, such as major drought events, and archaeological site abandonment.

Conclusions

Data currently available from observational, historical and tree-ring records in Southern Europe, South-West Asia and North Africa, while far from complete, suggest that high-frequency rainfall variability around the Mediterranean over the last 2–3 centuries has been linked to the North Atlantic Oscillation. This has involved a strong positive co-variance with high NAO index in Iberia-Maghreb and a somewhat weaker negative co-variance in the Levant-Egypt. The lack of longer highly-resolved climate records makes it difficult to know if such an east-west NAO-driven precipitation see-saw also applied during earlier periods of the Holocene. In contrast to high-frequency variability, there is evidence that over century-to-millennial timescales, there may have been greater spatial coherence in climate changes over the whole Mediterranean region. For the last millennium, records from both east and west Mediterranean appear to show phases of drought/flood broadly coeval with periods of colder/warmer climate in Europe (Little Ice Age/Medieval Warm Period). If so, this would be in anti-phase with precipitation oscillations in East Africa over the same time frame, where the period of the LIA was marked by generally wetter, rather than drier conditions (Verschuren, this volume). It would, however, broadly mirror Mediterranean climate changes over Milankovitch time-scales in which relatively dry climates prevailed during times of cold (glacial) climate, and wetter climates coincided with thermal maxima (Kallel et al., this volume; Magri et al., this volume).

At present, robust, well-dated climatic records are relatively deficient for the period from ca. 4000 to 1000 yr BP. This is the period during which human impact around the Mediterranean basin increased most markedly, and one for which it is can be particularly difficult to separate natural climate variability from anthropogenic landscape disturbance (e.g., in pollen records). GCM experiments have shown that changes over this time-period may have been further complicated by the feedback effects of human-induced land cover change on the regional climate system. None the less, Mediterranean palaeoclimate data indicate that centennial-scale climate changes did occur during proto-historic and early historic times, with suggestions in some records (e.g., Jordan rift (Heim et al. 1997)) of generally more arid phases during the later part of the first millennium AD and at least part of the first millennium BC. This is undoubtedly a time-period for which further palaeoclimate research work is required, with stable-isotope data from speleothems and lake sediments holding particular promise.

Prior to ca. 4000–5000 BP, most Mediterranean landscapes, although already impacted by early agricultural societies, were fundamentally different from those that have existed in historical and modern times. Much of this transformation is due to the impact of complex pre-industrial societies, but the weight of evidence also suggests that changes such as the extension northward in the area under xerophytic sclerophyll woodland and scrub, were partly climatically-controlled. Evidence for an overall decrease in moisture availability since the mid-Holocene in the Mediterranean comes from lake-level, speleothem and a range of other terrestrial archives, and is also in accord with marine core records, notably for the last (S_1) sapropel layer (Kallel et al., this volume). This period of overall more humid climate in the Mediterranean appears partly synchronous with a similar wetter phase in northern inter-tropical Africa, although the direct meteorological cause (viz. an expanded and intensified summer monsoon) seems unlikely to have applied to the Mediterranean. It should also be noted that not all terrestrial evidence from the Mediterranean is in accord with a wetter early Holocene climate, including records from Morocco and from some interior regions of South-West Asia (e.g., Griffiths et al. (2001)). The oscillating decline in precipitation in most Mediterranean areas during the 4th and 3rd millennia BC had significant but not predetermined impacts on human communities. While the short-term implications of drought events were often negative, in the longer term climatic instability appears to have acted as a trigger for social change and encouraged the adoption of new technologies (e.g., irrigation), and it is perhaps no coincidence that this period of overall climatic deterioration coincided with the emergence of the first complex societies around the Mediterranean basin.

Summary

Data sources on Holocene climatic change in the circum-Mediterranean region are particularly wide-ranging and include palaeobotany, palaeolimnology, palaeoceanography, speleothems, geomorphology/soils, archaeology, and written historical data. A key feature of the region is its role in linking and overlapping data sources between the tropical African zones of the PEP III transect and the temperate European sector. The region's flora is distinctive and adapted to both periodic desiccation and burning. Mediterranean lands have a long history of human use, which has led to long-standing debate about the relative importance of climate change and human impact in many proxy-data records.

In addition to exemplar records of climate change, we present a new regional synthesis for reconstructed vegetation (biomes) and climate at 6 kyr using modern analogue techniques on both pollen and pft transformations of pollen taxa. We also evaluate how climate variability has affected human activity and society over Holocene timescales, and assess how far climate change has been influential in the development of prehistoric and classical civilisations in the Mediterranean Basin. This is illustrated using a case example of cultural change during the proposed late 3rd millennium BC Bronze Age climatic crisis.

Instrumental climate records show a southeast-northwest Mediterranean precipitation see-saw over annual-decadal timescales linked to the NAO. By contrast, over longer periods most regions within the Mediterranean appear to show predominantly synchronous trends and directional changes. Over centennial timescales these include wet conditions broadly coincident with the Medieval Warm Period, followed by gradual shift towards drier conditions during the Little Ice Age. Over millennial timescales, they include evidence of generally wetter conditions during the early-mid Holocene, paralleling the formation of the Mediterranean S1 sapropel, followed by an oscillating trend towards drier conditions between 5 and 2–3 kyr. The reduced early-mid Holocene summer drought provides part of the explanation for the greater extent of forest biomes at 6 kyr, although the subsequent expansion of xerophytic scrubland and steppe biomes was also partly anthropogenic in origin. In fact, these two agencies may be partly related, as interactive climate modelling experiments show that land-cover change would have been sufficient to influence intensity of the summer drought in the Mediterranean region.

Acknowledgements

For their contributions to this chapter we thank W.J. Eastwood, J. Guiot, M. Jones, S. Juggins, I. Juhasz, M.J. Leng, O. Peyron, J. Reed, B. Rogers, ISOMED working group members, and to those who provided data to the European Pollen Database. GLCC data are distributed by the EROS Data Center Distributed Active Archive Center, Sioux Falls, South Dakota. This work includes work partly undertaken under funding from the UK NERC and a Newcastle University Lord Adams Fellowship.

References

Allen H. 2001. Mediterranean Ecogeography. Prentice-Hall, Harlow, 263 pp.

D'Arrigo R., Dalfes H.N., Cullen H., Jacoby G. and Onol B. 2001. Tree-ring records of hydrometeo-rological variability from Eastern Turkey. Abstracts of PAGES PEP III Conference "Past Climate Variability through Africa and Europe". Aix-en-Provence, pp. 73–74.

Barber K., Zolitschka B., Tarasov P. and Lotter A.F., this volume. Atlantic to Urals — the Holocene climatic record of Mid-Latitude Europe. In: Battarbee R.W., Gasse F. and Stickley C.E. (eds), Past Climate Variability through Europe and Africa. Kluwer Academic Publishers, Dordrecht, the Netherlands, pp. 417–442.

Bar-Matthews M. and Ayalon A., this volume. Speleothems as palaeoclimate indicators, a case study from Soreq Cave located in the Eastern Mediterranean Region, Israel. In: Battarbee R.W., Gasse F. and Stickley C.E. (eds), Past Climate Variability through Europe and Africa. Kluwer Academic Publishers, Dordrecht, the Netherlands, pp. 363–391.

Bar-Matthews M., Ayalon A. and Kaufman A. 1998. Middle to Late Holocene (6,500 Yr. Period) paleoclimate in the Eastern Mediterranean region from stable isotopic composition of speleothems from Soreq Cave, Israel. In: Issar A.S. and Brown N. (eds), Water, Environment and Society in Times of Climatic Change. Kluwer, Dordrecht, pp. 204–214.

Barriendos Vallvé M. and Martín-Vide J. 1998. Secular climatic oscillations as indicated by catas-trophic floods in the Spanish Mediterranean coastal area (14th–19th centuries). Climatic Change 38: 473–491.

Bottema S. 1995. Holocene vegetation of the Van area: palynological and chronological evidence from Söğütlü, Turkey. Veg. Hist. Archaeobot. 4: 187–193.

Bottema S., Entjes-Nieborg G. and van Zeist W. (eds) 1990. Man's Role in the Shaping of the Eastern Mediterranean Landscape. A.A. Balkema, Rotterdam.

Camuffo D. and Enzi S. 1992. Reconstructing the climate of Northern Italy from archive sources. In: Bradley R.S. and Jones P.D. (eds), Climate since AD 1500. Routledge, London/New York, pp. 143–154.

Carrión J.S. and Dupré M. 1996. Late Quaternary vegetational history at Navarrés, Eastern Spain. A two core approach. New Phytol. 134: 177–191.

Carroll R.P. 1986. Jeremiah. London, SCM.

Cheddadi R., Lamb H.F., Guiot J. and van der Kaars S. 1998. Holocene climatic change in Morocco: a quantitative reconstruction from pollen data. Clim. Dyn. 14: 883–890.

Cini Castagnoli G., Bernasconi S.M., Bonino G., Della Monica P. and Taricco C. 1999. 700 year record of the 11 year solar cycle by planktonic foraminifera of a shallow water Mediterranean core. Adv. Space Res. 24: 233–236.

Cullen H.M. and deMenocal P. 2000. North Atlantic influence on Tigris-Euphrates streamflow. Int. J. Climatol. 20: 853–864.

Cullen H.M., deMenocal P.D., Hemming S., Hemming G., Brown F.H., Guilderson T. and Sirocko F. 2000. Climate change and the collapse of the Akkadian empire: evidence from the deep sea. Geology 28: 379–82.

Dalfes H.N., Kukla G. and Weiss H. (eds) 1997. Third Millennium BC Climate Change and the Old World Collapse. Proceedings of NATO ASI Series I. vol. 49, Springer-Verlag.

Eastwood W.J., Roberts N. and Lamb H.F. 1998. Palaeoecological and archaeological evidence for human occupance in Southwest Turkey: the Beyşehir Occupation phase. Anat. Stud. 48: 69–86.

Frumkin A., Carmi I., Gopher A., Ford D.C., Schwarz P. and Tsuk T. 1999. A Holocene millennial-scale climatic cycle from a speleothem in Nathal Qanah Cave, Israel. The Holocene 9: 677–682.

Frumkin A., Magaritz M., Carmi I. and Zak I. 1991. The Holocene climatic record of the salt caves of Mount Sedom, Israel. The Holocene 1: 191–200.

Gates L.D. and Ließ S. 2001. Impacts of deforestation and afforestation in the Mediterranean region as simulated by the MPI atmospheric GCM. Glob. Planet. Chan. 30: 309–328.

Gasse F. and Battarbee R.W., this volume. Introduction. In: Battarbee R.W., Gasse F. and Stickley C.E. (eds), Past Climate Variability through Europe and Africa. Kluwer Academic Publishers, Dordrecht, the Netherlands, pp. 1–6.

Goodfriend G.A. 1999. Terrestrial stable isotope records of Late Quaternary paleoclimates in the eastern Mediterranean region. Quat. Sci. Rev. 18: 501–514.

Griffiths H.I., Schwalb A. and Stevens L.R. 2001. Environmental change in SW Iran: the Holocene ostracod fauna of Lake Mirabad. The Holocene 11: 757–764.

Grove A.T. and Rackham O. 2001. The Nature of Mediterranean Europe. An Ecological History. Yale University Press. New Haven and London.

Grove J.M. and Contiero A. 1995. The climate of Crete in the sixteenth and seventeenth centuries. Climatic Change 30: 223–247.

Harrison S.P. and Digerfeldt G. 1993. European lakes as palaeohydrological and palaeoclimatic indicators. Quat. Sci. Rev. 12: 233–248.

Heim C., Nowaczyk N.R., Negendank J.F.W., Leroy S.A.G. and Ben-Avraham Z. 1997. Near Eastern desertification: evidence from the Dead Sea. Naturwissenshaften 84: 398–401.

Huntley B. 1988. Europe. In: Huntley B. and Webb T. III (eds), Vegetation History. Kluwer Academic Publishers, Dordrecht, pp. 341–383.

Huntley B. and Birks H.J.B. 1983. An Atlas of Past and Present Pollen Maps for Europe: 0-13000 Years Ago. Cambridge University Press, Cambridge.

Jalut G., Esteban Amat A., Riera I., Mora S., Fontugne M., Mook R., Bonnet L. and Gauquelin T. 1997. Holocene climatic changes in the western Mediterranean: installation of the Mediterranean climate. C.R. Acad. Sci. 325: 327–334.

Jolly D., Harrison S.P., Damnati B. and Bonnefille R. 1998. Simulated climate and biomes of Africa during the Late Quaternary: comparison with pollen and lake status data. Quat. Sci. Rev. 17: 629–657.

Kallel N., Duplessy J., Labeyrie L., Fontugne M. and Paterne M., this volume. Mediterranean Sea palaeohydrology and pluvial periods during the Late Quaternary. In: Battarbee R.W., Gasse F. and Stickley C.E. (eds), Past Climate Variability through Europe and Africa. Kluwer Academic Publishers, Dordrecht, the Netherlands, pp. 307–324.

Kuniholm P.I., Kromer B., Manning S.W., Newton M., Latini C.E. and Bruce M.J. 1996. Anatolian tree rings and the absolute chronology of the eastern Mediterranean, 2220-718 BC. Nature 381: 780–783.

Lamb H.F. and van der Kaars S. 1995. Vegetational response to Holocene climatic change: pollen and palaeolimnological data from the Middle Atlas, Morocco. The Holocene 5: 400–408.

Lamb H.F., Gasse F., Benkaddour A., el-Hamouti N., van der Kaars S., Perkins W.T., Pearce N.J. and Roberts N. 1995. Relation between century-scale Holocene arid intervals in tropical and temperate zones. Nature 373: 134–137.

Lamb P.J. and Peppler R.A. 1991. West Africa. In: Glantz M.H., Katz R.W. and Nicholls N. (eds), Tele-connections between Worldwide Climate Anomalies. Cambridge University Press, Cambridge, pp. 121–89.

Mackay A. 1981. Climate and popular unrest in late Medieval Castille. In: Wigley T.M.L., Ingram M.J. and Farmer G. (eds), Climate and History: Studies in Past Climates and their Impact on Man. Cambridge University Press, Cambridge, pp. 356–375.

McDermott F., Frisia S., Huang Y.M., Longinelli A., Spiro B., Heaton T.H.E., Hawkesworth C.J., Borsato A., Keppens E., Fairchild I.J., van der Borg K., Verheyden S. and Selmo E. 1999. Holocene climate variability in Europe: evidence from delta O-18, textural and extension-rate variations in three speleothems. Quat. Sci. Rev. 18: 1021–1038.

McIntosh R.J., Tainter J.A. and McIntosh S.K. 2000. The Way the Wind Blows: Climate, History, and Human Action. Columbia University Press, New York.

Maheras P. and Kutiel H. 1999. Spatial and temporal variations in the temperature regime in the Mediterranean and their relationship with circulation during the last century. Int. J. Climatol. 19: 745–764.

Mann M.E. 2002. Large-scale climate variability and connections with the Middle East in past century. Climatic Change 55: 287–314.

Manning S.W., Kromer B., Kuniholm P.I. and Newton M.W. 2001. Anatolian tree rings and a new chronology for the East Mediterranean Bronze-Iron Ages. Science 294: 2532–2535.

Masson V., Cheddadi R., Braconnot P., Joussaume S., Texier D. and PMIP Participants. 1999. Mid-Holocene climate in Europe: what can we infer from PMIP model-data comparisons? Clim. Dyn. 15: 163–182.

Neumann K. 1992. The contribution of anthracology to the study of the late Quaternary vegetation history of the Mediterranean region and Africa. Bull. Soc. Bot. Fr. 139: 421–440.

Olson J.S. 1994. Global Ecosystem Framework. USGS EROS Data Center Internal Reports, Sioux Falls, SD.

Palutikof J.P., Conte M., Casimiro Mendes J., Goodess C.M. and Espirito Santo F. 1996. Climate and climatic change. In: Brandt C.J. and Thornes J.B. (eds), Mediterranean Desertification and Land Use. John Wiley, Chichester, pp. 43–86.

Peyron O., Guiot J., Cheddadi R., Tarasov P., Reille R., de Beaulieu J.-L., Bottema S. and Andrieu V. 1998. Climatic reconstruction in Europe for 18,000 yr BP from pollen data. Quat. Res. 49: 183–196.

Pons A. 1981. The history of the Mediterranean shrublands. In: di Castri F., Goodall D.W. and Specht R.L. (eds), Ecosystems of the World 11, Mediterranean-type Shrublands. Elsevier Scientific Publishing, Amsterdam, pp. 131–138.

Pons A. and Reille M. 1988. The Holocene and Upper Pleistocene pollen record from Padul (Granada, Spain). A new study. Palaeogeog. Palaeoclimatol. Palaeoecol. 66: 243–263.

Prentice I.C., Guiot J., Huntley B., Jolly D. and Cheddadi R. 1996. Reconstructing biomes from palaeoecological data: a general method and its application to European pollen data at 0 and 6 ka. Clim. Dyn. 12: 185–194.

Prentice I.C., Harrison S.P., Jolly D. and Guiot J. 1998. The climate and biomes of Europe at 6000 yr BP: comparison of model simulations and pollen-based reconstructions. Quat. Sci. Rev. 17: 659–668.

Prentice I.C., Jolly D. and Biome 6000 participants. 2000. Mid-Holocene and glacial-maximum vegetation geography of the northern continents and Africa. J. Biogeog. 27: 507–519.

Reale O. and Shukla J. 2000. Modeling the effects of vegetation on Mediterranean climate during the Roman Classical Period: Part II. Model simulation. Glob. Planet. Chan. 25: 185–214.

Reed J.M., Stevenson A.C. and Juggins S. 2001. A multi-proxy record of Holocene climate change in Southwestern Spain: the Laguna de Medina, Cádiz. The Holocene 11: 707–720.

Roberts N. 2002. Did prehistoric landscape management retard the post-glacial spread of woodland in Southwest Asia? Antiquity 76: 1002–10.

Roberts N., Meadows M.E. and Dodson J.R. (eds) 2001a. The Holocene history of mediterranean-type environments in the Eastern Hemisphere. The Holocene 11: 631–768.

Roberts N., Reed J., Leng M.J., Kuzucuoğlu C., Fontugne M., Bertaux J., Woldring H., Bottema S., Black S., Hunt E. and Karabıyıkoğlu M. 2001b. The tempo of Holocene climatic change in the eastern Mediterranean region: new high-resolution crater-lake sediment data from Central Turkey. The Holocene 11: 721–736.

Roberts N. and Wright H.E. Jr. 1993. Vegetational, lake-level and climatic history of the Near East and Southwest Asia. In: Wright H.E. Jr., Kutzbach J.E., Webb T. III, Ruddiman W.F., Street-Perrott F.A. and Bartlein P.J. (eds), Global Climates since the Last Glacial Maximum. University of Minnesota Press, Minneapolis, pp. 194–220.

Rodrigo F.S., Esteban-Parra M.J., Pozo-Vázquez D. and Castro-Diez Y. 2000. Rainfall variability in Southern Spain on decadal to centennial time scales. Int. J. Climatol. 20: 721–732.

Rosen A.M. 1991. Early Bronze Age Tel Erani: an environmental perspective. Tel Aviv 18: 192–204.

Rosen A.M. 1995. The social response to environmental change in Early Bronze Age Canaan. J. Anthrop. Arch. 14: 26–44.

Rosen A.M. and Rosen S.A. 2001. Determinist or not determinist? Climate, environment and archaeological explanation in the Levant. In: Wolff S.R. (ed.), Studies in the Archaeology of Israel and Neighboring Lands In Memory of Douglas L.Essel. Oriental Institute, Chicago, pp. 535–549.

Rossignol-Strick M. 1995. Sea-Land correlation of pollen records in the Eastern Mediterranean for the Glacial-Interglacial transition: biostratigraphy versus radiometric dating. Quat. Sci. Rev. 14: 893–915.

Sadori L. and Narcisi M. 2001. The Postglacial record of environmental history from Lago di Pergusa (Sicily). The Holocene 11: 655–672.

Schilman B., Bar-Matthews M., Almogi-Labin A. and Luz B. 2001. Global climate instability reflected by Eastern Mediterranean marine records during the late Holocene. Palaeogeog. Palaeoclimatol. Palaeoecol. 176: 157–176.

Scott L. and Lee-Thorp J.A. 2003. Holocene climatic trends and rhythms in Southern Africa. In: Battarbee R.W., Gasse F. and Stickley C.E. (eds), Past Climate Variability through Europe and Africa. Kluwer Academic Publishers, Dordrecht, the Netherlands, pp. 69–91.

Serre-Bachet F., Guiot J. and Tessier L. 1992. Dendroclimatic evidence from Southwestern Europe and Northwestern Africa. In: Bradley R.S. and Jones P.D. (eds), Climate since AD 1500. Routledge, London and New York, pp. 349–365.

Snowball I., Korhola A., Briffa K. and Koç N., this volume. Holocene climate dynamics in High-Latitude Europe and the North Atlantic. In: Battarbee R.W., Gasse F. and Stickley C.E. (eds), Past Climate Variability through Europe and Africa. Kluwer Academic Publishers, Dordrecht, the Netherlands, pp. 465–494.

Stevenson A.C. and Harrison R.J. 1992. Ancient forests in Spain: a model for land-use and dry forest management in South-West Spain from 4000 BC to 1900 AD. Proc. Prehist. Soc. 58: 227–247.

Tarasov P.E., Webb III T., Andreev A.A., Afanas'eva N.B., Berezina N.A., Bezusko L.G., Blyakharchuk T.A., Bolikhovskaya N.S., Cheddadi R., Chernavskaya M.M., Chernova G.M., Dorofeyuk N.I., Dirksen V.G., Elina G.A., Filimonova L.V., Glebov F.Z., Guiot J., Gunova V.S., Harrison S.P., Jolly D., Khomutova V.I., Kvavadze E.V., Osipova I.M., Panova N.K., Prentice I.C., Saarse L., Sevastyanov D.V., Volkova V.S. and Zernitskaya V.P. 1998a. Present-day and mid-Holocene biomes reconstructed from pollen and plant macrofossil data from the former Soviet Union and Mongolia. J. Biogeog. 25: 1029–1053.

Tarasov P.E., Cheddadi R., Guiot J., Bottema S., Peyron O., Belmonte J., Ruiz-Sanchez V., Saadi F. and Brewer S. 1998b. A method to determine warm and cool steppe biomes from pollen data; application to the Mediterranean and Kazakstan regions. J. Quat. Sci. 13: 335–344.

Till C. and Guiot J. 1990. Reconstruction of precipitation in Morocco since 1000 AD based on *Cedrus atlantica* tree-ring widths. Quat. Res. 33: 337–351.

Trincardi F., Cattaneo A., Asioli A., Correggiari A. and Langone L. 1996. Stratigraphy of the late-Quaternary deposits in the central Adriatic basin and the record of short-term climatic events. Mem. Ist. Ital. Idrobiol. 55: 39–70.

UNESCO-FAO 1970. Vegetation Map of the Mediterranean Zone. UNESCO, Paris.

Weiss H., Courty M.-A., Wetterstrom W., Guichard F., Senior L., Meadow R. and Curnow A. 1993. The genesis and collapse of Third Millennium North Mesopotamian civilization. Science 261: 995–1003.

Wick L., Lemcke G. and Sturm M. 2003. Evidence of Late-glacial and Holocene climatic change and human impact in Eastern Anatolia: high-resolution pollen, charcoal, isotopic and geochemical records from the laminated sediments of Lake Van, Eastern Turkey. The Holocene 13: 665–676.

Verschuren D., this volume. Decadal and century-scale climate variability in tropical Africa during the past 2000 years. In: Battarbee R.W., Gasse F. and Stickley C.E. (eds), Past Climate Variability through Europe and Africa. Kluwer Academic Publishers, Dordrecht, the Netherlands, pp. 139–158.

van Zeist W. and Bottema S. 1991. Late Quaternary Vegetation of the Near East. Beihefte Zum Tübinger Atlas des Vorderen Orients. Reihe A18, Dr L. Reichert Verlag, Wiesbaden, 156 pp.

Zolitschka B., Wulf S. and Negendank J.F.W. (eds) 2000. Mediterranean lacustrine records: a contribution to the ELDP. Quat. Int. 73/74: 1–144.

18. SPELEOTHEMS AS PALAEOCLIMATE INDICATORS, A CASE STUDY FROM SOREQ CAVE LOCATED IN THE EASTERN MEDITERRANEAN REGION, ISRAEL

MIRYAM BAR-MATTHEWS (matthews@mail.gsi.gov.il)
Geological Survey of Israel
30 Malchei Israel St.
Jerusalem, 95501
Israel

AVNER AYALON (ayalon@mail.gsi.gov.il)
Geological Survey of Israel
30 Malchei Israel St.
Jerusalem, 95501
Israel

Keywords: Speleothems, Eastern Mediterranean, Soreq cave, Palaeoclimate, δ^{18}O, δ^{13}C, Palaeo-rainfall

Introduction

The best available global climate information covering the last few thousand years is that obtained from the marine record, which averages worldwide effects of temperature and ice volume change, and is only marginally influenced by the localised climate changes that occur in areas less than continental in size. Consequently one of the key questions in palaeoclimate study is to understand climate changes occurring on land, and the sea-land relationships. Continental palaeoclimates have been studied using a variety of approaches such as geomorphological analyses, lake levels, archaeological studies, and pollen studies (e.g., Street and Grove (1976), Cerling et al. (1989), Gasse et al. (1990), Rossignol-Strick (1995), Weiss (2000), Magri et al. (this volume), Roberts et al. (this volume)). Recently there has been an increased interest in using the oxygen and carbon isotopic composition of speleothems as a climate proxy, because these isotopic compositions provide climate information, and U - series isotope dates give accurate chronological information. The study of the isotopic composition of speleothems has been performed throughout the world in different climatic zones in the northern hemisphere, the tropics, the Mediterranean, and the southern hemisphere (e.g., Dorale et al. (1992, 1998), Holmgren et al. (1995), Ayliffe et al. (1998), Burns et al. (1998, 2001), Bar-Matthews et al. (1999, 2000), Denniston et al. (1999), Frumkin et al. (1999), Lauritzen and Lundberg (1999), McDermott et al. (1999), Williams et al. (1999), Musgrove et al. (2001)).

R. W. Battarbee et al. (eds) 2004. *Past Climate Variability through Europe and Africa.*
Springer, Dordrecht, The Netherlands.

One of the main problems of speleothem studies is to understand how they acquire their $\delta^{18}O$ and $\delta^{13}C$ values. The $\delta^{18}O$ values of speleothems directly reflect the temperature of deposition and the $\delta^{18}O$ of the water from which they were deposited. In turn, the latter parameter depends on a number of factors including the isotopic composition of the source of clouds and the atmospheric and hydrological evolution of rainfall (e.g., Dansgaard (1964), Gat and Dansgaard (1972), Gat (1980), Jouzel (1980), Rozanski et al. (1993)). Variations in the $\delta^{13}C$ values of calcite speleothems reflect changes of the vegetation type in the vicinity of a cave and arise because of the differences in the photosynthetic pathways between C3 and C4 type vegetation. Enrichment in the ^{13}C of the speleothems' calcite usually reflects an increase in the contribution of C4 plants to the soil CO_2 (Cerling et al. 1991; Cerling and Quade 1993). Other factors controlling the carbon isotopic composition are the age of the soil organic matter, presence of soil cover, intensity of carbonate host-rock dissolution and interactions in the unsaturated zone involving open and closed system reactions between water, soil and rock (e.g., Hendy and Wilson (1968), Hendy (1971), Bar-Matthews et al. (1996), Ayalon et al. (1998), Genty et al. (2001)). Thus, the isotopic composition of the speleothems is site dependent, and it is necessary to calibrate in each cave site how the various parameters determine the isotopic compositions of the speleothems.

Another problem of speleothem studies is that speleothem growth can be discontinuous. Discontinuity can be climatic-related, such as in northern countries when speleothem growth ceased during glaciations (e.g., Schwarcz (1986), Gascoyne (1992)) or due to aridity, as in areas in Southeastern Australia (Ayliffe et al. 1998) or the desert area in Central Texas (Musgrove et al. 2001) Oman (Burns et al. 2001) and Israel (Vaks et al. 2003). Growth discontinuity can also occur because of changes in the routing of the water reaching a cave (Bar-Matthews et al. 1996), or due to tectonic activity (Forti 1998; Lemeille et al. 1999; Kagan et al. 2002). To determine whether hiatuses are climate- or tectonic-related it is necessary to study several speleothems at each cave site (Bar-Matthews et al. 1999; 2000).

A major advantage of speleothem research in the semi-arid type climate of the Eastern Mediterranean (EM) region in Israel is that the speleothems continuously grew throughout glacial and interglacial periods, making this area ideal for reconstructing climate change using their isotopic composition.

In this paper we integrate studies of the present-day parameters such as average cave and air temperature, rainfall amount and its isotopic composition, the isotopic composition of the sea-surface source, and of the host rock and soil, in order to understand how the isotopic composition of speleothems recorded palaeoclimate conditions during the last 185 kyr.

Geological and hydrological setting of the Soreq Cave (Israel)

The Soreq Cave is one of a series of karstic caves situated within the steep westward-dipping flank of the Judean Hill anticline and is located approximately 40 km inland and 400 m above sea level in Cenomanian dolomitic Weradim Formation (Fig. 1).

The location of the cave, on a steeply dipping flank results in a variation of its depth below the surface from less than 10 m at the western end to 40–50 m at the eastern end. The cave ceiling is crossed by two main fracture systems oriented in E-W and NW-SE respectively. The soil cover above the cave is about 30 cm thick and is composed of Terra Rosa and Rendsina soils, with typical Mediterranean C3 type vegetation. Access to the cave

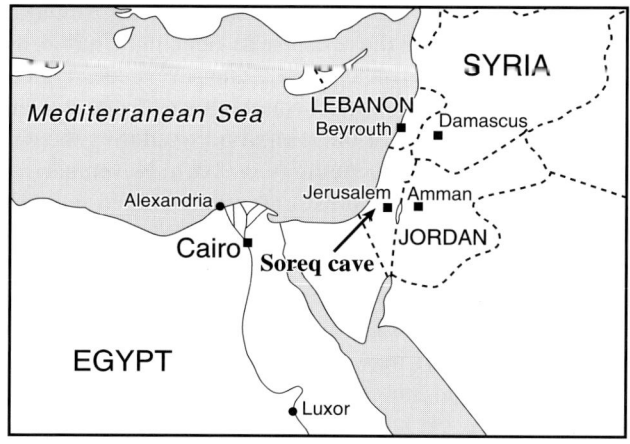

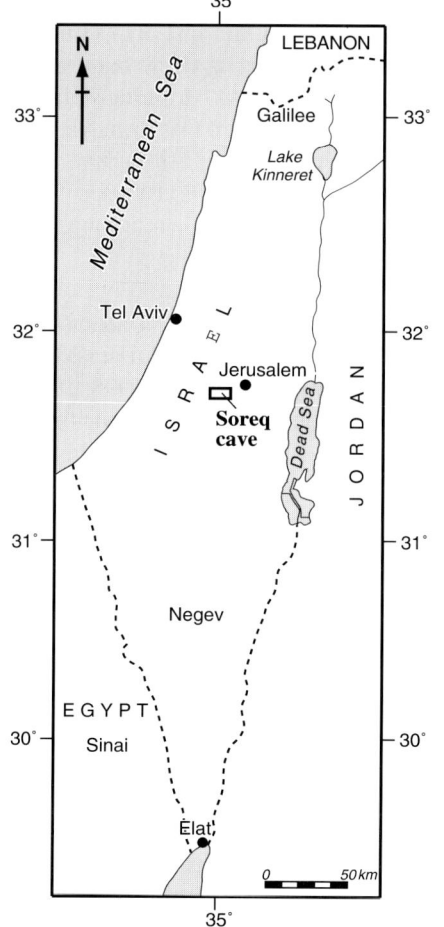

Figure 1. Map showing the location of Soreq cave, Israel. The cave is located approximately 40 km inland east of the Mediterranean Sea, and 400 m above sea level.

was made possible in 1968 as a result of quarrying and, since then, the interior conditions are controlled and kept almost identical to those before its opening. Until its discovery the cave was closed with no natural opening. The climate in the cave area is typical of Eastern Mediterranean (EM) semi-arid conditions, with mean winter temperatures of 14 °C, and mean summer temperatures of 26 °C. 70% of the rainfall occurs during the winter months (December - February), 10–20% during the autumn (October to November), and 10–20% during the spring (March to May). Most of the rainstorms are associated with Mediterranean fronts although some pass over the Red Sea. Average annual precipitation above the cave is 500 mm.

There are two major modes of water flowing in to the cave: (i) *vadose flow* (fast-drip water), flowing through large joints and fissure systems, and (ii) *vadose seepage* (slow drip water), slow migration of water through a network of small micro-fissures, their flow being mainly controlled by the structural and lithological nature of the host rock (e.g., Thrailkill (1968), Williams (1983), Choquette and James (1987), Hill (1987)). In the Soreq cave the fast-drip water runs into the cave during the winter months. The slow-drip water emerges out of the tips of stalactites, and occurs throughout the year. The quantitative balance of the water within the cave is dominated by the fast-drip sources. The dripping waters accumulate in pools, which vary in volume from a few hundred ml to about 400 litres. Some pools receive their water directly from slowly dripping water sources (stalactite-drip, flowstone); others receive most of their water from fast-drip. All waters are supersaturated with respect to calcium carbonate and capable of depositing low magnesium calcite (LMC) (Bar-Matthews et al. 1994).

Methods of study

In this paper we will not deal specifically with all the experimental methods that were applied to the study. Those are summarised in several publications (Bar-Matthews et al. 1991; 1996; 1997; 1998; 1999; 2000; Halicz et al. 1997; Ayalon et al. 1998; 1999; 2002; Kaufman et al. 1998; Matthews et al. 2000). What we wish to emphasise is the frequent monitoring of rain and cave water that has been carried out since 1990, and the high-resolution study of the isotopic composition of speleothems.

The collection of rainwater above the cave was made during all rain events in order to determine precisely the amount and distribution of the rainfall. The rainwater was quantitatively collected by allowing water to accumulate in a large funnel, and drip into a narrow-headed bottle. This type of bottle was used in order to reduce evaporation. The collecting bottles were changed several times during each rain event in order to monitor the change of the isotopic composition of the rainwater throughout the storm, and to minimise water loss due to evaporation. The average annual rainfall $\delta^{18}O$ and δD values were calculated from these measurements after weighing, to determine the proportion of the annual precipitation contributed by each rainfall event. The Meteorological Survey of Israel recorded temperatures continuously during all rainfall events.

Cave water was collected at various sites within the cave. The sampling locations were chosen in order to cover all types of cave water as well as different topographic levels and varying positions relative to the ceiling and the thickness of the rock cover. Sample collection was carried out at least 4–6 times per year with more frequent sampling during the rainy season.

Speleothem samples 60 to 250 mm in diameter and up to one metre high, from various locations within the cave were studied for their fossil record. The samples, mainly stalagmites and stalactites composed of LMC, were cut perpendicular or along their growth axis in order to expose the growth layers and permit a check for secondary alteration (Bar-Matthews et al. 1997). All samples were deposited in isotopic equilibrium according to the criteria of Hendy (1971), which is determined by the constancy of $\delta^{18}O$ and irregular $\delta^{13}C$ variations along a given layer of contemporaneous calcite growth (Bar-Matthews et al. 1993).

In order to perform a high-resolution study ^{230}Th-U dating was performed on a series of fine layers about 1.0 cm thick that were separated from each other with a 0.6 mm diamond saw. About 80% of the detrital material included in each sample was removed by physical methods using an ultrasonic probe as described in Bar-Matthews et al. (1997). The quality and the error of the ages were discussed in detail by Kaufman et al. (1998). The carbon and oxygen isotopic measurements were performed every 0.5 mm. A continuous isotopic record was obtained by assuming that the measured age represents the centre of the dated layer, and that the growth rate from the centre to the margin of each layer is constant.

To obtain maximum resolution of the isotopic record we chose the speleothems with the fastest growth rate. None therefore covered the entire age interval of 185 kyr. In order to obtain a continuous record we compared the $\delta^{18}O$ and $\delta^{13}C$ profiles of several speleothems covering similar time intervals, which enabled us to extend the isotopic record by matching the oldest layer of a younger speleothem with the youngest layer of an older speleothem. Because the isotopic record includes no significant time intervals in which ages are absent, we consider the record of the Soreq cave speleothems to be essentially continuous. The isotopic record was determined for 20 different stalagmite and stalactites from various locations within the cave, 95 Thermal Ionization Mass Spectrometric (TIMS) ages, and more than 2500 $\delta^{18}O$ and $\delta^{13}C$ analyses.

The isotopic composition of rain, cave water and modern speleothems

Rain

The pattern, amount, distribution, air temperature, and the isotopic composition of rain events vary within and between hydrological years. A hydrological year is determined from September, when first rain events may occur to the end of August (Ayalon et al. 1998). The main characteristics of the rain events are as follows: (i) the rainfall occurs as sporadic events lasting a few hours to few days and are separated by dry spells that extend from a few days to few weeks; (ii) the lowest $\delta^{18}O$ and δD values are associated with the most intense rain events (>15 mm), which usually occur during mid-winter (December to February) when the air temperature is between 5 and 10 °C. These events comprise more than 60% of the annual rainfall; (iii) the more isotopically enriched rains mostly occur during the early and late rainy season when air temperatures are usually above 15 °C; (iv) the isotopic composition during individual rain events varies by up to 5‰ (SMOW) in $\delta^{18}O$, and up to 50‰ (SMOW) in δD. During the whole rainy season $\delta^{18}O$ values vary by ~16‰ (−12‰ to +4‰) and δD by ~100‰ (−80‰ to +20‰); and (v) the cumulative $\delta^{18}O$ plots demonstrate a general stepwise decrease with time, which is dominated by extended plateaus and a steep drop between them (Fig. 2a-f).

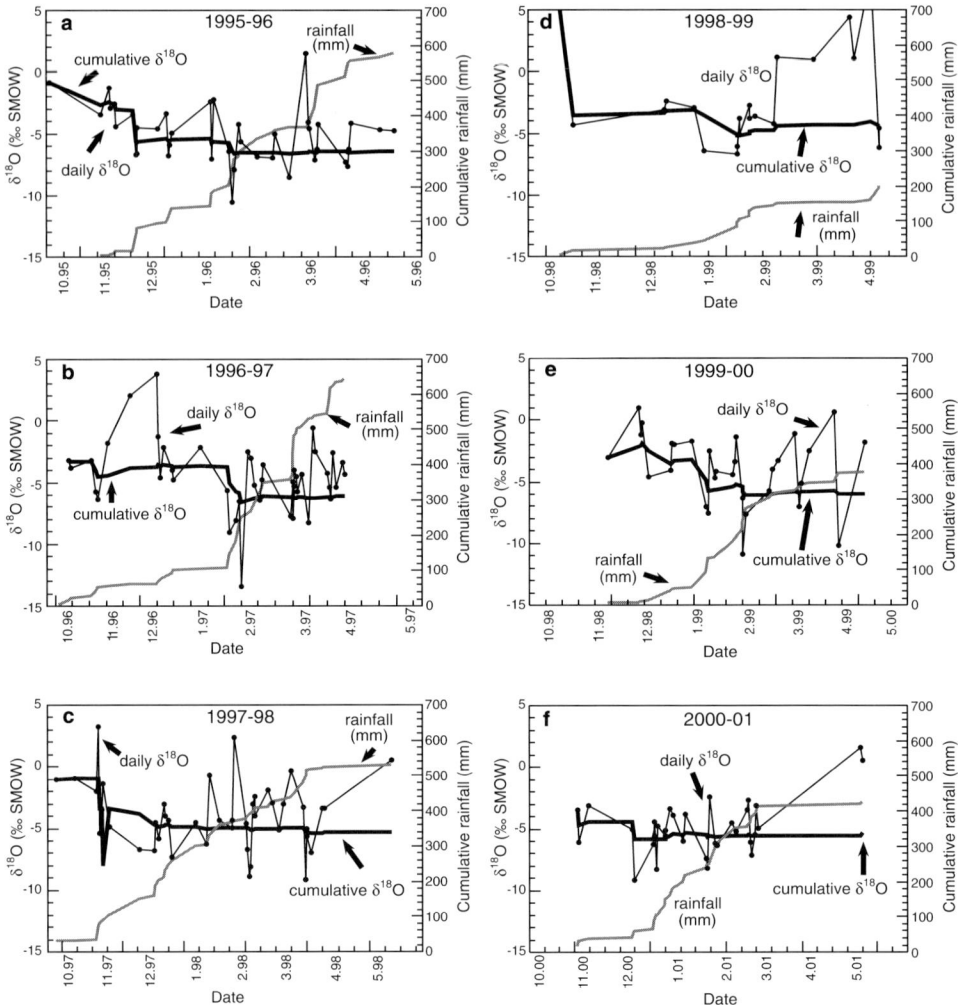

Figure 2. Daily and cumulative weighted $\delta^{18}O$ values of rainfall and the cumulative rainfall above the Soreq cave as a function of the timing of the rainfall in the years 1995–96 (a), 1996–97 (b), 1997–98 (c), 1998–99 (d), 1999–00 (e), 2000–01 (f).

The most intense rain events (>15 mm), follow the Mediterranean Meteoric Water Line (MMWL) trend with a slope of 8, and d-excess of 20 to 30‰ (Fig. 3), as defined by Gat and Dansgaard (1972), and Gat and Carmi (1970, 1987). This indicates that the rain in Soreq cave area is derived from the EM sea surface. As the amount of rainfall decreases to less than 15 mm, the rainwater become enriched in D and ^{18}O and the co-variations change with a slope of <8 and relatively lower d-excess, with some of the data intersecting and crossing the Meteoric Water line (MWL) (Fig. 3). The deviation from the MMWL of all events with less than 15 mm rainfall is ascribed to evaporation below the cloud (e.g., Gat and Carmi (1970, 1987), Gat (1980)).

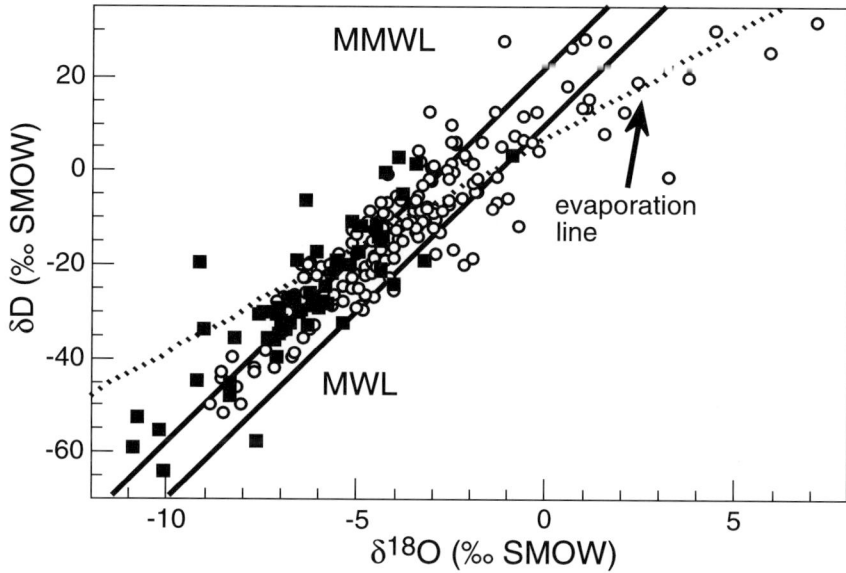

Figure 3. The δD vs. δ¹⁸O relationship of the rainwater at the Soreq cave area. Events with more than 15 mm rain (solid squares) fall on or above the MMWL. Events with less than 15 mm rain (open circles) fall on an evaporation line, which intersects the MMWL and the MWL. For a more detailed description of evaporation line trend see Gat (1996).

Eleven years monitoring of the relationship between the rainfall amount and its δ¹⁸O at the Soreq cave site are plotted in Figure 4. The results show an inverse linear relationship; with minimum δ¹⁸O values associated with rainy years and higher δ¹⁸O values with dry years. The variation of δ¹⁸O with rainfall amount given by a linear regression is $1.02 \pm 0.11‰$ per 200 mm rain.

Cave water

The rate of fast-drip water at different locations in the cave varies from 6 ml/min to more than 50 ml/min, depending on the rainfall amount, the rock cover thickness, and the fracture system. Between rainstorms, the drip rate slows down and ceases towards the end of the rainy season. There is a variation in both the timing of dripping relative to the onset of rain and in its isotopic composition as a function of the thickness of the rock-cover above the cave. Where the rock-cover exceeds 40–50 m, the time lag varies between several weeks and three months, when ~350–500 mm cumulative rainfall has fallen (Ayalon et al. 1998). The isotopic composition of this type of fast-drip varies only slightly from −6.0‰ to −5.6‰ in δ¹⁸O, and from −30‰ to −24‰ in δD. These values are closest to the average annual rainfall, thus the fast drip in the deep rock-cover produces the most accurate reflection of the yearly rainfall. The fast-drip immediately ceases after the end of the rainy season in areas with a thin rock-cover, whereas in areas with thick rock-cover, fast dripping may last up to two weeks and more.

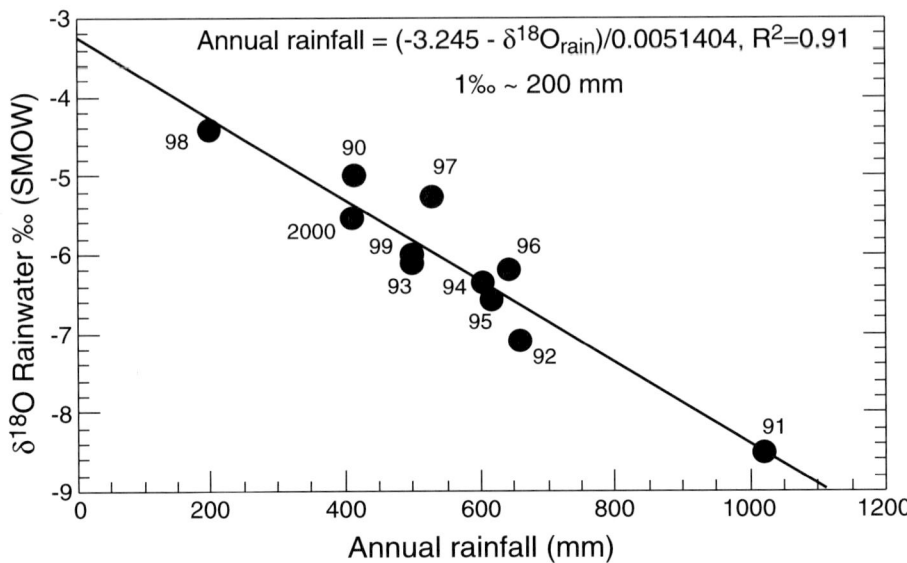

Figure 4. A plot of the average $\delta^{18}O$ values of rainfall against rainfall amount for 11 years sampling at the Soreq cave site, showing a linear regression of $1.02 \pm 0.11\%o$ per 200 mm rainfall ($R^2 = 0.91$).

The average $\delta^{18}O$ values of all fast-drips during the last 9 years are shown in Figure 5. The mean $\delta^{18}O$ values vary between years from $-5\%o$ to $-6\%o$ and δD variations are $-17\%o$ to $-28\%o$. The average annual $\delta^{18}O$ of fast-drip waters are almost always higher than those of the rainfall by $\sim 0.5\%o$ to $1.0\%o$ and by $\sim 4\%o$ in δD, apart for the driest year of 1998 (200 mm rain) when the fast-drip water are have slightly lower values than the rain, and during the wettest year of 1992 (1000 mm rain) the rain water have $\delta^{18}O$ and δD values much lower than the cave water.

The rate of slow-drip water from the tips of large numbers of straw stalactites is usually constant and slow during the dry seasons with only 0.5 ml/day. The $\delta^{18}O$ ($\sim -4.6\%o$) and δD ($\sim -14\%o$) values of slow-drip water are quite constant and vary only by less than $\sim 1\%o$ and less than $10\%o$ respectively between hydrological years (Fig. 5).

The fast-drip and the slow-drip waters are always isotopically higher than the average rainfall, because of partial loss of the isotopically lower intensive rain events by runoff, but also due to evapotranspiration and mixing process in the vadose zone. $\delta^{18}O$ and δD relations of both slow- and fast-drip waters follow the general trend of the MMWL (Fig. 6), showing that the relatively massive rain events are the dominant source of the cave water, though they are partially lost by runoff. The infiltrating water thermally equilibrates in the vadose zone and as a result, reaches the cave interior at an average air temperature of $\sim 18\,^{\circ}C$.

The $\delta^{13}C$ values of the Dissolved Inorganic Carbon (DIC) of the cave waters vary from $\sim -5\%o$ to $-14.5\%o$ and DIC concentrations range from 2 mM to 11.5 mM. Processes that could account for the large range in DIC concentrations and $\delta^{13}C$ values are the various modes of water infiltrating into the cave and Rayleigh distillation processes during CO_2 degassing and carbonate precipitation (Bar-Matthews et al. 1996). However, most of the $\delta^{13}C$ values of the DIC are between $-10\%o$ and $-14\%o$. These low values and the

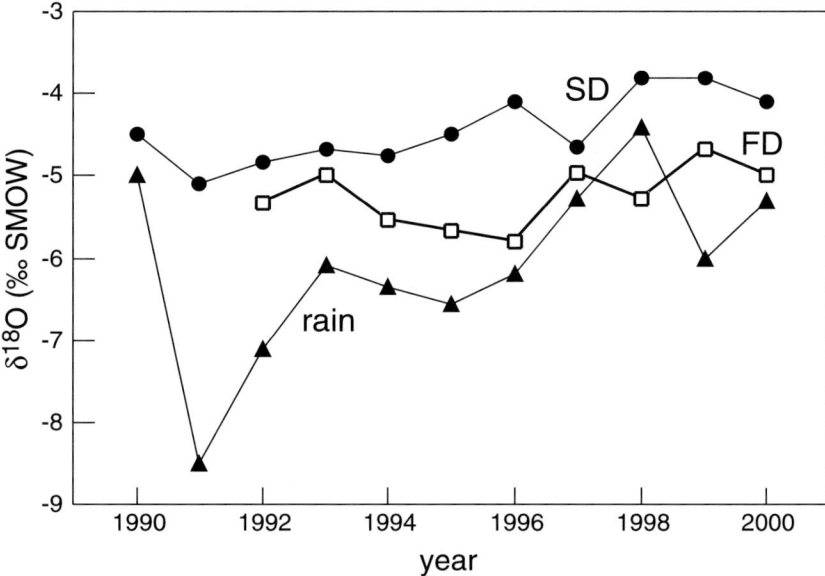

Figure 5. A plot showing the average annual $\delta^{18}O$ value of the rainwater above the Soreq cave (filled triangles), the cave waters: slow-drip (filled circles) and fast-drip (empty squares). The average annual $\delta^{18}O$ values of fast drip water are almost always higher than those of the rainwater by about 0.5‰ to 1.0‰. Fast-drip water was collected only since 1992.

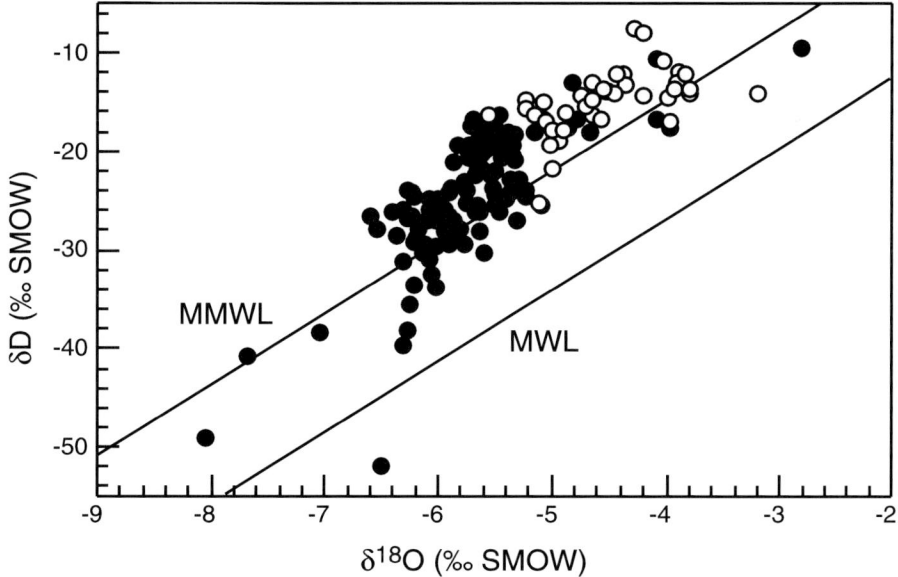

Figure 6. The δD versus $\delta^{18}O$ plot of fast-drip (filled circles) and slow-drip water (open circles) from the Soreq cave. The MMWL and global MWL are shown for reference. The δD and $\delta^{18}O$ values of the cave water follow the MMWL fall into two groups, the first in which the isotopically depleted values are associated with fast-drip water, and the second in which the most enriched values match slow-drip water.

supersaturation of the water with respect to calcium carbonate indicate that the seepage waters have dissolved both soil-CO_2 derived from C3 type vegetation and marine dolomitic host rock.

The residence time of the water in the vadose zone was determined by the study of tritium concentration of the cave water from different water sources within the cave, on samples collected between 1990 and 2000. The results were compared with the tritium concentrations of the annual rainfall from 1952 to 1998 at a station 35 km from the cave. The tritium in most of the cave waters is low (3–7 tritium units, TU), indicating that they originate from the last year's rain. However, in several of the cave waters tritium concentration was so high (up to 110 TU) that they must have originated mainly in the peak tritium rain years of 1964. Some of the other cave waters had tritium concentrations, which, though somewhat lower, must have originated largely from rain that fell between 1962 and 1966. The consequences of these tritium observations are that it generally takes 26–36 years for rainwater to percolate from the ground surface to the Soreq cave ceiling (Kaufman et al. 2003). This water resides longer within micro-fissures before reaching the cave.

Modern speleothems

Contemporary modern deposition of LMC samples is occurring at a number of sites: (i) pool deposits, consisting of the outermost LMC layer that are in direct contact with the pool water; (ii) surface rafts consisting of few mm thick LMC crusts floating on the surface of the pools; (iii) stalagmites, including those forming from fast-drip water that have started to grow on the concrete path built in the cave about 25 years ago, and newly-formed crusts on stalagmites located below present-day fast-drip and stalactite-drip sources; (iv) flowstones, consisting of newly-formed thin white crusts that are forming today where pools overflow after storm events; and (v) stalactites, consisting of the lower tips of straw stalactites that are in contact with drops of present-day stalactite-drip water.

The $\delta^{18}O$ values of the various types of modern LMC samples represented by 65 analyses are between –5.7‰ and –4.2‰ (PDB), but most of the modern speleothems (50 measurements) have a narrower range of $\delta^{18}O$ values of –5.6‰ to –5.2‰ with an average of –5.4±0.4‰. Using the O'Neil et al. (1969) calcite-water fractionation equation, it becomes clear that the speleothems are deposited under isotopic equilibrium with the average $\delta^{18}O$ value of the Soreq cave water (–4.7‰), at an average temperature of 18 °C.

$\delta^{13}C$ values vary from –11.6‰ to –7.8‰ (PDB). The higher values (−9.8‰ to –7.8‰) characterise the surface rafts due to CO_2 degassing, but most of the modern speleothems have $\delta^{13}C$ values of –10.7 ± 1.0‰. This value indicates that the water in the vadose zone reacts with soil-CO_2 derived from the C3 type vegetation and with the marine dolomitic host rock with a $\delta^{13}C$ value of ∼1–2‰ under equilibrium (Hendy 1971).

The isotopic composition of fossil speleothems

The isotopic profile of Soreq cave speleothems during the last 185 kyr is shown in Figure 7. This time period includes the last six marine isotopic stages (e.g., Imbrie et al. (1984), Martinson et al. (1987)), among them two interglacial stages, marine isotopic stage 5 and the Holocene. The main characteristic of the isotopic profile is that the $\delta^{18}O$ variations are

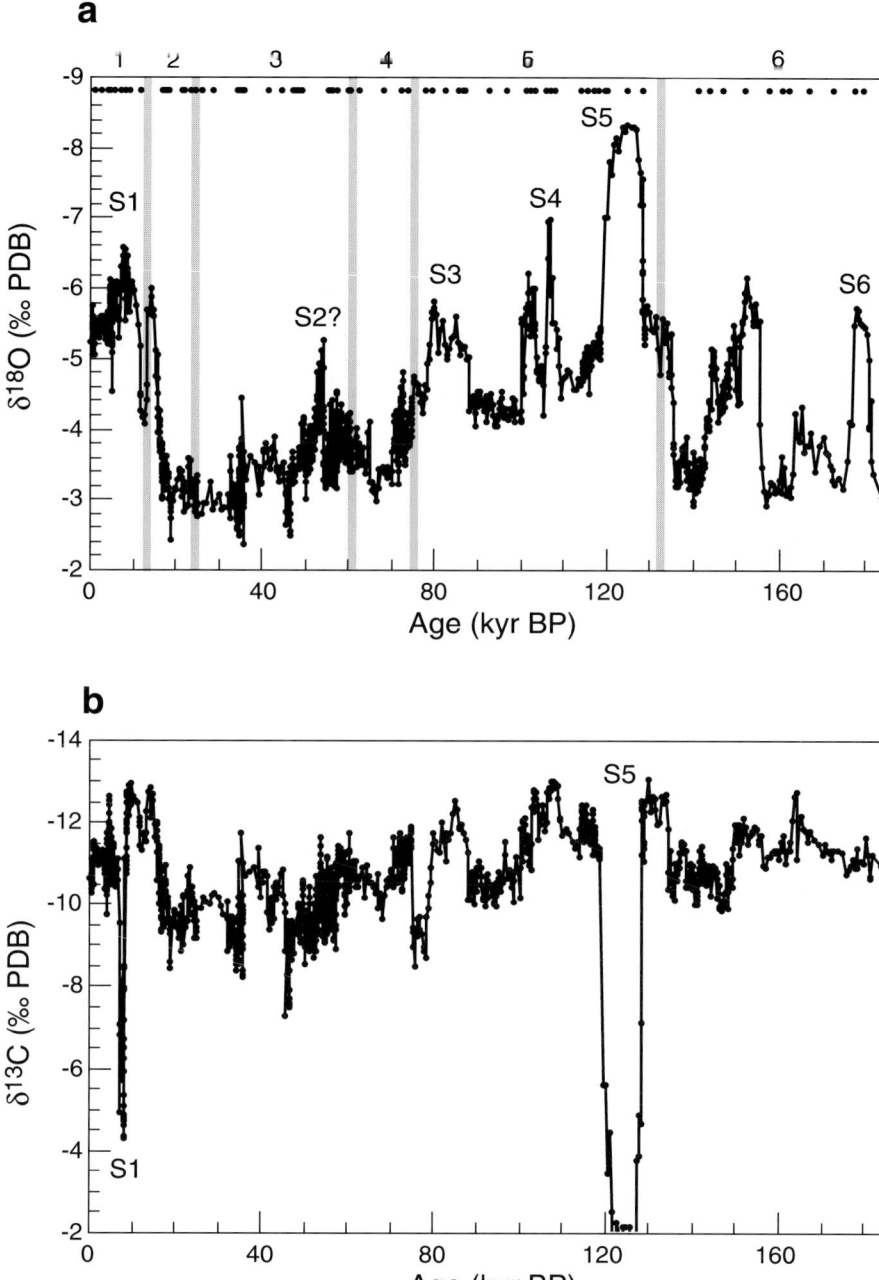

Figure 7. $\delta^{18}O$ (a) and $\delta^{13}C$ (b) values of Soreq cave speleothems deposited during the last 185 kyr. The circles at the top of A indicate the TIMS ages. Six marine isotopic stages 1–6 are marked and divided by vertical lines. The timing of sapropels S6 to S1 is indicated.

in order of 6‰ from −8.3‰ to −2.4‰ (Fig. 7a) with the highest $\delta^{18}O$ values occurring during glacial periods. The largest $\delta^{18}O$ change occurs during the transition from glacial to interglacial periods (marine isotopic stage 6 to stage 5 and between marine isotopic stage 2 and the Holocene). $\delta^{13}C$ values vary from −13‰ to −1.6‰, and the largest variation occurs during interglacial marine isotopic stages 5.5, and during the early Holocene (Fig. 7b). Superimposed on the long term isotopic fluctuations are smaller scale oscillations on millennial and centennial time-scales.

Marine isotopic glacial stages

The $\delta^{18}O$ profile during the marine isotopic stage 6 (185–130 kyr) time interval is shown in Figure 8a and the general pattern is of low amplitude $\delta^{18}O$ "oscillations" (less than 1.0‰) lasting for a few hundred years and of very high amplitude oscillations ranging from ∼−6‰ to ∼−3‰ and lasting for a longer time. High $\delta^{18}O$ values of ∼−3‰ occur at 185 kyr, between 175 kyr and 156 kyr (with a slight decrease to ∼−4‰ at 165 kyr) and between 142 kyr and 136 kyr. Minimum $\delta^{18}O$ values of ∼−6‰ occur at 180–178 kyr (peak #1) and at 155–152 kyr (peak #2, Fig. 8a).

$\delta^{13}C$ values of this stage vary only between −12.8‰ and −10.0‰ (Fig. 8b). However, the $\delta^{13}C$ peaks do not always follow the $\delta^{18}O$ peaks, and the general trend shows that the relatively lower $\delta^{13}C$ values from about −12.8‰ to −11‰ characterise the early marine isotope stage 6, from 185 kyr to 150 kyr, and relatively higher $\delta^{13}C$ values of −11‰ to −10‰ characterise the later period, from 150 kyr to 134 kyr (Fig. 8b).

The oxygen isotopic variations during the last glacial, i.e., marine isotopic stages 4, 3 and most of stage 2, from 74 kyr to 17 kyr, show a different pattern. Considerably larger $\delta^{18}O$ oscillations from −5.5‰ to −2.4‰, lasting a relatively long time, and low amplitude $\delta^{18}O$ "variation" (less than 1.0‰) lasting only for a few hundred years (Fig. 8c) is also observed during this time interval. However unlike marine isotopic glacial stage 6, during this period $\delta^{13}C$ values range from −12‰ to −7.5‰ and mimic the $\delta^{18}O$ trend (Fig. 8d).

$\delta^{18}O$ and $\delta^{13}C$ maxima, with the highest values up to about −2.4‰ and −7.5‰ to −8.5‰ respectively occur at 46 kyr, 35 kyr, 25 kyr and 19 kyr (peaks # 4, 6, 7 and 8 respectively), a minimum value of −5.5‰ occurs at 54 kyr and a fairly low value of −4.5‰ is found at 36 kyr (peaks #3 and 5 respectively, Fig. 8c). Analogous $\delta^{13}C$ minima (∼−11.0‰) are observed at the same time (Fig. 8d). A sharp drop of approximately 3‰ in both $\delta^{18}O$ (from −3‰ to −6‰) and $\delta^{13}C$ (from −9.5‰ to −13‰) occurs from 17 kyr to 14 kyr (equivalent to the period of deglaciation, peak #10, Fig. 8c). This drop is accompanied by a short-lived isotopic event at 16.5 kyr (peak #9, Fig. 8c), where both $\delta^{18}O$ and $\delta^{13}C$ show a sudden increase from −3.6‰ to −3‰ and from −10.5‰ to −9.5‰ respectively. At the end of marine isotopic stage 2 there is a relatively high value $\delta^{18}O$ event (−4.1‰) with its peak at 12.5 kyr (peak #11, Fig. 8c) that may be correlated with the Younger Dryas event (Berger and Jansen 1995).

Marine isotopic interglacial stages

During marine isotopic stage 5 (130–74 kyr) a number of low $\delta^{18}O$ events labeled #12, #13, and #14 can be identified (Fig. 9a). Event #12 lying between 128 and 119 kyr is characterised

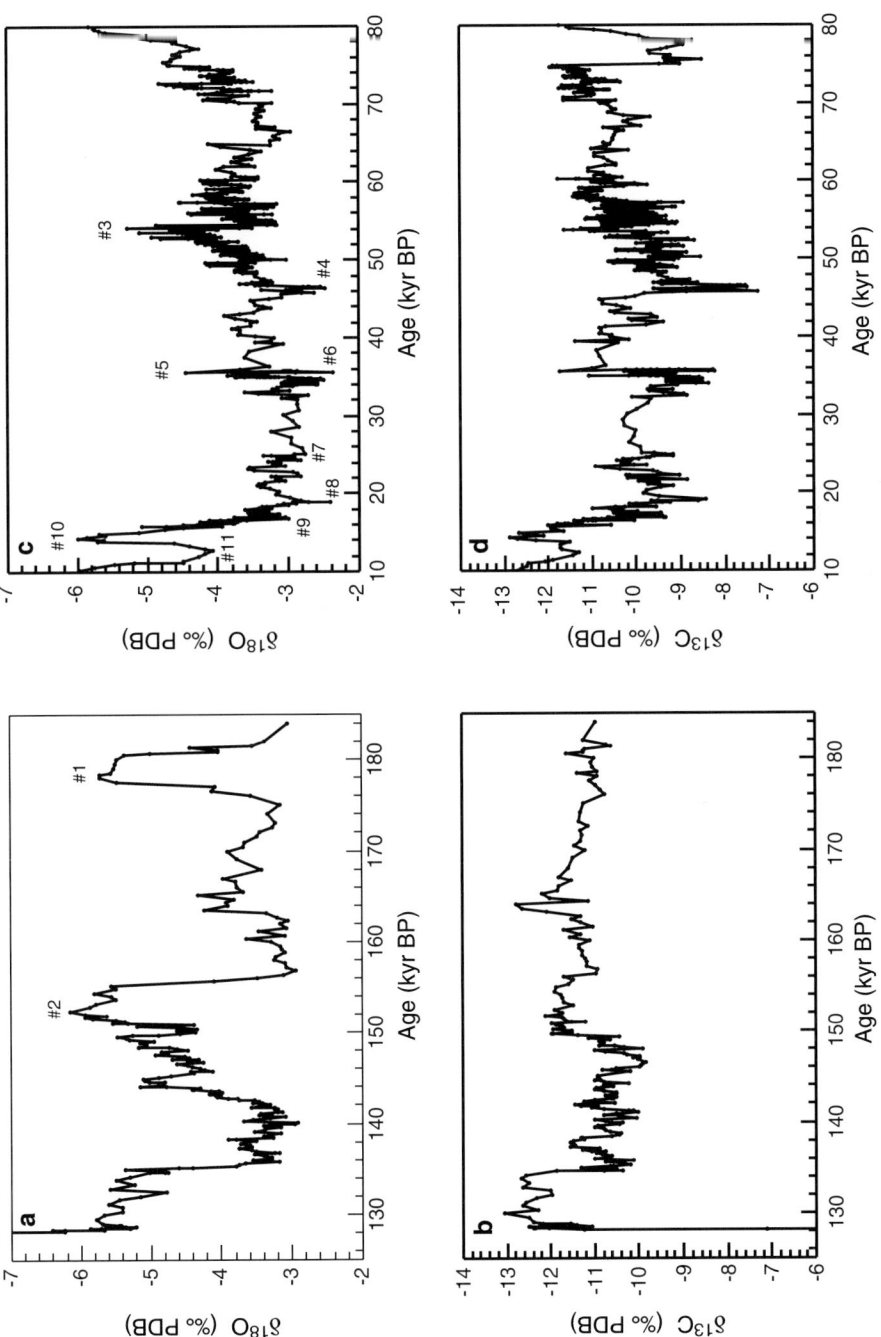

Figure 8. $\delta^{18}O$ and $\delta^{13}C$ values of Soreq cave speleothems deposited during marine isotopic glacial stages. The isotopic records during marine isotope stage 6 are shown in Figures 8a and 8b. Figures 8c and 8d show the $\delta^{18}O$ and $\delta^{13}C$ records of speleothems during marine isotope stages 4, 3 and 2. Low $\delta^{18}O$ events are defined at ~178 kyr and ~152 kyr (#1 and #2 in 8a); at 54, 36 and 14 kyr (#3, #5 and #10 in 8c). High $\delta^{18}O$ and $\delta^{13}C$ events are marked in 8c at 46 kyr, 35 kyr, 25 kyr 19 kyr, 16.5 kyr and 12.5 kyr (#4, #6, #7, #8, #9 and #11, respectively). The palaeoclimate significance of these events is described in the text.

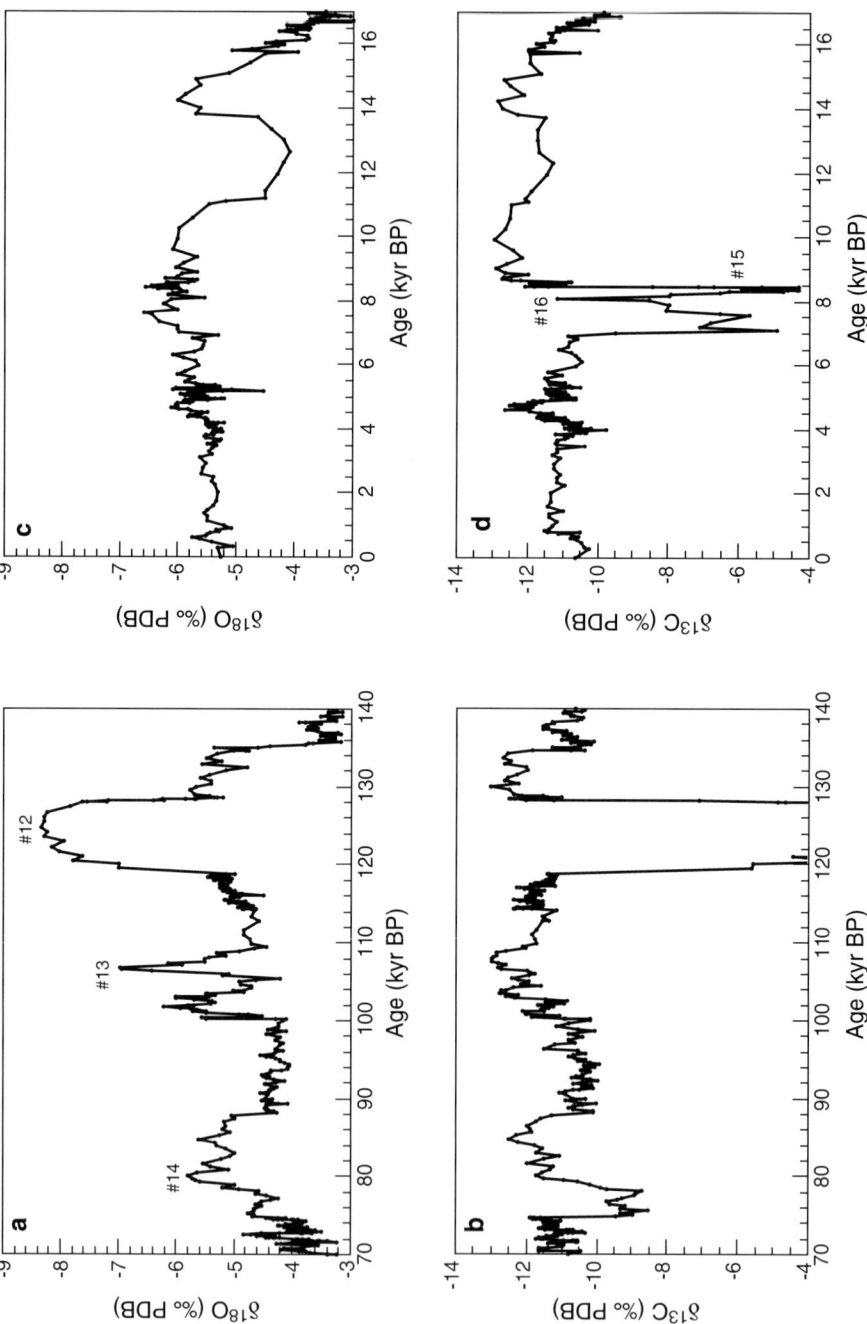

Figure 9. $\delta^{18}O$ and $\delta^{13}C$ values of Soreq cave speleothems deposited during interglacial marine isotope stages. Oxygen and carbon isotopic records during marine isotopic stage 5 are shown in Figures 9a and 9b respectively. Low $\delta^{18}O$ events coupled with very high $\delta^{13}C$ values are marked as #12 in Figure 9a. The other low $\delta^{18}O$ events that are coupled with low $\delta^{13}C$ values are marked as #13 and #14 in Figure 9a. The $\delta^{18}O$ and $\delta^{13}C$ record during the Holocene is shown in Figures 9c and 9d respectively, and the event of low $\delta^{18}O$ values coupled with high $\delta^{13}C$ values is marked as event #15 in Figure 9d. The 8.2-kyr event is marked as #16 in Figure 9d.

by the lowest $\delta^{18}O$ values ($-8.3‰$) and the highest $\delta^{13}C$ values of the studied time period, reaching about $-2‰$ (Fig. 9b). Event #12 was followed by two other isotopic events within marine isotopic stage 5. Event #13 lasted from 108 to 100 kyr and is characterised by the two minima at 107 kyr ($\delta^{18}O$ $-7‰$, and $\delta^{13}C$ $-13‰$), and at 101 kyr ($\delta^{18}O$ $-6.2‰$, and $\delta^{13}C$ $-12.5‰$). Event #14 lasted from 85 kyr to 79 kyr with minimum values of $\delta^{18}O$ ($-5.8‰$), and of $\delta^{13}C$ ($-12‰$) occurring at 80 kyr. Thus, unlike event #12, events #13 and #14 are characterised by low values of both $\delta^{18}O$ and $\delta^{13}C$.

The beginning of the Holocene is characterised by low $\delta^{18}O$ ($-6‰$) and low $\delta^{13}C$ values ($-12‰$), but immediately after, between 8.5 kyr and 7 kyr (peak #15, Fig. 9d) there is a combination of low oxygen isotopic values ($-6.5‰$) and very high carbon isotopic values ($-5.0‰$ to $-4.0‰$, similar to event #12). During this time interval another isotopic event is revealed at about 8.2 kyr, when $\delta^{18}O$ values increase only slightly (from $-6.5‰$ to $-5.7‰$) but $\delta^{13}C$ values decrease sharply to from $-4‰$ to $-11‰$ (event #16, Fig. 9d) and probably parallels the Holocene cooling event recorded in ice cores.

The $\delta^{18}O$ and $\delta^{13}C$ values of speleothems that were formed during the last 7000 years are shown in detail in Figure 10. Two major isotopic events are observed: the very short-lived high $\delta^{18}O$ ($-4.5‰$) value events at 5.2 kyr (peak #17) and the increase in $\delta^{18}O$ (from $-6.1‰$ to $-5.2‰$) and $\delta^{13}C$ (from $-12.7‰$ to $-9.7‰$) during a 400 year period, from 4.6 kyr to 4.0 kyr (peak #18). From 3.6 kyr to 1.5 kyr there is very little change in the isotopic composition of the speleothems, and during the last 1500 years $\delta^{18}O$ and $\delta^{13}C$ variations become more evident.

Discussion

Oxygen isotopic composition of fossil speleothems

Most of the rainstorms in the EM area originate in the North Atlantic and Northern Europe. However, the $\delta^{18}O$ - δD relationships of the rain in the EM region do not follow the global MWL but follow the MMWL which is also true for the rain in Soreq cave area (Fig. 3). This is due to the interaction of cold and dry fronts with warm and humid air above the EM Sea (Gat 1996). This implies that any changes in the isotopic composition of the EM Sea surface, which is directly related to sea surface temperatures (SST) and salinity, will directly affect the $\delta^{18}O$ of the rain. In addition, the isotopic composition of the rain is also determined by the rainfall amount as evident from the inverse linear relationships between the rainfall amount and its isotopic composition (Fig. 4). The waters that infiltrates into the cave also parallel the MMWL, indicating that they are derived from the rainwater (Fig. 6), although they are slightly more enriched in $\delta^{18}O$ (Fig. 5).

As shown above, the $\delta^{18}O$ value of modern speleothems reflect that they are formed under isotopic equilibrium from the cave water. The processes influencing the $\delta^{18}O$ of the cave water: rainfall amount, average cave temperature and the $\delta^{18}O$ value of the EM Sea source are all climate dependent. Thus, these processes are indirectly reflected in the $\delta^{18}O$ value of the speleothems.

The $\delta^{18}O$ profile of Soreq cave speleothems during the last 185 kyr shows a general correspondence with the stacked oxygen isotope record of Martinson et al. (1987) suggesting that global climatic changes are recorded by the Soreq cave speleothems. An even finer-scaled matching of peaks is obtained by the comparison of the Soreq cave speleothems

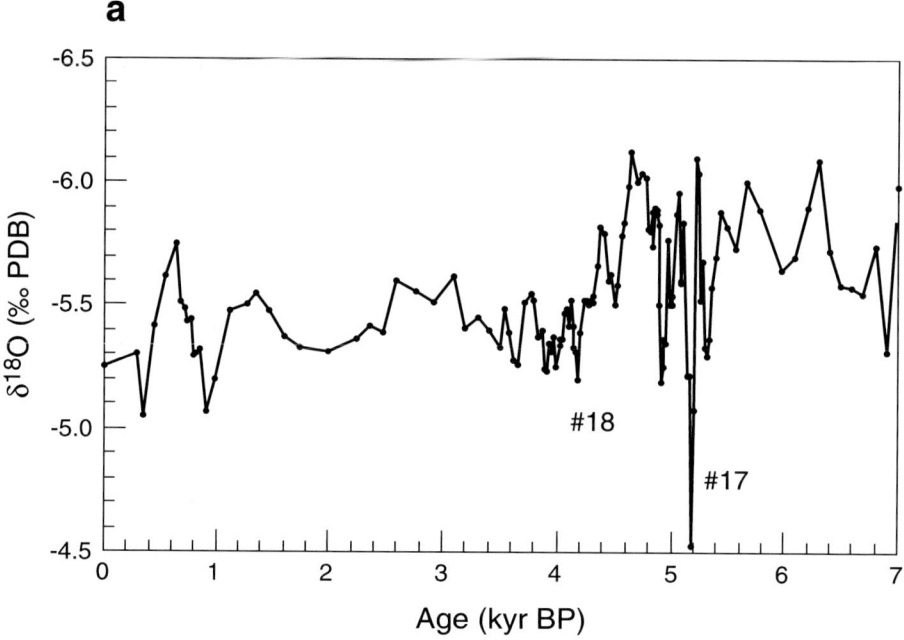

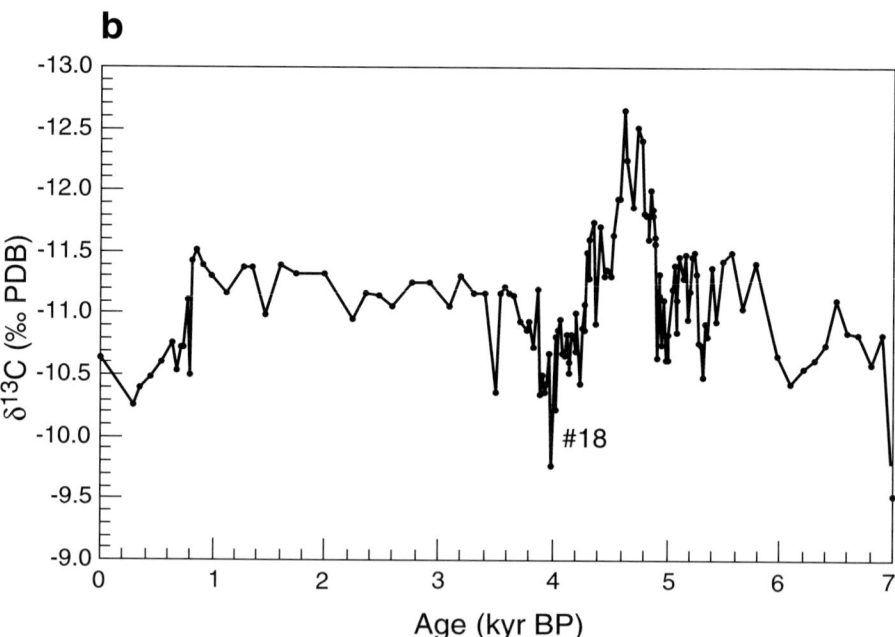

Figure 10. $\delta^{18}O$ (a) and $\delta^{13}C$ (b) values of Soreq cave speleothems deposited during the last 7000 years. The sharp increase in $\delta^{18}O$ values at 5.2 kyr is marked as #17 in (a), and the increase in $\delta^{18}O$ and $\delta^{13}C$ value from 4.6 kyr to 4.0 kyr is marked as #18 in (a) and (b).

isotopic record with the $\delta^{18}O$ record of *Globigerinoides ruber* (*G. ruber*) (Fig. 11) from studies by Vergnaud Grazzini et al. (1977) and Fontugne and Calvert (1992). *G. ruber* is a shallow epi-pelagic species, which thrives in the upper 100 m of the EM Sea (Reiss et al. 1999) and is considered to record sea-surface conditions (e.g., Fairbanks et al. (1982), Duplessy et al. (1981), Deuser (1987), van Os and Rohling (1993), Kallel et al. (2000), Kallel et al. (this volume)). The matching peaks and the amplitudes of the $\delta^{18}O$ changes are similar, suggesting that the climatic events in the EM Sea and on land have been linked throughout this period (Bar-Matthews et al. 2003). δD - $\delta^{18}O$ relationships of fluid inclusions extracted from Soreq cave speleothems plot along the trend of the MMWL also support this link. During cold glacial conditions, d-excess values of fluid inclusions in samples that were deposited during the coldest LGM event (event #8, Fig. 8c) are close to the global MWL. This shift away from the MMWL is due to the lower sea surface and atmospheric temperatures at the time of deposition (Matthews et al. 2000).

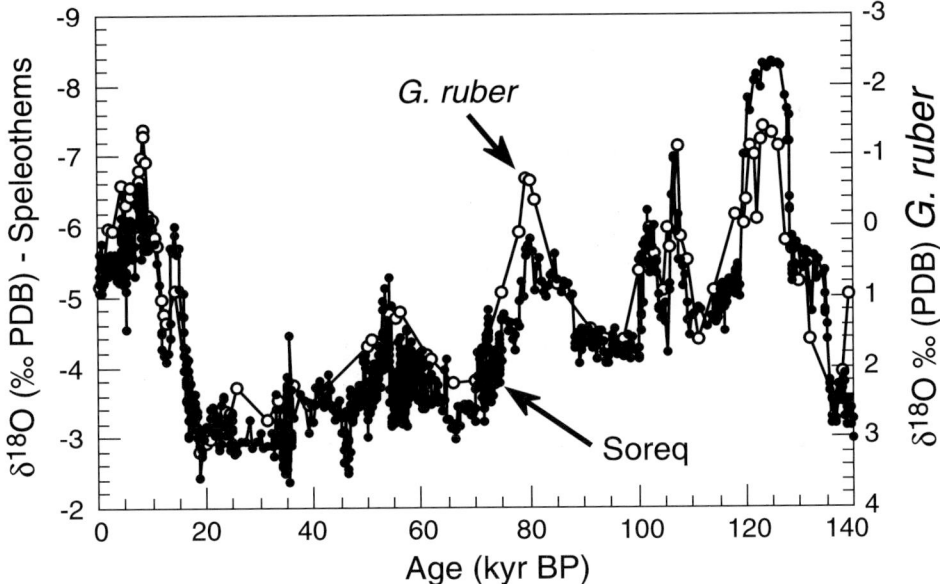

Figure 11. Comparison between the isotopic record of *G. ruber* (open circles) from core MD84651 located at 33°02'N, 32°38'E (Fontugne and Calvert 1992) and the oxygen isotopic record of Soreq cave speleothems (black closed circles) for the time period of 140 kyr to the present day. The isotopic profiles are plotted after correction of the mismatched *G. ruber* ages to the speleothem record. An excellent correspondence of the two isotopic records is observed

We will discuss first the low $\delta^{18}O$ values (#12, #13, #14, #15, Figs. 9a and 9d) of the most prominent speleothems because they can be better understood in the light of the conditions in the entire EM basin.

A distinctive feature of the Pleistocene and Holocene sediments of the EM Sea is the sapropels, which are thin, discrete, black, organic-rich layers that were deposited on a basin-wide scale as a marine response to low-latitude enhanced rainfall and fresh-water runoff from the continents. They contain abundant well-preserved planktonic microfossils and pollen and their origin has been extensively debated but not fully resolved (e.g.,

Vergnaud-Grazzini et al. (1977, 1986), Williams et al. (1978), Rossignol-Strick et al. (1982), Rossignol-Strick (1985), Rohling (1991, 1994), Fontugne and Calvert (1992), Cheddadi and Rossignol-Strick (1995a, b), Melières et al. (1997), Kallel et al. (1997, 2000, this volume)).

A total of five separate sapropel layers, referred to as S1, S3, S4, S5, S6 and possibly a sixth, S2, whose existence is controversial (Cita et al. 1977) were deposited in the EM basin during the last 185 kyr. These sapropels were formed both during interglacial and glacial interstadial isotopic stages (i.e., warm stages), and they correspond closely to the minima in the precession cycle and to periods of increased African monsoonal activity when the EM region experienced marked pluvial conditions (e.g., Rossignol-Strick et al. (1982), Rossignol-Strick (1983, 1985), Hilgen (1991), Cheddadi and Rossignol-Strick (1995a, b), Lourens et al. (1996), Melières et al. (1997)). The exact ages of most of the sapropels are still debated, although their approximate ages were established by biostratigraphy, oxygen isotope stratigraphy, tephrochronology, ^{14}C and [230]Th-U dating and paleomagnetism (e.g., Martinson et al. (1987), Tucholka et al. (1987), Fontugne et al. (1994), Severmann and Thompson (1998), Rossignol-Strick and Paterne (1999)).

The isotopic record of the Soreq cave speleothems show that four major minimum $\delta^{18}O$ events (#12, #13, #14 Fig. 9a and #15, Fig. 9d) correspond to sapropels S5, S4, S3, and S1 shown in Figure 7 (Bar-Matthews et al. 2000) and two of them (#1 and #3, Fig. 8a, 8c) correspond to glacial sapropels S6 (Ayalon et al. 2002) and possibly S2 (Fig. 7).

Based on the present-day relationships between rainfall amount and its isotopic composition (Fig. 4 and Bar-Matthews et al. (2003)) minimum speleothem $\delta^{18}O$ events are associated with increase rainfall on land. The formation of sapropels has also been taken to indicate that rainfall increased in the entire Mediterranean basin (Kallel et al. 1997; 2000). Thus two independent data show that the sapropels are related to high rainfall in the EM region.

In addition, several other lines of evidence indicate that the rainfall increased in the EM area during the early Holocene, among them the rise of the Dead Sea level (e.g., Frumkin (1997)), the rise of many lake levels in Africa and Arabia (e.g., COHMAP Members (1988)), and the oxygen and carbon isotopic composition of land snails in the Negev desert, Israel (Goodfriend 1990; 1991).

The $\delta^{18}O$ minimum events (#12, #13, #14, #15, Fig. 9a) during interglacial times also coincide with marine isotopic stages 5.5, 5.3, 5.1 and 1.1 when the EM Sea surface temperature increased (Emeis et al. 1998; 2000). This is a period when the global temperatures increased together with an increase in the atmospheric CO_2 and in sea-levels (e.g., Barnola et al. (1987), Shackleton (1987)) as evident from ice cores, deep-sea marine sediments and coral records. These observations emphasise the correlations between the global record to the EM Sea and land records.

The clear correlation between the Soreq cave speleothem $\delta^{18}O$ record and global events during marine isotopic glacial periods 4, 3, and 2, is evident from the correlation between the $\delta^{18}O$ maximum events with global events (Bar-Matthews et al. 1999). Some of these are marked in Figure 8c as events #4 to #11, and an additional event #16 during the Holocene is shown in Figure 9d. Peaks #4, #7 and #9 (Fig. 8c) correlate with Heinrich events, which are known to have occurred during periods of extreme cooling, reduced foraminifera fluxes and large decrease in the planktonic $\delta^{18}O$ as a result of sudden injections of large amounts of continental ice into the ocean at 46, 25, 16.5 kyr (e.g., Heinrich (1988), Bond et al. (1993)).

The $\delta^{18}O$ peak #8 which starts at 21 kyr and reaches its maximum at 19 kyr (Fig. 8c) is probably time equivalent to the last glacial maximum (LGM), which was characterised by a sudden cooling, with the ice sheets at their greatest extent over North Europe (e.g., Street and Grove (1979), COHMAP Members (1988)). The Younger Dryas event during which the North Atlantic sea surface temperature dropped and salinity decreased, corresponds with peak #11 in Figure 8c, and peak #16 in Figure 9d corresponds to the 8.2 kyr BP last major cooling event during the Holocene (Alley et al. 1997). The sharp drop in $\delta^{18}O$ starting at ~17.0 kyr is correlative with a time of temperature increase and ice-sheet melting during deglaciation (e.g., COHMAP Members (1988), Street-Perrott and Perrott (1990)).

The $\delta^{18}O$ values of the Soreq cave speleothems during the last 7000 years vary between −6‰ to −5‰ (apart for a very short-lasting peak at 5.2 kyr of −4.5‰). The isotopic fluctuations are not on the order of glacial-interglacial, or sapropel-non sapropel amplitudes, but they were significant enough to disturb late hunter-gatherer, and agriculture-based societies in the Levant (Weiss 2000). Among them are two major events: the sharpest, shortest event lasting less than 200 years at 5.2 kyr characterised by an increase in $\delta^{18}O$ from –6.1‰ to –4.5‰. This event corresponds with a period of population movement in the Levant (Weiss 2000). During the longer event lasting from 4.6 kyr to 4.0 kyr the $\delta^{18}O$ and $\delta^{13}C$ values increased from –6.3‰ to –5.3‰ and from –12.7‰ to –9.7‰, respectively. The only period with a major change in the $\delta^{13}C$ of the speleothems during the later part of the Holocene occurred during the latter event (Fig. 10). Such increases in the $\delta^{13}C$ values coinciding with increases of $\delta^{18}O$ most probably reflect an increase in the relative proportion of C4 type vegetation. This event corresponds with the collapse of the Akkadian Empire (Weiss et al. 1993; Weiss 2000), and the Egyptian Kingdom, possibly due to the sharp drop in the Nile water level (Hassan and Stucki 1987).

In order to estimate the change in the rainfall amount corresponding with these isotopic variations, we assume: (i) that the relationship between the annual rainfall amount and its $\delta^{18}O$ values (Fig. 4) was similar to the modern relationship. Such an assumption is probably justified for the Holocene because climate conditions are much closer to present-day than they were in glacial time intervals; and (ii) sea and land temperatures have not changed significantly during the last 7000 years. This assumption is supported by the slight SST changes measured in the EM sea surface (Emeis et al. 2000). With the assumption that the $\delta^{18}O$ value of cave water in the past was ~1‰ higher than the coeval rainfall, as in present-day conditions (Fig. 5 and Bar-Matthews et al. (1996)), we calculate the $\delta^{18}O$ of palaeo-rain water using the calcite-water fractionation equation of O'Neil et al. (1969). Using the relation between the annual amount of rainfall and its $\delta^{18}O$ (shown in Fig. 4), the variations in palaeo-rainfall during the last 7000 years can then be estimated. These calculations are plotted in Figure 12 and clearly show that at least 20% more rainfall occurred during the early Holocene in comparison with the 500 mm of today. For a very rapid short time interval, at 5.2 kyr, there was a very sharp drop in the rainfall amount, but conditions recovered immediately after. A major drop in the rainfall occurred from 4.6 kyr to 4.0 kyr, when the rainfall amount decreased from ~600 mm to ~400 mm, and from then remained below the present day average for a long time (Fig. 12). This drop in rainfall agrees well with the generally proposed trends towards aridity in North Africa and the Middle East, probably due to earth orbital changes (deMenocal et al. 2000). It is also associated with drops in lake levels, the deterioration of vegetation, and the increase in atmospheric CO_2 (e.g., Street and Grove (1976), Rossignol-Strick (1985), Petit-Maire and Guo (1996), Indermuhle et al. (1999), Schilman et al. (2001a, b)). The 5.2 kyr and 4.6–4.0 kyr events in the Soreq cave indicative of drier conditions coincides with archaeological disaster events, suggesting that dry conditions in the EM region during these times were widespread throughout the Levant and North Africa. The study of the isotopic profiles from a cave located in Ireland and receiving its rain from the North Atlantic, also show a sharp change in its $\delta^{18}O$ value about 4000 years ago (McDermott et al. 1999), suggesting that the latter event is also evident from speleothems in North Europe.

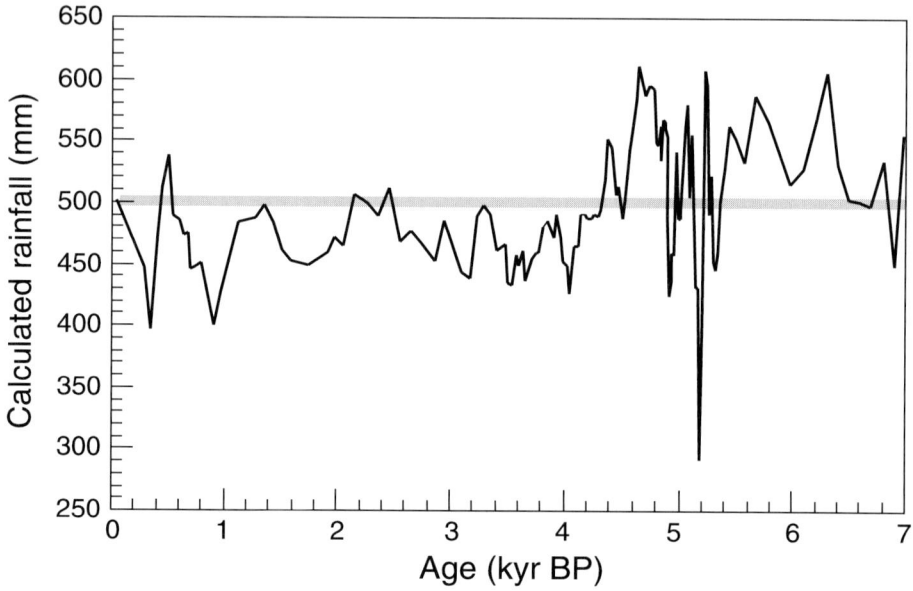

Figure 12. The calculated palaeo-rainfall amount during the last 7000 years. The horizontal line marks the present-day mean annual rainfall.

During glacial marine isotopic stage 6 between ~185 kyr and 135 kyr, the Soreq cave speleothem record correlates with events occurring in Northern Africa, rather than in the North Atlantic (Ayalon et al. 2002). During this stage, a large ice-sheet covered Northern Europe (e.g., Imbrie et al. (1984), Vostok Project Members (1995)), whereas the climate in the EM region is ambiguous because it includes a sapropel layer (S6), which differs from the interglacial sapropel layers. These sapropels contain pollen and faunal assemblages typical of glacial conditions (e.g., Rossignol-Strick (1983), Cheddadi and Rossignol-Strick (1995a, b)), and the SST deduced from alkenones show that temperatures were low (~15 °C; Emeis et al. (1998)). At the same time, conditions in the tropics were humid, with heavy African monsoons (e.g., Rossignol-Strick (1995), Melières et al. (1997), Rossignol-Strick and Paterne (1999)). Soreq cave speleothems show two prominent low $\delta^{18}O$ events (#1 and #2, Fig. 8a). The minimum $\delta^{18}O$ event at ~178 kyr is correlative with sapropel S6, and the other low $\delta^{18}O$ event at ~152 kyr is equivalent to low $\delta^{18}O$ values for *G. ruber* from EM marine cores (e.g., Vergnaud-Grazzini et al. (1977)), and is most probably equivalent also to the Monsoon Index Maxima at ~151 kyr (e.g., Melières et al. (1997)). However, there is no sapropel during this event probably because the increasing northern summer insolation was below the threshold for sapropel formation (Rossignol-Strick 1983). We therefore suggest that these two major low $\delta^{18}O$ events recorded in Soreq cave speleothems, record events occurring in low latitudes rather than in northern hemisphere.

Carbon isotopic composition of fossil speleothems

The average $\delta^{13}C$ value of modern speleothems is $-10.7 \pm 1.0\%o$. This value reflects interactions in a closed or in a mixed open-closed system between the percolating rain water, soil-CO_2 from C3 type vegetation, and host dolomite rock with $\delta^{13}C$ values of ~ 1–$2\%o$ (e.g., Hendy (1971), Bar-Matthews et al. (1996), Genty et al. (2001)). The $\delta^{13}C$ values of fossil speleothems are usually within the range of $-9\%o$ to $-13\%o$ for most of the last 185 kyr, excluding two episodes between 124 kyr and 119 kyr and between 8.5 kyr and 7.0 kyr when the $\delta^{13}C$ values are anomalously high (Fig. 7b). These events will be discussed separately.

However, there are also several short-term episodes during the last glacial when $\delta^{18}O$ values are high ($\sim -2.4\%o$, events #4, #6, #7, #8 and #9, Fig. 8c) and the $\delta^{13}C$ values are greater than $-9\%o$ (Fig. 8d). These episodes are considered to occur during cold climate conditions as discussed above. We therefore suggest that the combination of high $\delta^{18}O$ with relatively high $\delta^{13}C$ values reflects cold and dry periods. The coupling of such high $\delta^{18}O$ and $\delta^{13}C$ values only occur during the last glacial. During glacial marine isotopic stage 6, most of the $\delta^{13}C$ values are within the range of $-12\%o$ to $-10\%o$, even at times when $\delta^{18}O$ values are high as $-3\%o$ (Fig. 8b), suggesting that conditions during stage 6 were never as cold and dry as during the last glacial (Ayalon et al. 2002). During the early part of stage 6 (from 185 kyr to 150 kyr), the $\delta^{13}C$ values are relatively lower ($\sim -12\%o$ and $-11\%o$) compared with the later periods (from kyr 150 to 134 kyr) when $\delta^{13}C$ values are higher ($-11\%o$ to $-10\%o$, Fig. 8b), possibly because of a gradual increase in the proportion of C4 type vegetation.

During most of the interglacial marine isotopic stage 5, $\delta^{13}C$ values are between $-13\%o$ and $-10\%o$, but the trend of the $\delta^{13}C$ values follow that of the $\delta^{18}O$ values. The coupling of the lowest $\delta^{18}O$ and $\delta^{13}C$ values during interglacial conditions is taken to be indicative of warm and relatively wet periods with the dominance of C3 type vegetation (between 108 kyr and 101 kyr, and 85 kyr to 79 kyr). During interglacial periods when $\delta^{18}O$ and $\delta^{13}C$ values increase, conditions become drier (Fig. 9a, b). This trend is prominent during the late Holocene showing an increase in both $\delta^{18}O$ ($-6.2\%o$ to $-5.3\%o$) and $\delta^{13}C$ (from $-12.5\%o$ to $-10.5\%o$) from 4.6 kyr to 4.0 kyr (Fig. 10a, b), and indicates aridity (Bar-Matthews et al. 2003).

Anomalously high $\delta^{13}C$ values coupled with very low $\delta^{18}O$ values occur during S5 and S1 (i.e., between 124 kyr and 119 kyr and between 8.5 kyr and 7 kyr). Frumkin et al. (2000) argue that the sharp increase in $\delta^{13}C$ values reflect the transition to a warm and dry period. However, it is difficult to envisage that the high $\delta^{13}C$ values at these times reflect dry conditions, because there is a significant amount of evidence showing that the rainfall in the EM area increased (Kallel et al. 1997; 2000; this volume), Bar-Matthews et al. (2000, 2003) as previously discussed. There is also evidence for humid conditions from pollen studies indicative of the dominance of C3 type vegetation (e.g., Cheddadi and Rossignol-Strick (1995a, b)). We have analysed the $\delta^{13}C$ values of the organic material extracted from the speleothems deposited during the anomalously high $\delta^{13}C$ events, and they all are about -25%, indicating also that the dominant vegetation was C3 type. Thus we suggest that these episodes cannot reflect dry conditions, but they are indicative of deluge periods (Ayalon et al. 1999; Bar-Matthews et al. 2000; 2003), resulting in very fast infiltration of water. Because of the fast infiltration through the rock, the reactions between water, soil and rock were not in equilibrium, and the high $\delta^{13}C$ values reflect a higher proportion of rock signal and atmospheric CO_2. Kaufman et al. (1998) argued that the very sharp drop in $^{234}U/^{238}U$ of Soreq cave speleothems accompanying a sharp increase in $\delta^{13}C$ is due to the removal of soil. Ayalon et al. (1999) have shown that the $\delta^{13}C$ increase is marked by a sharp change in the petrography of the speleothems: thick, coarse-grained light-colored layer with large crystals showing preferred orientation (perpendicular to the growth layer) and parallel extinction, and containing less than 0.1% of detrital material, are followed by darker, smaller size, randomly oriented, equant LMC crystals, which are much richer in detrital material. Whereas the large calcite crystals must have formed from slowly dripping water, the fast flowing water conditions resulted in a larger input of detritus and oxides from the soil, and the formation of smaller crystal sizes. Coinciding with the petrographic and isotopic changes, a sharp drop occurred in the concentrations of Sr, Ba, and U, and in the ratios of $^{87}Sr/^{86}Sr$ and $^{234}U/^{238}U$ which reached minimum values during the wettest period, because of the larger input of the weathered dolomitic host-rock (Kaufman et al. 1998; Ayalon et al. 1999).

During the deluge period of the early Holocene, a sharp decrease in $\delta^{13}C$ values (from -4% to -11% over a period of 200 years) occurred at 8.2 kyr, and is equivalent to the global Holocene cooling event. In the Soreq cave speleothems, this event is unique because it is mostly expressed by the very sharp decrease in $\delta^{13}C$ values and by a less significant increase in $\delta^{18}O$ values from -6.5% to -5.7%. We suggest that because most of the entire period from 8.5 kyr to 7 kyr was a deluge period, the sharp cooling was associated with the drop in rainfall amount, resulting in the increased proportion in the $\delta^{13}C$ signal of the soil-derived CO_2.

Summary

The oxygen and carbon stable isotope compositions of speleothems from the Soreq cave, Central Israel, show that deposition continuously occurred during the last 185 kyr, and provide a powerful tool for understanding the EM palaeoclimate changes during glacial and interglacial conditions. In this review we examine the palaeoclimate history reflected in the Soreq cave speleothems in the light of the data that has been determined on modern speleothems, cave and rain waters.

Present-day $\delta^{18}O$ - δD relationships of the cave waters and rainwater above the Soreq cave, follow the trend of the MMWL, consistent with deriving their source from the EM sea surface. The $\delta^{18}O$ values of the rain also depend on the rainfall amount and there is a significant inverse linear relationship, showing that $\sim -1\%$ change in the $\delta^{18}O$ value of the rainwater is equivalent to an increase of 200 mm. $\delta^{13}C$ values of the dissolved inorganic carbon range between -10% and -14%, showing that the cave waters contain a dissolved component from soil-CO_2 derived from C3 type vegetation, and the marine dolomitic host rock. The average $\delta^{18}O$ of the cave water is -4.7% (SMOW) and the average $\delta^{18}O$ of most modern speleothems is –5.4‰ (PDB) indicating that they are formed in isotopic equilibrium with the cave water at an average temperature of 18 °C.

The variation of $\delta^{18}O$ values of Soreq cave speleothems is 6‰ (–2.4‰ to –8.3‰). The highest variation occurs during the transition from glacial to interglacial periods. Lower $\delta^{18}O$ values occur usually during interglacial marine isotopic periods. The time-trend isotopic profile shows a good correspondence with the stacked oxygen isotopes (Martinson et al. 1987) suggesting that the Soreq cave speleothems record global climatic changes. An even finer-scaled matching of peaks is obtained by comparison with the $\delta^{18}O$ record of the planktonic foraminifera G. ruber from the EM Sea. The matching peaks and the similar amplitudes of the $\delta^{18}O$ changes suggest that the climatic events on sea and land were linked throughout this period. The correlation between the global and the EM marine isotopic record with Soreq cave record are consistent with periods of low $\delta^{18}O$ values coinciding with warming, an increase in atmospheric CO_2, sea level rise and an increase in the rainfall amount in the entire Mediterranean basin, as also evident from the contemporaneous formation of sapropels. In contrast, high $\delta^{18}O$ value periods at 46 kyr, 35 kyr, 25 kyr and 19 kyr coincide with extreme global cooling episodes, such as Heinrich events and the Last Glacial Maximum.

$\delta^{13}C$ values vary largely from –1.6‰ to –13‰. Coupling of low $\delta^{18}O$ and $\delta^{13}C$ values is observed during interglacial conditions and is taken to be indicative of warm and wet periods with the dominance of C3 type vegetation (e.g., between 107 and 101 kyr, 85 kyr and 79 kyr). During interglacial periods, when $\delta^{18}O$ and $\delta^{13}C$ values increase, conditions become drier. Such an event is associated with the collapse of the Akkadian Empire and the Egyptian Kingdom and occurred from 4.6 kyr to 4.0 kyr. The coupling of anomalously high $\delta^{13}C$ values with low $\delta^{18}O$ values occurs during deluge periods, at times equivalent to the formation of sapropel S5 (124 kyr to 119 kyr) and S1 (8.5 kyr to 7 kyr). These anomalously high $\delta^{13}C$ values reflect an increased atmospheric CO_2 signal and dolomitic host rock contribution, at the expense of the soil CO_2 contribution. During the last glacial, several cold and dry episodes are characterised by both highest $\delta^{18}O$ and highest $\delta^{13}C$ values (above –9‰, at 46 kyr, 35 kyr, 25 kyr and 19 kyr).

Acknowledgments

This research was supported by The Israel Science Foundation (grants number 227/95-1, 151/98 – 13.0, 20/01-13.0). We express our gratitude to the Nature Protection Authority for their support throughout the study. We thank Dominique Genty and an anonymous reviewer for critically reviewing the manuscript and offering helpful comments for its improvement.

References

Alley R.B., Mayewski P.A., Sowers T., Stuiver M., Taylor K.C. and Clark P.U. 1997. Holocene climatic instability: A prominent, widespread event 8200 yr ago. Geology 25: 483–486.
Ayalon A., Bar-Matthews M. and Sass E. 1998. Rainfall-recharge relationships within a karstic terrain in the Eastern Mediterranean semi-arid region, Israel: $\delta^{18}O$ and δD characteristics. J. Hydrol. 207: 18–31.

Ayalon A., Bar-Matthews M. and Kaufman A. 1999. Petrography, trace elements (Ba, Sr, Mg and U) and isotope geochemistry of strontium and uranium in speleothems as paleoclimate proxies. Soreq Cave, Israel. Holocene 9: 715–722.

Ayalon A., Bar-Matthews M. and Kaufman A. 2002. Climatic conditions during marine isotopic stage 6 in the Eastern Mediterranean region as evident from the isotopic composition of speleothems. Soreq Cave, Israel. Geology 30: 303–306.

Ayliffe L.K., Marianelli P.C., Moriarty K.C., Wells R.T., McCulloch M.T., Motrimer G.E. and Hellstrom J.C. 1998. 500 ka precipitation record from Southeastern Australia: Evidence for interglacial relative aridity. Geology 26: 147–150.

Bar-Matthews M., Matthews A. and Ayalon A. 1991. Environmental controls of speleothems mineralogy in a karstic dolomitic terrain (Soreq Cave, Israel), J. of Geology 99: 189–207.

Bar-Matthews M., Ayalon A., Matthew A., Halicz L. and Sass E. 1993. The Soreq Cave speleothems as indicators of paleoclimate variations. Israel Geol. Surv. Curr. Res. 8: 1–3.

Bar-Matthews M., Ayalon A., Sass E. and Halicz L. 1994. Water behaviour in karstic system — Soreq Cave, Beit-Shemesh, summary of three years study. Israel Geol. Surv. Rep. # GSI/6/94.

Bar-Matthews M., Ayalon A., Matthews A., Sass E. and Halicz L. 1996. Carbon and oxygen isotope study of the active water-carbonate system in a karstic Mediterranean cave: implications for paleoclimate research in semiarid regions. Geochim. Cosmochim. Acta 60: 337–347.

Bar-Matthews M., Ayalon A. and Kaufman A. 1997. Late Quaternary palaeoclimate in the eastern Mediterranean region from stable isotope analysis of speleothems at Soreq Cave, Israel. Quat. Res. 47: 155–168.

Bar-Matthews M., Ayalon A. and Kaufman A. 1998. Middle to late Holocene (6500 Yr. period) paleoclimate in the eastern Mediterranean region from stable isotopic composition of speleothems from Soreq cave, Israel. In: Issar A.S. and Brown N. (eds), Water, Environment and Society in Time of Climate Change. Kluwer Academic Publishers, Dordrecht. pp. 203–214.

Bar-Matthews M., Ayalon A., Kaufman A. and Wasserburg G.J. 1999. The Eastern Mediterranean paleoclimate as a reflection of regional events: Soreq cave, Israel. Ear. Planet. Sci. Lett. 166: 85–95.

Bar-Matthews M., Ayalon A. and Kaufman A. 2000. Timing and hydrological conditions of Sapropel events in the Eastern Mediterranean, as evident from speleothems, Soreq cave, Israel. Chem. Geol. 169: 145–156.

Bar-Matthews M., Ayalon A., Gilmour M., Matthews A. and Hawkesworth C.J. 2003. Sea-land oxygen isotopic relationships from planktonic foraminifera and speleothems in the Eastern Mediterranean region and their implication for paleorainfall during interglacial intervals. Geochim. Cosmochim. Acta 67: 3181–3199.

Barnola J.M., Raynaud D., Korotkevich Y.S. and Lorius C. 1987. Vostok ice core provides 160,000-year record of atmospheric CO_2. Nature 329: 408–412.

Berger W.E. and Jansen E. 1995. Younger Dryas episode: ice collapse and super-fjord heat pump. In: Troelstra S.R., van Hinte J.E. and Ganssen G.M. (eds), Workshop R. Proc., pp. 61–105.

Bond G., Broecker W., Johnsen S., McManus J., Labeyrie L., Jouzel J. and Bonani G. 1993. Correlations between climate records from North Atlantic sediments and Greenland ice. Nature 365: 143–147.

Burns S.J., Matter A., Frank N. and Mangini A. 1998. Speleothem-based paleoclimate record from Northern Oman. Geology 26: 499–502.

Burns S.J., Fleitmann D., Matter A., Neff U. and Augusto M. 2001. Speleothem evidence from Oman for continental pluvial events during interglacial periods. Geology 29: 623–626.

Cerling T.E. and Quade J. 1993. Stable carbon and oxygen isotopes in soil carbonates. In: Swart P.K., Lohman K.C., McKenzie J. and Savin S. (eds), Climate Change in Continental Isotopic Records. Geoph. Monograph 78: 217–231.

Cerling T.E., Quade J., Wang Y. and Bowman J.R. 1989. Carbon isotopes in soils and palaeosols as ecology and palaeocology indicators. Nature 341: 138–139.

Cerling T.E., Quade J., Solomon D.K. and Bowman J.R. 1991. On the carbon isotopic composition of soil carbon dioxide. Geochim. Cosmochim. Acta 55: 3403–3405.

Cheddadi R. and Rossignol-Strick M. 1995a. Eastern Mediterranean Quaternary palaeoclimates from pollen and isotope records of marine cores in the Nile cone area. Paleoceanography 10: 291–300.

Cheddadi R. and Rossignol-Strick M. 1995b. Improved preservation of organic matter and pollen in eastern Mediterranean sapropels. Paleoceanography 10: 301–309.

Choquette P.W. and James N.P. 1987. Paleokarst, introduction. In: James N.P. and Choquette R.W. (eds), Paleokarst, Ch. 1, Springer-Verlag, New York, pp. 1–25.

Cita M.B., Vergnaud-Grazzini C., Robert C., Cahmley H., Ciaranfi N. and Donofrio S. 1977. Paleoclimatic record of a long deep-sea core from the Eastern Mediterranean. Quat. Res. 8: 205–235.

COHMAP Members 1988. Climatic changes of the last 18,000 years: observations and model simulations. Science 241: 1043–1052.

Dansgaard W. 1964. Stable isotopes in precipitation. Tellus 16: 436–468.

deMenocal P., Ortiz J., Guilderson T., Adkins J., Sarnthein M., Baker L. and Yarusinsky M. 2000. Abrupt onset and termination of the African Humid Period: rapid climate responses to gradual insolation forcing. Quat. Sci. Rev. 19: 347–361.

Denniston R.F., Gonzalez L.A., Baker R.G., Asmerom Y., Reagan M.K., Edwards R.L. and Alexander E.C. 1999. Speleothem evidence for Holocene fluctuations of the prairie-forest ecotone, North-Central USA. Holocene 9: 671–675.

Deuser W.G. 1987. Seasonal variations in isotopic composition and deep-water fluxes of the tests of perennially abundant planktonic foraminifera of the Saragasso Sea: Results from sediment-trap collections and their paleoceanographic significance. J. Foram. Res. 17: 14–27.

Dorale J.A., Gonzalez L.A., Reagan M.K., Pickett D.A., Murrell M.T. and Baker R.G. 1992. A high resolution record of Holocene climate change in speleothem calcite from Cold Water cave, Northeast Iowa, Science 258: 1626–1630.

Dorale J.A., Edwards R.L., Ito E. and Gonzalez L.A. 1998. Climate and vegetation history of the midcontinent from 75 to 25 ka: A speleothem record from Crevice Cave, Missouri, USA. Science 282: 1871–1874.

Duplessy L.C., Deliubrias G., Turon J.L., Pujol C. and Dupart J. 1981. Deglacial warming of the north-eastern Atlantic Ocean: correlation with the paleoclimatic evolution of the European continent. Palaeogeogr. Palaeoclimat. Palaoecol. 35: 121–144.

Emeis K.C., Schulz H.M., Struck U., Sakamoto T., Doose H., Erlenkeuser H., Howell M., Kroon D. and Paterne M. 1998. Stable isotope and alkenone temperature records of sapropels from sites 964 and 967: constraining the physical environment of sapropel formation in the eastern Mediterranean Sea. In: Robertson A.H.F., Emeis K.-C., Richter C. and Camerlenghi A. (eds), Proc. of the Ocean Drilling Prog., Ocean Drilling Prog., Sci. Res. 160: 309–331.

Emeis K.C., Struck U., Schulz H.M., Rosenberg R., Bernasconi S., Erlenkeuser H., Sakamoto T. and Martinez-Ruiz F. 2000. Temperature and salinity variations of Mediterranean Sea surface waters over the last 16,000 years from records of planktonic stable oxygen isotopes and alkenone unsaturation ratios. Palaeogeogr. Palaeoclimat. Palaeoecol. 158: 259–280.

Fairbanks R.G., Sverdlove M., Free R., Wiebe P.H. and Be' A.W.H. 1982. Vertical distribution of living planktonic foraminifera from the Panama Basin. Nature 298: 841–844.

Fontugne M. and Calvert S.E. 1992. Late Pleistocene variability of the carbon isotopic composition of organic matter in the Eastern Mediterranean: monitor of changes in carbon sources and atmosphere CO_2 concentrations. Paleoceanography 7: 1–20.

Fontugne M.R., Arnold M., Labeyrie L., Paterne M., Calvert S.E. and Duplessy J.-C. 1994. Palaeoenvironment, sapropel chronology and Nile river discharge during the last 20,000 years as indicated

388 BAR-MATTHEWS AND AYALON

by deep sea sediment records in the Eastern Mediterranean. In: Bar-Yosef O. and Kra R.S. (eds), Late Quaternary Chronology and Palaeoclimates of the Eastern Mediterranean. Radiocarbon, pp. 75–88.

Forti P. 1998. Seismotectonic and paleoseismic studies from speleothems: the state of the art, Han 98-Tectonique, karst et Seismws 79–81.

Frumkin A. 1997. The Holocene history of Dead Sea levels. In: Niemi T.M., Ben-Avraham Z. and Gat J.R. (eds), The Dead Sea: the Lake and its Setting. Oxford Monog. on Geol. and Geophys. 36: 237–248.

Frumkin A., Ford D.C. and Schwarcz H.P. 1999. Continental oxygen isotopic record of the last 170,000 years in Jerusalem. Quat. Res. 51: 317–327.

Frumkin A., Ford D.C. and Schwarcz H.P. 2000. Palaeoclimate and vegetation of the last glacial cycles in Jerusalem from a speleothem record. Global Biochem. Cyc. 14: 863–870.

Gascoyne M. 1992. Paleoclimatic determination from cave calcite deposits. Quat. Sci. Rev. 11: 609–632.

Gasse F., Tehet R., Durand A., Gibert E. and Fontes J.C. 1990. The arid-humid transition in the Sahara and the Sahel during the last deglaciation. Nature 346: 141–146.

Gat J.R. 1980. The isotopes of hydrogen and oxygen in precipitation. In: Fritz P. and Fontes Ch. (eds), Handbook of Environ. Isotope Geochem. 1, Elsevier, New York, pp. 21–47.

Gat J.R. 1996. Oxygen and hydrogen isotopes in the hydrologic cycle. Ann. Rev. Ear. Planet. Sci. 24: 225–262.

Gat J.R. and Carmi I. 1970. Evolution in the isotopic composition of atmospheric waters in the Mediterranean Sea area. J. Geophys. Res. 75: 3039–3048.

Gat J.R. and Carmi I. 1987. Effect of climate changes on the precipitation patterns and isotopic composition of water in a climate transition zone: case of the Eastern Mediterranean Sea area. The Influence of Climate Change and Climatic Variability on the Hydrologic Regime and Water Resources (Proc. of the Vancouver Symp., August 1987), IAHS Publ. 168: 513–523.

Gat J.R. and Dansgaard W. 1972. Stable isotope survey of the fresh water occurrences in Israel and the Jordan Rift Valley. J. Hydrol. 16: 177–211.

Genty D., Baker A., Massault M., Proctor C., Gilmour M., Pons-Branchu E. and Hamelin B. 2001. Dead carbon in stalagmites: Carbonate bedrock paleodissolution vs. ageing of soil organic matter. Implications for ^{13}C variations in speleothems. Geochim. Cosmochim. Acta 65: 3443–3457.

Goodfriend G.A. 1990. Rainfall in the Negev Desert during the middle Holocene, based on ^{13}C of organic matter in land snail shells. Quat. Res. 34: 186–197.

Goodfriend G.A. 1991. Holocene trends in ^{18}O in land snail shells from the Negev Desert and their implications for changes in rainfall source areas. Quat. Res. 35: 417–426.

Halicz L., Bar-Matthews M., Ayalon A. and Kaufman A. 1997. Determination of low concentrations of U and Th in Carbonate Rocks Using FI-ICP-MS. Atom. Specros. 18: 175–179.

Hassan F.A. and Stucki B.R. 1987. Nile floods and climatic change. In: Rampino M.R., Sanders J.E., Newman W.S. and Konigsson L.K. (eds), Climate: History, Periodicity and Predictability. Van Nostrand Reinhold, New York, pp. 37–46.

Heinrich H. 1988. Origin and consequences of cyclic ice rafting in the Northeast Atlantic Ocean during the past 130,000 years. Quat. Res. 29: 142–152.

Hendy C.H. 1971. The isotopic geochemistry of speleothems - I. The calculation of the effects of different modes of formation on the isotopic composition of speleothems and their applicability as paleoclimatic indicators. Geochim. Cosmochim. Acta 35: 801–824.

Hendy C.H. and Wilson A.T. 1968. Paleoclimatic data from speleothems. Nature 219: 48–51.

Hilgen F.J. 1991. Astronomical calibration of Gauss to Matuyama sapropels in the Mediterranean and implication for the geomagnetic polarity time scale. Ear. Planet Sci. Lett. 104: 226–244.

Hill C.A. 1987. Geology of Carlsbad Cavern and Other Caves in the Guadalupe Mountains, New Mexico and Texas. New Mexico Bureau of Mines and Miner. Resour. Bull. 117, 150 pp.

Holmgren K., Karlen W. and Shaw P.A. 1995. Paleoclimatic significance of the stable isotopic composition and petrology of a Late Pleistocene stalagmite from Botswana. Quat. Res. 43: 320–328.

Imbrie J., Hays J.D., McIntyre A., Mix A.C., Morley J.J., Pisias N.G., Prell W.L. and Shackleton N.G. 1984. The orbital theory of Pleistocene climate: Support from a revised chronology of the marine $\delta^{18}O$ record. In: Berger A., Imbrie J., Hays J. et al. (eds), Milankovich and Climate, Part 1. Reidel, Boston, pp. 269–305.

Indermuhle A., Stocker T.F., Joos F., Fisher H., Smith H.J., Wahlen M., Deck B., Mastroianni D., Tshumi J., Blunier T., Meyer R. and Stauffer B. 1999. Holocene carbon-cycle dynamics based on CO_2 trapped in ice at Taylor Dome, Antarctica. Nature 398: 121–126.

Jouzel J. 1980. Isotopes in cloud physics: multiphase and multistage condensation processes. In: Fritz P. and Fontes Ch. (eds), Handbook of Environ. Isotope Geochem. 1. Elsevier, New York, pp. 61–105.

Kagan E.J., Agnon A., Bar-Matthews M. and Ayalon A. 2002. Cave deposits as records of paleo-seismicity: a record from two caves located 60 km west of the Dead Sea Transform (Jerusalem, Israel). Environ. Catast. and Recov. in the Holocene. PAGES, Brunel University, p. 41.

Kallel N., Duplessy J.-C., Labeyrie L., Fontugne M., Paterne M. and Montacer M. 2000. Mediter-ranean pluvial periods and sapropel formation during the last 200,000 years. Palaeogeogr. Palaeoclimat. Palaoecol. 157: 45–58.

Kallel N., Paterne M., Duplessy J.-C., Vergnaud-Grazzini C., Pujol C., Labeyrie L., Arnold M., Fontugne M. and Pierre C. 1997. Enhanced rainfall in the Mediterranean region during the last sapropel event. Oceanol. Acta 20: 697–712.

Kallel N., Duplessy J., Labeyrie L., Fontugne M. and Paterne M., this volume. Mediterranean Sea palaeohydrology and pluvial periods during the Late Quaternary. In: Battarbee R.W., Gasse F. and Stickley C.E. (eds), Past Climate Variability through Europe and Africa. Kluwer Academic Publishers, Dordrecht, the Netherlands, pp. 307–324.

Kaufman A., Wasserburg G.J., Porcelli D., Bar-Matthews M., Ayalon A. and Halicz L. 1998. U-Th isotope systematics from the Soreq Cave Israel and climatic correlations. Ear. Planet. Sci. Lett. 156: 141–155.

Kaufman A., Bar-Matthews M., Ayalon A. and Carmi I. 2003. The vadose flow above Soreq Cave, Israel: a tritium study of the cave waters. J. Hydrol. 273: 155–163.

Lauritzen S.E. and Lundberg J. 1999. Calibration of the speleothem data function: an absolute temperature record for the Holocene in Northern Norway. Holocene 9: 659–669.

Lemeille F., Cushing M., Carbon D., Grellet B., Bitterli Th., Flehoc Ch. and Innocent Ch. 1999. Co-seismic ruptures and deformations recorded by speleothems in the epicentral zone of the Basel earthquake. Geodinam. Acta 12: 3-4, pp. 179–191.

Lourens L.J., Antonarakou A., Hilgren F.J., van Hoof A.A.M., Vergnaud-Grazzini C. and Zachariasse W.J. 1996. Evaluation of the Plio-Pleistocene astronomical timescale. Paleoceanography 11: 391–413.

Magri D., Kallel N. and Narcisi B., this volume. Palaeoenvironmental changes in the Mediterranean re-gion 250–10 kyr BP. In: Battarbee R.W., Gasse F. and Stickley C.E. (eds), Past Climate Variability through Europe and Africa. Kluwer Academic Publishers, Dordrecht, pp. 325–341.

Martinson D.G., Pisias N., Hays J.D., Imbrie J., Moore T.C. Jr. and Shackleton N.J. 1987. Age dating and the orbital theory of the ice ages: development of a high-resolution 0-300,000-year chronostratigraphy. Quat. Res. 27: 1–29.

Matthews A., Ayalon A. and Bar-Matthews M. 2000. D/H ratios of fluid inclusions of Soreq cave (Israel) speleothems as a guide to the Eastern Mediterranean Meteoric Line relationships in the last 120 ky. Chem. Geol. 166: 183–191.

McDermott F., Frisia S., Huang Y., Longinelli A., Spiro B., Heaton T.H.E., Hawkesworth C.J., Borsato A., Keppens E., Fairchild I.J., Borg K., Verheyden S. and Selmo E. 1999. Holocene climate variability in Europe: Evidence from $\delta^{18}O$, textural and extension rate variations in three speleothems. Quat. Sci. Rev. 18: 1021–1038.

Melières M.A., Rossignol-Strick M. and Malaize B. 1997. Relation between low latitude insolation and $\delta^{18}O$ change of atmospheric oxygen for the last 200 kyrs, as revealed by Mediterranean sapropels. Geophys. Res. Lett. 24: 1235–1238.

Musgrove M.L., Banner J.L., Mack L.E., Combs D.M., James E.W., Cheng H. and Edwards R.L. 2001. Geochronology of late Pleistocene to Holocene speleothems from Central Texas: Implications for regional palaeoclimate. GSA Bull. 113: 1532–1543.

O'Neil J.R., Clayton R.N. and Mayeda T.K. 1969. Oxygen isotope fractionation of divalent metal carbonates. J. Chem. Physics. 30: 5547–5558.

Petit-Maire N. and Guo Z. 1996. Mise en evidence de variations climatiques Holocenes rapides, en phase dans les deserts actuels de Chine et de Nord de l'Afrique. Sci. de la Terre et des Planet. 322: 847–851.

Reiss Z., Halicz E. and Luz B. 1999. Late-Holocene foraminifera from the SE Levantine Basin. Israel J. Earth Sci. 48: 1–27.

Roberts N., Stevenson A.C., Davies B., Cheddadi R., Brewer S. and Rosen A., this volume. Holocene climate, environment and cultural change in the circum-Mediterranean region. In: Battarbee R.W., Gasse F. and Stickley C.E. (eds), Past Climate Variability through Europe and Africa. Kluwer Academic Publishers, Dordrecht, pp. 343–362.

Rohling E.J. 1991. Shoaling of the Eastern Mediterranean pycnoline due to reduction of excess evaporation: implication for sapropel formation. Paleoceanography 6: 747–753.

Rohling E.J. 1994. Review and new aspects concerning the formation of eastern Mediterranean sapropels. Mar. Geol. 122: 1–28.

Rossignol-Strick M. 1983. African monsoons, as immediate climate response to orbital insolation. Nature 304: 46–48.

Rossignol-Strick M. 1985. Mediterranean Quaternary sapropels, an immediate response of the African monsoon to variation of insolation. Palaeogeogr. Palaeoclimatol. Palaoecol. 49: 237–263.

Rossignol-Strick M. 1995. Sea-land correlation of pollen records in the eastern Mediterranean for the glacial-interglacial transition: Biostratigraphy versus radiometric time-scale. Quat. Sci. Rev. 14: 893–915.

Rossignol-Strick M. and Paterne M. 1999. A synthetic pollen record of the Eastern Mediterranean sapropels of the last 1 Ma: implications for the time-scale and formation of sapropels. Marine Geol. 153: 221–237.

Rossignol-Strick M., Nesteroff W., Olive P. and Vergnaud-Grazzini C. 1982. After the deluge: Mediterranean stagnation and sapropel formation. Nature 295: 105–110.

Rozanski K., Araguas-Araguas L. and Gonfiantini R. 1993. Isotopic patterns in modern global precipitation. In: Swart P.K., Lohman K.C., McKenzie J. and Savin S. (eds), Climate Change in Continental Isotopic Record. Geoph. Monograph 78, Am. Geophy. Union, Washington, DC, pp. 1–37.

Schilman B., Almogi-Labin A., Bar-Matthews M., Labeyrie L., Paterne M. and Luz B. 2001a. Long- and short-term carbon fluctuations in the Eastern Mediterranean during the late Holocene. Geology 29: 1099–1102.

Schilman B., Bar-Matthews M., Almogi-Labin A. and Luz B. 2001b. Global climate instability reflected by Eastern Mediterranean marine records during the late Holocene, Palaeogeogr. Palaeoclimatol. Palaeoecol. 176: 157–176.

Schwarcz H.P. 1986. Geochronology and isotopic geochemistry of speleothems. In: Fritz P. and Fontes J.C. (eds), Handbook of Environ. Isotope Geochem., The Terrestrial Environment B, Elsevier, Amsterdam, pp. 271–303.

Severmann S. and Thomson J. 1998. Investigation of the intergrowth of radioactive daughters of ^{238}U in Mediterranean sapropels as a potential dating tool. Chem. Geol. 150: 317–330.

Shackleton N.J. 1987. Oxygen isotopes, ice volume and sea level. Quat. Sci. Rev. 6: 183–190.

Street F.A. and Grove A.T. 1976. Environmental and climatic implication of late Quaternary lake level fluctuation in Africa. Nature 261: 385–390,

Street F.A. and Grove A.T. 1979. Global maps of lake-level fluctuations since 30,000 yr BP. Quat. Res. 12: 83–118.

Street-Perrott F.A. and Perrott R.A. 1990. Abrupt climatic fluctuations in the tropics: the influence of Atlantic Ocean circulations. Nature 34: 607–612.

Thrailkill J. 1968. Dolomite cave deposits from Carlsbad Caverns, J. Sedim. Petrol. 38: 141–145.

Tucholka P., Fontugne M., Guichard F. and Paterne M. 1987. The Blake magnetic polarity episode in cores from the Mediterranean Sea. Ear. Planet. Sci. Lett. 86: 320–326.

Vaks A., Bar-Matthews M., Ayalon A., Schilman B., Frumkin A., Kaufman A., Matthews A., Gilmour M. and Hawkesworth C.J. 2003. Paleoclimate reconstruction based on the timing of speleothem growth, oxygen and carbon isotope composition from a cave located in the 'rain shadow', Israel. Quat. Res. 59: 182–193.

van Os B.J.H. and Rohling E.J. 1993. Oxygen isotope depletions in Eastern Mediterranean sapropels exclude estuarine circulation in primary and diagenetic signals in Mediterranean sapropels and North Atlantic turbidites. Origin and fate of trace metals and paleo-proxies. Geol. Utraiectina 109: 3–12.

Vergnaud-Grazzini C., Ryan W.B.F. and Cita M.B. 1977. Stable isotope fractionation, climate change and episodic stagnation in the Eastern Mediterranean during the Late Quaternary. Marine Micropaleont. 2: 353–370.

Vergnaud-Grazzini C., Devaux M. and Znaid J. 1986. Stable isotope "anomalies" in Mediterranean Pleistocene records. Marine Micropaleontol. 10: 35–69.

Vostok Project Members 1995. International effort helps decipher mysteries of Palaeoclimate from Antarctic ice cores. EOS 76: 169–170.

Weiss H. 2000. Beyond the Younger Dryas collapse as adaptation to abrupt climate change in ancient West Asia and the Eastern Mediterranean. In: Bawden G. and Reycraft R. (eds), Confronting Natural Disaster: Engaging the Past to Understand the Future. University of New Mexico Press, Albuquerque, pp. 75–98.

Weiss H., Corty M.A., Wetterstorm W., Guichard F., Senior L., Meadow R. and Curnow A. 1993. The genesis and collapse of third millennium north Mesopotamian civilization. Science 261: 995–1004.

Williams D.F., Thunell R.C. and Kennett J.P. 1978. Periodic freshwater flooding and stagnation of the Eastern Mediterranean Sea during the Late Quaternary. Science 201: 252–254.

Williams P.W. 1983. The role of the subcutaneous zone in karst hydrology. J. Hydrol. 61: 45–67.

Williams P.W., Marshall A., Ford D.C. and Jenkinson A.V. 1999. Palaeoclimatic interpretation of stable isotope data from Holocene speleothems of the Waitomo District, North Island, New Zealand. Holocene 9: 649–657.

19. CLIMATIC AND ENVIRONMENTAL VARIABILITY IN THE MID-LATITUDE EUROPE SECTOR DURING THE LAST INTERGLACIAL-GLACIAL CYCLE

JEF VANDENBERGHE (jef.vandenberghe@falw.vu.nl)
Faculty of Earth and Life Sciences
Vrije Universiteit
Amsterdam
The Netherlands

JOHN LOWE (j.lowe@rhul.ac.uk)
Department of Geography
Royal Holloway
University of London
Egham
Surrey, TW20 0EX
UK

RUSSELL COOPE (r.coope@rhul.ac.uk)
Department of Geography
Royal Holloway
University of London
Egham
Surrey, TW20 0EX
UK

THOMAS LITT (t.litt@uni-bonn.de)
Institute for Palaeontology
University of Bonn
Bonn
Germany

LUDWIG ZÖLLER (Ludwig.Zoeller@uni-bayreuth.de)
Geographisches Institut
University of Bonn
(at present University of Bayreuth)
Germany

Keywords: Quantified and semi-quantified palaeoclimate estimates, Mid-Latitude Europe, Eemian, Weichselian Middle Pleniglacial, Last Glacial Maximum, Weichselian Lateglacial, Geochronology

393

R. W. Battarbee et al. (eds) 2004. *Past Climate Variability through Europe and Africa.*
Springer, Dordrecht, The Netherlands.

Introduction

Key criteria when constructing regional summaries of the sequence and patterns of climate change during the last interglacial-glacial cycle are: (i) well-dated, high resolution palaeo-environmental data-sets based on multi-disciplinary investigations; (ii) the construction of sophisticated databases able to store, and facilitate analysis of, such complex data-sets; and (iii) the development of closer links between the 'palaeo-data' and global climate modelling communities, in order to test ideas about the nature and causes of abrupt climate changes. In this review we examine the nature of the palaeo-environmental data generated by studies conducted within the Mid-Latitude Europe sector of PEP III. The review is restricted to Time Stream 2 only (Gasse and Battarbee, this volume).

The Mid-Latitude Europe sector of the PEP III transect approximates to the broad band of generally low-lying land that lay between the southern edge of the northern ice sheet and the southern limit of permafrost during the last cold stage (Fig. 1). A rich array of palaeo-environmental data has been obtained from numerous sites in this region during the past 50 years or so. The cumulative results indicate that the region endured a predominantly periglacial environment during the last cold stage, but also that a number of pronounced climatic and environmental fluctuations took place. Frost-thaw cycles varied in intensity, and for brief periods the ground thawed, and boreal forests were able to migrate quite far north. Long periods of frozen ground and loess deposition were therefore interrupted by short-lived interstadial periods, during which soils formed. By contrast, environmental variability during the last interglacial stage appears to have been much less pronounced.

The western sector of Mid-Latitude Europe lies adjacent to the North Atlantic, and it has long been assumed that sea-surface temperature variations in the North Atlantic constituted the prime driver of abrupt climate changes in Europe during the last glacial cycle (Ruddiman and McIntyre 1973; Ruddiman et al. 1977). Two key questions therefore are: (i) how did changes in ice-sheet extent, North Atlantic sea-surface temperatures (SST) and thermohaline circulation modify North European climate and hydrology? And (ii) what were the magnitudes of climate changes, and of west-east climate gradients, in Mid-Latitude Europe during successive episodes of advance and retreat of the northern ice-sheets?

Since the publication of the Greenland ice-core records from the beginning of the 1990s onward (e.g., Dansgaard et al. (1993)), ideas about the frequency, magnitude and rapidity of the climate changes that affected temperate Europe during the last glacial cycle have been radically revised. Furthermore, subsequent high-resolution records from the North Atlantic (Bond et al. 1993; 1997; Cortijo et al. 2000) show remarkable similarities to the Greenland ice-core records for the last glacial cycle, suggesting a close coupling in the behaviour of the North Atlantic and the Greenland ice sheet. This new perspective has led to a re-framing of the questions being posed about the nature of ice-land-sea interactions, the degree of climate instability experienced in Europe, and the extent to which palaeoclimatic records from Europe match those obtained from the Greenland ice cores and adjacent seas (e.g., Maslin et al. (2001)). Two important questions that have recently emerged are: (i) were *all* of the climatic and environmental variations that affected Mid-Latitude Europe during the last interglacial-glacial cycle driven by North Atlantic thermohaline changes? And (ii) how quickly did the climate of Europe respond to N. Atlantic forcing during, for example, successive Dansgaard-Oeschger and Bond cycles?

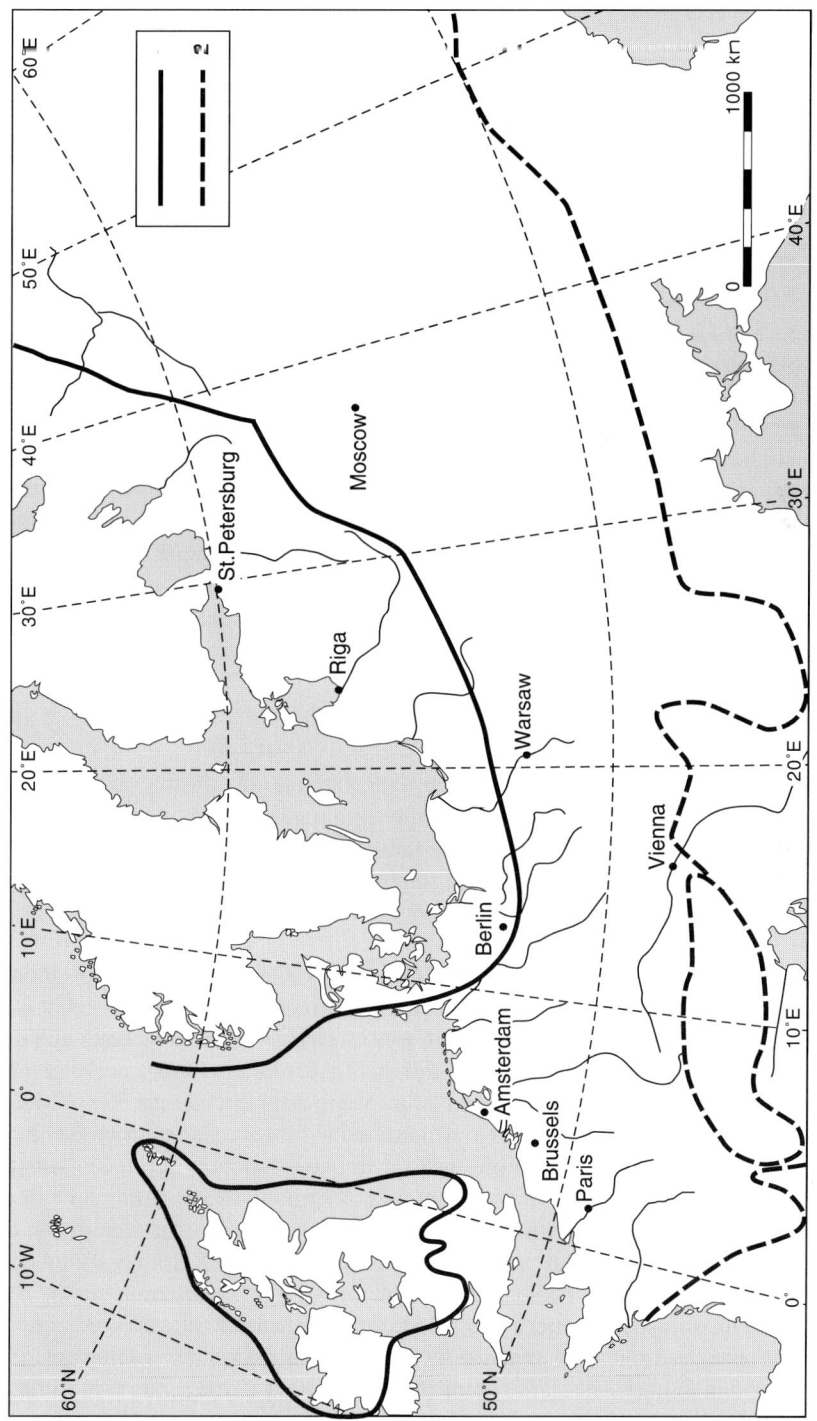

Figure 1. The Mid-Latitude Europe sector, corresponding to the area that lies between the Weichselian ice sheets of Northern Europe (full lines: 1) and the southern margin of the permafrost zone during the Weichselian LGM (broken lines: 2). Note: small central and southern European glacier extents are not represented.

Variations in climate in Mid-Latitude Europe during the last interglacial-glacial cycle

Records which span the last interglacial-glacial cycle as a continuous sequence are rare in the mid-latitude sector of Europe, due to the influences of the last (Weichselian) ice sheet and the periglacial zone which fringed its southern edge. The climate record for this long interval is therefore based on a composite model, an amalgam of the results of investigations of widely-dispersed sites that contain sediment records that span part of the cycle only, and which frequently have been dated only crudely. Exceptions may be some of the more continuous loess sequences, some of which span most of the last glacial cycle (e.g., in the Ukraine), and a few long lake sequences, such as those reported from the sites of Les Echets and La Grande Pile. These show evidence for a series of stadial-interstadial oscillations, but the chronology of the sequences is imprecise, and they do not match the Greenland ice-core records in terms of the frequency of climate oscillations inferred.

In this section of the paper we summarise the state-of-play concerning interpretations of the palaeoclimatic record obtained for 4 subdivisions of the last interglacial-glacial cycle: the last interglacial (corresponding to Marine Isotope Stage 5e - or MIS 5e), the Middle Pleniglacial (corresponding to MIS 3), the Late Pleniglacial (corresponding to MIS 2) including the Last Glacial Maximum, and the Weichselian Lateglacial (or Last Termination). This will illustrate the current problems of comparing palaeoclimatic data with, for example, the ice-core records, problems which are more severe for the earlier of these episodes than for the later ones.

Climate variations during the Eemian (MIS 5e)

The magnitude and frequency of possible climatic fluctuations during the last (Eemian) interglacial is a topic of widespread interest at present, in view of the fact that the Eemian is the most recent episode of fully interglacial conditions which was not affected by human-induced influences on climate. Hence, stratigraphical records for this period offer the best opportunity for establishing the natural sensitivity of global climate during a full interglacial period.

Some of the evidence obtained from the GRIP ice core (Greenland), which is thought to extend back to the Last Interglacial, suggests that high amplitude temperature fluctuations characterise isotope stage 5e (Dansgaard et al. 1993). However, these features do not have close parallels in the GISP2 ice core at Summit (Taylor et al. 1993), nor in a number of high-resolution isotope records obtained from North Atlantic marine cores (McManus et al. 1994). Palaeobotanical records from continental Europe serve to cloud the issue further. Field et al. (1994), for example, used a transfer function approach, interpreted palaeobotanical records from the site of Bispingen, Germany as indicating an episode of marked winter cooling, though without substantial changes in summer conditions, during the mid-Eemian. Their interpretation may be supported by records from Southern Europe (e.g., Thouveny et al. (1994)) and some marine records (e.g., Seidenkrantz et al. (1995)). On the other hand, some researchers interpret Eemian palaeobotanical records, including the Bispingen site, as indicating that the Eemian was a period of uninterrupted warm climatic conditions (e.g., Menke and Tynni (1984), Frenzel (1991), Zagwijn (1996), Litt et al. (1996), Boettger et al. (2000), Drescher-Schneider (2000)). The results obtained by

Field et al. (1994) using the transfer function approach contrast with those obtained using an indicator species method (Fig. 2). The latter suggest that the early part of the last interglacial was influenced by sub-continental conditions, while the middle part (*Carpinus* phase, zone 5) was characterised by a markedly more oceanic climatic regime, as seems to be indicated by the persistent records of *Hedera*, *Ilex* and *Buxus* in the mid-Eemian (see Frenzel (1991), Zagwijn (1996), Litt et al. (1996), Aalbersberg and Litt (1998)). The probability density function (*pdf*) method (Kühl et al. 2002; Litt et al. 2001) has been applied, involving the definition of a 'climate-space' in much the same way as has been developed for analysis of fossil beetle assemblages (Atkinson et al. 1987). It supports the interpretations based on the indicator species method, for they suggest that the most probable January and July temperatures during the Eemian were little different to those prevailing in Mid-Latitude Europe today (Fig. 3).

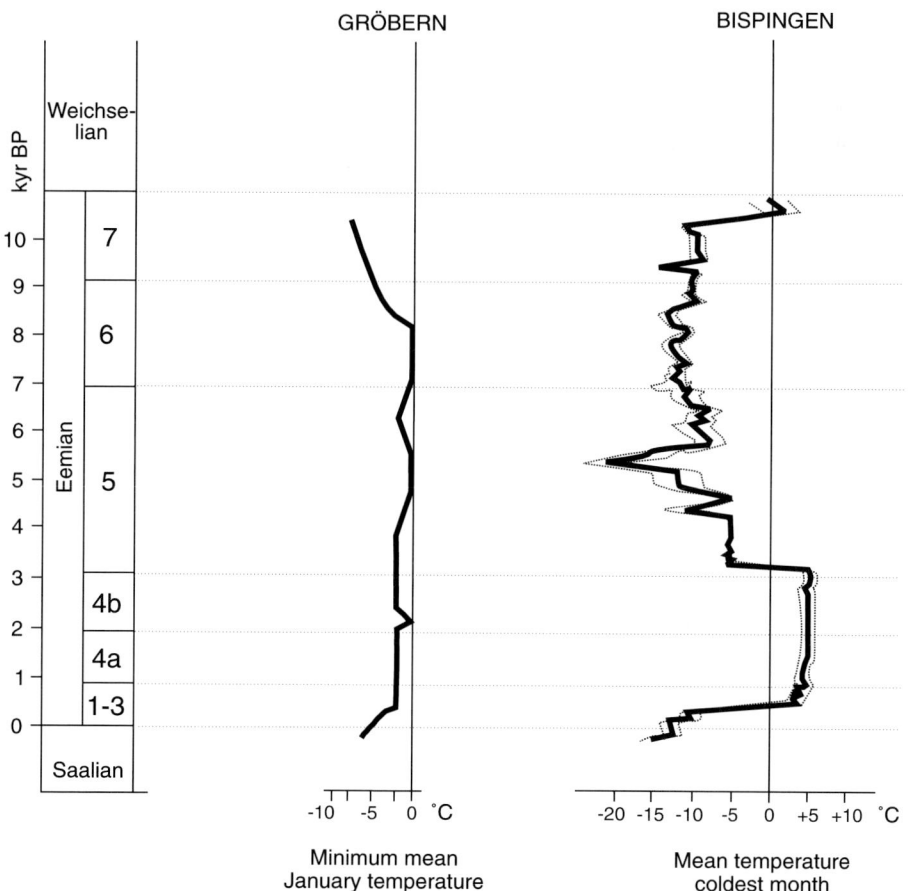

Figure 2. Reconstructed January temperatures for: (a) the Eemian record at Gröbern, Germany, based on an indicator species approach (Litt et al. (1996); see also Fig. 1): and (b) the Eemian record at Bispingen, Germany, based on a transfer function approach (Field et al. 1994). The left-hand column shows the Eemian pollen assemblage zones.

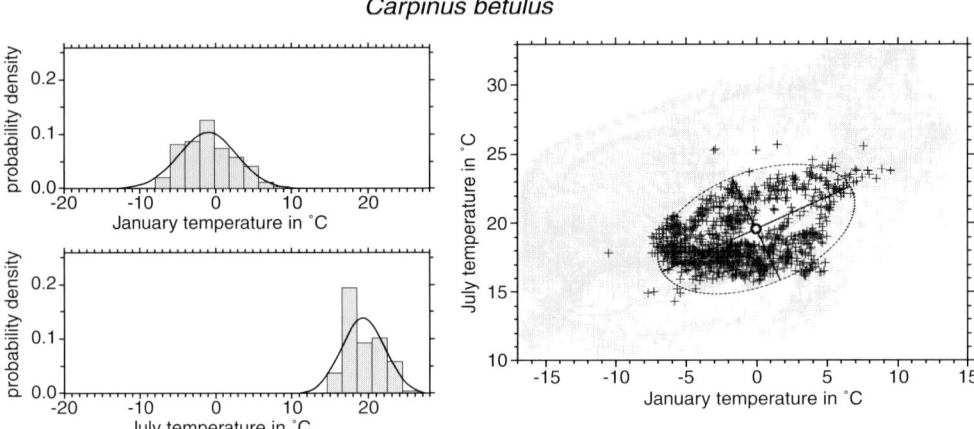

Figure 3. The probability density function (pdf) method applied to a fossil assemblage record from the Gröbern site, consisting of remains of *Acer, Carpinus betulus, Corylus avellana, Fraxinus excelsior, Hedera helix, Ilex aquifolium, Picea abies, Quercus* spp. and *Taxus baccata*. The results support the interpretation based on the indicator species method, in suggesting that the most likely mean January temperature during the Eemian *Carpinus* phase (pollen zone 5) was close to 0 °C (small circle). The thick ellipse denotes the 90% uncertainty range for the combined pdf data. The present-day mean January value is represented by a cross.

The history of climate development in Europe during the Eemian remains controversial; whether interglacial periods are prone to sudden climate fluctuations remains one of the key questions to be resolved by the palaeoclimate science community (see also Rioual et al. (2001)). Some further light may be shed on this question if the investigations could be widened to include other methods and proxy types, such as, for example, the generation of beetle Mutual Climatic Range (MCR) data which uses the mutual overlap of modern climate envelopes for several species that co-occur in a fossil assemblage (Atkinson et al. 1987; Coope et al. 1998). This was attempted for the La Grande Pile sequence, though the results were of too limited a temporal resolution (see Ponel (1995)). In addition, the transfer function approach may be used, for example, for fossil chironomid assemblages. The latter method is relatively new and enables quantitative estimates of summer palaeotempera-tures. It is based on measuring the relationships between the present-day distributions of chironomid taxa and modern climate gradients (Brooks and Birks 2000).

The Weichselian Middle Pleniglacial (MIS 3)

At one time it was considered that the environment of Europe during the Middle Pleniglacial (MIS 3, ca. 60–28 kyr BP) was characterised by the occurrence of only a few interstadials (Van der Hammen et al. 1967; Behre and Lade 1986). It was subsequently argued that these so-called 'interstadials' did not represent warmer conditions (Kolstrup and Wijmstra 1977; Vandenberghe 1985), which led to the suggestion, at one time widely believed, that the whole of the Middle Pleniglacial was characterised by a rather monotonous succession of tundra vegetation, uninterrupted by significant warming events (Vandenberghe 1992). This view was, however, based almost entirely upon the results of pollen-stratigraphical studies of the sediment units dating to this period. Because there was no evidence of significant changes in arboreal pollen percentages in these records, it was concluded that this indicated that there had been no significant changes in temperature.

Studies of plant macrofossil and beetle records, however, suggest that there was a series of short-lived climatic warmings throughout MIS-3. Not all of these can be found in stratigraphical succession at a single site, and they cannot yet be precisely dated. Except for the Hengelo Interstadial, there is no correlation with the 'interstadials' previously defined on the basis of pollen stratigraphy. It is therefore difficult to establish the full effect on Europe of each of these warming events.

Certain characteristic periglacial phenomena and landforms are diagnostic of continuous and discontinuous permafrost. Many of them are often well preserved in the geological record (Huijzer and Isarin 1997), the most widespread being thermal contraction wedges, large cryoturbation structures and perennial frost mounds (in permafrost regions), and small cryoturbations and frost fissures (the latter can also occur under conditions of deep seasonal frost). If these phenomena can be dated precisely, then they provide an indication of the severity (especially in terms of mean annual air temperatures) of ground freezing conditions (e.g., Vandenberghe and Pissart (1993)). Caution is required, however, when interpreting data derived using this approach (Van Huissteden et al. 2003).

Some important advances are beginning to be made. For example, the Hengelo Interstadial (ca. 39–36 ^{14}C kyr BP), originally identified by Zagwijn (1974), has been recognised more widely in Mid-Latitude Europe and there is increasingly robust proxy data to reconstruct the climatic conditions that prevailed during that event (Van Huissteden 1990; Kasse et al. 1995). It was preceded by a distinctly colder phase, the Hasselo Stadial (ca. 40–38.7 ^{14}C kyr BP), during which permafrost developed in The Netherlands and in Germany (Van Huissteden 1990; Kasse et al. 2003). Just prior to this Hasselo Stadial, distinctly warmer conditions were recognised in Europe by Bos et al. (2001) from evidence in East Germany, an event which probably correlates with the Upton Warren Interstadial of the British Isles (Coope 1977). In both the latter event and in the Hengelo Interstadial, summer temperatures rose by ca. 2 °C to 3 °C, in marked contrast to the bitterly cold conditions that had generally prevailed in the region during the intervening stadial (Fig. 4). In fluvial sediment successions the evidence is more difficult to interpret, since the disappearance of permafrost could also be induced by river flooding (Kasse et al. 1995). Following the Hasselo Stadial, permafrost conditions appear to have persisted up until the Late Pleniglacial only on loess subsoils in Western Europe and in more eastern regions (Haesaerts 1974; Van Vliet-Lanoë 1989; Vandenberghe et al. 1998a; Vandenberghe and Nugteren 2001; Kasse et al. 2003). The loess record too points to frequent climatic oscillations on a millennial time scale (e.g., Antoine et al. (2001), Rousseau et al. (1998), Schirmer (2000), Weidenfeller and Zöller (1999)). In this respect, certain key "marker loesses", originally defined in the Czech Republic and Slovakia, promise to provide important stratigraphic links (e.g., Rousseau et al. (2001)).

One technique that offers much potential for the dating and correlation of loess sequences is the analysis of $\delta^{13}C$ variations in the organic matter contained within loess deposits. Hatté et al. (2001) have shown that $\delta^{13}C$ variations in a loess succession of last glacial age from the upper Rhine area can be correlated with the GRIP/GISP Greenland

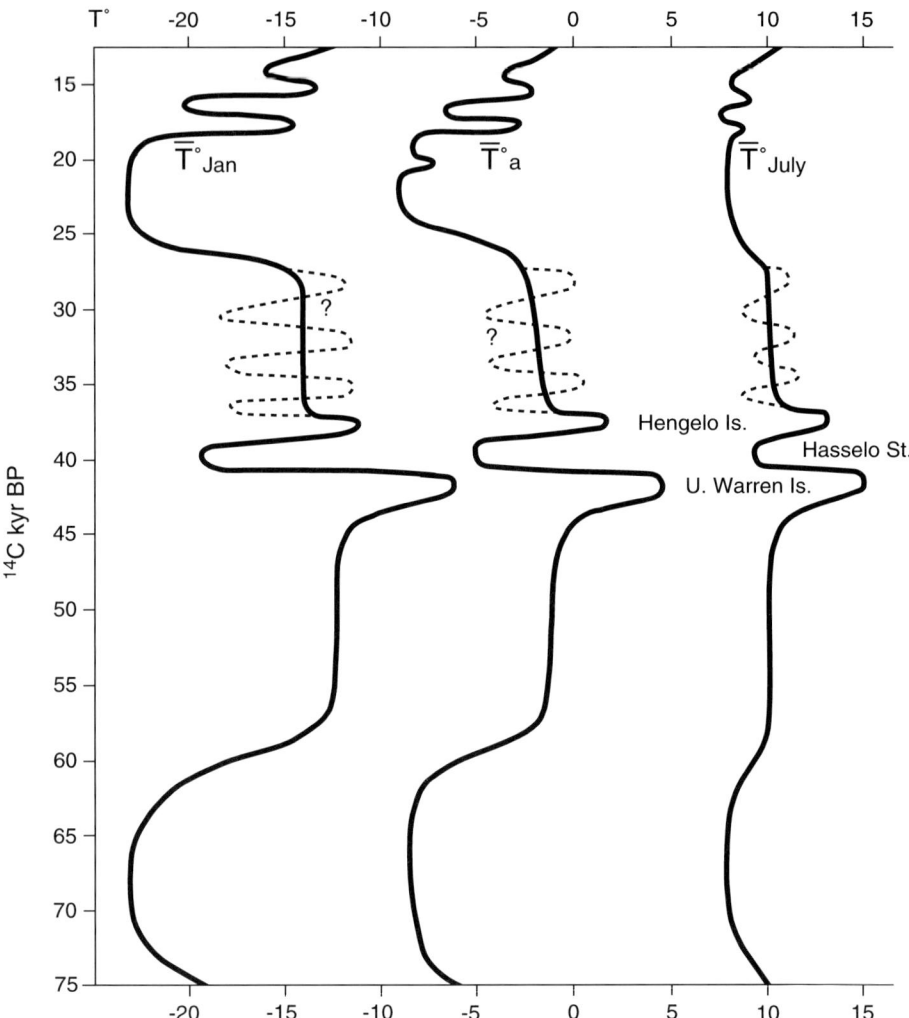

Figure 4. Variations in mean annual temperature (T°a) and in mean temperatures of the warmest (T°July) and coldest (T°Jan) months during the Weichselian Pleniglacial in Western Europe, as reconstructed using a multi-proxy palaeoclimate approach.

ice isotopic record. These results indicate that the loess of the Rhine area during the last glacial stage was not deposited in a dry steppe environment, as previously assumed, but instead under a humid forest steppe or tundra environment, while palaeo-precipitation reconstructions point to annual precipitation values similar to those of the present day, with the exceptions of drier phases towards the end of MIS 2 and MIS 4.

A state-of-the-art model of changing summer, winter and mean annual temperatures during the Weichselian Pleniglacial in Western Europe is presented in Figure 4, though much remains to be refined, in terms of quantifying the magnitude of the climate oscillations as well as establishing the precise ages of the principal climate shifts. The latter is particularly problematic because of serious discrepancies between different models used to calibrate radiocarbon dates for this period (e.g., Kitagawa and van der Plicht (1998a, 1998b), Beck et al. (2001)). The scheme in Figure 4 remains rather tentative, and comparisons with detailed ice-core and marine records for MIS 3 cannot yet be made with a reasonable precision.

The Last Glacial Maximum (MIS 2)

The LGM is a very distinct signal in records from mid-latitude Europe. Many records can be assigned to this time period, and the prevailing climatic conditions can be reconstructed in quite some detail. Accordingly, relatively robust palaeoclimatic data can be provided to palaeoclimatic modellers for global reconstructions that focus on the LGM (Velichko 1984; Van Vliet-Lanoë 1989; Vandenberghe and Pissart 1993; Huijzer and Vandenberghe 1998).

Indications for continuous permafrost are very widespread in Mid-Latitude Europe during the coldest phase of the LGM. There are, however, also some indications that ice wedges degraded temporarily, but resumed their activity shortly afterwards (Van Vliet-Lanoë 1992; Van Huissteden et al. 2000). It is not certain that such (partial) permafrost degradation had a climatic origin, but the possibility is not discounted. Continuous permafrost disappeared finally around 19–20 cal kyr BP in the coversand region (Bateman and Van Huissteden 1999). Within the loess substratum in Belgium, and in the Southern Netherlands, ice-wedge formation appears to have ceased at about the same time, but then resumed shortly afterwards, with final degradation not taking place until just before 17 cal kyr BP (Fig. 5; Van den haute et al. (1998), Renssen and Vandenberghe (2003)). The dating of the coldest phase of the LGM in Mid-Latitude Europe to between 19 and 23 cal kyr BP (continuous permafrost, MAAT $<-8\,°C$) coincides roughly with the age of the lowest ocean level at the LGM (EPILOG: Mix et al. (2001)). The Scandinavian ice sheet, however, reached its maximum position in the NW Russian plain slightly later, i.e., around 18 cal kyr BP (Lunkka et al. 2001; Saarnisto and Lunkka, this volume).

A synthesis of the available coleopteran and palaeobotanical evidence for the LGM in NW Europe suggests that the mean temperature of the warmest month was approximately $8\,°C$. However, in the north the mean temperature of the warmest month was probably no more than $4\,°C$, so that a north to south gradient is evident in the data.

The $-4\,°C$ isotherm of the mean annual air temperature (representing the boundary between continuous and discontinuous permafrost, and based on such evidence as ice-wedge casts) was situated near the French-Belgian border during the LGM, and a north-south temperature gradient is implied. Nevertheless, isolated parts of Central and Northern France may have developed a more continuous permafrost because of increased altitude or because the local substrates consisted of fine-grained soils (cf. Van Vliet-Lanoë (1989)). These overall conclusions are supported by independent fossil coleopteran data.

The mean temperature of the coldest month during the LGM is based on the evidence of ice-wedge casts which indicate values below $-20\,°C$. Coleopteran evidence points to a mean temperature of the coldest month between about $-25\,°C$ and $-18\,°C$. A combination of both kinds of proxy data suggests a mean temperature of the coldest month

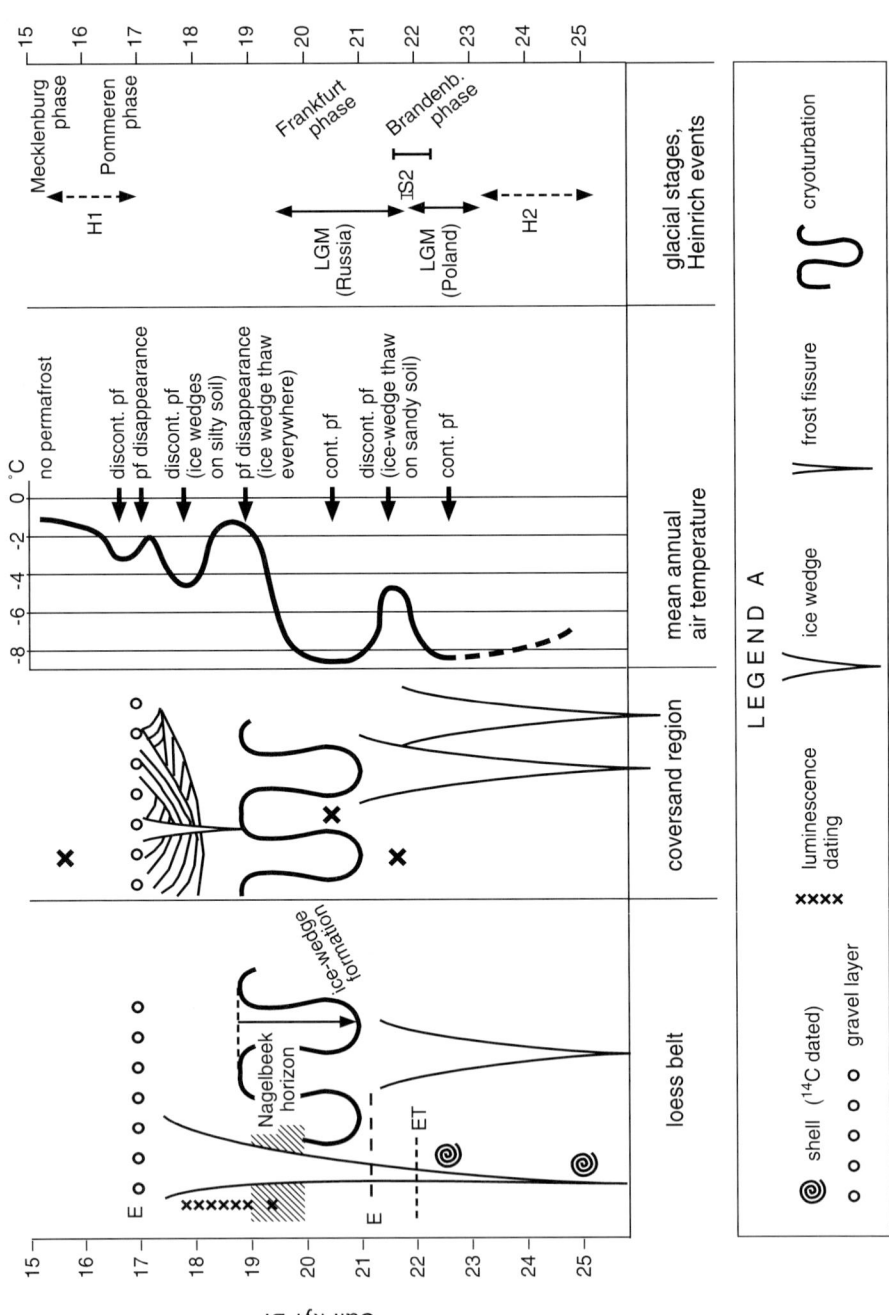

Figure 5. Summary of developments in the Dutch-Belgian region during the Weichselian Late Pleniglacial. Represented schematically are evidence for permafrost development, mean annual temperature variations, and principle episodes of glacial advance and timing of some Heinrich Events (H$_1$, H$_2$). ET = Eltville Tuff (volcanic ash layer from the Eiffel Mountains). E = erosion levels, and 'IS' = isotope stage. LGM = Last Glacial maximum (after Renssen and Vandenberghe (2003)).

somewhere between −25 °C and −20 °C. The temperature of the coldest month may also be approximated from influences of the minimum mean temperature of the warmest month and the maximum mean annual temperature, which give similar results. As a consequence, the annual temperature amplitude during the LGM was probably of the order of 28 °C to 33 °C, indicating a high degree of continentality at that time.

From this climate overview it seems likely that the north to south thermal gradient over North-Western Europe during the LGM was much stronger than that of the present-day. This is reflected especially in the reconstructions of mean annual and winter temperatures, but is less pronounced in reconstructions of mean summer temperature.

There is ample evidence for widespread aeolian activity in the European lowlands during the LGM. Extensive accumulation of loess and sand suggests predominantly arid conditions, although there was sufficient water available for reworking of the aeolian deposits. In addition, the contrast between an extensive sand belt in the north (in Denmark, The Netherlands, Belgium, Northern Germany) and the finer loess deposits in the south (Southern Belgium, Southern part of The Netherlands, Northern France) suggests a dominant wind direction from the northern quadrant, leading to along-track sorting of particle size. However, there is evidence to suggest that wind patterns may have been more complex than this during the LGM (Vandenberghe et al. 1999).

An important phase of fluvial incision took place at the transition from the Middle to the Late Pleniglacial substages (i.e., at the start of the LGM), as recorded in many catchments in Western and Central Europe (references in Mol (1997)). At some locations the subsequent aggradation led to a change from anastomosing to braided-river systems. In general, these river pattern changes around 27 kyr are linked to increased peak discharges and a relatively high supply of sediments into the river valleys. From the collective geomorphological and sedimentary evidence, relatively high precipitation is inferred for the beginning of this time interval. The high peak discharges probably reflect the spring thawing of snow that accumulated during winter. Following this early phase, aeolian activity increased in importance, suggesting that drier conditions prevailed towards the end of the Late Pleniglacial.

The Weichselian Lateglacial (Last Termination)

A fuller understanding of the mode of operation of abrupt climatic changes requires analysis of the *spatial variations* in the palaeo-data. The availability of a large number of detailed and diverse palaeo-environmental records also offers the opportunity to apply quality screening measures to individual records (see e.g., Huijzer and Isarin (1997)). NW Europe probably has the highest density of sites with published records spanning the Last Termination/ Weichselian Lateglacial of any comparable region in the world. Over 400 palynological studies of part or all of the Late-glacial have been undertaken in The Netherlands alone (Hoek 1997). In theory, therefore, fairly sophisticated palaeoclimatic reconstructions ought to be possible, based on this archive of information. Indeed, attempts have been made to synthesise the collective proxy data, and to compare the results with model simulations (see below). There is a huge potential to generate valuable palaeoclimatic syntheses, and some tentative steps in this direction have already been made. There are, however, three significant constraints that affect the available data-base, which limit its potential to meet PEP III objectives.

Data quality

Not all of the records have been studied at a high temporal resolution, and many are based on a single proxy, or on a limited selection of proxy types. It is only in the last decade or so that truly multi-disciplinary investigations, involving large scientific teams, have become common-place (Huijzer and Isarin 1997; Vandenberghe et al. 1998b). Multi-proxy investigations of sequences that span the Last Termination and early Holocene are in progress at a number of key sites located across the Mid-Latitude Europe transect, though there is no comprehensive 'register' of these activities, and there appears to be a relative paucity of such studies in the extreme parts of the transect (Ireland and Eastern Europe).

This trend towards the organisation of large collaborative projects, designed to generate palaeo-environmental records based on multi-proxy investigations, reflects the growing awareness within the palaeo-data community that not all proxy indicators reflect climate influences directly, while some may show time-lagged responses to climate signals (e.g., Lotter et al. (2000)). The diversity of proxy indicators utilised in site investigations also appears to be growing, with some recent studies combining analyses of fossil assemblages (usually several of the following: pollen, plant macrofossil, cladoceran, chironomid, coleopteran, diatom and molluscan assemblages) with stable isotope analyses of both the sediment matrix and of selected fossil types (e.g., Lotter et al. (1997), Ammann et al. (2000), Von Grafenstein et al. (2000), Mayle et al. (1999)). This diversity of palaeo-environmental records is to be welcomed, as it may provide information on different climatic controls affecting the records, as well as independent checks on interpretations based upon different proxies.

Quantification of climatic reconstructions

Until recently, very few of the published palaeoclimatic reconstructions expressed the magnitude of inferred climate change in quantified terms. Exceptions were the palaeo-temperature curves based on analysis of fossil coleopteran data (e.g., Coope and Lemdahl (1995), Coope et al. (1998)) and the climate indicator species approach applied to palaeobotanical data (Aalbersberg and Litt 1998; Isarin and Bohncke 1999), though the latter were not, until very recently, based on investigations of high temporal resolution. More recently, techniques have been developed that enable quantified inferences of mean summer temperatures to be obtained from pollen and cladoceran stratigraphy (see above and Lotter et al. (2000)) as well as fossil chironomid assemblages (Brooks and Birks 2000) using modern calibration sets. These methods hold out great promise for reconstructing the sequence and pattern of climate changes at a higher temporal resolution than hitherto. The number of sites from which such detailed records have been obtained remains relatively low, however, and their geographical spread is patchy.

Geochronological precision

One of the most serious obstacles to the successful synthesis of palaeoclimatic data for the Last Termination is the level of uncertainty in the dating and correlation methods employed. The INTIMATE[1] Group has recently reviewed the limitations affecting age estimates based

---

[1] **INTIMATE** (INTegration of Ice-core, MArine and TErrestrial records of the Last Termination) is a core programme of the International Quaternary Union (INQUA) Palaeoclimate Commission (INTIMATE @ http://www.geog.uu.nl/fg/palaeoclimate/intimate).

on radiocarbon dating, ice-layer counting, varve chronology (Lowe et al. 2001). The over-riding problem is essentially one of temporal resolution: climate transitions, according to the Greenland ice core records, could be affected within just a few decades, while some of the climate events (e.g., the short oscillations of Greenland Interstadial 1, or the 'Bølling - Allerød phase) lasted only some 100 to 300 ice-core years. However, age estimates for events within the Last Termination obtained using ice-layer counting, radiocarbon dating and varve chronology are frequently in excess of 200 years at 1 σ (Blockley et al. pers. comm.). Clearly, it is difficult to date and correlate the various high-resolution records with the required degree of precision, and records can only be considered 'synchronous' within the rather wide uncertainty limits of the dating methods employed.

Despite these problems, some palaeoclimatic reconstructions which span the Last Termination obtained from sites in Mid-Latitude Europe show strong resemblances to the pattern of climate variations reflected in the Greenland ice-core records. Examples are inferred temperature records based on fossil beetle assemblages (Lowe et al. 1999) and on fossil chironomid assemblages (Brooks and Birks 2000) from sites in the UK, and stable oxygen isotope variations from sequences in Germany (Von Grafenstein et al. 1999), Switzerland (Schwander et al. 2000) and The Netherlands (Hoek and Bohncke 2001). There are, however, differences in detail between these records, and they are all subject to significant dating uncertainties. In none of the reconstructions are the statistical uncertainties in the chronologies represented.

There is some diversity of opinion as to whether the climate events represented in high-resolution sequences in Europe were synchronous with those represented in the Greenland ice-core and in marine records. Lowe et al. (1995) suggested a time-lag of ca. 200 to 300 years between the timing of events in the UK and of equivalent events in the GISP record. Others have suggested that climate events are synchronous between the two regions. The problem is that it is not possible to test these claims satisfactorily at present, given the inadequacies of the dating methods currently employed.

A further possible complication that needs to be borne in mind is that climate changes during the Last Termination may have been diachronous even within the confines of Europe, for there is evidence to suggest that there was a significant delay in warming at the start of the Last Termination in the north, particularly in Southern Scandinavia, by comparison with farther south (Coope and Lemdahl 1995; Witte et al. 1998). This may reflect the influence of the Scandinavian ice sheet, which may have cooled the areas in its immediate periphery. Some reconstructions therefore suggest that considerable thermal gradients prevailed across Europe at certain times during the Last Termination, much steeper than those that prevail there at the present time. Coope et al. (1998) have produced palaeoclimate maps for 8 time-slices within the Last Termination-early Holocene time-span, based on quantified palaeotemperature records from 77 sites (cf. Plate 8). These may give a useful insight into the complex evolution of the climate of Europe at the close of the last cold stage. It is the most detailed reconstruction yet attempted for Europe, which is based entirely on quantified palaeoclimate estimates.

Renssen and Isarin (1998) have employed an even more extensive palaeoclimatic data-base based on 300 site records for the 'Younger Dryas' (GS-1) period to construct palaeoclimate maps for NW Europe. The records are heterogeneous, in that a variety of proxy methods have been used to generate the palaeoclimate inferences, but an attempt has been made to derive quantified climate estimates for each record. From these data they

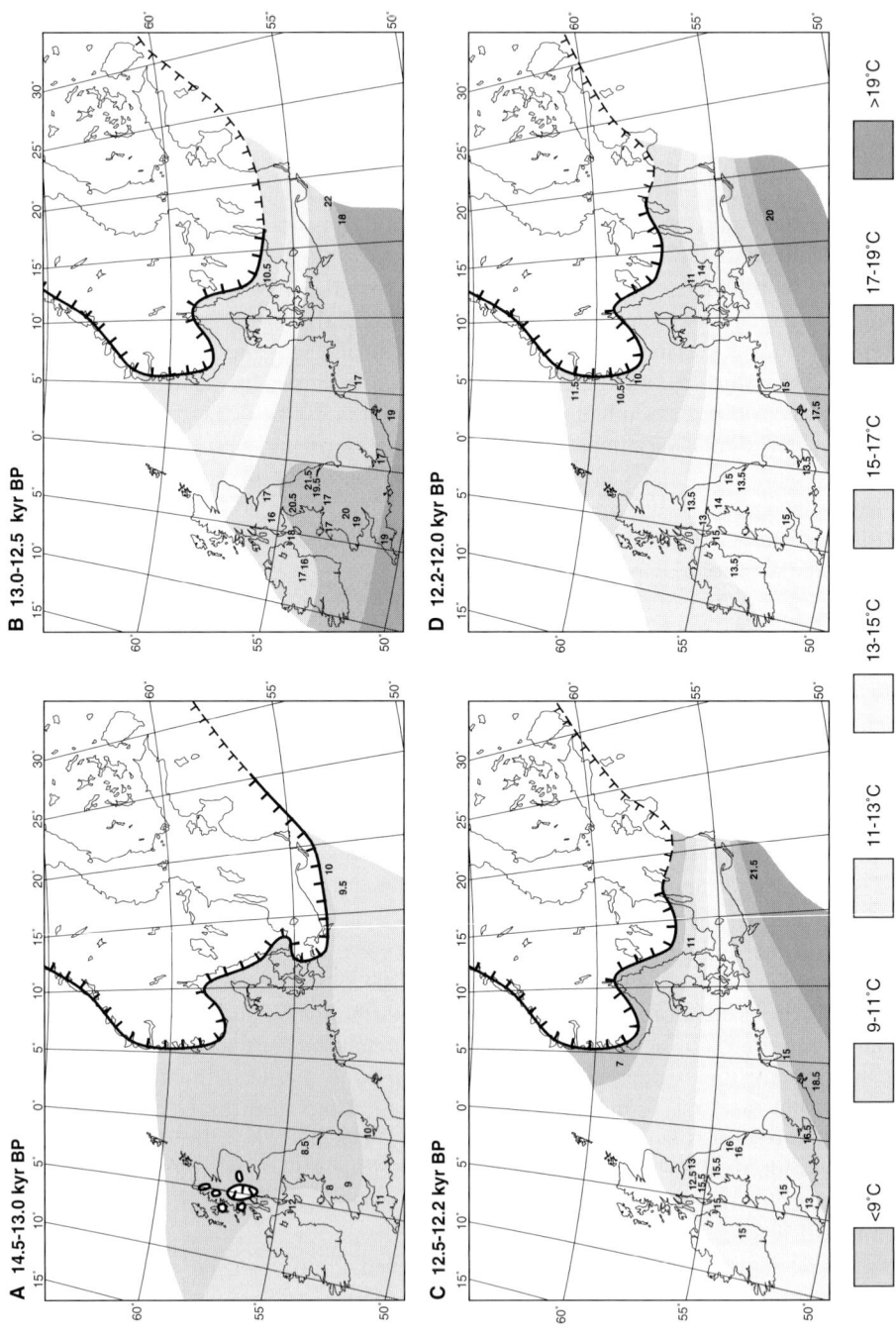

Plate 8. Generalised isotherms for the late-glacial period in Europe, based on beetle MCR interpretations. The diagram shows four of the eight time-slices for which such reconstructions were provided in Coope et al. 1998. Colour version of this Plate can be found in Appendix, p.635

have generated maps of mean winter, summer and annual temperature, with interpolated isotherms summarising climate gradients across the region during the 'Younger Dryas' (Fig. 6).

New developments and potentials

Climate modelling

Comparisons of the results of AGCM (Atmospheric General Circulation Model) simulation experiments with reconstructions of synoptic palaeoclimatic patterns in mid-latitude Europe based on proxy palaeo-data include those attempted for MIS 3 and the Younger Dryas (Isarin et al. 1997; Renssen and Isarin 1998; Isarin and Renssen 1999; Van Huissteden et al. 2003). The most important boundary conditions during those episodes, according to these experiments, were ocean surface conditions (sea-surface temperature and extent of sea ice-cover), the extent and elevation of the major ice sheets, insolation and atmospheric concentration of CO_2. Reconstructions of Younger Dryas winter temperatures are consistent with simulated (AGCM) winter conditions in North Europe but deviate from them in Southern Europe, where the simulated temperatures are 10 °C warmer than those suggested by the palaeo-data. If the palaeo-data are reliable, then they suggest that the N. Atlantic was significantly colder during the YD than is prescribed for the AGCM (Renssen and Isarin 2001). Similar results emerged from an experiment concerned with conditions during MIS 3. In both cases, sensitivity experiments focused especially on the relative contribution of sea-surface temperature, the extent of sea ice and the influence of permafrost and vegetation. Both studies strongly suggest that it is especially the winter sea-ice boundary in the northern Atlantic Ocean that controls the development and extent of permafrost conditions on the continent. Setting the sea-ice boundary too far north in the AGCM might be the reason for the discrepancy between the model outputs and the reconstructions based on palaeo-data.

Tephrochronology

Davies et al. (2002) have reviewed the potential of applying tephrochronology to the dating of Weichselian Late-glacial sequences in Europe. Altogether some 33 different tephras have been reported from sediment sequences in Europe and the NE Atlantic region which date to the period between 18.5 and 8.0 ^{14}C yr BP, though some of these are in micro-tephra form only (invisible to the naked eye). Because tephra layers are deposited virtually instantaneously (in geological terms), they effectively represent time-parallel marker horizons within stratigraphical sequences (Turney and Lowe 2001). In theory, therefore, all 33 tephras could prove valuable for correlation purposes, either in a local context or, in those cases where the ash deposits have been widely dispersed, between regional type-sequences. Davies et al. (2002) illustrate how the wider use of tephrochronology could provide a more robust geochronological framework for the dating of European sequences and of palaeoclimatic events.

It appears that there is also much potential for the application of tephrochronology to the dating and correlation of earlier events, as Wastegård and Rasmussen (2001) have

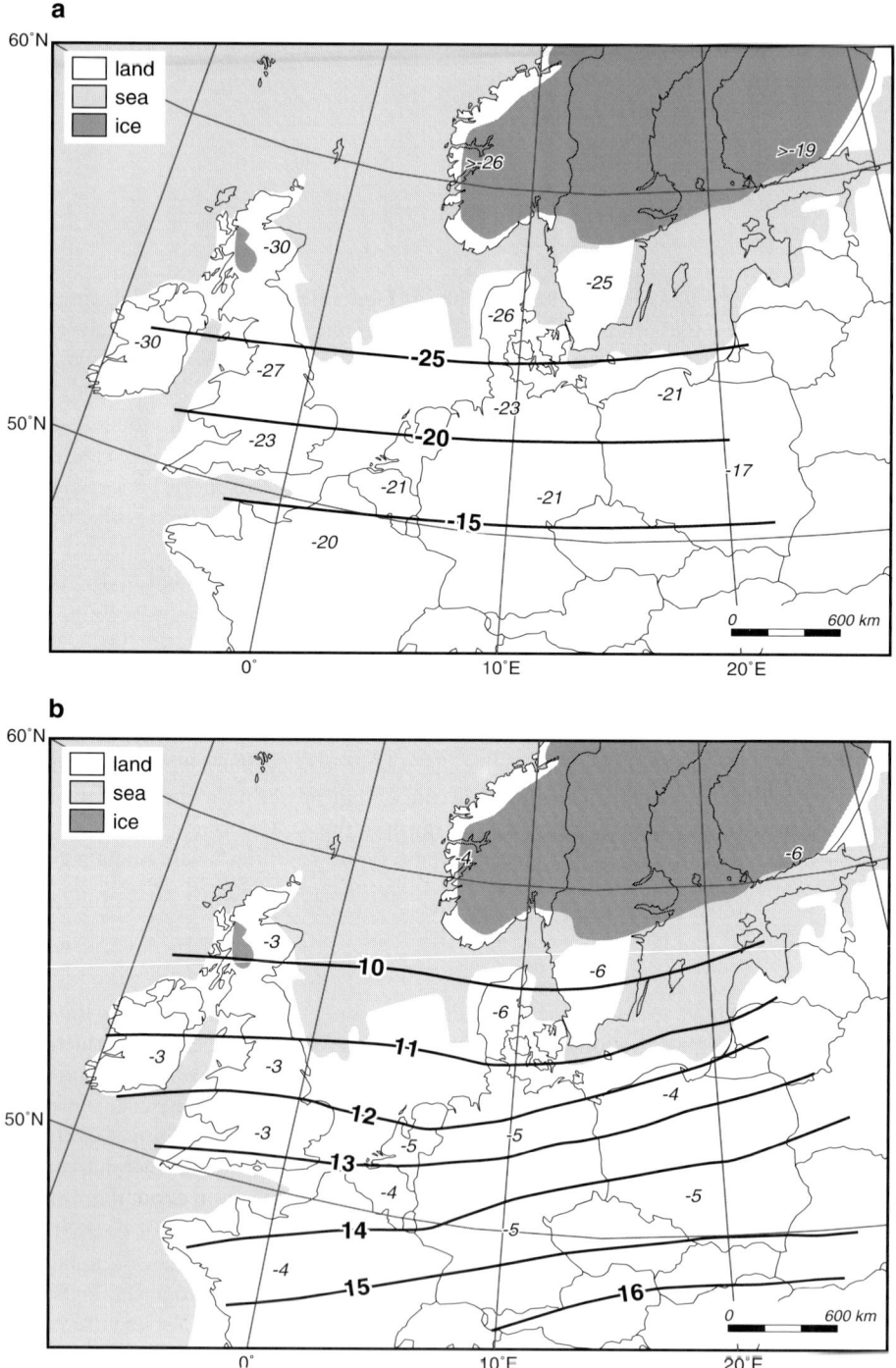

Figure 6. Reconstructed temperatures for sea-level in NW Europe for the Younger Dryas period: (a) winter (maximum mean, coldest month); and (b) summer (min, mean, warmest month) isotherms (°C) in bold and deviations from present values (°C) in italics (after Isarin et al. (1998)).

discovered tephra horizons of Icelandic province and of MIS-5 age from two cores recovered from the North Atlantic. They suggest that at least one of the tephras is likely to be widespread, and that the potential exists for its wider detection in terrestrial sequences on mainland Europe as well as in Greenland ice core records.

Annually-laminated sediments

Annually-laminated lacustrine sediments offer great potential for the detailed chronology and correlation of late Quaternary sequences, and hence for models of migration of biota, time-transgressive climate changes and other palaeo-environmental reconstructions. Varved sediments that are continuous over long intervals are comparatively rare in Europe. During the last few years, however, the Central Europe working group of the *European Lake Drilling Project* (ELDP), an international collaborative initiative funded by the European Science Foundation, has located a number of varved sediment sequences that span the Weichselian Late-glacial and has synchronised these records (Litt et al. 2001). The sites synchronised by ELDP members occur on a west-east transect from western (Eifel Maar region: Brauer et al. (1999, 2001), Litt and Stebich (1999)) and Northern Germany (Hämelsee: Merkt and Müller (1999)) to central (Lake Gościąż: Ralska-Jasiewiczowa et al. (1998)) and Eastern Poland (Lake Perespilno: Goslar et al. (1999)). Correlation of the records is based on a combination of varve counting, pollen stratigraphy, tephrochronology and stable isotope stratigraphy.

The results suggest the Younger Dryas/Preboreal transition to be quasi-synchronous in the region, and to be dated to 11,530–11,590 varve years BP. The Younger Dryas Stadial (YD) is estimated to have lasted about 1100 varve years (lakes Gościąż and Perespilno, Meerfelder Maar). The age of the Laacher See Tephra (LST), an important time marker in these sequences, has been determined to about 12,900 yr BP (varve counting at Meerfelder Maar and Ar/Ar dating), which is nearly 200 years older than the age estimated for the Allerød/YD transition (Hämelsee, Meerfelder Maar). The duration of the Allerød biozone, as defined in Jutland and in Northern Germany, was about 625–670 varve years (Hämelsee, Meerfelder Maar), whereas the older Lateglacial biozones can only be clearly defined by varve chronology in the Meerfelder Maar profile.

Regional chronologies based on annually-laminated sediments, such as those now available from Germany and Poland, are vital for providing independently-dated palaeoclimatic reconstructions that can be compared with marine and ice-core reconstructions. Precise correlations between terrestrial, marine and ice-core reconstructions are difficult to make at present, because of uncertainties with radiocarbon data-sets and with radiocarbon calibration procedures (Lowe et al. 2001). Robust age models based upon annually-laminated sediments enable the ages of climatic events in mid-latitude Europe to be dated independently from, for example, events dated using the Greenland ice-core chronologies. Hence the idea that climatic signals in mid-latitude Europe were synchronous with climatic changes in Greenland can be tested, rather than assumed, the latter having been the general practice hitherto.

It is only by adopting such an approach that regional responses to global or North Atlantic climate signals can be established objectively.

Luminescence dating

There has been striking progress in the luminescence dating of aeolian and fluvio-aeolian sands and of loess, which have accumulated in Northwest Europe during the last ten thousand years (Fig. 5; Frechen et al. (2001), Bateman and van Huissteden (1999)). These results hold out much promise for improved dating of sedimentary sequences spanning the last glacial cycle, and thereby the climatic events that can be inferred from them.

Challenges for future work

While enormous strides forward have been taken during the past two decades in the con-struction of more robust models of the climate history of Mid-Latitude Europe during the last inter-glacial-glacial cycle, much remains to be done.

As intimated in earlier sections of this paper, the impact of the N-Atlantic Ocean on European climate is not yet fully understood. There is a need for much more detailed comparisons between terrestrial and ocean records, which should be based on quantified palaeoclimate reconstructions, and which will need to be dated much more precisely than has been the case hitherto.

The wider use of tephrochronology, varve counting and improved approaches to lu-minescence dating will undoubtedly improve geochronological precision, though new approaches to radiocarbon dating are also required if the precision of age estimates for the younger events (e.g., the Weichselian Late-glacial) is to be significantly improved (cf. Lowe and Walker (2000)). The INTIMATE group has recommended a protocol for improved precision in the dating of events that fall within the Last Termination (Lowe et al. 2001).

There is a need to develop methods that will provide reconstructions of past precipitation levels. There are several methods that can be used to generate palaeotemperature variations and former wind patterns and strength, but former precipitation patterns are notoriously difficult to reconstruct since moisture is rarely the limiting factor for vegetation development and modern analogues tend to span a wide range of precipitation estimates.

There is a need to rationalise data-base facilities in Europe, to establish coherent links between them, and to expand the facilities to cope with more comprehensive data-sets. Some preliminary data-bases have been developed, but they do not cover the entire area of mid-latitude Europe (e.g., the EU-funded EPECC project at Amsterdam), or they do not have a multi-proxy remit (e.g., the European Pollen Data-base, based in Arles), or they serve as repositories of site records from many different contexts and periods (e.g., PANGAEA, based in Germany).

In view of the potential for increased clarity in how the global climate system works, a number of national initiatives have been launched to stimulate greater interaction between the palaeo-data and climate modelling communities, such as NOClim (Norway) and the UK's Rapid Climate Change thematic programme (NERC).

Summary

This paper summarises what is currently known about abrupt palaeoclimatic events and prevailing patterns of climate in the mid-latitude sector of Europe during the last interglacial-glacial cycle. It also addresses the question of whether the data that are available are adequate to meet the scientific goals of PEP III. Significant climatic and environmental changes in this sector are reflected in (i) marked shifts in the position of vegetation belts and the distribution of fauna, some areas having experienced extreme shifts between temperate and boreal forests, and polar deserts, (ii) alternations between episodes of loess deposition and of soil formation, (iii) major modifications of river patterns and of fluvial processes, and (iv) the periodic development of permafrost. From a methodological point of view, considerable steps forward have been made in the quantification of inferred past palaeo-environmental conditions. The paper illustrates how records from key sites offer the best potential for reconstructing the sequence of abrupt climatic changes during the last interglacial-glacial cycle, but also why spatial reconstructions are required for a fuller understanding of the modes and regional effects of climatic events. The use of an array of palaeoclimatic proxy data obtained from high-resolution records from key sites, combined with assessments of the regional climatic patterns that prevailed over Europe during selected time windows, seems the best way forward for developing robust palaeoclimatic reconstructions that will be of use to the climate modelling community. We illustrate this by reference to the evidence available for the following periods: the Eemian, the Weichselian Middle Pleniglacial, the Last Glacial Maximum and the Weichselian Lateglacial.

Acknowledgments

The authors wish to thank D.-D. Rousseau and M. Magny for their constructive comments on an earlier draft.

References

Aalbersberg G. and Litt T. 1998. Multi-proxy climate reconstructions for the Eemian and Early Weichselian. J. Quat. Sci. 13: 367–390.

Ammann B., Birks H.J.B., Brooks S.J., Eicher U., von Grafenstein U., Hofmann W., Lemdahl G., Schwander J., Tobolski K. and Wick L. 2000. Quantification of biotic responses to rapid climatic changes around the Younger Dryas — a synthesis. Palaeogeogr. Palaeoclim. Palaeoecol. 159: 313–348.

Antoine P., Rousseau D.-D., Zöller L., Lang A., Munaut A.V., Hatté C. and Fontugne M. 2001. High resolution record of the last interglacial-glacial cycle in the Nussloch loess paleosol sequences, Upper Rhine Area, Germany. Quat. Int. 76/77: 211–229.

Atkinson C.T., Briffa K.R. and Coope G.R. 1987. Seasonal Temperatures in Britain during the past 22,000 years, reconstructed using beetle remains. Nature 325: 587–592.

Bateman M.D. and Van Huissteden J. 1999. The timing of last glacial periglacial and aeolian events, Twente, Eastern Netherlands. J. Quat. Sci. 14: 277–283.

Beck J.W., Richard D.E., Edwards L., Silverman B.W., Smart P.L., Donahue D.J., Herrar-Osterheld S., Burr G.S., Calsoyas L., Timothy Jull A.J. and Biddulph D. 2001. Extremely large variations of atmospheric ^{14}C concentration during the last glacial period. Science 292: 2453–2458.

Behre K.-E. and Lade U. 1986. Eine Folge von Eem und vier Weichsel-Interstadialen in Oerel/Niedersachsen und ihr Vegetationsablauf. Eiszeitalter und Gegenwart 36: 11–36.

Boettger T., Junge F.W. and Litt T. 2000. Stable climatic conditions in Central Germany during the last interglacial. J. Quat. Sci. 15: 469–473.

Bond G., Showers W., Cheseby M., Lotti R., Almasi P., deMenocal P., Priore P., Cullen H., Hajdas I. and Bonani G. 1997. A pervasive millennial-scale cycle in North Atlantic Holocene and glacial climates. Science 278: 1257–1265.

Bond G., Broecker W., Johnsen S., Mcmanus J., Labeyrie J. and Bonani G. 1993. Correlations between climate records from North Atlantic sediments and Greenland ice. Nature 365: 143–147.

Bos J.A.A., Bohncke S., Kasse C. and Vandenberghe J. 2001. Vegetation and climate during the Weichselian Early Glacial and Pleniglacial in the Nierderlausitz, Eastern Germany — macrofossil and pollen evidence. J. Quat. Sci. 16: 269–289.

Brauer A., Endres Ch. and Negendank J.F.W. 1999. Lateglacial calendar year chronology based on annually laminated sediments from lake Meerfelder Maar, Germany. Quat. Int. 61: 17–25.

Brauer A., Litt T., Negendank J.F.W. and Zolitschka B. 2001. Lateglacial varve chronology and biostratigraphy of lakes Holzmaar and Meerfelder Maar, Germany. Boreas 30: 83–88.

Brooks S.J. and Birks H.J.B. 2000. Chironomid-inferred Late-glacial air temperatures at Whitrig Bog, South-East Scotland. J. Quat. Sci. 15: 759–764.

Coope G.R. 1977. Fossil coleopteran assemblages as sensitive indicators of climatic changes during the Devensian (last) cold stage. Phil. Trans. r. Soc., Lond. B 280: 313–340.

Coope G.R. and Lemdahl G. 1995. Regional differences in the Lateglacial climate of Northern Europe based on coleopteran analysis. J. Quat. Sci. 10: 391–395.

Coope G.R., Lemdahl G., Lowe J.J. and Walkling A. 1998. Temperature gradients in Northern Europe during the last glacial-Holocene transition (14–9 ^{14}C ka BP) interpreted from coleopteran assemblages. J. Quat. Sci. 13: 419–34.

Cortijo E., Labeyrie L., Elliot M., Balbon E. and Tisnerat N. 2000. Rapid climate variability of the North Atlantic Ocean and global climate: a focus of the IMAGES program. Quat. Sci. Rev. 19: 227–41.

Dansgaard W., Johnsen S.J., Clausen H.B., Dahl-Jensen D., Gundestrup N.S., Hammer C.U., Hvidberg C.S., Steffensen J.P., Sveinbjörnsdottir A.E., Jouzel J. and Bond G. 1993. Evidence for general instability of past climate from a 250-kyr ice-core record. Nature 364: 218–220.

Davies S.M., Branch N.P., Lowe J.J. and Turney C.S.M. 2002. Towards a European tephrochronological framework for Termination 1 and the early Holocene. Phil. Trans. r. Soc., Lond. A. (in press).

Drescher-Schneider R. 2000. The Riss-Würm interglacial from West to East in the Alps: an overview of the vegetational succession and climate development. Geologie en Mijnbouw 79: 233–239.

Field M.H., Huntley B. and Müller H. 1994. Eemian climate fluctuations observed in a European pollen record. Nature 371: 779–783.

Frechen M., Vanneste K., Verbeeck K., Paulissen E. and Camelbeeck T. 2001. The deposition history of the coversands along the Bree Fault Escarpment, NE Belgium. Netherlands Journal of Geosciences/ Geologie en Mijnbouw 80: 171–185.

Frenzel B. 1991. Das Klima des letzten Interglazials in Europa. In: Frenzel B. (ed.), Klimageschichtliche Probleme der letzten 130 000 Jahre. Fischer, Stuttgart, pp. 51–78.

Gasse F. and Battarbee R.W., this volume. Introduction. In: Battarbee R.W., Gasse F. and Stickley C.E. (eds), Past Climate Variability through Europe and Africa. Kluwer Academic Publishers, Dordrecht, the Netherlands, pp. 1–6.

Goslar T., Bałaga K., Arnold M., Tisnerat N., Starnawska E., Kuźniarski M., Chróst L., Walanus A. and Więckowski K. 1999. Climate-related variations in the composition of the Lateglacial and Early Holocene sediments of Lake Perespilno (Eastern Poland). Quat. Sci. Rev. 18: 899–911.

Haesaerts P. 1974. Séquence paléoclimatique du Pleistocène Supérieur du Bassin de la Haine (Belgique). Annales de la Société Géologique de Belgique 97: 105–137.

Hatté C., Antoine P., Fontugne M., Lang A., Rousseau D.-D. and Zöller L. 2001. δ^{13}C of Loess Organic Matter as a Potential Proxy for Paleoprecipitation. Quat. Res. 55: 33–38.

Hoek W. 1997 Palaeogeography of Lateglacial vegetations. Ph.D. Thesis, Vrije Universiteit, Amsterdam, 147 pp.

Hoek W. and Bohncke S. 2001. Oxygen-isotope wiggle-matching as a tool for synchronising ice-core and terrestrial records over Termination 1. Quat. Sci. Rev. 20: 1251–1264.

Huijzer A.S. and Isarin R.F.B. 1997. The multi-proxy approach to the reconstruction of past climates with an example of the Weichselian Pleniglacial in North-Western and Central Europe. Quat. Sci. Rev. 16: 513–533.

Huijzer A.S. and Vandenberghe J. 1998. Climatic reconstruction of the Weichselian Pleniglacial in North-Western and Central Europe. J. Quat. Sci. 13: 391–417.

Isarin R.F.B., Renssen H. and Koster E.A. 1997. Surface wind climate during the Younger Dryas in Europe as inferred from aeolian records and model simulations. Palaeogeogr. Palaeoclim. Palaeoecol. 134: 127–148.

Isarin R.F.B., Renssen H. and Vandenberghe J. 1998. The impact of the North Atlantic Ocean on the Younger Dryas climate in Northwestern and Central Europe. J. Quat. Sci. 13: 447–453.

Isarin R.F.B. and Bohncke S. 1999. Mean July temperatures during the Younger Dryas in Northwestern and Central Europe as inferred from climate indicator species. Quat. Res. 51: 158–173.

Isarin R.F.B. and Renssen H. 1999. Reconstructing and modelling Late Weichselian climates: the Younger Dryas in Europe as a case study. Earth Sci. Rev. 48: 1–38.

Kasse C., Bohncke S. and Vandenberghe J. 1995. Fluvial periglacial environments, climate and vegetation during the Middle Pleniglacial with special reference to the Hengelo Interstadial. Mededelingen Rijks Geologische Dienst 52: 387–414.

Kasse C., Vandenberghe J., Van Huissteden J., Bohncke S.J.P. and Bos J.J.A. 2003. Sensitivity of Weichselian fluvial systems to climate change (Nochten mine, Eastern Germany). Quat. Sci. Rev. (in press).

Kitagawa H. and Van der Plicht J. 1998a. Atmospheric radiocarbon calibration to 45,000 yr BP: late glacial fluctuations and cosmogenic isotope production. Science 279: 1187–90.

Kitagawa H. and Van der Plicht J. 1998b. A 40,000 year varve chronology from Lake Suigetsu, Japan: extension of the ^{14}C calibration curve. Radiocarbon 40: 505–515.

Kolstrup E. and Wijmstra T.A. 1977. A palynological investigation of the Moershoofd, Hengelo, and Denekamp Interstadials in The Netherlands. Geologie en Mijnbouw 56: 85–102.

Kühl N., Gebhardt C., Litt T. and Hense A. 2002. Probability density functions as botanical. climatological transfer functions for climate reconstruction. Quat. Res. 58 (3): 381–392.

Kukla G. 1977. Pleistocene Land-Sea Correlations. I. Europe. Ear. Sci. Rev. 13: 307–374.

Litt T. and Stebich M. 1999. Bio- and chrono-stratigraphy of the Lateglacial in the Eifel region, Germany. Quat. Int. 61: 5–16.

Litt T., Brauer A., Goslar T., Merkt J., Balaga K., Müller H., Ralska-Jasiewiczowa M., Stebich M. and Negendank J.F.W. 2001. Correlation and synchronisation of Lateglacial continental sequences in Northern Central Europe based on annually-laminated lacustrine sediments. Quat. Sci. Rev. 20: 1233–1249.

Litt T., Junge F. and Böttger B. 1996. Climate during the Eemian in North-Central Europe — a critical review of the palaeobotanical and stable isotope data from Central Germany. Vegetation History and Archaeobotany 5: 247–256.

Lotter A.F., Birks H.J.B., Eicher U., Hofmann W., Schwander J. and Wick L. 2000. Younger Dryas and Allerød summer temperatures at Gerzensee (Switzerland) inferred from fossil pollen and cladoceran assemblages. Palaeogeogr. Palaeoclim. Palaeoecol. 159: 349–362.

Lotter A.F., Birks H.J.B., Hofmann W. and Marchetto A. 1997. Modern diatom, cladocera, chironomid and chrysophyte cyst assemblages as quantitative indicators for the reconstruction of past environmental conditions in the Alps. I. Climate. J. Paleolim. 18: 395–420.

Lowe J.J., Birks H.H., Brooks S.J., Coope G.R., Harkness D.D., Mayle F.E., Sheldrick C., Turney C.S.M. and Walker M.J.C. 1999. The chronology of palaeoenvironmental changes during the last glacial-Holocene Transition: towards an event stratigraphy for the British Isles. Quat. J. Geol. Soc., Lond. 156: 397–410.

Lowe J.J. and Walker M.J.C. 2000. Radiocarbon dating the last glacial-interglacial transition (ca. 14–9 ^{14}C ka BP) in terrestrial and marine records: the need for new quality assurance protocols. Radiocarbon 42: 53–68.

Lowe J.J., Coope G.R., Harkness D.D., Sheldrick C. and Walker M.J.C. 1995. Direct comparison of UK temperatures and Greenland snow accumulation rates, 15–12,000 calendar years ago. J. Quat. Sci. 10: 175–180.

Lowe J.J., Hoek W. and INTIMATE Group 2001. Inter-regional correlation of palaeoclimatic records for the Last Glacial-Interglacial Transition: a protocol for improved precision recommended by the INTIMATE project group. Quat. Sci. Rev. 20: 1175–1188.

Lunkka J.P., Saarnisto M., Gey V., Demidov I. and Kiselova V. 2001. Extent and age of the Last lacial Maximum in the southeastern sector of the Scandinavian Ice Sheet. Glob. Plan. Chan. 31: 407–425.

Maslin M., Seidov D. and Lowe J.J. 2001. Synthesis of the nature and causes of rapid climate transitions during the Quaternary. In: Seidov D., Maslin M. and Haupt B.J. (eds), The Oceans and Rapid Climatic Change: Past, Present and Future. American Geophysical Union, Geophysical Monograph 126, pp. 9–52.

Mayle F.E., Bell M., Birks H.H., Brooks S.J., Coope G.R., Lowe J.J., Sheldrick C., Turney C.S.M. and Walker M.J.C. 1999. Response of lake biota and lake sedimentation processes in Britain to variations in climate during the last glacial-Holocene transition. J. Geol. Soc., Lond. 156: 411–23.

McManus J.F., Bond G.C., Broecker W.S., Johnsen S., Labeyrie L. and Higgins S. 1994. High resolution climate records from the North Atlantic during the last Interglacial. Nature 371: 326–329.

Menke B. and Tynni R. 1984. Das Eeminterglazial und das Weichselfrühglazial von Rederstall/Dithmarschen und ihre Bedeutung für die mitteleuropäische Jungpleistozän-Gliederung. Geologisches Jahrbuch A76: 3–120.

Merkt J. and Müller H. 1999. Varve chronology of Lateglacial in Northwest Germany from lacustrine sediments of the Hämelsee/Lower Saxony. Quat. Int. 61: 41–59.

Mix A.C., Bard E. and Schneider R. 2001. Environmental processes of the ice age: land, ocean, glaciers (EPILOG). Quat. Sci. Rev. 20: 627–657.

Mol J. 1997. Fluvial response to Weichselian climate changes in the Niederlausitz (Germany). Journal of Quaternary Science 12: 43–60.

Ponel P. 1995. Rissian, Eemian and Würmian Coleoptera assemblages from La Grande Pile (Vosges, France). Palaeogeogr. Palaeoclim. Palaeoecol. 114: 1–41.

Ralska-Jasiewiczowa M., Goslar T., Madeyska T. and Starkel L. (eds) 1998. Lake Gościąż, Central Poland, A Monographic Study. Part I., W. Szafer Institute of Botany, Kraków, 340 pp.

Renssen H. and Isarin R.F.B. 2001. The two major warming phases of the last deglaciation at ~14.7 and ~11.5 kyr cal BP in Europe: climate reconstructions and AGCM experiments. Glob. Plan. Chan. 30: 117–153.

Renssen H. and Isarin R.F.B. 1998. Surface temperature in NW Europe during the Younger Dryas: AGCM simulation compared with temperature reconstructions. Clim. Dyn. 14: 33–44.

Renssen H. and Vandenberghe J. 2003. Investigation of the relationship between permafrost distribution in NW Europe and extensive winter sea-ice cover in the North Atlantic Ocean during the cold phases of the Last Glaciation. Quat. Sci. Rev. 22: 209–223.

Rioual P., Andrieu-Ponel V., Rietti-Shati M., Battarbee R.W., de Beaulieu J.-L., Cheddadi R., Reille M., Svobodova H. and Shemesh A. 2001. High-resolution record of climate stability in France during the last interglacial period. Nature 413: 293–296.

Rousseau D.-D., Zöller L. and Valet J.-P. 1998. Late Pleistocene Climatic Variations at Achenheim, France, Based on Magnetic Susceptibility and TL Chronology of Loess. Quat. Res. 49: 255–263.

Rousseau D.-D., Gerasimenko N., Matviischina Z. and Kukla G. 2001. Late Pleistocene Environments of the Central Ukraine. Quat. Res. 56: 349 356

Ruddiman W.F. and McIntyre A. 1973. Time-transgressive deglacial retreat of polar waters from the North Atlantic. Quat. Res. 3: 117–30.

Ruddiman W.F., Sancetta C.D. and McIntyre A. 1977. Glacial/interglacial response rate of subpolar North Atlantic waters to climatic change: the record left in deep-sea sediments. Phil. Trans. r. Soc., Lond. B 280: 119–142.

Saarnisto M. and Lunkka J.P., this volume. Climate variability during the last interglacial-glacial cycle in NW Eurasia. In: Battarbee R.W., Gasse F. and Stickley C.E. (eds), Past Climate Variability through Europe and Africa. Kluwer Academic Publishers, Dordrecht, the Netherlands, pp. 443–464.

Schirmer W. 2000. Eine Klimakurve des Oberpleistozäns aus dem rheinischen Löss. Eiszeitalter und Gegenwart 50: 25–49.

Schwander J., Eicher U. and Ammann B. 2000. Oxygen isotopes of lake marl at Gerzensee and Leysin (Switzerland), covering the Younger Dryas and two minor oscillations, and their correlation to the GRIP ice core. Palaeogeogr. Palaeoclim. Palaeoecol. 159: 203–214.

Seidenkrantz M.-S., Kristensen P. and Knudsen K.L. 1995. Marine evidence for climatic instability during the last interglacial in shelf records from Northwest Europe. J. Quat. Sci. 10: 77–82.

Taylor K.C., Hammer C.U., Alley R.B., Clausen H.B., Dahl-Jensen D., Gow A.J., Gundestrup N.S., Kipfstuhl J., Moore J.C. and Waddington E.D. 1993. Electrical conductivity measurements from the GISP2 and GRIP Greenland ice cores. Nature 366: 549–552.

Thouveny N., De Beaulieu J.-L., Bonifay E., Creer K.M., Guiot J., Icole M., Johnsen S., Reille M., Williams T. and Williamson D. 1994. Climatic variations in Europe over the past 140 kyr deduced from rock magnetism. Nature 371: 503–506.

Turney C.S.M. and Lowe J.J. 2001. Tephrochronology. In: Last W.M. and Smol J.P. (eds), Tracking Environmental Change Using Lake Sediments: Physical and Chemical Techniques. Kluwer Academic Publishers, Dordrecht.

Van den haute P., Vancraeynest L. and De Corte F. 1998. The Late Pleistocene loess deposits and palaeosols of Eastern Belgium: new TL age constraints. J. Quat. Sci. 13: 487–497.

Van der Hammen T., Maarleveld G.C., Vogel J.C. and Zagwijn W. 1967. Stratigraphy Climatic succession and radiocarbon dating of the last glacial in the Netherlands. Geologie en Mijnbouw 4: 79–95.

Van Huissteden J. 1990. Tundra Rivers of the Last Glacial: sedimentation and geomorphological processes during the Middle Pleniglacial (Eastern Netherlands). Mededelingen Rijks Geologische Dienst 44-3: 1–138.

Van Huissteden J., Vandenberghe J., Van der Hammen T. and Laan W. 2000. Fluvial and eolian interaction under permafrost conditions: Weichselian Late Pleniglacial, Twente, Eastern Netherlands. Catena 40: 307–321.

Van Huissteden J., Vandenberghe J. and Pollard D. 2003. Palaeotemperature reconstructions of the European permafrost zone during oxygen isotope stage 3 compared with climate model results. J. Quat. Sci. (in press).

Van Vliet-Lanoë B. 1989. Dynamics and extent of the Weichselian permafrost in Western Europe (substage 5e to stage 1). Quat. Int. 3-4: 109–113.

Van Vliet-Lanoë B. 1992. Le niveau à langues de Kesselt, horizon repère de la stratigraphie du Weichselien supérieur européen: signification paléoenvironnementale et paléoclimatique. Mémoires de la Société géologique de France n.s. 160: 35–44.

Vandenberghe J. 1985. Paleoenvironment and stratigraphy during the Last Glacial in the Belgium-Dutch border region. Quat. Res. 24: 23–38.

Vandenberghe J. 1992. Geomorphology and climate of the cool oxygen isotope stage 3 in comparison with the cold stages 2 and 4 in The Netherlands. Zeitschrift für Geomorphologie, Suppl. Bd. 86: 65–75.

Vandenberghe J. and Pissart A. 1993. Permafrost changes in Europe during the last glacial. Permafrost and Periglacial Processes 4: 121–135.

Vandenberghe J., Huijzer A.S., Mücher H. and Laan W. 1998a. Short climatic oscillations in a western European loess sequence (Kesselt, Belgium). J. Quat. Sci. 13: 471–485.

Vandenberghe J., Coope G.R. and Kasse C. 1998b. Quantitative reconstructions of palaeoclimates during the last interglacial-glacial in Western and Central Europe: an introduction. J. Quat. Sci. 13: 361–366.

Vandenberghe J., Isarin R.F.B. and Renssen H. 1999. Comments on 'Windpolished boulders as indicators of a Late Weichselian wind regime in Denmark in relation to neighbouring areas' by Christiansen and Svensson [9(1): 1–21, 1998]. Permafrost and Periglacial Processes 10: 199–201.

Vandenberghe J. and Nugteren G. 2001. Abrupt climatic changes recorded in loess sequences. Glob. Plan. Chan. 28: 1–9.

Velichko A.A. (ed.) 1984. Late Quaternary Environments of the Soviet Union. Longman, London, 327 pp.

Von Grafenstein U., Erlenkauser H., Brauer A., Jouzel J. and Johnsen S.J. 1999. A mid-European decadal isotope-climate record from 15,500 to 5,000 years BP. Science 284: 1654–7.

Von Grafenstein U., Eicher U., Erlenkauser H., Ruch P., Schwander J. and Ammann B. 2000. Isotope signature of the Younger Dryas and two minor oscillations at Gerzensee (Switzerland): palaeoclimatic and palaeolimnological interpretation based on bulk and biogenic carbonates. Palaeogeogr. Palaeoclimat. Palaeoecol. 159: 215–229.

Wastegård S. and Rasmussen T.L. 2001. New tephra horizons from Oxygen Isotope Stage 5 in the North Atlantic: correlation potential for terrestrial, marine and ice-core archives. Quat. Sci. Rev. 20: 1587–1593.

Weidenfeller M. and Zöller L. (eds) 1999. Loess in the Middle and Upper Rhine Area, Loessfest 1999 Field Guide. Geologisches Landesamt Rheinland-Pfalz, Mainz, 83 pp.

Witte H.J.L., Coope G.R., Lemdahl G. and Lowe J.J. 1998. Regression coefficients of thermal gradients in Northwestern Europe during the last glacial-Holocene transition using beetle MCR data. J. Quat. Sci. 13: 435–446.

Zagwijn W. 1974. Vegetation, climate and radiocarbon datings in the Late Pleistocene of The Netherlands. Part II: Middle Weichselian. Mededelingen Rijks Geologische Dienst 25: 101–110.

Zagwijn W. 1996. An analysis of Eemian climate in Western and Central Europe. Quat. Sci. Rev. 15: 451–469.

Zöller L., Oches E.A. and McCoy W.D. 1994. Towards a revised chronostratigraphy of loess in Austria with respect to key sections in the Czech Republic and in Hungary. Quaternary Geochronology (Quaternary Science Reviews) 13: 465–472.

20. ATLANTIC TO URALS — THE HOLOCENE CLIMATIC RECORD OF MID-LATITUDE EUROPE

KEITH BARBER (keith.barber@soton.ac.uk)
Palaeoecology Laboratory
Department of Geography
University of Southampton
Southampton, SO17 1BJ
UK

BERND ZOLITSCHKA (zoli@uni-bremen.de)
GEOPOLAR
Institut für Geographie
Universität Bremen
FB 8 Celsiusstraße FVG-M
28359, Bremen
Germany

PAVEL TARASOV (paveltarasov@hotmail.com)
Department of Geography
Moscow State University
Vorobievy Gory, 119899 Moscow
Russia

ANDRÉ F. LOTTER (a.lotter@bio.uu.nl)
Laboratory of Palaeobotany and Palynology
University of Utrecht
Budapestlaan 4, 3584 CD Utrecht
The Netherlands

Keywords: Holocene proxy records, Peatlands, European Pollen Database, Laminated lake sediments, Tree rings, Speleothems, Alpine treelines.

Introduction and rationale

The Mid-Latitude belt of Europe, broadly between 45° and 65 °N, is probably the most intensively studied area of the PEP III transect, but providing a synthesis of Holocene climatic change over this large and varied area is not easy. Stretching from the Atlantic coast of Ireland to the Ural Mountains of Russia (Fig. 1), it takes in the United Kingdom,

R. W. Battarbee et al. (eds) 2004. *Past Climate Variability through Europe and Africa.*
Springer, Dordrecht, The Netherlands.

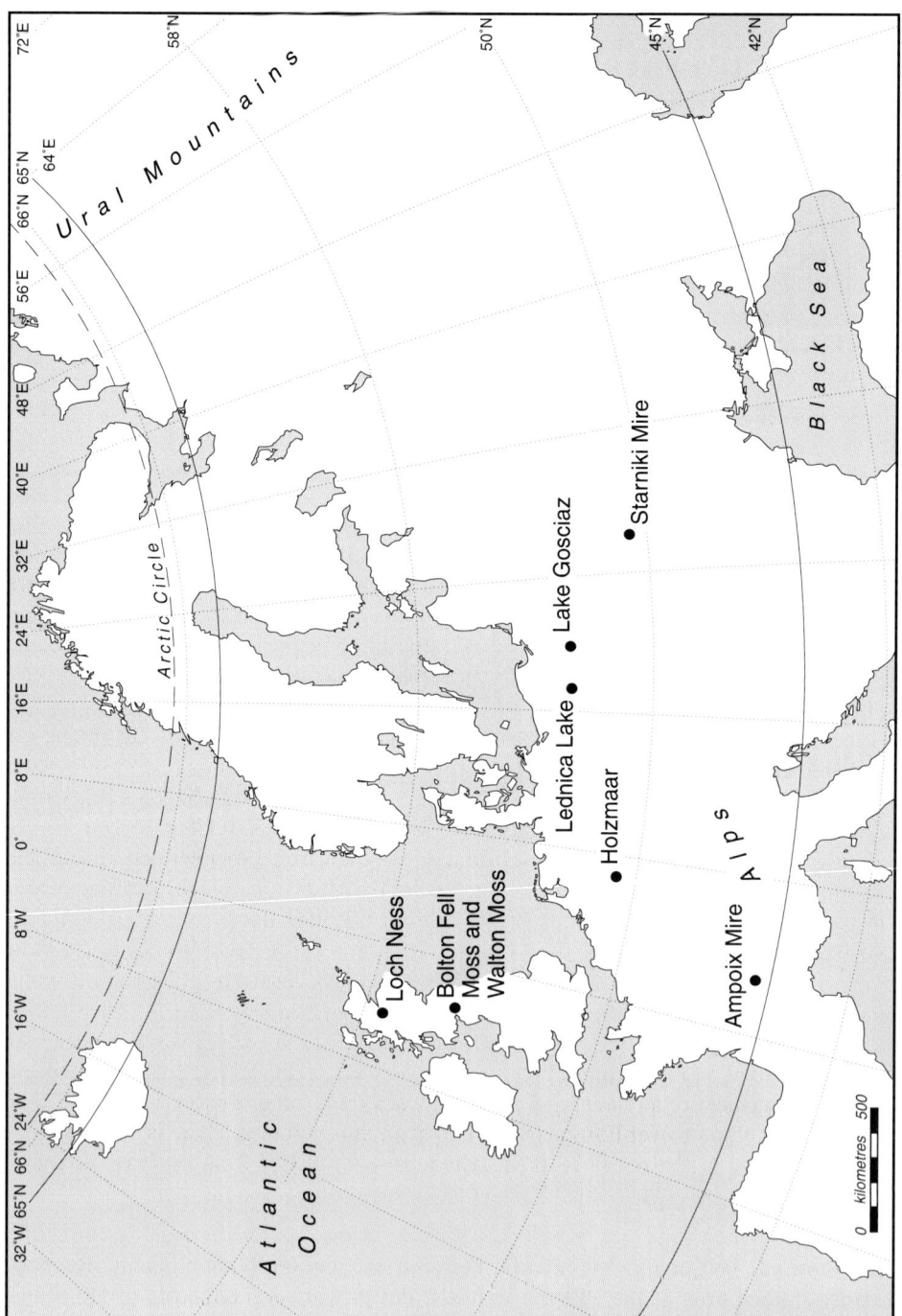

Figure 1. Map of the European Mid-Latitude transect showing key sites referred to in the text.

France north of the Mediterranean coastal zone, the Low Countries, Germany, the Alpine countries, Central and Eastern Europe north of the Alps, and the lowlands of the former USSR. Climate varies along the transect from hyper-oceanic in the west, to very continental in the east (Crawford 2000).

In this chapter we outline the nature and range of the records, and the clarity and timing of the climatic signal. After brief consideration of the background record of change furnished by the ice and ocean-core records we deal with two of the oldest terrestrial sources of proxy climate records from our transect area, peat and pollen data, and then consider annually-resolved records from lakes, tree rings and speleothems, before summarising changes from the Alps. We highlight evidence from key sites with high quality proxy records, rather than attempting to synthesise a Europe-wide picture, which would be premature, and needs further refinement of site chronologies. Existing compilations of palaeoecological data for Europe, such as Berglund et al. (1996), illustrate the size of such a task.

The nature of the records and the major climatic phases and events

Ocean and ice core records

Ocean and ice-core records provide a background scale of change to what was happening on the continent, but the tremendous events of the late-glacial in these records have tended to obscure the importance of Holocene climatic fluctuations until recently. Temperature variations in the order of 1–2 °C may appear as minor variations in an ice-core record but such changes had effects on glaciers, lakes, treelines and bogs, and on people. The direct effects on humanity are moderated by the adaptability of societies, but there must have been impacts, especially on farming.

The two major Holocene fluctuations recorded in the $\delta^{18}O$ records of Greenland ice cores are the temperature depressions related to the Preboreal Oscillation (ca. 10,000–9500 BP) and the 8200-year event (Alley et al. 1997; Meese et al. 1994; O'Brien et al. 1995). Both of these events register for only about 200 years in the ice record, but the response of terrestrial ecosystems may lag behind the initial impulse and be a prolonged response, or the ecosystem may be insensitive to the forcing. Examples of all three states are given below, and it is clear that the linkages between ocean and/or ice-driven climatic events are complex.

According to Bond et al. (1997, 2001), North Atlantic ocean cores of Holocene age show eight abrupt cooling events, besides the Little Ice Age (LIA), caused by polar and sub-polar water moving further south and causing deposition of ice-rafted debris (IRD). These events occur with a cyclicity of around 1470 years with a variability of ca. 500 years, and interact with the thermohaline circulation of the North Atlantic (Vidal and Arz, this volume) to influence sea-surface temperatures, depression tracks and other factors which are then expressed in the climate over Europe (Snowball et al., this volume). There is also evidence of solar variability in the ocean records, which may be reflected in sensitive terrestrial ecosystems (Bond et al. 2001; Chambers and Blackford 2001; Mauquoy et al. 2002).

Peatland records of climatic change

The stratigraphy of European peat bogs was one of the first proxy climate records, and was used in sub-dividing the Holocene (Godwin 1975). Due to a number of mistaken ideas on bog growth the potential of the record was not properly exploited until the work of Aaby (1976), van Geel (1978) and Barber (1981). Over the last two decades the work of Dupont (1986), Svensson (1988a, 1988b), Blackford and Chambers (1991, 1995), Barber et al. (1994), Chambers et al. (1997), van Geel et al. (1996) Charman et al. (1999), Mauquoy and Barber (1999), Ellis and Tallis (2000) and Chiverrell (2001), amongst others, has shown the important role of climate as an allogenic forcing factor in ombrotrophic (rain-fed) bog growth and therefore of peat stratigraphy as a proxy climate record. Recent developments of more quantified analyses (Barber et al. 1994; Blackford and Chambers 1993; Charman et al. 1999; Barber et al. 2000) have revived the usefulness of the peat archive, and correlations between bog surface wetness (BSW) and chironomid-derived temperature reconstructions from a lowland lake support the hypothesis that changes in BSW are driven primarily by summer temperature (Barber and Langdon 2001). Charman and Hendon (2000) and Charman et al. (2001) have also demonstrated that changes in BSW can be correlated to the GISP2 ice core record and sea surface temperature.

Haslam (1987) undertook an extensive survey of the Main Humification Change (=*Grenzhorizont*) in bogs across Europe in a transect from oceanic Ireland to continental Poland, including sites in the UK, the Netherlands, Germany, Denmark and Southern Sweden. This showed that bogs in the oceanic climate of Ireland and Northern England reacted sensitively to changes in effective precipitation (rain and snowfall minus evapotranspiration) with frequent changes in humification and plant assemblages, whereas bogs in a more continental climate such as inland Northern Germany showed infrequent sudden changes as if forced over a climatic threshold. In the much more continental climate of Eastern Poland only one minor shift to wetter conditions in an otherwise dry stratigraphy was shown during the last 4000 years (Barber 1993). It appears therefore that away from the blanket and raised bogs of Western Europe the peatland record of Eastern Europe is not suitable for detailed climatic reconstructions; indeed many of the peatlands are not ombrotrophic (Moore 1984).

At the key sites of Bolton Fell Moss and Walton Moss, adjacent to each other in Cumbria, Northern England (Fig. 1), the climatically sensitive record extends back into the early Holocene with an average accumulation rate of 10 yrs/cm. Both bogs have been studied intensively (Barber 1981; Barber et al. 1994; 1998; Hughes et al. 2000; Mauquoy et al. 2002; Barber et al. 2003) and detailed palaeoclimate records have been extracted using three independent proxies: plant macrofossils, humification and testate amoebae. Both sites became ombrotrophic through the development of *Eriophorum / Calluna* peat above fen peat early in the Holocene at about 9600 cal. BP, and the first wet shift is recorded at both sites around 7800 cal. BP. This is particularly prominent in the Detrended Correspondence Analysis (DCA) of the macrofossil data from Walton Moss (Fig. 2; Hughes et al. (2000)) and may well be a lagged terrestrial response to the 8200 BP event; at both sites there is some evidence of a change to wetter conditions before the major wet-shift at 7800 cal. BP. After this the BSW records imply a generally dry and / or warm climate until ca. 4400 cal. BP, but with wet episodes recorded at, for example, Bolton Fell Moss at around 6200, 5700, 5420 and 5250 cal. BP. As shown in Table 1, (amended from Hughes et al. (2000), with some new data), similar changes can be found in bogs from The Netherlands and Denmark, and it must be noted that not all bogs would be at an ombrotrophic stage during this period or may only have started growth later, such as the South Cumbrian bogs on estuarine flats investigated by Wimble (1986).

Table 1. Dates of wet-shifts from selected European bogs. NIM = Not in Model — the top of each peat profile was not included in the age / depth model for this study.

Site	Reference												
Abbeyknockmoy Co. Galway, Ireland	Barber et al. 2003	NIM						2200	2750	3150	4000 4250		
Mongan Bog Co. Offaly Ireland	Barber et al. 2003	NIM	450 600 850			1600 1800		2250	2350 2450 2750	3200			
Kentra Moss Western Scotland	Ellis and Tallis 2000	330	600 880	1150	1400			2150	2550	3250			
Talla Moss, Borders, Scotland	Chambers et al. 1997	540		1100		1700	1930	2270	2600	3460			
Border mires Northern England	Mauquoy and Barber 1999ab	180 550	850	1030	1400	1740	1980	2130	2540 2710				
Bolton Fell Moss, Cumbria, England	Barber et al. 2003	NIM	620 720	1000	1400			2200	2350 2440 2580 2900	3020 3200 3600 3750	4020 4280 4420 4620	5250 5420 5700 6200	7500 7800
Bolton Fell Moss	Barber 1981, et al. 1994	210 500		1000 1170	1300			2200	2900	3600	4350		
Walton Moss Cumbria, England	Hughes et al. 2000	100 350	–	–	1450	1750		2320-2040	3170-2860	3500	4410-3990	5300 5900	7800
South Cumbrian bogs, England	Wimble 1986		600 800	1050	1350 1500	1700		2250 2250	2900 2900 2900	3400 3400 3500	4300 3800		
British and Irish blanket bogs	Blackford and Chambers 1991, 1995	490		1150	1310 1330 1310		1910						
Engbertsdijksveen The Netherlands	van Geel 1978 van Geel et al. 1996								2850 3020 2750-2450	3750	4350	5450 5850 6450	6800 7150
Bourtangerveen The Netherlands	Dupont 1986						1950			3300-3650	4450	5300	
Draved Mose Jutland, Denmark	Aaby 1976	450	660 860		1500	1700		2250	3000	3400	4000 4300 4600	4850 5050 5400	
NW European raised bogs	Haslam 1987			1150		1850			2550 3050		4200		

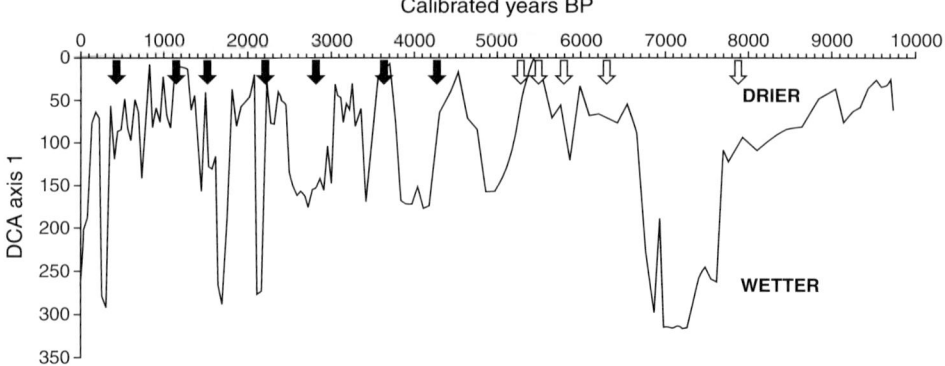

Figure 2. Reconstruction of bog surface wetness (BSW) at Walton Moss, Cumbria, England. Axis 1 scores from the DCA of macrofossil results from core WLM 11 plotted against calibrated years BP, with periods of increased BSW at the nearby Bolton Fell Moss indicated by arrows. Open arrows = minor changes; filled arrows = major changes. Redrawn from Hughes et al. (2000) with other data from Barber et al. (2003).

Further climatic shifts can be seen in the Walton Moss record (Fig. 2) and these are mirrored by similar changes in other western European bogs. The dates in Table 1 are of prominent wet-shifts at selected sites from Ireland across to Sweden, the dates from Haslam (1987) being the most prominent changes at a large number of sites surveyed from Ireland, Britain, the Netherlands, Germany, Denmark and Sweden. Many sites have been omitted. This selection is to illustrate the point, and all of them have at least reasonable dating control. The wet-shifts have been grouped subjectively into clusters and have been calibrated and rounded to the nearest decade. However, it must be noted that the dates are those given by each individual author to the beginning of the wet-shift, and as each author has interpreted their own data slightly differently they should be regarded as having an error term of ±50 years, and comparisons between sites made with some caution. It is however apparent that there are clusters of dates at sites across this broad swathe of Mid-Latitude Europe, especially in the late-Holocene. In particular, wet and / or cold conditions were widespread at around 4400–4000, 2800–2200, 1800–1700, 1400–1300, and 1100–1000 cal. BP. After cal. 1000 BP, (AD 950) there exists an increasing amount of documentary evidence (Lamb 1977; 1995), which allows us to link and validate proxy climate signals from the peat record with independent evidence of climatic change. Unfortunately this is also the time when peat cutting for fuel begins to become significant over much of Britain and Ireland, and even more so in The Netherlands. Where the upper peat still exists the two phases of the LIA are often very marked, between AD 1300–1500 and especially AD 1650–1800 (Barber 1981; Barber et al. 2000; Mauquoy et al. 2002).

Periodicities of ca. 1100, 800, 600 and 200 years, amongst others, have been recognised from the peat bog record, and these may be linked to solar forcing and oceanic changes (Barber et al. 1994; Chambers et al. 1997; van Geel and Renssen 1998; Hughes et al. 2000; Chambers and Blackford 2001; Langdon and Barber 2001). It is clear that the teleconnections in climate changes over the last 2000 years between sites some 300 km apart and at different altitudes (Barber et al. 2000) can also be found during earlier times between sites across a wide area of oceanic Mid-Latitude Europe. Considering the rather small differences in the present-day climate of the area (Crawford 2000) this is not too surprising, and it does show that there is a good climate "signal" over and above any biological "noise" In BSW changes. The climate mechanism of most importance could be the effects of summer water balances on the growing plants (Mauquoy and Barber 1999; 2002), rather than, for example, winter precipitation. The variations in climate most likely to be behind the recorded changes could be driven by fluctuations in the position of the Polar Front, which could bring cooler and damper weather further south; the position of the front depends on the thermohaline circulation of the North Atlantic and possibly solar variations (van Geel and Renssen 1998; Bond et al. 2001; Mauquoy et al. 2002). BSW records are an integration of temperature and precipitation, and progress is being made in disentangling these two climatic parameters, with the evidence so far perhaps favouring temperature as the main forcing factor (Barber and Langdon 2001; Charman and Hendon 2000), which is plausible since it is spatially coherent over wide areas of Europe, whereas precipitation can vary markedly over small distances (Barber et al. 2000). These peat data constitute valuable lowland records of change, reacting sensitively to climatic forcing in a way that established forests did not.

Pollen-derived climatic reconstructions along the regional transect

Pollen assemblages reflect not only climatically-induced vegetation changes but can also be influenced by human disturbance and other factors. However, pollen records from Europe have been used to derive quantitative climate reconstructions by different non-statistical (e.g. Grichuk 1969; Savina and Khotinsky 1984; Frenzel et al. 1992) and statistical approaches (e.g., Klimanov 1984; Guiot et al. 1993; Cheddadi et al. 1997; Tarasov et al. 1999a, 1999b). Prentice et al. (1992, 1996), Guiot et al. (1993) and Sykes et al. (1996) demonstrated that forest vegetation in Europe is controlled by the "bioclimatic" variables of (i) mean temperature of the coldest month (MTCO); (ii) growing degree days above 5 °C (GDD5), and (iii) ratio of actual to equilibrium evapotranspiration (α). These three variables have been chosen for this reconstruction using the best modern analogue approach.

The details of the standard best modern analogues method were first described by Guiot (1990). A chord distance (Euclidian metric between two points in the n-dimension space defined by the square root of the pollen percentages) is used to determine the similarity between each fossil pollen spectrum and the reference modern data. In the present study the reference modern pollen data-set is that compiled by Tarasov et al. (1999a, 1999b) from Northern Eurasia and North-West USA and Canada. It includes 1245 surface spectra for which taxa percentages were calculated based on the sum of 64 terrestrial pollen taxa. The same taxa were selected in the fossil records. For each analysed fossil spectrum 10 modern spectra which have the smallest chord distance were considered as the best modern analogues following Guiot (1990) and the results were used for the construction of Figure 3. The three climate variables estimated at the sites of the modern analogues were averaged by a weighting inverse to the chord distance. The average thus obtained gives the value of the reconstructed climate parameter for each fossil spectrum. In the best modern analogues method the error bars for the reconstructed values are defined by the climate variability among the chosen number of analogues. However, the small error bars that can be obtained in this way are often underestimated, especially when analogues are situated close to each other.

Mid-Latitude Europe is well represented in the European (EPD) and in the Global Pollen Database (GPD), with over 300 sites between 5W-60 °E and 45–60 °N, but the

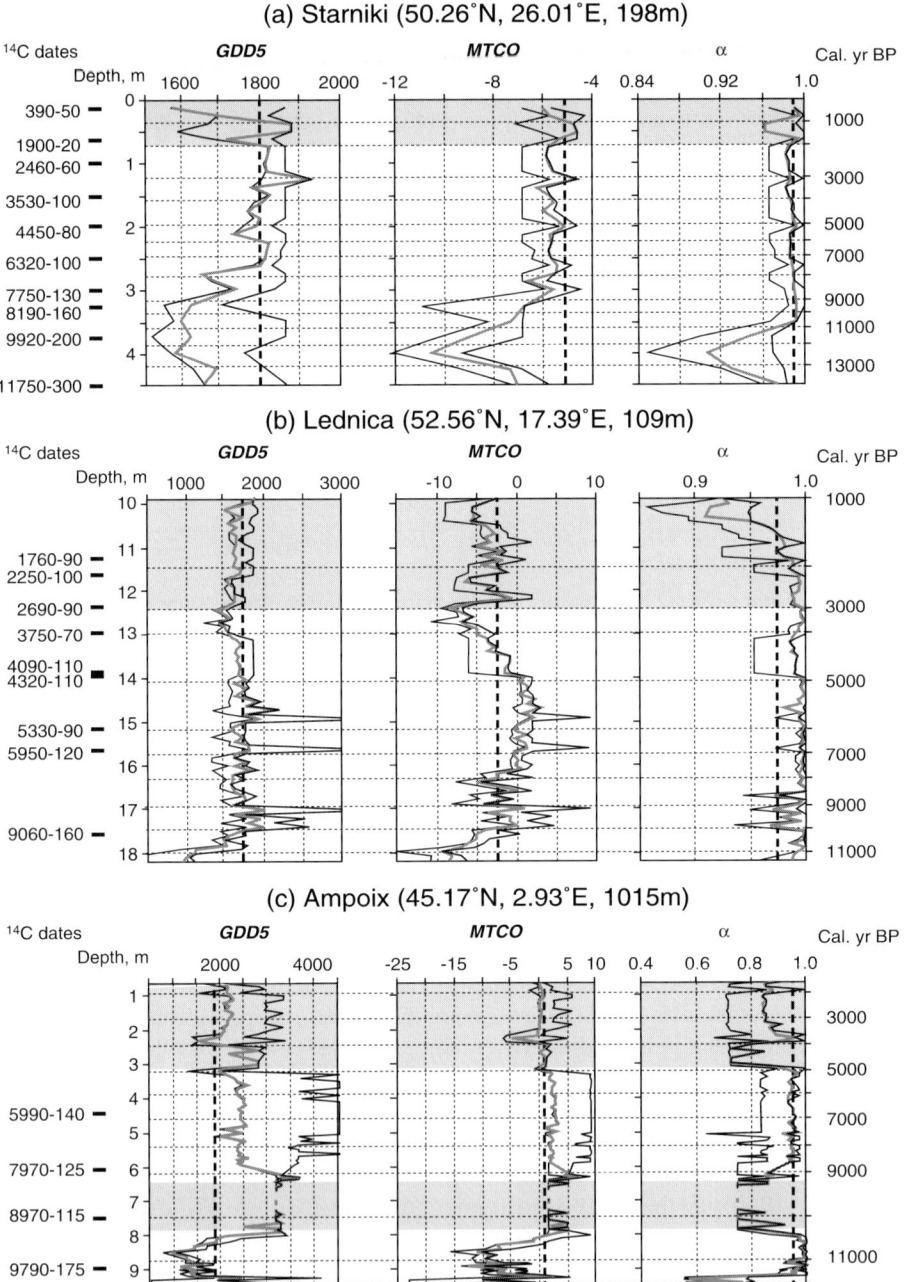

Figure 3. Reconstructed bioclimatic variables with standard deviations for three pollen records from Mid-Latitude Europe: (a) Starniki; (b) Lednica; and (c) Ampoix. Sum of growing degree days above 5 °C (*GDD5*), mean temperature of the coldest month (*MTCO*, in °C) and actual to equilibrium evapotranspiration (α). Thick dashed line shows modern values. Grey colour used to indicate zones where reconstruction is affected by increasing human disturbance of the vegetation or by limited number of modern analogues. Note that latitude and longitude are given as decimal coordinates.

number of sites and the data quality varies between the west and east; the region east of 25 °E has only 10 to 20 poorly dated sites. We selected three well dated pollen records, representing different parts of Mid-Latitude Europe (Fig. 1). The easternmost site, Starniki mire (50.26 °N, 26.01 °E, 198 m.a.s.l: Bezusko et al. (1985)) is situated in the Ukraine Polessie at the margin between the cool mixed and temperate deciduous forest belts. Lednica lake (52.56 °N, 17.39 °E, 109 m.a.s.l: Makohonienko 1991) lies in the agricultural area of the Polish Lowland, where the natural vegetation is represented by the patches of hornbeam-oak forest. Ampoix mire (45.17 °N, 2.93 °E, 1015 m.a.s.l: Beaulieu and Goeury 1987) occupies an ancient maar basin in the Massif Central, France.

Data from Starniki (Fig. 3a) suggest that reconstructed MTCO values were significantly lower than modern values from ca. 13,700 to about 9000 cal. BP. MTCO seems to be slightly lower than present during the middle and late Holocene. However, we assume that pollen-based reconstruction systematically underestimates MTCO values. This assumption is based on the test with modern pollen spectra collected in the vicinity of Starniki mire. Temperatures reconstructed from the modern spectra are about 1–1.5 °C lower than actual temperatures in the study area. This is due to a persistent input of *Picea* pollen transported from the Carpathian mountains by westerly winds. Reconstructed changes in both GDD5 and α are small compared to the error bars.

Data from Lednica lake (Fig. 3b) show that MTCO was similar to present by about 10,000 cal. BP. Winter temperatures fluctuate around the modern values between 9000 and 8000 cal. BP, becoming 1–3 °C warmer than present at 7500–5000 cal. BP, giving a distinct mid-Holocene climatic optimum. This is followed by marked falls in the MTCO reconstructed between 4500 and 3000 cal. BP, ca. 2500 cal BP, and before 1000 cal. BP. In contrast to the warmer than present mid-Holocene winters, the curves of GDD5 and α do not show distinct differences from today at that time. However, a slight decrease in accumulated summer warmth is reconstructed between 4000 and 3000 cal. BP. After that time the reconstruction can be taken to be affected by human disturbance of the vegetation, as evidenced by the appearance of *Secale cereale* pollen (Makohonienko 1991).

The results from Ampoix (Fig. 3c) show that Preboreal time (between ca. 11,500 and 10,500 cal. BP) was characterised by the spread of *Pinus* and *Betula* forest patches in the region (Beaulieu and Goeury 1987), followed by warmer winters than now during the mid-Holocene (10,500–5000 cal. BP), while reconstructed GDD5 values were greater than present between 10,500 and 2000 cal. BP. There are distinct minima in the α curves between 10,000 and 9000 and between 5000 and 4000 cal. BP, but the reconstructions at these intervals are based on a limited number of analogues. In all three records, climatic reconstructions for the last 2000–3000 cal. years are affected by being situated in regions with high human populations and longstanding agricultural activity.

It is apparent that the reconstructed patterns of climate are similar in all three records. However, Holocene warming appeared in the eastern-most Starniki record only around 9000 cal. BP, 1000 years later than in Lednica and 1500 years later than in the Ampoix record. According to the BIOME1 vegetation model (Prentice et al. 1992) based on plant ecological studies, the principal climate variables for definition of the different forest types in Mid-Latitude Europe are MTCO and GDD5. The limit between forest and non-forest PFTs is given by $\alpha = 65\%$. However, our reconstruction (Fig. 3) suggests that changes in this parameter might have influenced vegetation in humid temperate regions of Europe only during the relatively dry climate phase before the onset of the Holocene. It is tempting to speculate that the oscillations in the MTCO curve sometime before 8000 cal. BP in the Starniki and Lednica records, while no changes appear in Ampoix record, is a result of the 8200 cal. BP event, but the relatively coarse pollen and time resolutions of these records cannot confirm this. These same problems of coverage and chronology limit our ability to do precise inter-regional correlations and to discuss short-term climate events.

This is the first time that GDD5 and α have been reconstructed in Mid-Latitude Europe, and therefore comparisons with other studies are only available using reconstructed MTCO. Frenzel et al. (1992) presented maps of mean January temperatures for the Mid-Holocene Climatic Optimum (MHO: 6000–5500 uncal. BP), and Mid-Latitude Europe is represented by 32 sites. The results show that MHO winters were about 1 to 3 °C warmer than present across the area, with lower values (1 to 2 °C) between 20 and 40 °E and slightly more than 3 °C in the Volga-Ural watershed area.

Cheddadi et al. (1997) used modern pollen analogues constrained by lake status data to reconstruct climate in Europe at 6000 BP. However, they found that GDD5 and MTCO anomalies (6000 BP - present) for most of Mid-Latitude Europe were not significant. Similar results were obtained with the plant-functional-type method for the central part of the East European Plain (Tarasov et al. 1999a). Positive α anomalies (0 to 0.08) are shown as significant in Mid-Latitude Europe, excluding areas of the British Isles and Denmark (Cheddadi et al. 1997).

Velichko et al. (1997) presented climate reconstructions at four locations in the central part of the East-European Plain and showed that MTCO was lower than present until ca. 9500–9300 uncal. BP in Central Belarus (54 °N, 27 °E) and around Moscow (54–57 °N, 37°–39 °E), but that further east at Bashkiriya (55 °N, 57 °E) it did not reach modern values until 8000 uncal. BP, and was 3–5 °C lower than present between 9000 and 8000 uncal. BP, and then 1–2 °C warmer between 8000 and 5000 uncal. BP. At all sites the reconstructions show decreases in MTCO below modern levels at about 4500, 2500 and after 1000 uncal. BP, with the amplitude of the cooling increasing from west to east, with values of -0.5 / -1 °C in Belarus to -1 / -2 °C in Bashkiriya.

Annual palaeoenvironmental records from lakes

Investigations of lake sediments provide an integrated view of palaeoenvironmental conditions both in the lakes and their catchment areas (Oldfield 1977; Haworth and Lund 1984; Smith et al. 1991; Gierlowski-Kordesch and Kelts 1994), including records of human impact. Understanding of the different individual and interacting roles of climate and human impact on process-response mechanisms during the past is rather complex and calls for interdisciplinary studies (Dearing and Zolitschka 1999). Some lake sediments are laminated, facilitating high resolution sampling and dating, sometimes up to annual resolution (Hicks et al. 1994). Such records correlate particularly well with ice-core records (Brauer et al. 2000; Goslar et al. 1995), and reconstructions become available on the scale of human generations, such as the case studies of Elk Lake, Minnesota, USA (Bradbury and Dean 1993), Lake Gosciaz, Poland (Ralska-Jasiewiczowa et al. 1998) and Baldeggersee, Switzerland (Wehrli 1997).

Lakes are unevenly distributed between the Atlantic and the Urals, being concentrated in areas of the last glaciation such as Fennoscandia (Snowball et al., this volume), the North-Eastern European Lowlands and the Alps (see below). Quite a few non-glacial lakes exist in-between these areas, but the number of sites with good time control and high-resolution interdisciplinary sedimentological investigations is rather limited. For reasons of comparison with other sites and types of archives we focus on three well-dated and annually laminated sediment records: Loch Ness in Scotland, Lake Holzmaar in Western Germany and Lake Gosciaz in Central Poland.

The laminated sediments from Loch Ness span almost the entire Holocene and consist of clastic varves: the pale and silt-rich part of the couplet is deposited during the winter and spring runoff-season, whereas the dark and clay-rich laminae are deposited throughout the rest of the year (Cooper and O'Sullivan 1998). Variations in annual laminae thickness are related to changes in atmospheric circulation over the North Atlantic. A high North Atlantic Oscillation (NAO) index causes enhanced westerly airflow bringing more cyclonic rainfall to Scotland, thus increasing runoff and sediment transfer into Loch Ness. Signal analyses of the varve record for the last century reveal a correlation to the NAO index. The most recent, as well as the Holocene data, correlate with the duration of the sea-ice cover off Iceland as well as to the 11-year sun spot and the 88-year Gleissberg cycle (Cooper et al. 2000). As increased varve thickness is also related to a high sun-spot number, the latter can be interpreted as being associated with more precipitation (Cooper et al. 2000). This demonstrated correlation is so far restricted to the Atlantic region of Scotland and needs to be verified by other high resolution records on the European mainland.

Sun spot and Gleissberg cycles have also been recognised in the annually laminated record from Lake Holzmaar, Germany (Vos et al. 1997). At this site the annually laminated sediments are organic varves consisting of spring to autumn diatom blooms followed by a winter to spring lamina of organic and minerogenic detritus. The record is cross-dated by several independent dating methods, e.g., AMS ^{14}C dating on terrestrial macrofossils, tephrochronology, optically stimulated luminescence and thermoluminescence dating (Zolitschka et al. 2000). This archive provides insights into Holocene environmental variations mainly driven by climate as indicated by a comparison of the modern sediment pattern (AD 1952–1990) with meteorological data (Zolitschka 1996). Thus it could be demonstrated that annual deposition or varve thickness (sediment accumulation rate) is mainly controlled by increased runoff and sediment transfer to the lake during the winter (Zolitschka 1996; Zolitschka and Negendank 1998). Increased varve thickness is therefore attributed to either colder winters, with increased snow-melt runoff events, or to wetter winters with an overall increase in rainfall related runoff.

In addition to the natural forcing of accumulation rates, increased human activities in the catchment may lead to a decrease in plant cover and transpiration and thus to a surplus in runoff which would also modify the climate signal. However, according to the non-arboreal pollen record (Litt and Kubitz 2000), intensified human activities in the catchment area started not earlier than ca. 2800 cal. BP with the onset of the Iron Age, and ended with the re-afforestation in the 19th century. Most of the Holocene record from Lake Holzmaar can therefore be regarded as responding to climatic forcing, but during the last 3000 years it is very difficult, if not impossible, to differentiate human from climatic forcing.

Nine varve thickness maxima at Holzmaar, interpreted as cool or more humid periods (Fig. 4), have been recognised at the following periods (all cal. years BP): 10,500–10,200 (Boreal Oscillation), 9500–8600, 8100–7800 (8200-year event?), 7400–7000, 6400–6000, 5300–4800, 4100–4000, 3500–3200 and 2750–2450 cal. BP. These events agree in timing with climatic variations as reconstructed from Alpine glacier fluctuations (Haas et al. 1998);

five of them are coincident with ice-rafted debris (IRD) events in the North Atlantic (Bond et al. 1997), and three are correlated with increased amounts of dust in the Summit ice-core from Greenland (O'Brien et al. 1995). Additionally, the most prominent fluctuations of the global radiocarbon record around 11,200, 10,200 and 2800 cal. BP (Stuiver et al. 1998) correspond to cooling periods at Lake Holzmaar. However, the major Holocene climatic event recorded in Greenland ice-cores at 8200 cal. BP (Alley et al. 1997) is not as pronounced at Lake Holzmaar as could be expected (Fig. 4). It is well-known from oxygen isotope measurements on ostracod shells from Ammersee in Southern Germany, which record a 200-year climatic deterioration with a temperature decrease of 1.7 °C (von Grafenstein et al. 1998; 1999a). The event is also evident in pollen studies at Schleinsee, also in Southern Germany, as well as at Soppensee in Switzerland (Tinner and Lotter 2001), and Lake Gosciaz, Central Poland (Ralska-Jasiewiczowa et al. 1998). The fact that the event is of minor importance at Holzmaar might be related to a relatively stable environmental system with a high buffering capacity during this time period.

Varve thickness is a complex response to a number of different environmental changes, of which climatic forcing is only one option. Since 2800 cal. BP human impact in the Holzmaar catchment becomes a second factor that can cause a sediment signal similar to climatic forcing. The good correlation of the Holzmaar record with other continental and global climatic records makes it likely that climate was the main influence on most of the fluctuations recorded in the sediment, especially for Roman and Medieval times. However, the climate-sediment response function is now changed completely. Instead of cool and humid climatic periods being responsible for an increased accumulation rate, increased varve thickness in Roman and Medieval times is related to warm climatic conditions that favour human activities in the catchment area and thus deforestation with increased runoff and sediment transfer to the lake. The degree to which the presence and activities of prehistoric and historic cultures, and their influence on lake systems, have been controlled by climatic fluctuations has to remain an open question (Merkt and Müller 1994).

At Lake Gosciaz the annually laminated sediments are mainly organic varves with a high proportion of autochthonous precipitated calcite - pale calcite-rich layers alternate with dark layers composed of minerogenic and organic detritus (Ralska-Jasiewiczowa et al. 1998). Four periods of higher water level have been determined for Lake Gosciaz by higher rates of lacustrine deposition and other evidence in the catchment area. Periods of increased flood frequency were reconstructed for 9450–8450, 6350–6300, 4825–4775 and 3230–1950 cal. BP (Ralska-Jasiewiczowa et al. 1998; Starkel et al. 1996). These data show reasonable correlations with Alpine glacier fluctuations and with the Holzmaar sediment record. However, human impact has been recognised in the sediments since the Linear Pottery Culture of the early Neolithic (Ralska-Jasiewiczowa and van Geel 1992), a fact that makes interpretation, especially for the youngest period, ambiguous.

Annual records from tree-rings and speleothems

Tree-ring and speleothem records represent valuable high-frequency, calendar-dated proxy-climate records, but they are few in number across the Mid-Latitude transect and the full exploitation of their potential demands more research.

Baillie (1995) sets out the problems encountered in using tree-rings from the Mid-Latitude region to reconstruct past environmental conditions. He points out (Baillie 1995)

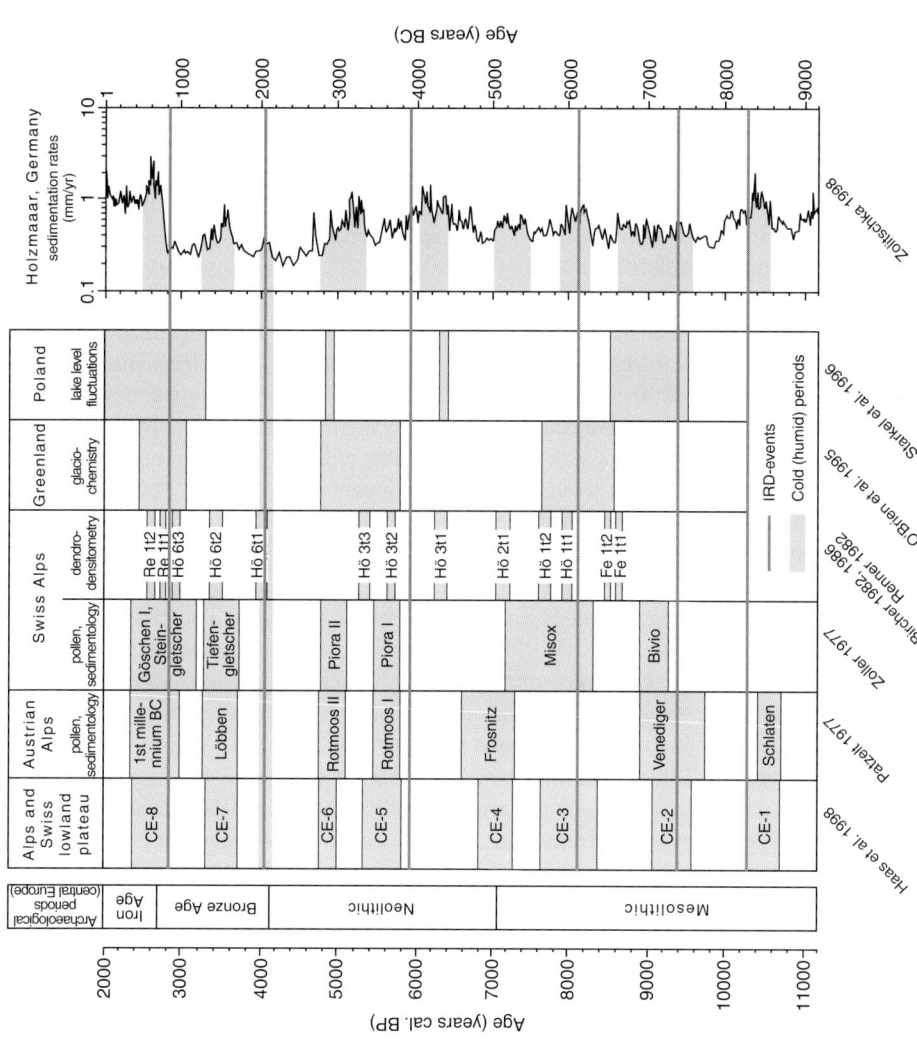

Figure 4. Holocene climatic fluctuations as recorded in the Swiss and Austrian Alps, on the Swiss Lowland Plateau, in Polish and German lakes and in Greenland ice cores. Archaeological periods for Central Europe are indicated, and ice-rafted debris (IRD) events according to Bond et al. (1997) are marked. Left hand timescale is in years cal. BP with the reference year of AD 1950; the right hand scale is in years BC. Redrawn from Haas et al. (1998) with other data from Zolitschka (1998).

that there has only been limited success in reconstructing climate change since the completion of the long oak chronologics in Ireland, England and Germany, although there are many examples of the use of tree-rings to infer information on growing conditions both of the trees and therefore of crop plants, and to provide archaeologists with proxy-climate information. It seems, however, as though the record of tree-rings in the Mid-Latitude area does not provide as clear a climate signal as in high latitudes (Baillie pers. comm., Briffa et al. 1990, 1999; Briffa 1999; Snowball et al. this volume) but some periods stand out in the more southerly record such as "…a pronounced decrease in oak growth in the middle 6th century AD that may be linked to very cool conditions seen in the high-northern trees in AD 536" (Briffa 1999), and a century of poor growth between 6270–6040 BC, probably again a reflection of the 8200-yr event.

Research into the palaeoclimate record of speleothems is very active (Lauritzen and Lundberg 1999) and in Mid-Latitude Europe there have been a number of recent advances (Baker et al. 2000; Proctor et al. 2002). McDermott et al. (1999), in a comparative study of speleothems from Western Ireland, SE France and NW Italy, found that the two southerly records were out of phase with that from the Atlantic seaboard; an interesting insight into climatic contrasts between Mediterranean and Atlantic climates. The Irish site record showed cool conditions at 10,000 cal. BP, warming between 9000–6000 cal. BP, but with a cooling trend from 7800–3500 cal. BP, and finally a warming trend since then. McDermott et al. (2001) have also reported that a "…high-resolution oxygen isotope record from a speleothem in Southwestern Ireland provides evidence for centennial-scale $\delta^{18}O$ variations that correlate with subtle $\delta^{18}O$ changes in the Greenland ice cores, indicating regionally coherent variability in the early Holocene".

Northwest Scotland has proved to be fruitful area for speleothem research, with the added bonus that the cave records can be linked to peatland archives. The cave at Uam an Tartair, at 58 °N and 5 °W, has provided stalagmites that show detailed changes over the last 2500 years. Baker et al. (1999) compared records of speleothem luminescence and peat humification and demonstrated a cyclicity in bog surface wetness of 90–100 years, arguing that this is related to rainfall variations generated by shifts in tracks of Atlantic depressions. Proctor et al. (2000) studied the last 1100 years in another stalagmite from the same cave and related changes in the annual luminescent banding to precipitation and thus to the NAO. Interestingly, they reconstruct high precipitation through much of the Medieval Warm Period (MWP, AD 1080–1330), implying a persistently high NAO index. This is in accord with the evidence for wet conditions in the Cairngorms for the same period (Barber et al. 2000). In another recent study in this area Charman et al. (2001) concentrated on replicating data from the peats near the cave and then compared the humification changes of three profiles to the speleothem record of Baker et al. (1999). They concluded that this record is more strongly influenced by precipitation than by temperature, and that it again reflects southward movements of depression tracks.

Clearly both of these types of annual record have great potential for future research in tracking climatic change on meridional transects and in answering key questions involving the eastern extent of the influence of the Atlantic Ocean.

Climate records from the Alps

The treeline represents the major ecotone in the Alps that reacts to climatic change. Tentative

late-glacial and Holocene treeline reconstructions based on palaeobotanical data have been presented for different parts of the Alps (Wegmüller 1966; Welten 1982; Schneider 1985; Burga 1987; Ammann 1993). It is, however, most important to validate pollen-derived treeline limits by plant macrofossil analyses. There is only a small set of such combined investigations available from higher elevations in the Alps where a reliable estimation of past treelines is possible (e.g. Wegmüller and Lotter 1990; Ponel et al. 1992; Ammann and Wick 1993; Wick and Tinner 1997), and only a few studies are dated well enough to allow correlations of climatic events between sites. Based on the study of two well-dated sites in the Central Alps, Wick and Tinner (1997) could correlate Holocene treeline fluctuations with glacier advances (Patzelt 1977), solifluction phases (Gamper 1993; Tinner et al. 1996), and dendroclimatic data (Renner 1982; Bircher 1982, 1986; see also Kaiser 1991). Heiri (2001) derived Holocene treeline fluctuations from two well-dated sites in the Swiss Alps using pollen, plant macrofossil data as well as chironomids, and revealed a record of warmer summers during the early to mid-Holocene, as well as a close correlation between six treeline depressions and chironomid-inferred decreases in mean July temperature. Four of these cooling phases, at 10,500–10,400, 9200–9100, 8200–7700, and 6000–5800 cal. BP, show a close temporal agreement to ocean current changes reported from the North Atlantic (Bond et al. 1997) and to air circulation variations inferred from the Greenland ice core record (O'Brien et al. 1995).

According to several authors (e.g. Wick and Tinner 1997; Burga and Perret 1998) the range of climatically-induced treeline fluctuations during the Holocene was not more than 100–150 m. Given the modern air temperature lapse rates in the Alps of 6–7 °C km^{-1} this implies that Holocene climatic fluctuations during the warm season had an amplitude of between ± 0.5 and ± 1 °C.

Several studies on fossil biota or stable isotopes from lowland sites in Central Europe suggest a temperature increase at the onset of the Holocene in the order of 2–6 °C (Ammann 1989; Eicher 1991; von Grafenstein et al. 1999b; Isarin and Bohncke 1999; Lotter et al. 2000). After this glacial / interglacial transition it becomes more difficult to detect the low amplitude climatic fluctuations of the Holocene in the lowland ecosystems in Mid-Latitude Europe, the notable exception being peat bogs (see above). However, using treeline fluctuations in the Alps and different palaeoecological reconstructions of climate from sites on the Swiss Plateau, Haas et al. (1998) identified eight synchronous pre-Roman cold phases (at 9600–9200, 8600–8150, 7550–6900, 6600–6200, 5350–4900, 4600–4400, 3500–3200, and 2600–2350 conventional radiocarbon years BP; Fig. 4) and suggested that there is a 1000-year cyclicity in the data. In the Jura mountains Magny (1998) found several lake-level changes during the Holocene. As these changes coincided with fluctuations in δ^{14}C and glacier fluctuations in the Alps he concluded that they were forced by changes in ocean circulation and by changes in solar activity.

Holocene climatic changes have also been detected from fluctuations in the oxygen isotopes of ostracod valves from a pre-Alpine lake in Southern Germany (von Grafenstein et al. 1999a), including the 8200 cal BP event (von Grafenstein et al. 1998). These fluctuations are interpreted as being caused by a short weakening of the thermohaline circulation through episodic freshwater input into the North Atlantic. Tinner and Lotter (2001) have also found a striking coincidence between the 8200 cal. BP event and the onset of changes in the vegetation composition in two annually laminated Central European lakes which they interpreted as the onset of moister conditions due to changed air-mass trajectories.

Inner Alpine moraine stages imply very fast late-glacial ice decay (Maisch 1995), with an ice retreat from the lowlands into the Alpine valleys before 13,000 cal. BP. At the onset of the Holocene most Alpine glaciers retreated to dimensions comparable to the ones documented for their historical advances (Maisch 1992). Several phases of Holocene glacier advances have been described and attributed to cooler and moister climatic conditions. These advances were, however, within the extent of, or only slightly larger than the ones documented for the LIA (e.g., Patzelt and Bortenschlager (1973)). During phases of warmer and drier Holocene climate the dimension of the Alpine glaciers may have been smaller than today (e.g. Porter and Orombelli 1985; Burga 1991; Nicolussi and Patzelt 2000; Hormes et al. 2001), and Leemann and Niessen (1994), in a study of varves at Lake Silvaplana in the Eastern Swiss Alps, have suggested that there were very few or no glaciers at all in the catchment between 9400 and 3300 BP. Neoglaciation began around 3300 BP and the highest glacier-induced sediment accumulation, that also points to the maximum Holocene glacier extent, occurred during the LIA. For the LIA several well-documented glacier advances are evidenced in the Alps (Wanner et al. 2000). A first advance occurred in the mid-14th century AD and was followed by several phases of advance that in most cases culminated in the greatest extension of the glaciers around 1850 AD. Inferred meteorological conditions from historical records (e.g., Pfister 1992) as well as tree-ring records from the Alps (Schweingruber et al. 1979), show good agreement with the glacier oscillations for this 500-year period of climate favourable for glacier growth. In a high-resolution, multi-proxy study Hausmann et al. (2002) give an example of how LIA climatic oscillations influenced treeline ecosystems through changes in land-use and pasturing patterns in the Alps, whereas Lotter et al. (2000b) showed the direct impact of LIA climate change on aquatic and terrestrial biota above treeline.

Discussion

It is clear from the studies quoted above that changes in the Atlantic Ocean can be detected in the climate over the European landmass, both in individual events such as that at 8200 cal. BP, and in cycles of change that may reflect IRD events and changes in the thermohaline circulation. In the highest frequency records, the annually-resolved speleothem, tree-ring and laminated lake sediment records, there is also scope for the recognition of the North Atlantic Oscillation (e.g., Proctor et al. 2000).

Changes in oceanicity / continentality are more difficult to detect. A recent review makes clear that vegetation responses to oceanity are complex and that "…oceanicity describes a multi-faceted situation. Species found in proximity to the ocean may live there for a variety of ecological, physiological and historical reasons" (Crawford 2000). Furthermore, oceanicity and continentality are usually measured by mean annual temperature range, adjusted for latitude, and it is by no means clear how the various proxies used in climatic reconstruction are affected by this. However, potential summer water deficits are reduced in oceanic climates and this undoubtedly affects peat bogs, as mentioned above. The pollen-based reconstructions also show greater winter cooling further to the east.

A growing number of cycles are being recognised in various proxy data-series. As noted above there are cycles of 200, 600, 800 and 1100 years in peat stratigraphic data, and these have been found at a number of sites since the original detection of a cyclicity of 260 years in Danish bog data by Aaby (1976). As the number and precision of radiocarbon-dated profiles has increased, these periodicities have become more well-founded and the search for causes has been stimulated (Chambers and Blackford 2001). It is plausible that the bog cycles are being driven by oceanic changes, as originally suggested by Barber et al. (1994), the mechanism being changes in summer temperatures and therefore water deficits. The identification of Holocene oceanic cycles of 550 and 1100 years by Chapman and Shackleton (2000) are a close match to the cycles of 600 and 1100 years from Walton Moss (Hughes et al. 2000), and the distinct cycle (in two independent proxies) of 1100 years from Temple Hill Moss in Southeast Scotland (Langdon et al. 2003). Solar forcing of individual events, such as the abrupt climate change around 2650 BP, (van Geel et al. 1996; 1998; van Geel and Renssen 1998), and the search for solar cycles in other data such as radiocarbon fluctuations (Stuiver and Brazunias 1993), ice core records (O'Brien et al. 1995; Stuiver et al. 1998), glacier and treeline fluctuations (Karlen and Kuylenstierna 1996), and lake varves (Negendank et al. 1999) testify to the growing importance attached to solar forcing; see also review and references in Beer et al. (2000) and Mauquoy et al. (2002).

The MWP and the LIA are well-expressed and well-dated in a number of different proxy records, especially in peat stratigraphy both visually in the field (Barber 1981) and in detailed analyses (Barber et al. 2000; Mauquoy et al. 2002). In some records however, both periods are poorly expressed if at all. For example, the LIA may be found only as a period of 80 years (AD 1570–1650) in tree rings in Fennoscandia and there is little evidence of a MWP in the same record (Briffa et al. 1990). The reality of the LIA, signalled by the dramatic expansion of glaciers and recorded in a number of other proxies, can hardly be doubted in the Alps, and there are of course documentary accounts, in various forms, of the LIA and to a lesser extent the MWP (Lamb 1995). Deriving palaeoclimatic trends from long tree-ring chronologies is not as straightforward as is sometimes assumed (LaMarche 1974; Cook et al. 1995; Cook and Peters 1997), and contrasts with the low frequency peat records which can be viewed as acting as a low-pass biological filter on the climate signal.

The behaviour of natural ecosystems and human societies in response to climate variation is, of course, a huge and complex question, especially given the biological tolerances of species within ecosystems and also the adaptability of human societies. Nevertheless, climatic changes of the Holocene are reflected in ecosystems which are in marginal situations, such as the Alpine treeline, and those lowland ecosystems which are especially sensitive to climate, such as peat bogs. Early in the Holocene, before the establishment of closed forest cover, then some climatic oscillations such as the 8200-yr event, may be recorded by lake ecosystems, and where lakes are situated in remote montane areas, with minimal human impact, then the climatic signal may also be clear (Battarbee et al. 2001).

Human societies are subject to numerous influences, ranging from the environmental to the spiritual. The extent to which climatic change may impact on humans has recently been reviewed by Messerli et al. (2000) and will not be rehearsed here; suffice it to say that the changes outlined above must have had some impact on agrarian based societies, especially those exploiting marginal land, and may have led to population movements as postulated by van Geel et al. (1996).

Summary

The record of climatic changes in Mid-Latitude Europe is a rich and varied one, with many records covering the whole of the Holocene and many showing the dramatic transition from the late-glacial. Individual records often demonstrate coherent changes within a region, as demonstrated by the bog record and by the lake sediment record, but progress in some areas is limited because the linking of different types of record demands high-resolution time control, and even then there may be lags and non-linear responses in the biological proxies being compared.

In many of the records, however, it is possible to see some common periods of change, even bearing in mind the chronological problems. The Preboreal Oscillation and the 8200-yr event are clearly shown in many records, especially lake sediments (Magny et al. 2001). There is then a period of general stability until wet shifts appear in bogs in Northern England and the Netherlands around 6400–5900 cal. BP. A major and fairly widespread change to cooler and probably wetter conditions sets in around 4400 cal. BP, with further deteriorations clustering around 3500, 2900 and especially 2800 cal. BP, the latter date being the beginning of the still recognised Sub-Atlantic period of Blytt and Sernander. Amongst late-Holocene climatic changes those at about 1400 cal. BP and 1000 cal. BP stand out, but the major climatic downturn is that of the LIA beginning in many records around 650 cal. BP / 1300 AD with a first trough around 1450 AD, a slight amelioration around 1500 to 1580 AD, and then the coldest period of the last 2000 or more years between about 1600 and 1850 AD.

Future research needs to move to linking high quality data from key sites through more precise time control, as well as refining and standardising methods.

Acknowledgments

The work reported on peat bogs has benefitted from funding from the UK Natural Environment Research Council.

References

Aaby B. 1976. Cyclic climatic variations in climate over the past 5500 years reflected in raised bogs. Nature 263: 281–284.

Alley R.B., Mayewski P.A., Sowers T., Stuiver M., Taylor K.C. and Clark P.U. 1997. Holocene climatic instability: a prominent, widespread event 8200 yr ago. Geology 25: 483–486.

Ammann B. 1989. Response times in bio- and isotope-stratigraphies to Late-Glacial climatic shifts — an example from lake deposits. Ecol. geol. Helvet. 82: 183–190.

Ammann B. 1993. Flora und Vegetation im Paläolithikum und Mesolithikum der Schweiz. In: Die Schweiz vom Paläolithikum bis zum frühen Mittelalter. Verlag Schweizerische Gesellschaft für Ur- und Frühgeschichte, Basel, pp. 66–84.

Ammann B. and Wick L. 1993. Analysis of fossil stomata of conifers as indicators of the alpine tree line fluctuations during the Holocene. In: Frenzel B. (ed.), Oscillations of the Alpine and Polar Tree Limits in the Holocene. Paläoklimaforschung. Gustav Fischer Verlag, Stuttgart, pp. 175–185.

Baillie M.G.L. 1995. A Slice through Time: dendrochronology and precision dating. Batsford, London.

Baker A., Bolton L., Brunsdon C., Charlton M. and McDermott F. 2000. Visualisation of luminescence excitation-emission timeseries: palaeoclimate implications from a 10,000 year stalagmite record from Ireland. Geophys. Res. Lett. 27: 2145–2148.

Baker A., Caseldine C.J., Gilmour M.A., Charman D., Proctor C.J., Hawkesworth C.J. and Phillips N. 1999. Stalagmite luminescence and peat humification records of palaeomoisture for the last 2500 years. Earth Planet. Sci. Lett. 165: 157–162.

Barber K.E. 1981. Peat Stratigraphy and Climatic Change: A Palaeoecological Test of the Theory of Cyclic Peat Bog Regeneration. Balkema, Rotterdam, Netherlands, 219 pp.

Barber K.E. 1993. Peatlands as scientific archives of biodiversity. Biodivers. Conserv. 2: 474–489.

Barber K.E., Chambers F.M., Maddy D., Stoneman R.E. and Brew J.S. 1994. A sensitive high-resolution record of late Holocene climatic change from a raised bog in Northern England. The Holocene 4: 198–205.

Barber K.E., Dumayne-Peaty L., Hughes P.D.M., Mauquoy D. and Scaife R.G. 1998. Replicability and variability of the recent macrofossil and proxy-climate record from raised bogs: field stratigraphy and macrofossil data from Bolton Fell Moss and Walton Moss, Cumbria, England. J. Quat. Sci. 13: 515–528.

Barber K.E., Maddy D., Rose N., Stevenson A.C., Stoneman R.E. and Thompson R. 2000. Replicated proxy-climate signals over the last 2000 years from two distant UK peat bogs: new evidence for regional palaeoclimate teleconnections. Quat. Sci. Rev. 18: 471–479.

Barber K.E. and Langdon P.G. 2001. Testing the palaeoclimatic signal from peat bogs — temperature or precipitation forcing? Abstracts, PAGES-PEPIII / ESF-HOLIVAR International Conference: Past Climate Variability Through Europe and Africa, ECRC / CEREGE, pp. 58–59.

Barber K.E., Chambers F.M. and Maddy D. 2003. Holocene palaeoclimates from peat stratigraphy: macrofossil proxy-climate records from three oceanic raised bogs in Britain and Ireland. Quat. Sci. Rev. 22: 521–539.

Battarbee R.W., Cameron N.G., Golding P., Brooks S.J., Switsur R., Harkness D., Appleby P.G., Oldfield F., Thompson R., Monteith D.T. and McGovern A. 2001. Evidence for Holocene climate variability from the sediments of a Scottish remote mountain lake. J. Quat. Sci. 16: 339–346.

Beaulieu J.-L. de and Goeury C. 1987. Zonation automatique appliquée à l'analyse pollinique: exemple de la narse d'Ampoix (Puy de Dome, France). Bull. Assoc. fr. Quat. 1: 49–61.

Beer J., Mende W. and Stellmacher R. 2000. The role of the sun in climate forcing. Quat. Sci. Rev. 19: 403–415.

Berglund B.E., Birks H.J.B., Ralska-Jasiewiczowa M. and Wright H.E. 1996. Palaeoecological Events during the last 15,000 years: Regional Syntheses of Palaeoecological Studies of Lakes and Mires in Europe. Wiley, Chichester, 764 pp.

Bezusko L.G., Kajutkina T.M., Kovalukh N.N. and Artjushenko A.T. 1985. Paleobotanical and radiological studies of deposits from bog Starniki (Maloe Polessie). Ukr. Botan. Zhurn. 42: 27–30.

Bircher W. 1982. Zur Gletscher- und Klimageschichte des Saastales. Glazialmorphologische und dendroklimatologische Untersuchungen. Phys. Geog. 9: 1–233.

Bircher W. 1986. Dendrochronology applied in mountain regions. In: Berglund B.E. (ed.), Handbook of Holocene Palaeoecology and Palaeohydrology. Wiley, Chichester, pp. 387–403.

Blackford J.J. and Chambers F.M. 1991. Proxy records of climate from blanket mires: evidence for a Dark Age (1400 BP) climatic deterioration in the British Isles. The Holocene 1: 63–67.

Blackford J.J. and Chambers F.M. 1993. Determining the degree of peat decomposition for peat based palaeoclimatic studies. Int. Peat J. 5: 7–24.

Blackford J.J. and Chambers F.M. 1995. Proxy climate record for the last 1000 years from Irish blanket peat and a possible link to solar variability. Earth Planet. Sci. Lett. 133: 145–150.

Bond G., Showers W., Cheseby M., Lotti R., Almasi P., deMenocal P., Priore P., Cullen H., Hajdas I. and Bonani G. 1997. A pervasive millennial-scale cycle in North Atlantic Holocene and glacial climates. Science 278: 1257–1266.

Bond G., Kromer B., Beer J., Muscheler R., Evans M.N., Showers W., Hoffmann S., Lotti-Bond R., Hajdas I. and Bonani G. 2001. Persistent solar influence on North Atlantic climate during the Holocene. Science 294: 2130–2136.

Bradbury J.P. and Dean W.E. (eds) 1993. Elk Lake, Minnesota: Evidence for Rapid Climate Change in the North-Central United States. Geol. Soc. Am., Boulder, Colorado. Special Paper 276, 336 pp.

Brauer A., Günter C., Johnsen S.J. and Negendank J.F.W. 2000. Land-ice teleconnections of cold climatic periods during the last Glacial/Interglacial transition. Clim. Dyn. 16: 229–239.

Briffa K.R. 1999. Analysis of dendrochronological variability and associated natural climates — the last 10000 years (ADVANCE-10 K). PAGES Newslett. 7: 6–8.

Briffa K.R., Bartholin T.S., Eckstein D., Jones P.D., Karlen W., Schweingruber F.H. and Zetterberg P. 1990. A 1,400-year tree-ring record of summer temperatures in Fennoscandia. Nature 346: 434–439.

Briffa K.R., Jones P.D., Vogel R.B., Schweingruber F.H., Baillie M.G.L., Shiyatov S.G. and Vaganov E.A. 1999. European tree rings and climate in the 16th century. Clim. Change 43: 151–168.

Burga C.A. 1987. Vegetationsgeschichte der Schweiz seit der Späteiszeit. Geog. Helvet. 42: 71–77.

Burga C.A. 1991. Vegetation history and paleoclimatology of the Middle Holocene: pollen analysis of alpine peat bog sediments, covered formerly by the Rutor Glacier, 2510 m (Aosta Valley, Italy). Global Ecol. Biogeog. Lett. 1: 143–150.

Burga C.A. and Perret R. 1998. Vegetation und Klima der Schweiz seit dem jüngeren Eiszeitalter. Ott Verlag, Thun.

Chambers F.M., Barber K.E., Maddy D. and Brew J. 1997. A 5500-year proxy-climate and vegetation record from blanket mire at Talla Moss, Borders, Scotland. The Holocene 7: 391–399.

Chambers F.M. and Blackford J.J. 2001. Mid- and Late-Holocene climatic changes: a test of periodicity and solar forcing in proxy-climate data from blanket peat bogs. J. Quat. Sci. 16: 329–338.

Chapman M.R. and Shackleton N.J. 2000. Evidence of 550-year and 1000-year cyclicities in North Atlantic circulation patterns during the Holocene. The Holocene 10: 287–291.

Charman D.J., Hendon D. and Packman S. 1999. Multi-proxy surface wetness records from replicate cores on an ombrotrophic mire: implications for Holocene palaeoclimate records. J. Quat. Sci. 14: 451–463.

Charman D.J. and Hendon D. 2000. Long-term changes in soil water tables over the past 4500 years: relationships with climate and North Atlantic atmospheric circulation and sea surface temperature. Clim. Change 47: 45–59.

Charman D.J., Caseldine C., Baker A., Gearey B., Hatton J. and Proctor C. 2001. Paleohydrological records from peat profiles and speleothems in Sutherland, Northwest Scotland. Quat. Res. 55: 223–234.

Cheddadi R., Yu G., Guiot J., Harrison S.P. and Prentice I.C. 1997. The climate of Europe 6000 years ago. Clim. Dyn. 13: 1–9.

Chiverrell R.C. 2001. A proxy record of late Holocene climate change from May Moss, Northeast England. J. Quat. Sci. 16: 9–29.

Cooper M.C. and O'Sullivan P.E. 1998. The laminated sediments of Loch Ness, Scotland: Preliminary report on the construction of a chronology of sedimentation and its potential use in assessing Holocene climatic variability. Palaeogeogr. Palaeoclim. Palaeoecol. 140: 23–31.

Cooper M.C., O'Sullivan P.E. and Shine A.J. 2000. Climate and solar variability recorded in Holocene laminated sediments — a preliminary assessment. Quat. Int. 68-71: 363–371.

Cook E.R., Briffa K.R., Meko D.M., Graybill D.A. and Funkhouser G. 1995. The 'segment length curse' in long tree-ring chronology development for palaeoclimatic studies. The Holocene 5: 229–237.

Cook E.R. and Peters K. 1997. Calculating unbiased tree-ring indices for the study of climatic and environmental change. The Holocene 7: 361–370.

Crawford R.M.M. 2000. Ecological hazards of oceanic environments. New Phytol. 147: 257–281.

Dearing J.A. and Zolitschka B. 1999. System dynamics and environmental change: an exploratory study of Holocene lake sediments at Holzmaar, Germany. The Holocene 9: 531–540.

Dupont L.M. 1986. Temperature and rainfall variations in the Holocene based on comparative palaeoecology and isotope geology of a hummock and a hollow (Bourtangerveen, The Netherlands). Rev. Palaeobot. Palynol. 48: 71–159.

Eicher U. 1991. Oxygen isotope studies in lacustrine carbonate sediments. In: Frenzel B., Pons A. and Gläser B. (eds), Evaluation of Climate Proxy Data in Relation to the European Holocene. Paläoklimaforschung. Gustav Fischer Verlag, Stuttgart, pp. 171–173.

Ellis C.J. and Tallis J.H. 2000. Climatic control of blanket mire development at Kentra Moss, North-West Scotland. J. Ecol. 88: 869–889.

Frenzel B., Pecsi M. and Velichko A.A. 1992. Atlas of Paleoclimates and Paleoenvironments of the Northern Hemisphere. Gustav Fischer Verlag, Stuttgart, 153 pp.

Gamper M. 1993. Holocene solifluction in the Swiss Alps: dating and climatic implications. In: Frenzel B. (ed.), Solifluction and Climatic Variation in the Holocene. Paläoklimaforschung. Gustav Fischer Verlag, Stuttgart, pp. 1–9.

Gierlowski-Kordesch E. and Kelts K. 1994. Global Geological Record of Lake Basins; Volume 1. Cambridge, Cambridge University Press.

Godwin H. 1975. History of the British Flora. Cambridge University Press, Cambridge.

Goslar T., Arnold M. and Pazdur M.F. 1995. The Younger Dryas cold event — was it synchronous over the North Atlantic region? Radiocarbon 37: 63–70.

Grichuk V.P. 1969. An experiment in reconstructing some characteristics of climate in the Northern Hemisphere during the Atlantic period of Holocene. In: Neishtadt M.I. (ed.), Golotsen Nauka, Moscow, pp. 41–57.

Guiot J. 1990. Methodology of palaeoclimatic reconstruction from pollen in France. Palaeogeogr. Palaeoclim. Palaeoecol. 80: 49–69.

Guiot J., Harrison S.P. and Prentice I.C. 1993. Reconstruction of Holocene precipitation patterns in Europe using pollen and lake-level data. Quat. Res. 40: 139–149.

Haas J.N., Richoz I., Tinner W. and Wick L. 1998. Synchronous Holocene climatic oscillations recorded on the Swiss Plateau and at timberline in the Alps. The Holocene 8: 301–309.

Haslam C.J. 1987. Late Holocene peat stratigraphy and climate change — a macrofossil investigation from the raised mires of Western Europe. Ph.D. thesis, University of Southampton.

Hausmann S., Lotter A.F., van Leeuwen J.F.N., Ohlendorf C., Lemcke G., Grönlund E. and Sturm M. 2002. Interactions of climate and land use documented in the varved sediments of Seebergsee in the Swiss Alps. The Holocene 12: 279–289.

Haworth E.Y. and Lund J.W.G. (eds) 1984. Lake Sediments and Environmental History. Leicester University Press, Leicester, 411 pp.

Heiri O. 2001. Holocene palaeolimnology of Swiss mountain lakes reconstructed using subfossil chironomid remains: past climate and prehistoric human impact on lake ecosystems. Ph.D. thesis, University of Bern.

Hicks S., Miller U. and Saarnisto M. (eds) 1994. Laminated Sediments. Rixenart, PACT Belgium, 148 pp.

Hormes A., Müller B.U. and Schlüchter C. 2001. The Alps with little ice: evidence for eight Holocene phases of reduced glacier extent in the Central Swiss Alps. The Holocene 11: 255–265.

Hughes P.D.M., Mauquoy D., Barber K.E. and Langdon P.G. 2000. Mire development pathways and palaeoclimatic records from a full Holocene peat archive at Walton Moss, Cumbria, England. The Holocene 10: 465–479.

Isarin R.F.B. and Bohncke S.J.P. 1999. Mean July temperatures during the Younger Dryas in the Northwestern and Central Europe as inferred from climate indicator plant species. Quat. Res. 51: 158–173.

Kaiser K.F. 1991. Tree-rings in Switzerland and other mountain regions: late glacial through Holocene. In: Frenzel B., Pons A. and Gläser B. (eds), Evaluation of Climate Proxy Data in relation to the European Holocene. Paläoklimaforschung, Gustav Fischer Verlag, Stuttgart pp. 119–131.

Karlen W. and Kuylenstierna J. 1996. On solar forcing of Holocene climate: evidence from Scandinavia. The Holocene 6: 359–365.

Klimanov V.A. 1984. Paleoclimatic reconstruction based on the information statistical method. In: Velichko A.A., Wright H.E. and Barnosky C.W. (eds), Late Quaternary Environments of the Soviet Union. University of Minnesota Press, Minneapolis, pp. 297–303.

LaMarche V.C. Jr. 1974. Paleoclimatic inferences from long tree ring records. Science 183: 1043–1048.

Lamb H.H. 1977. Climate: Present, Past and Future. Volume 2, Climatic History and the Future. Methuen, London. 835 pp.

Lamb H.H. 1995. Climate, History and the Modern World. (2nd Edition.) Routledge, London. 433 pp.

Langdon P.G. and Barber K.E. 2001. New Holocene tephras and a proxy climate record from a blanket mire in northern Skye, Scotland. J. Quat. Sci. 16: 753–759.

Langdon P.G., Barber K.E. and Hughes P.D.M. 2003. A 7500 year peat-based palaeoclimatic reconstruction and evidence for an 1100 year cyclicity in mire surface wetness from Temple Hill Moss, Pentland Hills, Southeast Scotland. Quat. Sci. Rev. 22: 259–274.

Lauritzen S.-E. and Lundberg J. (eds) 1999. Speleothems as high-resolution palaeoclimatic archives: Holocene Special Issue. The Holocene 9: 643–742.

Leemann A. and Niessen F. 1994. Holocene glacial activity and climatic variations in the Swiss Alps: reconstructing a continuous record from proglacial lake sediments. The Holocene 4: 259–268.

Litt T. and Kubitz B. 2000. Anthropogenic indicators in pollen diagrams based on varved sediments from lakes Holzmaar and Meerfelder Maar, Germany. Terra Nostra 2000/7: 50–54.

Lotter A.F., Birks H.J.B., Eicher U., Hofmann W., Schwander J. and Wick L. 2000a. Younger Dryas and Alleröd summer temperatures at Gerzensee (Switzerland) inferred from fossil pollen and cladoceran assemblages. Palaeogeogr. Palaeoclim. Palaeoecol. 159: 349–361.

Lotter A.F., Hofmann W., Kamenik C., Lami A., Ohlendorf C., Sturm M., van der Knaap W.O. and van Leeuwen J.F.N. 2000b. Sedimentological and biostratigraphical analyses of short sediment cores from Hagelseewli (2339 m a.s.l.) in the Swiss Alps. J. Limnol. 59: 53–64.

Magny M. 1998. Reconstruction of Holocene lake-level changes in the French Jura: methods and results. In: Harrison S.P., Frenzel B., Huckriede U. and Weiss M.M. (eds), Palaeohydrology as Reflected in Lake-Level Changes as Climatic Evidence for Holocene Times. Paläoklimaforschung. Gustav Fischer Verlag, Stuttgart, pp. 67–85.

Magny M., Guiot J. and Schoellammer P. 2001. Quantitative reconstruction of Younger Dryas to mid-Holocene paleoclimates at Le Locle, Swiss Jura, using pollen and lake-level data. Quat. Res. 56: 170–180.

Maisch M. 1992. Die Gletscher Graubündens. Rekonstruktion und Auswertung der Gletscher und deren Veränderungen seit dem Hochstand von 1850 im Gebiet der östlichen Schweizer Alpen (Bünderland und angrenzende Regionen). Geographisches Institut der Universität Zürich, Zürich.

Maisch M. 1995. Gletscherschwundphasen im Zeitraum des ausgehenden Spätglazials (Egesen-Stadium) und seit dem Hochstand von 1850 sowie Prognosen zum künftigen Eisrückgang in den Alpen. In: Gletscher im ständigen Wandel. Publikation der SANW. vdf. pp. 81–100.

Makohonienko M. 1991. Materialy do postglacjalnej historii roslinnosci okolic Lednicy. Czesc II. Badania palinologiczne osadow Jeziora Lednickiego — rdzen I/86 i Wal/87. In: Tobolski K. (ed.), Wstep do paleoekologii Lednickiego Parku Krajobrazowego. Wydawnictwo Naukowe UAM: 63–70.

Mauquoy D. and Barber K.E. 1999. A replicated 3000 yr proxy-climate record from Coom Rigg Moss and Felecia Moss, the Border Mires, Northern England. J. Quat. Sci. 14: 263–275.

Mauquoy D. and Barber K.E. 2002. Testing the sensitivity of the palaeoclimatic signal from four paired ombrotrophic peat bogs in Northern England and the Scottish Borders. Rev. Paleobot. Palynol. 119: 219–240.

Mauquoy D., van Geel B., Blaauw M. and van der Plicht J. 2002. Evidence from northwest European bogs shows 'Little Ice Age' climatic changes driven by variations in solar activity. The Holocene 12: 1–6.

McDermott F., Frisia S., Huang Y.M., Longinelli A., Spiro B., Heaton T.H.E., Hawkesworth C.J., Borsato A., Keppens E., Fairchild I.J., van der Borg K., Verheyden S. and Selmo E. 1999. Holocene climate variability in Europe: evidence from δ^{18}O, textural and extension-rate variations in three speleothems. Quat. Sci. Rev. 18: 1021–1038.

McDermott F., Mattey D.P. and Hawkesworth C.J. 2001. Centennial-scale Holocene climate variability revealed by a high-resolution speleothem δ^{18}O record from SW Ireland. Science 294: 1328–1331.

Meese D.A., Gow A.J., Grootes P., Mayewski P.A., Ram M., Stuiver M., Taylor K.C., Waddington E.D. and Zielinski G.A. 1994. The accumulation record from the GISP2 core as an indicator of climate change throughout the Holocene. Science 266: 1680–1682.

Merkt J. and Müller H. 1994. Laminated sediments in South Germany from the Neolithic to the Hallstatt period. In: Hicks S., Miller U. and Saarnisto M. (eds), Laminated Sediments. Rixenart, PACT Belgium, pp. 101–116.

Messerli B., Grosjean M., Hofer T., Núñez L. and Pfister C. 2000. From nature-dominated to human-dominated environmental changes. Quat. Sci. Rev. 19: 459–479.

Moore P.D. 1984. European Mires. Academic Press, London, 367 pp.

Negendank J.F.W., Zolitschka B., Rein B., Brauer A., Bruch-Mann C., Sanchez A. and Vos H. 1999. Varves and solar variability (Lake Holzmaar, Eifel, Germany). Bull. Soc. Belg. Geol. 106: 53–61.

Nicolussi K. and Patzelt G. 2000. Discovery of early-Holocene wood and peat on the forefield of the Pasterze Glacier, Eastern Alps, Austria. The Holocene 10: 191–199.

O'Brien S.R., Mayewski P.A., Meeker L.D., Meese D.A., Twickler M.S. and Whitlow S.I. 1995. Complexity of Holocene climate as reconstructed from a Greenland ice core. Science 270: 1962–1964.

Oldfield F. 1977. Lakes and their drainage basins as units of sediment based ecological study. Prog. Phys. Geog. 1: 460–504.

Patzelt G. 1977. Der zeitliche Ablauf und das Ausmass postglazialer Klimaschwankungen in den Alpen. In: Frenzel B. (ed.), Dendrochronologie und Postglaziale Klimaschwankungen in Europa. F. Steiner Verlag, Wiesbaden, pp. 248–259.

Patzelt G. and Bortenschlager S. 1973. Die postglazialen Gletscher- und Klimaschwankungen in der Venedigergruppe (Hohe Tauern, Ostalpen). Zeitschrift für Geomorphologie N.F. Suppl. 16: 25–72.

Pfister C. 1992. Monthly temperature and precipitation in Central Europe 1525–1979: quantifying documentary evidence on weather and its effects. In: Bradley R.S. and Jones P.D. (eds), Climate since AD 1500. Routledge, London, pp. 118–142.

Ponel P., de Beaulieu J.L. and Tobolski K. 1992. Holocene paleoenvironment at the timberline in the Taillefer Massif: pollen analysis, study of plant and insect macrofossils. The Holocene 2: 117–130.

Porter S.C. and Orombelli G. 1985. Glacier contraction during the middle Holocene in the western Italian Alps: evidence and implications. Geology 13: 296–298.

Prentice I.C., Cramer W., Harrison S.P., Leemans R., Monserud R.A. and Solomon A.M. 1992. A global biome model based on plant physiology and dominance, soil properties and climate. J. Biogeogr. 19: 117–134.

Prentice I.C., Guiot J., Huntley B., Jolly D. and Cheddadi R. 1996. Reconstructing biomes from palaeoecological data: a general method and its application to European pollen data at 0 and 6 ka. Clim. Dyn. 12: 185–194.

Proctor C.J., Baker A., Barnes W.L. and Gilmour R.A. 2000. A thousand year speleothem proxy record of North Atlantic climate from Scotland. Clim. Dyn. 16: 815–820.

Proctor C.J., Baker A. and Barnes W.L. 2002. A three thousand year record of North Atlantic climate. Clim. Dyn. 19: 449–454.

Ralska-Jasiewiczowa M. and van Geel B. 1992. Early human disturbance of the natural environment recorded in annually laminated sediments of Lake Gosciaz, Central Poland. Veg. Hist. Archaeobot. 1: 33–42.

Ralska-Jasiewiczowa M., Goslar T., Madeyska T. and Starkel L. (eds) 1998. Lake Gosciaz, Central Poland — A Monographic Study, Part 1. Krakow, W. Szafer Institute of Botany, 340 pp.

Renner F. 1982. Beiträge zur Gletschergeschichte des Gotthardgebietes und dendroklimatologische Analysen an fossilen Hölzern. Phys. Geog. 8: 1–180.

Savina S.S. and Khotinsky N.A. 1984. Holocene paleoclimatic reconstructions based on the zonal method. In: Velichko A.A., Wright H.E. and Barnosky C.W. (eds), Late Quaternary Environments of the Soviet Union. University of Minnesota Press, Minneapolis, 289–296.

Schneider R. 1985. Palynologic research in the southern and southeastern Alps between Torino and Trieste. Dissertationes Botanicae 87: 83–103.

Schweingruber F.H., Braker O.U. and Schar E. 1979. Dendroclimatic studies on conifers from Central Europe and Great Britain. Boreas 8: 427–452.

Smith J.P., Appleby P.G., Battarbee R.W., Dearing J.A., Flower R., Haworth E.Y., Oldfield F. and O'Sullivan P.E. (eds) 1991. Environmental History and Palaeolimnology. Developments in Hydrobiology 67: Dordrecht, Kluwer, 382 pp.

Snowball I., Korhola A., Briffa K.R. and Koç N., this volume. Holocene climate dynamics in Fennoscandia and the North Atlantic. In: Battarbee R.W., Gasse F. and Stickley C.E. (eds), Past Climate Variability through Europe and Africa. Kluwer Academic Publishers, Dordrecht, the Netherlands, pp. 465–494.

Starkel L., Pazdur A., Pazdur M., Wicik B. and Wieckowski K. 1996. Lake-level and groundwater-level changes in the Lake Gosciaz area, Poland: palaeoclimatic implications. The Holocene 6: 213–224.

Stuiver M. and Brazunias T.F. 1993. Sun, ocean, climate and atmospheric $^{14}CO_2$: an evaluation of causal and spectral relationships. The Holocene 3: 89–305.

Stuiver M., Reimer P.J., Bard E., Beck J.W., Burr G.S., Hughen K.A., Kromer B., McCormac G., van der Plicht J. and Spurk M. 1998. INTCAL98 radiocarbon age calibration, 24,000-0 cal. BP. Radiocarbon 40: 1041–1083.

Svensson G. 1988a. Fossil plant communities and regeneration patterns on a raised bog in South Sweden. J. Ecol. 76: 41–59.

Svensson G. 1988b. Bog development and environmental conditions as shown by the stratigraphy of Store Mosse mire in Southern Sweden. Boreas 17: 89–111.

Sykes M.T., Prentice I.C. and Cramer W. 1996. A bioclimatic model for the potential distributions of north European tree species under present and future climates. J. Biogeogr. 23: 203–233.

Tarasov P.E., Guiot J., Cheddadi R., Andreev A.A., Bezusko L.G., Blyakharchuk T.A., Dorofeyuk N.I., Filimonova L.V., Volkova V.S. and Zernitskaya V.P. 1999a. Climate in Northern Eurasia 6000 years ago reconstructed from pollen data. Earth Planet. Sci. Lett. 171: 635–645.

Tarasov P.E., Peyron O., Guiot J., Brewer S., Volkova V.S., Bezusko L.G., Dorofeyuk N.I., Kvavadze E.V., Osipova I.M. and Panova N.K. 1999b. Last Glacial Maximum climate of the Former Soviet Union and Mongolia reconstructed from pollen and plant macrofossil data. Clim. Dyn. 14: 227–240.

Tinner W. and Lotter A.F. 2001. Central European vegetation response to abrupt climate change at 8.2 ka. Geology 29: 551–554.

Tinner W., Ammann B. and Germann P. 1996. Treeline fluctuations recorded for 12,500 years by soil profiles, pollen, and plant macrofossils in the central Swiss Alps. Arct. Alp. Res. 28: 131–147.

van Geel B. 1978. A palaeoecological study of Holocene peat bog sections in Germany and the Netherlands, based on the analysis of pollen, spores and macro- and microscopic remains of fungi, algae, cormophytes and animals. Rev. Paleobot. Palynol. 25: 1–120.

van Geel B. and Renssen H. 1998 Abrupt climate change around 2,650 BP in North-West Europe: evidence for climatic teleconnections and a tentative explanation. In: Issar A.S. and Brown N. (eds), Water, Environment and Society in Times of Climatic Change. Kluwer, Dordrecht, pp. 21–41.

van Geel B., Buurman J. and Waterbolk H.T. 1996. Archaeological and palaeoecological indications of an abrupt climate change in The Netherlands, and evidence for climatological teleconnections around 2650 BP. J. Quat. Sci. 11: 451–460.

van Geel B., van der Plicht J., Kilian M.R., Klaver E.R., Kouwenberg J.H.M., Renssen H., Reynaud-Farrera I. and Waterbolk H.T. 1998. The sharp rise of ^{14}C ca. 800 cal BC: possible causes, related climatic teleconnections of human environments. Radiocarbon 40: 535–550.

Velichko A.A., Andreev A.A. and Klimanov V.A. 1997. The dynamics of climate and vegetation in the tundra and forest zone during the Late Glacial and Holocene. Quat. Int. 41-42: 71–96.

Vidal L. and Arz H., this volume. Oceanic climate variability at millennial time-scales: modes of climate connections. In: Battarbee R.W., Gasse F. and Stickley C.E. (eds), Past Climate Variability through Europe and Africa. Kluwer Academic Publishers, Dordrecht, the Netherlands, pp. 31–44.

von Grafenstein U., Erlenkeuser H., Müller J., Jouzel J. and Johnsen S. 1998. The cold event 8200 years ago documented in oxygen isotope records of precipitation in Europe and Greenland. Clim. Dyn. 14: 73–81.

von Grafenstein U., Erlenkeuser H., Brauer A., Jouzel J. and Johnsen S.J. 1999a. A Mid-European decadal isotope-climate record from 15,500 to 5000 years BP. Science 284: 1654–1657.

von Grafenstein U., Erlernkeuser H. and Trimborn P. 1999b. Oxygen and carbon isotopes in modern fresh-water ostracod valves: assessing vital offsets and autecological effects of interest for palaeoclimate studies. Palaeogeogr. Palaeoclim. Palaeoecol. 148: 133–152.

Vos H., Sanchez A., Zolitschka B., Brauer A. and Negendank J.F.W. 1997. Solar activity variations recorded in varved sediments from the crater lake of Holzmaar — a maar lake in the Westeifel Volcanic Field, Germany. Surv. Geophys. 18: 163–182.

Wanner H., Holzhauser H., Pfister C. and Zumbühl H. 2000. Interannual to century scale climate variability in the European Alps. Erdkunde 54: 62–69.

Wegmüller S. 1966. Über die spät- und postglaziale Vegetationsgeschichte des südwestlichen Juras. Beitr. Geobotan. Landes. Schweiz 48: 1–143.

Wegmüller S. and Lotter A.F. 1990. Palynostratigraphische Untersuchungen zur spät- und postglazialen Vegetationsgeschichte der nordwestlichen Kalkvoralpen. Bot. Helvet. 100: 37–73.

Wehrli B. 1997. High resolution varve studies in Baldeggersee. Aquat. Sci. 59: 283–375.

Welten M. 1982. Vegetationsgeschichtliche Untersuchungen in den westlichen Schweizer Alpen: Bern-Wallis. Denkschrift. Schweiz. Natur. Gesell. 95: 1–104.

Wick L. and Tinner W. 1997. Vegetation changes and timberline fluctuations in the Central Alps as indicators of Holocene climatic oscillations. Arct. Alp. Res. 29: 445–458.

Wimble G.A. 1986. The palaeoecology of lowland coastal raised mires of South Cumbria. Ph.D. thesis, University of Wales (Cardiff).

Zolitschka B. 1996. High resolution lacustrine sediments and their potential for palaeoclimatic reconstruction. In: Jones P.D., Bradley R.S. and Jouzel J. (eds), Climatic Variations and Forcing Mechanisms of the Last 2000 Years. NATO ASI Series I 41., Springer, Berlin and Heidelberg, pp. 453–478.

Zolitschka B. 1998. A 14,000 year sediment yield record from Western Germany based on annually-laminated lake sediments. Geomorphology 22: 1–17.

Zolitschka B. and Negendank J.F.W. 1998. A high resolution record of Holocene palaeohydrological changes from Lake Holzmaar, Germany. Paläoklimaforschung 25 ESF Special Issue 17: 37–52.

Zolitschka B., Brauer A., Negendank J.F.W., Stockhausen H. and Lang A. 2000. Annually dated late Weichselian continental paleoclimate record from the Eifel, Germany. Geology 28: 783–786.

21. CLIMATE VARIABILITY DURING THE LAST INTERGLACIAL-GLACIAL CYCLE IN NW EURASIA

MATTI SAARNISTO (matti.saarnisto@gsf.fi)
Geological Survey of Finland
P.O. Box 96, FIN-02150 Espoo
Finland

JUHA P. LUNKKA (juha.pekka.lunkka@oulu.fi)
Institute of Geosciences
P.O. Box 3000, 90014 University of Oulu
Finland

Keywords: Late Quaternary, Eemian interglacial, Weichselian glaciation, Scandinavian Ice Sheet, Interglacial, Interstadial, Pollen stratigraphy, Palaeohydrology, Palaeoclimate, NW Russia

Introduction

The history of the Eurasian continental ice sheets (Scandinavian and Barents and Kara Sea Ice Sheets) during the last interglacial-glacial cycle has been the subject of several national and international projects in recent years. Great efforts have been made to study the Russian part of the ice sheets in particular, and to evaluate in the field the interpretations presented in the extensive Russian literature. One of the tasks of the European Science Foundation's initiative QUEEN (Quaternary Environment of the Eurasian North) was to establish the extent of glaciers and focus on the climatic conditions that existed at the Last Glacial Maximum (LGM, ca. 21,000–18,000 years ago) in Eurasia. Much of the work has been undertaken within an EU-funded project 'Eurasian Ice Sheets' (cf. Larsen et al. (1999), Thiede et al. (2001)).

New data have been collected from terrestrial and marine sequences, and an existing picture of the timing of the growth of the ice sheet, its extent and its deglaciation chronology have been considerably revised. Svendsen et al. (1999 and references therein) have shown that the extent and timing of the Kara and Barents Ice Sheets during the Valdai (Weichselian) stage differed from that presented by Grosswald (1980, 1998). Recent results on the glacial cover over vast areas of Eurasia suggest that Grosswald's model over-estimates the size of the ice-cover at the Last Glacial Maximum (LGM) by more than 50%. The Kara and Barents Ice Sheets are considered to be much more restricted in size in the most recent reconstructions (Svendsen et al. 1999).

The interplay of the Scandinavian, Kara and Barents Ice Sheets during the Valdai cold stage is of vital importance in understanding the palaeohydrology of Northern Eurasia. It

443

R. W. Battarbee et al. (eds) 2004. *Past Climate Variability through Europe and Africa.*
Springer, Dordrecht, The Netherlands.

is particularly important to know the directions in which the drainage of the continental and glacial waters were directed at various stages of the glaciation: whether towards the North Atlantic, or towards the Caspian-Black-Mediterranean Sea, since this must have had a huge impact on Atlantic circulation patterns and consequently upon regional climate.

The glacier fluctuations can be directly linked with climate fluctuations and the most detailed picture has emerged from analysis of continuous marine sediment cores obtained from localities adjacent to the Scandinavian Ice Sheet (Baumann et al. 1995). The picture obtained from terrestrial sequences is, by comparison, much more fragmentary.

In this chapter special reference will be made to the south-eastern sector of the Scandinavian Ice Sheet, which extends from Finland to the Valdai-Vologda area of the Russian Plain, into which region the ice advanced during the Late Weichselian. This is the area where much new data have been collected in recent years from terrestrial sequences, and where the history of glacier fluctuations during the Valdai glaciation can now be reconstructed in considerably more detail than hitherto (Lunkka et al. 2001; Saarnisto and Saarinen 2001). Data obtained from the SE sector of the Scandinavian Ice Sheet provide more precise dates on the growth and decay of ice better than are available from other parts of Eurasia.

The aim of the present chapter is: (i) to describe the fluctuations of Eurasian glaciers since the last (i.e., Eemian/Mikulino) interglacial and assess their influence on palaeohydrology; (ii) to summarise biostratigraphical data available for the last interglacial/glacial cycle in that area; and (iii) to compare these data with other proxy data on climate.

Dating

As most of the history of the last cold stage lies beyond the range of radiocarbon dating, the chronology of events remains the principal difficulty. The sequence of events can be established using lithostratigraphical and biostratigraphical methods, whereas the timing of events is based on the most probable correlation with the marine oxygen isotope scheme. Direct 'absolute' dates have been obtained using the OSL technique, the principal dating tool adopted for the 'Eurasian Ice Sheet'-project. All dates generated during the operation of the project were measured at the Nordic Laboratory for Luminescence Dating, Risø National Laboratory, Denmark by Dr. Andrew Murray. Measurements of quartz grains were made using the Single Aliquot Regenerated dose protocol (e.g., Murray and Wintle (1999)). At its best, this method can at present only group individual age estimates into last interglacial age or into substages of the last glaciation. Hence the accuracy of the method is of the order of thousands of years. Only where a number of compatible dates from suitable sediments for OSL dating are obtained from the same stadial/interstadial stage, can the dating be considered reliable.

The growth and decay of the Eurasian Ice Sheets to and from their largest extent in the Late Weichselian is well within the effective range of radiocarbon dating. The combination of varved clay records, palaeomagnetic measurements and calibrated radiocarbon dates have resulted in a number of considerable revisions to the deglaciation chronology of the last Scandinavian Ice Sheet. As will be argued below, the maximum ice extent was reached at a later time in the east, while deglaciation was more rapid and occurred ca. 1000 years earlier in the east than was previously thought. The revisions of the Swedish varve chronology that dates the deglaciation phases in Sweden adds about 900 more varve years to the conventional

Swedish deglaciation chronology (Andrén et al. 1999). In its revised form, the Swedish deglaciation chronology is consistent with the new deglaciation history of the south-eastern sector of the ice sheet in NW Russia and Finland (Saarnisto and Saarinen 2001). The dating of the major climate events during deglaciation are now in harmony, within 100 years, to chronologies based on oxygen isotope records from ice cores, tree-ring data and annually laminated lake sediments in Central Europe (cf. Litt et al. (1999), Saarnisto and Saarinen (2001)). This agreement also includes the Younger Dryas cold event, which is now dated to between 12,600 and 11,500 calendar years BP (Litt et al. 1999).

Annually-laminated lake sediment sequences from sites located beyond the maximal limits of the last ice-sheet offer a great opportunity to date precisely the changes in the terrestrial environment during the last glacial stage. Such sites include maar lakes in Central Europe which contain sediments that cover the time since the Last Glacial Maximum (Zolitschka and Negendank 1999) and the records from Italy which cover the entire last glacial cycle (e.g., Allen et al. (1999)). Long lake sediment sequences located outside the maximal extent of the Scandinavian Ice Sheet are also known in NW Russia. Semenenko et al. (1981) have studied a core from the former Tatischevo Lake (The First of May Factory), which lie some 200 km north of Moscow. This core contains a nearly continuous, annually-laminated sequence that covers the Valdai/Weichselian cold stage and the Eemian. A detailed pollen diagram from this site shows vegetation and climate fluctuations over the past 120,000 years. However, in view of the difficulties of varve identification and of counting long varve sequences, the chronology may need to be reconfirmed. In any case, this sediment sequence requires further attention, using more modern stratigraphical techniques in order to test the chronology and palaeoclimatic reconstructions. In conclusion, so long as the dating is based on correlation with the deep-sea oxygen isotope scheme a risk of circular reasoning remains.

Eurasian glacier fluctuations during the Weichselian Stage with special emphasis on fluctuations of the Scandinavian Ice Sheet

Early Weichselian

In different parts of Eurasia ice started to build up after the Eemian. In general, the geographical centres of glaciation migrated from east to west during the Weichselian as the moisture source moved towards the west as a result of progressive cooling of North Atlantic and Siberian coastal waters (Karabanov et al. 1998). The major ice build up in NW Siberia took place during MIS 5d (Karabanov et al. 1998), while the Kara-Barents ice sheets were at their largest during the Early and Middle Weichselian but were considerably more restricted during the LGM. In contrast, the Scandinavian ice sheet was relatively small during the Early and Middle Weichselian but attained its largest extent during the Late Weichselian (Mangerud et al. 1999; Astakhov et al. 1999, Fig. 1).

Northern Fennoscandia became covered by ice during MIS 5d and MIS 5b in the Early Weichselian, ca. 110,000 and 90,000 years ago, respectively (Saarnisto and Salonen 1995). These sub-stages were separated by the Brørup interstadial (MIS 5c) and the stadial MIS 5b was followed by the Odderade interstadial (MIS 5a). Lithostratigraphical observations combined with OSL results suggest that there was a long ice-free period after the Saalian

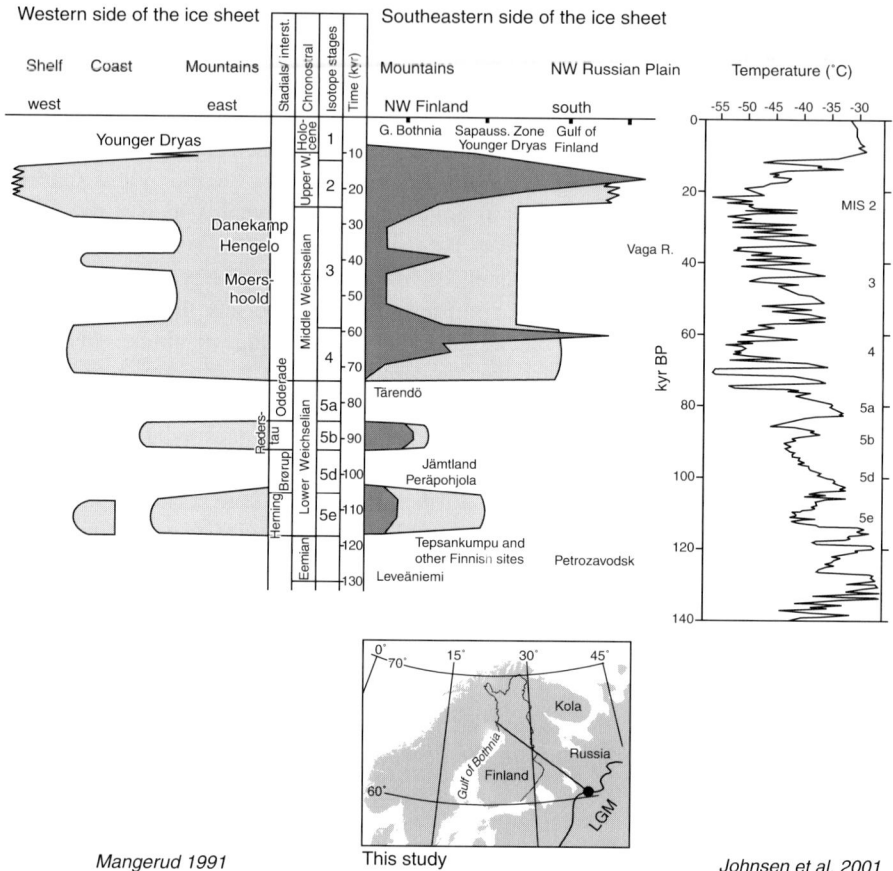

Figure 1. Time-distance diagram from the Gulf of Bothnia to the Vologda area, NW Russia (see inset map) showing the growth and decay of the Scandinavian Ice Sheet during the Weichselian in the south-eastern sector (dark shading) compared to the previously established glaciation curve (lighter shading) of Mangerud (1991). Greenland ice core temperature reconstructions (according to Johnsen et al. (2001)) are also shown for comparison.

deglaciation of Western Finland that continued most probably throughout the Early Weichselian (cf. Hütt et al. (1993), Nenonen (1995)). However, parts of Northern and Westernmost Finland, as well as the Scandinavian Mountain range, were covered by ice between the Early Weichselian interstadials (Fig. 2; Hirvas 1991; Lundqvist 1992).

Middle Weichselian

It is generally believed that during MIS 4, at the beginning of the Middle Weichselian, (ca. 70,000–60,000 years ago), the whole of Fennoscandia became covered by the continental ice sheet, with perhaps limited ice free areas on the west coasts of Norway and in Sweden (Lundqvist 1994). In fact, little is known about the Middle Weichselian ice cover in most of Sweden, Finland and Russian Karelia. However, it is known that the ice margin

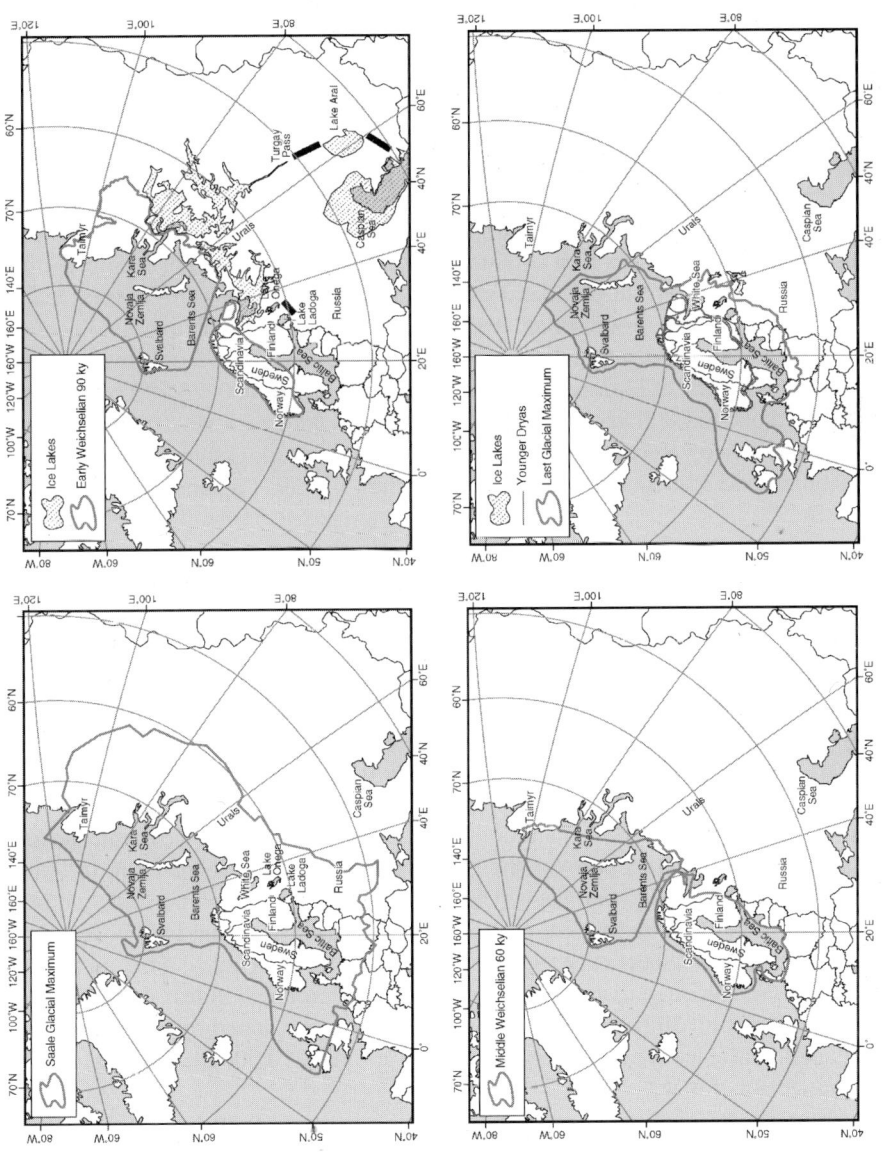

Figure 2. The greatest extent of the ice sheet in Northern Eurasia during the Saalian, the Early/ Middle Weichselian (90,000 and 60,000 years ago) and in the Late Weichselian modified after Svendsen et al. (1999 and references therein). Ice-dammed lakes 90,000 years ago are shown according to Mangerud et al. (2001). Comparable lakes existed most probably also 60,000 years ago. The maximum extent in the west (at over 20 kyr) is considerably older compared to the maximum extent in the east (at around 17 000–18 000 years ago) as also discussed in text. Ice-dammed lakes are from Lunkka et al. (2001; see Fig. 7). The Younger Dryas ice margin of the Scandinavian Ice Sheet is also shown according to Andersen et al. (1995).

fluctuated a great deal on the Atlantic coast of Norway and on Svalbard during this period and that several ice-free periods alternated with periods of rapid ice growth (cf. Mangerud (1991), Mangerud et al. (1996), Olsen (1997)). The Scandinavian Ice Sheet, at its maximum around 60,000 years ago, covered the whole of Fennoscandia, part of Russian Karelia and the Kola Peninsula (Fig. 2).

The detailed history of the Scandinavian Ice Sheet extent in its eastern part during the Middle Weichselian (MIS 4 and MIS 3), a very long time period spanning more than 50,000 yr, is still poorly known, but a general picture of this time interval is now emerging. The evidence for ice sheet extent is based on lithostratigraphy, mainly till stratigraphy and OSL ages. The data indicate that at around 60,000 years ago, the Scandinavian Ice Sheet covered the whole territory of Finland and perhaps also the most western parts of NW Russia as well as the northern part of Estonia. However, studies of till stratigraphy combined with OSL dates in Western Finland suggest that there were possibly several ice-free periods during the Middle Weichselian, even in Western Finland which was adjacent to the glaciation centre in the Scandinavian Mountains (cf. Nenonen (1995)). It seems that the southern and central parts of Finland were free of ice between 40,000–25,000 years ago and also between 50,000–55,000 years ago (Nenonen 1995; Ukkonen et al. 1999; Lunkka and Saarnisto, unpublished). Recently Helmens et al. (2001) also suggested that there was an ice-free period around 42,000 years ago, even in eastern Finnish Lapland.

Late Weichselian

The Scandinavian Ice Sheet reached its maximum extent during the Late Weichselian in all sectors, although the maximum positions achieved by different margins were not synchronous (Fig. 3). The maximal extent of the ice margin had already been attained by ca. 25,000 years ago on the SW coast of Norway (Sejrup et al. 1994), whilst in the eastern sector it was not reached until 17,000 to 18,000 years ago (Larsen et al. 1999; Lunkka et al. 2001).

According to dates obtained from mammoth bones found in glacigenic sediments in Finland, the southern and central parts of the country were ice-free at least for 10,000 years until about 25,000 calendar years ago (Ukkonen et al. 1999; calibration scheme of Kitagawa and Plicht 1998). After that ice advanced to its maximal extent in NW Russia (Lunkka et al. 2001). The ice sheet expanded from Northern Sweden across Finland to its maximal extent in NW Russia, a distance more than 1000 kilometres, in less than 10,000 years, (Fig. 1, Lunkka et al. 2001). The maximum position was reached by about 17,000 to 18,000 years ago, while ice retreat was completed in less than 10,000 years, suggesting very marked changes in climate.

The ice advance to its maximum position across the extreme south of Sweden and Eastern Denmark was also a late, short event that took place roughly 20,000 radiocarbon years ago (Lundqvist 1994), i.e., about 22,000 calendar years ago. The maximum position of the eastern sector of the Scandinavian Ice Sheet may well have extended from east of the White Sea to the western Kanin Peninsula, across the Pechora Sea to the Kara Sea basin and perhaps to the north-western coast of the Taimyr Peninsula (Svendsen et al. 1999). It followed the edge of the continental shelf around the Barents Sea and west of Norway. It was in contact with the British Isles ice mass before 25,000 years ago, but not during

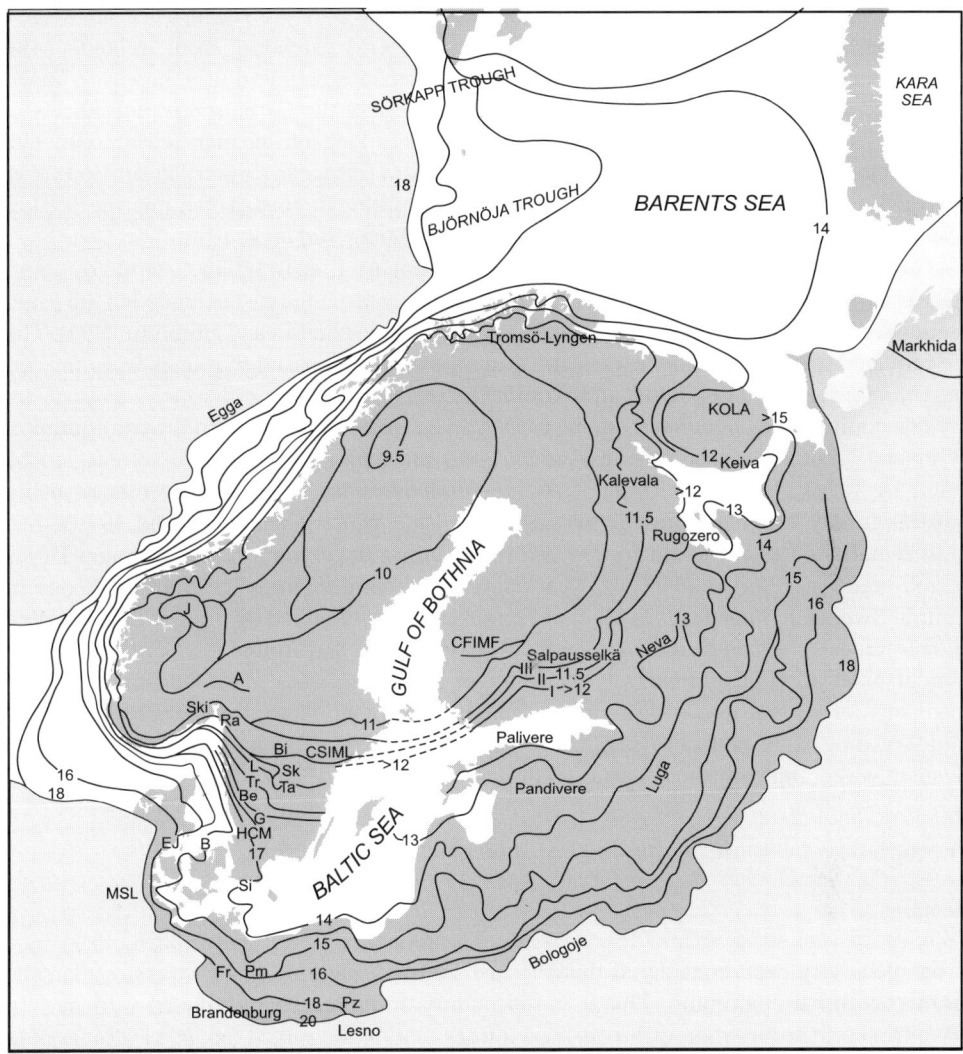

Figure 3. Retreat stages of the Scandinavian ice-sheet from the Last Glacial Maximum as summarised in the results of IGCP Project 253, 'Termination of the Pleistocene', by Lundqvist and Saarnisto (1995). Dates are in calendar years according to current interpretations (discussed in text). Modifications to the 1995 model in the NE sector are according to Svendsen et al. (1999), see Figure 2, and by the QUEEN project members, for Southern Sweden according to Lundqvist and Wohlfartth (2001). Note that the northern Russian mainland remained ice-free during the Late Weichselian.

its later advance, dated to between 19,000 and 15,000 radiocarbon years BP (Sejrup et al. 1994), when the ice in the east was in general still growing.

During the LGM the mainland of northern European Russia was free of ice because of the limited extent of the Barents-Kara Ice Sheets, which were centered over the shelf areas only. Only valley glaciers were present in the Ural Mountains (Astakhov et al. 1999). The ice-sheet reconstruction outlined above contradicts some previous reconstructions, which envisaged a huge continuous ice-cover during the LGM extending from Ireland to the Taimyr Peninsula (Grosswald 1980; 1998). The latter interpretation was, however, based on limited stratigraphical field data. The restricted size of the Barents-Kara Ice Sheets at the LGM is further supported by discoveries of traces of Palaeolithic man at Mamontovaya Kurya at the Arctic circle near the polar Urals in northern European Russia dated to 36,000 yr BP (Pavlov et al. 2001), together with the Upper Palaeolithic Byzovaya dwelling site 300 km SW of Mamontovaya Kurya dated to 28,000 yr BP (Mangerud et al. 1999).

Deglaciation from the LGM limit, the Main Stationary Line in Jylland, started by about 22,000 calendar years ago, but a major Baltic readvance from the east extended to the Main Stationary Line once again at 18,000 years ago (Houmark-Nielsen and Humlum 1994). The ice retreated to the Swedish west coast, where the Halland coastal moraines were formed ca. 16,000 years ago (Lundqvist and Wohlfarth 2001).

The course of deglaciation from the LGM can be traced by analysis of the configuration of glacial landforms, especially eskers and end moraines, and stages of retreat can be dated by varve chronology (Fig. 3). Larger end moraines have been correlated using biostratigraphy and with climatic events of the Late Weichselian (Berglund 1979). The most remarkable end moraine zone was formed during the extremely cold Younger Dryas period between 12,600 and 11,500 years ago. The Younger Dryas Ice-Marginal Zone in Central Sweden has correlatives all over Fennoscandia, in Finland (the major Salpausselkä end moraines), Russian Karelia, across the Kola Peninsula and around the Norwegian coast (Andersen et al. 1995). In Norway the Younger Dryas moraines mark the limit of an ice readvance of several tens of kilometres, whereas for other sectors of the ice sheet, only minor advances are generally suggested. The final ice retreat to the interior of Northern Sweden was completed shortly after 10,000 BP (Lundqvist 1994).

Vegetation and climate during the last interglacial-glacial cycle

Eemian

It has generally been considered that the last Eemian interglacial was warmer than the Holocene climatic optimum. This is based mainly on pollen stratigraphical evidence. In NW Russia, the proportion of broad-leaved trees mainly the mixed oak (QM) component of *Quercus, Tilia* and *Ulmus* and especially *Carpinus* were more abundant with higher *Corylus* frequencies during the Eemian climatic optimum compared with the Holocene values (Ikonen and Ekman 2001; Gey et al. 2001). This is reflected in the Mikulino/Eemian pollen diagram from Petrozavodsk in Russian Karelia (Fig. 4). Spruce was an important element during most of the Eemian in NW Russia, and only during the climatic optimum was it virtually absent. High frequences of steppe elements, together with spruce, suggest a continental climate. In Northern Finland, high spruce and alder pollen values, together with frequent occurrences of QM, suggest a climate warmer than during the Holocene climatic optimum, as shown in the Tepsankumpu pollen diagram which covers most of the interglacial (Fig. 5) (Saarnisto et al. 1999).

Zagwijn (1996) analysed selected climatic indicator species recorded in 31 pollen diagrams from sites in Western and Central Europe, and concluded that the Amsterdam area had mean temperatures 3 °C higher in January and 2 °C higher in July than at present.

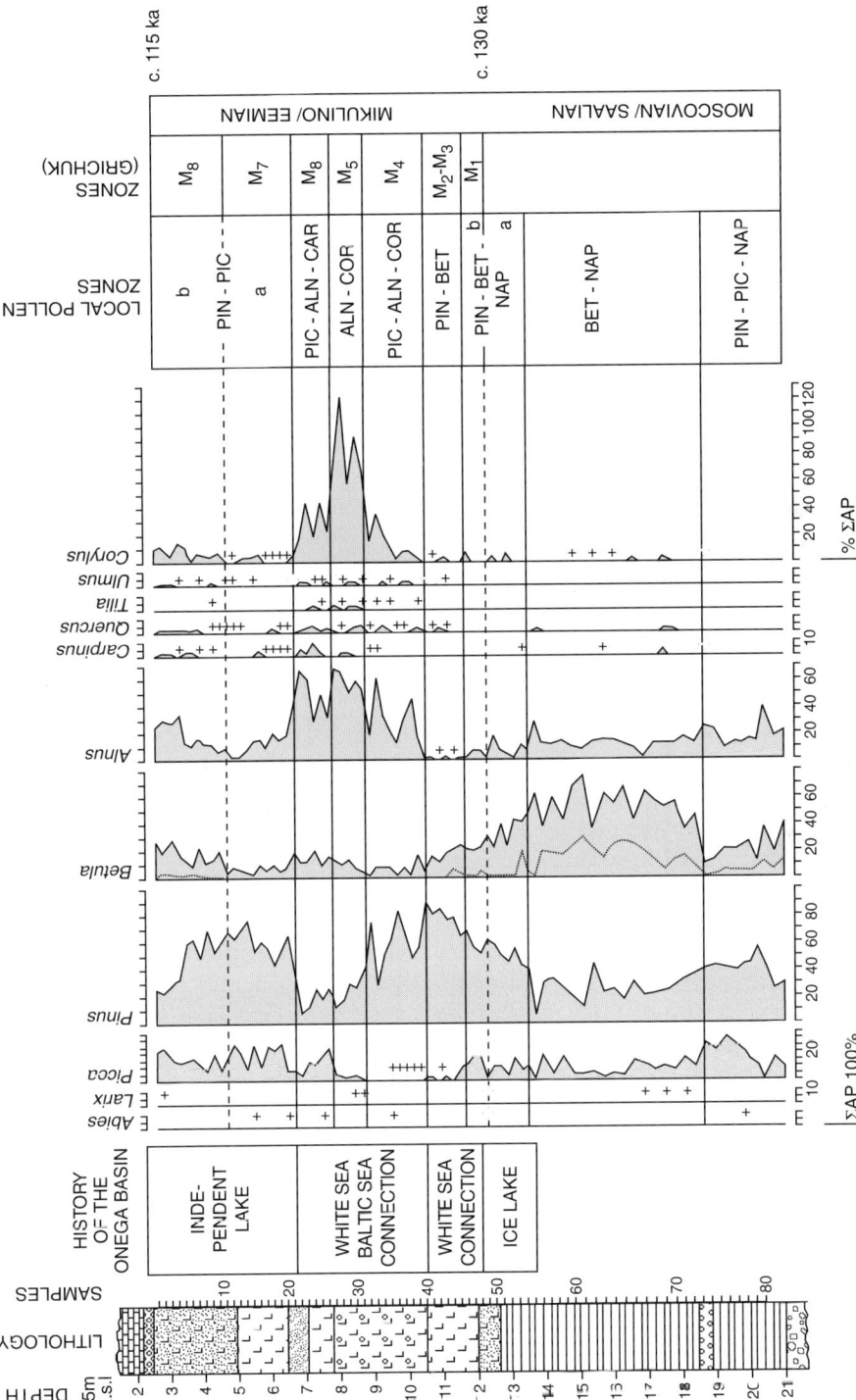

Figure 4. Selected pollen diagram from an Eemian/Mikulino sequence from Petrozavodsk, Russian Karelia. The time of the marine submergence between the pollen zones M_1-M_8 is indicated by broken lines. Local pollen zones and their correlation with Grichuk regional pollen zones are also shown. Modified after Ikonen and Ekman 2001.

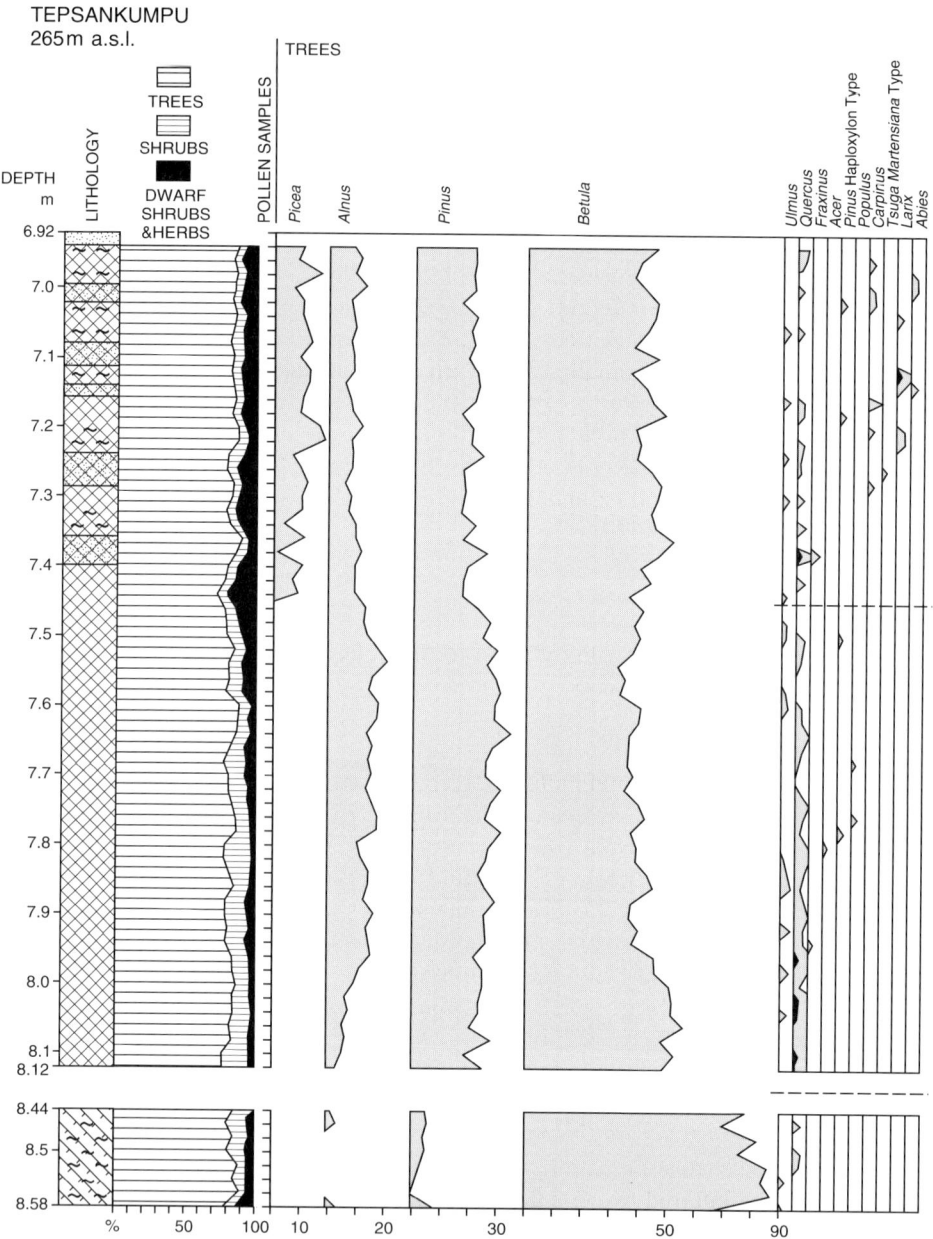

Figure 5. Selected pollen diagram from the Eemian deposits of Tepsankumpu Finnish Lapland, which lies close to the Weichselian glaciation centre, showing a tree pollen sequence resembling that of the Holocene. Modified after Saarnisto et al. (1999).

Aalbersberg and Litt (1998) presented their multi-proxy climate reconstructions from an east-west transect across Europe between 50 °N and 60 °N from the Baltic states in the east to Ireland in the west. They concluded that the July minimum mean temperatures were between 16 °C–20 °C, while the January minimum mean temperatures were between −2 °C–1 °C during the Eemian intergalacial. Moreover, the Eemian climate became progressively more oceanic in character, which was also the conclusion reached from evidence obtained from Finland (Eriksson 1993). Pollen diagrams from Central Europe as well as from Finland suggest (Saarnisto et al. 1999) stable Eemian climates with minor fluctuations. The Greenland ice cores indicate that temperatures were 3 °C higher during the Eemian compared with the Holocene climate optimum (Johnsen et al. 2001).

The pollen evidence from Northern Finland and Sweden suggests that the early Eemian climate became warm very rapidly, with birch forest advancing soon after deglaciation (Robertsson 1997; Saarnisto et al. 1999), which is compatible with the ice-core record (e.g., Johnsen et al. (2001)). By contrast, the termination of the Eemian seems to have been gradual according to the pollen evidence from NW Russia (Gey et al. 2001) and Northern Central Europe (Caspers and Freund 2001).

Early and Middle Weichselian

Climatic conditions during the Early and Middle Weichselian fluctuated between glacial and interstadial stages. Oxygen isotope records from the Greenland ice cores (Johnsen et al. 2001) indicate that the fluctuations were within 15 °C and that there was a general trend towards lower temperatures. This is also indicated by climate reconstructions constructed by Aalbersberg and Litt (1998) and pollen diagrams from Northern Germany (Behre 1989), which suggest average July temperature of nearly 15 °C during the Early Weichselian Brørup and Odderade interstadials. The vegetation during these two interstadials was characterised by forests (e.g., Zagwijn (1989)). Similarly, in NW Finland, which lies closer to the glaciation centre, the Early Weichselian interstadials were forested, the data indicating a succession from birch-dominated to pine-dominated woodland (Donner 1983; Peltoniemi et al. 1989). This suggests that July temperatures were above 10 °C, which is the temperature at the present tree limit in Northern Fennoscandia (Hyvärinen 1975). The Early Weichselian interstadial deposits discovered in Finland are often correlated with the Brørup interstadial (MIS 5c) but some of them may represent a somewhat cooler interstadial, the Odderade (MIS 5a), as suggested by Nenonen (1995). The Early Weichselian stadials MIS 5d and MIS 5a in Central Finland were characterised by arctic tundra conditions (Hütt et al. 1993).

The Early Weichselian climate in the area around the Late Weichselian ice margin in NW Russia can be inferred from numerous pollen diagrams, although their age control is poor. However, in this area the Early Weichselian landscape appears to have been open birch forest with spruce and with fluctuating NAP values. Towards the end of the Early Weichselian, birch and alder increased suggesting a warmer climate (Gey et al. (2001), see also Gey and Malakhovsky (1998)). Unfortunately, biostratigraphical correlation of the Early Weichselian cores in NW Russia with established schemes in Western Europe is not yet possible.

The Middle Weichselian sub-stage (MIS 4 and MIS 3) started ca. 74,000 years ago with a rapid cooling. The Scandinavian Ice Sheet expanded to its most extensive position so far

at around 60,000 years ago, to cover most of Fennoscandia. The prominent climate fluctuations, known as Dansgaard / Oeschger cycles (D/O), commenced during this substage,
according to the Greenland ice-core record (cf. Johnsen et al. (2001)).

Pollen evidence from the NW Russian Plain suggests that the local vegetation alternated between spruce-dominated forests and nearly treeless tundra. The glaciation curve
constructed for the SE sector of the Scandinavian Ice Sheet (Fig. 1), and based on a number
of new OSL dates from Finland (Lunkka and Saarnisto, unpublished), suggest substantial
and rapid fluctuations of the ice margin. The major ice advances took place at ca. 60,000
years ago and at around ca. 40,000 years ago and are comparable in scale with glacier
fluctuations in the Atlantic sector of the ice sheet (Mangerud 1991).

The Middle Weichselian interstadials in Central Europe were not forested, i.e., the
July temperature was below +10 °C (Behre 1989; Caspers and Freund 2001). The Middle
Weichselian pollen records from Northern Scandinavia, which was adjacent to the glaciation
centre, are poor. Tundra-type *Artemisia*-dominated pollen assemblages have been recently
discovered in a Middle Weichselian (ca. 50,000 yr ago) sequence at Ruunaa, Eastern Finland (Lunkka et al., unpublished). In addition, Helmens et al. (2000) interpret the pollen
assemblages in the Middle Weichselian sediments at Sokli, eastern Finnish Lapland, north
of the Arctic Circle, as representing a shrub tundra environment.

The upper Middle Weichselian pollen data from NW Russia are fragmentary. The
Palaeolithic man record from Mammontovaya Kura in northern European Russia, dated
to 36,000 radiocarbon years ago, coincides with evidence for a nearly treeless steppe
environment, when the climate was colder and more continental than at present. Correlation
with the Hengelo interstadial has been suggested by Pavlov et al. (2001). A pollen profile
from organic-bearing waterlain sediments (Fig. 6) exposed in the bank of the Vaga River,
north of Velsk (61 °N Lat and 42 °E Long), more than 500 km south of Mammontovaya
Kura, indicates an open spruce-dominated forest at 35,200 radiocarbon years ago. Today
the area is covered by closed boreal taiga forests.

Late Weichselian

During the Late Weichselian, at around 20–17 kyr, the climate was extremely cold and dry
east of the Scandinavian Ice Sheet. Polar deserts dominated in the Siberian coastal areas,
whereas further south, in NW Russia, precipitation was adequate to support ice-dammed
lakes. Ice melted by ablation and not only by sublimation (Lunkka et al. 2001). Aeolian
sand deposits are widespread outside the ice margin. Periglacial activity is indicated by
fossil ice-wedges, which increase in size towards the north, when observed along the Late
Weichselian ice margin from the Valdai upland area to the Vaga river valley (Lunkka et al.
2001). Palaeobotanical evidence, although limited, suggests that winter temperatures were
at least 20 °C, and summer temperatures 5–11 °C lower, than at present in this region, while
precipitation was reduced to one third of present values (Tarasov et al. 1999).

The deglaciation from the Late Weichselain maximum ice-limits commenced as early
as 17,000 years ago and by the latest at 15,000 years ago, when distinct warming had begun
(Larsen et al. 1999; Lunkka et al. 2001). In the north, the deglaciation of the Barents Sea
commenced soon after that (Polyak et al. 1995; Lundqvist and Saarnisto 1995). The generally warming trend was interrupted between 12,650 and 11,500 years ago by the Younger

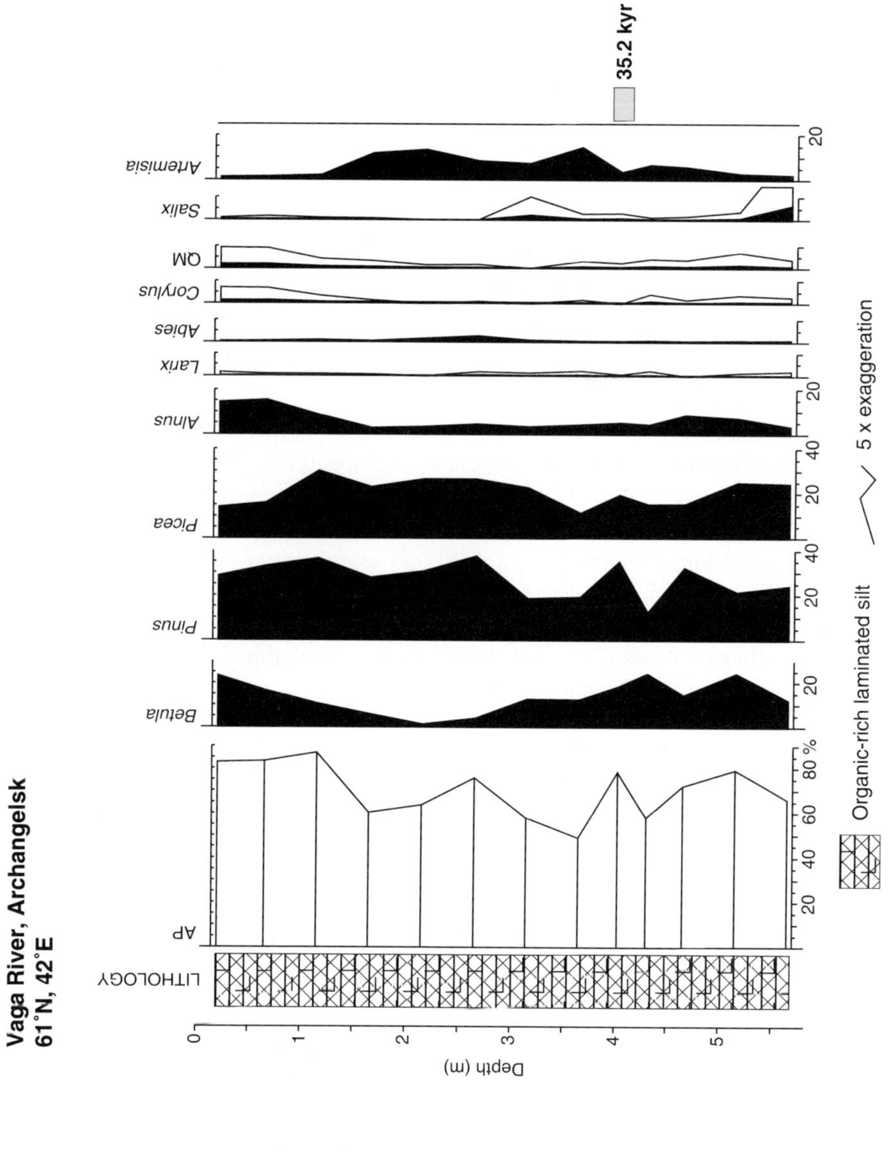

Vaga River, Archangelsk
61°N, 42°E

Figure 6. Selected pollen diagram from Late Middle Weichselian fluvial deposits on the Vaga River (north of the town of Velsk, Archangelsk District, Russia) indicating open spruce-dominated forest at ca. 35 kyr BP i.e., predating the Late Weichselian cold substage. QM = mixed oak (*Quercus, Tilia, Ulmus* and *Carpinus*).

Dryas stadial, when distinct end moraines were formed around Fennoscandia (Fig. 2). Fossil ice wedges found in the Salpaussselkä moraines in Southern Finland suggest mean annual temperatures during the Younger Dryas below −6 °C (cf. Péwé et al. (1969), Vandenberghe and Pissart (1993)), which is at least 10 °C less than present-day temperatures. No fossil ice-wedges are known from deposits in Finland that post-dates the Younger Dryas. The cold climate of the Younger Dryas is indicated by *Artemisia*-dominated tundra vegetation outside the ice margin (Hyvärinen 1973; Donner 1995) and increased minerogenic deposition in lake sediments. Abrupt warming at the end of the Younger Dryas stadial (in Greenland more than 10 °C, see Alley et al. (1993)) resulted in a rapid expansion of birch forests in South-East Finland and Russian Karelia (Hyvärinen 1973; Elina 1985).

Palaeohydrology

The drainage of the northern Eurasian rivers has altered dramatically during the last interglacial-glacial cycle as the continental ice-sheets have repeatedly blocked exits into the Arctic Ocean and the Baltic Sea. The growth and decay of the continental ice sheets caused eustatic fluctuations of sea level over a range of 150 metres, between the Eemian high sea levels (5–10 m above the present sea level) to the levels attained at the LGM, 18,000–21,000 years ago, of 120 to 140 metres below present sea level (Waelbroek et al. 2002). Eustatically low sea levels exposed extensive shelf areas along the northern Eurasian coast. After the Early Middle Weichselian, some 70,000 years ago, the Bering Strait was dry land until the Late Weichselian (e.g., Sher (1997)) and the White Sea basin was occupied by a fresh-water lake during ice-free periods (Lunkka and Saarnisto, unpublished). All this must have had an influence on coastal currents and ocean circulation in general.

Eemian molluscs found in sediment sequences indicate that warm Atlantic waters penetrated along the North European Russian and Siberian coasts much further east than during the Holocene, indicating limited sea ice-cover during the Eemian (e.g., Karabanov et al. (1998), Mangerud et al. (1999), Larsen et al. (1999), Raukas (1991)). There was an open passage between the Barents Sea and the Baltic Sea over the White Sea, and the Onega and Ladoga basins for the major part of the Eemian (Ikonen and Ekman 2001 and references therein). This passage undoubtedly had a significant impact on ocean circulation around Fennoscandia (Mangerud et al. 1999).

The growth of the Barents-Kara Ice Sheets to their maximum Weichselian positions during the Early Weichselian at around 90,000 years ago resulted in the damming of rivers that flowed to the Arctic Ocean, and huge ice-dammed lakes formed in front of the ice margin (cf. Mangerud et al. (2001)). In Siberia, the Ob and Jenisei rivers were blocked by the Kara Ice Sheet, and an ice-dammed lake occupied the entire western Siberian plain (Arkhipov et al. 1995 and references therein). The ice-lake drained via the Aral Sea to the Caspian Sea and further to the Black Sea and Mediterranean Sea and finally to the mid-latitude Atlantic, presumably having a significant impact on ocean conditions and circulation, and hence climate.

The first Weichselian ice advances of the Barents-Kara Ice Sheets onto the mainland (dated at around 90,000 years ago) dammed the north-bound rivers and created one of the most extensive Eurasian ice lakes, the recently discovered Komi Lake (Astakhov et al. 1999; Maslenikova and Mangerud 2001; Mangerud et al. 2001). This lake extended from the Urals to the White Sea basin and drained via the Baltic basin to the Atlantic. Its present altitude lies at 100 metres above sea level. The drainage of the lake was eventually via the Ojat river valley between the Onega and Ladoga basins (Maslenikova and Mangerud 2001; Saarnisto and Lunkka, unpublished). Very little is known about the termination of these lakes, although the only possible drainage was towards the Barents and Kara seas. It is highly probable that, for example, the ice-lake that occupied the White Sea Basin emptied via the straight between Eastern Kola and the mainland. There is no direct evidence, so far, of any catastrophic outburst events that could have brought large quantities of fresh-water into the North Atlantic Ocean.

The palaeohydrology during the Late Weichselian was different from that of the Early and Middle Weichselian. As discussed above, the Barents-Kara Ice Sheets did not penetrate into the mainland of Russia during that sub-stage, and consequently the drainage of the northbound Russian rivers was not blocked by ice. The Scandinavian Ice Sheet, however, attained its greatest extent during the Late Weichselian ca. 18,000–17,000 years ago and relatively extensive water bodies were formed in front of the ice margin. In the south-eastern sector of the ice sheet, in the area of the Mologa - Sheksna and Sukhona basins, glacial lakes extended some 500 kilometres along the ice margin (Fig. 7). These lakes, the highest shorelines of which are now at 130 m.a.s.l., belonged to the Volga catchment area and drained eventually into the Caspian Sea (Lunkka et al. 2001).

When the Scandinavian Ice Sheet started to retreat from its Late Weichselian maximum position ca. 15,000–17,000 years ago, the drainage of the glacial meltwaters was directed towards the Baltic Basin, and a series of short-lived ice lakes were formed at the retreating ice margin (Kvasov 1979). An ice-lake in the Onega basin existed between 14,400 and 12,900 years ago (Saarnisto and Saarinen 2001). Its outlet was across the Onega - Ladoga isthmus into the Baltic and the ice-lake phase terminated when it started to drain to the White Sea basin following the retreat of the ice margin.

In the Baltic Basin, the Baltic Ice Lake was in contact with the receding ice margin from ca. 14,000 years ago. It was most extensive during the Younger Dryas stadial when glacial meltwaters built the extensive marginal deltas of the Salpausselkä moraines in Finland up to the level of the Baltic Ice Lake. It also occupied the Lake Ladoga basin. Its water level dropped nearly 30 metres to the contemporaneous level of the ocean at the end of the Younger Dryas stadial (ca. 11,590 years ago).

It is assumed that the drainage of enormous volumes of water from glacial lakes, for example, Lake Agassiz into the North Atlantic at around 11,000 radiocarbon years ago i.e., ca. 12,500 cal BP (Broecker et al. 1989) had a major influence on the thermohaline circulation of the ocean and hence on climate. This has been suggested as the main reason for the abrupt Younger Dryas cooling (Broecker et al. 1988), although other explanations, such as the role of the sun, have been also suggested (Goslar et al. 1995). The repeated fluctuations of the Eurasian and North American drainage patterns during the Weichselian may be the explanation of the major climate fluctuations, the Dansgaard/Oeschger events, recognised in the Greenland ice cores, and dated to between ca. 80,000 and 15,000 years ago. The rapid temperature shifts at this time are estimated to have been up to 12 °C to 15 °C (Bond et al. 1997; Johnsen et al. 2001). Although there seems to be a direct connection between D-O events found in ice-core records and so-called Heinrich events discovered in marine cores (cf. Bond and Lotti (1995)), these relatively short, though intensive fluctuations in temperature are difficult to identify, and even if identified they are also difficult to date in terrestrial records, for example in NW Russia.

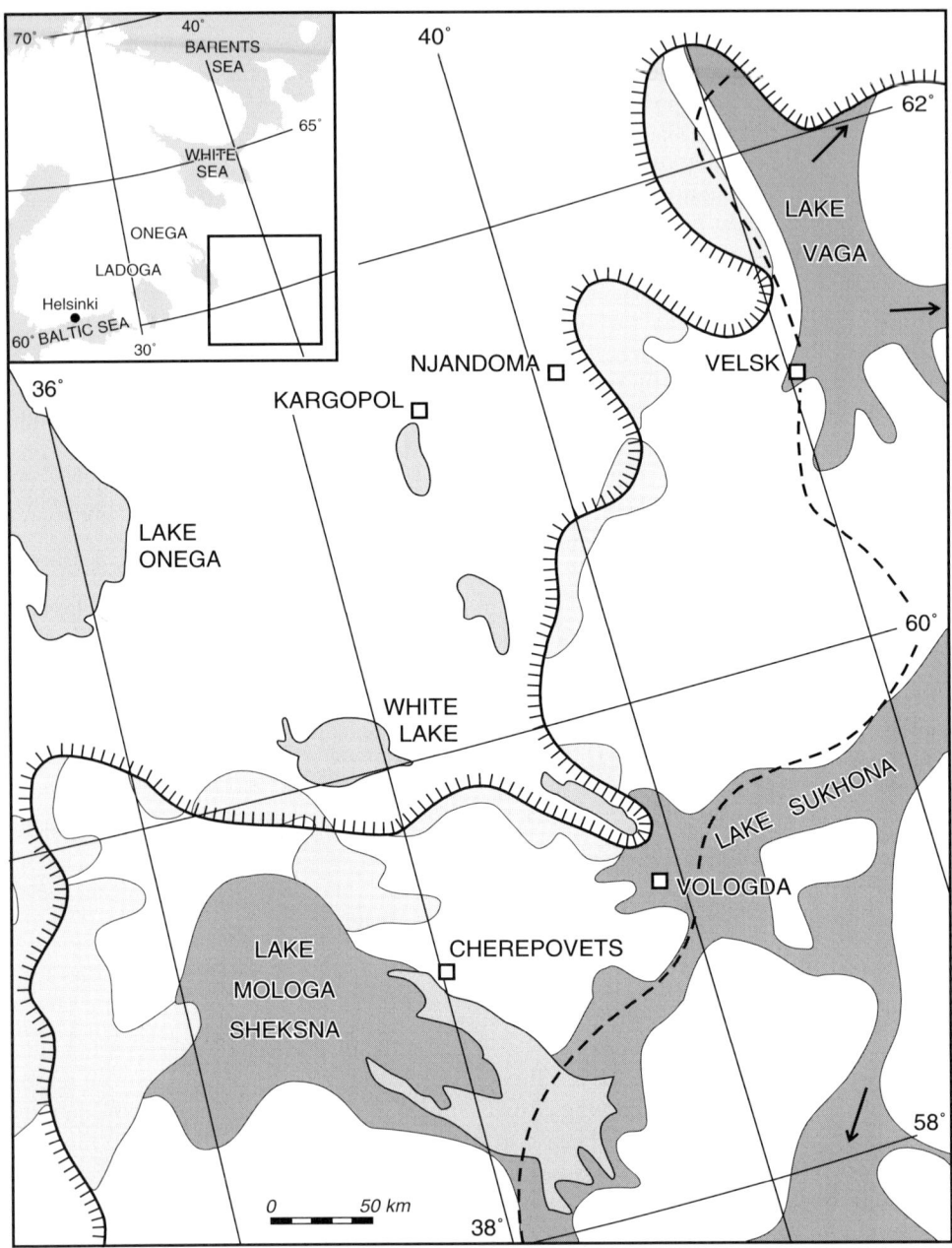

Figure 7. Ice-dammed lakes in the SE sector of the Scandinavian Ice Sheet, dated approximately at 18 kyr BP, drainage at that time was southwards to the Volga river system. From Lunkka et al. (2001).

Discussion

Although the terrestrial record of the Weichselian sediments is far from being continuous and the dating resolution is relatively poor, the behaviour of the Eurasian ice sheets is more complex than previously thought. The results suggest that the Kara and Barents ice-sheets were at their greatest during the Early (at around ca. 90,000 years ago) and Middle Weichselian (at around 60,000 years ago) while the Scandinavian ice sheet was greatest in the Late Weichselian. At present it is impossible to correlate the terrestrial record with the ice-core and marine records. However, based on stratigraphical evidence and optically stimulated luminescence (OSL) and ^{14}C AMS age determinations, it is possible for the first time to reconstruct a glaciation curve for the eastern flank of the Scandinavian ice sheet. The data from this area clearly indicate one major ice advance at ca. 60,000 and another minor advance at around 40,000 years ago, and the greatest advance during the LGM. These advances are comparable to those found in the western flank of the ice sheet (cf. Mangerud (1991)).

Contrary to previous assumptions, recent research suggests that Southern Finland and NW Russia were ice-free most of the Weichselain cold stage. During the Early Weichselian interstadials (Brørup and Odderade) pollen evidence indicates that even areas close to the glaciation centre and to the south-east of it were forested, in turn indicating July temperatures above 10 °C. Arctic tundra conditions persisted in SW Finland during the Early Weichselian stadials, unlike in NW Russia where the landscape was open birch forest with spruce, and with relatively high NAP values at times. During the ice-free periods of the Middle Weichselian, tundra type vegetation persisted in Finland while further south-east in Russia the landscape was forested. Finally, climate started to deteriorate at ca. 30,000 years ago, towards the Late Weichselian, and the climate was extremely cold at around 18,000 years ago. Palaeobotanical evidence, although limited, suggests that the winter temperatures were at least 20 °C and summer temperatures 5–11 °C lower than today in the NW Russian Plain.

Huge ice-lakes formed in front of the Kara and Barents ice-sheets at around 90,000 years ago and possibly also at around 60,000 years ago when the northbound Russian rivers were blocked. Similarly ice-lakes of smaller extent were formed in front of the Scandinavian ice-sheet at around 18,000 years ago. It has been discussed in several papers that large volumes of melt water into the Mid-Atlantic during the glacial phases and the outbursts of the Eurasian ice-dammed lakes into the North Atlantic Ocean would have had a major impact on ocean circulation and hence climate. In order to further test this hypothesis, it is vital to obtain more precise dates for these meltwater events, to map the drainage routes in more detail and to obtain detailed information on those events from marine cores from the Arctic Ocean.

Summary

The growth and decay patterns of the Eurasian ice sheets and their influence on paleo-hydrology and regional climate during the Weichselian cold stage have been intensively studied by several international projects during the past decade. Here we present a summary on the behaviour of the Eurasian Ice Sheets, and their influence on paleoenvironments and climate during the Valdai cold stage.

Recent investigations indicate that the Barents and Kara Sea ice sheets, centred on the present shelf areas of the Kara and Barents Seas, attained their greatest extent during the Early and Middle Weichselian. In contrast, the Scandinavian Ice Sheet was relatively small during the Early and Middle Weichselian but attained its greatest extent during the Late Weichselian. In this article, special reference is made to the south-eastern sector of the Scandinavian Ice Sheet in which area we have collected new data. As a result of this the history of glacier fluctuations during the Valdai glaciation can now be reconstructed in considerably more detail than hitherto.

Pollen evidence from Central Europe suggests that the Eemian climate was 2 °C–3 °C warmer than during the Holocene climate optimum and rather stable, although the climate became progressively more oceanic in character. During the Early and Middle Weichselian climate fluctuated between glacial and interglacial conditions and there was a general trend towards lower temperatures in the Middle Weichselian. Pollen evidence indicates that, during two distinct Early Weichselian interstadials (Brørup and Odderade), even areas close to the glaciation centre and to the south-east of it were forested, indicating July temperatures above 10 °C. During the Early Weichselian stadials arctic tundra conditions persisted in SW Finland. In NW Russia, the landscape was open birch forest with spruce, and with relatively high NAP values (40%–50%) at times. There is evidence of the Middle Weichselian ice-free periods in Finland that indicate tundra type vegetation while further south-east in Russia the landscape was forested. Climate started to deteriorate at ca. 30,000 years ago towards the Late Weichselian, and climate was extremely cold in NW Russia. Palaeobotanical evidence, although limited, suggests that the winter temperatures at around the last glacial maximum were at least 20 °C and summer temperatures 5–11 °C lower than today in the NW Russian plain, and precipitation was reduced to one third of present values.

Fluctuations of the Eurasian ice sheets caused drastic changes in palaeohydrology and climate during the Weichselian. The Kara and Barents ice sheets blocked the north-bound Russian rivers during the Early and Middle Weichselian, and huge ice-marginal lakes were formed. Similarly, ice lakes of smaller extent formed in front of the Scandinavian Ice Sheet at around 19,000–17,000 years ago.

References

Aalbersberg G. and Litt T. 1998. Multiproxy climate reconstructions for the Eemian and Early Weichselian. J. Quat. Sci. 13: 367–390.

Allen J.R.M., Brandt U., Brauer A., Hubberten H.W., Huntley B., Keller J., Kraml M., Mingram J., Negendank J.F.W., Nowaczyk N.R., Oberhänsli H., Watts W.A., Wulf S. and Zolitschka B. 1999. Evidence of rapid last glacial environmental fluctuations from Southern Europe. Nature 400: 740–743.

Alley R.B, Meese D.A., Shuman C.A., Gow A.J., Taylor K.C., Grootes P.M., White J.W.C., Ram M., Waddington E.D., Mayewski P.A. and Zielinski G.A. 1993. Abrupt increase in Greenland's snow accumulation at the end of the Younger Dryas event. Nature 362: 527–529.

Andersen B.G., Lundqvist J. and Saarnisto M. 1995. The Younger Dryas margin of the Scandinavian Ice Sheet — An Introduction. IGCP 253 — Termination of the Pleistocene — Final Report. Quat. Int. 28: 145–146.

Andrén T., Björck J. and Johnsen S. 1999. Correlation of the Swedish glacial varves with the Greenland (GRIP) oxygen isotope stratigraphy. J. Quat. Sci. 14: 361–371.

Arkhipov S.A., Ehlers J., Johnson R.G. and Wright H.E. Jr. 1995. Glacial drainage towards the Mediterranean during the Middle and Late Pleistocene. Boreas 24: 196–206.

Astakhov V., Svendsen J.I., Matiouchkov A., Maslenikova O. and Tveranger J. 1999. Marginal formations of the last Kara and Barents ice sheets in Northern Russia. Boreas 28: 23–45.

Baumann K.-H., Lackschewitz K., Mangerud J., Spielhagen R., Wolf-Welling T., Rüdiger H. and Kassens H. 1995. Reflection of Scandinavian ice sheet fluctuations in Norwegian Sea sediments during the past 150,000 years. Quat. Res. 43: 185–187.

Behre K.-E. 1989. Biostratigraphy of the last glacial period in Europe. Quat. Sci. Rev. 8: 25–44.

Berglund B.E. 1979. The deglaciation of Southern Sweden 13,500–10,000 BP. Boreas 8: 89–118.

Bond G. and Lotti R. 1995. Iceberg discharges into the North Atlantic on millennial time scales during the last glaciation. Science 267: 1005–1010.

Bond G., Showers W., Cheseby M., Lotti R., Almasi P., de Menocal P., Priore P., Cullen H., Hajdas I. and Bonani G. 1997. A pervasive millennial-scale cycle in North Atlantic Holocene and glacial climates. Science 278: 1257–1266.

Broecker W., Andree M., Wolfli W., Oeschger H., Bonani G., Kennett J. and Peteet D. 1988. The chronology of the last deglaciation; implications to the cause of the Younger Dryas event. Paleoceanography 3: 1–19.

Broecker W., Kennett J., Flower B., Teller J., Trumbore S., Bonani G. and Wolfli W. 1989. Routing of meltwater from the Laurentide Ice Sheet during the Younger Dryas cold episode. Nature 341: 318–320.

Caspers G. and Freund H. 2001. Vegetation and climate in the Early- and Pleni-Weichselian in Northern Central Europe. J. Quat. Sci. 16: 31–48.

Donner J. 1983. The Identification of Eemian Interglacial and Weichselian Interstadial Deposits in Finland. Ann. Acad. Scient. Fennicae A III 135, 38 pp.

Donner J.J. 1995. The Quaternary History of Scandinavia. Cambridge University Press, Cambridge, 200 pp.

Elina G. 1985. The history of vegetation in Eastern Karelia (USSR) during the Holocene. Aquilo, Ser. Botanica 22: 1–36.

Eriksson B. 1993. The Eemian Pollen Stratigraphy and Vegetational History in Ostrobothnia, Finland. Bull. Geol. Surv. Finland 352, Helsinki, 36 pp.

Gey V. and Malakhovsky D. 1998. On the age and extension of the maximum upper Pleistocene glaciation in western Vologda area. Izvestija Russkogo Geographicheskogo Obschestva 1: 43–53.

Gey V., Saarnisto M., Lunkka J.P. and Demidov I. 2001. Mikulino and Valdai palaeoenvironments in the Vologda area, NW Russia. Glob. Plan. Chan. 31: 347–366.

Goslar T., Arnold M., Bard E., Kuc T., Pazdur M.F., Ralska-Jasiewiczowa M., Rozanski K., Tisnerat N., Walanus A., Wicik B. and Wieckowski K. 1995. High concentration of atmospheric (super 14) C during the Younger Dryas cold episode. Nature 377: 414–417.

Grosswald M.G. 1980. Late Weichselian ice sheet of Northern Eurasia. Quat. Res. 13: 1–32.

Grosswald M.G. 1998. Late-Weichselian ice sheets in Arctic and Pacific Siberia. Quat. Int. 45/46: 3–18.

Helmens K., Räsänen M., Johanssson P., Jungner H. and Korjonen K. 2000. The Last Interglacial — Glacial cycle in NE Fennoscandia: a nearly continuous record from Sokli (Finnish Lapland). Quat. Sci. Rev. 19: 1605–1623.

Hirvas H. 1991. Pleistocene Stratigraphy of Finnish Lapland. Bull. Geol. Surv. Finland 354. 123 pp.

Houmark-Nielsen M. and Humlum O. 1994. High deglaciation rates in Denmark during the Late Weichselian — implications for the palaeoenvironment. Geografisk Tidsskrift, Danish Journal of Geography 94: 26–27.

Hütt G., Jungner H., Kujansuu R. and Saarnisto M. 1993. OSL and TL dating of buried podsols and overlying sands in Ostrobothnia, Western Finland. J. Quat. Sci. 8: 125–132.

Hyvärinen H. 1973. The deglaciation history of Eastern Fennoscandia — recent data from Finland. Boreas 2: 85–102.

Hyvärinen H. 1975. Absolute and Relative Pollen Diagrams from Northernmost Fennoscandia. Fennia 142, Helsinki, 22 pp.

Ikonen L. and Ekman I. 2001. Biostratigraphy of the Mikulino Interglacial Sediments in NW Russia: the Petrozavodsk Site and a Literature Review. Ann. Acad. Scient. Fennicae, Geologica-Geographica 161, Helsinki, 88 pp.

Johnsen S., Dahl-Jensen D., Gundestrup N., Steffensen J., Clausen H., Miller H., Masson-Delmotte V., Sveinbjörnsdottir A. and White J. 2001. Oxygen isotope and palaeotemperature records from six Greenland ice-core stations: Camp Century, Dye-3, GRIP, GISP 2, Renland and North GRIP. J. Quat. Sci. 16: 299–307.

Karabanov E., Prokopenko A., Williams D. and Colman S. 1998. Evidence from Lake Baikal for Siberian Glaciation during Oxygen-Isotobe Substage 5d. Quat. Res. 50: 46–55.

Kitagawa H. and van der Plicht J. 1998. A 40 000-year varve chronology from Lake Suigetsu, Japan: extension of the ^{14}C calibration curve. Radiocarbon 40: 505–515.

Kvasov D. 1979. The Late-Quaternary History of Large Lakes and Inland Seas of Eastern Europe. Ann. Acad. Sci. Fennicae A III 127, 71 pp.

Larsen E., Funder S. and Thiede J. (eds) 1999. Late Quaternary History of Northern Russia and Adjacent Shelves. Boreas 28, 242 pp.

Larsen E., Lyså A., Demidov I., Funder S., Houmark-Nielsen M., Kjær K.H. and Murray A.S. 1999. Age and extent of the Scandinavian ice sheet in Northwest Russia. Boreas 28: 115–132.

Litt T., Brauer A., Goslar T., Merkt J., Balaga K., Muller H., Ralska-Jasiewiczowa M., Stebish M. and Negendank J. 1999. Correlation and synchronisation of Lateglacial continental sequences in Northern Central Europe based on varved limnic sediments. Terra Nostra 88/19: 58–63.

Lundqvist J. 1992. Glacial stratigraphy in Sweden. Geol. Surv. Finland, Special Paper 15: 43–59.

Lundqvist J. 1994. The deglaciation. In: Fredén C. (ed.), Geology. National Atlas from Sweden, Stockholm, pp. 124–133.

Lundqvist J. and Saarnisto M. 1995. Summary of Project IGCP-253. IGCP 253 — Termination of the Pleistocene — Final Report. Quat. Int. 28: 9–18.

Lundqvist J. and Wohlfarth B. 2001. Timing and east-west correlation of South Swedish ice marginal lines during the Late Weichselian. Quat. Sci. Rev. 20: 1127–1148.

Lunkka J.P., Saarnisto M., Gey V., Demidov I. and Kiselova V. 2001. Extent and age of the Last Glacial Maximum in the south-eastern sector of the Scandinavian Ice Sheet. Glob. Plan. Chan. 31: 407–425.

Mangerud J. 1991. The Scandinavian Ice Sheet through the last interglacial/glacial cycle. In: Frenzel B. (ed.), Klimageschichtliche Probleme der Letzten 130,000 Jahre. G. Fisher, Stuttgart, New York, pp. 307–330.

Mangerud J., Astakhov V., Jakobsson M. and Svendsen J.I. 2001. Huge Ice-age lakes in Russia. J. Quat. Sci. 16: 773–777.

Mangerud J., Jansen E. and Landvik J.Y. 1996. Late Cenozoic history of the Scandinavian and Barents Sea ice sheets. Glob. Plan. Chan. 12: 11–26.

Mangerud J., Svendsen J.I. and Astakhov V.I. 1999. Age and extent of the Barents and Kara ice sheets in Northern Russia. Boreas 28: 46–80.

Maslenikova O. and Mangerud J. 2001. Where was the outlet of the ice-dammed Lake Komi, Northern Russia? Glob. Plan. Chan. 31: 337–345.

Murray A.S. and Wintle A.G. 1999. Luminescence dating of quartz using an improved single aliquot regenerative-dose protocol. Radiation Measurements 32: 57–73.

Nenonen K. 1995. Pleistocene Stratigraphy and Reference Sections in Southern and Western Finland. Geological Survey of Finland, Regional Office for Mid-Finland, Helsinki, 94 pp.

Olsen L. 1997. Rapid shifts in glacial extension characterise a new conceptual model for glacial variations during the Mid and Late Weichselian in Norway. NGU-Bulletin 433: 54–55.

Pavlov P., Svendsen J.I. and Indrelid S. 2001. Human presence in the European Arctic nearly 40,000 years ago. Nature 413: 64–67.

Peltoniemi H., Eriksson B., Grönlund T. and Saarnisto M. 1989. Marjamurto, an interstadial site in a till covered esker area of central Ostrobothnia, Western Finland. Bull. Geol. Soc. Finland 43: 209–237.

Péwé T.L., Church R.E. and Andersen M.J. 1969. Origin and Paleoclimatic Significance of Large Scale Patterned Ground in the Donelly Dome Area, Alaska. Geol. Soc. Am., Bolder, Colorado, USA, Special Paper 103, 87 pp.

Polyak L., Lehman S., Gataullin V. and Jull A.J. 1995. Two-step deglaciation southeastern Barents Sea. Geology 23: 567–571.

Raukas A. 1991. Eemian interglacial record in the northwestern European part of the Soviet Union. Quat. Int. 10-12: 183–189.

Robertsson A.-M. 1997. Reinvestigation of the interglacial pollen flora at Leveäniemi, Swedish Lapland. Boreas 26: 81–89.

Saarnisto M., Eriksson B. and Hirvas H. 1999. Tepsankumpu revisited -pollen evidence of stable Eemian climates in Finnish Lapland. Boreas 28: 12–22.

Saarnisto M. and Saarinen T. 2001. Deglaciation chronology of the Scandinavian ice sheet from East of Lake Onega basin to the Salpausselkä end moraines. Glob. Plan. Chan. 31: 387–405.

Saarnisto M. and Salonen V.P. 1995. Glacial history of Finland. In: Ehlers J., Kozarski S. and Gibbard P. (eds), Glacial Deposits in North-East Europe. A.A. Balkema, Rotterdam, pp. 3–10.

Sejrup H.P., Haflidason H., Aarseth I., Forsberg C.F., King E., Long D. and Rokoengen K. 1994. Late Weichselian glaciation history of the northern North Sea. Boreas 23: 1–13.

Semenenko L.T., Aleshinskaya Z.V., Arslanov Kh.A., Valuyeva M.N. and Krasnovskaya F.I. 1981. Opprrnyi razrez verkhnego pleistotsena u fabriki 'Pervoe Maya' Dmitrovskogo raiona Moskovskoi oblasti (otlozheniya drevnego Tatischevskogo ozera). Upper Plestocene Section near 'First of May' Factory in Dmitrov district, Moscow region (sediments from the former Tatischevskoe Lake). XI INQUA Congress, Moscow 1982, p. 121–135.

Sher A. 1997. Late-Quaternary extinction of large mammals in Northern Eurasia: A new look at the Siberian contribution. In: Huntley B., Wolfgang C., Morgan A.V., Prentice H.C. and Allen J.R.M. (eds), Past and Future Rapid Environmental Changes: The Spatial and Evolutionary Responses of Terrestrial Biota. NATO ASI Series, Vol. 147. Springer-Verlag, Berlin-Heidelberg, pp. 319–339.

Svendsen J.I., Astakhov V.I., Bolshiyanov D.Yu., Demidov I., Dowdeswell J.A., Gataullin V., Hjort C., Hubberten H.W., Larsen E., Mangerud J., Melles M., Möller P., Saarnisto M. and Siegert M.J. 1999. Maximum extent of the Eurasian ice sheets in the Barents and Kara Sea region during the Weichselian. Boreas 28: 234–242.

Tarasov P.E., Peyron O., Guiot J., Brewer S., Volokova V.S., Bezusko L.G., Dorofeyuk N.I., Kvavadze E.V., Osipova I.M. and Panova N.K. 1999. Last Glacial Maxium climate of the former Soviet Union and Mongolia reconstructed from pollen and plant macrofossil data. Clim. Dyn. 15: 227–240.

Thiede J., Bauch H., Hjort C. and Mangerud J. 2001. Editorial. The late Quaternary stratigraphy and environments of Northern Eurasia and the adjacent Arctic seas — new contributions from QUEEN. Glob. Plan. Chan. 31: vii-x.

Ukkonen P., Lunkka J.P., Jungner H. and Donner J. 1999. New radiocarbon dates from Finnish mammoths indicating large ice-free areas in Fennoscandia during the Middle Weichselian. J. Quat. Sci.14: 711–714.

Waelbroek C., Labeyrie L., Michel E., Duplessy J., McManus J., Lambeck K., Balbon E. and Labracherie M. 2002. Sea-level and deep water temperature changes derived from benthic *Foraminifera* isotopic records. Quat. Sci. Rev. 21: 295–306.

Vandenberghe J. and Pissart A. 1993. Permafrost changes in Europe during the Last Glacial. Permafrost and periglacial processes 4: 121–135.

Zagwijn W.H. 1989. Vegetation and climate during warmer intervals of the Late Pleistocene of Western and Central Europe. Quat. Int. 3/4: 57–67.

Zagwijn W.H. 1996. An analysis of Eemian climate in Western and Central Europe. Quat. Sci. Rev. 15: 451–469.

Zolitschka B. and Negendank J.F.W. 1999. High-resolution records from European lakes. Quat. Sci. Rev. 18: 885–888.

22. HOLOCENE CLIMATE DYNAMICS IN FENNOSCANDIA AND THE NORTH ATLANTIC

IAN SNOWBALL (ian.snowball@geol.lu.se)
Geobiosphere Science Centre
Quaternary Sciences
Lund University
Sölvegatan 12, SE-223 62 Lund
Sweden

ATTE KORHOLA (atte.korhola@helsinki.fi)
Department of Ecology and Systematics
Division of Hydrobiology
University of Helsinki
P.O. Box 17 (Arkadiankatu 7), FIN-0014
Finland

KEITH R. BRIFFA (k.briffa@uea.ac.uk)
Climatic Research Unit
University of East Anglia
Norwich, NR4 7TJ
UK

NALAN KOÇ (nalan.koc@npolar.no)
Norwegian Polar Institute
N-9296 Tromsø
Norway

Keywords: Holocene, Climate dynamics, North Atlantic, Fennoscandia, Thermohaline circulation

Introduction

Set against the high amplitude climatic variations that characterised the Pleistocene, the majority of the Holocene epoch was once considered to be comparatively stable. Indeed, observed post-19th century "global" warming is frequently thought of as an anomaly. However, the Holocene in the high latitude terrestrial environment of the PAGES PEP III transect has not been climatically stable, as was indicated by 19th century studies of Scandinavian and Scottish peat bogs by Blytt (1876) and Sernander (1908). Taken from Andersson's early 20th century review of Late-Quaternary Swedish climate research, the

R. W. Battarbee et al. (eds) 2004. *Past Climate Variability through Europe and Africa.*
Springer, Dordrecht, The Netherlands.

following quote puts into perspective the advances we have made during the last nine decades towards understanding the complexities of Holocene climate dynamics in the northern PEP III region.

"…in the whole of Scandinavia, from the most southerly to the most northerly parts, there are found, on land and in the sea, traces of a warmer period in post-glacial time during which the time of vegetation was considerably longer than now, and with about 2.5 °C mean temperature higher, while the winters were presumably about the same as now or inconsiderably warmer." Andersson (1909, p. 65).

Andersson's (1909) summary was made without the help of modern geochronology and the modern statistical techniques that are now applied to quantify past geological data in terms of meteorological parameters. However, it was uncannily akin to more recently acquired quantified estimates of climate change, which we review here.

Over the last nine decades the northern geographical region that includes Fennoscandia and the North Atlantic has probably been the most intensively studied domain of all the PEP transects. It is arguably unrivalled in its collection of natural "palaeo-climate" archives and access to meteorological observations and historical annals. As will be discussed more fully in the following sections, ideal reconstructions of climate dynamics require that *rates of change* are established, which in turn demands that the most accurate and precise chronologies provided by nature are exploited. In addition to ice-cores, this region contains an abundance of alternative annually-resolved archives of environmental change in the form of tree-ring series and varved sediment sequences. Example archives, which are discussed later, are shown in Figure 1. Within the confines of these pages this review cannot refer to all the work undertaken by the large number of scholars who have improved our understanding of Holocene climate change. Huntley et al. (2002) provide an extensive review of the development of North-West Europe during the Holocene. In this review, we focus on a selected range of reconstructions obtained from numerical techniques and approaches, which have recently produced quantitative palaeoclimatic data with a relatively high precision (e.g., Koç et al. (1993), Dahl and Nesje (1996), Rosén et al. (2001), Seppä and Birks (2001), Bigler et al. (2002), Korhola et al. (2002)). The added-value and limitations of these quantitative reconstructions are discussed in the context of potential dynamic links between long-term orbital forcing, the circulation of the North Atlantic and the response of terrestrial ecosystems on neighbouring landmasses.

The Holocene epoch was essentially subdivided into five chronozones by Mangerud et al. (1974) with the transition between the Younger Dryas (YD) and the Preboreal placed at 10,000 uncal. ^{14}C yr BP. Björck et al. (1998) proposed a scheme where the calendar year dated GRIP Greenland ice-core record formed an event stratigraphy for the North Atlantic region. In this scheme the boundary between the YD (alternatively denoted as Greenland Stadial 1 or GS-1) and the Holocene is placed at 11,550 cal BP. Ages are subsequently expressed in calendar years before present (cal BP), where the "present year" is classified as AD 1950.

Climate forcing

The Earth's orbital path promoted high summer solar insolation in the northern hemisphere (Fig. 2) during the early Holocene. The climate of the northern PEP III region was, however,

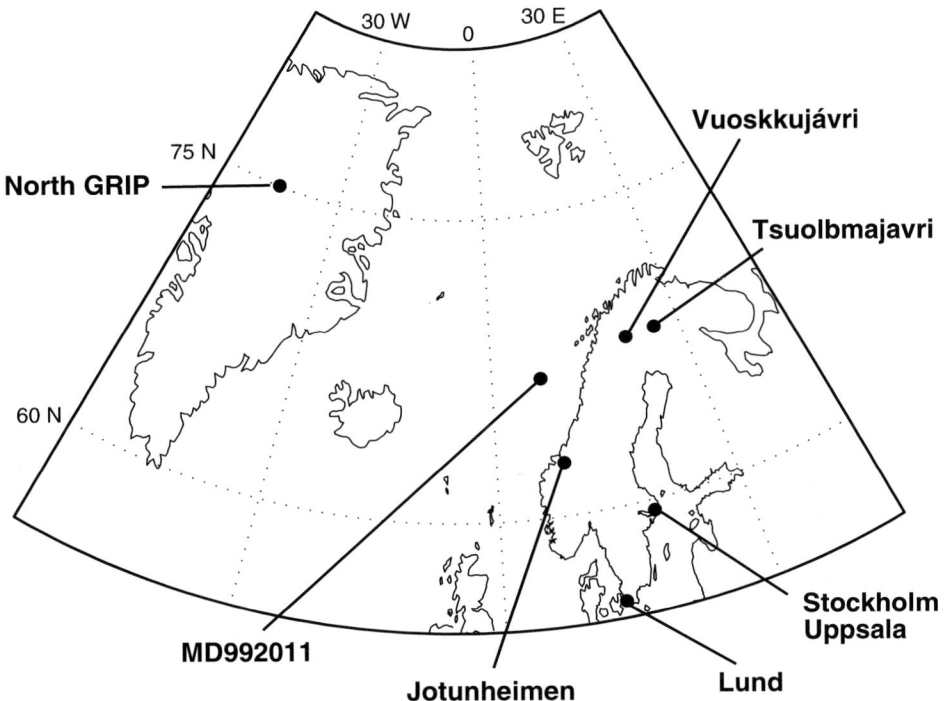

Figure 1. Map of the PEP III high latitude region and selected archives of climate change discussed in the text. Lund, Stockholm and Uppsala represent instrumental data-sets.

undoubtedly influenced by the hysteresis associated with the delayed melting of the Laurentide and Fennoscandian ice-sheets, and pulses of meltwater discharge caused extremely rapid changes in the physical characteristics of the North Atlantic and the temperature of overlying air masses. It has long been assumed that the stable oxygen isotope composition of precipitation over Greenland reflects the temperature of cloud vapour and, therefore, that $\delta^{18}O$ measurements of ice-cores reveal palaeo-temperatures (Johnsen et al. 2001). The assumption that first order variability in stable isotope measurements is temperature related is to a certain degree supported by the similar trend of directly measured borehole temperatures (Dahl-Jensen et al. 1998). Figure 2 shows the recently acquired oxygen isotope record from the NorthGRIP core (Johnsen et al. 2001). The ca. 3500 year period between the start of the Holocene and 8000 cal BP is characterised by a general warming trend, although there are significant deviations from this trend, which may be related to the early Holocene ice-rafting events (Bond et al. 1997). These cold reversals provide testimony to the non-linear nature of the general climate system, although it should also be noted that the general warming trend ground to a halt at ca. 9500 cal BP, when the summer insolation in the northern hemisphere reached its peak and started to decline. A distinct increase in sea salt and terrestrial dust concentrations, plus changes in the chemical composition of soluble impurities in the ice during the period 8800 to 7800 cal BP suggest an expansion of the north polar vortex or intensification of meridional airflow, and thus climatic cooling on Greenland (O'Brien et al. 1995). A long-term cooling trend since ca. 8000 cal BP is

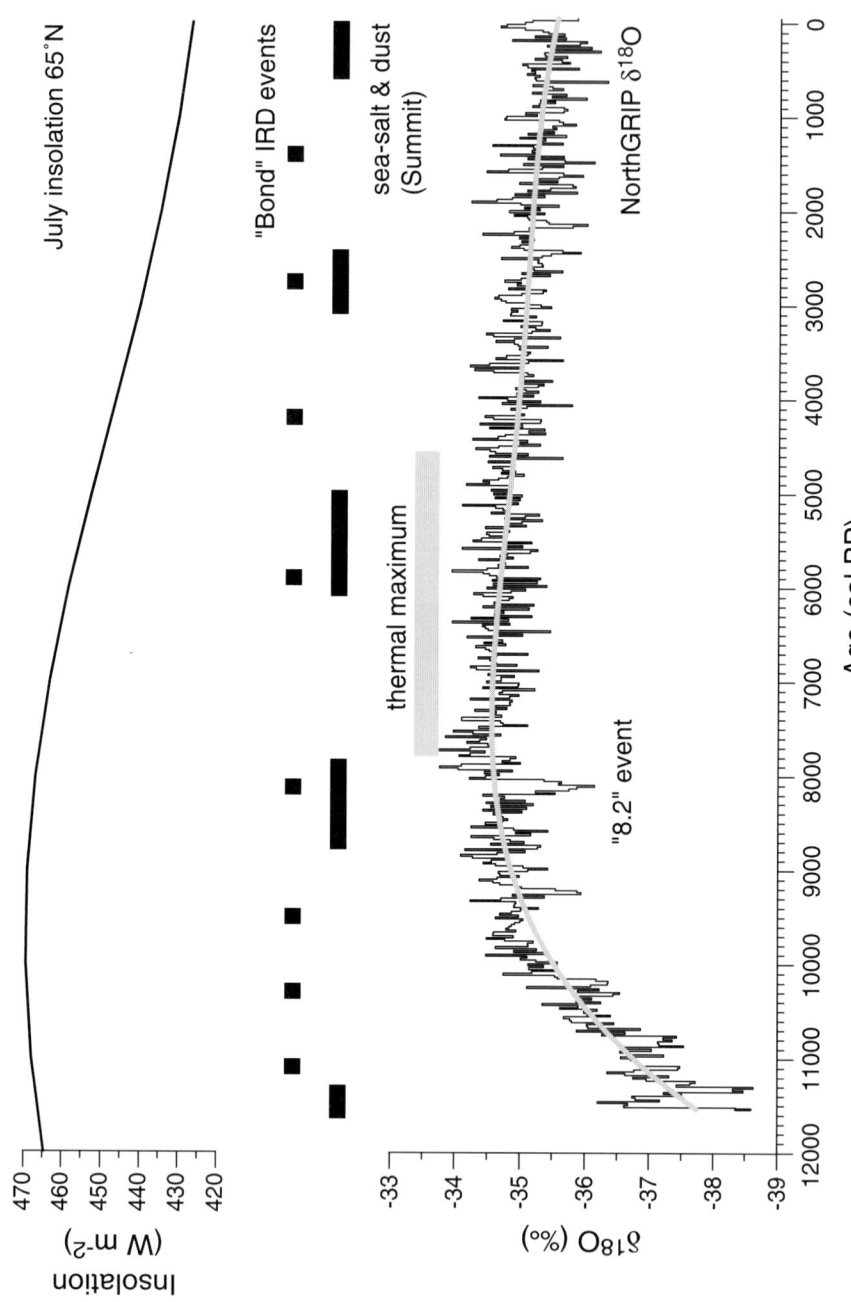

Figure 2. NorthGRIP δ^{18}O (Johnsen et al. 2001) and July insolation values at 65°N (Berger and Loutre 1991), which demonstrate the initial post-glacial warming and a thermal maximum between 8000 and 4000 cal BP. Higher-frequency cooling events detected in marine sediments (from Bond et al. (1997)) that have no direct correlation to the insolation flux are marked by filled squares. Filled bars show periods of increased dust and sea-salts in the Greenland ice-cores, which are interpreted as evidence of expansions of the northern polar vortex, or increased meridional airflow (from O'Brien et al. (1995)).

characteristic of the NorthGRIP ice-core and likely reflects the long term energy balance associated with orbital forcing. However, the earlier obtained Summit GRIP ice-core lacks a distinct Holocene thermal maximum (HTM), and significant differences in the stable isotope composition between the NorthGRIP and GRIP records may also be interpreted in terms of atmospheric circulation patterns (Johnsen et al. 2001).

Several other lines of evidence also suggest that both the long-term and short-term physical states of the North Atlantic, the Nordic seas and Greenland during the Holocene have been determined, directly and indirectly, by the strength of the northern hemisphere summer insolation (Koç et al. 1993; Koç and Jansen 2002). On the other hand, there is also evidence that relatively high-frequency variations in solar activity have forced climatic changes in the high northern latitudes, but only through amplification of their effects via changes in North Atlantic circulation (Mikalsen 1999; Bond et al. 2001; Björck et al. 2001). These recent investigations, plus those referred to by Huntley et al. (2002) suggest that a variety of forcing mechanisms, which include (at least) solar activity, ocean circulation, the atmospheric concentrations of greenhouse gases and aerosols emitted as a consequence of volcanic eruptions and even impacts with cometary debris (Baillie 1999) have combined to mould Holocene climate development in the northern PEP III region. This plethora of potential climate forcings leads one to realise that no single factor dominated climate change during the foregone part of the Holocene, as must be the case at present and in the future.

The northern instrumental climate record

Not only do instrumental records provide precisely dated and quantified climatic data; they are also required for the calibration of proxy-derived parameters. In addition to excellent geographical coverage, the instrumental data-base for Fennoscandia and Western Russia contains a number of excellent long instrumental climate records, such as the daily temperature series for Uppsala (Bergström and Moberg 2002) and Stockholm (Moberg et al. 2002) and the precipitation series from Uppsala and Lund (Tabony 1981), all of which reach back into the 18th century. These series are reviewed in the context of other long European records in Jones (2001). For those who use these records as calibration series it is fortunate that temperature variability across Northern Europe is highly coherent on inter-annual and decadal timescales over distances of many hundreds of kilometres (Briffa and Jones 1993; Jones and Briffa 1996). It is valid, therefore, to represent the large-scale thermal history of the northern PEP III region by amalgamating a network of local records. Figure 3 illustrates just such an amalgamation: the average of monthly mean temperatures at Uppsala, Stockholm, St. Petersburg, Trondheim, Vardo and Archangel'sk (Jones et al. 2002), here shown as separate annual, summer (JJA) and winter (DJF) series.

A positive trend over the 250 years is clearly evident in the mean annual data, as is the general warmth over most of the 20th century. Much of the warming appears as a relatively rapid change between about 1900 and 1920. A particularly pronounced increase in mean annual temperature, of about 1.5 °C, occurred in Swedish and Finnish Lapland between the late 1880s and the mid-1930s (Eriksson and Alexandersson 1990; Sorvari et al. 2002). The seasonal data, however, indicate that the winter warming was more gradual and began in the later half of the 19th century. This warming trend is muted in the annual data by a contemporaneous run of cool summers in the late 19th and early 20th centuries and the

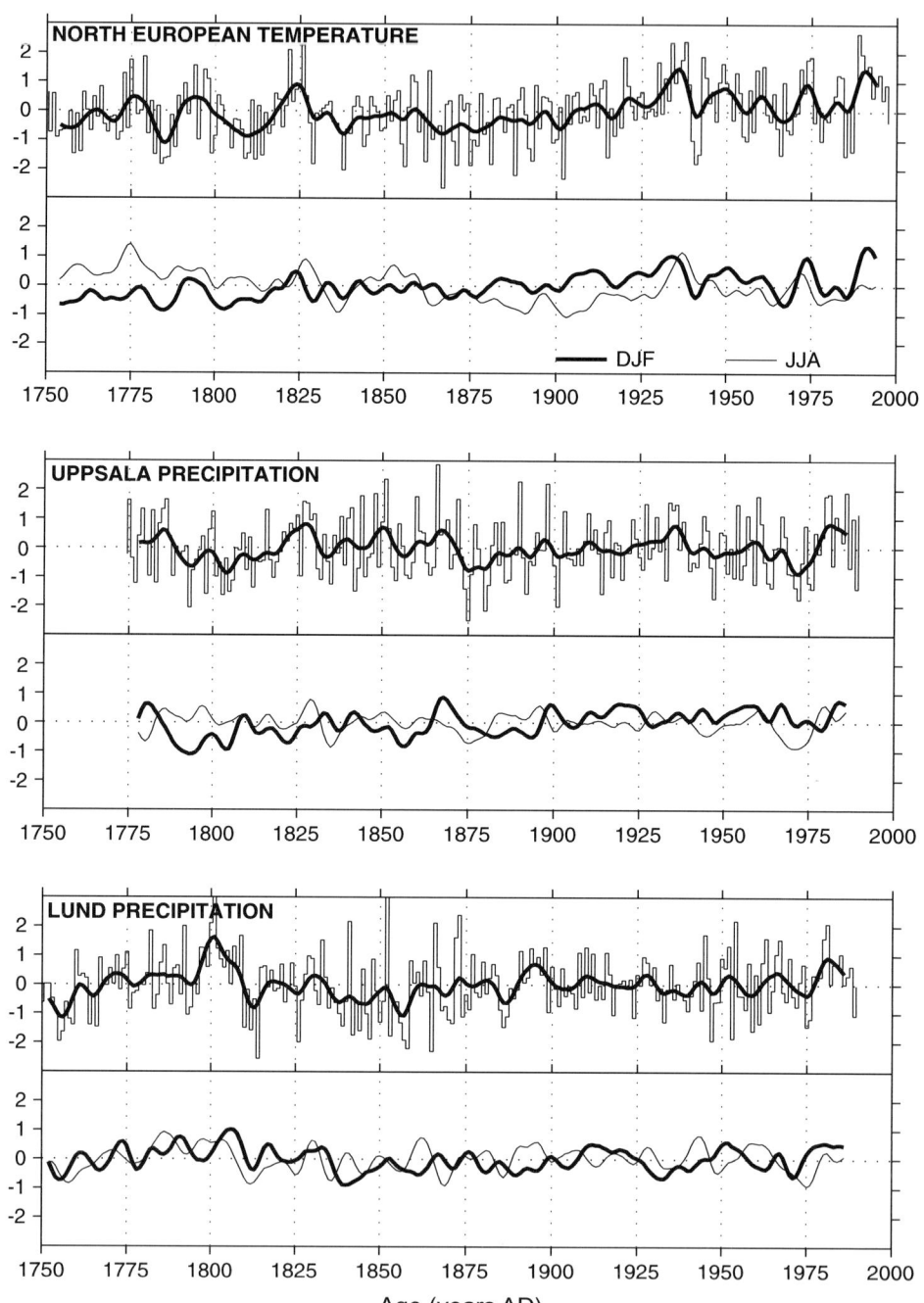

Figure 3. Representative temperature and precipitation records for the last 250 years from North West Europe plotted as standardised departures from the common base period (AD 1961-90). Northern European temperatures are the average of data from six northern stations, the mean annual data shown as a histogram with 10-year smoothing superimposed and separate summer (JJA) and winter (DJF) decadally-smoothed data shown below.

unusual warmth since the early 1990s is predominantly a winter phenomenon. What is also remarkable is that the summers were apparently as warm, or warmer, during the second half of the 18th century as in the 20th century, with the possible exception of the very warm 1930s.

However, a warm bias in the early Uppsala and Stockholm temperature records, due to thermometer exposure problems before 1860 (Moberg et al. 2002) almost certainly means that the summer 'Northern European' series may be several tenths of a degree Celsius too warm at this time. The correlation between the North European temperature record and the equivalent seasonal averages of the mean northern hemisphere (land and marine) data (Jones et al. 1999) averages are low and insignificant at inter-annual and decadal timescales (Briffa and Jones 1993), but there is much gross similarity in the multi-decadal trends. The northern hemisphere mean data show very similar trends in the summer and annual average data: relative cooling from 1860–1910; warming from 1910–1935; stable during 1935–1960; cooling to 1975 and subsequent warming. These trends are qualitatively similar to the post-1860 annual and summer North European trends in Figure 3: cooling up until 1905; warming until 1935; cooling until 1965. The strong post-1980 warming is not apparent in the North European summer record, but is identifiable to some extent in the winter and spring data. The winter (and annual) North European temperature, however, warmed consistently between 1860 and 1930, and the 1930s, in both summer and winter, were as warm as the last two decades. According to Tuomenvirta et al. (2000), however, the 1988–95 warm winters surpassed even the warm 1930s. The spring has warmed in the Arctic part of Europe throughout the twentieth century (Tuomenvirta et al. 2000). This increase is especially important as it forms part of an even longer warming trend in springtime, with a steady rise in spring temperatures since 1800 at an average rate of 0.005 °C yr^{-1} (Sorvari et al. 2002). The increase in spring temperatures is also manifested by earlier ice cover break-up dates and later freeze dates in the long data series of lakes and rivers in Northern Europe (Magnuson et al. 2000; Sorvari et al. 2002). Despite the focus on contemporary "global warming" that does apply to the northern hemisphere as a whole, it must be stressed that Northern Europe clearly cooled during the 1990's.

The northern record also displays a change in character between about 1860 and 1920, at a time of relatively cool conditions and distinct inter-annual fluctuations. Before and following this period, the variance was more pronounced at the longer decadal-timescale, particularly in winter. It is tempting to equate this with a similar pattern of changing persistence in the variability of the winter North Atlantic Oscillation that represents the strength of westerly air flow across the eastern North Atlantic (Visbeck 2002). However, while it is known that southern Scandinavian winter (but not summer) temperatures are strongly correlated with the local zonal geostrophic wind there is no significant correlation between the North European winter temperature record and West European zonal wind strength, on either inter-annual or decadal timescales (Jones et al. 2002).

Figure 3 also shows two examples of long Fennoscandian precipitation records, from Uppsala and Lund. Even allowing for some uncertainty in these records, especially in the 1870's, there are no strong overall trends in these data and little correlation between them. Seasonal differences at each location are essentially random, but stations do show mutually high levels during the last 20 years. As shown by studies of precipitation patterns in Southern Sweden (Linderson 2002) there is little co-variance of precipitation events over large distances and as such, reconstructions of high-frequency precipitation events based on geographically restricted proxy data may have only local or sub-regional significance. Nevertheless, according to Parry (2000) precipitation in Northern Europe has increased by 10–40% during the last 100 years and it can be anticipated that centennial to multi-millennial scale changes in precipitation amounts can be reconstructed from suitable proxies (e.g., Sander et al. (2002)).

Dendroclimatic studies in Northwest Eurasia — the last two millennia

Dendroclimatic studies in Northern Europe date back to the early part of the 20th century (e.g., Erlandsson (1936), Mikola (1956), Høeg (1956)). Major advances, however, in climate reconstruction using tree-ring data stem mainly from the 1960s (e.g., Sirén (1961)) with a major expansion of work occurring from the 1980s onwards (Bartholin and Karlén 1983; Aniol and Eckstein 1984; Briffa et al. 1988; Schweingruber et al. 1988).

As with the instrumental data, tree-ring derived records display strong coherence across wide areas of northern Fennoscandia and similar tree-growth responses to summer temperature forcing have been demonstrated in numerous studies (Briffa et al. 1990; Kalela-Brundin 1998; Lindholm et al. 2000; Lindholm and Eronen 2000; Kirchhefer 2001). The particular strength of dendrochronology lies in its representation of climatic variability at sub-centennial timescales: no other, even ice-core, data can provide evidence of high-frequency, even seasonally specific, events with such accurate dating control, though as with all proxies the interpretation in terms of forcing, magnitude and spatial representativeness must be viewed cautiously. Even extremely narrow or very low-density tree rings, evident at particular sites, may not represent climate extremes experienced over wide regions, but with shrewd data integration across wider areas, strong and objective evidence of major climate anomalies can be provided. These anomalies can be linked with possible forcing mechanisms such as the likely volcanic cooling in years such as AD 1641, 1601 and 536. The last of these is particularly notable as it marks the start of a period of continental-wide cooling across Northern Eurasia that lasted for over a decade and that may have had profound societal consequences across a very large area of the globe (Baillie 1999; Keys 1999). Though presently based only on one localised tree-ring collection (from the Yamal peninsula), recent work by Hantemirov et al. (2000) has shown that anomalous wood anatomical features (especially frost rings) may indicate abrupt freezing in cool springs or late summers. These provide new evidence for very cold events to which can be assigned precise dates, such as AD 1109, 1278, 1466, 1601, 1783 and 1882. Detailed descriptions of the general high-frequency (i.e., inter-annual to multi-decadal timescale) variability in summer temperatures in different parts of the northern PEP III region are to be found in the various papers cited above. However, most focus on recent centuries only.

Kalela-Brundin (1998) constructed a 600-year pine chronology representing summer temperature change over much of South-Eastern Norway, though with reduced reliability prior to 1500. She provides a comparison of decadal timescale temperature variability in this region with other, more northerly, reconstructed data. Relative warmth occurred in the 1750s and 1760s; at around 1830; during the 1890s; during the 1930s and 1940s; and in recent decades. The first half of 16th century and the very early 15th century were also likely to have been warm. Noticeably cool summers clearly occurred between 1780 and 1820; at around 1710 and 1740; in the early decade of the 20th century; and probably at about 1380 and between 1450 and 1470. More northern records (e.g., Lindholm and Eronen (2000), Kirchhefer

(2001)) show some subtle differences across northern Fennoscandia, but there is widespread evidence of relatively cool summers in the 540s and 550s; the early decades of the 17th century; the 1230s and 1240s; and the 1460s and 1470s. Relative warmth is registered in the 1920s and 1930s; 1650s and 1660s (in Northern Finland); and in the 1420s and 1430s.

A representative summary of current tree-ring evidence for longer timescale (i.e., multi-decadal to century) changes in past temperatures in the Northern European part of the PEP III transect is provided in Figure 4b. This figure focuses on larger-scale changes in the western (Scandinavia) and more eastern (West Siberia) areas of the northern PEP III region during the last two millennia. Wide regional evidence has been extracted from networks of tree-ring density chronologies in each area (Schweingruber et al. 1991; Vaganov et al. 1996; Briffa et al. 2001; 2002), but these are of restricted length, whereas more local and much more sparse regional ring-width data extend back much further (Briffa 2000). Comparison of these independent records demonstrates the general representativeness of the extended Northwest Eurasian curve (Fig. 4b) and demonstrates the extent to which the 20th century was warm, probably the warmest century of the last two millennia. The extended record, however, displays a multi-centennial oscillation with other periods of warmth clearly evident in the late 10th and early 11th centuries and in the 15th and early 1st centuries AD. The cool summers of the 17th and 18th centuries are also clearly expressed, but again they are preceded by other, perhaps less persistent, periods of cool summers in the early 4th, early 6th, mid 7th, and early 9th centuries. The very considerable variability in northern summers known to occur on inter-annual and decadal time-scales is, therefore, superimposed on an underlying periodic variability with an oscillation occurring on a timescale of several centuries (about 300 years in the 1st millennium and 500 in the 2nd millennium AD).

Ongoing work continues to extend the length of the tree-ring records in a number of locations in Northern Europe and Siberia. There are now continuous chronologies that combine local living, historical and older subfossil records together stretching over more than 7 millennia in Sweden (Grudd et al. 2002), Finland (Eronen et al. 2002; Helama et al. 2002) and West and Central Siberia (Hantemirov and Shiyatov 2002; Naurzbaev et al. 2002). It is notable that only two periods in the northern Swedish and Finnish series, at 600 and 250 BC, appear to have exceeded the general level of warmth attained during the 20th century (though both are associated with large uncertainty because of the relative scarcity of data that go to make up the series at these times).

Dendroclimatology has, therefore, provided evidence for very variable high-latitude climate with considerable year-to-year, decadal, and century-time-scale changes. All the chronologies, however, require the incorporation of further sample data to enhance the robustness of the longer-term statistical signals expressed in them, particularly changes on time-scales longer than a century. None of the long tree-ring data-series indicate any distinct multi-millennial trends in climate parameters such as those shown in tree-line data (e.g., Hantemirov and Shiyatov (2002)) and other proxy climate data obtained for the northern PEP III region, which we now turn to.

Glacier dynamics: has the North always been frozen?

Any review of Fennoscandian Holocene climate development must mention work undertaken to reconstruct glacier variations (reviewed by Karlén (1988)). Karlén's pioneering study of Swedish glacier and tree-limit variations in Swedish Lapland showed that the

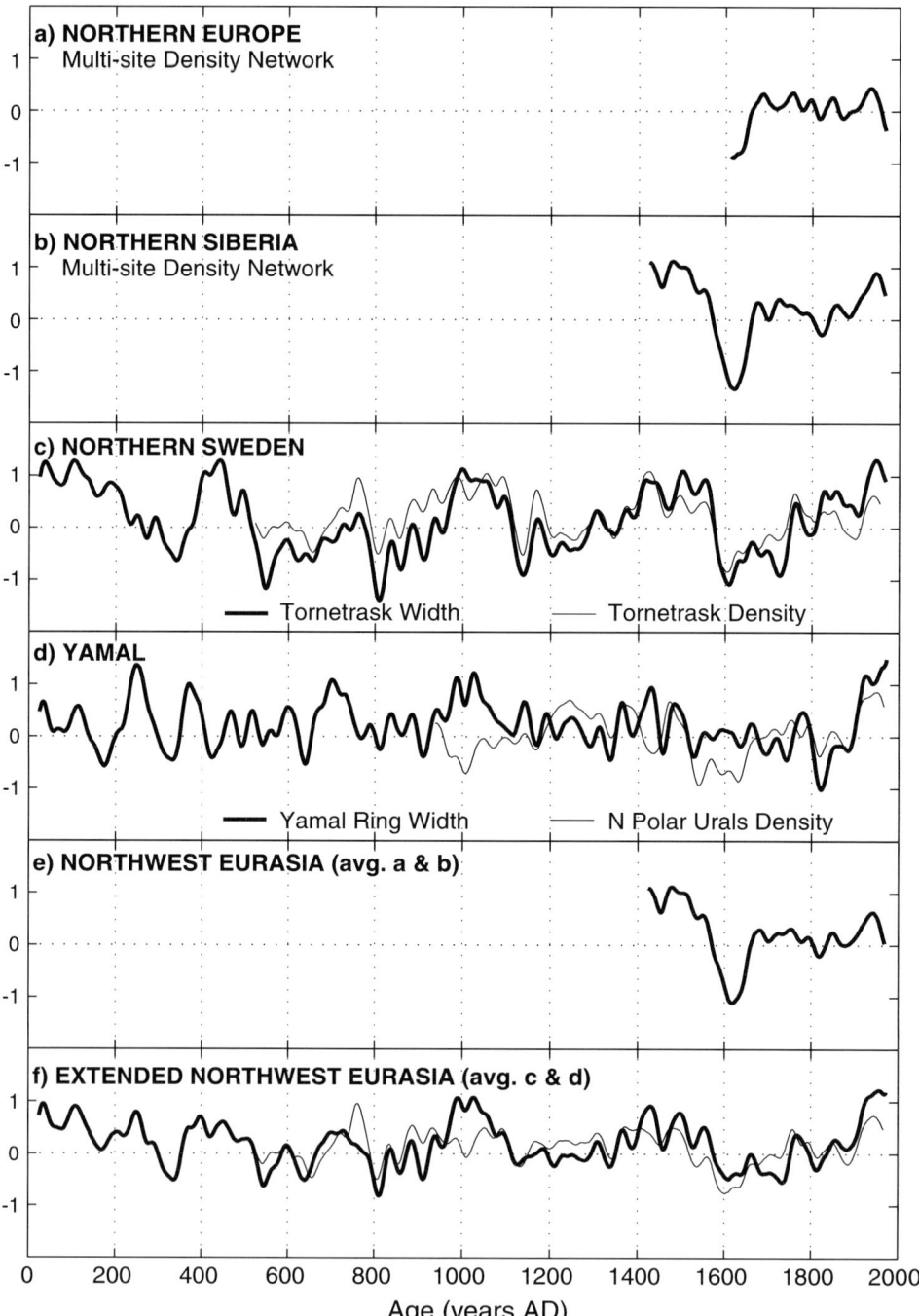

Figure 4. Selected North European tree-ring curves representing summer season temperatures in Northern Scandinavia and Western Russia over the last 2000 years. The various series are all plotted here as decadally-smooth series and as standardised departures from the common base period.

Holocene was characterised by high-frequency fluctuations superimposed upon a general picture of initial warming and reversion to cooling (Karlén 1976). Of course, there exists a considerable climatic contrast between the locations of glaciers situated near the maritime Norwegian coast and those on the more continental, eastwards, side of the Scandinavian mountain chain. This contrast causes the mass balance of continental glaciers, such as those initially studied by Karlén, to be determined by summer temperature; while the mass balance of maritime glaciers in Norway is largely driven by winter precipitation. A number of further studies have now produced a clear picture of glacier variations in Fennoscandia (e.g., Dahl and Nesje (1994), Nesje et al. (1994), Snowball and Sandgren (1996), Nesje et al. (2001)). An extensive study of glacio-lacustrine sediment cores was undertaken by Matthews et al. (2000) to reconstruct Holocene glacier variations in Central Jotunheimen, Southern Norway. Their reconstruction is presented in Figure 5 as a representative example of the climatic signals revealed by other studies of glacier variations in Fennoscandia. In agreement with the Greenland ice-core records, these studies indicate an unstable early Holocene. Multiple early Holocene advances and retreats occurred prior to ca. 8000 cal BP, which was followed by the rapid retreat of glaciers and the complete disappearance of many during a long term mid-Holocene that was characterised by markedly low glacial activity. Significant glacier advances and reformations, which form the so-called Neoglaciation, began as early as 5000 cal BP and culminated in the major advances of the last few centuries (commonly referred to as the Little Ice Age, LIA).

Quantitative reconstructions of climate parameters

Instrumental records are not long enough to capture the full range of climate variability and, therefore, the geological record must be quantified in terms of physical climate parameters. The northern PEP III region has effectively formed a testing ground for quantitative palaeo-environmental reconstructions (Birks 1995) and palaeo-climatology, which are now applied to both marine and terrestrial ecosystems. Although transfer functions have been constructed primarily for the calibration and quantification of biological data-sets such as tree-rings (reviewed by Briffa (2000)) and microfossil assemblages (see Table 1 for a list of recently constructed terrestrial microfossil-based transfer functions) the approach has also been used to calibrate physical data-sets, e.g., stable isotope ratios (Lauritzen and Lundberg 1999), glacier mass balance (Dahl and Nesje 1996) and stream discharge (Sander et al. 2002). Despite reservations concerned with non-analogue situations, quantified reconstructions are likely to pave the way for inter-regional comparisons of palaeo-climates and, eventually, to understanding long-term transfers of energy within the Earth system. Figure 6 shows a selection of quantitative temperature reconstructions, including those based on different proxies from the same sediment cores. These reconstructions are shown as examples of the many other studies that we refer to in the following sections.

Oceanic circulation — sea surface temperature studies

Koç and Jansen (2002) reviewed the Holocene climate evolution of the North Atlantic Ocean and the Nordic Seas, primarily with respect to sea-surface temperatures (SST's). Climatic

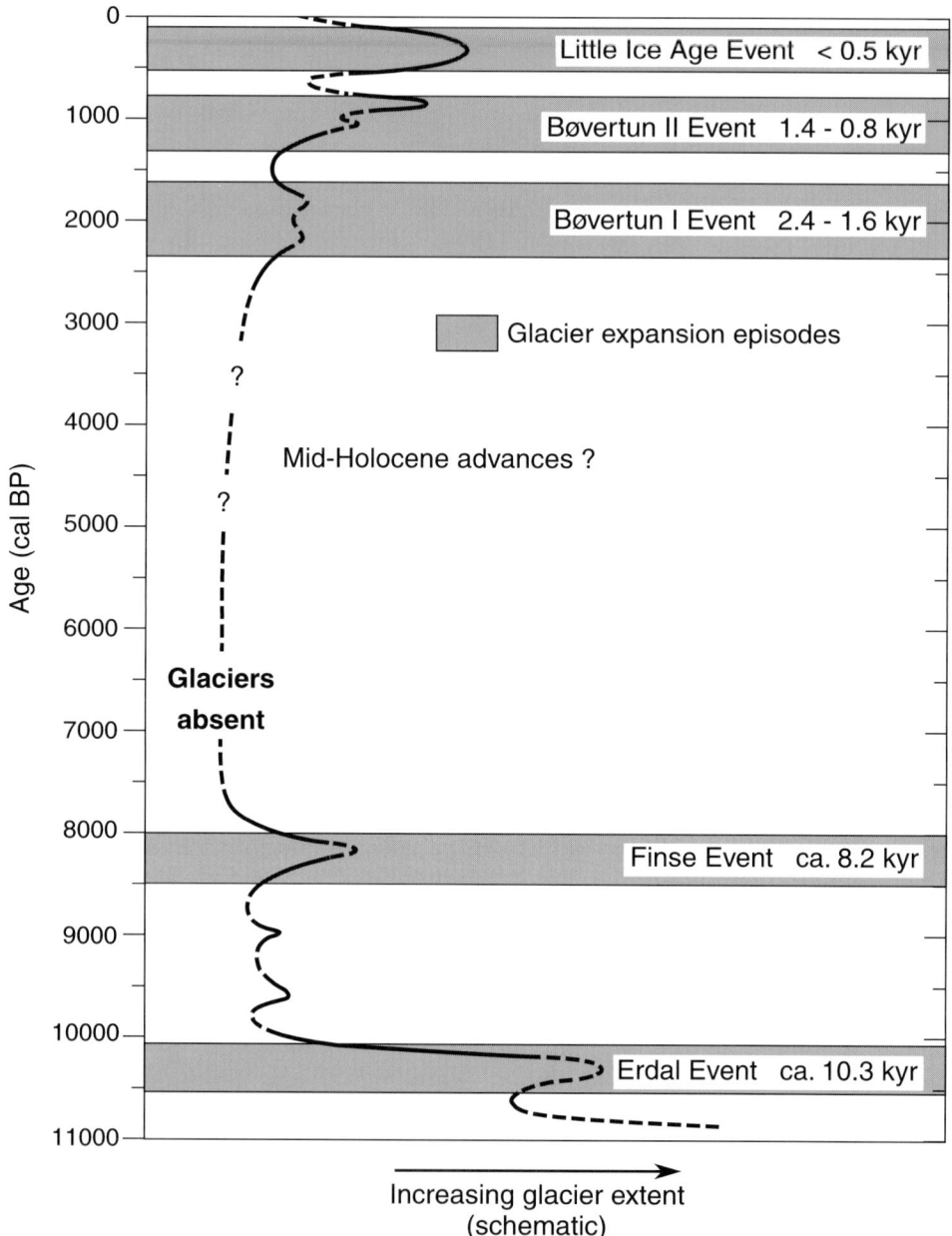

Figure 5. Adapted from Matthews et al. (2000) the intensity of glacial activity in Central Jotunheimen is a representative example of several studies of glacier dynamics in Fennoscandia.

Table 1. Performance statistics for weighted averaging with inverse deshrinking (WA-inv), WA partial least squares (WA-PLS), partial least squares (PLS), and Bayesian (Finnish chironomid data only) calibration models for mean July air temperature and the regional diatom, Cladocera, pollen and chironomid data-sets from Finland, Sweden, Norway, and Fennoscandia (combined data set of Norway, Finland & Sweden for pollen). The root mean square error of prediction (RMSEP) and r^2 values between predicted and observed temperatures are all based on the results of leave-one-out cross-validation (jackknifing). (n = number of samples).

Local calibration data-set	Temperature range (°C)	Model	$r^2_{(jack)}$	RMSEP (°C)	Reference
Finland					
Diatoms (n = 38)	7.9–14.9	WA-PLS, 2 components	0.78	0.89	Korhola et al. 2000.
Diatoms (n = 64)	7.9–14.9	WA-PLS, 2 components	0.67	0.95	Weckström and Korhola 2001.
Cladocera (n = 36)	9.3–15.0*	PLS, 1 component	0.30	1.19	Korhola 1999.
Pollen (n = 113)	10.9–17.1	WA-PLS, 2 components	0.84	0.60	Seppä and Birks 2001.
Chironomids (n = 53)	8.5–14.9	WA-PLS, 2 components	0.61	0.95	Olander et al. 1999.
Chironomids (n = 62)	8.5–14.9	Bayesian model	0.74	0.80	Vasko et al. 2000.
Chironomids (n = 62), improved taxonomy	8.5–14.9	WA-PLS, 2 components	0.77	0.73	Seppä et al. 2002.
Sweden					
Diatoms (Sarek/Abisko) (n = 60)	7.0–14.3	WA-PLS, 3 components	0.74	0.88	Rosén et al. 2001.
Diatoms (Abisko) (n = 100)	7.0–14.7	WA-PLS, 2 components	0.75	0.96	Bigler and Hall 2000.
Pollen (Sarek) (n = 55)	7.5–15.0	WA-inv	0.33	1.20	Rosén et al. 2001.
Chironomids (Sarek) (n = 40)	7.5–13.0	WA-inv	0.44	1.02	Rosén et al. 2001.
Chironomids (Abisko)	7.0–14.7	WA-PLS, 2 components	0.65	1.13	Larocque et al. 2001.
Norway					
Pollen (n = 191)	7.7–16.4	WA-PLS, 3 components	0.54	1.03	Birks et al., unpub.
Chironomids (n = 44)	5.7–14.1	WA-PLS, 1 component	0.69	1.13	Brooks and Birks 2000a.
Chironomids (n = 109)	3.5–15.6	WA-PLS, 3 component	0.94	0.93	Brooks and Birks 2000b.
Chironomids (n = 153)	3.5–16.0	WA-PLS, 3 component	0.91	1.01	Brooks and Birks, unpub.
Fennoscandia (combined)					
Pollen (n = 304)	7.7–17.1	WA-PLS, 2 components	0.71	0.99	Seppä and Birks 2001.

*Water temperature

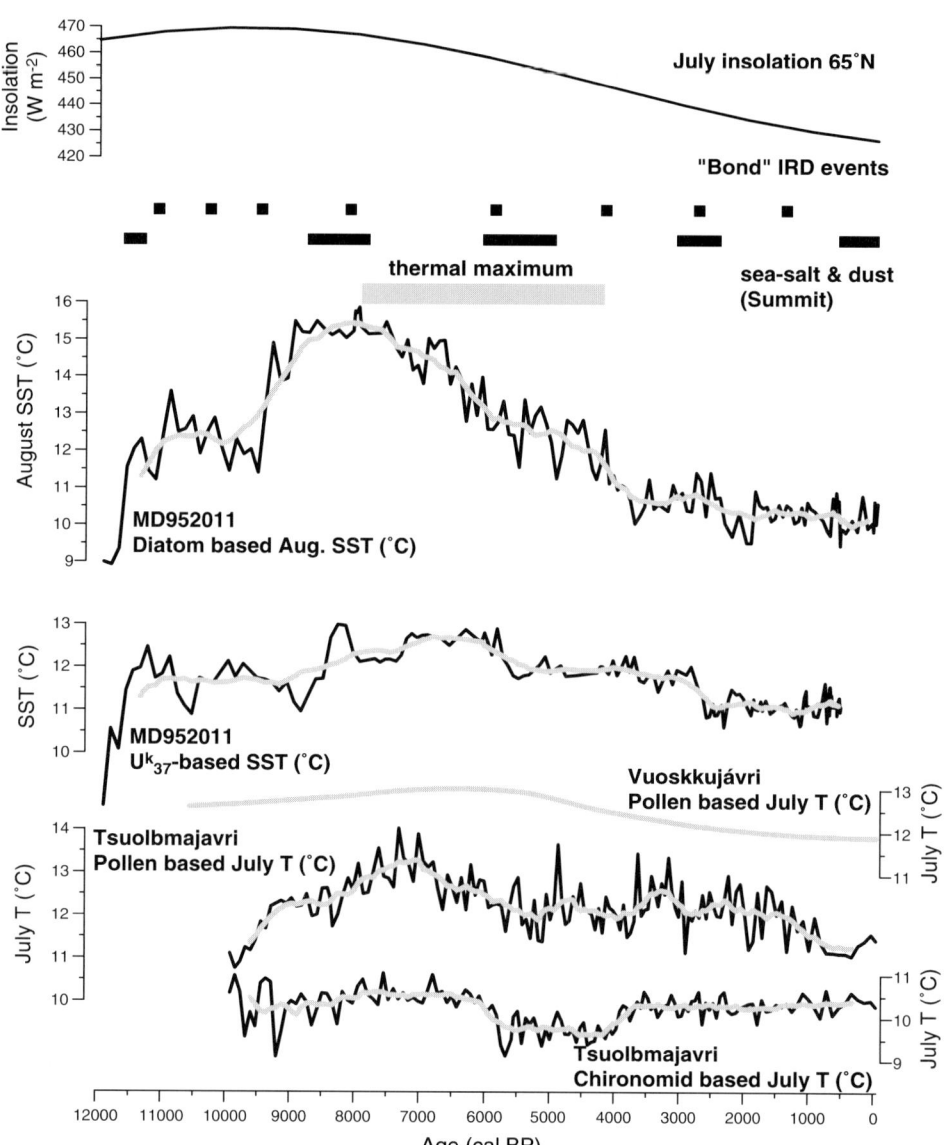

Figure 6. Selected temperature reconstructions for the Holocene obtained using transfer function techniques, which are plotted on the same y-(temperature) scales. These may be compared to the July insolation (Berger and Loutre 1991), ice-rafting events (Bond et al. 1997) and sea salt-dust concentrations in the Greenland ice cheet (O'Brien et al. 1995) as shown in Figure 1. The upper two temperature plots show diatom- (Birks and Koç 2002) and alkenone- (Calvo et al. 2002) based reconstructions from core MD952011 in the Norwegian Sea. The central curve is a terrestrial pollen based July temperature estimation for northernmost Sweden from Bigler et al. (2002) obtained using a LOESS smoother. The two lower curves are pollen- and chironomid-based July temperatures obtained from sediments in Lake Tsuolbmajarvi, northernmost Finland (Korhola et al. 2000).

conditions drastically changed in the Nordic Seas during the onset of the Holocene, when the sea-surface temperature rose ~9 °C within <50 years (Koç et al. 1993; Hald and Aspeli 1997; Andruleit and Baumann 1998). The strong influx of warm Atlantic waters caused the north-westerly retreat of the sea-ice margin and the polar front, which had been positioned in the eastern Nordic Seas during the deglaciation period, to positions off Greenland (Koç et al. 1993). The transition into the Holocene appears dramatic at northern high latitudes because the YD cooling was so marked (at least ~6 °C); in the mid-latitude North Atlantic the YD cooling was only about 1–2 °C (Koç et al. 1996).

After the initial Preboral warming the sea-surface temperatures cooled (~2 °C in the southeastern Norwegian Sea and ~0.5 °C in mid-latitude North Atlantic) at ca. 11,300 cal BP (a cooling known as the Preboreal oscillation, PBO). The PBO corresponds in time to the second global meltwater pulse (or peak - MWP 1b), although no clear evidence of a causal linkage exists (Koç et al. 1993; Hald and Hagen 1998). Marine data also testify to the effect of the short cold snap at ca. 8200 cal BP, which is represented as a SST decline of 2–3 °C (Klitgaard-Kristensen et al. 1998, B. Risebrobakken, pers. comm.). Barber et al. (1999) suggest that catastrophic drainage of freshwater glacial lakes Agassiz and Ojibway pertubated thermohaline circulation (THC) at 8470 cal BP. However, the freshwater forcing of this event on the North Atlantic remains to be identified and it must be noted that there is no evidence of the "8.2 kyr" event in the diatom- and alkenone-based temperature reconstructions presented by Birks and Koç (2002) and Calvo et al. (2001).

Despite the intermittent cold snaps caused by the final melting of the northern Hemi-sphere ice-sheets, it would appear that the North Atlantic and the Nordic Seas were warmest during the first half of the Holocene, particularly between 9000 and 6000 cal BP (Koç et al. 1993; 1996; Fronval and Jansen 1997; Hald and Hagen 1998). However, on the western Iceland Plateau, SSTs started to decrease immediately after ca. 8000 cal BP. Thus, the duration of the HTM was shorter in the north and at the margins of the Nordic Seas. It is also notable that maximum SSTs existed in the eastern basin between 9000 and 7000 cal BP, while the western basin was warmest between 11,500 and 9500 cal BP (L. Labeyrie, unpublished). This difference may suggest a possible anti-phase relationship between the eastern and western basins of the North Atlantic.

A general increase in the extent of seasonal sea-ice cover, possibly since 7000 cal BP, occurred in the western Nordic Seas (Koç et al. 1996). It is possible that, in the manner foreseen by Imbrie et al. (1992), the spread of sea-ice cover increased the albedo of the region and amplified the orbital cooling effect. On the other hand, in the eastern part of the Nordic Seas, which was still influenced by the advection of warm Atlantic water, the thermal maximum lasted until ~5000 cal BP. Since then the surface waters have cooled to modern values.

In the late Holocene, there is a significant cooling associated with the relatively short LIA. Compared with contemporary temperatures, Core MD 2011/JM97 from the Vøring Plateau documents a general cooling of 1–2 °C during the LIA (Jansen and Koç 2000). Higher temperatures in the same record may represent the Medieval Warm Period (MWP), which has also been called the Medieval Climate Anomaly (MCA). It should be appreciated, however, that none of these periods was monotonously warm or cold, for example, quite variable climatic conditions with frequent intervals of severe cold characterise a LIA-type interval in Nansen Fjord, Eastern Greenland (Jennings and Weiner 1996).

Terrestrial reconstructions

In addition to the evidence from the cryosphere, the broad-scale patterns of climatic development during the Holocene in Fennoscandia have traditionally been reconstructed from
pollen, macrofossil and tree-megafossil analyses. Due to today's prevailing oceanic climate,
mountain birch (*Betula pubescens* ssp. *tortuosa*) forms the dominant tree-line species in
northern Fennoscandia. However, changes in the latitudinal and altitudinal limits of pine (*Pinus sylvestris*, L.) have constituted a valuable source of palaeoclimatic data in Fennoscandia.
Although birch woods were the dominant vegetation communities in Fennoscandia during
the early Holocene, pine also spread rapidly northwards after the deglaciation and reached
its highest limits in southern Fennoscandia between 10,000 and 9000 cal BP (Huntley and
Birks 1983). Megafossil records show that pine trees were growing 200–300 m above their
present limit in Southern Norway and on the southern Swedish Scandes at 9000 cal BP
(Aas and Faarlund 1988; Kullman 1993). Stratigraphic records of macrofossil abundance
(Barnekow 2000) and megafossils of mountain birch found well above present tree line
in the northern Swedish Scandes (Kullman 1999) show that the tree limit was 300–400 m
higher than today prior to ca. 5000 cal BP. By comparison with the position of the modern
tree line and accounting for isostatic land uplift the summer temperature is suggested to
have been about 1.5 °C higher than today during the early Holocene (Barnekow 2000).

In agreement with the ice-core records, palaeolimnological records suggest highly
unstable climatic conditions during the early Holocene (e.g., Korhola et al. (2000) and
(2002), Rosén et al. (2001)). The long term trends in the quantitative temperature reconstructions shown in Figure 6 clearly demonstrate the early to mid-HTM, which possibly
ended rather abruptly at ca. 5800 cal BP (Korhola et al. 2002). These statistically robust
estimates point towards summer (July) temperatures that were 1–3 °C warmer than today,
which are in close agreement with the more qualified estimates of Andersson (1909) and
the SST reconstructions. The most intensive warmth occurred according to these records in
northern Fennoscandia between 8000 and 5500 cal BP, with a peak in many records around
6000 cal BP. Again, it should be pointed out that no distinct "8.2 kyr" cold event can be
seen in the quantitative reconstructions, although some qualitative records suggest a period
of significant climate cooling near this time (e.g., Karlén (1976), Snowball et al. (2002)).
It must also be stressed that little significance is attached to the high-frequency variability
of quantitative reconstructions in the sub-millennial timescale, which may be an artefact
of the numerical methods (e.g., Seppä and Birks (2001)).

Proxy data sets do point to regional differences in the age and duration of the Holocene
thermal maximum. For example, pollen assemblages and pine megafossil data indicate that
the Holocene thermal maximum occurred in Finnish Lapland between 7000 and 5000 cal
BP (Seppä 1996; Eronen et al. 1999; Seppä and Birks 2001) and the pine limit had reached
a maximum and already started to decrease prior to 6000 cal BP in Central and Southern
Scandinavia (Dahl and Nesje 1996). However, based on similar evidence, the HTM in
Northern Sweden occurred somewhat later, about 6000–4500 cal BP (Berglund et al. 1996;
Barnekow 2000). Of course, this marked discrepancy in the behaviour of pine forests in
southern and northern Fennoscandia, which is frequently interpreted in terms of climate
change, may at least in part reflect varying migration patterns.

Despite the reservations associated with the geographical representativeness of quantified precipitation reconstructions, several biological and physical studies do point to

a long term Holocene reduction in precipitation in northern Fennoscandia, which may be connected to the southwards migration of Arctic air masses since about 7000 cal BP (Hammarlund et al. 2002; Bigler et al. 2002). According to the reconstruction of Nesje et al. (2001) the maritime Jostedalsbreen region in Western Norway experienced millennial scale Holocene winter precipitation variations between 28 and 160% relative to the calibration period (AD 1961–1990). The very earliest Holocene, prior to ca. 9500 cal BP, experienced the driest winters. Despite a general increase in winter precipitation between 9500 and 6500 cal BP, the equilibrium line altitude (ELA) was high due to elevated annual temperatures and thus glaciers did not expand. For recent years, enhanced solid winter precipitation occurred during periods of positive North Atlantic Oscillation (NAO) index (Nesje et al. 2000). Nesje et al. (2001) invoked a millennial-scale correlation between cold/dry periods and periods of enhanced ice-rafting in the North Atlantic reconstructed by Bond et al. (1997) and shown in Figure 1, which have subsequently been linked to solar variability (Bond et al. 2001). Careful examination of the proposed correlation (Nesje et al. (2001), their Fig. 13) indicates that the ice-rafted-debris (IRD) events occurred at the onset of warm/wet periods rather than during the coldest/driest times. This correlation would suggest that high SST's could promote the enhanced melting of ice-sheet margins, as proposed by Broecker et al. (1990) for the last glacial maximum (LGM) and by Moros et al. (2002) for the build up to the LGM.

Synthesis

Based on both relative and quantified proxies derived from the different realms reviewed above, a relatively coherent story emerges if we attempt a regional synthesis of Holocene climate dynamics in the northern PEP III region.

Unstable early Holocene: 11,550–7500 cal BP

The current evidence suggests that the northern PEP III region experienced a quite variable climate during the early Holocene. Due to peak solar insolation, SSTs and atmospheric temperatures may have reached their Holocene maximum almost immediately after the YD, which through an intimately coupled ocean-atmosphere system may explain the sporadic finds of individual tree megafossils of comparable age at anomalously high altitudes and in favourable micro-climates in the Scandinavian mountains (Kullman 1995; 1999). However, the gradual intensification of North Atlantic THC was frequently interrupted by discharges of meltwater from the still diminishing northern hemisphere ice-sheets. Although the triggers and precise origins of these discharges remain to be identified, three cooling events were superimposed upon the general warming trend that took place between the end of the YD and ca. 8000 cal BP, specially the Preboreal oscillation at 11,300–11,150 cal BP (Björck et al. 1997), a second event at 10,300 cal BP (Björck et al. 2001) and the "8.2 kyr" event (Alley et al. 1997). The latter cold event around 8200 cal BP was probably the most striking, although the dynamics of climate change associated with this event remain to be quantified and dating uncertainties have made it problematical to identify its trigger and the feedback mechanisms that subsequently amplified or dampened its effects (Snowball et al. 2002).

This review must stress that temperature variability is only one aspect of a dynamic climate system. Qualitative and quantitative pollen-based temperature and precipitation records from northern Fennoscandia (Berglund et al. 1996; Seppä and Birks 2001) indicate that climate in the early Holocene in Northern Europe had a more oceanic character than at present, with the increased efficiency of moisture transport across the Scandinavian mountains (Hammarlund et al. 2002). With the additional contribution from SST studies we can conclude that enhanced cyclonic activity in the early Holocene was promoted by a large temperature difference between the warming Atlantic surface waters, which originated from lower latitudes, and the colder air masses circulating the northern hemisphere. This difference promoted the westerly transport of relatively mild and moist laden air over Northern Europe until the advent of the last cold snap at 8200 cal BP. The temperature difference between the ocean and atmosphere gradually became smaller and lead to a gradual reduction in cyclonic activity (Hammarlund et al. 2003). Thus, in northern Fennoscandia the general rise of summer temperatures in the early Holocene was accompanied by a trend towards a drier, more continental climate.

Thermal maximum: 7500–5500 cal BP

The marine reconstructions provide clear evidence for a distinct HTM, although the resolution of these records and their ages can be refined. Physical and biological data from varved lake-sediments in Northern Sweden would suggest that THC intensified significantly within 75 years at the end of the final cold snap at ca. 7500 cal BP (Snowball et al. 2002). Following this relatively rapid warming a period ca. 2 °C warmer than present occurred between approximately 7500 and 5500 cal BP. In particular, the period between 7500 and 6300 cal BP was associated either with a distinct northward shift of polar vortex contraction or a weaker meridional circulation. This period was most likely characterised by the development of dominant anti-cyclonic summer conditions over Scandinavia and averaged centennial summer temperatures reached their Holocene maxima, which favoured primary production. Given the widespread evidence for glacial activity lower than today, there is no doubt that this part of the Holocene in the northern PEP III region was warmer than the 20th century AD. It was also warmer than the conservative estimates for regional warming that is predicted to have taken place by the middle of the 22nd century. This extended warmth allowed forests composed of thermophilous trees and other temperature sensitive plant communities to spread further north (Donner 1995) and to rise altitudinally.

The Fennoscandian climate during the HTM was not only warm, but also dry, which led to the starvation of glaciers and low lake levels. This effect was notable in Finland, where lake-level reconstructions from small closed basins in the Finnish tree-line areas indicate that the levels during this period were 3–4 m lower than at present (Hyvärinen and Alhonen 1994; Korhola and Rautio 2001). On the basis of pollen influx values it can be suggested that the driest period in Northwestern Finland dates to 6000–5000 cal BP (Seppä and Hammarlund 2000; Seppä et al. 2002). A similar pattern can be found in Southern Sweden too, where a major period of increased dryness, recorded as decreasing lake levels, begun ca. 7500 cal BP and culminated at ca. 5600–4500 cal BP (Digerfeldt 1988).

Abrupt mid-Holocene cooling: between 6000 and 5500 cal BP

Despite the evidence of a gradual lowering of sea-surface temperatures other records indicate a rapid end to the thermal maximum and the possible breakdown of a still dominant anti-cyclonic weather pattern at ca. 6000 cal BP (Fig. 6). A record from a tree-line lake in NW Finnish Lapland suggests that a distinct cooling phase started around 5800 cal BP (Korhola et al. 2002). This cooling coincides with one of the major ice-rafting episodes in the North Atlantic identified by Bond et al. (1997).

With the onset of this mid-Holocene cooling, certain glaciers in Northern Sweden became more active again (Karlén 1988; Karlén and Kuylenstierna 1996). There are also indications of increased inter-annual variability in the Finnish tree-ring records from ca. 5800 cal BP, suggesting a shift towards increased climatic instability (Zetterberg et al. 1996; Eronen et al. 1999). A temperature drop is further supported by data of glacier variations and relative equilibrium line altitude (ELA) fluctuations of maritime glaciers in Western Norway that show a shift from mild/wet winters towards cold/dry winters around 6000 cal BP (Nesje et al. 2001). In Holocene pollen records, this is the period when there was an elm (*Ulmus*) decline in Northwestern Europe (about 5700 cal BP). The elm decline has been assigned to specific pathogen attacks (Peglar and Birks 1993), yet it may now be worth considering the effect of climate cooling at least on the spread of epidemics. Again, the non-linear response of environmental systems to external forcing may prove to be responsible for the some of the apparently abrupt climate changes reconstructed from proxy records.

Towards late Holocene climatic instability: 5500 cal BP - present

As already discussed, the late Holocene trend of decreasing temperature most likely reflects a regional response to lower summer insolation. However, many short-term transitions were superimposed on the long term trend, particularly during the past four millennia when the pine tree-limit descended (Kullman 1995; Barnekow 2000), glaciers advanced (Nesje and Kvamme 1991) and erosion increased (Snowball et al. 1999). It would appear that, via both reduced THC intensity and solar insolation, the dominant anti-cyclonic pattern of the mid-Holocene was replaced by a more variable climatic regime, possibly one in a transition mode. The increase in lake levels can be linked to the end of the dry period at ca. 5000 cal BP and to the subsequent development of moister climate (Hyvärinen and Alhonen 1994; Korhola and Rautio 2001). The increase of moisture was probably a result of decreased summer temperatures and associated reduction of summer evaporation. Superimposed on this trend has been the spread of both alpine and boreal peatlands, which is indicated by increasing pollen influx values of *Sphagnum* and vascular plants associated with moist meadows in the north (Hyvärinen 1975; Seppä 1996; Seppä and Weckström 1999) and extensive paludification of originally less wet mineral soils and increased peatland carbon accumulation rates in the south (Korhola 1995; Korhola et al. 1996).

A particularly rapid transition to a more variable climatic regime at 3700 cal BP was indicated by study of a varved lake sediment sequence in Northern Sweden by Snowball et al. (1999), where periods of variable erosion since 3700 cal BP were coupled to the accumulation of snow, which was subsequently released as spring meltwater discharge. Single site studies may, however, reflect the delayed response of a fluvial system to external forcing and the eventual breach of a system threshold. Despite the potential for delayed response, a mid-Holocene transition has corroborated by isotopic studies on lacustrine carbonates in Southern Sweden (Hammarlund et al. 2003) and in Central Norway (Nesje et al. 2001). Thus, at least in western Fennoscandia, the period between 4500 and 3500 Cal BP saw a remarkable change towards a significantly more variable climate regime.

A generally colder climate is marked in the Finnish dendrochronological data series around 2200 cal BP by very few subfossil pines (Eronen et al. 1999). However, this period is not directly shown in the available temperature curves, although certain changes in the species composition can be found around this time (Rosén et al. 2001; Korhola et al. 2001).

Temperatures during the MWP (ca. AD 700–1300) were ca. 0.5–0.8 °C higher than those of today, while it was almost 1 °C cooler during the Little Ice Age. Historical records from all over Northern Europe and Greenland attest to the reality of both events, and their profound impact on human society. For example, the colonisation of Greenland by the Vikings early in the 2nd millennium AD was only possible because of the medieval warmth. At the onset of the LIA in the middle of the 14th century, which was possibly marked a series of particularly cold winters and summer between AD 1351 and 1355, Viking settlements in Western Greenland were abandoned (Grove 2002).

Climatic cycles and periodicities

Repeated millennial-scale climate shifts at high northern latitudes during the Holocene were first proposed by Denton and Karlén (1973). As increasingly higher resolution proxy-climate records are being generated, the temptation to search for periodicities/quasi-periodicities using time-series analyses cannot be resisted, even if true time-series are rare in a geological context (with the possible exception of annually resolved records). Aaby (1976) detected millennial-scale cycles in Danish peat bogs and Chapman and Shackleton (2000) found evidence of 550-year and 1000-year cycles in North Atlantic circulation patterns during the Holocene. In addition, a pervasive millennial-scale cycle close to ~1500 years (though not strictly periodic) in both the glacial and the Holocene North Atlantic sediments was suggested by Bond et al. (1997). This cycle has been interpreted as southward shifts of ice-bearing surface waters into the sub-polar North Atlantic, which were triggered by solar variability (Bond et al. 2001). Based on a sortable-silt grain-size study of Holocene sediments from the Gardar drift south of Iceland (core NEAP-15K) Bianchi and McCave (1999) similarly documented a quasi-periodicity of ~1500 years in the flow speed of the Iceland-Scotland Overflow Water (ISOW). These latter two records suggest a possible existence of a quasi-periodicity of ~1500 years in the physical behaviour of North Atlantic surface- and deep water masses. Spectral analysis of high-resolution grey-scale data from the same core indicates variations in the delivery of terrigeneous sediment to the oceans as a result of iceberg rafting and/or changes in surface productivity, reveals cyclicities at 1650, 1000 and 450 years (Chapman et al. 1999). High-resolution sea surface temperature reconstructions from the Vøring Plateau (core MD952011) indicate variability in the order of 1–2 °C during the last 2000 years, with the presence of a ~220 year periodicity (Koç et al., unpublished). Oxygen isotope records of benthic foraminifera in a core from a fjord in Western Norway show cyclic changes with strong periodicities of 380 and 206 years, which are interpreted to reflect bottom water temperature changes of up to 3–4 °C through the last 2500 years (Mikalsen 1999). Based on a good correlation between the 206 year cycle in their record and in the atmospheric $\delta^{14}C$ record these authors also suggest that variations in solar radiation may have forced oceanic surface temperature changes and THC through the last 2500 years. Solar forcing of climate has been suggested from instrumental data (Reid 1987; Lean et al. 1992; 1995) and modelling experiments (Tett et al. 1999; Wood et al. 1999). At present, however, it is uncertain if the controversial evidence of periodicities in various palaeo-climate data-sets is the result of dating interpolation and statistical methods, or of different and poorly understood forcing mechanisms and system thresholds.

Lessons learnt from the northern experience and future challenges

One difficulty encountered when trying to reconstruct quantitatively Holocene climatic fluctuations is that they were far less pronounced than during the late-glacial times. Despite undeniable qualitative evidence for major and often abrupt shifts in climate the reconstructed oscillations frequently remain within the prediction errors of the inference models. For example, the current high-latitude calibration data sets have a predictive precision for mean July air temperature of $\sim$0.9 °C (diatoms), 0.8–1.1 °C (chironomids) and 0.6–1.2 °C (pollen). A similar precision is also provided by diatom based SST reconstructions. The present array of transfer functions is capable, therefore, of providing information on broad-scale Holocene climate fluctuations and trends, but it is probably not sensitive enough for the statistically significant detection of higher frequency oscillations (with the possible exception of dendro-climatology).

Even so, the evidence from various terrestrial and marine archives in the high North Atlantic and Fennoscandia point to a long-term, post-glacial Holocene cooling trend, which reflects the dominant role of orbital forcing and summer insolation. However, a synthesis of palaeo-records shows that the long-term trend was possibly stepwise and punctuated by rapid climate shifts, predominantly as cooler periods. In fact, only one millennial-scale period of uninterrupted warmth occurred, between ca. 7500 and 6300 cal BP.

There is no serious doubt that the northwards transfer of heat, via the North Atlantic THC, has regulated climate in high latitude Europe on a wide range of timescales, from decades to millennia. Yet, the errors associated with the majority of geo-chronometers force us to admit that we have difficulties to separate temporally records of the triggers (causes) of rapid climate change from their effects on ecosystems: the "8.2 kyr event" is an excellent example, despite its apparently widespread impact. Here we must mention the exciting role that tephrochronology can play in the northern PEP III region. Tephra particles emitted from Icelandic volcanoes have produced a series of isochronous marker horizons across the North Atlantic and Northern Europe. In most cases the pre-historical tephra layers have been dated by the radiocarbon method and their ages include the uncertainties associated with this method (e.g., Eríksson et al. (2000), Boygle (1998)). Mid-Holocene tephra layers (H-3, Kebister and H4) have been identified in varved lake-sediments in Central Southern Sweden and allocated calendar year ages (Zillen et al. 2002). Such tephra layers, when positively identified in ice-cores, marine sequences and further terrestrial deposits will improve our ability to synchronise palaeo-climate records and make robust connections between the high- and mid-latitudes (see Barber et al., this volume).

Statistical analyses of palaeo-data can be carried out on a wide range of well-dated archives in the high latitudes. Annually layered ice-cores, tree-rings and varved sediments are the closest analogues of true time series that nature can provide and analyses of these archives will provide more accurate tests of hypotheses suggesting that (i) repeated cooling events occurred with a recurrence interval of ca. 1500 years through the Holocene and (ii) solar forcing may lie behind repeated centennial-scale and millennial-scale climate variability (Bond et al. 2001). Synchronisation between ice-cores, high-resolution marine sediments and terrestrial archives via independent, accurate and precise dating techniques will lead to important advances in our understanding of Holocene climate development in high latitude Europe, where ecosystems are already responding to the observed post-19th century AD warming (Sorvari et al. 2002).

Summary

At the same time as the average global temperature was observed to rise during the 20th century AD, intensive efforts were made by geologists working in Fennoscandia, Greenland and the North Atlantic to recover natural archives of climate change. Even the earliest studies of terrestrial plant remains found in peat deposits concluded that average annual temperatures higher than those that prevailed during the 19th century AD must have previously existed in Fennoscandia during the Holocene. More recent investigations, which could benefit from the application of an array of dating techniques, indicate that the period between 11,550 and ca. 8000 cal BP was characterised by relatively rapid warming, although this trend was punctuated by a series of cold reversals caused by the lingering effects of melting northern hemisphere ice-sheets. Through the quantitative analysis of marine sediments recovered from the North Atlantic, the Greenland ice cores and highly resolved continental records it is now clear that temperatures ca. $2\,^{\circ}C$ higher than today prevailed during an extended early "Holocene thermal maximum" between ca. 8000 and 6000 cal BP, which was most likely a consequence of the peak Holocene flux of solar radiation during the summer season at northern high latitudes. Regional discrepancies do occur between the apparent ages of the period of absolute maximum temperatures, which may reflect dating problems, the occurrence of regional differences in climate development or the delayed response of ecosystems to external forcing. Evidence provided by the studies of tree-lines, ice-cores and SST's indicate a general long-term Holocene cooling trend since the thermal maximum, which is consistent with a general reduction in insolation. Stable isotope and lake-level data also suggest a shift from oceanic to more continental conditions from ca. 9000 to 5000 cal BP. Following the mid-Holocene warmth and dryness, precipitation again increased from ca. 4000 cal BP and there is no doubt that the overall frequency and amplitude of climate variability increased, probably in connection with changing atmospheric circulation patterns and the possible development of a sensitive climate regime in a transition mode. The limited synthesis of high-latitude records available for the last 2000 years indicate some diversity in the timing and magnitudes of regional anomalies (e.g., the LIA and MWP), but they show clear evidence of overall 20th century warmth and suggest a pattern of prior swings between relatively warm and cool periods; warm ca. AD 400–500; 1000–1100; and 1400–1500, and cool in the 6th century; 1200–1300; and markedly between 1600 and 1750, and in the early 19th century.

The precise causes of the relatively abrupt climate changes, which are superimposed on the long term trend, remain to be positively identified, but significant influences of solar irradiance change and volcanic eruptions are certainly implicated. Advances in our knowledge of the individual and combined roles of such influences are likely to result from the production of precisely synchronised high-resolution proxy records of climatic forcing and response.

Acknowledgments

All those engaged in studies of Holocene climate variability in the northern PEP III region have contributed to this review. We particularly grateful to, in alphabetical order of surname, the following people for discussions and critique: Björn Berglund, Christian Bigler, John Birks, Svante Björck, Dan Hammarlund, Sigfus Johnsen, Isabelle Larocque, Matthias Moros, Atle Nesje, Sylvia Peglar, Tine Rasmussen, Ingemar Renberg, Peter Rosén, Mats Rundgren, Heikki Seppä, Stefan Wastegård, Jan Weckström and Barbara Wohlfarth. The final version benefited from the critical reviews of Vivienne Jones and Matti Saarnisto.

References

Aaby B. 1976. Cyclic climatic variations in climate over the past 5,500 yr reflected in raised bogs. Nature 263: 281–284.

Aas B. and Faarlund T. 1988. Postglaciale skoggrenser i sentrale sørnorske fjelltrakter. [14]C datering av subfossile furu- og bjørkrester. (Postglacial forest limits in Central South Norwegian mountains. Radiocarbon datings of subfossil pine and birch specimens). Norsk Geografisk Tidsskrift 42: 25–61.

Alley R.B., Mayewski P.A., Sowers T., Stuiver M., Taylor K.C. and Clark P.U. 1997. Holocene climatic instability: A prominent, widespread event 8200 yr ago. Geology 25: 483–486.

Aniol R.W. and Eckstein D. 1984. Dendroclimatological studies at the northern timberline. In: Mörner N.-A. and Karlén W. (eds), Climatic Changes on a Yearly to Millennial Basis. Reidel, Dordrecht, pp. 273–279.

Andersson G. 1909. The Climate of Sweden in the Late-Quaternary Period. Sveriges Geologiska Undersökning, Series C, No. 218, 88 pp.

Andruleit H.A. and Baumann K.H. 1998. History of the last deglaciation and Holocene in the Nordic seas as revealed by coccolithophore assemblages. Mar. Micropaleontol. 35: 179–201.

Baillie M.G.L. 1999. Exodus to Arthur: Catastrophic Encounters With Comets. Batsford, London, 272 pp.

Barber D.C., Dyke A., Hillaire-Marcel C., Jennings A.E., Andrews J.T., Kerwin M.W., Bilodeau G., McNeely R., Southon J., Morehead M.D. and Gagnon J.-M. 1999. Forcing of the cold event of 8,200 years ago by catastrophic drainage of Laurentide lakes. Nature 400: 344–348.

Barker P., Talbot M.R., Street-Perrott F.A., Marret J., Scourse J. and Odada E., this volume. Late Quaternary climatic variability in intertropical Africa. In: Battarbee R.W., Gasse F. and Stickley C.E. (eds), Past Climate Variability through Europe and Africa. Kluwer Academic Publishers, Dordrecht, the Netherlands, pp. 117–138.

Barnekow L. 2000. Holocene regional and local vegetation history and lake-level changes in the Torneträsk area, Northern Sweden. J. Paleolim. 23: 399–420.

Bartholin T.S. and Karlén W. 1983. Dendrokronologi I Lappland. Dendrokronologiska Sällskapet, meddelende 5: 3–16.

Berger A. and Loutre M.F. 1991. Insolation values for the climate of the last 10000000 years. Quat. Sci. Rev. 10: 297–317.

Berglund B.E., Barnekow L., Hammarlund D., Sandgren P. and Snowball I.F. 1996. Holocene forest dynamics and climate changes in the Abisko area, North Sweden — the Sonesson model of vegetation history reconsidered and confirmed. Ecol. Bull. 45: 15–30.

Bergström H. and Moberg A. 2002. Daily air temperature and pressure series for Uppsala (1722–1998). Clim. Chan. 53: 213–252.

Bianchi G.G. and McCave N. 1999. Holocene periodicity in North Atlantic climate and deep-ocean flow south of Iceland. Nature 397: 515–517.

Bigler C. and Hall R.I. 2000. Diatoms as indicators of climatic and limnological change in Swedish Lapland: a 100-lake calibration set and its validation for paleoecological reconstructions. J. Paleolim. 27: 97–115.

Bigler C., Larocque I., Peglar S.M., Birks H.J.B. and Hall R.I. 2002. Quantitative multiproxy assessment of long-term patterns of Holocene environmental change from a small lake near Abisko, Northern Sweden. The Holocene 12: 481–596.

Birks C.J.A. and Koç N. 2002. A high-resolution diatom record of late-Quaternary sea-surface temperatures and oceanographic conditions from the eastern Norwegian Sea. Boreas 31: 323–344.

Birks H.J.B. 1995. Quantitative palaeoenvironmental reconstructions. In: Maddy D. and Brew J.S. (eds), Statistical Modelling of Quaternary Science Data. Quaternary Science Association, Cambridge, pp. 161–254.

Björck S., Walker M.J.C., Cwynar L.C., Johnsen S., Knudsen K.-L., Lowe J., Wohlfarth B. and INTIMATE members. 1998. An event stratigraphy for the Last Termination in the North Atlantic region based on the Greenland ice-core record: a proposal by the INTIMATE group. J. Quat. Sci. 13: 283–292.

Björck S., Muscheler R., Kromer B., Andresen C.S., Heinemier J., Johnsen S.J., Conley D., Koç N., Spurk M. and Veski S. 2001. High-resolution analyses of an early Holocene climate event may imply decreased solar forcing as an important climate trigger. Geology 29: 1107–1110.

Blytt A. 1876. Forsög til en theori om invandringen av Norges Flora under vexlende regnfulde og törre tider. Nyt Mag. F. Naturv. Vol. 21.

Briffa K.R. 2000. Annual variability in the Holocene: interpreting the message of ancient trees. Quat. Sci. Rev. 19: 87–105.

Briffa K.R., Jones P.D., Pilcher J.R. and Hughes M.K. 1988. Reconstructing summer temperatures in northern Fennoscandia back to AD 1700 using tree-ring data from Scots pine. Arct. Alp. Res. 20: 385–394.

Briffa K.R., Bartholin T., Eckstien D., Jones P.D., Schweingruber F.H. and Zetterberg P. 1990. A 1,400-year tree-ring record of summer temperatures in Fennoscandia. Nature 306: 434–439.

Briffa K.R. and Jones P.D. 1993. Global surface air temperature variations during the twentieth century: Part 2, implications for large-scale high-frequency palaeoclimatic studies. The Holocene 3: 77–88.

Briffa K.R., Osborn T.J., Schweingruber F.H., Harris I.C., Jones P.D., Shiyatov S.G. and Vaganov E.A. 2001. Low-frequency temperature variations from a northern tree-ring-density network. J. Geophys. Res. 106: 2929–2941.

Briffa K.R., Osborn T.J., Schweingruber F.H., Jones P.D., Shiyatov S.G. and Vaganov E.A. 2002. Tree-ring width and density around the Northern Hemisphere: Part 2, spatio-temporal variability and associated climate patters. The Holocene 12: 759–789.

Brooks S.J. and Birks H.J.B. 2000a. Chironomid-inferred Late-glacial air temperatures at Whitrig Bog, Southeast Scotland. J. Quat. Sci. 15: 759–764.

Brooks S.J. and Birks H.J.B. 2000b. Chironomid-inferred late-glacial and early-Holocene mean July air temperatures for Kråkenes Lake, Western Norway. J. Paleolim. 23: 77–89.

Bond G., Showers W., Cheseby M., Lotti R., Almasi P., DeMenocal P., Priore P., Cullen H., Hajdas I. and Bonani G. 1997. A pervasive millennial-scale cycle in North Atlantic Holocene and glacial climates. Science 278: 1257–1266.

Bond G., Kromer B., Beer B., Muscheler R., Evans M.N., Showers W., Hoffman S., Lotti-Bond R., Hajdas I. and Bonani G. 2001. Persistent solar influence on North Atlantic climate during the Holocene. Science 294: 2130–2136.

Boygle J. 1998. A little goes a long way: discovery of a new mid-Holocene tephra in Sweden. Boreas 27: 195–199.

Broecker W.S., Bond G., Klas M., Bonani G. and Wolfli W. 1990. A salt oscillator in the glacial Atlantic? 1. The concept. Paleoceanography 5: 469–477.

Calvo E., Grimalt J. and Jansen E. 2002. High resolution U^K_{37} sea surface temperature reconstruction in the Norwegian sea during the Holocene. Quat. Sci. Rev. 21: 1385–1394.

Chapman M.R., Shackleton N.J., Bianchi G.B. and McCave I.N. 1999. Millenial-scale climate fluctuations in the subpolar North Atlantic during the Holocene. Terra Nova 11: 169.

Chapman M.R. and Shackleton N.J. 2000. Evidence of 550-year and 1000-year cyclicities in North Atlantic circulation patterns during the Holocene. The Holocene 10: 287–291.

Dahl S.O. and Nesje A. 1994. Holocene glacier flucuations at Hardangerøkulen, Central-Southern Norway: a high resolution composite chronology from lacustrine and terrestrial deposits. The Holocene 4: 269–277.

Dahl S.O. and Nesje A. 1996. Approach to calculating Holocene winter precipitation by combining glacier equilibrium-line altitudes and pine-tree-limits: a case study from Hardangerjøkulen, Central Southern Norway. The Holocene 6: 391–398.

Dahl-Jensen D., Mosegaard K., Gundestrup N., Clow G.D., Johnsen S.J., Hansen A.W. and Balling N. 1998. Past temperatures directly from the Greenland Ice Sheet. Science 282: 268–271.

Denton G.H. and Karlén W. 1973. Holocene climatic variations — their pattern and possible cause. Quat. Res. 3: 155–205.

Digerfeldt G. 1988. Reconstruction and regional correlation of Holocene lake-level fluctuations in Lake Hysjn, South Sweden. Boreas 17: 162–182.

Donner J.J. 1995. The Quaternary History of Scandinavia. Cambridge University Press, Cambridge, 200 pp.

Eriksson B. and Alexandersson H. 1990. Our Changing Climate. Agric. Forest Meteorol. 50: 55–64.

Eiríksson J., Knudsen K.-L., Haflidason H. and Heinemeier J. 2000. Chronology of late Holocene climatic events in the northern North Atlantic based on AMS ^{14}C dates and tephra markers from the volcano Hekla, Iceland. J. Quat. Sci. 15: 573–580.

Erlandsson S. 1936. Dendro-Chronological Studies. Data Från Stockholm Högskolas. GeoKronologiska Institute 23: Stockholm, 116 pp.

Eronen M., Hyvarinen H. and Zetterberg P. 1999. Holocene humidity changes in northern Finnish Lapland inferred from lake sediments and submerged Scots pines dated by tree-rings. The Holocene 9: 569–580.

Eronen M., Zetterberg P., Briffa K.R., Lindholm M., Meriläinen J. and Timonen M. 2002. The supralong scots pine tree-ring record for Finnish Lapland: Part 1, chronology construction and initial inferences. The Holocene 12: 673–680.

Fronval T. and Jansen E. 1997. Eemian and early Weichselian (140–60 ka) paleoceanography and paleoclimate in the Nordic seas with comparisons to Holocene conditions. Paleoceanography 12: 443–462.

Grove J.M. 2002. Climatic change in Northern Europe over the last two thousand years and its possible influence on human activity. In: Wefer G., Berger W., Behre K.-E. and Jansen E. (eds), Climate and History in the North Atlantic Realm. Springer-Verlag Berlin, Heidelberg, pp. 313–326.

Grudd H., Briffa K.R., Karlén W., Bartholin T.S., Jones P.D. and Kromer B. 2002. A 7400-year tree-ring chronology in northern Swedish Lapland: natural climatic variability expressed on annual to millennial timescales. The Holocene 12: 657–665.

Hald M. and Aspeli R. 1997. Rapid climatic shifts of the northern Norwegian Sea during the last deglaciation and the Holocene. Boreas 26: 15–28.

Hald M. and Hagen S. 1998. Early preboreal cooling in the Nordic seas region triggered by meltwater. Geology 26: 615–618.

Hammarlund D., Barnekow L., Birks H.J.B., Buchardt B. and Edwards T.W.D. 2002. Holocene changes in atmospheric circulation recorded in oxygen-isotope stratigraphy of lacustrine carbonates from Northern Sweden. The Holocene 12: 339–351.

Hammarlund D., Björck S., Buchardt B., Israelson C. and Thomsen C. 2003. Rapid hydrological changes during the Holocene revealed by stable isotope records of lacustrine carbonates from Lake Igelsjön, Southern Sweden. Quat. Sci. Rev. 22: 353–370.

Hantemirov R.M. 2000. The 4309-year tree-ring chronology from Yamal Peninsula and its application to the reconstruction of the climate of the past in the northern part of West Siberia. In: Problems of Ecological Monitoring and Ecosystem Modelling 17, Institute of Global Climate and Ecology, Gidrometeoizdat, St. Petersburg, pp. 287–301.

Hantemirov R.M., Shiyatov S.G. and Gorlanova L.A. 2000. Dendroclimatic potential of *Juniperus sibirica*. Burgsd. Lesovedine 6: 33–38 (in Russian).

Hantemirov R.M. and Shiyatov S.G. 2002. A continuous multimillennial ring-width chronology in Yamal, Northwestern Siberia. The Holocene 12: 717–726.

Helama S., Lindholm M., Timonen M., Meriläinen J. and Eronen M. 2002. The supra-long scots pine tree-ring record for Finnish Lapland: Part 2, interannual to centennial variability in summer temperature for 7500 years. The Holocene 12: 681–687.

Huntley B. and Birks H.J.B. 1983. An Atlas of past and present pollen maps for Europe: 0-13000 BP. Cambridge University Press, Cambridge, 667 pp.

Huntley B., Baillie M., Grove J.M., Hammer C.U., Harrison S.P., Jacomet S., Jansen E., Karlen W., Koc N., Luterbacher J., Negendank J. and Schibler J. 2002. Holocene palaeoenvironmental changes in North-West Europe: Climatic implications and the human dimension. In: Wefer G., Berger W., Behre K.-E. and Jansen E. (eds), Climate and History in the North Atlantic Realm. Springer-Verlag Berlin, Heidelberg, pp. 258–298.

Hyvärinen H. 1975. Absolute and relative pollen diagrams from northernmost Fennoscandia. Fennia 142: 1–23.

Hyvärinen H. and Alhonen P. 1994. Holocene lake-level changes in the Fennoscandian tree-line region, western Finnish Lapland: diatom and cladoceran evidence. The Holocene 4: 251–258.

Høeg O.A. 1956. Growth ring research in Norway. Tree-Ring Bull. 21: 2–15.

Imbrie J., Boyle F.A., Clemens S.C., Duffy A., Howard W.R., Kukla G., Kutzbach J., Martinson D.G., McIntyre A., Mix A.C., Molfino B., Morley J.J., Peterson L.C., Pisias N.G., Prell W.L., Raymo M.E., Shackleton N.J. and Toggweiler J.R. 1992. On the structure and origin of major glaciation cycles 1. Linear responses to Milankovitch forcing. Paleoceanography 7: 701–738.

Jansen E. and Koç N. 2000. Century to decadal scale records of Norwegian Sea surface temperature variations of the past 2 millennia. PAGES/CLIVAR Newsletter 8 (1): 13–14.

Jennings A.E. and Weiner N.J. 1996. Environmental change in Eastern Greenland during the last 1300 years: evidence from foraminifera and lithofacies in Nansen Fjord, 68 °N. The Holocene 6: 179–191.

Johnsen S.J., Dahl-Jensen D., Gundestrup N., Steffesen P., Clausen H.B., Miller H., Masson-Delmotte V., Sveinbjörnsdottir A.E. and White J. 2001. Oxygen isotope and palaeotemperature records from six Greenland ice-core stations: Camp Century, Dye-3, GRIP, GISP2, Renland and North GRIP. J. Quat. Sci. 16: 299–307.

Jones P.D. 2001. Early European instrumental records. In: Davies T.D. and Briffa K.R. (eds), History and Climate: Memories of the Future. Kluwer Academic/Plenum, New York, pp. 1–8.

Jones P.D. and Briffa K.R. 1996. What can the instrumental record tell us about longer timescale palaeoclimate reconstructions? In: Jones P.D., Bradley R.S. and Jouzel J. (eds), Climate Variations and Forcing Mechanisms of the Last 2000 Years. NATO ASI Series 41, Springer-Verlag, Berlin, pp. 9–41.

Jones P.D., New M., Parker D.E., Martin S. and Rigar L.G. 1999. Surface air temperature and its variation over the last 150 years. Rev. Geophysics 37: 173–199.

Jones P.D., Briffa K.R., Osborn T.J., Moberg A. and Bergström H. 2002. Relationships between circulation strength and the variability of growing season and cold season climate in Northern and Central Europe. The Holocene 12: 643–656.

Kalela-Brundin M. 1998. Climatic information from tree rings of *Pinus sylvestris* L. and a reconstruction of summer temperatures back to AD 1500 in Femundsmarka, Eastern Norway, using partial least squares regression (PLS). The Holocene 9: 59–77.

Keys D. 1999. Catastrophe: An Investigation Into the Origins of the Modern World. Ballantine Books, New York, 343 pp.

Karlén W. 1976. Lacustrine sediments and tree-limit variations as indicators of Holocene climatic fluctuations in Lappland, Northern Sweden. Geogra. Ann. A58: 1–34.

Karlén W. 1988. Scandinavian glacial and climatic fluctuations during the Holocene. Quat. Sci. Rev. 7: 199–209.

Karlén W. and Kuylenstierna J. 1996. On the solar forcing of Holocene climate: evidence from Scandinavia. The Holocene 6: 359–365.

Kirchhefer A.J. 2001. Reconstruction of summer temperature from tree-rings of Scots pine (*Pinus sylvestris* L.) in coastal Northern Norway. The Holocene 11: 41–52.

Klitgaard-Kristensen D., Sejrup H.P., Haflidason H., Johnsen S.J. and Spurk M.A. 1998. A regional 8200 cal. yr BP cooling event in NW Europe. J. Quat. Sci. 13: 165–169.

Korhola A. 1995. Holocene climatic variations in Southern Finland reconstructed from peat initiation data. The Holocene 5: 43–58.

Korhola A. 1999. Distribution patterns of Cladocera in subarctic Fennoscandian lakes and their potential in environmental reconstruction. Ecography 22: 357–373.

Korhola A., Alm J., Tolonen K., Turunen J. and Junger H. 1996. Three-dimensional reconstruction of carbon accumulation and CH4 emission during nine millennia in a raised mire. J. Quat. Sci. 11: 161–165.

Korhola A., Weckstrom J., Lasse H. and Panu E. 2000. Quantitative Holocene climatic record from diatoms in northern Fennoscandia. Quat. Res. 2: 284–294.

Korhola A. and Rautio M. 2001. Cladocera and other branchiopod crustaceans. In: Smol J.P., Birks H.J.B. and Last W.M. (eds), Tracking Environmental Change Using Lake Sediments. Volume 4: Zoological Indicators. Kluwer Academic Publishers, Dordrecht, pp. 5–42.

Korhola A., Vasko K., Toivonen H.T.T. and Olander H. 2002. Holocene temperature changes in northern Fennoscandia reconstructed from chironomids using Bayesian modelling. Quat. Sci. Rev. 21: 1841–1860.

Koç N., Jansen E. and Haflidason H. 1993. Paleoceanographic reconstructions of surface ocean conditions in the Greenland, Iceland and Norwegian seas through the last 14 ka based on diatoms. Quat. Sci. Rev. 12: 115–140.

Koç N., Jansen E., Hald E. and Labeyrie L. 1996. Lateglacial-Holocene sea-surface temperatures and gradients between the North Atlantic and the Norwegian Sea: Implications for the Nordic heat pump. In: Andrews J.T., Austin W.E.N. and Labeyrie L. (eds), The Late Glacial Paleoceanography of the North Atlantic Margins. Geological Society, London, Special Publication No. 111, 177–185.

Koç N. and Jansen E. 2002. Holocene climate evolution of the North Atlantic Ocean and the Nordic Seas — a synthesis of new results. In: Wefer G., Berger W., Behre K.-E. and Jansen E. (eds), Climate and History in the North Atlantic Realm. Springer-Verlag Berlin, Heidelberg, pp. 165–173.

Kullman L. 1993. Holocene thermal trend inferred from tree-limit history in the Scandes Mountains. Global Ecol. Biogeogr. Lett. 2: 181–188.

Kullman L. 1995. Holocene tree-limit and climate history from the Scandes Mountains. Ecology 76: 2490–2502.

Kullman L. 1999. Early Holocene tree growth at a high elevation site in the northermost Scandes of Sweden (Lapland). A palaeobiogeographical case study based on megafossil evidence. Geogra. Ann. A81: 63–74.

Lauritzen S.E. and Lundberg J. 1999. Calibration of the speleothem delta function: an absolute temperature record for the Holocene in Northern Norway. The Holocene 9: 659–669.

Larocque I., Hall R.I. and Grahn E. 2001. Chironomids as indicators of climate change: a 100-lake training set from a subarctic region of Northern Sweden (Lapland). J. Paleolim. 26: 307–322.

Lean J., Beer J. and Bradley R. 1995. Reconstruction of solar irradience since 1610: Implications for climate change. Geophys. Res. Lett. 22: 3195–3198.

Lean J., Skumanich A. and White O. 1992. Estimating the sun's radiative output during the Maunder Minimum. Geophys. Res. Lett. 19: 1591–1594.

Linderson M.-J.F. 2002. The spatial distribution of precpitation in Scania, Southern Sweden. Ph.D. thesis, Department of Physical Geography and Ecosystems Analysis, Lund University, 44 pp.

Lindholm M. and Eronen M. 2000. A reconstruction of mid-summer temperatures from ring-widths of scots pine since AD 50 in northern Fennoscandia. Geogra. Ann. A82: 527–535.

Lindholm M., Lehtonen H., Kolström T., Meritainen J., Eronen M. and Timonen M. 2000. Climate signals extracted from ring-width chronologies of scots pines from the northern, middle and southern parts of the boreal forest belt in Finland. Silva Fennica 34: 317–330.

Magnuson J.J., Robertson D.M., Benson B.J., Wynne R.H., Livingstone D.M., Arai T., Assel A.A., Barry R.G., Card V., Kuusisto E., Granin N.G., Prowse T.D., Stewart K.M. and Vuglinski V.S. 2000. Historical trends in lake and river ice cover in the Northern Hemisphere. Science 289: 1743–1746.

Mangerud J., Anderson S.T., Berglund B.E. and Donner J. 1974. Quaternary stratigraphy of Norden, a proposal for terminology and classification. Boreas 3: 109–129.

Matthews J.A., Dahl S.O., Nesje A., Berrisford M.S. and Andersson C. 2000. Holocene glacier variations in central Jotunheimen, Southern Norway based on distal glaciolacustrine sediment cores. Quaternary Sci. Rev. 19: 1625–1647.

Mikalsen G. 1999. Western Norwegian fjords; recent benthic foraminifera, stable isotope composition of fjord and river water, and Late Holocene variability in basin water characteristics. Ph.D. thesis, University ersityof Bergen.

Mikola P. 1956. Tree-ring research in Finland. Tree-Ring Bull. 21: 16–20.

Moberg A., Tuomenvirta H. and Nordli Ø. 1993. Recent climate trends. In Seppälä M. (ed.), The Physical Geography of Fennoscandia. Oxford Regional Environmental Series, Oxford University Press, Oxford, 200 pp.

Moberg A., Bergström H., Krigsman J. and Svanered O. 2002. Daily air temperature and pressure series for Stockholm (1756–1998). Climatic Change 53: 171–212.

Moros M., Kuijpers A., Snowball I., Lassen S., Bäckström D., Gingele F. and McManus J. 2002. Were glacial iceberg surges in the North Atlantic triggered by Climatic warming? Mar. Geol. 192: 393–417.

Naurzbaev N.M., Vaganov E.A., Siderova O.V. and Schweingruber F.H. 2002. Summer temperatures in eastern Taimyr inferred from a 2427-year late-Holocene tree-ring chronology and earlier floating series. The Holocene 12: 727–736.

Nesje A. and Kvamme M. 1991. Holocene glacier and climate variations in Western Norway: Evidence for early Holocene glacier demise and multiple Neoglacial events. Geology 19: 610–612.

Nesje A., Dahl S.O., Løvlie R. and Sulebak J.R. 1994. Holocene glacier activity at the southwestern part of Harangerjøkulen, Central-Southern Norway: evidence from lacustrine sediments. The Holocene 4: 377–382.

Nesje A., Lie Ø. and Dahl S.O. 2000. Is the North Atlantic Oscillation reflected in Scandinavian glacier mass balance records? J. Quat. Sci. 15: 587–601.

Nesje A., Matthews J.A., Dahl S.O., Berrisford M.S. and Andersson C. 2001. Holocene glacier fluctuations of Flatebreen and winter-precipitation changes in the Jostedalsbreen region, Western Norway, based on glaciolacustrine sediment records. The Holocene 11: 267–280.

O'Brien S.R., Meyewski P.A., Meeker L.D., Meese D.A., Twickler M.S. and Whitlow S.I. 1995. Complexity of Holocene climate as reconstructed from a Greenland ice core. Science 270: 1962–1964.

Olander H., Birks H.J.B., Korhola A. and Blom T. 1999. An expanded calibration model for inferring lakewater and air temperatures from fossil chironomid assemblages in northern Fennoscandia. The Holocene 9: 279–294.

Parry M.L. (ed.) 2000. Assessment of Potential Effects and Adaptations for Climate Change in Europe: Summary and Conclusions. Jackson Environment Institute, University of East Anglia, Norwich, 24 pp.

Peglar S.M. and Birks H.J.B. 1993. The mid-Holocene Ulmus fall at Diss Mere, South-East England — disease and human impact? Veget. Hist. Archaeobot. 2: 61–68.

Reid G.C. 1987. Influence of solar variability on global sea surface temperatures. Nature 329: 142–143.

Rosén P., Segerström U., Eriksson L., Renberg I. and Birks H.J.B. 2001. Holocene climate change reconstructed from diatoms, chironomids, pollen and near-infrared spectroscopy at an alpine lake (Sjuodjijaure) in Northern Sweden. The Holocene 11: 551–562.

Sander M., Bengtsson L., Holmquist B. and Wohlfarth B. 2002. The relationship between annual varve thickness and maximum annual discharge (1901–1971 AD). J. Hydrol. 263: 23–35.

Schweingruber F.H., Bartholin T., Schär E. and Briffa K.R. 1988. Radiodensitometric-dendroclimatological conifer chronologies from Lapland (Scandinavia) and the Alps (Switzerland). Boreas 17: 559–566.

Schweingruber F.H., Briffa K.R. and Jones P.D. 1991. Yearly maps of summer temperatures in Western Europe from AD 1750 to 1975 and Western North America from 1600 to 1982: Results of a radiodensotometrical study on tree rings. Vegetation 92: 5–71.

Seppä H. 1996. Post-glacial dynamics of vegetation and tree-lines in the far north of Fennoscandia. Fennia 174: 1–96.

Seppä H. and Weckström J. 1999. Holocene vegetational and limnological changes in the Fennoscandian tree-line area as documented by pollen and diatom records from Lake Tsuolbmajarvi, Finland. Ecoscience 6: 621–635.

Seppä H. and Hammarlund D. 2000. Pollen-stratigraphical evidence of Holocene hydrological change in northern Fennoscandia supported by independent isotopic data. J. Paleolim. 24: 69–79.

Seppä H. and Birks H.J.B. 2001. July mean temperature and annual precipitation trends during the Holocene in the Fennoscandian tree-line area: pollen based climate reconstructions. The Holocene 11: 527–539.

Seppä H., Nyman M., Korhola A. and Weckström J. 2002. Changes of the tree-lines and alpine vegetation in relation to post-glacial climate dynamics in northern Fennoscandia based on pollen and chironimid records. J. Quat. Sci. 17: 287–301.

Sernander R. 1908. On the evidences of postglacial changes of climate furnished by the peat-mosses of Northern Europe. Geol. Fören. Stock. For. 30.

Sirén G. 1961. Skogsgränstallen som indikator för klimafluktuationerna i norra Fennoskandien under historisk tid. Communicationes Instituti Forestatis Fenniae 54: 66 pp.

Snowball I.F. and Sandgren P. 1996. Lake sediment studies of Holocene glacial activity in the Kårsa valley, Northern Sweden: contrasting opinions. The Holocene 6: 367–372.

Snowball I.F., Sandgren P. and Petterson G. 1999. The mineral magnetic properties of an annually laminated Holocene lake sediment sequence in Northern Sweden. The Holocene 9: 353–362.

Snowball I.F., Zillén L. and Gaillard M.-J. 2002. Rapid early Holocene environmental changes in Northern Sweden based on studies of two varved lake sediment sequences. The Holocene 12: 7–16.

Sorvari S., Korhola A. and Thompson R. 2002. Lake diatom response to recent Arctic warming in Finnish Lapland. Global Change Biol. 8: 171–181.

Tabony R.C. 1981. A principal component and spectral analysis of European rainfall. J. Climatol. 1: 283–294.

Tett S.F.B., Stott P.A., Allen M.R., Ingram W.J. and Mitchell J.F.B. 1999. Causes of twentieth-century temperature change near the Earth's surface. Nature 399: 569–572.

Tuomenvirta H., Alexandersson H., Drebs A., Frich P. and Nordli P.O. 2000. Trends in Nordic and Arctic temperature extremes and ranges. J. Climate 13: 977–990.

Vaganov E.A., Shiyatov S.G. and Mazepa Y.S. 1996. Dendroclimatic Study in Ural-Siberian Subarctic. Novasiberisk: Nauka Siberian Publishing Forum RAS 24, 245 pp. (in Russian).

Vasko K., Toivonen H. and Korhola A. 2000. A Bayesian multinomial Gaussian response model for organism-based environmental reconstruction. J. Paleomin. 24: 243–250.

Visbeck M. 2002. The ocean's role in Atlantic climate variability. Science 297: 2223–2224.

Weckström J. and Korhola A. 2001. Patterns in the distribution, composition and diversity of diatom assemblages in relation to ecoclimatic factors in Arctic Lapland. J. Biogeography 28: 31–46.

Wood R.A., Keen A.B., Mitchell J.F.B. and Gregory J.M. 1999. Changing spatial structure of the thermohaline circulation in response to atmospheric CO_2 forcing in a climate model. Nature 399: 572–575.

Zetterberg P., Eronen M. and Lindholm M. 1996. The mid-Holocene climatic change around 3800 BC: tree-ring evidence from Northern Scandinavia. Paläeoklimaforschung 20: 135–146.

Zillén L.M., Wastegård S. and Snowball I.F. 2002. Calendar year ages of three mid-Holocene tephra layers identified in varved lake sediments in West Central Sweden. Quat. Sci. Rev. 21: 1583–1591.

23. RECENT DEVELOPMENTS IN HOLOCENE CLIMATE MODELLING

HANS RENSSEN (hans.renssen@geo.falw.vu.nl)

Faculty of Earth and Life Sciences
Vrije Universiteit Amsterdam
De Boelelaan 1085, NL-1081 HV Amsterdam
The Netherlands

PASCALE BRACONNOT (pasb@lsce.saclay.cea.fr)

Laboratoire des Sciences du Climat et de l' Environnement
UMR CEA-CNRS 1572
CEA Saclay bat 709
91191 Gif sur Yvette cedex
France

SIMON F.B. TETT (simon.tett@metoffice.com)

Hadley Centre for Climate Prediction and Research
(Reading Unit)
Met Office
Meteorology Building, University of Reading
Reading, RG6 5615
UK

HANS VON STORCH (hans.von.storch@gkss.de)

GKSS-Forschungszentrum Geesthacht GmbH
Max-Planck-Straße
D-21502 Geesthacht
Germany

S.L. WEBER (weber@knmi.nl)

Royal Netherlands Meteorological Institute KNMI
P.O. Box 201
NL-3730 AE De Bilt
The Netherlands

Keywords: Recent developments, Climate models, Holocene, General circulation models, Earth system models of intermediate complexity, Internal variability, 20th century, Glacier response

R. W. Battarbee et al. (eds) 2004. *Past Climate Variability through Europe and Africa*.
Springer, Dordrecht, The Netherlands.

Introduction

To improve our understanding of climate variability on decadal to centennial time-scales, it is crucial to use a hierarchy of climate models in addition to palaeoclimate reconstructions based on proxy data. Climate models give a physically consistent overview of the global climate on all time-scales. They are useful tools in palaeoclimatology, since: (i) they can be used to test hypotheses that have been inferred from palaeo-data; and (ii) they can provide plausible explanations of observed phenomena (e.g., Isarin and Renssen (1999), Kohfeld and Harrison (2000)). In recent years, considerable progress in palaeoclimate modelling has been made with the extensive use of models that consider the coupling of the different components of the climate system (atmosphere, ocean, sea-ice, vegetation). The aim of this paper is to inform the palaeo-data community on recent developments in palaeoclimate modelling, with special reference to the Holocene climate. In the first section, different model types and experiments are discussed, together with a short overview of Holocene climate modelling studies and differences between models and palaeo-data. In the second section, three important issues are further illustrated by discussing in detail three studies that use state-of-the-art models.

Different types of models

Different types of numerical models of the coupled atmosphere-ocean system vary in complexity. For an extensive overview of the history of numerical climate models, the reader is referred to McGuffie and Henderson-Sellers (2001). It suffices here to summarise the main current model types. It is convenient to make a distinction between complex climate models, simple climate models, and models of intermediate complexity.

Complex models: General Circulation Models

The standard complex models that are currently operational, are so-called atmosphere-ocean general circulation models (or AOGCMs, see e.g., McAvaney et al. (2001), IPCC (2001)). These models have been developed in the last 15 years, based on the combination of atmospheric and oceanic general circulation models (AGCMs and OGCMs, respectively) that have been developed since the 1970s. AOGCMs can be thought of consisting of a three-dimensional matrix with a global coverage, in which the dynamics of atmosphere and ocean are simulated based on physical laws. In the atmosphere, the models simulate the weather, including climatic parameters such as temperature, wind and humidity. In the oceanic part, ocean currents, temperature and salinity are computed. Furthermore, a sea-ice model is included that simulates sea-ice thermodynamics and dynamics. At the atmosphere-ocean interface, AOGCMs calculate the vertical exchange of momentum, heat and fresh water. Most AOGCMs include the simulation of both an annual and diurnal cycle. The typical horizontal resolution is 125 to 250 km, whereas the vertical resolution is generally about 20 layers for both the atmosphere and ocean (IPCC 2001). Processes that take place on smaller spatial scales (so-called sub-grid scale processes, e.g., cloud formation) cannot be computed, so that their effects have to be parameterized (i.e., described in a simple way based on the relationship with large-scale variables).

AOGCMs require the definition of some upper and lower boundary conditions. The upper boundary condition is the amount of solar radiation received at the top of the atmosphere, which varies as a function of the seasonal cycle. Together with changes in the concentration of atmospheric trace gases (such as CO_2 and CH_4), this variation in solar radiation can be considered as the driving force of the model. Lower boundary conditions are related to the characteristics of the earth's surface, such as bathymetry and land topography, but also land surface characteristics such as albedo and soil type. Recently developed AOGCMs incorporate complex land-surface schemes, some of which include a dynamical vegetation component (i.e., AOVGCMs, e.g., Cox et al. (2000)). In these AOVGCMs, the potential vegetation types (or biomes) are calculated based on bioclimatic parameters that determine their distribution, such as precipitation and the length of the growing season. Subsequently, the relevant characteristics of the biomes, such as surface albedo and roughness, are fed back into the atmospheric component. Consequently, in AOVGCMs it is no longer necessary to prescribe the vegetation characteristics as in AOGCMs. Similarly, in AOGCMs the temperatures of the ocean surface are calculated and not prescribed as in AGCMs. Thus, the lower boundary conditions that have to be prescribed depend on the components of the climate system that are included in the climate model. A drawback of AOGCMs is that their complexity requires time-consuming calculations on super-computers, which makes their application expensive, and transient experiments are limited to 500 to 1000 years around specific time periods in the past.

Simple models: Energy Balance Models

At the other end of the complexity spectrum are simple climate models. In this family of models, energy balance models (EBMs) are the most widely applied category in palaeoclimatology. Generally, latitude is the only dimension that is varying in EBMs. As their name suggests, EBMs calculate the energy balance for latitudinal zones, followed by a prediction of the surface temperature. Atmospheric dynamics are not included in EBMs, as heat transport is calculated on the basis of diffusion. Typically, the grids used are much coarser that applied in GCMs. As a consequence of the simplified calculation and lower resolution, EBMs run on simpler computers and can be applied on much longer time-scales. EBMs have been widely applied to study climate change at a hemispheric-to-global scale (e.g., Crowley (2000)).

Earth System Models of Intermediate Complexity

A recent line in model development is focused on building earth system models of intermediate complexity (or EMICs, e.g., Claussen et al. (2002)). EMICs incorporate climatic sub-systems (i.e., atmosphere, ocean, biosphere, land ice) in a simplified but efficient manner, while retaining those properties of the climate system that are relevant for specific research purposes (Pethoukov et al. 2000). Within the family of EMICs, large differences in complexity exist. Some EMICs include a two dimensional description of the climate system, with an EBM as atmospheric component. On the other hand, some EMICs are very close to GCMs in the sense that they contain a three-dimensional description of atmospheric dynamics on time-scales ranging from synoptic to millennial (e.g., Opsteegh et al.

(1998)). EMICs have in common that they incorporate a large number of parameterizations compared to GCMs. As a result of the reduced complexity, EMICs are relatively fast and are ideal to study the interaction between various components on centennial-to-millennial time-scales. GCMs and EMICs have been recently expanded with process-based forward models (transfer equations) that simulate the behaviour of proxy indicators themselves (e.g., biomes, glaciers, isotopes; see example in this chapter). With the inexorable increase in computing power through Moore's law, an EMIC may be an earlier and lower resolution version of a current AOGCM.

Different types of experiments

Equilibrium experiments

For a proper understanding of palaeoclimate modelling studies, it is important to know the difference between equilibrium and transient experiments. In equilibrium experiments, the boundary conditions are changed and kept time invariant (except for the annual and diurnal cycles). Where the focus is on the equilibrium response, the normal experimental design is to use a simplified and non-dynamical ocean. It is possible to perform such an equilibrium experiment with forcings for the present-day, and also for a period in the past (e.g., a snapshot experiment for 6 kyr BP). Thus an equilibrium experiment shows the equilibrium response of the climate system to the forcings at a given moment in time, giving insight into the 'background' internal variability of the model at various time-scales. Often an equilibrium experiment with present-day boundary conditions is called a "control" experiment, which is used as a reference experiment to which simulations with different boundary conditions are compared to analyse the model's sensitivity to a particular forcing. Later in this chapter an example of a control run is discussed.

Transient experiments

In transient experiments, time-dependant forcings are used. These forcings may be changes in solar radiation and greenhouse-gas concentrations, but also "slow" factors like orbital parameters if the interest is in climate change at relatively long time-scales (centuries to millennia). This implies that the model adjusts constantly to the 'external' changes. As will be shown in an example later on, a comparison of a forced run with the control run provides valuable information on the effect of forcings on the dynamics of the coupled system. Transient experiments on "palaeoclimatic" time-scales (multi-millennial) are only feasible with simple models and EMICs, as coupled GCMs are still too expensive to make such long runs.

Data assimilation

A third type of model experiment exists that is likely to play an important role in palaeo-climate modelling in the near future. In this type of simulation, the model is driven by data, as is commonly applied in modern weather forecasting. In so-called data assimilation, meteorological observations are used from selected sites (i.e., stations) to drive (or 'nudge') the climate model toward the state presented by the data. This method is not yet commonly applied in palaeoclimatology, but in the near future it will be used for past climates by introducing proxy data into the model (see von Storch et al. (2000)).

Holocene climate modelling studies

Climate model experiments on Holocene climate have been performed since the 1980's (e.g., Kutzbach (1981), Kutzbach and Otto-Bliesner (1982)). In the early stages, AGCMs were used to study the effects of orbital variations on the climate of the early-to-mid Holocene by performing snapshot experiments for 9, 6 and 3 kyr BP. The latter simulations were part of a series that started at the last glacial maximum (21 kyr BP) and went forward at 3 kyr intervals. They were done primarily within the Cooperative Holocene Mapping Project (COHMAP, e.g., Kutzbach and Street-Perrott (1985), Kutzbach and Guetter (1986), COHMAP members (1988), Wright et al. (1993)). The simulations for the early Holocene showed clearly the strengthening of the summer monsoons in response to orbital forcing, which is in agreement with reconstructed high lake levels in North Africa and Southeast Asia (e.g., Kutzbach and Street-Perrott (1985)).

A few years later, AGCMs were coupled to simplified ocean models (so-called mixed layer models) to simulate the 9 kyr BP climate, thereby taking the exchange between atmosphere and ocean into consideration (Mitchell et al. 1988; Kutzbach and Gallimore 1988). In the 1990's, the Palaeoclimate Modelling Intercomparison Project (PMIP, e.g., Joussaume and Taylor (1995)) aimed at comparing the performance of GCMs for climatic conditions different from today, by setting up equilibrium palaeoclimate simulations with identical sets of boundary conditions. They focussed at the 21 and 6 kyr BP time-slices. For the 6 kyr BP experiments it was assumed that the orbital forcing was the primary forcing factor affecting the climate system. Therefore, all other boundary conditions were kept at present-day values. The results of PMIP made it clear that all AGCMs capture the large scale features of the 6 kyr BP climate. However, some systematic discrepancies were found compared to data reconstructions. In Northern Africa, for instance, AGCMs clearly underestimated the strengthening of the summer monsoons as indicated by lake-level studies and palaeo-vegetation reconstructions (Joussaume et al. 1999). Further studies showed that part of this model-data mismatch can be attributed to the lack of oceanic and land-surface feedbacks in the atmospheric models (e.g., Kutzbach and Liu (1997), Broström et al. (1998), Harrison et al. (1998), Braconnot et al. (1999, 2000a, 2000b), DeNoblet et al. (2000), Texier et al. (1997, 2000)). However, even fully coupled AOGCMs and EMICs (including ocean and vegetation components) underestimate the African monsoon amplification at 6 kyr BP (Ganopolski et al. 1998; Hewitt and Mitchell 1998; McAvaney et al. 2001).

The 6 kyr BP simulation results for Europe showed that most PMIP models were able to reproduce the summer warming reconstructed using pollen data (Masson et al. 1999). Also, the reconstructed drier conditions in Northwestern Europe and wetter conditions in Southern Europe were simulated by some models. However, the characteristic winter temperature pattern of warming in Northeastern Europe was poorly reproduced, probaly due to different sea surface temperature patterns in the Atlantic Ocean at 6 kyr BP that were not included in the PMIP simulations (Masson et al. 1999). The reader is referred to Braconnot et al. (this volume) for an overview of the PMIP 6 kyr BP simulations carried out with coupled atmosphere-ocean models.

In addition to the PMIP-type equilibrium simulations, transient experiments have recently been completed, in which time-dependent forcings are applied to simulate the evolution of the Holocene climate. Fully coupled AOGCMs have been used to look at the last few hundred years (e.g., Cubasch et al. (1997), Tett et al. (1999)), whereas EMICs have been used to take the entire Holocene into account (Claussen et al. 1999; Weber 2001; Brovkin et al. 2002; Crucifix et al. 2002). Furthermore, EMICs are now used to study specific Holocene climate events in transient experiments, such as the 8.2 kyr BP cooling event (Renssen et al. 2001a; 2002) and the termination of the African Humid Period at ~6 kyr BP (Claussen et al. 1999; Renssen et al. 2003).

Scale differences between models and palaeo-data

If we want to make a useful comparison of palaeo-data with palaeoclimate model simulations, it is important that both represent similar spatial scales. In AOGCMs, the skill with which mean climate and climate variability is reliably simulated depends on the spatial scale (von Storch 1995). Large-scale features, reflecting global differential heating, the impact of continental-scale patterns of mountain ranges and land-sea contrasts are usually well reproduced, whereas smaller scales are not well described because of insufficient resolution of the physiographic details, in particular secondary mountain ranges (like the Alps), marginal seas (like the North Sea) or specifics of land use. The Third Assessment Report of the IPCC (IPCC 2001) found that AOGCMs fare realistically on spatial scales on 10^7 km^2 and more, while smaller scales should be considered less reliably simulated in global climate models (Giorgi et al. 2001). This limitation of climate models represents a significant obstacle for co-operation between palaeoclimatologists and climate modelers, as the former collect data and generate knowledge mostly on scales much smaller than 10^7 km^2, although there are exceptions to this. Thus comparison of climate model data with palaeoclimatic evidence is methodically difficult.

A solution to this dilemma is given by the "downscaling" concept. According to this concept, the state and statistics of smaller scales are a function of the state and statistics of the larger scales. The details of the "downscaling" function are determined by the physiographic details of the considered region or locality.

Two main classes of downscaling are in use (for an overview, refer to Giorgi et al. (2001)). First, for the "dynamical method" (Giorgi 1990) a regional climate model with a high spatial resolution is "nested" in a GCM with a relatively low resolution. The regional model is driven at the boundary, and possibly in the interior, by large-scale information taken from the global model. The skill of this approach was recently demonstrated by the "Big Brother" experiment of Denis et al. (2002). This technique has matured in the past few years (e.g., Whetton et al. (2001)) and has been applied to palaeoclimates (e.g., Hostetler et al. (1994), Renssen et al. (2001b)). Second, the "empirical method" (Wigley et al. 1990) assumes that an empirically determined link between the large-scales and the regional or local scales remains valid also under changing conditions. Then the link is used to interpret the large-scale output of a climate model. The method has been applied to various parameters, both meteorological (like precipitation) and proxy (like phenological dates) data.

The inverse of these downscaling methods is used for the reconstruction of relevant atmospheric patterns with the help of local proxy-data (e.g., Appenzeller et al. (1998),

Crueger and von Storch (2001)). This method, named "upscaling", is also a key tool in attempts to force climate models to incorporate specific proxy-data based information (data-driven experiments, see above, von Storch et al. (2000)).

Recent examples

In this section, some recent developments in climate modelling are discussed that are relevant to palaeoclimatology. First, natural variability, as simulated in an AOGCM control run, is discussed. Second, the potential of performing a series of transient experiments with varying forcings is considered, using an example of AOGCM simulations for the 20th century. Third, forward modelling is further explained by discussing the simulation of Holocene glacier length variations driven by a transient EMIC simulation.

Example 1: internal climate variability in a control run

As already mentioned, the purpose of control runs is to simulate an equilibrium climate, given the repeated cycles of radiation and the constant values of greenhouse-gas concentrations, and its variability. Control runs have demonstrated the skill of state-of the-art AOGCMs to reproduce reliably the large-scale aspects of contemporary climate. Control runs prepared with different climate models produce rather similar results on large scales, whereas on regional scales there may be marked differences (McAvaney et al. 2001). The important point, often overlooked by geoscientists, is that the climate is varying for internal reasons independent of any time dependent forcing (except for the prescribed but constant annual insolation cycle). In fact, the proper way to conceptualise climate as a process is to think about it as a random process, whose characteristics are conditioned by certain external factors. When the latter are fixed, the characteristics are fixed, when they are time dependent (or transient), the characteristics are time dependent as well.

The random character of the climate system in state-of-the-art climate models is demonstrated in Figure 1, which displays a 1000-years time series of air temperature at the ground averaged over different regions. The model used is ECHO-G, an AOGCM developed in Hamburg (Legutke and Voss 1999). Apart from a slight downward trend, the time-series is characterised by irregular variations with significant deviations from the "normal" (the long-term mean). For instance, in the 6th century, prolonged warming appears in Europe, with maximum anomalies of more than 0.5 K. Comparison of the time-series for Europe and the Northern Hemisphere also shows that the amplitude of climate variability decreases when averaging over a larger area.

The dynamical background of these seemingly spontaneous variations lies mainly in the chaotic dynamics of the atmosphere. That non-linear systems can develop such irregular variations has been known for many decades. The difference from, for instance, Lorenz' famous 3-component system (Lorenz 1976) is that the dynamics in the real atmosphere and, to a lesser degree, in state-of-the-art climate models has very many degrees of freedom, with many of them behaving chaotically. The sum of these many chaotic processes is not a "simple" pattern of variations, as non-linear dynamical theorists like to show, but the truly "chaotic" variations as shown in Figure 1. Being in character multiple non-linear, the

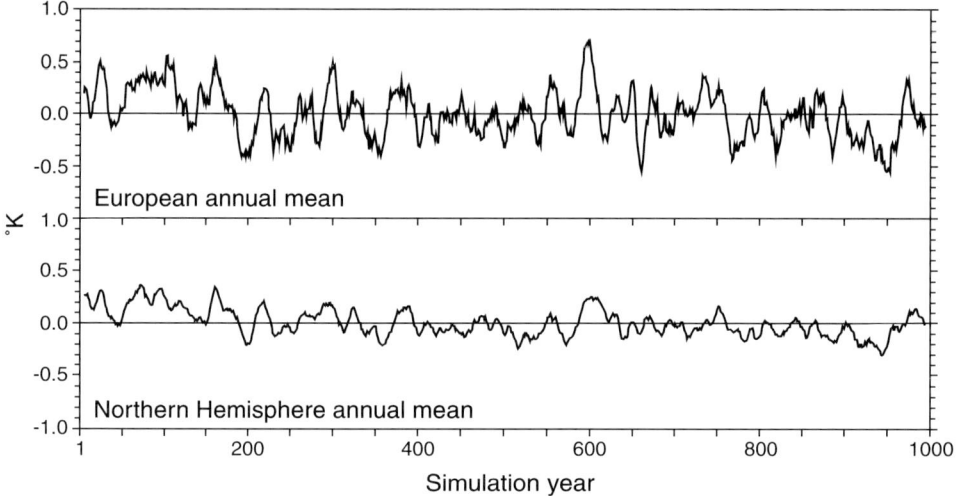

Figure 1. Time-series of area-averaged air temperature near the ground in a 1000 year control run (Zorita, pers. comm.) simulated by an AOGCM.

phenomena can consistently and efficiently be described with the mathematical concept of randomness (von Storch et al. 2001).

The "internally generated" variability, as simulated in control runs, makes the analysis of climate change signals difficult, as these signals are often masked by the noise. This is one of the problems that are faced in the detection of natural and anthropogenic signals in climate (Hasselmann 1979), in particular for assessing the reality of global warming (e.g., Hegerl et al. (1997)). For palaeoclimatology it is equally important to obtain a good signal-to-noise ratio in palaeoclimate simulations and climate reconstructions based on proxy records.

Example 2: simulating the 20th century using transient ensemble experiments

In this section we give an example of transient simulations carried out with a state-of-the art AOGCM. It is important for the palaeo-data community to recognise the difference between this type of experiment compared to the 'standard' PMIP-type equilibrium experiment, as it is expected that transient simulations will become more important in palaeoclimatology as computer power increases.

The numerical experiments were carried out at the Hadley Centre with the aim of understanding 20th century temperature change. The model used for these simulations was HadCM3 (Pope et al. 2000; Gordon et al. 2000); the third generation Hadley Centre coupled model (AOGCM). Four ensembles of experiments were run. Each ensemble used the same forcing but started from different initial conditions. These initial conditions represent the state of the coupled ocean-atmosphere system in the model at the start of the simulation. It is necessary to perform multiple experiments (i.e., ensembles) with slight changes in these initial conditions, because it is useful to consider the range of internally generated year-to-year variability in the model (see example 1). As a coupled climate model is sensitive to the initial state, each member of the ensemble gives a different, but equally valid, solution. Thus ensembles are needed to average out the noise

The average of each ensemble is used to determine the response to the forcing. Each ensemble was for the period 1860 to 1999. The four ensembles are:

- **GHG** – forcing with well-mixed greenhouse gases alone. Changes in CO_2, CH_4, NO and several (H)CFC's are included in these simulations.

- **Anthro** – forcing with changes in well-mixed greenhouse gases, anthropogenic sulphate aerosols and ozone. Both the direct scattering effect of aerosols and their indirect effect of increasing cloud albedo are included in the simulations. Note that the poorly understood indirect effect is the dominant negative forcing in these simulations. Ozone changes include both tropospheric increases and, since 1974, stratospheric decreases.

- **Natural** – forcing with changes in volcanic aerosol and solar irradiance changes.

- **All** – forcing with both natural and anthropogenic forcing.

More details of the experiments and their analysis can be found in Stott et al. (2000) and Tett et al. (2002). These experiments are compared with estimates of near-surface temperature changes from Parker et al. (1994, updated to 1999). The **All** ensemble average performs well in reproducing observed global-mean temperature changes (Fig. 2). None of the other simulations manages to reproduce the observed changes. The **All** simulation also captures the large-scale patterns of temperature change (not shown but see Stott et al. (2000)). Thus both natural and anthropogenic forcings are required to explain changes in 20th century temperatures. Closer examination of the **Anthro** simulations shows that very little net anthropogenic warming happened prior to the mid 1960s (some warming then cooling). By comparing this with the results from the **GHG** ensemble this seems to be due to the impacts of the other anthropogenic forcings and in particular the cooling effect of sulphate aerosol on climate. In the **Natural** ensemble the climate warms until the 1960s when it cools again. Over the 20th century as a whole natural forcings cause no net warming.

We extend the analysis by making the simple assumption that the observations are a linear combination of the simulated signals plus climate noise:

$$\mathbf{y} = \beta\mathbf{X} + \mathbf{u}$$

where $\mathbf{y}$ is the observations, $\mathbf{X}$ a matrix of simulated signals, $\mathbf{u}$ a realisation of climate noise and β a vector of amplitudes applied to the simulated signals. Climate noise is estimated from simulated climate variability and β is estimated using the observations and the ensemble average signals. We carry out an analysis of temperature change for the period 1897 to 1997 using decadal mean observed and simulated temperature anomalies. We considered only three signals — **GHG**, **Anthro** and **Natural**. The **GHG** and **Anthro** signals were transformed to give **G** (the effect of greenhouse gases alone) and **SO** (the effect of sulphates and ozone). We then estimate amplitudes and uncertainty ranges (5– 95%) for these signals. Uncertainty estimates are computed from a control simulation (see above). From the amplitudes and uncertainty estimates we can estimate the linear-trend in temperature from different forcings. Figure 3 shows the results of this analysis for the entire century and for two fifty-year periods. Note that all trends, including their uncertainties,

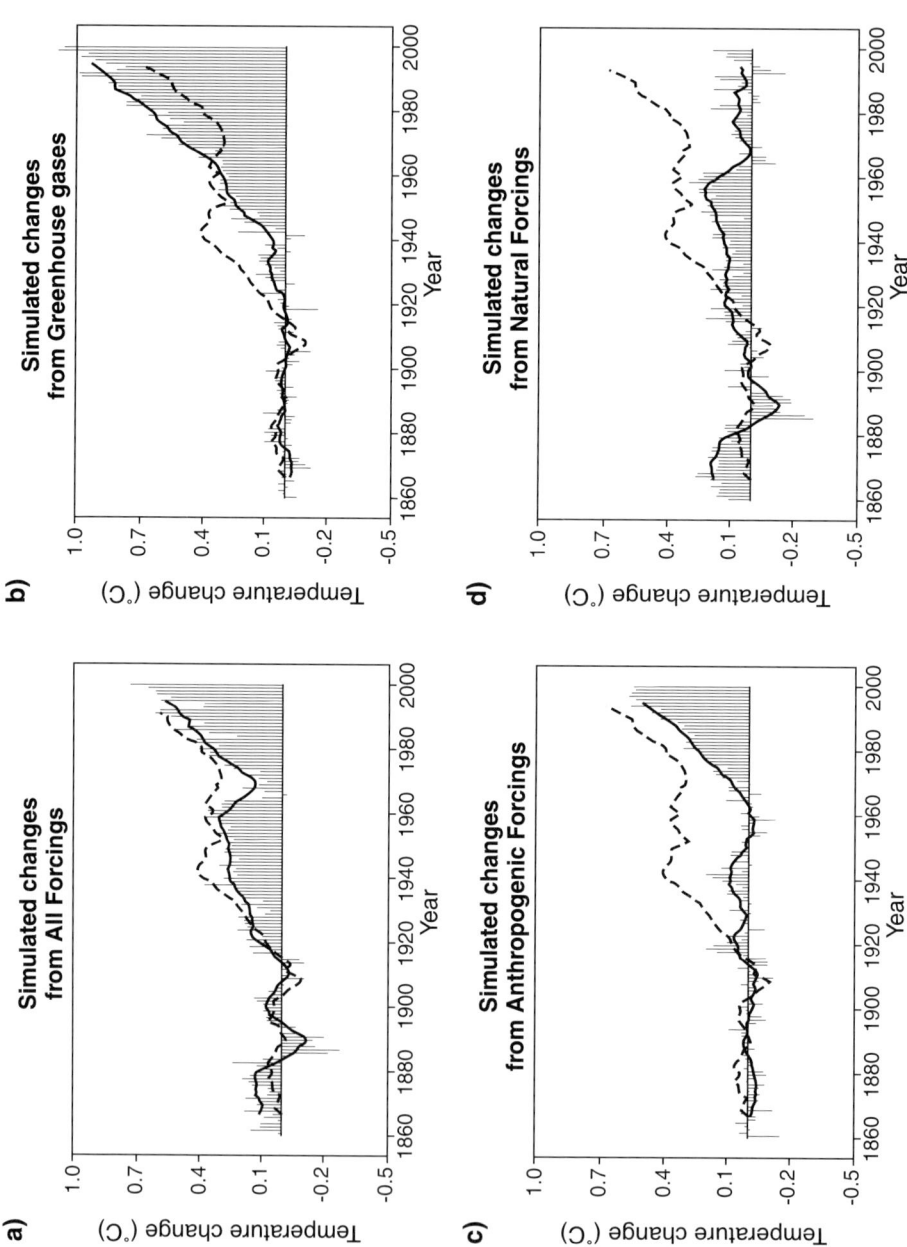

Figure 2. Time-series of simulated and observed global-mean temperatures. a) **All**, b) **GHG**, c) **Anthro**, d) **Natural** ensembles. All values are anomalies relative to 1890–1919 and the simulated values had data discarded where there were no observed data. The bars are annual-averages while the thick line shows a 10-year running mean. The observations are shown as a dashed line.

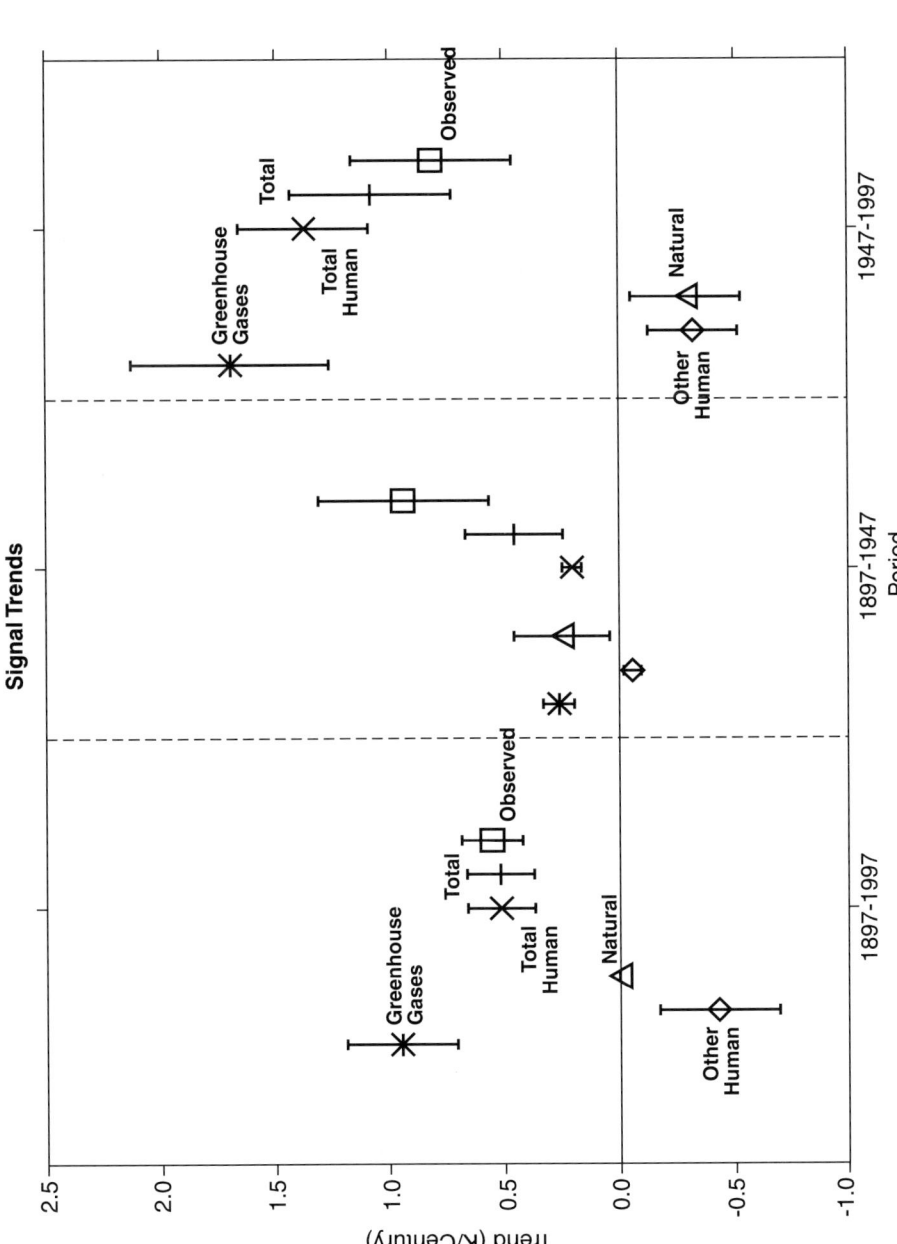

Figure 3. Best-estimate linear trend and uncertainty ranges (K/century) for G (left error bar with diamond), Natural (right error bar with triangle). The error bars show the 5 to 95% uncertainty ranges. The best-estimate trend is shown as a symbol at the centre of the bar. Also shown is the total anthropogenic trend (x), total trend (+) and observed trends (square). Uncertainty in the observations is estimated from internal climate variability simulated by the **Control** simulation.

exclude zero and are of the expected sign — all signals are detected. Thus we have positive evidence that all the signals we consider are present in the observations.

Over the 20th century anthropogenic forcings cause a warming trend of 0.5 ± 0.15 K/century ("total human" in Fig. 3). The trend due to greenhouse gases is 0.9 ± 0.24 K/century while remaining anthropogenic factors cool at a rate of 0.4 ± 0.26 K/century ("other human" in Fig. 3). The uncertainty in the total anthropogenic warming trend is less than the uncertainties in the individual trends as they are correlated with one another. Over the century natural forcings contribute little to the observed trend ("natural" in Fig. 3).

Our analysis considers only uncertainty in the amplitude of the simulated response and neglects uncertainty in the time-dependence of the forcing and in the spatial patterns of response, as well as uncertainties in the observations. However our best estimates are consistent with the observations. Furthermore in a single ensemble of simulations forced with both natural and anthropogenic forcings changes in simulated near-surface temperature are consistent with those observed (see Fig. 2) suggesting that those uncertainties may not be too great.

During the first half of the 20th century, greenhouse gases and natural forcings cause warming trends of about 0.2 to 0.3 K/century, while other anthropogenic factors produce negligible cooling trends (Fig. 3). Over the last half of the century, greenhouse gases warm the climate at a rate of 1.7 ± 0.43 K/century with natural forcings (largely volcanic aerosol) and other anthropogenic factors (mainly the indirect effect of sulphate aerosols) both causing an estimated cooling trend of about 0.3 ± 0.2 K/century. Thus, since 1947, changes in aerosol concentrations (anthropogenic and natural) have offset about a third of the greenhouse gas warming. Results presented here all rely on simulated internal variability to assess consistency or to estimate uncertainty ranges. We find that amplifying simulated internal variability by a factor of two still leads to a detection of the effects of greenhouse gases but no detection of other anthropogenic effects or those of natural forcings.

The above analysis shows how multiple transient simulations can be used to quantify the contribution of various forcing factors to observed climate change. It is expected that similar transient simulations will be performed in the near future for past centuries, e.g., the last 1000 years, including the Little Ice Age, or even earlier millennia. Consequently, this type of analysis will enable palaeoclimatologists to obtain a better understanding of the mechanisms behind climatic changes observed in proxy data.

Example 3: forward modelling

Clearly it is crucial to develop an understanding of proxy data in order to be able to interpret past climatic fluctuations registered by proxy data. To achieve this aim, models of proxy data are being developed and tested. These so-called "forward models" are process-based environmental models (physical, biological, chemical, or empirical), which are driven by climate model output to simulate a synthetic proxy record or time series, which can be directly compared with the actual proxy data (Weber and Von Storch 1999). Some forward models are dynamically incorporated within climate models, so that their results feed back into the atmospheric component and thus can have an effect on the simulated climate. Examples are the simulation of biomes in AOVGCMs, (Harrison et al. 1998), the inclusion of large lakes (Hostetler et al. 1993) and the incorporation of dust uptake, transport and deposition (e.g., Mahowald et al. (1999)). Other forward models operate off-line and are only fed by the climate model output (one-way procedure). Recent applications of this type are e.g., wetlands, lakes and rivers (Coe 1995; 1997), stable water isotopes (Jouzel et al. 2000; Werner et al. 2000), ocean sediment cores (Heinze 2001), glaciers (Reichert et al. 2001) and local sea level (van der Schrier et al. 2002). Such forward models consist of two components: one, a process-based model of the proxy parameter itself and two, a component transferring the climatic conditions to a forcing term driving the first component. Here we give an example where synthetic glacier length records are generated for the Holocene epoch using a glacier model coupled to the intermediate-complexity climate model ECBilt (Weber and Oerlemans 2003). The glacier model consists of a mass-balance component and an ice-flow component.

The climate model ECBilt is a dynamic, three-dimensional EMIC. It is computationally fast because its atmospheric component has a coarse vertical resolution (i.e., 3 layers) and simplified parameterisations. The Holocene climate is simulated in a 10,000-year transient experiment driven by orbital insolation changes (Weber 2001). All other boundary conditions (orography, concentration of trace gases and surface characteristics) are set to their present-day values. It is thus an idealised experiment, aimed at a better understanding of the glacier response to orbital forcing. The simulated climatic signal can be compared directly to proxy-based reconstructions of the climate during, for example, the mid-Holocene climate at 6 kyr BP. In addition, the implied glacier length changes provide an indirect validation of the simulated climatic signal.

We consider three glaciers, ranging from maritime to continental. The simulated glacier lengths are shown in Figure 4. The long-term trends in glacier length are associated with the orbital forcing. Nigardsbreen (Southern Norway) shows a phase of rapid expansion during the mid-Holocene, followed by more gradual growth. Here, the long-term trend in the annual mass balance is primarily determined by high summer temperatures in the early Holocene. There is a smaller contribution from enhanced winter precipitation. Both the temperature signal (Cheddadi et al. 1997) and the precipitation signal (Dahl and Nesje 1996) seem consistent with proxy data. At Rhonegletscher (the Swiss Alps) there is a maximum glacier extent at 3–5 kyr BP, which seems unlikely. ECBilt simulates warmer and wetter summers in the early Holocene. Proxy data confirm the temperature signal (Cheddadi et al. 1997), but they are not conclusive with respect to the precipitation signal. Earlier time-slice model experiments also differ widely in the simulated water budget response (Masson et al. 1999). The present results suggest that the simulated precipitation response is not realistic.

The climatic response at the Abramov glacier (Kirghizia) is characterised by enhanced summer precipitation, associated with a northward extension of the Indian monsoon reaching its maximum at 6 kyr BP. This signal is well known (Joussaume et al. 1999). The timing of the maximum, which is consistent with lake level data (Harrison et al. 1996), can be understood from the lagged response of the monsoon system to the orbital forcing. September precipitation, which dominates the transient signal, peaks around 6 kyr BP following August insolation. The simulated glacier length reaches a pronounced postglacial maximum at that time. This result indicates that the northward extension of the monsoon is coincident with glacial advance, which should be recognisable in the field by the remains of end moraines.

There is considerable variability on time-scales shorter than millennial in the simulated glacier length records. These length variations are due to internal climatic variability. They

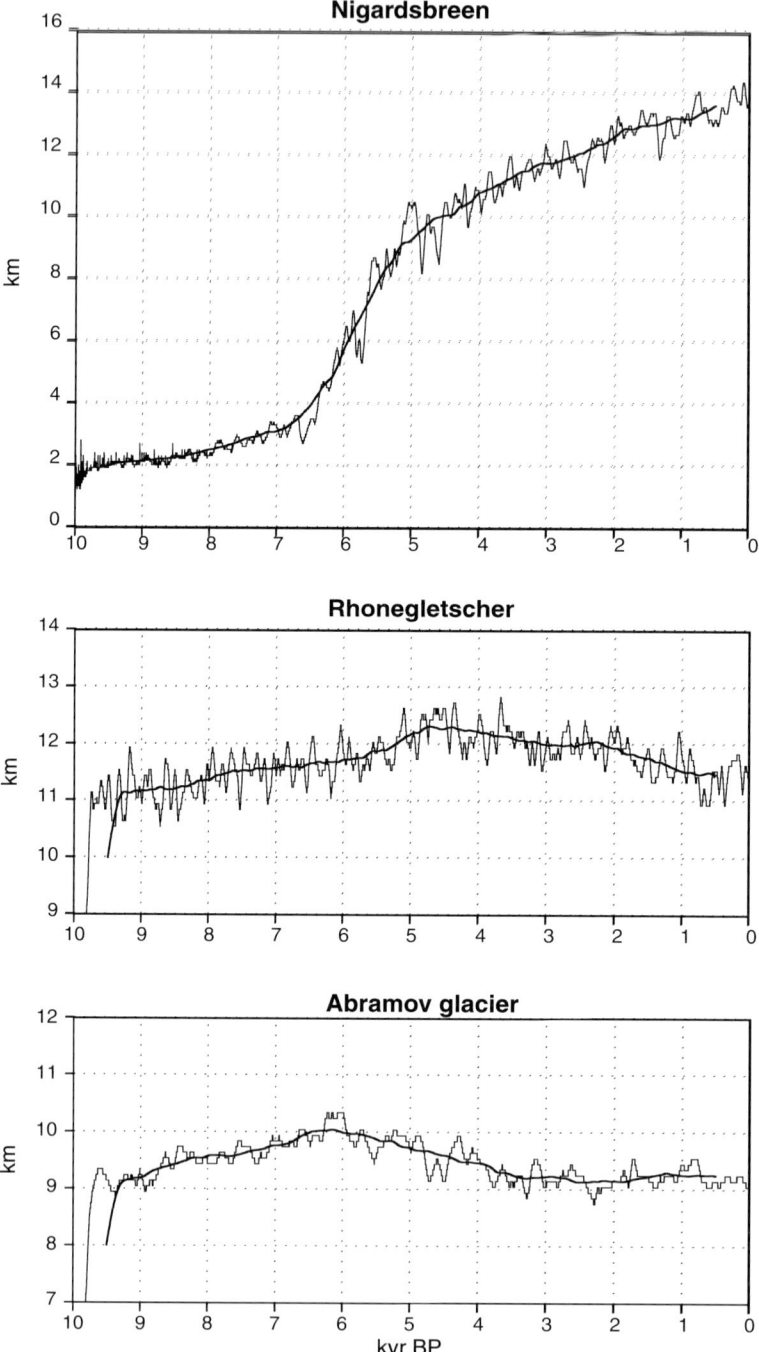

Figure 4. The glacier length (in km) as simulated by ECBilt for the three glaciers Nigardsbreen, Rhonegletscher and Abramov glacier as a function of time (in kyr BP). The long-term trends, as given by the 999-yr running mean, are also shown.

are typically asynchronous among the three different glaciers. Length variations can be shown to behave as a lagged moving-average process, with a glacier-specific memory.

Conclusion

This paper has discussed recent developments in Holocene climate modelling that are important for narrowing the gap between climate modelling and palaeoclimatology. The evolution of climate models is rapid; a process that is partly governed by the development of increasingly fast computers and partly by the needs of the climate community. Standard general circulation models are now describing the coupled atmosphere-ocean system, and dynamical vegetation models are presently being incorporated. This means that, at present, the dynamics of the climate system can be simulated far more completely than by atmospheric GCMs which were the standard models until recently. A more complete incorporation of the interaction of several components of the climate system is also the aim of recently developed EMICs. The use of coupled models also enables the performance of more realistic time-dependent (or transient) experiments that are driven by time-varying forcings, as opposed to the equilibrium experiments with AGCMs that have been dominating palaeoclimate research until recently. Transient experiments produce time-varying climate variability that can be compared with observations. Another development relevant to palaeoclimatologists is forward modelling. Forward models use climatic input from climate models to calculate the dynamics of the proxy data, so that it becomes possible to make a more direct data-model comparison.

Summary

The aim of this paper is to inform the palaeo-data community about recent developments in palaeoclimate modelling, with special reference to the Holocene climate. First, these developments are discussed in some detail, and subsequently the main issues are illustrated with the aid of examples. In a first example, natural variability in a coupled climate model simulation without a change in forcing is discussed. A second example discusses the response of a coupled climate model to transient forcings, thus clarifying the difference between natural "background" climatic noise and forced climate variability. A third example discusses the forward modelling approach. In particular, the simulation of glacier length variations for the last 10,000 years is shown in detail.

Acknowledgments

The valuable comments of R.W. Battarbee, S.P. Harrison and an anonymous referee are acknowledged. The authors would like to thank the European Science Foundation for supporting the 'Climate Modelling' workshop at the PEPIII conference in Aix-en Provence, August 2001. SFBT was funded by the UK Government Meteorological Research Program.

References

Appenzeller C., Stocker T.F. and Anklin M. 1998. North Atlantic Oscillation dynamics recorded in Greenland ice cores. Science 282: 446–449.

Braconnot P., Harrison S.P., Joussaume S., Hewitt C.D., Kitoh A., Kutzbach J.E., Liu Z., Otto-Bliesner B., Syktus J. and Weber S.L., this volume. Evaluation of PMIP coupled ocean-atmosphere simulations of the mid-Holocene. In: Battarbee R.W., Gasse F. and Stickley C.E. (eds), Past Climate Variability through Europe and Africa. Kluwer Academic Publishers, Dordrecht, the Netherlands, pp. 515–533.

Braconnot P., Joussaume S., de Noblet N., Ramstein G. and PMIP participating groups. 2000a. Mid-Holocene and Last Glacial Maximum African monsoon changes as simulated within the Paleoclimate Modelling Intercomparison Project. Global and Planetary Change 26: 51–66.

Braconnot P., Marti O., Joussaume S. and Leclainche Y. 2000b. Ocean feedback in response to 6 kyr BP insolation. J. Climate 13: 1537–1553.

Braconnot P., Joussaume S., Marti O. and de Noblet N. 1999. Synergistic feedbacks from ocean and vegetation on the African monsoon response to mid-Holocene insolation. Geophys. Res. Lett. 26: 2481–2484.

Broström A., Coe M., Harrison S.P., Gallimore R., Kutzbach J.E., Foley J., Prentice I.C. and Bartlein P.J. 1998. Land surface feedbacks and paleomonsoons in Northern Africa. Geophys. Res. Lett. 25: 3615–3618.

Brovkin V., Bendtsen J., Claussen M., Ganopolski A., Kubatzki C., Petoukhov V. and Andreev A. 2002. Carbon cycle, vegetation and climate dynamics in the Holocene: experiments with the CLIMBER-2 model. Global Biogeochem. Cycles 16: DOI: 10.1029/2001GB001662.

Cheddadi R., Yu G., Guiot J., Harrison S.P. and Prentice I.C. 1997. The climate of Europe 6000 years ago. Clim. Dyn. 13: 1–9.

Claussen M., Kubatzki C., Brovkin V., Ganopolski A., Hoelzmann P. and Pachur H.J. 1999. Simulation of an abrupt change in Saharan vegetation in the mid-Holocene. Geophys. Res. Lett. 26: 2037–2040.

Claussen M., Mysak L.A., Weaver A.J., Crucifix M., Fichefet T., Loutre M.F., Alexeev V.A., Berger A., Ganopolski A., Goosse H., Lohman G., Lunkeit F., Mohkov I., Petoukhov V., Stone P., Wang W. and Weber S.L. 2002. Earth system models of intermediate complexity: closing the gap in the spectrum of climate system models. Clim. Dyn. 18: 579–586.

Coe M.T. 1995. The hydrologic cycle of major continental drainage and ocean basins: a simulation of the modern and Mid-Holocene conditions and a comparison with observations. J. Clim. 8: 535–543.

Coe M.T. 1997. Simulating continental surface waters: an application to Holocene North Africa. J. Clim. 10: 1680–1689.

COHMAP members. 1988. Climatic changes of the last 18,000 years: observations and model simulations. Science 241: 1043–1052.

Cox P.M., Betts R.A., Jones C.D., Spall S.A. and Totterdell I.J. 2000. Acceleration of global warming due to carbon-cycle feedbacks in a coupled climate model. Nature 408: 184–187.

Crowley T.J. 2000. Causes of climate change over the past 1000 years. Science 289: 270–277.

Crucifix M., Loutre M.F., Tulkens P., Fichefet T. and Berger A. 2002. Climate evolution during the Holocene: a study with an Earth system model of intermediate complexity. Clim. Dyn. 19: 43–60.

Crueger T. and von Storch H. 2001. Creation of "Artificial Ice Core"; Accumulation from Large-Scale GCM Data: Description of the downscaling method and application to one North Greenland Ice Core. Clim. Res. 20: 141–151.

Cubasch U., Voss R., Hegerl G., Waskewitz J. and Crowley T.J. 1997. Simulation of the influence of solar radiation variations on the global climate with an ocean-atmosphere general circulation model. Clim. Dyn. 13: 757–767.

Dahl S.O. and Nesje A. 1996. A new approach to calculating Holocene winter precipitation by combining glacier ELA and pine-tree limits: a case study from Hardangerjokulen, Central South Norway. Holocene 6: 381–398.

Denis B., Laprise R., Caya D. and Cote J. 2002. Downscaling ability of one-way nested regional climate models: The Big brother experiment. Clim. Dyn. 18: 627–646.

DeNoblet-Decoudré N., Claussen M. and Prentice I.C. 2000. Mid-Holocene greening of the Sahara: first results of the GAIM 6000 yr BP experiment with two asynchronously coupled atmosphere/biome models. Clim. Dyn. 16: 643–659.

Ganopolski A., Kubatzki C., Claussen M., Brovkin V. and Petoukhov V. 1998. The influence of vegetation-atmosphere-ocean interaction on climate during the Mid-Holocene. Science 280: 1916–1919.

Giorgi F. 1990. Simulations of regional climate using limited-models nested in a general circulation models. J. Climate 3: 941–963.

Giorgi F., Hewitson B., Christensen J., Julme M., von Storch H., Whetton P., Jones R., Mearns L. and Fu C. 2001. Regional climate information — evaluation and projections. In: Houghton J.T., Ding Y., Griggs D.J., Noguer M., van der Linden P.J., Dai X., Maskell K. and Johnson C.A. (eds), Climate Change 2001: The Scientific Basis. Contributions of Working Group I to the Third Assessment Report of The IPCC, Cambridge University Press, Cambridge, pp. 583–638.

Gordon C., Cooper C., Senior C.A., Banks H., Gregory J.M., Johns T.C., Mitchell J.F.B. and Wood R.A. 2000. The simulation of SST, sea ice extents and ocean heat transports in a version of the Hadley Centre coupled model without flux adjustments. Clim. Dyn. 16: 147–168.

Harrison S.P., Yu G. and Tarasov P.E. 1996. Late Quaternary lake-level record from Northern Eurasia. Quat. Res. 45: 138–159.

Harrison S.P., Jolly D., Laarif F., Abe-Ouchi A., Dong B., Herterich K., Hewitt C., Joussaume S., Kutzbach J.E., Mitchell J., De Noblet N. and Valdes P. 1998. Intercomparison of simulated global vegetation distributions in response to 6 kyr BP orbital forcing. J. Climate 11: 2721–2742.

Hasselmann K. 1979. On the signal-to-noise problem in atmospheric response studies. In: Shaw B.D. (ed.), Meteorology Over the Tropical Oceans. Royal Meteorological Society, Bracknell, Berkshire, pp. 251–259.

Hegerl G.C., Hasselmann K.H., Cubasch U., Mitchell J.F.B., Roeckner E., Voss R. and Waszkewitz J. 1997. Multi-fingerprint detection and attribution analysis of greenhouse gas, greenhouse gas-plus-aerosol and solar forced climate change. Clim. Dyn. 13: 613–634.

Heinze C. 2001. Towards the time dependent modeling of sediment core data on a global basis. Geophys. Res. Lett. 28: 4211–4214.

Hewitt C.D. and Mitchell J.F.B. 1998. A fully coupled GCM simulation of the climate of the mid-Holocene. Geophys. Res. Lett. 25: 361–364.

Hostetler S.W., Bates G.T. and Giorgi F. 1993. Interactive coupling of a lake thermal-model with a regional climate model. J. Geophys. Res. 98: 5045–5057.

Hostetler S.W., Giorgi F., Bates G.T. and Bartlein P.J. 1994. Lake-atmosphere feedbacks associated with Paleolakes Bonneville and Lahontan. Science 263: 665–668.

Houghton J.T., Ding Y., Griggs D.J., Noguer M., van der Linden P.J., Dai X., Maskell K. and Johnson C.A. (eds) 2001. Climate Change 2001: The Scientific Basis. Contributions of Working Group I to the Third Assessment Report of the Intergovernmental Panel on Climate Change (IPCC). Cambridge University Press, Cambridge, 881 pp.

Isarin R.F.B. and Renssen H. 1999. Reconstructing and modelling late Weichselian climates: the Younger Dryas in Europe as a case study. Earth Sci. Rev. 48: 1–48.

Joussaume S. and Taylor K.E. 1995. Status of the Paleoclimate Modeling Intercomparison Project, Proceedings of the First International AMIP Scientific Conference, Monterey, USA, pp. 425–430.

Joussaume S., Taylor K.E., Braconnot P., Mitchell J.F.B., Kutzbach J.E., Harrison S.P., Prentice I.C., Broccoli A.J., Abe-Ouchi A., Bartlein P.J., Bonfils C., Dong B., Guiot J., Herterich K.,

Hewitt C.D., Jolly D., Kim J.W., Kislov A., Kitoh A., Loutre M.F., Masson V., McAvaney B., McFarlane N., de Noblet N., Peltier W.R., Peterschmitt J.Y., Pollard D., Rind D., Royer J.F., Schlesinger M.E., Syktus J., Thompson S., Valdes P., Vettoretti G., Webb R.S. and Wyputta U. 1999. Monsoon changes for 6000 years ago: results of 18 simulations from the Paleoclimate Modeling Intercomparision Project (PMIP). Geophys. Res. Lett. 26: 859–862.

Jouzel J., Hoffmann G., Koster R.D. and Masson V. 2000. Water isotopes in precipitation: data/model comparison for present-day and past climates. Quat. Sci. Rev. 19: 363–379.

Katz R.W. and Parlange M.B. 1996. Mixtures of stochastic processes: applications to statistical downscaling. Clim. Res. 7: 185–193.

Kohfeld K.E. and Harrison S.P. 2000. How well can we simulate past climates? Evaluating the models using palaeoenvironmental datasets. Quat. Sci. Rev. 19: 321–346.

Kutzbach J.E. 1981. Monsoon climate of the early Holocene: Climate experiment with the earth's orbital parameters for 9000 years ago. Science 214: 59–61.

Kutzbach J.E. and Gallimore R.G. 1988. Sensitivity of a coupled atmosphere/mixed ocean model to changes in orbital forcing at 9000 years BP. J. Geophys. Res. 93: 803–821.

Kutzbach J.E. and Guetter P.J. 1986. The influence of changing orbital parameters and surface boundary conditions on climate simulations for the past 18000 years. J. Atmos. Sci. 43: 1726–1759.

Kutzbach J.E. and Liu Z. 1997. Response of the African monsoon to orbital forcing and ocean feedbacks in the Middle Holocene. Science 278: 440–443.

Kutzbach J.E. and Otto-Bliesner B.L. 1982. The sensitivity of the African-Asian monsoonal climate to orbital parameter changes for 9000 yr BP in a low-resolution general circulation model. J. Atmos. Sci. 39: 1177–1188.

Kutzbach J.E. and Street-Perrott F.A. 1985. Milankovitch forcing in the level of tropical lakes from 18 to 0 kyr BP. Nature 317: 130–134.

Legutke S. and Voss R. 1999. The Hamburg Atmosphere-Ocean Coupled Circulation Model ECHO-G, Technical Report No. 18, DKRZ, Hamburg, 62 pp.

Lorenz E.N. 1976. Nondeterministic theories of climatic change. Quat. Res. 6: 495–506.

Mahowald N., Kohfeld K.E., Hansson M., Balkanski Y., Harrison S.P., Prentice I.C., Schulz M. and Rodhe H. 1999. Dust sources and deposition in the last glacial maximum and current climate: A comparison of model results with paleodata from ice cores and marine sediments. J. Geophys. Res. 104: 15,895–15,916.

Masson V., Cheddadi R., Braconnot P., Joussaume S., Texier S. and PMIP participating groups. 1999. Mid-Holocene climate in Europe: What can we infer from PMIP model data comparisons? Clim. Dyn. 15: 163–182.

McAvaney B.J., Covey C., Joussaume S., Kattsov V., Kitoh A., Ogana W., Pitman A.J., Weaver A.J., Woord R.A. and Zhao Z.-C. 2001. Model evaluation. In: Houghton J.T., Ding Y., Griggs D.J., Noguer M., van der Linden P.J., Dai X., Maskell K. and Johnson C.A. (eds), Climate Change 2001: The Scientific Basis. Contributions of Working Group I to the Third Assessment Report of the Intergovernmental Panel on Climate Change (IPCC). Cambridge University Press, Cambridge, pp. 471–523.

McGuffie K. and Henderson-Sellers A. 2001. Forty years of numerical climate modelling, Int. J. Climatol. 21: 1067–1109.

Mitchell J.F.B., Grahame N.S. and Needham K.J. 1988. Climate simulations for 9000 years before present: seasonal variations and effects of the Laurentide Ice Sheet. J. Geophys. Res. 93: 8283–8303.

Opsteegh J.D., Haarsma R.J., Selten F.M. and Kattenberg A. 1998. ECBILT: a dynamical alternative to mixed boundary conditions in ocean models. Tellus 50A: 348–367.

Parker D.E., Jones P.D., Folland C.K. and Bevan A. 1994. Interdecadal changes of surface temperature since the late nineteenth century. J. Geophys. Res. 99: 14373–14399.

Pethoukhov V., Ganopolski A., Brovkin V., Claussen M., Eliseev A., Kubatzki C. and Rahmstorf S. 2000. CLIMBER-2: a climate system model of intermediate complexity. Part I: model description and performance for present climate. Clim. Dyn. 16: 1–17.

Pope V.D., Gallani M.L., Rowntree P.R. and Stratton R.A. 2000. The impact of new physical parametrizations in the Hadley Centre climate model — HadAM3. Clim. Dyn. 16: 123–146.

Reichert B.K., Bengtsson L. and Oerlemans J. 2001. Midlatitude forcing mechanisms for glacier mass balance investigated using general circulation models. J. Climate 14: 3767–3784.

Renssen H., Goosse H., Fichefet T. and Campin J.M. 2001a. The 8.2 kyr BP event simulated by a global atmosphere–sea-ice–ocean model. Geophys. Res. Lett. 28: 1567–1570.

Renssen H., Isarin R.F.B., Jacob D., Podzun R. and Vandenberghe J. 2001b. Simulation of the Younger Dryas climate in Europe using a regional climate model nested in an AGCM: preliminary results. Glob. Plan. Chan. 30: 41–57.

Renssen H., Goosse H. and Fichefet T. 2002. Modeling the effect of freshwater pulses on the early Holocene climate: the influence of high frequency climate variability. Paleoceanography 17, 1020, DOI 10.1029/2001PA000649.

Renssen H., Brovkin V., Fichefet T. and Goosse H. 2003. Holocene climate instability during the termination of the African Humid Period. Geophys. Res. Lett. 30, 1184, DOI 10.1029/2002GL016636.

Stott P.A., Tett S.F.B., Jones G.S., Allen M.R., Mitchell J.F.B. and Jenkins G.J. 2000. External control of twentieth century temperature by natural and anthropogenic forcings. Science 290: 2133–2137.

Tett S.F.B., Stott P.A., Allen M.R., Ingram W.J. and Mitchell J.F.B. 1999. Causes of twentieth-century temperature change near the Earth's surface. Nature 399: 569–572.

Tett S.F.B., Jones G.S., Stott P.A., Hill D.C., Mitchell J.F.B., Allen M.R., Ingram W.J., Johns T.C., Johnson C.E., Jones A., Roberts D.L., Sexton D.M.H. and Woodage M.J. 2002. Estimation of natural and anthropogenic contributions to 20th century temperature change. J. Geophys. Res. 107.

Texier D., de Noblet N. and Braconnot P. 2000. Sensitivity of the African and Asian monsoons to mid-Holocene insolation and data-inferred surface changes. J. Climate 13: 164–181.

Texier D., de Noblet N., Harrison S.P., Haxeltine A., Jolly D., Joussaume S., Laarif F., Prentice I.C. and Tarasov P. 1997. Quantifying the role of biosphere-atmosphere feedbacks in climate change: coupled model simulations for 6000 years BP and comparison with paleodata for Northern Eurasia and Northern Africa. Clim. Dyn. 13: 865–882.

van der Schrier G., Weber S.L. and Drijfhout S.S. 2002. Sea level changes in the North Atlantic by solar forcing and internal variability. Clim. Dyn. 19: 435–447.

von Storch H. 1995. Inconsistencies at the interface of climate impact studies and global climate research. Meteorol. Zeitschrift 4 NF: 72–80.

von Storch H., von Storch J.-S. and Müller P. 2001. Noise in the climate system — ubiquitous, constitutive and concealing. In: Engquist B. and Schmid W. (eds), Mathematics Unlimited - 2001 and Beyond. Part II. Springer-Verlag, Heidelberg, pp. 1179–1194.

von Storch H., Cubasch U., González-Ruoco J., Jones J.M., Widmann M. and Zorita E. 2000. Combining paleoclimatic evidence and GCMs by means of Data Assimilation Through Upscaling and Nudging (DATUN). 11th Symposium on Global Change Studies, American Meteorological Society, Washington D.C., pp. 28–31.

von Storch J.-S., Kharin V., Cubasch U., Hegerl G., Schriever D., von Storch H. and Zorita E. 1997. A description of a 1260 year control integration with the coupled ECHAM1/LSG general circulation model. J. Climate 10: 1526–1543.

Weber S.L. and von Storch H. 1999. Simulating climatic millennial timescales and combining paleoclimatic evidence and GCM-based dynamical knowledge. EOS 80: 380.

Weber S.L. 2001. The impact of orbital forcing on the climate of an intermediate-complexity coupled model. Glob. Plan. Chan. 30: 7–12.

Weber S.L. and Oerlemans J. 2003. Holocene glacier variability: three case studies using an intermediate-complexity climate model. The Holocene 13: 353–363.

Werner M., Mikolajewicz U., Heimann M. and Hoffmann G. 2000. Borehole versus isotope temperatures on Greenland: seasonality matters. Geophys. Res. Lett. 27: 723–726.

Whetton P.H., Katzfey J.J., Hennessy K.J., Wu X., McGregor L.J. and Nguyen K. 2001. Developing scenarios of climate change for Southeastern Australia: an example using regional climate model output. Clim. Res. 16: 181–201.

Wigley T.M.L., Jones P.D., Briffa K.R. and Smith G. 1990. Obtaining sub-grid-scale information from coarse-resolution general circulation model output. J. Geophys. Res. 95: 1943–1953.

Wright H.E., Kutzbach J.E., Webb III T., Ruddiman W.F., Street-Perrott F.A. and Bartlein P.J. (eds) 1993. Global Climates Since the Last Glacial Maximum. University of Minnesota Press, Minneapolis, 569 pp.

24. EVALUATION OF PMIP COUPLED OCEAN-ATMOSPHERE SIMULATIONS OF THE MID-HOLOCENE

PASCALE BRACONNOT (pasb@lsce.saclay.cea.fr)
Laboratoire des Sciences du Climat et de l' Environnement
UMR CEA-CNRS 1572
CEA Saclay bât 709
91191 Gif sur Yvette cedex
France

SANDY P. HARRISON (sharris@bgc-jena.mpg.de)
Max-Planck-Institut für Biogeochemie
Postfach 100164, 07701 Jena
Germany
Currently at
Department of geography
Bristol University
Bristol
UK

SYLVIE JOUSSAUME (joussaume@cea.fr)
Laboratoire des Sciences du Climat et de l' Environnement
UMR CEA-CNRS 1572
CEA Saclay bât 709
91191 Gif sur Yvette cedex
France

CHRIS D. HEWITT (chris.hewitt@metoffice.com)
Hadley Centre for Climate Prediction and Research
Met Office
FitzRoy Road, Exceter
Devon, EX1 3BP
UK

AKIO KITOH (kitoh@mri-jma.go.jp)
Climate Research Department
Meteorological Research Institute
Nagamine 1-1, Tsukuba
Ibaraki, 305-0052
Japan

R. W. Battabee et al. (eds) 2004. *Past Climate Variability through Europe and Africa.*
Springer, Dordrecht, The Netherlands.

JOHN E. KUTZBACH (jek@facstaff.wisc.edu)
Department of Atmospheric and Oceanic Sciences
University of Wisconsin - Madison
1225 West Dayton Street
Madison
Wisconsin 53706
USA

ZHENGYU LIU (zliu3@facstaff.wisc.edu)
Department of Atmospheric and Oceanic Sciences
University of Wisconsin - Madison
1225 West Dayton Street
Madison
Wisconsin 53706
USA

BETTE OTTO-BLIESNER (ottobli@ucar.edu)
Climate Change Research
National Center for Atmospheric Research
1850 Table Mesa Drive
P.O. Box 3000, Boulder
Colorado 80307
USA

JOZEF SYKTUS (jis@dar.csiro.au)
CSIRO Division of Atmospheric Research
Private Bag No 1
Aspendale 3195
Victoria
Australia

S.L. WEBER (weber@knmi.nl)
Royal Netherlands Meteorological Institute KNMI
P.O. Box 201
NL-3730 AE De Bilt
The Netherlands

Keywords: Climate, Model, Proxy data, Biomes, Atmosphere, Ocean, Land-surface, Monsoon

Introduction

Physically based models provide unique means to predict the probable future impact of anthropogenic changes in atmospheric composition and land use. These models are continually improving their ability to simulate the major features of today's climate (IPCC 2001). The accurate simulation of current climate is an important benchmark but does not guarantee that a model will correctly simulate climatic conditions very different from today. It is difficult to evaluate model performance solely on the basis of the instrumental record because the changes in climate since the middle of the last century have been relatively modest (e.g., Mann et al. (1995), Tett et al. (1999)). Evaluating model performance under the extreme climatic conditions that occur in the distant past provides an opportunity to evaluate how models respond to larger changes in forcing, and ultimately provides a credibility test for modelling the future.

The Paleoclimate Modeling Intercomparison Project (PMIP), coordinated by S. Joussaume (CNRS, France) and K. Taylor (Lawrence Livermore, USA), was initiated in order to coordinate and encourage the systematic study of climate models for key periods in the past (Joussaume and Taylor 1995; PMIP 2000). The PMIP effort developed out of a NATO Advanced Research Workshop in 1991. The workshop participants agreed to focus initially on two specific periods: the last glacial maximum, 21000 calendar years before present (yr BP), and the mid-Holocene (6000 yr BP). The last glacial maximum provides an opportunity to assess the models' ability to simulate extreme cold conditions, and to study the feedbacks associated with both a decrease in the atmospheric CO_2 concentration and the presence of 2–3 km thick ice-sheets over North America and Northern Europe. The simulation of the mid-Holocene was chosen to test the response of the climate system to a change in the seasonal contrast of the incoming solar radiation at the top of the atmosphere (insolation).

PMIP is an international project involving 18 climate modelling groups around the world. It is endorsed by both the International Geosphere Biospere Project (under PAGES, Past Global Changes) and the World Climate Research Program (within the working group on coupled models).

Model evaluation is crucially dependent on the existence of spatially explicit data sets that can be compared with outputs from the model simulations. Thus, one goal of PMIP has been to foster the creation of well-documented, spatially explicit data-sets designed for use in model evaluation. Although the construction of palaeoenvironmental data-sets for model evaluation began prior to PMIP (e.g., Street and Grove (1976), Wright et al. (1993)), PMIP has played a key role in stimulating the continued development and improvement of such data-sets and has been instrumental in the creation of two new data sets: the BIOME 6000 data set (Prentice and Webb III 1998; Prentice et al. 2000) and the 21 ka Tropical Terrestrial data synthesis (Farrera et al. 1999).

Basic PMIP experiments were designed to test atmospheric general circulation models (AGCMs). Model-model and model-data comparisons were conducted within sub-projects led by a scientific coordinator. Results from the different sub-projects are summarised in PMIP (2000). However, in order to understand the basic PMIP experiments better and to enhance our knowledge of the climate system, a number of complementary experiments were also performed by individual modelling groups. These complementary experiments explored, for example, the role of land-surface and ocean feedbacks on climate. Simulations in which vegetation changes were prescribed to the atmospheric model (Broström et al. 1998; Texier et al. 2000) or computed using coupled atmosphere-vegetation models (Claussen and Gayler 1997; de Noblet-Ducoudré et al. 2000; Doherty et al. 2000; Texier et al. 1997) have shown that vegetation enhances the orbitally-induced monsoon changes over Northern Africa. Ocean feedbacks on the African monsoon during the mid-Holocene have been explored using coupled ocean-atmosphere models (Braconnot et al.

1999, Braconnot et al. 2000, Hewitt and Mitchell 1998, Kitoh and Shigenori pers. comm., Kutzbach and Liu 1997, Otto-Bliesner 1999, Voss and Mikolajewicz 2001, Weber 2001).

Coupled ocean-atmosphere models (OAGCMs) have become the basic tool for projection of future climate change (IPCC 2001). This makes it important to evaluate such models under the radically different climate conditions of the past. A working group on coupled ocean-atmosphere simulations has thus emerged within PMIP, with the aim of documenting:

• the robust differences between OAGCM and AGCM simulations;

• the role of the ocean response in the timing of the changes in the seasonal cycle, and more specifically the role of ocean dynamics in the mid-Holocene enhancement of the northern hemisphere monsoons (i.e., the Asian, African and North American monsoons);

• how the simulated climates compare with palaeoenvironmental data, focussing on regions such as Northern Africa, Europe and the high northern latitudes where standard data-model comparisons have been developed and used for the evaluation of the basic PMIP simulations.

In this paper, we will present preliminary results from the working group on coupled experiments. We first describe the boundary conditions and the models (section 2) and some aspects of the simulated mid-Holocene changes (section 3) as shown by both the basic PMIP experiments and the coupled OAGCM experiments. In sections 4 and 5, we present the data-sets and the methodology used for the model-data comparisons. We then focus on the evaluation of the coupled simulations over Northern Africa (section 6) and compare them with the basic PMIP AGCM simulations. Finally we discuss the implications of these results for the future.

PMIP experimental design and coupled models

At 6000 yr BP, the main change in insolation is due to the displacement of the longitude of the perihelion; changes in other orbital parameters are small (Table 1). Compared to modern conditions, this orbital configuration intensifies (weakens) the seasonal distribution of insolation in the northern (southern) hemisphere by about 5% (Fig. 1). The insolation forcing is not at a maximum at 6000 yr BP, but the impact of insolation changes during the early Holocene are more difficult to isolate because the climate is also affected by the remains of the Laurentide ice sheet.

In atmosphere-only simulations, sea-surface temperatures (SSTs), land-surface conditions, and atmospheric trace gas concentrations have to be defined (prescribed). In the basic PMIP mid-Holocene AGCM experiments, SSTs were kept at modern values. Except at high northern latitudes and a few coastal regions, differences in SST between 6000 yr BP and today are small and within the error bars associated with the reconstruction methods (Koç Karpuz and Schrader 1990; Ruddiman and Mix 1993). In the absence of a global reconstruction of mid-Holocene SST patterns, it therefore seems reasonable to prescribe modern SSTs in the mid-Holocene experiment. Mid-Holocene pollen data record large vegetation changes, especially in the African monsoon region (Jolly et al. 1998; Prentice et al. 2000; Elenga et al., this volume; Hoelzmann et al., this volume). However, partly because the lack of pollen data from some regions makes it difficult to construct a global vegetation map and partly through a desire to keep the experimental design relatively

Table 1. Description of the boundary conditions used in the PMIP experiments for the mid-Holocene (6000 yr BP). Orbital parameters are derived from Berger (Berger 1978) and CO_2 from Raynaud et al. (1993).

Boundary conditions	Modern	6000 y BP
Sea surface temperature	Control run or PMIP data	No change
Ice cover	set	
CO_2	345 ppm or Ctrl run: Ccont	280 ppm or (280/345)* Ccont
Insolation		
• solar constant	1365 Wm-2 or Ctrl run	No change
• orbital parameters		
Eccentricity	0,016724	0,018682
Axial tilt	23,446	24,105
ω-180°	102,04	0,87

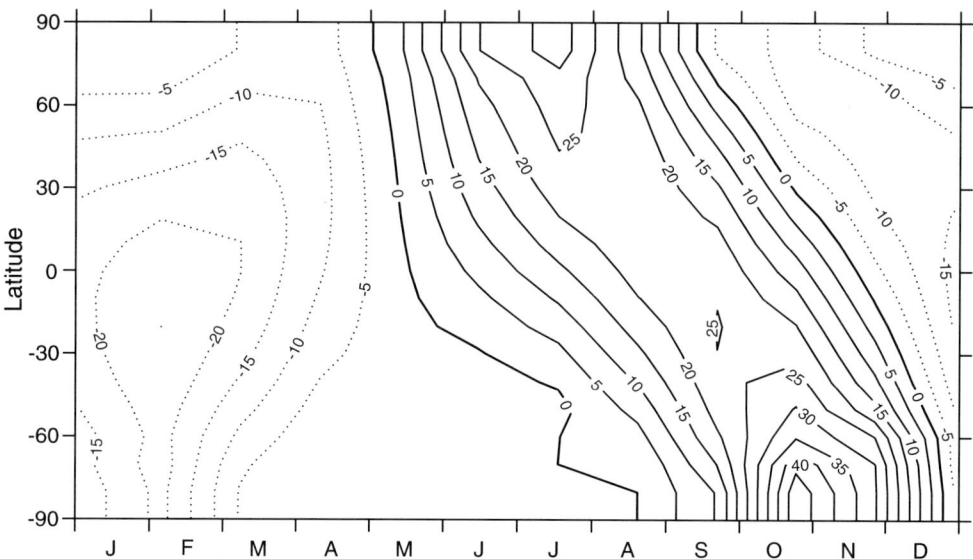

Figure 1. Insolation changes at 6000 year BP: latitude-month distribution of the changes in incoming solar radiation at the top of the atmosphere. Isolines at every 5 W/m².

simple, it was decided not to change land-surface characteristics in the mid-Holocene PMIP experiment. The CO_2 concentration was prescribed at its pre-industrial value of 280 ppm (Raynaud et al. 1993) in the mid-Holocene experiment and at 345 ppm in the modern day control experiment. The vernal equinox was set at March 21, but the definition of the seasons was kept as present. The changes in the length of the seasons are small at 6000 yr BP and do not have a major effect on simulated climate means when the models are forced by modern SSTs (Joussaume and Braconnot 1997).

The OAGCM experiments described here were made with models that couple an atmospheric GCM with an oceanic GCM. The two components exchange information about the conditions at the air-sea interface (sea-surface temperature, sea-ice cover, momentum, heat and fresh water fluxes) once a day. The coupled simulations were performed with models of different resolution and complexity (Table 2). All of the simulations follow the PMIP protocol for insolation changes, including the date of the vernal equinox and the definition of the seasons. Similarly, they follow the original protocol in keeping land-surface conditions the same in the modern (control) and mid-Holocene experiments. However, unlike the basic PMIP AGCM experiments, all except two (CSIRO, ECHAM/LSG) of the simulations prescribe the CO_2 level at 6000 yr BP to be the same as in the modern simulation. One model (CSM) was used to make two simulations, one in which CO_2 was changed (ΔCSM) and one in which it was not changed between the control and mid-Holocene experiments. The actual CO_2 concentration used differs from model to model (Table 2). By keeping the CO_2 concentration unchanged, we are able to test the sensitivity of the model to orbital forcing alone. A further motivation for not changing the CO_2 concentration is that most of these simulations are not run to equilibrium, which would require running the coupled model for many hundreds of years. The global energetics of the model is only slightly affected by changes in the orbital parameters alone, and thus the changes in the mean seasonal cycle can be analysed with some confidence over a short period of time. Changes in CO_2, however, have a larger impact on the global energetics and thus longer runs would be required to have confidence in the results. Some of the coupled models used a so-called "flux-correction" at the air-sea interface to prevent model drift.

Table 2. Characteristics of the coupled simulations. Long, lat, and lev refer respectively to longitude, latitude and vertical levels. For the spectral model, the type of truncation is indicated, whereas for grid point models, the number of grid points is indicated.

MODEL	RESOLUTION		FLUX CORRECTION	CO_2	
	ATM Long × lat (levels)	OCEAN Long × lat (levels)		CTRL	6000 yr BP
CSM1.2	T31 (18)	102 × 116 (25)	none	280	280
CSM1.2Δ	T31 (18)	102 × 116 (25)	none	355	280
CSIRO	64 × 36 (9)	64 × 36 (12)	SST, SSS, τ	330	280
UKMO (HADCM2)	96 × 73 (19)	96 × 73 (20)	SST, SSS	323	323
IPSL-CM1	64 × 50 (11)	92 × 76 (31)	none	345	345
MRI2	72 × 46 (15)	144 × 111 (23)	SST, SSS	345	345
ECHAM3/LSG	T21	64 × 32	SST, SSS	345	280
ECBILT	T21 (3)	64 × 32 (12)	none	345	345
FOAM	R15	128 × 128	none	330	330

The analyses of the coupled OAGCM simulations presented here are preliminary in nature. For purely pragmatic reasons resulting from the time at which individual runs were completed and made available, the different analyses (e.g., Fig. 3, Plate 11) were performed using different subsets of the experiments. In the future, and in particular when these experiments are re-run using a standard protocol, the analyses will need to be repeated.

Large scale changes in simulated climate

Following the 6000 yr BP insolation forcing, all the PMIP AGCM simulations produce an increased seasonal cycle of surface temperature over the continents of the northern hemisphere. The ensemble mean of the PMIP simulations show that the maximum summer warming ($>2\,^{\circ}$C) occurs between $40\,^{\circ}$N and $50\,^{\circ}$N, whereas the maximum winter cooling occurs in the tropics (Fig. 2). During summer, all the models produce a pronounced warming over Eurasia, which deepens the summer thermal low over the continents and thus intensifies the pressure gradient between land and ocean. The flux of moisture from the ocean to the continent is enhanced, resulting in an increase in African and Asian monsoon activity. This monsoon enhancement is marked by a northward extension of the rains in Northern Africa and increased inland penetration of the monsoon in Asia (Fig. 2). Although the mechanism is common to all simulations, individual models show significant differences in both the magnitude and the spatial patterns of the simulated changes in the surface climate of the monsoon region (Braconnot et al. 2000; de Noblet-Ducoudré et al. 2000; Guiot et al. 1999; Harrison et al. 1998; Joussaume et al. 1999; Masson et al. 1999; Yu and Harrison 1996).

Results from the coupled OAGCM simulations show basically the same features as the basic PMIP AGCM simulations. All of the coupled models show an increase in the magnitude of the seasonal cycle of temperature averaged over the northern hemisphere compared to the atmosphere-only simulations (Fig. 3). The CSIRO simulation is systematically colder in all seasons than the other models because of the lower CO_2 imposed in this simulation. The shape of the curve is nevertheless comparable to the other simulations. Braconnot et al. (2000) have shown that the changes in the seasonal cycle of temperature over land is of smaller magnitude in a coupled simulation, thus the differences between the coupled simulations and the basic PMIP experiments result from amplification of the seasonal cycle of SSTs.

The ocean causes a subtle shift in the timing of the response to insolation forcing (Fig. 3). In the basic PMIP AGCM simulations, orbitally-induced warming starts in May and persists through into August. In the coupled OAGCM simulations, the ocean remains relatively cold in the spring and orbitally-induced warming is not registered until July. However, warmer conditions persist longer into the autumn because the warmer ocean delays the onset of orbitally-induced winter cooling.

These ocean-induced changes in the seasonal cycle of temperature have implications for the response of the African monsoon to mid-Holocene orbital forcing. Ocean feedbacks enhance the African monsoon and cause a northward expansion of the monsoon precipitation belt compared to the basic PMIP simulations. The monsoon season is also lengthened in the coupled simulations. Cold Atlantic SSTs in spring, when the land surface is already beginning to warm, favour an early initialisation of monsoon flow from the ocean to the

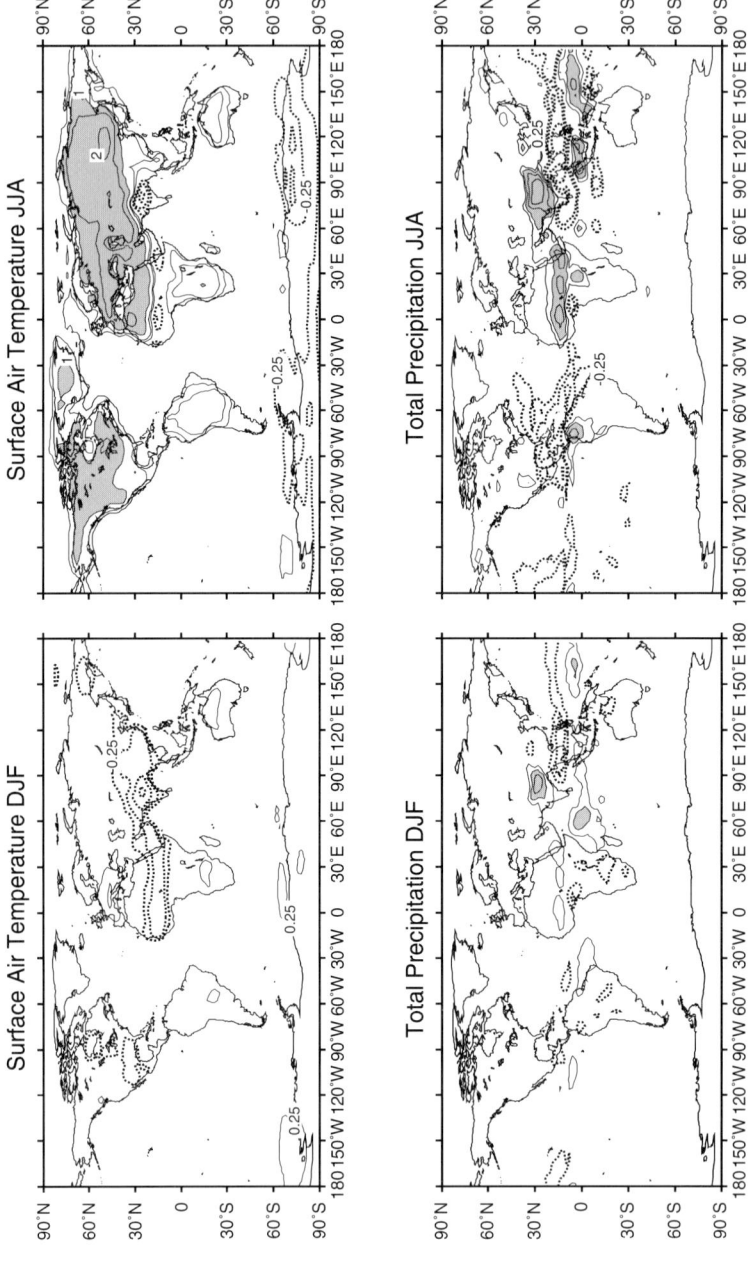

Figure 2. Winter (DJF) and summer (JJA) changes in a) surface air temperature and b) precipitation, averaged for 17 of the 18 PMIP simulations (the MSU model results are excluded here because of the low resolution of this model). Isolines at ±0.25, ±0.5, and then at every 0.5 °C with grey shading above 1 °C. Isolines ±0.25, ±0.5, ±1, ±2 mm/day with grey shading above 0.5 mm/day for precipitation changes.

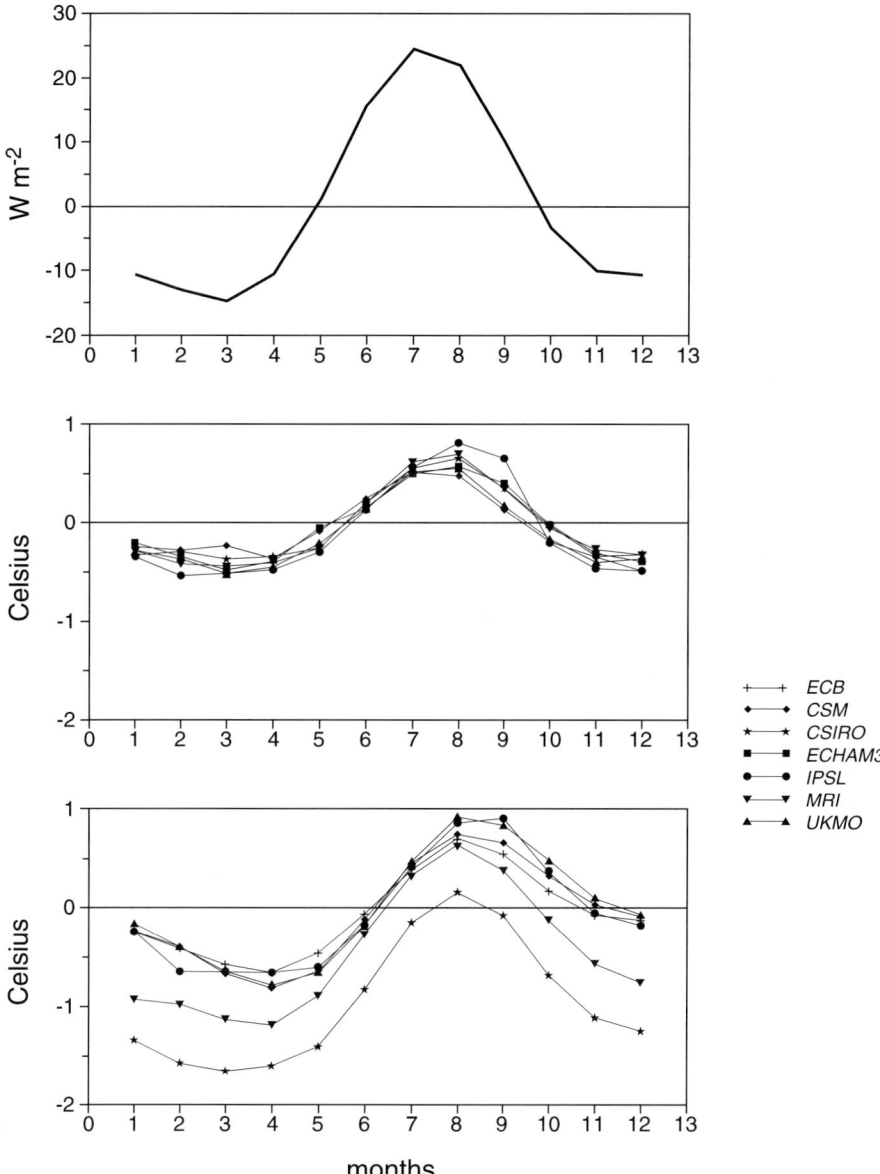

Figure 3. Mid-Holocene change in insolation (W/m² top), surface air temperature (°C) for PMIP simulations performed with the atmospheric component (middle) and the coupled ocean-atmosphere simulations (bottom) averaged over the northern hemisphere and plotted as a function of months.

continent (Hewitt and Mitchell 1998). In the late summer, most of the models exhibit a strong gradient in SSTs at ca 10 °N, with cooler SSTs than today to the south and warmer SSTs to the north. This structure helps to maintain the ITCZ in a more northerly position over the Atlantic and Western Africa (Braconnot et al. 2000; Kutzbach and Liu 1997), thus encouraging the inland penetration of monsoon rains. A similar dipole over the Atlantic occurs today during years when Sahelian rainfall is above average (Folland et al. 1986; Fontaine and Janicot 1996; Palmer 1986).

Global mid-holocene palaeoenvironmental data sets

Two sources of data have been widely used for evaluation of the mid-Holocene PMIP simulations: the Global Lake Status Data Base and the BIOME 6000 data set.

The Global Lake Status Data Base (GLSDB: Kohfeld and Harrison (2000), Qin et al. (1998)) is a long-standing international effort to compile the geomorphic and biostrati-graphic data for changes in lake level, area, or volume (collectively referred to as lake status), in order to document changes in regional water balance during the last 30,000 years. Developed with data-model comparisons as a primary objective, the GLSDB builds on the earlier Oxford Lake Level Data Base (Street and Grove 1976) and contains data both from closed-basin lakes in now-arid regions and from currently overflowing lakes in temperate and wet tropical regions (Jolly et al. 1998; Tarasov et al. 1996; Yu and Harrison 1996).

Lake status data from the GLSDB for 6000 yr BP show that conditions were wetter than today across Northern Africa, the Arabian Peninsula, Northern India, and Southwest China (Plate 9), indicating expansion of the Afro-Asian summer monsoons. Conditions were slightly wetter than today in Central America and SW USA, reflecting expansion of the North American monsoon (Harrison et al. 2003). In Central Eurasia, lake records show conditions similar to or slightly wetter than today, while the limited evidence from the mid-latitudes of the southern hemisphere and in the high northern latitudes suggests conditions were also wetter than today. The only regions where lakes show conditions were drier than today are interior North America and Western Europe.

The Palaeovegetation Mapping Project (known as BIOME 6000: Prentice and Webb III (1998)) has developed global palaeovegetation data-sets for the LGM and the mid-Holocene. Broadscale vegetation types (biomes) are reconstructed from pollen or plant-macrofossil data using a standardised, objective method (biomisation) based on plant functional types (PFTs: Prentice et al. (1996)). Plant taxa are first assigned to PFTs, and then the set of PFTs that can occur in each biome is specified. The allocation of pollen or plant-macrofossil assemblages to biomes is made on the basis of an affinity-score procedure that takes into account both the diversity and the abundance of taxa belonging to each PFT in the sample. Extensive tests using modern surface samples have shown that the method is capable of reproducing natural vegetation patterns even in regions heavily impacted by human activities (Plate 10a).

The BIOME 6000 data set for 6000 yr BP (Plate 10b) shows that the Arctic forest limit was north of its present position in the Mackenzie Delta region (Edwards et al., in press), Europe (Prentice et al. 1996) and Western and Central Siberia (Texier et al. 1997), and south of its present position in Quebec-Labrador (Williams et al. 2000). The northward expansion of northern temperate forest zones was more dramatic than the relatively modest change in the Arctic forest limit. Warmer winters (as well as summers) are required to explain some of these shifts in northern temperate forests (Prentice et al. 2000). Temperate deciduous forests were greatly extended in Europe, southwards into the Mediterranean region as well

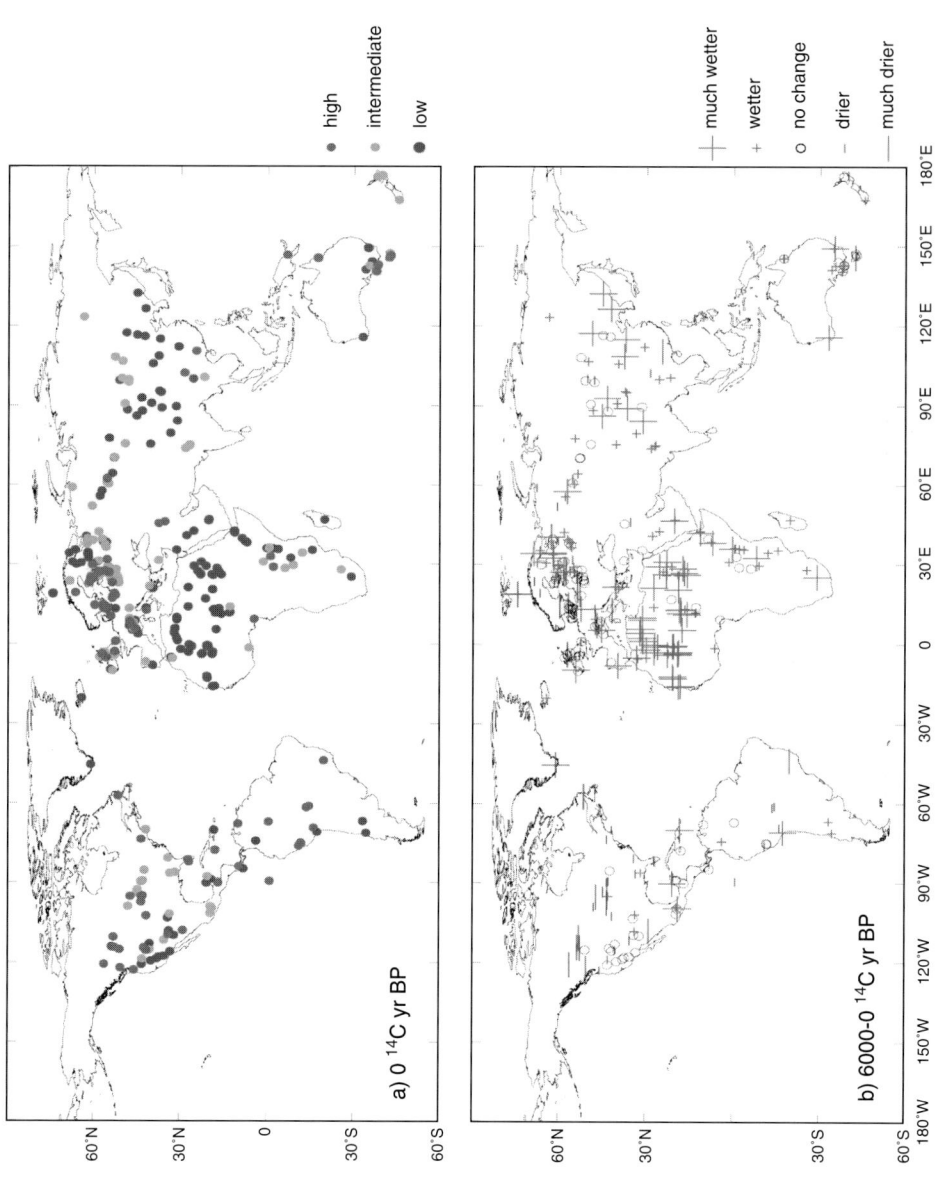

Plate 9. Lake status (a) today and (b) the change in lake status at 6000 ^{14}C yr BP compared to present. The data are derived from the Global Lake Status Data Base (Kohfeld and Harrison 2000; Yu et al. 2001; Harrison et al. 2003). Only sites with dating control ≤4 are included. Colour version of this Plate can be found in Appendix, p.636

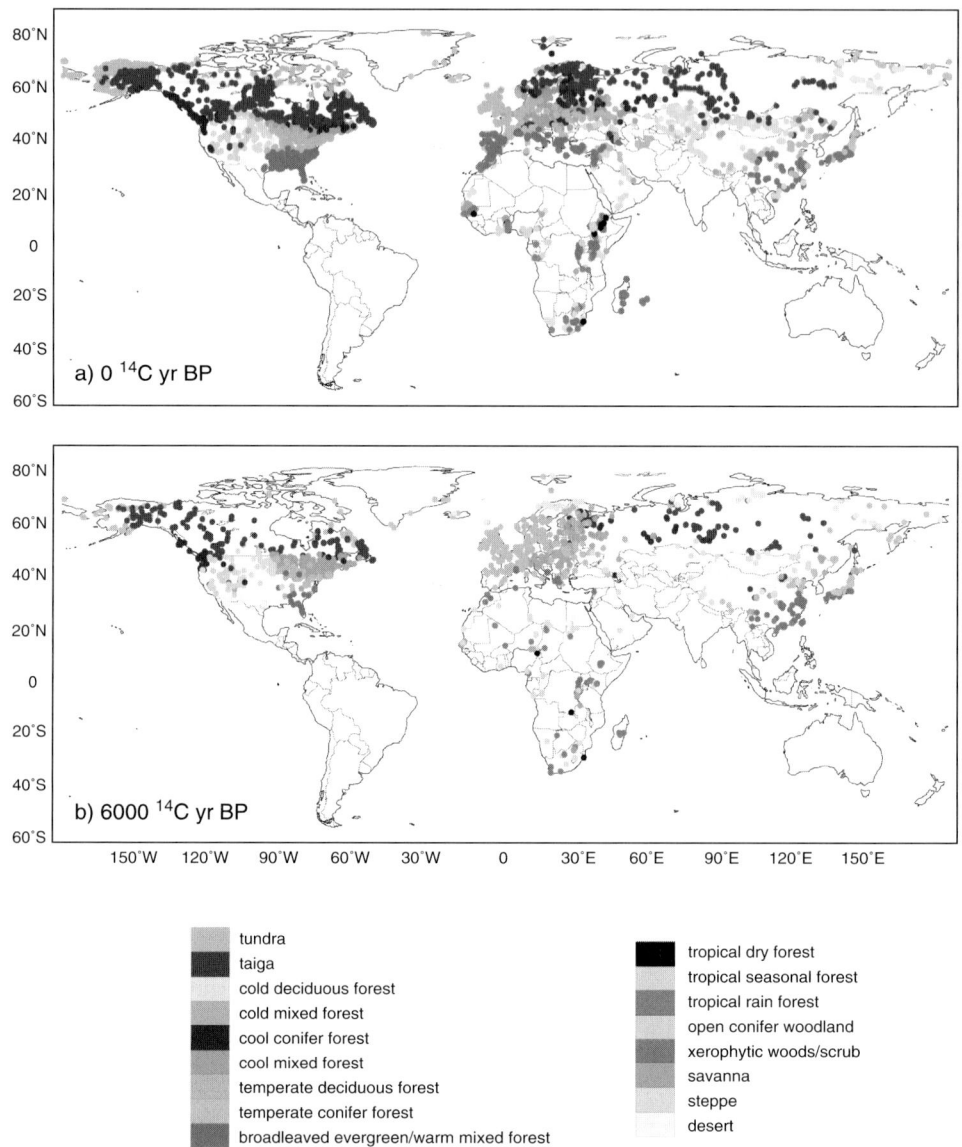

Plate 10. Biome reconstructions for (a) today and (b) 6000 ^{14}C yr BP (from Prentice et al. (2000)). Colour version of this Plate can be found in Appendix, p.637

as to the north (Roberts et al., this volume). Steppe vegetation occurred in areas occupied today by forests in North America in response to drier conditions (Williams et al. 2000), but forest biomes encroached on the present-day steppe in Southeastern Europe and Central Asia (Tarasov et al. 1998). Enhanced monsoons extended forest biomes inland in China (Yu et al. 1998) and Sahelian vegetation into the Sahara, while the African rainforest was reduced (Jolly et al. 1998) consistent with a more seasonal climate in the equatorial zone.

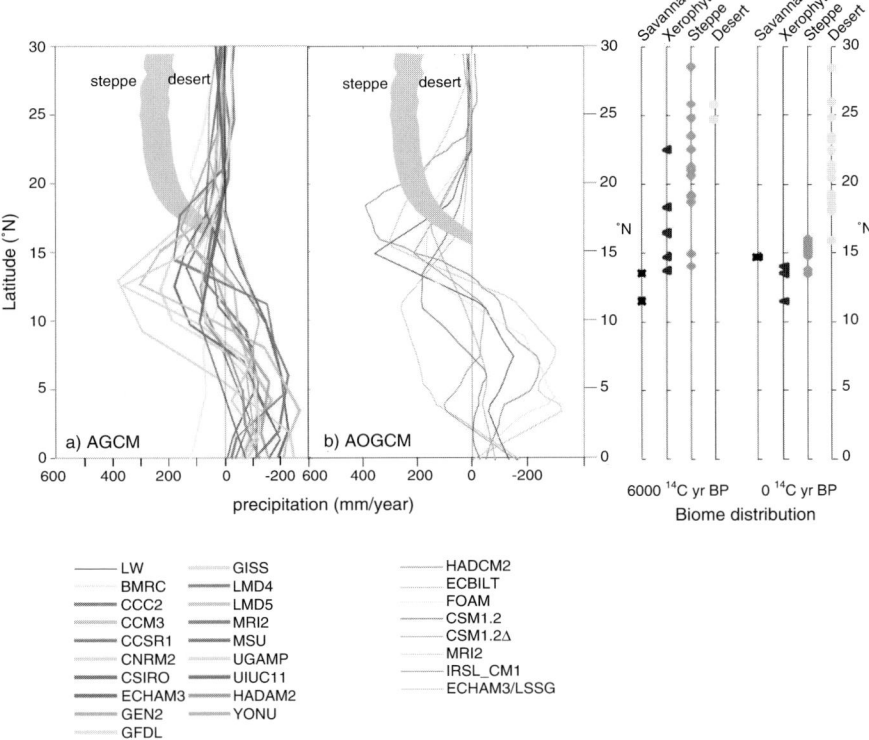

Plate 11. Zonally-averaged simulated annual precipitation anomalies (6000 yr BP minus control) over northern African land grid cells (20 °W–30 °E) vs. latitude (from 0° to 30 °N) as simulated (a) by the 18 AGCMs participating in the PMIP basic experiment, and (b) in a suite of 9 OAGCM experiments. The grey shaded area in both diagrams represents upper and low estimates of the precipitation needed in excess of the modern precipitation to support steppe vegetation at each latitude (Joussaume et al. 1999). The plots at the right of the figure (c) show the latitudinal distribution of vegetation at 6000 ^{14}C yr BP compared to today, and show that steppe and xerophytic vegetation were extended much further north than today at 6000 ^{14}C yr BP. Colour version of this Plate can be found in Appendix, p.638

Methodology for model-data comparisons

Two complementary approaches to comparing observations and simulations have been used within PMIP: inverse techniques and forward-modelling techniques. Inverse methods are particularly useful when the geological data are abundant. The forward modelling approach maximises the use of relatively sparse data sets for model evaluation.

In the forward modelling approach, process-based models are used to predict the response of palaeoenvironemental indicators (vegetation, hydrology) to the simulated climate. Although the use of a second model can introduce a source of uncertainties about the cause of the mismatches between simulations and observations, forward modelling has been used within the PMIP project to facilitate comparisons with terrestrial vegetation data and with lake data. Thus, we have used terrestrial biosphere models from the BIOME family (e.g.,

Prentice et al. (1992)) for direct comparison with the BIOME 6000 data set (see Harrison et al. (1998)). Topographically-explicit terrestrial hydrological models, which predict the surface area of lakes and wetlands, and river discharge, from simulated runoff, precipitation and evaporation (e.g., HYDRA Coe (1998)) have been used in a similar manner and directly compared with palaeo-lake area (see Coe and Harrison (2002)).

In the inverse approach, palaeodata are translated into climatic parameters using statistical algorithms (e.g., transfer functions, modern analogs). Initially, inverse techniques were used to reconstruct standard climate parameters (e.g., mean July temperature, mean January temperature, mean annual precipitation) from vegetation. These reconstructions will have large error bars if the vegetation is not directly controlled by aspects of the climate that are well correlated with the standard climate parameters. In order to avoid over-simplifications of the climate-vegetation relationship, PMIP has strongly encouraged reconstructions of non-standard climatic variables that are more closely related to the underlying controls on specific palaeoenvironmental indicators, such as the coldest month temperature, the accumulated temperature sum during the growing season, or a moisture index. The need for extensive data arrays means that the use of quantitative reconstructions to evaluate PMIP simulations has largely been confined to Europe and North America (e.g., Cheddadi et al. 1997; Bonfils et al. 2000; Guiot et al. 1999; Masson et al. 1999). However, inverse techniques have been used to quantify the amount of precipitation required to maintain the steppe vegetation that characterised Northern Africa during the mid-Holocene (Joussaume et al. 1999).

Evaluation of AGCM and OAGCM simulations of the african monsoon using palaeoenvironmental data

The expansion of the area influenced by the Afro-Asian summer monsoon at 6000 yr BP is one of the most striking features shown by palaeoenvironmental data, and thus this region has become one of the major foci for model evaluation in PMIP. Comparisons of the simulated precipitation-evaporation (P-E) fields with lake data from the GLSDB indicate that the basic PMIP simulations consistently underestimate the northward shift of the monsoon front (Yu and Harrison 1996). Similarly, BIOME3 simulations made with outputs from the PMIP simulations consistently fail to reproduce the observed northward shift in the Sahara/Sahel boundary (Harrison et al. 1998). The precipitation required to generate the observed latitudinal distribution of steppe (grassland) in Northern Africa at 6000 yr BP has been estimated using a combination of forward-modelling and inverse techniques. Joussaume et al. (1999) showed that the basic PMIP simulations underestimate the required precipitation at ca 23 °N by at least 100 mm (Plate 11a). When output from the PMIP experiments is used to simulate the extent of lakes across Northern Africa using the HYDRA model, the maximum extent of Lake Chad is <30% of the "observed" area (570,000 km^2) of this lake at 6000 yr BP (Coe and Harrison 2002). Thus, data-model comparisons show that the PMIP simulations consistently underestimate both (i) the northward shift in the monsoon belt shown by palaeoenvironmental data, and (ii) the magnitude of the precipitation required to produce the observed lake and vegetation changes in Northern Africa.

Results of the coupled simulations (Plate 11b) show that the ocean feedbacks help to enhance the African monsoon and to shift the belt of maximum precipitation further north than in the basic PMIP simulations. In the basic experiments, the belt of maximum precipitation is located between ca 10–15 °N; in the coupled simulations this belt lies between ca 10–18 °N (Plate 11b). As shown for the PMIP simulations (Braconnot et al. 2000; Joussaume et al. 1999), the location of the main precipitation belt is influenced by the position of the rain belt in the control simulation. This helps to explain why the northward shift in precipitation is smaller in some models (e.g., UKMO HADCM3) than in others. The precipitation increase induced by the combined effect of orbital forcing and ocean feedbacks is still not sufficient to maintain steppe vegetation in Northern Africa.

Implications for the future

The PMIP community, through systematic comparisons of model simulations against benchmarks provided by regional or global syntheses of palaeoenvironmental data, has demonstrated that the observed large changes in mid-Holocene climates cannot be simulated without explicitly considering ocean- and land-surface feedbacks, and the synergies between them (see Braconnot et al. (1999), Ganopolski et al. (1998)). However, the preliminary analyses of a suite of coupled OAGCMs presented here show that inter-model differences are larger when the ocean is explicitly simulated than in atmosphere-only experiments. It is imperative to understand how these differences arise. In the future, PMIP will seek to address this question through the analysis of simulations of the mid-Holocene climate using both OAGCMs and fully coupled ocean-atmosphere-vegetation (OAVGCM) models. These simulations will need to be rigorously benchmarked against palaeoenvironmental data. Coupled models, whether OAGCMs or OAVGCMs, will make it possible to investigate the changes in interannual to multi-decadal variability during specific time intervals. Analyses of two of the coupled OAGCM simulations presented here suggest that the inter-annual variability of surface temperature was enhanced over Eurasia and reduced in the tropics during the mid-Holocene. Otto-Bliesner (1999) also reported changes in the characteristics of the El-Niño phenomenon. Again, these model results need to be systematically compared with high-resolution palaeoenvironmental records in order to determine whether the coupled models are capable of simulating short-term climate variability in a realistic fashion. The PMIP project, therefore, remains committed to improving existing data sets, developing better analytical tools and encouraging additional syntheses of palaeoenvironmental data in order to be able to evaluate the models that will subsequently be used to simulate potential future climate changes.

Acknowledgments

The PMIP project is endorsed by both the IGBP, through its programme element Past Global Changes (PAGES), and the World Climate Research Programme, through the working group on coupled simulations. The GLSDB is sponsored by IGBP through PAGES Palaeo-Mapping Project (PMAP). BIOME 6000 is sponsored by IGBP through its programme elements Global Analysis, Intercomparison and Modelling (GAIM), the Data and Information System (DIS), Global Change and Terrestrial Ecosystems (GCTE) and PAGES. PMIP archive and website are maintained at PCMDI (http://www-pcmdi.llnl.gov/pmip). Selected results from the coupled OAGCM simulations are archived at LSCE (Paris) and MPI-BGC (Jena).

References

Berger A. 1978. Long-term variations of caloric solar radiation resulting from the earth's orbital elements. Quat. Res. h 9: 139–167.

Bonfils C., de Noblet N., Guiot J., Bartlein P.J. et al. 2000. New method for comparing models and data: Application to European climate 6 kyr BP. In: Bracconot P. (ed.), Paleoclimate Modeling Intercomparison Project (PMIP), Proceedings of the Third PMIP Workshop. WCRP-111,WMO/TD-No. 1007:271: 95–98.

Braconnot P., Joussaume S., Marti O. and de Noblet N. 1999. Synergistic feedbacks from ocean and vegetation on the African monsoon response to mid-Holocene insolation. Geophys. Res. Lett. 26: 2481–2484.

Braconnot P., Marti O., Joussaume S. and Leclainche Y. 2000. Ocean feedback in response to 6 kyr Before Present Insolation. Journal of Climate 13: 1537–1553.

Braconnot P., Joussaume S., de Noblet N., Ramstein G. and PMIP participating groups. 2000. Mid Holocene and last glacial maximum African monsoon changes as simulated within the Paleoclimate Modeling Intercomparison Project. Glob. Plan. Chan. 26: 51–66.

Broström A., Coe M.T., Harrison S.P., Gallimore R.G., Kutzbach J.E., Foley J., Prentice C.I. and Behling P. 1998. Land surface feedbacks and paleomonsoons in Northern Africa. Geophys. Res. Lett. 25: 3615–3618.

Cheddadi R., Yu G., Guiot J., Harrison S.P. and Prentice C.I. 1997. The climate of Europe 6000 years ago. Clim. Dyn. 13: 1–9.

Claussen M. and Gayler V. 1997. The greening of the Sahara during the mid-Holocene: results of an interactive atmosphere-biome model Global Ecology and Biogeography Letters 6: 369–377.

Coe M.T. 1998. A linked global model of terrestrial hydrologic processes: simulation of modern rivers, lakes, and wetlands. J. Geophys. Res. 103: 8885–8899.

de Noblet-Ducoudré N., Claussen M. and Prentice C. 2000. Mid-Holocene greening of the Sahara: First results of the GAIM 6000 yr BP experiment with two asynchronously coupled atmosphere/biome models. Clim. Dyn. 16: 643–659.

Doherty R., Kutzbach J., Foley J. and Pollard D. 2000. Fully coupled climate/dynamical vegetation model simulations over Northern Africa during the mid-Holocene. Clim. Dyn. 16: 561–573.

Edwards M.E., Anderson P.M., Brubaker L.B., Ager T., Andreev A.A., Bigelow N.H., Cwynar L.C., Eisner W.R., Harrison S.P., Hu F.-S., Jolly D., Lozhkin A.V., MacDonald G.M., Mock C.J., Ritchie J.C., Sher A.V., Spear R.W., Williams J. and Yu G. 2000. Pollen-based biomes for Beringia 18,000, 6000 and 0 ^{14}C yr BP. J. Biogeogr 27(3): 521–554.

Elenga H., Maley J., Vincens A. and Farrera I., this volume. Palaeoenvironments, palaeoclimates and landscape development in Atlantic Equatorial Africa: a review of major terrestrial sites covering the last 25 kyrs. In: Battarbee R.W., Gasse F. and Stickley C.E. (eds), Past Climate Variability in Europe and Africa. Kluwer Academic, Dordrecht, The Netherlands, pp. 181–198.

Farrera I., Harrison S.P., Prentice I.C., Ramstein G., Guiot J., Bartlein P.J., Bonnefille R., Bush M., Cramer W., von Grafenstein U., Holmgren K., Hooghiemstra H., Hope G., Jolly D., Lauritzen S.-E., Ono Y., Pinot S., Stute M. and Yu G. 1999. Tropical climates at the last glacial maximum: a new synthesis of terrestrial palaeoclimate data. I. Vegetation, lake-levels and geochemistry. Clim. Dyn. 15: 823–856.

Folland C.K., Palmer T.N. and Parker D.E. 1986. Sahel rainfall and world wide sea surface temperature. Nature 320: 602–607.

Fontaine B. and Janicot S. 1996. Sea Surface Temperature Fields Associated with West African Rainfall Anomaly Types. Journal of Climate 9: 2935–2940.

Ganopolski A., Kubatzki C., Claussen M., Brovkin V. and Petoukhov V. 1998. The influence of Vegetation-atmosphere-Ocean Interaction on Climate During the Mid-Holocene. Science 280: 1916–1919.

Guiot J., Boreux J.J., Braconnot P., Torre F. and PMIP-participating-groups 1999. Data-models comparison using fuzzy logic in palaeoclimatology. Clim. Dyn. 15: 569–581.

Harrison S.P., Jolly D., Laarif F., Abe-Ouchi A., Dong B., Herterich K., Hewitt C., Joussaume S., Kutzbach J.E., Mitchell J., de Noblet N. and Valdes P. 1998. Intercomparison of Simulated Global Vegetation Distributions in Response to 6 kyr BP Orbital Forcing. J. Clim. 11: 2721–2742.

Harrison S., Kutzbach K.E., Liu Z., Barthlein P., Otto-Bliesner B., Muhs D., Prentice I.C., Thomson R.S. 2003. Mid-Holocene climates of the Americas: a dynamical response to changed seasonality. Clim. Dyn. 20: 663–668.

Hewitt C.D. and Mitchell J.F.B. 1998. A fully coupled GCM simulation of the climate of the mid-Holocene. Geophys. Res. Lett. 25: 361–364.

Hoelzmann P., Gasse F., Dupont L.M., Salzmann U., Staubwasser M., Leuscher D.C. and Sirocko F., this volume. Palaeoenvironmental changes in the arid and subarid belt (Sahara-Sahel-Arabian Peninsula) from 150 ka to present. In: Battarbee R.W., Gasse F. and Stickley C.E. (eds), Past Climate Variability in Europe and Africa. Kluwer Academic, Dordrecht, The Netherlands, pp. 219–256.

Jolly D., Harrison S.P., Damnati B. and Bonnefille R. 1998. Simulated climate and biomes of Africa during the Late Quaternary: comparison with pollen and lake status data. Quat. Sci. Rev. 17: 629–657.

Jolly D., Prentice I.C., Bonnefille R., Ballouche A., Bengo M., Brenac P., Buchet G., Burney D., Cazet J.-P., Cheddadi R., Edohr T., Elenga H., Elmoutaki S., Guiot J., Laarif F., Lamb H., Lezine A.-M., Maley J., Mbenza M., Peyron O., Reille M., Reynaud-Ferrera I., Riollet G., Ritchie J.C., Roche E., Scott L., Ssemmanda I., Straka H., Umer M., Van Campo E., Vilimumbala S., Vincens A. and Waller M. 1998. Biome reconstruction from pollen and plant macrofossil data for Africa and the Arabian peninsula at 0 and 6 ka. J. Biogeogr. 25: 1007–1028.

Joussaume S. and Taylor K.E. 1995. Status of the Paleoclimate Modeling Intercomparison Project in Proceedings of the first international AMIP scientific conference, WCRP-92, Monterey, USA: pp. 425–430.

Joussaume S. and Braconnot P. 1997. Sensitivity of paleoclimate simulation results to season definitions J. Geophys. Res. 102: 1943–1956.

Joussaume S., Taylor K.E., Braconnot P., Mitchell J.F.B., Kutzbach J., Harrison S.P., Prentice I.C., Broccoli A.J., Abe-Ouchi A., Bartlein P.J., Bonfils C., Dong B., Guiot J., Herterich K., Hewitt C.D., Jolly D., Kim J.W., Kislov A., Masson V., McAvaney B., McFarlane N., de Noblet N., Peltier W.R., Peterschmitt J.-Y., Pollard D., Rind D., Royer J.-F., Schlesinger M.E., Syktus J., Thompson S., Valdes P., Vettoretti G., Webb R.S. and Wyputta U. 1999. Monsoon changes for 6000 years ago: results of 18 simulations from the Paleoclimate Modeling Intercomparison Project (PMIP). Geophys. Res. Lett. 26: 859–862.

Koç N. and Schrader H. 1990. Surface sediment distribution and Holocene paleotemperature variations in the Greenland, Iceland and Norwegian Sea. Paleoceanography 5: 557–580.

Kohfeld K.E. and Harrison S. 2000. How well can we simulated past climates? Evaluating the models using global palaeoenvironmental data sets. Quat. Sci. Rev. 19: 321–346.

Kutzbach J.E. and Liu Z. 1997. Response of the African monsoon to orbital forcing and ocean feedbacks in the middle Holocene. Science 278: 440–443.

Mann M.E., Park J. and Bradley R.S. 1995. Global interdecadal and century-scale climate oscillations during the past five centuries. Nature 378: 266–270.

Masson V., Cheddadi R., Braconnot P., Joussaume S., Texier S. and PMIP participating groups. 1999. Mid-Holocene climate in Europe: what can we infer from PMIP model-data comparisons? Clim. Dyn. 15: 163–182.

Otto-Bliesner B.L. 1999. El Niño/La niña and Sahel precipitation during the middle Holocene. Geophys. Res. Lett. 26: 87–90.

Palmer T.N. 1986. Influence of the Atlantic, Pacific and Indian Oceans on Sahel rainfall. Nature 322: 251–253.

Prentice I.C. and Webb III T. 1998. BIOME 6000: reconstructing global mid-Holocene vegetation patterns from palaeoecological records. J. Biogeogr. 25: 997–1005.

Prentice I.C., Jolly D. et al. 2000. Mid-Holocene and glacial-maximum vegetation geography of the northern continents and Africa. J. Biogeogr. 27(3): 507–519.

Prentice I.C., Guiot J., Huntley B., Jolly D. and Cheddadi R. 1996. Reconstructing biomes from palaeoecological data: a general method and its application to European pollen data at 0 and 6 ka. Clim. Dyn. 12: 185–194.

Prentice I.C., Cramer W., Harrison S.P., Leemans R., Monserud R.A. and Solomon A.M. 1992. A global biome model based on plant physiology and dominance, soil properties and climate. J. Biogeogr. 19: 117–134.

Qin B., Harrison S. and Kutzbach J. 1998. Evaluation of modelled regional water balance using lake status data: a comparison of 6 ka simulations with the NCAR CCM. Quat. Sci. Rev. 17: 535–548.

Raynaud D., Jouzel J., Barnola J.M., Chappellaz J., Delmas R.J. and Lorius C. 1993. The ice record of greenhouse gases. Science 259: 926–934.

Roberts N., Stevenson A.C., Davis B., Cheddadi R., Brewer S. and Rosen A., this volume. Holocene climate, environment and cultural change in the circum-Mediterranean region. In: Battarbee R.W., Gasse F. and Stickley C.E. (eds), Past Climate Variability in Europe and Africa. Kluwer Academic, Dordrecht, The Netherlands, pp. 343–362.

Ruddiman W.F. and Mix A.C. 1993. The north and equatorial Atlantic at 9000 and 6000 yr BP. In: Wright H.E. Jnr., Kutzbach J.E., Webb III T., Ruddiman W.F., Street-Perrott F.A. and Bartlein P.J. (eds), Global Climates Since the Last Glacial Maximum. University of Minnesota Press, Minneapolis, pp. 94–124.

Street F.A. and Grove A.T. 1976. Environmental and climatic implications of late Quaternary lake-level fluctuations in Africa. Nature 261: 385–390.

Tarasov P., Pushenko M.Y., Harrison S.P., Saarse L., Andreev A.A., Aleshinskaya Z.V., Davydova N.N., Dorofeyuk N.I., Efremov Y.V., Elina G.A., Elovicheva Y., Filimonova L.V., Gunova V.S., Khomutova V.I., Kvadaze E.V., Neustreuva I., Pisareva V.V., Sevastyanov D.V., Shelekhova T.S., Uspenskaya O.N. and Zernitskaya V.P. 1996. Lake Status Record from the Former Soviet Union and Mongolia: Documentation of the Second Version of the Database. NOAA Paleoclimatology Publications Series Report 5: 224 pp.

Tarasov P., Webb III T., Andreev A.A., Afanas'eva N.B., Berezina N.A., Bezusko L.G., Chernova G.M., Dorofeyuk N.I., Diksen V.G., Elina G.A., Filimonova L.V., Glebov F.Z., Guiot J., Gunova V.S., Harrison S.P., Jolly D., Khomutova V.I., Kvavadze E.V., Osipova I.M., Ponova N.K., Prentice C.I., Saarse L., Sevastyanov D.V., Volkova V.S. and Zernitskaya V.P. 1998. Present-day and mid-Holocee biomes reconstructed from pollen and plant macrofossil data from the former Soviet Union and Mongolia. J. Biogeogr. 25: 1029–1053.

Tett S.F.B., Stott P.A., Allen M.R., Ingram W.J. and Mitchell J.F.B. 1999. Causes of twentieth-centry temperature change near the Earth's surface. Nature 399: 569–572.

Texier D., de Noblet N. and Braconnot P. 2000. Sensitivity of the African and Asian monsoons mid-Holocene insolation and data-inferred surface changes. J. Clim. 13: 164–181.

Texier D., de Noblet N., Harrison S.P., Haxeltine A., Jolly D., Joussaume S., Laarif F., Prentice I.C. and Tarasov P. 1997. Quantifying the role of biosphere-atmosphere feedbacks in climate change: coupled model simulations for 6000 years BP and comparison with paleodata for Northern Eurasia and Northern Africa. Clim. Dyn. 13: 865–882.

Voss R. and Mikolajewicz U. 2001. The climate 6000 years BP in near-equilibrium simulations with a coupled AOGCM. Geophys. Res. Lett. 28: 2213–2216.

Weber S.L. 2001. The impact of orbital forcing on the climate of an intermediate-complexity coupled model. Glob. Plan. Chan. 30 (1-2): 7–12.

Williams J.W., Webb III T., Richard P.J.H. and Newby P. 2000. Late Quaternary biomes of Canada and the Eastern United States. J. Biogeogr. 27(3): 585–607.

Wright H.E. Jnr., Kutzbach J.E., Webb III T., Ruddiman W.F., Street-Perrott F.A. and Bartlein P.J. 1993. Global Climates since the Last Glacial Maximum. University of Minnesota Press, Minneapolis, 544 pp.

Yu G. and Harrison S.P. 1996. An evaluation of the simulated water balance of Northern Eurasia at 6000 yr BP using lake status data. Clim. Dyn. 12: 723–735.

Yu G., Prentice C.I., Harrison S.P. and Sun X. 1998. Pollen-based reconstructions for China for 0 ka and 6 ka. J. Biogeogr. 25: 1055–1069.

25. FAMINE, CLIMATE AND CRISIS IN WESTERN UGANDA

PETER ROBERTSHAW (proberts@csusb.edu)
Department of Anthropology
California State University
San Bernardino, CA 92407-2397
USA

DAVID TAYLOR (taylord@tcd.ie)
Department of Geography
Trinity College
University of Dublin
Dublin 2
Ireland

SHANE DOYLE (s.d.doyle@leeds.ac.uk)
School of History
University of Leeds
Leeds LS2 9JT
UK

ROBERT MARCHANT (marchant@tcd.ie)
Department of Botany
Trinity College
University of Dublin
Dublin 2
Ireland

Keywords: Famine, Climate change, History, Uganda, Environmental determinism

Introduction

That human societies must adapt to crises arising from climate variability is self-evident. Yet the nature of that adaptation is often understood either within a geographically extensive framework or within the narrow confines of environmental determinism. Some of the contributors to *Chronology, Migration and Drought in Interlacustrine Africa* (Webster 1979a) utilise both approaches in linking references to drought and famine in oral histories from across the Interlacustrine region of inter-tropical Africa. The historian James McCann has recently criticised this volume for its "embarrassingly uncritical use of oral sources

R. W. Battarbee et al. (eds) 2004. *Past Climate Variability through Europe and Africa.*
Springer, Dordrecht, The Netherlands.

to reconstruct East African climate and demographic history in a way that was at best overoptimistic and at worst grossly incompetent" (McCann (1999), p. 265). In addition to misgivings concerning the use of oral histories to reconstruct and date past environmental and human catastrophes, doubts exist as to whether references to droughts in oral histories should be interpreted as referring literally to a single or series of catastrophic events, as Webster (1979b) contends, or whether they should be seen as metaphors for periods of food insecurity, some of which may have been caused by factors other than shortages of rainfall (e.g., Henige (1982)).

An alternative approach to understanding the human dimensions of climate variability in the past is called for, founded on the premises that: (i) adaptation occurs at the local level and is particular to specific cultural and temporal contexts; and (ii) adaptation is never directly determined by climate, as any response is always mediated through social, political, economic and other cultural filters. One cannot assume that people will respond in predictable ways to climate change. For example, we argue below that different episodes of drier climatic conditions promoted different responses in terms of the relative importance of agriculture and cattle-keeping in our study area because of contingent historical factors. The great importance of these historical factors is clearly illustrated by our account of events in the modern period. Therefore, any approach to exploring the relationships between climate and human history must be based upon relevant archaeological, ethnological and historical research, in addition to palaeoclimatic reconstructions, and is likely to involve in-depth and multidisciplinary case studies of small areas that are to some degree culturally homogenous.

In this paper we provide an example of the alternative approach, based on the Banyoro people in the humid, Interlacustrine region of Western Uganda known historically as Kitara, the region roughly centred on Mubende and encompassing much of what was regarded as Bunyoro, Toro, Nkore, and perhaps western parts of Buganda in the late nineteenth century (Sutton 1993). Kitara has the advantage of being a locus for a range of relevant, multi-disciplinary information, sources of which include rich oral histories, archaeological excavations at a comparatively advanced state and sediment-based, palaeoenvironmental studies.

Study area

The study area lies between around 1150 and 1350 m.a.s.l. and is centred upon Munsa Earthworks (ca. 1300 m.a.s.l., N 0°49′30″; E 31°18′00″), an archaeological site investigated by Robertshaw (1997) located in the climatically wetter and topographically undulating, southern part of Bunyoro. Rivers in Bunyoro drain into the White Nile, either directly or via lakes Albert and Victoria (Fig. 1). The main basement rocks are granite intrusions and argillites and quartzites of the Pre-Cambrian Bunyoro-Toro system (Harrop 1970).

Rainfall is closely related to the passage of the Inter-Tropical Convergence Zone (ITCZ), with the onset, intensity and duration of the two wetter periods during the year determined by the annual cycle of circulation over the Indian Ocean (Hastenrath et al. 1993). In addition to this dominant climatic system, rainfall is influenced by irregularly occurring El Niño Southern Oscillation (ENSO)-related phenomena (Phillips and McIntyre 2000). According to rainfall data from the meteorological site at Masindi (N 1° 41′; E 31° 43′, 1147 m.a.s.l., WMO station number: 636540; the closest reliable meteorological station in the same climatic zone as Munsa (Atlas of Uganda 1967)) and based on years for which an entire

Figure 1. Map of the study area, showing locations of places mentioned in the text. Boundaries of pre-colonial kingdoms are approximate and based upon Figure 5 in Posnansky (1963). The eastern shoreline of Lake Albert and the southern bank of the (White) Nile River effectively form the western and northern boundaries of Bunyoro.

12-month dataset is available (a total of 68 years of data), mean annual rainfall for the period 1908–1999 was 1330.5 mm (Botanical, Forestry and Scientific Department (1909, 1910), British East African Meteorological Service (1932 *et sequ*), Uganda Meteorological Observations (1909–1938, 1943–1948), and data supplied by Uganda Meteorological Services, Kampala). Of the two wetter seasons affecting the study area, the later one in the year (August-November) tends to be associated with higher levels of rainfall than the earlier one (March-May).

A patchwork of small farms, interspersed with remnants of Medium Altitude Semi-Deciduous Forest (*sensu* Langdale-Brown et al. (1964)), characterises the sides of flat-topped hills, with intervening swampy, papyrus-filled valley bottoms. Dale (1954) suggests that forest cover declined in extent during the pre-colonial period, before partially recovering during the colonial period. This recovery has continued into post-colonial times, according to information obtained during interviews with elderly male and female Banyoro farmers in July 2001. Patches of tall elephant grass (*Pennisetum purpureum*) commonly encountered today may represent abandoned farmland, while areas that are cultivated are occupied by a range of perennial crops, such as banana and coffee, and annuals, e.g., beans and grains. The planting of annual crops coincides with the beginning of each wetter season, when the soils are easiest to cultivate and water availability for germination is high, with final maturity taking place during the following dry season. Livestock, though valued, do not make a major contribution to food production, as the main cattle-rearing areas are to the southwest and north of the study area.

Climate variability and societal change in Bunyoro

The history of crises presented here is divided into pre-modern (the pre-colonial period, or pre-1900 AD) and modern (the colonial and post-colonial periods). Lake and swamp sediment records provide an important source of information on the pre-modern, although many of these records are characterised by poor sampling and temporal resolution. Levels of Lake Victoria are also a proxy of effective precipitation over a large region that includes the study area, with the lake's general shallowness amplifying the effect of variations in the balance between rainfall and evaporation. Some documentary evidence of former levels of Lake Victoria is available for the late pre-modern period (Nicholson 1998; Nicholson and Yin 2001), particularly from 1850 AD (Fig. 2). Levels of Lake Victoria and rainfall pre-1850 AD can be approximated from records of Nile River minima during the Northern Hemisphere late spring/early summer (Nicholson and Yin 2001), when Nile River levels are most strongly influenced by discharge from the White Nile. Records of minima and maxima levels are available from the Nilometer at Rodah for most of the second millennium AD (Hassan 1981; Nicholson 1996; 1998) (Fig. 3).

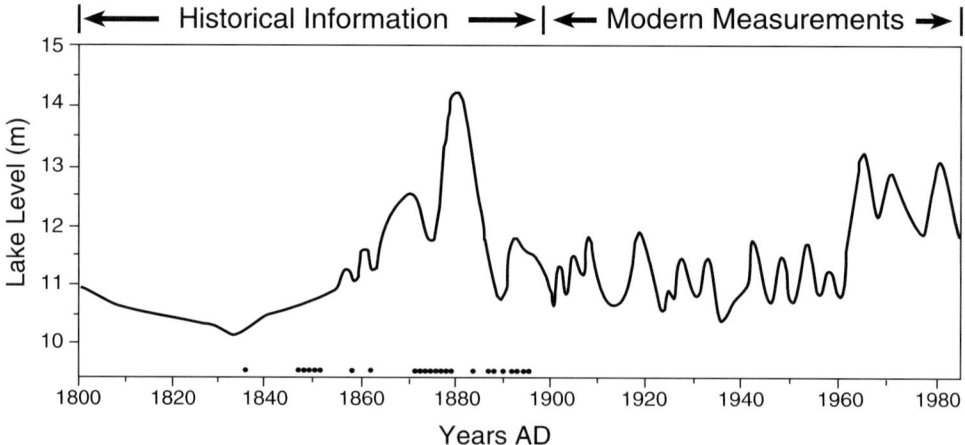

Figure 2. Fluctuations in water levels in Lake Victoria from 1800 AD to the 1980s, based on historical and instrumental data (after Nicholson (1998)).

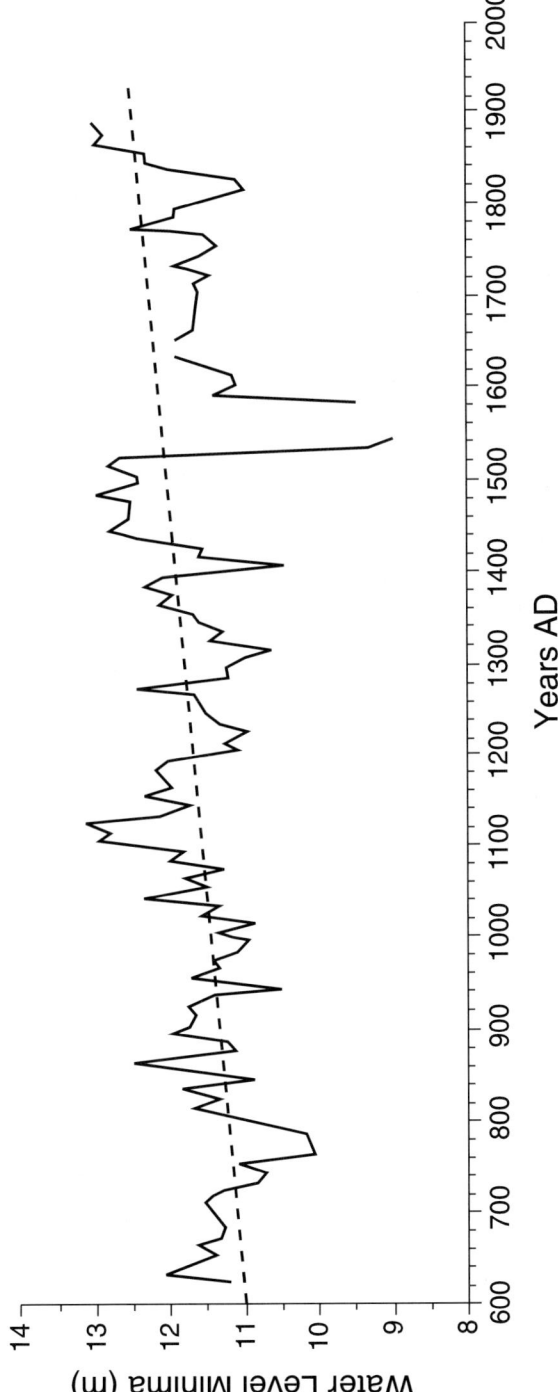

Figure 3. Rodah Nilometer readings showing fluctuations (scale on left of figure, in m) in levels of (NH summer) Nile minima since @ AD 650, based upon decadal averages (after Hassan (1981), Nicholson (1996, 1998)).

The resolution of information is much improved for the modern period and includes gauge records of rainfall and lake levels, documents from colonial archives and published materials. Oral traditions and oral histories are also much more specific as living informants can recall specific droughts, famines and other calamities from early in their own lives or from stories passed down from relatives. Some of these histories may spill over into the late pre-modern, specifically to events in the reign of Kabaleega (1870–1899), the last independent *mukama* (king) of Bunyoro who fought a fierce war against the imposition of colonial rule.

Pre-modern

At least four major shifts in climate may have occurred in Bunyoro and adjacent regions during the 1000 years to the beginning of the colonial period, according to indicators of past rainfall. While the precise timing of each transition remains open to debate, each seemingly occurred around the same time as significant socio-economic changes (see Robertshaw and Taylor (2000) and Taylor et al. (2000) for a detailed exposition of these).

Humid conditions around the beginning of the last millennium, evident in high Nile minima levels and possibly correlated with the Medieval Warm Period of temperate latitudes, lasted until the end of the 12th century AD and witnessed movements of people into Kitara from the Western Rift and Lake Victoria shores (Schoenbrun 1998). While historical linguistic evidence indicates that cattle flourished in the grasslands south of the Katonga River in the first centuries of the second millennium (Schoenbrun 1993), archaeological excavations, particularly at the site of Ntusi, revealed that cattle keeping was in fact fully integrated with cereal agriculture (Reid 1996). Humid conditions in the context of a relatively under-populated, internal frontier region (see Kopytoff (1987) for discussion of the "internal African frontier") where there was little or no competition for land promoted a diversified subsistence base that could exploit the full range of available micro-environments while providing insurance against any calamities, such as livestock disease, that would affect only one component of the economy.

Subsequent, gradual changes in settlement and subsistence patterns may have coincided with increased aridity from around the end of the 12th century. The period from about AD 1180 to 1400 witnessed an increase in the number and size of agricultural settlements in the relatively more humid regions north of the Katonga River (Robertshaw 1994). Perhaps reduced agricultural productivity in the relatively drier areas surrounding Ntusi prompted some families that were more dependent upon agriculture than livestock to migrate northwards. This period also witnessed the first episodes of occupation at sites that are later identified with Cwezi religious cults (Robertshaw and Taylor 2000). Towards the end of the period, the first earthworks with an extensive system of 4 m-deep ditches may have been constructed at Bigo on the south bank of the Katonga River (Posnansky 1969). It would be foolhardy to attribute these events entirely to climatic circumstances. While there is a danger of over-interpretation of sparse archaeological data, it is nevertheless tempting to see these centuries as a period of uncertainty and social unrest, hence the rise of the Cwezi cults, which, *inter alia*, attempted to control nature, particularly rainfall, through religious practices (see Schoenbrun (1998)). Socio-political processes, as individuals and factions jockeyed for power in different areas of the region within the context of an increasing human population, were likely the primary agents of change.

A return to relatively humid climatic conditions from around AD 1400 coincides with extensive forest clearance north of the Katonga River, as is evident from pollen and charcoal in a sediment core at Kabata Swamp (Taylor et al. 1999). This forest clearance, which may well have begun in the preceding drier episode, is indicative of the importance of agriculture, an interpretation that is confirmed by abundant archaeological evidence (Robertshaw and Taylor 2000). This was a period of political consolidation that culminated in the construction of major earthworks at Munsa and Kibengo. It was presumably a continuation and outcome of the competitive politics operating prior to AD 1400. Power at the major centres was rooted in a variety of political strategies, including military might expressed in raids on neighbours for cattle and women, possible control of the production and distribution of iron and prestige goods, control of surplus food production, and monopolies of forms of ritual and religious authority (Robertshaw 1999). Insofar as climate played a role in any of these events, reliable and relatively high rainfall would have permitted the production of agricultural surpluses that were then appropriated by political elites to finance the construction of earthworks.

A shift to a regime of lower rainfall levels beginning early in the 16th century and possibly associated with the Little Ice Age of higher latitudes may actually have been a series of major climate fluctuations (Verschuren et al. 2000). In Kitara, a period of general aridity appears to have set in from around AD 1520. Given that the major ditch systems at Munsa and Kibengo were probably constructed in the 15th or 16th centuries, their use is likely to have been associated in some manner with increased aridity beginning in the early AD 1500s. While we argued above that the earthworks were most likely constructed towards the end of the preceding period of higher rainfall, the available AMS radiocarbon dates do not offer sufficient chronological precision to rule out construction early in the succeeding drier period. However, it seems reasonable to suggest that while the period of higher rainfall promoted conditions suitable for the production of an agricultural surplus sufficient to finance the excavation of the ditch systems, the mode shift to drier conditions in the early 16th century enhanced the power and economic security of those living within the earthworks. If drier conditions were a factor promoting insecurity in subsistence and, by extension, in the political arena, then those living in or near earthworks may have been at an advantage. For it appears that the outer ditches may have protected agricultural crops from elephant predations, while the inner ditches offered inhabitants protection against human raiders. Nevertheless, the earthworks at Munsa appear to have been abandoned by perhaps the beginning of the 18th century; an event that may be enshrined in oral tradition as the story of the killing of chief Kateboha inside his earthworks by his disgruntled peasantry (Lanning 1959).

It seems that the settlement pattern that characterised Bunyoro in the 18th and 19th centuries was one of a dispersed rural population dominated by a large peripatetic royal capital and local chiefly courts. Judging from 19th century evidence there seems to be three main reasons why Nyoro royal capitals were peripatetic. Royal capitals could be relocated for strategic purposes, either to organise punitive military action against local rebels or to seek refuge from a foreign aggressor. Capitals may also have been moved because of the public health implications of accumulated human waste resulting from a large, settled population. Finally the crucial ritual and nutritional importance of the royal cattle herd may have compelled the royal court to shift to new grazing grounds in times of localised drought or due to the exhaustion of local grasslands.

Repeated episodes of prolonged drought could have influenced the availability of water and fodder and thus encouraged dispersed and peripatetic settlements. Moreover, the drier climate may have lead to a greater emphasis upon cattle, rather than cereal agriculture, for subsistence, at least among the nobility. Nicholson and Yin (2001) maintain that the first few decades of the 19th century were marked throughout Africa by drought that was most extreme during the 1820s and 1830s (cf. Verschuren et al. (2000)). A range of oral histories from across East Africa strengthens the evidence for severe drought during this period (Cohen 1977; Hartwig 1979; Koponen 1988). Evidence also exists for drought during the preceding century: Nile River minima are relatively low for much of the 1700s, rising to a peak in 1785, while Laws et al. (1975) argue that changes in forest composition in Bunyoro around 1780–1820 were due to a prolonged drought that caused Bunyoro's huge elephant population to forage in forested areas. While Laws et al.'s estimates of the age of trees is open to question, as those researchers themselves admit, it is possible that these records and reports of low rainfall form part of a much longer period of anomalously low precipitation that commenced during the 16th century.

In addition to shortfalls in rainfall, changes in population levels, political relations and in the nature of conflict are also likely to have been crucial. The kingdom of Bunyoro, once the most powerful state in the region, declined significantly in territory and power from the beginning of the 19th century. Here too climate may have played a part by helping to shift the balance of power in the region towards the Baganda. Thus, Médard (2000) states, "les périodes de mauvaises récoltes peuvent ... être bénéfiques pour le Buganda." The same extended dry period may also have facilitated the secession of Tooro and land bordering Buganda, while repeated invasions disrupted trade, denuded Bunyoro of livestock and reduced food security. War is still remembered by Banyoro informants as a major cause of famine during the late 19th century. One informant in the present study stated that famine occurred because "the men had gone for war and then a drought set in; so women couldn't grow enough food; they only had small gardens of millet", while another described the destruction of crops stored in grain pits by the Baganda.

Bunyoro experienced a revival under king Kabaleega from 1870 to 1893. Territory was regained, the country was restocked with cattle and a great expansion of trade brought new wealth to the kingdom. These improvements partly coincided with high Nile River minima and water levels in lakes Naivasha, Tanganyika and Victoria (Nicholson 1996; 1998; Nicholson and Yin 2001) and presumably relatively humid climatic conditions. Falling and low lake levels during the last two decades of the 19th century (Hastenrath 2001; Nicholson and Yin 2001) indicate a return to a regime of low rainfall. A severe famine struck Bunyoro in 1898–1900, by the end of which the kingdom was almost completely depopulated (Doyle 1998). Low levels of rainfall are unlikely to have been the only factor in food insecurity, however, as Bunyoro was also in the midst of the most prolonged and destructive war of colonial conquest experienced in the region. The war facilitated the spread of epidemic disease, as populations were both unusually mobile and uncommonly closely settled as a consequence of conflict, and the complete breakdown of anti-famine measures, such as underground grain stores and networks of trade. Scorched earth and hunger were two of the main weapons employed in the war; British-led colonial forces systematically attempted to pacify local populations by emptying the granaries and destroying the crops and livestock of villages that were suspected of disloyalty (Doyle 1998). Imperial cattle raiding resulted in the complete de-stocking of Bunyoro, which undermined not only the food security of

the kingdom, but also its systems of exchange and patronage. Market systems and social networks broke down due to worsening poverty and insecurity as the war progressed. Perhaps of greatest significance to social stability, Bunyoro's traditional rulers were unable to serve as sources of assistance in times of crises; even the newly installed king of Bunyoro was reduced to begging for food from European missionaries.[1]

Modern

According to the data available from Masindi for 1908–1999, great inter-annual variability in rainfall levels is apparent, with 1912–1913, early 1919 and 1925–1929 years of below average levels of rainfall during the early part of the modern period (Fig. 4). This variability is also clearly evident in recorded fluctuations in levels of Lake Victoria, with the early 1910s and 1920s periods of low stands (Nicholson and Yin 2001). That rainfall data measured at Masindi appear to co-vary with recorded levels of Lake Victoria is also borne out by the 1961 "event" (Hastenrath 2001). Lake Victoria levels show a dramatic rise (of 2 m) during the 1960s, with a rise of 1 m occurring in 1961, while data from Masindi indicate rainfall in 1961 at 1,628.9 mm was more than 20% above the mean for the period 1908–1999.

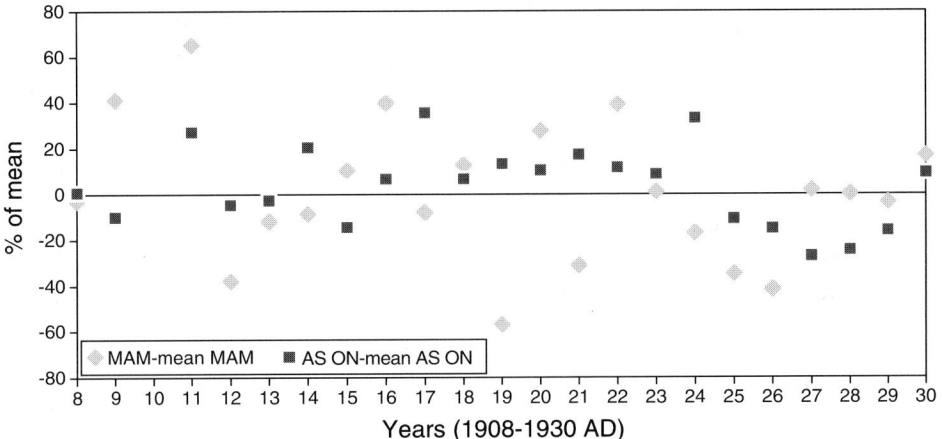

Figure 4. Rainfall data from Masindi for the period 1908–1930, showing deviations from the means for levels of rainfall in the earlier (MAM) and later (ASON) annual wetter seasons. Twenty-two years of data; no data are available for 1910.

Bunyoro experienced famine or serious food shortages in 24 of the 63 years of the colonial period, which ended in 1962. In almost every case, a combination of government food imports, reliance on locally maintained famine reserves, food purchases at market and traditional famine avoidance strategies ensured that large-scale mortality did not occur. Serious mortality did occur however during the famines of 1907 and 1917–18 (known locally as, respectively, *kiromere* (widespread) and *kabakuli* (small dish)). Mortality was also

---

[1] A. Fisher to H. Fox, 30 Nov. 1905, Birmingham, Church Missionary Society (C.M.S.) archive, Unofficial Papers Acc. 84, F3, Book XVI.

associated with a famine referred to locally as *zimya etaala* (blow out the candle) and dated to 1914–15 (Byaruhanga-Akiiki 1971; Bazaara 1988), as were drought-induced reductions in growth rates in forest plantations (Botanical, Forestry and Scientific Department 1914). However, this famine is not as well documented as the ones in 1907 and 1917–18, possibly because colonial government officers and staff were pre-occupied with the onset of the First World War, aside from reports insisting that chiefs maintain famine reserves in the face of recurrent food shortages.[2]

Accounts of severe drought and progressively depleted food reserves can be found in colonial records in the years leading up to 1907.[3] Government response to an impending crisis was muted, however.[4] Probably the most important reason for the colonial government's poor response was that it continued to face serious civil unrest in Bunyoro, as local leaders protested against the growing power of foreign chiefs in their country.[5] The severe famine in 1917–1918 was encouraged by the disruption brought by terrible disease mortality, due mainly to meningitis, smallpox and influenza.[6] The rainfall data from Masindi do not indicate a prolonged drought during this period, although more localised shortfalls in precipitation, unrecorded by the meteorological station, may have placed internal strain on food reserves that was compounded by the colonial administration's decision in 1917 to supply government troops with food from Bunyoro's famine stores.[7] The contrast with the 1907 famine was in the government response to the developing crisis. This was the first famine in Uganda when a recognisably modern famine relief programme was implemented. Food was trucked into Bunyoro and distributed from a series of depots through the local administration[8] and mortality was accordingly relatively limited compared to 1907.[9]

Undoubtedly the capacity of the colonial administration to source foodstuffs from other countries and to transport them very rapidly made food relief in a crisis one of the areas where British rule in Uganda was more effective than pre-colonial regimes. This effectiveness came at a high price, however. The total cost of the relief food trucked in during 1917 and 1918 had to be repaid immediately by a special tax of two rupees, and the price of famine relief was therefore severe poverty in succeeding years.[10] A high incidence of food shortages during the colonial period, even though population levels were generally low (Doyle 2000) and soils fertile, may have been partly caused by a return to an emphasis on sedentary means of producing food, a switch to cash crops such as coffee and a disruption of trading links following the establishment of international borders. Thus,

---

[2] Eden, Northern Province monthly reports, June-July 1914, Entebbe, Uganda National Archives (U.N.A), A46/791.

[3] H. Ladbury, 1907 annual letter, 12 Oct. 1907, C.M.S. annual letters, p. 237, Birmingham, C.M.S. archives; Masindi reports, Jan. and June 1907, U.N.A., A43/73.

[4] Mrs. Ladbury, Journals (2 vols, unpublished ms, Kampala, Makerere University Library), I, 26 May 1907.

[5] Speke, Masindi annual report, 1907–08, 15 Apr. 1908, U.N.A., A44/195; Knowles' comments on western province report, Mar. 1907, U.N.A., A43/74; Sub-commissioner Knowles' comments on Masindi report, Mar. 1907, U.N.A., A43/73.

[6] Northern province annual report, 1915–16, U.N.A., A46/809; Northern province reports, Jan.-Dec. 1917, U.N.A., A46/794; Medical department annual report 1917, 9, London, Public Record Office (P.R.O.), C.O./685/3; Medical department annual report 1919, 9, P.R.O., C.O./685/3; Medical report 1919, 8 July 1920, P.R.O., C.O./536/101.

[7] Northern province annual report, 1918–19, U.N.A., A46/811.

[8] Hence the famine being called 'kabakuli' after the ration bowls used in food distribution.

[9] Bowers, annual letter 1918, 30 Nov. 1918, Birmingham, C.M.S. archives, G3/AL/1917–1934.

[10] Medical department annual report 1917, 9, P.R.O., C.O./685/3.

while certain actions and policies of the colonial administration reduced the likelihood of major famine, others led to increased levels of food insecurity. One example of the latter was colonial hunting restrictions, which were enforced with particular rigour in Bunyoro (Doyle 1998). Enforcement not only removed one means of famine relief (i.e., hunting for game), it also led to the abandonment of traditional techniques of limiting crop raiding, resulting in increased losses of food crops. Elephant populations increased enormously in the thirty years following colonial invasion in 1893, as hunting pressure almost completely disappeared. A policy of culling was introduced in 1925 but it was only in the 1940s that elephant populations began to decline. Animals other than elephants, such as baboons, chimpanzees and wild pigs, may have been even more destructive overall.

Discussion

There is little doubt that a long-established relationship exists in Bunyoro between rainfall anomalies and food security. However, documented evidence for the late pre-modern and early modern periods suggests that major famines occurred when a combination of human and physical agents undermined systems designed to mediate variations in the availability of food. It is difficult to argue against famines in the more distant past being contingent upon the convergence of a number of factors rather than simply determined by shortfalls in rainfall. In the case of the period of transition to colonial government in Bunyoro, surplus agricultural capacity, the basis for a system of food reserves, was drained away through unusually heavy taxation, labour extraction, the requirement to provide food for government troops and crops for cash, and progressively degraded soils, while government policies and administrative inaction in other areas of responsibility ensured that famine, when it did finally occur, had a maximum impact. Only later in the colonial period did increased administrative and technical ability help circumvent the occurrence of major famines. That famine is contingent upon a number of environmental and human agents is hardly a revolutionary new finding; Watts (1983) argued the same in the early 1980s. However it is one that is probably worth considering before presumptions are made regarding the impacts on humans of past and future climatic variations.

Fieldwork around Munsa during July 2001 found that the names of famines could act as a valuable trigger of historical memory, releasing useful information about food crises in general and how people responded to them in the past. However there was marked disagreement among informants about the dates of major, named famines of the 20th century and between the informants' and documented dates for the famines. In some cases, the disagreement between the dates given by informants and those documented was more than 30 years. This confusion presumably stems from the fact that the major famines occurred just long enough ago to mean that informants would at best be remembering events from their childhood, or more likely recalling stories passed on from their parents. In addition, there is the possibility that oral historical research will uncover local, rather than collective, memories of famines, and that different local communities will have given different names to the same famine, or even the same name to different famines. Many of the names for famines that have been recorded in Bunyoro referred to the nature of the particular period of hunger or to objects associated with it (such as the small bowls used to distribute relief during the *kabakuli* famine), rather than to a particular event that coincided with the famine. Thus it is perhaps more likely that the years when famines occurred will be misremembered. All these problems are likely to be exaggerated when attempts are made to correlate social memories across large distances and between very different environments and cultural groups. Famine names therefore probably provide a more reliable means of eliciting information on responses to food shortages in the past than of dating them (contra Webster (1979b)).

This paper commenced with a criticism of attempts to correlate, date and interpret in deterministic ways references to major famines and droughts in oral histories from across the Interlacustrine region. Dubious interpretations of oral historical data are rendered no more reliable when they are simplistically correlated with palaeoclimatic data. While our own data vary in quantity and quality, and thus offer possibilities for alternative interpretations, our discussion of both the pre-modern and modern periods in Bunyoro nevertheless amply demonstrates the complex interplay of historical, political, and economic factors in determining both long-term human responses to climate change and the varied short-term consequences of droughts. Moreover, our excellent historical data from the modern period show that there is little correlation between the severity of a drought, as recorded in rainfall records, and its impact in terms of famine.

Our current goal is to obtain more data from Bunyoro, particularly for the pre-modern period. As part of an effort to understand the human dimensions of past variations in precipitation and other environmental variables, series of continuous cores of sediments were extracted from small sedimentary basins in Bunyoro and adjacent parts of Kitara during fieldwork in July 2001. Each of the basins cored contains within its immediate catchment a major former occupation site that has recently been the subject of archaeological excavations (the sites at Bigo, Kasunga, Munsa and Ntusi). Analyses of the sediment cores when combined with the results of the excavations may yield direct evidence of adaptation to major droughts in the form of, for example, adjustments in the mode of food production and the quality of crops grown and even the abandonment of settled sites, while on-going archaeological, ethnological and historical research should provide recent and contemporary contexts.

Summary

Attempts to understand the human dimensions of climate change are susceptible to post hoc explanations without thoughtful consideration of the likely mechanisms through which people and societies might have adapted to changed conditions. This paper describes a multidisciplinary approach to understanding human-environment interrelationships during the second millennium AD, focussing on the ancient Interlacustrine kingdom of Bunyoro in Western Uganda. The period of interest incorporates several major changes in climate and the first appearance of hierarchical societies (in the form of chiefdoms) with mixed economies centred upon large, apparently permanent settlements. There follows a transition in the 18th century AD to dispersed homesteads and a more narrowly based economy in which pastoralism had greater prominence than before. All of the major climate changes appear to coincide with significant changes in socio-economic conditions in the study area. The dangers of linking the two sets of variables in simple deterministic ways are illustrated through an examination of the documented history of famine and climate change in Bunyoro since late pre-modern (late pre-colonial) times.

Acknowledgments

We would like to thank the following, without whose help the research that underpins this paper would not have been possible: The UNCST, President's Office, Government of Uganda for research permission (reference number: EC487); The NGS, US, through a research grant (reference number: 6950-01); The ESRC for a PhD scholarship to Shane Doyle; Dr. Ephraim Kamuhangire and Professor Remigius Ziraba-Bukenya for useful advice and for facilitating the process of obtaining research permission; Julius Bunny, Moses Mufabi and Dismas Ongwen for assistance in the field, and the British Institute in Eastern Africa for provision of field equipment and a Landrover and driver; and Graham Bartlett at the National Meteorological Library, Bracknall, UK. For our most recent visit to Uganda (July 2001) we would especially like to thank Nathan Lubega, Laura Tindimubona and Nanny Carder.

References

Bazaara N. 1988. The food question in colonial Bunyoro-Kitara: capital penetration and peasant response. Unpublished M.A. dissertation, Makerere University, Kampala, Uganda.

Botanical, Forestry and Scientific Department 1909,1910. Annual Report. Government Printer, Entebbe, Uganda.

British East African Meteorological Service 1932 *et sequ.* Summary of Rainfall in Uganda Protectorate. Government Printer, Nairobi, Kenya.

Byaruhanga-Akiiki A. 1971. Religion in Bunyoro. Unpublished Ph.D. thesis, Makerere University, Kampala, Uganda.

Cohen D.W. 1977. Womunafu's Bunafu: a Study of Authority in a Nineteenth-Century African Community. Princeton University Press, Princeton, USA, 216 pp.

Dale I.R. 1954. Forest spread and climatic change in Uganda during the Christian era. The Empire Forestry Review 33: 23–29.

Department of Land and Surveys. 1967. Atlas of Uganda 1967. 2nd Edition. Government Printer, Entebbe, Uganda, 81 pp.

Doyle S. 1998. An environmental history of the kingdom of Bunyoro in Western Uganda, from c. 1860 to 1940. Unpublished Ph.D. thesis, University of Cambridge, UK.

Doyle S. 2000. Population decline and delayed recovery in Bunyoro, 1860–1960. Journal of African History 41: 429–458.

Harrop J.F. 1970. Climate. In: Jameson J.D. (ed.), Agriculture in Uganda. 2nd Edition. Oxford University Press, UK, pp. 24–29.

Hartwig G.W. and Patterson K.D. (eds) 1978. Disease in African History: An Introductory Survey and Case Studies. Duke University Press, Durham, USA, 258 pp.

Hassan F.A. 1981. Historical Nile floods and their implications for climatic change. Science 212: 1142–1145.

Hastenrath S. 2001. Variations of East African climates during the past two centuries. Climatic Change 50: 209–217.

Hastenrath S., Nicklis A. and Greischar L. 1993. Atmospheric-hydrospheric mechanisms of climate anomalies in the western equatorial Indian Ocean. J. Geophys. Res.-Oceans 98 (C11): 20219–20235.

Henige D.P. 1980. Ganda and Nyoro kinglists in a newly literate world. In: Miller J.C. (ed.), The African Past Speaks. Folkestone Press, UK, pp. 240–261.

Henige D.P. 1982. Oral Historiography. Longman, New York, USA, 150 pp.

Koponen J. 1988. People and Production in Late Precolonial Tanzania: History and Structures. Finnish Society of Development Studies #2, Helsinki, Finland, 434 pp.

Kopytoff I. 1987. The internal African frontier: the making of African political culture. In: Kopytoff I. (ed.), The African Frontier: The Reproduction of Traditional African Societies. Indiana University Press, Bloomington, USA, pp. 3–84.

Langdale-Brown I., Osmaston H.A. and Wilson J.G. 1964. The Vegetation of Uganda and its Bearing on Land-use. Uganda Government Printer, Entebbe, Uganda, 159 pp.

Lanning E.C. 1959. The death of chieftain Kateboha. Uganda Journal 24: 183–196.

McCann J.C. 1999. Climate and causation in African history. International Journal of African Historical Studies 32: 242–262.

Laws R.M., Parker I.S.C. and Johnstone R.C.B. 1975. Elephants and Their Habitats: The Ecology of Elephants in North Bunyoro, Uganda. Clarendon Press, Oxford, UK, 376 pp.

Médard H. 2000. Croissance et Crises de la Royauté de Buganda au XIXe Siècle. Unpublished Ph.D. thesis, University of Paris 1.

Nicholson S.E. 1996. Environmental change within the historical period. In: Adams W.A., Goudie A.S. and Orme A.R. (eds), Physical Geography of Africa, Oxford University Press, Oxford, pp. 60–87.

Nicholson S.E. 1998. Historical fluctuations of Lake Victoria and other lakes in the northern Rift valley of East Africa. In: Lehman J.T. (ed.), Environmental Change and Response in East African Lakes, Kluwer Academic, Dordrecht, The Netherlands, pp. 7–35.

Nicholson S.E. and Yin X. 2001. Rainfall conditions in equatorial East Africa during the nineteenth century as inferred from the record of Lake Victoria. Climatic Change 48: 387–398.

Phillips J. and McIntyre B. 2000. ENSO and interannual rainfall variability in Uganda: implications for agricultural management. International Journal of Climatology 20: 171–182.

Posnansky M. 1963. Towards an historical geography of Uganda. East African Geographical Review 1: 7–20.

Posnansky M. 1969. Bigo bya Mugenyi. Uganda Journal 33:125–50.

Reid D.A.M. 1996. Early settlement and social organisation in the interlacustrine region. Azania 29-30: 303–313.

Robertshaw P. 1994. Archaeological survey, ceramic analysis and state formation in Western Uganda. African Archaeological Review 12: 105–31.

Robertshaw P. 1997. Munsa earthworks: a preliminary report on recent excavations. Azania 32: 1–20.

Robertshaw P. 1999. Woman, labour and state formation in Western Uganda. In: Bacus E.A. and Lucero L.J. (eds), Complex Polities in the Ancient Tropical World. Archaeological Papers of the American Anthropological Association Number 9, Arlington, Virginia, pp. 51–65.

Robertshaw P. and Taylor D. 2000. Climate change and the rise of political complexity in Western Uganda. Journal of African History 41: 1–28.

Schmidt P.R. 1997. Archaeological views on a history of landscape change in East Africa. Journal of African History 38: 393–421.

Schoenbrun D.L. 1993. Cattle herds and banana gardens: the historical geography of the western Great Lakes region, ca. AD 800–1500. African Archaeological Review 11: 39–72.

Schoenbrun D.L. 1998. A Green Place, A Good Place. James Currey, Oxford, 301 pp.

Sutton J.E. 1993. The antecedents of the interlacustrine kingdoms. Journal of African History 34:33–64.

Tantala R.L. 1989. The Early History of Kitara in Western Uganda: Process Models of Religious and Political Change. Unpublished Ph.D. thesis, University of Wisconsin, Madison, USA.

Taylor D., Marchant R. and Robertshaw P. 1999. Late glacial-Holocene history of lowland rain forest in Central Africa: a record from Kabata Swamp, Ndale volcanic field, Uganda. J. Ecol. 87: 303–315.

Taylor D., Robertshaw P. and Marchant R.A. 2000. Environmental change and political-economic upheaval in precolonial Western Uganda. The Holocene 10: 527–536.

Uganda Meteorological Observations 1909–1938, 1943–1948. Annual Reports. Government Printer, Entebbe, Uganda.

Verschuren D., Laird K.R. and Cumming B.R. 2000. Rainfall and Drought in Equatorial East Africa During the Past 1,100 Years. Nature 403: 410–414.

Watts M. 1983. Silent Violence: Food, Famine and Peasantry in Northern Nigeria. University of California Press, Berkeley, 687 pp.

Webster J.B. (ed.) 1979a. Chronology, Migration and Drought in Interlacustrine Africa. Longman and Dalhousie University Press, London, 345 pp.

Webster J.B. 1979b. Noi! Noi! Famines as an aid to Interlacustrine chronology. In: Webster J.B. (ed.), Chronology, Migration and Drought in Interlacustrine Africa. Longman and Dalhousie University Press, London, pp. 1–38.

26. PALAEO-RESEARCH IN AFRICA: RELEVANCE TO SUSTAINABLE ENVIRONMENTAL MANAGEMENT AND SIGNIFICANCE FOR THE FUTURE

DANIEL O. OLAGO (dolago@unobi.ac.ke)
Department of Geology
University of Nairobi
PO Box 30197
Nairobi
Kenya

ERIC O. ODADA (eodada@uonbi.ac.ke)
Department of Geology
University of Nairobi
PO Box 30197
Nairobi
Kenya

Keywords: Palaeo-research, Climate variability, Freshwater, Land cover, Environment, Sustainability, Management, Vulnerability, Adaptation, Biodiversity

Introduction

Africa's environment is closely linked with its climate, so that climatic constraints have been a major force in the development of vegetation, soils, agriculture and general livelihood (Nicholson 2001). The African continent, one of the most vulnerable regions to climate change, is subject to frequent droughts and famine. These events reflect the large range of climatic variability that envelops mean trends in the major climatic parameters such as temperature and precipitation.

The livelihoods of most Africans are largely dependent on utilisation of land-based resources, as well as on freshwater lacustrine and riverine systems as sources of potable water, fish, transport etc. There has been increasing awareness of the dependence of various economic activities on climate fluctuations and on the implications of long-term change (Ottichilo et al. 1991). For example, crop and livestock production are major employers and make significant contributions to GDP and export earnings (IPCC 2001). The predominance of rain-fed subsistence agriculture and, across Southern Africa, over-dependence on (water-demanding) maize has helped ensure that food security for most of the continent is inextricably linked to the quality of each rainy season (IPCC 2001). Global warming will lead to higher temperatures estimated to be between 0.2 and 0.5 °C per decade for

R. W. Battarbee et al. (eds) 2004. *Past Climate Variability through Europe and Africa.*
Springer, Dordrecht, The Netherlands.

Africa (Hulme et al. 2001). Climate studies and modelling experiments indicate that the anthropogenically-driven rise in global temperatures and land-use changes may adversely affect climatic, hydrological and environmental parameters.

Africa will experience the effects of the human-induced changes in climate, but much work remains to be done in trying to isolate those aspects of African climate variability that are natural from those that are related to human influences (Hulme et al. 2001). However, the current economic decline caused by high rate of population growth, inefficient resource use, weak institutional capacity, inadequate human resources, low levels of investment and savings and a general decline in income and living standards is expected to impair Africa's capacity to respond effectively to disruptions emanating from climate change (Ottichilo et al. 1991). Nonetheless, effective responses to climate change impacts can only evolve from an understanding of the driving forces of climate change and from effective prediction of the future state, not only on seasonal or annual time-scales, but also on inter-annual and decadal timescales. The short instrumental record in Africa (mainly from the late 1880s to present) does not provide an adequately long time-series to capture and understand fully the range of natural climate variability, nor the frequency and intensity of unique events such as El Niño. It is an impediment to the recognition and understanding of the complex workings and interactions of long-term (decadal, inter-decadal or centennial scale) features of the climate system. For this, we have to turn to the palaeo-records archived within the continent.

Palaeo-data covering the last glacial through the Holocene to the present day from various parts of the continent show that Africa has experienced natural, large and sometimes abrupt fluctuations in climate, hydrology, and environment. Although orbitally-induced changes in monsoon strength account for a large part of long-term climatic changes in tropical Africa, the late Pleistocene-Holocene hydrological fluctuations rather appear to have been a series of abrupt events that reflect complex interactions between orbital forcing, atmosphere, ocean and land surface conditions (Gasse 2000). More generally, the climatic, hydrological and environmental oscillations of the low-latitude regions during the Holocene are linked to changes in earth-surface temperatures, sea-surface temperatures (SSTs), ocean and atmospheric circulation patterns, regional topography, land-surface albedo etc. The relative importance of these forcing factors, and the extent of the linkages between them are still unclear, but the data suggest that the climate and hydrology of tropical regions may be adversely affected by the anthropogenically driven rise in global temperatures and land cover change.

Role and significance of palaeo-research in Africa

Palaeo-environmental and palaeoclimatic research in Africa is of great importance for several reasons. It provides a historical perspective on past variability due to natural and human causes, and thus provides a baseline for efficient long-term management of natural resources; this is essential in a poor continent subject to frequent droughts and famine. Meteorological records and written observations are limited to the very recent past (often only the past few decades); thus data on longer term cyclical fluctuations is very limited, as is our understanding of how these impact on regional environments and human societies, or how these various components interact. It is noted, for example, that during the late Holocene when natural forcings and boundary conditions were similar to today, climate variability often exceeded anything that is seen in modern instrumental records (Oldfield and Alverson 2003). Knowledge of long-term climate change, therefore, is necessary in order to assess the significance of historically documented, and modern-day climate change. Palaeo-research also enables us to estimate better the range or 'envelope' of natural climate variability under boundary conditions similar to the present, and also to discriminate between natural and anthropogenic perturbations of the climate system. It enables us to recognise locally and regionally significant human impacts, and is critical in the development and testing of models which can then be used to simulate future climate change and trends.

It is abundantly clear from palaeoclimate records that large and sometimes abrupt changes have occurred in the global climate system at certain times in the past. Apparently, non-linear responses have occurred as critical thresholds were passed. Our understanding of what these thresholds are is completely inadequate. We cannot be certain that anthropogenic changes in the climate system will not lead us, inexorably, across such thresholds, beyond which may lie a dramatically different future climate state. Only by careful attention to such episodes in the past, can we hope to fully comprehend the potential danger of future global changes due to human-induced effects on the climate system. Predictions about the possible significance, trends and consequences of natural interactions in the earth-climate system and the possible effects on people and environment, and modulation of the environment and climate by anthropogenic activity, rely essentially on the ability of earth-climate system models to realistically reproduce the complex changes observed in the past. Palaeoclimatic evidence provides the essential perspective on climate system variability, its relationship to forcing mechanisms and to feedbacks that may amplify or reduce the direct consequences of particular forcings. Particularly in the tropics, such a perspective cannot be provided by the very limited set of instrumental data at our disposal.

Palaeoclimatic data provide a critical test of general circulation models used to simulate future climates; if they can accurately simulate climatic conditions that are known to have existed in the past, confidence in their ability to predict future climatic conditions will be enhanced. Palaeoclimatic records also provide evidence of how biological and environmental systems have responded in the past to changes in climate. High-resolution, well-dated cores are particularly required from the tropics to better understand the long- and short-term periodicities that characterise the climate of the region. The mechanisms underlying these cycles need to be elucidated by use of high-performance regional and global models validated with reliable palaeo-datasets.

Relevance to present and future sustainable environmental management

Climate variability

It has been observed that in Africa, changes in precipitation have a much larger envi-ronmental impact than changes in temperature. Temperature has a much wider diurnal range as compared to its annual range which remains relatively constant from year to year except in the highest latitudes (Nicholson 2001). Using both systematic rainfall records and proxy information concerning lakes and rivers and the occurrence of famine and drought, Nicholson (2001) observed that the most significant climatic change that has occurred in Africa over the past two centuries has been a long-term reduction in rainfall in the semi-arid regions of West Africa. In the 1990s, however, the rainfall situation in the Sahel has ranged from normal to above normal (Amani 2001). More generally, over the past 30 years or so, unusually severe and/or prolonged droughts in African drylands have seriously affected agriculture and wildlife and caused many deaths and severe malnutrition (IPCC 2001).

Examples from the palaeo-record

High resolution palaeo-records have to varying degrees and in specific regions, extended our knowledge of inter-annual, decadal and inter-decadal climate variability. For example, a 1,100 yr record describes the hydrological response of Lake Naivasha, to a succession of decade-scale fluctuations in the regional balance of rainfall and evaporation, providing an excellent record of rainfall and drought in equatorial East Africa (Verschuren et al. 2000; this volume). The data indicate that, over the past millennium, equatorial East Africa has alternated between contrasting climatic conditions, with significantly drier climate than today during the Medieval Warm Period (ca. AD 1000 to 1270) and a relatively wet climate during the Little Ice Age (AD 1270–1850) which was interrupted by three prolonged dry episodes. The arid periods or drought events were broadly coeval with phases of high solar radiation, and intervening periods of increased moisture were coeval with phases of low solar radiation (Verschuren et al. 2000). The drought periods matched oral historical records of famine, political unrest and large-scale migration of indigenous peoples, while the wet periods were prosperity years.

Palaeo-records also reflect El Niño Southern Oscillation (ENSO)-induced changes in precipitation in Eastern Africa, showing that this phenomenon is a long-lived feature of the climate system. The palaeo-records are also sufficiently long to enable a study of its range of variability and intensity. Time-series analysis of a varved core from northern Lake Malawi (Pilskaln and Johnson 1991; Johnson 1996) shows periodicities at 2.6 and 3.5 years, very similar to periodicities in rainfall anomalies for this part of Africa (cf. Nicholson (1996)). These types of cyclicity, including multiples of ENSO occurrence, are probably archived in several other lakes, but, as is the case for Lake Turkana, the present uncertainty in the calibration of the radiocarbon timescale does not allow for confirmation of periodicities that are less than one century (Johnson 1996). High-resolution, well-dated cores are, therefore, particularly required from the tropics to better understand the long- and short-term periodicities that underlie the climate of the region.

Implications for long-term sustainable development and management

Verschuren et al. (2000) note, from the 1,100 yr Naivasha record, that the magnitude of natural decade-scale rainfall variability in sub-humid East Africa implies that sustainable development and protection of food security will require agricultural management strategies adjusted to major long-term variation in water-resource availability, irrespective of any future effects of anthropogenic climate change on the hydrological cycle. The fact that persistent droughts, well beyond the range of those recently experienced, have been common in the past, suggest that there is a high possibility of their occurrence in the future (Oldfield and Alverson 2003). These are important outcomes from palaeo-research that support the need for the institution of informed and long-term sustainable development policies to cope with and adapt to future climate change impacts.

Current evidence also suggests that inter-annual and inter-decadal climate variability have a direct influence on the epidemiology of vector-borne diseases (Githeko et al. 2000). For example, climatic anomalies associated with ENSO have been linked to outbreaks of

malaria in Africa, Asia and South America (Githeko et al. 2000). IPCC conclusions show that most (90%) global mortality due to malaria occurs in Africa (IPCC 2001). In recent years the number of epidemics of this disease have increased in the East African region with devastating effects. In the two warming periods in the 1930s to 1940s and the late 1980s, malaria epidemics were observed in the East African region (Roberts 1964; Githeko and Ndegwa 2001). Githeko and Ndegwa (2001) noted an association between malaria, rainfall, and unusually high maximum temperatures, and produced a model to predict malaria epidemics in the Kenyan Highlands. With predicted higher mean temperatures for this century and an increase in the amplitude of temperature variability, the gap between mean and critical "threshold" maximum temperatures that mark the onset of epidemics would be reduced, with possible increases in the frequency of malaria epidemics (cf. Patz et al. (2002)). Although there are no palaeo-records available on the long-term relationship between climate variability and the incidence of disease, this example illustrates how knowledge of the range of past climate variability can be incorporated to model the future spatial and temporal spread of climate-related diseases.

Freshwater resources

The major effects of climate change on African water systems will be through changes in the hydrological cycle, the balance of temperature, and rainfall (IPCC 2001). Global climate variability and change will affect the water supply (Hulme 1996) although non-climatic changes such as water policy and management practice may have significant effects. Freshwater resources and changes in those resources are perhaps the most important global change issues in sub-Saharan Africa (Gash et al. 2001), as water supply undoubtedly is a most important resource for Africa's social, economic, and environmental well-being (IPCC 2001). Many African countries are today experiencing water stress, and it is projected that many more will shift from a water surplus state to a water scarce state by 2025 due to changes in population alone (IPCC 2001). Some of the basic problems with water as a resource in Africa are not necessarily the lack of rainfall, but rather the very high potential evaporation which occurs throughout the year and is in excess of 2000 mm per annum over large tracts, very high aridity indices, and a generally low conversion of rainfall to runoff (Schulze 2001). These factors are further compounded by an often very concentrated seasonality of rainfall, and hence runoff, a strong response to the ENSO signal and thus generally high inter-annual coefficient of variability of rainfall, and an amplification of the inter-annual coefficient of variability of rainfall by the hydrological cycle (Schulze 2001). Other problems that affect the quantity, quality and availability of freshwater include increasing population pressure and pollution of water resources, land-use leading to enhanced erosion/siltation, and possible ecological consequences of land-use change on the hydrological cycle.

Examples from the palaeo-record
During the Holocene period when temperature has been more or less similar to today, several large lakes registered abrupt transgressions and recessions of magnitudes far larger than any witnessed in recent times. There is a large body of evidence on lacustrine extensions in the Sahelian and Saharan subtropical latitudes in the early Holocene (Lézine 1989; Street-Perrott et al. 1989). By 9000 yr BP, a belt of high lake levels extended from 4 °S to 33 °N,

suggesting that large areas now arid were regularly receiving substantial tropical rainfall. During this period there were regression events of extremely large magnitude recorded in some lakes of Eastern Africa. In the Ziway-Shala basin of Ethiopia, for example, a lake lowering of 50 m is recorded between 8000 to 6500 yr BP (Gasse and Street 1978), with lowest levels between 7800 and 7000 yr BP (Lézine 1982; Gillespie et al. 1983). It generally correlates with a desiccation event in many other African lakes between 8000 and 7500 yr BP (Street-Perrott et al. 1985). The Holocene wet phase lasted until about 4000 yr BP when drier conditions set in (e.g., Gasse and Street (1978), Hoelzmann (2002), Hoelzmann et al., this volume). As recently as several hundred years ago, Lake Malawi is believed to have been 50 m shallower than it has been during the last 150 years (Owen et al. 1990).

    Palaeohydrological studies of Sahelian-Saharan groundwaters have shown that the waters are thousands of years old, recharged in wetter climates than today (most of Sahara receives less than 100 mm rainfall today) (Edmunds et al., this volume). Most of the groundwater recharge of the regional and local Saharan and Sahelian aquifers occurred (during periods wetter and cooler compared to today) prior to, and immediately after, the Last Glacial Maximum (e.g., Thorweihe and Heinl (2002)). The Nubian Aquifer System in the eastern Sahara covers about 2 million square kilometres. Radiocarbon dating indicates that the groundwater recharge took place during two wet phases: prior to 20,000 yr BP and between 14,000 and 4000 yr BP (Thorweihe and Heinl 2002).

Implications for long-term sustainable development and management
The great spatial and temporal variability of rainfall and evaporation over Africa presents a significant problem for sustainable development of water and land resources; these variations have neither been predictable nor uniform over the continent (Soliman 2001). Although the severity of the impacts of climate change depend primarily on the magnitude of change, the different hydrological sensitivities of the river basins are also important. The Nile and Zambezi are especially sensitive to climate warming as runoff decreases in these basins even when precipitation increases, due to the large hydrological role played by evaporation (IPCC 2001). GCM scenarios provide widely diverging pictures of possible future river flows, from a 30% increase to a 78% decrease; the large uncertainty in climate-change projections makes it very hard for basin managers to adopt any response policy (IPCC 2001). This reflects a poor understanding of the hydrological variability of the river basins, partly due to the lack of long-term climate series predating 1900 that would indicate the extremes of variability. If lake-level regressions such as those that occurred in the past (under boundary conditions similar to today) were to occur today, the impact on human society and environment would be disastrous. Possible scenarios include freshwater shortage and conflict over its limited supply, disruption of infrastructure, reduced output or complete shut-down in hydroelectric power supply, collapse of agriculture, etc.

    Groundwater systems monitored over many years in different countries show steady decline. It has been shown that the present groundwater recharge of the Nubian Aquifer System is negligible, and that the groundwater extraction is in fact mining of a non-renewable resource, but large amounts of groundwater (150,000 km^3) allow for restricted extraction in limited areas (Thorweihe and Heinl 2002). Indeed, much of the Sahara/Sahelian palaeo-groundwaters can be considered as a non-renewable resource that needs to be diligently used. It is clear that if abstraction of these palaeo-waters continues the resource will be eventually exhausted, with serious implications for the communities of this region. The impact would be further exacerbated by the threat of reduced Nile River flow due to destruction of catchment areas in the Ethiopian Highlands and Lake Victoria basin.

Land-cover change

Land-cover changes, both natural and human-related are significant in modifying regional climates (Xue 1997). Presently, existing precipitation in the Sahel savannah, Western Africa, could support a much richer flora and fauna, were it not for human activities (Cloudsley-Thompson 1974). Increased albedo, atmospheric dust, and reduced soil moisture resulting from anthropogenic deforestation, enhances drought in the region (Cloudsley-Thompson 1993). Episodic conflicts have generated large movements of refugees that have led to marginal land-cover changes in some places. Savannah fires are widespread and have significant ecological impacts on a wider region given the long-range transport of atmospheric emissions. The increased demand for fuelwood due to population pressure has exacerbated the deforestation trend in Africa (Ottichilo et al. 1990). For example, forest species richness and tree density has declined in the West Africa Sahel over the last half of the 20th century and the rural population exceeded the 1993 carrying capacity for firewood from shrubs (Gonzalez 2001).

In Southern Africa, the natural vegetation has been highly modified by crop cultivation, urbanisation, overgrazing (Schulze 2000) and biomass burning (Scholes and Andreae 2000). Up to 91% of rainfall in Southern Africa is returned to the atmosphere via evapotranspiration (Gondwe and Jury 1997) compared to a global average of 65–75% (Martyn 1992). This implies that the depletion of vegetation, particularly over the eastern highlands north of 25 °S, could alter the feedback budget with negative implications (Gondwe and Jury 1997). In addition, the amount of CO_2 that is exchanged with the atmosphere annually owing to vegetation fires in Southern Africa is very large, about 20% of net primary production, thus a relatively small perturbation to the fire regime could have significant consequences for the net global carbon budget (Scholes and Andreae 2000).

All these have undermined Africa's bioproductive systems and its economic base while contributing to the greenhouse effects (cf. Ottichilo et al. (1990)). Many African countries have implemented economic recovery programmes aimed at halting the decline of the productive sector and enhancing its sustained growth (Ottichilo et al. 1990). Such efforts, if successful, are expected to enhance the continent's capability to manage the effects of climate change. However, the models under the economic recovery programmes have often had adverse social and environmental impacts on the economies of the countries concerned (Ottichilo et al. 1990). This is partly due to the fact that they were launched without sufficient understanding of the scientific basis of the problem and of the interactions and feedbacks between human activities and the environment.

Examples from the palaeo-record
The landscape response to climate change in the Holocene has not been as abrupt as that of water resources, but the magnitude has been large. The major trend is from wet/moist vegetation in the early Holocene to drier vegetation from the middle Holocene to present. The starkest example of regional-scale vegetation change from palaeoenvironmental studies comes from the Sahara-Sahel region, where there was a rapid extension of humid vegetation

to the north from 9000 yr BP, correlated with the intensification of the Atlantic monsoon (Lézine 1989). Tropical savannahs shifted 500 to 700 km northwards of their present range between 7000 and 6500 yr BP, receding slightly after 6000 yr BP to 300–400 km north of their present range (Neumann 1991). Following the termination of the late Pleistocene, the biomes of Southern Africa began to reflect modern conditions by 7000 yr BP (Scott et al. 1997; this volume). Generally, the advent of moister conditions in Southern Africa is recorded earlier (7500 to 6500 yr BP) in the north at ca. 26 °S than around 31 °S (5,000 yr BP), and this is provisionally associated with a relative shift in seasonality from a predominance of all-season precipitation to a greater proportion of summer rainfall (Scott 1993). Related patterns of change have been found in the southern Kalahari (Scott 1990). Micro-mammalian evidence suggests that the Namib desert was grassier at ca. 6,500 yr BP (Brain and Brain 1977) and moister conditions during the last 2,000 years are supported by pollen data from hyrax middens from the Kuiseb River (Scott et al. 1997).

Thus, the latitudinal march of vegetation belts has been shown from palaeo-research to be dependent primarily on changes in precipitation, but this relationship is now being modified by human impact on land.

Implications for long-term sustainable development and management
The findings in the Sahel and Southern Africa region indicate that better understanding is required of the long-term interactions between humans, vegetation changes, climate variability and the role of fire, particularly over the past 2000 years in order to better understand present and future climate change impacts on land cover. While general circulation models simulate changes of African climate as a result of increased greenhouse gas concentrations, two potentially important drivers of African climate, ENSO and land-cover change, are poorly or not at all (respectively) represented in the models (Hulme et al. 2001). Problems of soil erosion and deforestation may be exacerbated by increased precipitation in some regions and increased droughts in others, respectively, while shifts in climatic zones may create serious problems in human settlements, agricultural land-use and wildlife management under the existing land tenure systems (Ottichilo et al. 1991). Climate - land-cover interactions, therefore, require much more investigation, as the impacts of anthropogenically-driven land-cover change can, potentially, significantly alter regional climate boundary characteristics in the short-term, and may feedback positively into the global climate system as a whole in the long-term.

If models are able to simulate accurately the natural biome changes that occurred in the past, then an understanding of the processes and feedbacks that drive natural changes will emerge. This will enable a de-linking of natural and anthropogenically driven change, leading to the generation of more robust predictive models that can be used for analysis and formulation of long-term sustainable development options in areas such as food security and land management.

Ecosystem stability and biodiversity in the African Great Lakes

The large lakes of the East African Rift Valley are amongst the oldest on Earth and are vital resources (e.g., for transportation, water supply, fisheries, waste disposal, recreation and tourism) for the indigenous populations that inhabit their basins. The lakes are unique in many ways: they are sensitive to climatic change; their circulation dynamics, water column chemistry and biological complexity differ from large lakes at higher altitude; they have long, continuous, high-resolution records of past climatic change; and they have rich and diverse populations of endemic organisms.

The African large lakes such as Victoria, Tanganyika and Malawi as well as many smaller freshwater bodies (including wetlands and rivers) in the region are under considerable pressure from a variety of inter-linked human activities. Overfishing, siltation, erosion of deforested watersheds, species introductions, industrial pollution, eutrophication and climate change are all contributing to a host of rapidly evolving changes occurring in these lakes that seriously threaten both their ecosystem function and overall diversity (Hecky 1993; Cohen et al. 1996; Twongo 1996). For example, the extinction of several hundred species of haplochromine cichlid fish in Lake Victoria following the introduction of Nile Perch, a large voracious predatory species, ranks as the largest single recorded vertebrate extinction attributable to specific human actions on earth (Johnson et al. 1996). These lakes, as earlier mentioned, have experienced large water level fluctuations in the past, with as yet poorly known consequences on aquatic ecosystem change and biodiversity that would result from changes in pH, salinity, geography, stratification, sediment and nutrient load changes and fluxes.

Lakes Malawi/Nyasa, Victoria and Tanganyika are famous for their endemic species flocks of cichlid fishes. Lake Malawi hosts a large flock, estimated to include 700+ cichlid fish species (Snoeks 2000). Before the introduction of the predatory Nile Perch, the Lake Victoria cichlid fish species flock included 500+ species (Seehausen 1996). Lake Tanganyika hosts 250+ cichlid species parsed between several sub-flocks (Snoeks et al. 1994). The African cichlid fish are the largest and most diverse radiation of vertebrates on earth. Lake Tanganyika, with more than 2000 species of plants and animals, is among the richest freshwater ecosystems in the world. With their great number of species, including endemic species, genera and families, the African Great Lakes make an important contribution to global biodiversity.

Examples from the palaeo-record

The substantial hydrological changes that occurred in lakes during the Holocene (see above) resulted in a modification of their geography, hydrology, limnology and habitats, and are believed to have had major consequences for speciation and extinction particularly in the African Great Lakes (Johnson et al. 1996; Plisnier 2002).

It is thought that the proto Lake Tanganyika was colonised by organisms from the ancient Zaire River system (which pre-dates the lake), and these pioneer species evolved and radiated within the lake basin, creating Tanganyika's great diversity (Coulter 1994). Palaeo-research findings have shown that Lake Victoria dried up completely during the Lake Pleistocene, before 12,400 yr BP, implying that the rate of speciation of cichlid species (500+; Seehausen (1996)) has been very rapid and, barring any satellite lakes in the basin where the fish could seek refuge during the arid period, is the fastest ever recorded for such a large number of vertebrate species (Johnson et al. 1996). Palaeo-research has also provided evidence of increased productivity in Lake Victoria since 1900, reflected by increases in the cyanobacteria population (Lipiatou et al. 1996). Eutrophication-induced loss of deep-water oxygen started in the early 1960s, and may have contributed to the 1980s collapse of indigenous fish stocks by eliminating suitable habitats for certain deep-water cichlids (Verschuren et al. 2002).

Implications for long-term sustainable development and management
Natural changes in lake level of similar or greater magnitude as those modelled for the future have occurred in the past (e.g., Owen et al. (1990)). Palaeo-research would help to elucidate the mechanisms of change by availing long, continuous records. Opportunities also exist to study the evolution of biodiversity and ecology of the lakes. For example, there is a considerable redundancy of species in ancient lakes such as Tanganyika, and it is argued that redundancy can be a buffering mechanism against species loss in (more or less predictable) fluctuating environments, since a large number of species per functional guild will allow species loss without ecosystem collapse (Martens 2002). If the palaeo-perspective bears this out, then, for example, conservation programmes should not focus exclusively on keystone taxa and function, but should rather manage for redundancy as a buffer for ecosystem resilience to both climatic and human induced disturbances (Martens 2002).

Conclusions

Long-term records of change are required to assess the significance of historically docu-mented and modern-day change in Africa. Anthropogenic impacts on the earth system are becoming increasingly evident, leading to globally important environmental changes. These include changes in climate, reduction in freshwater availability, decreased food security, increase in aridification etc. and which are negatively impacting upon the aspects of the socio-economic structures in various regions of the world. Given the rapidity of human impact on the environment, and the associated uncertainties regarding the impact on global climate and environment, as well as the modes of feedback and degree of interaction between the various components/factors of change, there is a need to understand better the natural climate and environmental variability. Rapid progress is critical, particularly in Africa where there is a relative dearth of palaeo-information. The mechanisms underlying the abrupt, large-scale climatic events in the Holocene need to be understood as they occurred during a period with similar climatic boundaries to today. It is evident that more, high-resolution proxy records with wide spatial coverage are required in order to differentiate between local and regional impacts, and between human-induced and natural change.

The palaeo-research will, in addition to answering fundamental science questions, be important to the better and holistic understanding of the natural cycles and interactions of the various components of the earth system. It needs to address the issues in such a manner that the findings will also be relevant to, and inform on, current and future sustainable environmental development, and policy formulation for the global society at large.

Summary

The livelihoods of most Africans are largely dependent on utilisation of land-based re-sources, as well as on freshwater lacustrine and riverine systems as sources of potable water, fish and transport. Global warming and land-use changes may adversely affect the services provided by these climate-sensitive resources, in turn placing greater stress upon the societies that are currently experiencing economic decline, widespread poverty and high population growth rates. Many African countries are today experiencing water stress, and it is projected that many more will shift from a water surplus state to a water scarce state by 2025 due to changes in population alone (IPCC 2001). Increasing population pressure, pollution of water resources, land-use leading to enhanced erosion/siltation, and possible ecological consequences of land-use change on the hydrological cycle are contributing to reduced surface water and groundwater supplies. The predominance of rain-fed subsistence agriculture has helped ensure that food security for most of the continent is inextricably linked to the quality of each rainy season (IPCC 2001). In addition, the increased demand for fuelwood due to population pressure has exacerbated the deforestation trend in Africa (Ottichilo et al. 1990; Gonzalez 2001). Changes of African climate are not, however, well understood. For example, while general circulation models simulate changes of African climate as a result of increased greenhouse gas concentrations, two potentially important drivers of African climate, ENSO and land-cover change are poorly or not at all (respectively) represented in the models (Hulme et al. 2001).

Palaeoclimate records show that large and sometimes abrupt changes, far exceeding anything that has been instrumentally recorded, have occurred in the past under boundary conditions similar to today. The short instrumental record in Africa (mainly from the late 1880s to present) does not provide an adequately long time series to capture and understand fully the range of natural climate variability. As a result, the long-term (decadal, inter-decadal or centennial scale) features of the climate system are only poorly understood, and consequently, predictions of possible future climate impacts are highly uncertain. African societies may, therefore, not be able to adequately prepare for, cope with, or adapt to, climate change impacts. Effective responses need to be devised: these can only evolve from an understanding of the driving forces of climate change and from effective prediction of the future state, not only on seasonal or annual time-scales, but also on inter-annual and decadal timescales.

Palaeo-records can, and do, provide temporally long climate change data that is necessary for assessing the significance of historically documented, and modern-day climate change. The African palaeo-records show, for example, that there have been sustained and persistent droughts (decadal to inter-decadal), well beyond the range of those recently experienced. The drought periods in sub-humid East Africa over the past millennium matched oral historical records of famine, political unrest and large-scale migration of indigenous peoples (Verschuren et al. 2000). If lake-level regressions such as those that occurred in the African lakes in the past (under boundary conditions similar to today) were to occur today, the impact on human society and environment would be disastrous. Possible scenarios include freshwater shortage and conflict over its limited supply, disruption of infrastructure, reduced output or complete shut-down in hydroelectric power supply, collapse of agriculture, loss of biodiversity, etc. The palaeo-records thus enable us to: estimate better the range or 'envelope' of natural climate variability under boundary conditions similar to the present; discriminate between natural and anthropogenic perturbations of the climate system; recognise locally and regionally significant human impacts; develop and test realistic models which can then be used to simulate future climate change and trends. Such information is critical to the generation of robust predictive models that can be used for analysis and formulation of long-term sustainable development options in areas such as food security, sustainable water use, land management and conservation of biodiversity.

Acknowledgments

We would like to recognise the efforts of the following organisations for supporting research and capacity building in African palaeoclimatology and limnology: IDEAL programme, START, PAGES, MacArthur Foundation, NSF and NORAD.

References

Amani A. 2001. Rainfall and water resources variability in the Sahelian region: a review. In: Gash J.H.C., Odada E.O., Oyebande L. and Schulze R.E. (eds), Freshwater Resources in Africa: Proceedings of a Workshop, Nairobi, Kenya, October 1999. BAHC International Project Office, Potsdam, pp. 59–64.

Brain C.K. and Brain V. 1977. Microfaunal remains from Mirabeb: some evidence of palaeoecological changes in the Namib. Madoqua 10: 285–305.

Cloudsley-Thompson J.L. 1974. The expanding Sahara. Environmental Conservation 1 (1): 5–13.

Cloudsley-Thompson J.L. 1993. The future of the Sahara. Environmental Conservation 20 (4): 335–338.

Cohen A.S., Kaufman L. and Ogutu-Ohwayo R. 1996. Anthropogenic threats, impacts and conservation strategies in the African Great Lakes: a review. In: Johnson T.C. and Odada E.O. (eds), The Limnology, Climatology and Paleoclimatology of the East African Lakes. Gordon and Breach Publishers, Australia, pp. 575–624.

Coulter G.W. 1994. Lake Tanganyika. In: Martens K., Goddeeris B. and Coulter G. (eds), Speciation in Ancient Lakes. Arch. Hydrobiol. 44: 13–38.

Edmunds W.M., Dodo A., Djoret D., Gasse F., Gaye C.B., Goni I.B., Travi Y., Zouari K. and Zuppi G.M., this volume. Groundwater as an archive of climatic and environmental change. In: Battarbee R.W., Gasse F. and Stickley C.E. (eds), Past Climate Variability through Europe and Africa. Kluwer Academic Publishers, Dordrecht, the Netherlands, pp. 279–306.

Gash J.H.C., Fosberg M., Odada E.O., Oyebande L. and Schulze R.E. 2001. Freshwater resources research in Africa. In: Gash J.H.C., Odada E.O., Oyebande L. and Schulze R.E. (eds), Freshwater Resources in Africa: Proceedings of a Workshop, Nairobi, Kenya, October 1999. BAHC International Project Office, Potsdam, pp. 3–6.

Gasse F. 2000. Hydrological changes in the African tropics since the Last Glacial Maximum. Quat. Sci. Rev. 19: 189–211.

Gasse F. and Street F.A. 1978. Late Quaternary lake-level fluctuations and environments of the northern Rift Valley and Afar region (Ethiopia and Djibouti). Palaeogeogr. Palaeoclim. Palaeoecol. 24: 279–325.

Gillespie R., Street-Perrott F.A. and Switsur R. 1983. Post-glacial arid episodes in Ethiopia have implications for climate prediction. Nature 306: 680–683.

Githeko A.K., Lindsay S.W., Confalonieri U. and Patz J. 2000. Climate Change and Vector borne diseases: A regional analysis. Bull. World Health Org. 78: 1136–1147.

Githeko A.K. and Ndegwa W. 2001. Predicting malaria epidemics in the Kenyan Highlands using climate data: a tool for decision makers. Global Change and Human Health 2: 54–63.

Gondwe M.P. and Jury M.R. 1997. Sensitivity of vegetation (NDVI) to climate over Southern Africa: relationships with summer rainfall and OLR. South African Geographical Journal 79 (1): 52–60.

Gonzalez P. 2001. Desertification and a shift of forest species in the West Africa Sahel. Clim. Res. 17: 217–228.

Hecky R.E. 1993. The eutrophication of Lake Victoria. Proc. Int. Ass. Theor. Appl. Limnol. 25: 39–48.

Hoelzmann P. 2002. Lacustrine sediments as key indicators of climate change during the Late Quaternary in Western Nubia (Eastern Sahara). In: Lenssen-Erz T., Tegtmeier U., Kröpelin S., Berke H., Eichhorn B., Herb M., Jesse F., Keding B., Kindermann K., Linstädter J., Nußbaum S., Reimer H., Schuck W. and Vogelsang R. (eds), Tides of the Desert. Monographs on African Archaeology and Environment, Africa Prehistorica 14: 375–388.

Hoelzmann P., Gasse F., Dupont L.M., Salzmann U., Staubwasser M., Leuschner D.C. and Sirocko F., this volume. Palaeoenvironmental changes in the arid and subarid belt (Sahara-Sahel-Arabian Peninsula) from 150 kyr to present. In: Battarbee R.W., Gasse F. and Stickley C.E. (eds), Past Climate Variability through Europe and Africa. Kluwer Academic Publishers, Dordrecht, the Netherlands, pp. 219–256.

Hulme M. 1996. Climate Change and Southern Africa: An Exploration of Some Potential Impacts and Implications in the SADC Region. Climate Research Unit, University of East Anglia, Norwich, UK, 104 pp.

Hulme M., Doherty R.M., Ngara T., New M.G. and Lister D. 2001. African climate change: 1900–2100. Clim. Res. 17: 145–168.

IPCC 2001. Special Report on The Regional Impacts of Climate Change: An Assessment of Vulnerability. Intergovernmental Panel on Climate Change, WMO/UNEP, 27 pp.

Johnson T.C. 1996. Sedimentary processes and signals of past climatic change in the large lakes of the East African Rift Valley. In: Johnson T.C. and Odada E.O. (eds), The Limnology, Climatology and Paleoclimatology of the East African Lakes. Gordon and Breach Publishers, Australia, pp. 367–412.

Johnson T.C., Scholz C.A., Talbot M.R., Kelts K., Ricketts R.D., Ngobi G., Beuning K., Ssemanda I. and McGill J.W. 1996. Late Pleistocene desiccation of Lake Victoria and rapid evolution of cichlid fishes. Science 273: 1091–1093.

Lézine A.M. 1989. Vegetational palaeoenvironments of northwest tropical Africa since 12,000 yr BP: pollen analysis of continental sedimentary sequences (Senegal-Mauritania). Palaeoecol. Afr. 20: 187–188.

Lipiatou E., Hecky R.E., Eisenreich S.J., Lockhart L., Muir D. and Wilkinson P. 1996. Recent ecosystem changes in Lake Victoria reflected in sedimentary natural and anthropogenic compounds. In: Johnson T.C. and Odada E.O. (eds), The Limnology, Climatology and Paleoclimatology of the East African Lakes. Gordon and Breach Publishers, Australia, pp. 524–542.

Martens K. 2002. Redundancy and ecosystems stability in the fluctuating environments of long-lived lakes. In: Odada E.O. and Olago D.O. (eds), The East African Great Lakes: Limnology, Palaeolimnology and Biodiversity. Kluwer Academic Publishers, Dordrecht, The Netherlands, pp. 309–319.

Martyn D. 1992. Climates of the World. Developments in Atmospheric Science 18, Polish Scientific Publishers, Warsaw, 435 pp.

Neumann K. 1991. In search for the green Sahara: palynology and botanical remains. Palaeoecol. Afr. 22: 203–212.

Nicholson S.E. 1996. A review of climate dynamics and climate variability in Eastern Africa. In: Johnson T.C. and Odada E.O. (eds), The Limnology, Climatology and Paleoclimatology of the East African Lakes. Gordon and Breach Publishers, Australia, pp. 25–56.

Nicholson S.E. 2001. Climatic and environmental change in Africa during the last two centuries. Clim. Res. 17: 123–144.

Oldfield F. and Alverson K. 2003. The societal relevance of palaeoenvironmental research. In: Alverson K.D., Bradley R.S. and Pedersen T.F. (eds), Paleoclimate, Global Change and the Future. Springer-Verlag Berlin, pp. 1–11.

Ottichilo W.K., Kinuthia J.H., Ratego P.O. and Nasubo G. 1991. Weathering the Storm: Climate Change and Investment in Kenya. ACTS Press, Nairobi, 90 pp.

Owen R.B., Crossley R., Johnson T.C., Tweddle D., Kornfield I., Davison S., Eccles D.H. and Engstrom D.E. 1990. Major low levels of Lake Malawi and implications for speciation rates in cichlid fishes. Proc. Roy. Soc. Lond. 240: 519–553.

Patz J.A., Hulme M., Rosenzweig C., Mitchell T.D., Goldberg R.A., Githeko A.K., Lele S., McMichael A.J. and Le Sueur D. 2002. Regional warming and malaria resurgence. Nature 420: 627–628.

Pilskaln C. and Johnson T.C. 1991. Seasonal signals in Lake Malawi sediments. Limnol. Oceanogr. 36: 544–557.

Plisnier P.-D. 2002. Limnological profiles and their variability in Lake Tanganyika. In: Odada E.O. and Olago D.O. (eds), The East African Great Lakes: Limnology, Palaeolimnology and Biodiversity. Kluwer Academic Publishers, Dordrecht, The Netherlands, pp. 349–366.

Roberts J.M.D. 1964. Control of epidemic malaria in the highlands of Western Kenya, Part III. After the campaign. Journal of Tropical Medicine and Hygiene 67: 230–237.

Schulze R.E. 2000. Modelling hydrological responses to land use and climate change: a southern African perspective. Ambio 29: 12–22.

Schulze R.E. 2001. Managing water as a resource in Africa: are we asking the right questions in the quest for solutions? In: Gash J.H.C., Odada E.O., Oyebande L. and Schulze R.E. (eds), Freshwater Resources in Africa: Proceedings of a Workshop, Nairobi, Kenya, October 1999. BAHC International Project Office, Potsdam, pp. 9–14.

Scholes M. and Andreae M.O. 2000. Biogenic and pyrogenic emissions from Africa and their impact on the global atmosphere. Ambio 29: 23–29.

Scott L. 1990. Palynological evidence for late Quaternary environmental change in Southern Africa. Palaeoecol. Afr. 21: 259–268.

Scott L. 1993. Palynological evidence for late Quaternary warming episodes in Southern Africa. Palaeogeogr. Palaeoclim. Palaeoecol. 101: 229–235.

Scott L., Anderson H.M. and Anderson J.M. 1997. Vegetation history. In: Cowling R.M., Richardson D.M. and Pierce S.M. (eds), Vegetation of Southern Africa. Cambridge University Press, Cambridge, pp. 62–84.

Scott L. and Lee-Thorp J.A., this volume. Holocene climatic trends and rhythms in Southern Africa. In: Battarbee R.W., Gasse F. and Stickley C.E. (eds), Past Climate Variability through Europe and Africa. Kluwer Academic Publishers, Dordrecht, the Netherlands, pp. 69–91.

Seehausen O. 1996. Lake Victoria Rock Cichlids: Taxonomy, Ecology and Distribution. Verduijn Press, Zevenhuizen, The Netherlands.

Snoeks J. 2000. How well known is the ichthyodiversity of the large East African Lakes? In: Rossiter A. and Kawanabe H. (eds), Ancient Lakes: Biodiversity, Ecology and Evolution. Adv. ecol. Res. 31: 17–38.

Snoeks J., Ruber L. and Verheyen E. 1994. The Tanganyika problem: comments on the taxonomy and distribution patterns of its cichlid fauna. In: Martens K., Goddeeris B. and Coulter G. (eds), Speciation in Ancient Lakes. Arch. Hydrobiol. 44: 355–372.

Soliman W.R. 2001. The African challenge for the millennium: focus on the African Development Bank Water Resources Management. In: Gash J.H.C., Odada E.O., Oyebande L. and Schulze R.E. (eds), Freshwater Resources in Africa: Proceedings of a Workshop, Nairobi, Kenya, October 1999. BAHC International Project Office, Potsdam, pp. 21–25.

Street-Perrott F.A., Roberts N. and Metcalfe S. 1985. Geomorphic implications of late Quaternary hydrological and climatic changes in the Northern Hemispheric tropics. In: Douglas I. and Spencer T. (eds), Environmental Change and Tropical Geomorphology. George Allen and Union, London, pp. 165–183.

Street-Perrott F.A., Marchand D.S., Roberts N. and Harrison S.P. 1989. Global Lake-Level Variations from 18,000 to 0 Years Ago: A Palaeoclimatic Analysis. United States Department of Energy, Washington, DC. 213 pp.

Stuiver M. and Reimer P.J. 1993. Extended ^{14}C data base and revised CALIB 3.0 ^{14}C age calibration program. Radiocarbon 35, 1: 215–230.

Thorweihe U. and Heinl M. 2002. Groundwater Resources of the Nubian Aquifer System, NE Africa: Synthesis. Observatoire du Sahara et du Sahel, OSS, Paris, 24 pp.

Twongo T. 1996. Growing impact of water hyacinth on near shore environments of Lakes Victoria and Kyoga (East Africa). In: Johnson T.C. and Odada E.O. (eds), The Limnology, Climatology and Paleoclimatology of the East African Lakes. Gordon and Breach Publishers, Australia, pp. 633–642.

Verschuren D., this volume. Decadal and century-scale climate variability in tropical Africa during the past 2000 years. In: Battarbee R.W., Gasse F. and Stickley C.E. (eds), Past Climate Variability through Europe and Africa. Kluwer Academic Publishers, Dordrecht, the Netherlands, pp. 139–158.

Verschuren D., Kathleen R.L. and Cumming B.F. 2000. Rainfall and drought in equatorial East Africa during the past 1,100 years. Nature 403: 410–413.

Xue Y. 1997. Biosphere feedback on regional climate in tropical North Africa. Q. J. R. Meteorol. Soc. 123: 1483–1515.

27. CLIMATE VARIABILITY IN EUROPE AND AFRICA: A PAGES - PEP III TIME STREAM I SYNTHESIS

DIRK VERSCHUREN (dirk.verschuren@rug.ac.be)
Department of Biology
Ghent University
Ledeganckstraat 35
B-9000 Gent
Belgium

KEITH R. BRIFFA (k.briffa@uea.ac.uk)
Climatic Research Unit
University of East Anglia
Norwich, NR4 7TJ
UK

PHILIPP HOELZMANN
(phillip.hoelzmann@bgc-jena.mpg.de)
Max-Planck-Institut für Biogeochemie
Postfach 100164, 07701 Jena
Germany

KEITH BARBER (keith.barber@soton.ac.uk)
Palaeoecology Laboratory
Department of Geography
University of Southampton
Southampton, SO17 1BJ
UK

PHILIP BARKER (p.barker@lancaster.ac.uk)
Department of Geography
Lancaster University
Lancaster, LA1 4YB
UK

LOUIS SCOTT (scottl@sci.uovs.ac.za)
Department of Botany and Genetics
University of the Free State
P.O. Box 339, Bloemfontein 9300
South Africa

R. W. Battarbee et al. (eds) 2004. *Past Climate Variability through Europe and Africa.*
Springer, Dordrecht, The Netherlands.

IAN SNOWBALL (ian.snowball@geol.lu.se)
Quaternary Geology
Geobiosphere Science Centre
Lund University
Solvegatan 12, SE-223 62 Lund
Sweden

NEIL ROBERTS (cnroberts@plymouth.ac.uk)
School of Geography
University of Plymouth
Plymouth, PL4 8AA
UK

RICHARD W. BATTARBEE (rbattarbee@geog.ucl.ac.uk)
Environmental Change Research Centre
Department of Geography
University College London
26 Bedford Way
London, WC1H 0AP
UK

Keywords: Holocene climate variability, Milankovitch climate forcing, Solar climate forcing, Little Ice Age, Medieval Warm Period, Europe, Africa, Regional climate linkages

Introduction

In each of the six major study regions constituting the PEP III Europe-Africa transect, the principal trend of Holocene climate change reflects the global climate forcing that is exerted by variation in the seasonal distribution of solar insolation received at the Earth surface, due to precession of the Earth's orbit around the Sun. In the Northern Hemisphere, summer insolation was above-average between about 15 and 5 kyr (kyr here denotes 10^3 calendar years ago), peaked at 10 kyr, and is currently near its minimum; in the Southern Hemisphere, summer insolation has been above average for the past 5 kyr, and is currently near its peak. However, the regional expression of this forcing is modulated by a multitude of amplification, damping, and feedback processes involving all four components of the climate system: atmosphere, ocean, continents, and cryosphere. Consequently, regional histories of Holocene temperature and rainfall change often deviate significantly from the perfect hemispheric anti-phasing which orbital forcing would predict. In addition, superimposed on the long-term climate trends attributed to orbital forcing are various modes of Holocene climate variability operating at inter-annual to millennial time scales. Some of these can be linked to other external forcing mechanisms, such as volcanic eruptions and variations in the radiation output of the Sun, but others appear to result from poorly understood periodicity in the internal dynamics of the climate system. These processes have generated tremendous complexity in Holocene climate history at both regional and continental scales, so that the challenge to document this complexity in enough detail to elucidate the exact mechanisms involved constantly strains the possibilities of available climate-reconstruction methods.

This Time Stream 1 synthesis chapter serves two purposes. First, a summary of the main patterns of Holocene climate change across the PEP III transect aims to draw attention to the (real or apparent) synchrony and time-lags between distinct climatic anomalies in the different regions, in the hope that it may help generate the sequence of speculation, modelling, and more detailed reconstruction efforts which typically yield exciting new insights. Second, it aims to draw attention to some of the more persistent problems in high-resolution climate reconstruction, in the hope that efforts to address these problems face-on may lead to an improved methodology of climate reconstruction which, when applied to both new and existing data, will advance our understanding of exactly how the world's climate system operates.

Patterns of Holocene climate change in Europe and Africa

Long-term patterns: orbital (Milankovitch) insolation forcing

When after the Younger Dryas - Preboreal transition (YD - PB) sea-ice retreated to the northwestern Nordic Sea off Greenland, and the North Atlantic Ocean assumed a near-modern thermohaline circulation (THC) regime, strong influx of warm water from the Equatorial Atlantic increased SST off Southern Norway 9 °C within ~50 years (Snowball et al., this volume). A Holocene thermal maximum of ~15 °C summer SST was reached already by 10.8 kyr, after a ~500-yr relapse to slightly cooler conditions known as the Preboreal Oscillation (PBO). This two-step increase reflects the slowing-down of the THC (and hence meridional heat transport) due to intermittent melt-water discharge from the retreating Northern Hemisphere ice-sheets into the North Atlantic. High-latitude continental areas were also influenced by lingering effects of the melting Laurentide and Fennoscandian ice sheets. Initially, climate in northern Europe was more oceanic than today, as high SSTs combined with relatively cold air masses enhanced cyclonic activity, and brought mild, moist weather inland. As the continent itself started heating up, cyclonic incursions decreased and climate became drier, starving mountain glaciers and lowering lake levels. Fossil lake biota indicate that northern Europe experienced peak Holocene temperatures, possibly 1–1.5 °C warmer than at present, between 8 and 6 kyr, with warm and dry summers especially 7–6 kyr ago. This early Holocene warming allowed northward expansion of vegetation zones, with pine reaching its furthest position north of the present-day tree line between 10 and 9 kyr. However, vegetation response to the early Holocene thermal maximum was only completed by 7 kyr in Finland, and 6 kyr in Northern Sweden.

Decreasing Northern Hemisphere insolation from 10 kyr onwards led to greater seasonal ice cover and lower SSTs in the western Nordic Sea as early as 7 kyr, but in the eastern North Atlantic, continued advection of warm water seems to have maintained high SSTs until 5 kyr (Snowball et al., this volume). A speleothem record from northern Norway suggests that on land, gradual cooling set in around 7 kyr; according to fossil lake biota, cooling set in rather abruptly at 5.8 kyr. Maritime glaciers, which received sufficient moisture, benefited from this cooling and experienced significant advances, but glaciers farther inland did not because continental climate remained dry until ~4.5 kyr, as evidenced by low lake levels and limited boreal peat accumulation. A significant vegetation response to the end of the Holocene optimum started at 5–4.5 kyr, with the retreat of pine and mountain birch from their former northern limits in Finland and Sweden. Reduced summer evaporation due to cooling then improved the continental water balance, causing lake levels to rise and peat-bog surfaces to become wetter from 4.5 kyr onwards, and many Scandinavian glaciers to expand from 3 kyr. Expansion of boreal and alpine peatlands world-wide over the past 4–5 kyr is considered the principal cause of late-Holocene increases in atmospheric methane concentration, from ~600 ppm at 5 to 3 kyr to its pre-industrial (~AD 1850) level of 700 ppm.

The mid-latitude North Atlantic Ocean adjacent to temperate Europe warmed only 1–2 °C at the YD-PB transition, adding perhaps another 0.5 °C after the PBO to reach peak summer SSTs by 10.5 kyr (Snowball et al., this volume). Also in the mid-latitude Atlantic the marine thermal maximum appears to have lasted until ~5 kyr. On land, Irish speleothems suggest that the Holocene optimum lasted from 9 to 6 kyr, and that the ensuing cooling trend started as early as 7.8 kyr (Barber et al., this volume). Pollen records of vegetation response to this orbitally-induced warming and cooling indicate that the Holocene optimum of 1–3 °C above modern values was characterised by an enhanced continentality gradient across Europe, with very little winter warming compared to today in the easternmost areas. In Central Europe, early Holocene warming was sufficient to significantly reduce Alpine glaciation and possibly melt away most or all glaciers in the eastern Swiss Alps between 9.4 and 3.3 ^{14}C years BP.

In the Mediterranean region, a pollen-based reconstruction from the Atlas Mountains in Northwest Africa suggests that Holocene climate was warmer than at present both in summer (1 °C) and winter (~2.5 °C) from before 11 kyr to ~7 kyr, and wetter than at present between 6.3 and ~2.5 kyr; the early Holocene was mostly dry, except for a somewhat moister episode between 10.3 and 8.2 kyr (Roberts et al., this volume). Persistence of high lake levels up to the present time in Northwest Africa suggests a favourable water balance here for much of the past 7 kyr, but during the last 3 kyr this has possibly depended on reduced evaporation by virtue of the cooler winters. Elsewhere in the Mediterranean, from Spain through Italy to Turkey and the Near East, a variety of continental proxy climate indicators suggest that favourable water-balance conditions started and ended earlier, at ~10 kyr and 4 kyr respectively. This period also includes the formation of the last Mediterranean Sea sapropel layer (Kallel et al., this volume). It appears largely synchronous with the time of humid conditions in sub-tropical Africa south of the Sahara desert, although it is unlikely that the Mediterranean benefited directly from an expanded and intensified monsoon circulation. Early to mid-Holocene vegetation in much of the Mediterranean was dominated by temperate deciduous woodland, followed by late-Holocene replacement by drought- and fire-adapted trees and shrubs. Modelling studies suggest that the albedo changes resulting from large-scale anthropogenic forest clearance could by itself have sufficiently diminished summer rainfall levels to cause this vegetation shift. The weight of paleo-environmental evidence, however, suggests that the northward spread of xerophytic woodland and scrub, which started before 4 kyr, was at least partly climate-controlled.

In the southern half of the PEP III transect, i.e., the tropical and subtropical regions of Africa in both hemispheres plus the Arabian Peninsula, Holocene climate variability mostly involved changes in water balance, including temperature effects on evaporation, rather than temperature change *per se*. Again, the main long-term climate trends resulted from control of the Milankovitch precessional cycle on the strength and latitudinal reach of

tropical monsoon dynamics, through changes in relative summer insolation over Northern and Southern Hemisphere areas. Temporal patterns of lake-level fluctuation across the northern arid/sub-arid belt and northern inter-tropical and equatorial Africa provide strong support for orbital forcing as a dominant climate mechanism, but it is less obvious to what extent southern inter-tropical and subtropical Africa manifested the opposing temporal patterns which this mechanism would predict (Barker et al., this volume). Secondly, a marked west-to-east rainfall gradient extending from northern sub-tropical to southern inter-tropical Africa, maintained by moist south-westerly air flow from the tropical Atlantic Ocean, created potential for regional patterns of climate change between continental areas differently affected by moisture coming from the Atlantic and Indian Oceans.

Along the northern margin of the Sahara desert, rainfall today results from southward displacement of mid-latitude westerlies during winter, whereas along its southern margin the rain is of monsoonal origin and comes during summer, governed by seasonal migration of the Inter-tropical Convergence Zone (ITCZ) into the Northern Hemisphere sub-tropics. Today, there is almost no overlap or interaction between these summer and winter rainfall regimes, and consequently a true arid zone exists between $\sim 20°$ and $\sim 30°$N. During the late Glacial and early Holocene, insolation-driven intensification of African and Indian monsoon dynamics established wet conditions in the arid/sub-arid belt of North Africa in two abrupt steps dated to 15 ± 0.5 kyr and 11.5–10.8 kyr, separated by an arid interval centered at ~ 12.4 kyr corresponding to the YD chronozone in the North Atlantic region (Hoelzmann et al., this volume). The second step is thus coeval with the PB warming seen in northern boreal and temperate regions, exemplifying the strong climatic linkage between the tropical Atlantic and Indian Oceans and high-latitude regions of the Northern Hemisphere. Intensified monsoon circulation peaked at ~ 9.4 kyr, as evidenced in maximum Holocene lake levels and the furthest spread of vegetation into the Sahara desert; not much from the north, although Mediterranean woodland did encroach upon *Artemisia* semi-desert, but mostly from the south, where wooded and grass-savanna greatly expanded to cover the present Sahel region, and steppe covered much of the present hyper-arid desert. These spectacular landscape changes imply rainfall increases of up to 350 mm above modern, and as a result the area of true desert in the Saharo-Arabian zone was greatly reduced in extent, if it existed at all. The early-Holocene boundary between winter and summer rainfall regions in North Africa is thought to have been located at 21–23 °N, and at times the spatial domains of the two rainfall regimes may have overlapped. Some upland regions and parts of the eastern Mediterranean may even have experienced the kind of bi-modal rainfall distribution found today in Northern Mexico and the US Southwest, which lies at a comparable latitude to the Sahara and Arabia within the PAGES-PEP I transect. This landscape transformation must have involved significant positive feedback between the ocean, atmosphere, and vegetation itself.

On the Arabian Peninsula, the early Holocene humid period probably came into existence only around 9.9 kyr, and ended by 6.2–6.1 kyr. Some records suggest that the humid period ended already at ~ 8.1 kyr in those places, roughly coeval with the temporary reduction in monsoon strength registered in many lacustrine and vegetational records from arid North Africa. In the north of the peninsula, cyclonic activity originating above the Mediterranean helped limit evaporation of monsoonal moisture during winter, allowing relatively wet conditions to persist until 6.1 kyr, while in the South there is evidence that humid conditions lasted until 6.2 kyr, in good correspondence with lake-level records from the Sahara. The sedimentary record of aeolian sediment deposition in the sub-tropical North Atlantic, thought to reflect the crossing of a threshold in vegetation cover on the adjacent African continent, places the end of the African Humid Period at 5.5 kyr, when 20 °N summer insolation fell below 4.2% greater than present. Pollen records of vegetation change in the arid/sub-arid belt show drying to have started by 7.8–6.8 kyr, accelerating between 4.2 and 2.4 kyr. In the Sahara, arid to hyper-arid conditions prevailed since 3.2 kyr, except that ~1.5 kyr ago there was a slight shift to somewhat wetter but more variable conditions.

The general expression and timing of the Holocene humid period in northern inter-tropical and equatorial Africa were similar to that in the arid/sub-arid belt to the north, and at least on the longest time scale displayed the patterns predicted by orbital forcing of summer insolation in the Northern Hemisphere. In inter-tropical Africa this forcing can explain enhancement of monsoon rainfall by 35–45%, sufficient to re-establish positive water balances after the markedly dry YD, fill many of the large and smaller Rift lakes to overflow levels, and increase connectivity between Africa's major drainage systems (Barker et al., this volume). This 'northern' pattern of insolation forcing dominated equatorial Africa down to 9 °S, and exerted influence down to 15 °S at least. Sub-tropical southern Africa, in contrast, was dry (and cool) between 11 and 7.5 kyr, only then experiencing a wetting (and warming) trend which culminated in a distinctly wet phase dated to 6.5–5.1 kyr. In adjacent southern inter-tropical Africa very few sites show this 'southern' pattern, and in those that do it was not an equal and opposite response, rather a more muted version of the 'northern' Holocene humid period. The northern-most site currently suggesting some conformity to the expected 'southern' pattern of Holocene climate change is Lake Malawi (10–14 °S), where some proxy climate indicators suggest drier conditions than today between 10.5 and 5 kyr, although other indicators suggest the lake did remain close to overflow level. In the small winter-rain region of south-western South Africa, pronounced cooling during the Southern Hemisphere insolation minimum at ~10 kyr seems to have maintained relatively moist conditions in the earliest Holocene (Scott and Lee-Thorp, this volume). Soon after, this area became drier, and stayed dry also after 7.5 kyr while the summer-rain portion of southern Africa started benefiting from the intensified monsoon associated with southward displacement of the ITCZ. However, as early as 5 kyr the monsoon appears to have weakened again. Drought in the Southwest strengthened and culminated in widespread drought by ~2 kyr, despite the peaking of Southern-Hemisphere summer insolation around this time. Thus, only to some degree did broad-scale Holocene rainfall patterns in monsoon-influenced tropical and sub-tropical Africa reflect the anti-phase behaviour of precession-induced hemispheric variation in solar insolation: influence of northern-hemisphere forcing extended to 15 °S, and in sub-tropical southern Africa the predicted increase in rainfall over the past 7–8 kyr was cut short after 5 kyr. Comparison between the Mediterranean and the climatically similar Cape region of southern Africa also fails to show a clear synchronic or anti-phase relationship between the Northern and Southern Hemispheres in long-term Holocene climate changes.

Century-scale (sub-Milankovitch) climate variability

In all areas of the PEP III transect, suitable high-resolution climate-proxy records display evidence of climatic instability at time scales from individual years to decades to millennia.

Apart from threshold responses of the global climate system to gradual orbital insolation forcing (e.g., the effect of a sudden THC slow-down due to an ice-sheet melt-water pulse the timing of which is controlled by topography), this sub-Milankovitch climatic variability has its origin in two other external climate-forcing mechanisms, namely variation in the radiation output of the Sun and the heat-shielding effect of volcanic eruptions; in cyclic processes within the world's climate system, such as El Niño-Southern Oscillation (ENSO); or some as yet poorly understood resonance between the external forcing and internal climate dynamics. The direct forcing of these mechanisms being relatively modest and short-lived, response amplification by feed-back processes between the four components of the climate system is less pronounced than that enjoyed by orbital forcing, so that their signatures in climate-proxy archives can often be rather indistinct, and clouded by lagged indirect effects and/or chronological uncertainty. Still, in recent years increasing evidence has been found for the prominent influence of variations in solar radiation on Holocene climates, at decadal to millennial time scales. Often the case for solar forcing of local or regional climate is built on the observation of cyclic climate fluctuations with frequencies approaching those found in the record of solar activity. This is somewhat problematic on the longer time-scales, because the reconstruction of atmospheric ^{14}C production that is often used as a proxy record for solar activity variation is complicated by interference from variable rates of ^{14}C cycling through marine and terrestrial reservoirs. In other cases, distinct and well-dated climatic anomalies have been established to covary directly with a known forcing function, such as cooling in Europe during the Maunder minimum in solar activity of AD 1650–1715, or short-lived improvement of hydrological conditions in semi-arid Turkey following the ca. 1650 BC eruption of the volcanic island of Thera (Santorini) in the Aegean Sea. As with the Milankovitch pace-makers of Ice-Age climate, consideration of the geographical distribution and expression of individual climate events can yield clues to the physical mechanisms that link cause and effect, and so enhance understanding of how the climate system operates at those time-scales.

In boreal and cold-temperate Europe, one of the most prominent climatic anomalies of the Holocene is a century-scale cooling episode at 8.2 kyr (Snowball et al., this volume). It involved up to 5 °C cooling in Central Greenland and a 2–3 °C depression of North Atlantic SST, which reverberated over Northwest Europe as an abrupt 1.5–2 °C cooling, significant enough to trigger intermittent advance of some Scandinavian glaciers (the Finse Event) and to shift Swiss and German vegetation in favour of drought-sensitive trees (Barber et al., this volume). The 8.2 kyr cooling event is thought of as the last of a series of interruptions in post-glacial THC intensification caused by periodic melt-water discharge events associated with Northern Hemisphere deglaciation, starting with the YD and PBO. In some Scandinavian lakes, fossil aquatic biota recorded at least two such cooling events between the PBO and 8.2 kyr. These century-scale climatic anomalies coincide with the so-called Bond cycles recorded in North Atlantic sediments, which reflect the quasi-periodic drifting of ice-berg 'armadas' far southward into the North Atlantic, slowing down Atlantic THC with a mean periodicity of 1500 years throughout the Holocene. Tentatively linked to long-term variation in production rates of ^{14}C and ^{10}Be, these Bond cycles suggest that solar radiation output must have had a marked influence on Holocene climate variation in all regions influenced by changes in Atlantic Ocean circulation, including tropical Africa. Solar forcing of tropical climate variability may also have been more direct, as suggested by excellent correlation between the ^{14}C production record and a speleothem record of early-Holocene (9.6 to 6.1 kyr) monsoon intensity from the Arabian Peninsula (Hoelzmann et al., this volume).

Multi-proxy studies from various sites in the Alps show close temporal agreement between the reconstructed changes in Atlantic Ocean THC and the local sequence of century-scale climate variability. Episodes of cooler, wetter climate are clearly manifested in intermittent timberline depression and glacier advance, even though temperature anomalies there probably amounted to only 0.5–1 °C. Similar sequences of about eight distinct Holocene cooling events at more or less regular time intervals are also documented in lake records from Germany, Poland, the Faroe Islands, in the peat records of British and Irish bogs, and in the records of glacial advance and retreat in Scandinavia. Periodicities of c. 1100, 800, 600 and 200 years, among others, have been recognised from the peat-bog record, and these may be linked to solar forcing and oceanic THC changes. The connection with solar activity variation is particularly striking in the case of the Subboreal/Subatlantic transition at c. 850 cal yr BC, a distinct shift to wetter conditions evidenced in peat bogs throughout the North Atlantic region that coincides with a major increase in atmospheric $\delta^{14}C$ (Barber et al., this volume).

In most areas of the arid/sub-arid belt as well as northern inter-tropical and equatorial Africa, the early-Holocene humid period was punctuated by a severe dry spell correlative with the 8.2 kyr event, usually dated to 8.4–8.0 kyr. Only in the Eastern Sahara has it so far not been found in lake-level records, probably due to the strong continentality of this region and the buffering effect of fully recharged aquifers (Hoelzmann et al., this volume). In contrast with the marine sediment record, which suggests the early-Holocene humid period to have ended in a single episode of aridification at 5.5 kyr, lake-level histories from the Sahara and Sahel suggest that it ended in a succession of two pronounced dry spells at 6.7–5.5 and 4.0–3.6 kyr. In the Horn of Africa the first dry spell is situated at 7.0–6.5 kyr, earlier than elsewhere, while the major drop in lake levels seems to have occurred at 5.5–5 kyr (Umer et al., this volume). Inter-tropical Africa also experienced marked climatic change around 8.2 kyr, but a ~6 kyr event is less well expressed than in the drier regions to the north. Widespread lake regression is there centered on 4.2–4 kyr, causing a drastic decline in the base flow of the Nile which severely impacted on Egyptian society during the Old Kingdom. A pulse of dust deposition into the Arabian Sea at ~4.2 kyr reflects a century-scale drought event in Mesopotamia, thought to be coincident with collapse of the Akkadian Empire (Weiss 2000; deMenocal 2001).

Also in other parts of the PEP III transect, the period around 4 kyr appears to have been characterised by climatic upheaval. Boreal tree-ring records and a speleothem record from Northern Norway show that the main, orbitally-driven Holocene cooling trend was punctuated by one particularly abrupt anomaly, dated in the trees to 4.4 kyr. This abrupt event is also found in many mid-latitude bogs (Barber et al., this volume). Most temperature reconstructions covering this period in some detail indicate that late-Holocene cooling resumed after this event, but some pollen records from mid-latitude Europe show strong mid-Holocene cooling to 2–3 °C below modern values between 4.5 and 3 kyr, after which climate appears to have recovered towards modern values. Also Irish speleothem records and a lake-temperature reconstruction from northern Fennoscandia appear to indicate that Northern Hemisphere cooling did not continue until the present day, but that some recovery from mid-Holocene cooling occurred in the last 4–3.5 kyr. This evidence is at odds with what orbital theory would predict, and also difficult to explain using the link between solar radiation output and Atlantic Ocean circulation. One possibility is that late-Holocene temperature inferences from these sites are affected by an as yet unrecognised hydrological response of the climate archives.

Without doubt the best known climatic anomalies of the past 2000 years are the Little Ice Age (LIA; AD 1250–1800) and the Medieval Warm Period (MWP; AD 900–1250) or Medieval Climatic Optimum (MCO), recently often referred to as the Medieval Climatic Anomaly (MCA) due to apparent complexity in its regional expression. From a North Atlantic perspective, the LIA is just the most recent Bond cycle of enhanced ice-berg formation and drift which started around 0.8 kyr, separated from the previous one (dated to 2–1.2 kyr) by a period of sea-ice retreat, the MWP. However, LIA glacier advances in the Alps and Scandinavia often overrode most of the earlier Holocene advances, suggesting that this last Bond event was the most serious cold/wet episode since the onset of the Holocene. The bracketing ages used to define the LIA and MWP are derived from the residual ^{14}C record, which shows mostly above-average solar radiation between AD 900 and 1250 (only interrupted by the Oort Minimum, AD 1020–1100), followed by a 550-yr period of mostly below-average solar radiation comprised of the Wolf, Spörer, and Maunder Minima. In Europe, the term LIA is often reserved for the distinctly cold 17th and 18th centuries, although LIA-related glacier advance in the Alps started ~AD 1250 and was well underway by the mid-14th century.

In the mid-latitude Atlantic Ocean, the MWP - LIA transition involved an SST depression of 1.5–2 °C. In much of mid- and high-latitude Europe, MWP warming appears to have been 0.5–0.8 °C above, and peak LIA cooling ~1 °C below, mid-19th century (pre-industrial) values. The available high-resolution proxy records show substantial individuality in the exact timing and evolution of MWP- and LIA-related climatic anomalies, but given the relatively modest climate changes involved, a significant fraction of the apparent differences may be due to archive- or site-specific responses of the proxy indicator rather than regional differences in climate history. In fact, temperature variability across Northern Europe is highly coherent at least on inter-annual and decadal time-scales, and this also seems to be true for Northwest Eurasian tree-ring records, thus justifying the amalgamation of regional data networks into a single master curve. Tree-ring data imply extended warm periods particularly in the 10th, 11th, and 15th centuries, and cool summers in the 17th and 18th centuries (Snowball et al., this volume). The record is also punctuated by temperature depressions immediately following major volcanic eruptions, e.g., in AD 536 (of an unidentified volcano), 1601 (Huaynaputina, Peru), and 1816 (Tambora, Indonesia). Although relatively short-lived, the societal impact of some of these episodes is well-documented, both negative (in regions with marginal growing-degree days for important food crops) and positive (beneficial for the water balance of regions with endemic summer drought).

In the peat record of oceanic Europe, the MWP and LIA are clearly evident in drier bog surfaces during 1.3–0.7 kyr, and two wet/cold phases dated to AD 1300–1500 and AD 1600–1850 (Barber et al., this volume). In the Mediterranean region, the LIA appears to have started with increased wetness around AD 1100–1400, followed by decline towards cooler, drier conditions AD ~1700–1850, coincident with the coldest phase of the LIA in mid-latitude Europe (Roberts et al., this volume). Because Mediterranean winter rainfall is largely cyclonic and of Atlantic origin, inter-annual rainfall variability there is strongly linked to rainfall patterns in temperate Europe and shows similarly strong covariance with the NAO index. Further, a 'Mediterranean Oscillation' or West-East precipitation see-saw is evident both at this inter-annual time scale and in a more long-term, 20th-century wetting trend in the northern and western Mediterranean relative to the eastern Mediterranean. Century-scale climate variability during the MWP and LIA appears to have overridden this West-East contrast, however, as it shows a fairly uniform sequence of events across the region.

Knowledge of the last 1,000 years of climate history in the arid/sub-arid belt of North Africa is fragmentary, with chronological uncertainty in the available proxy records precluding the distillation of any regional patterns at this time (Hoelzmann et al., this volume). Somewhat less fragmentary proxy data from inter-tropical Africa, then, appear to indicate that much of the region was drier than today during the MWP-equivalent period ~AD 900–1250, and that the ~AD 1250 onset of the LIA involved a transition to moister conditions (Verschuren, this volume). Drought was also widespread in the late 18th and early 19th centuries. For three centuries before this drought, i.e., the period coincident with greatest LIA cooling in Europe, rainfall in equatorial Africa and southern inter-tropical Africa appears to have been inversely correlated, showing generally moist conditions (albeit interrupted by two severe droughts) persisting to the late 18th century in equatorial East Africa, whereas further to the south a (fairly gradual) return to drier conditions already started in the 16th century. To what extent high LIA lake levels in equatorial Africa were sustained by reduced evaporation due to cooling, rather than higher rainfall alone, is unclear. Reconstructed rainfall variations in sub-tropical Southern Africa are broadly consistent with those in southern inter-tropical Africa, and some pollen and speleothem evidence appears to suggest that there at least, the LIA was both drier and colder than today.

Challenges of high-resolution climate reconstruction

Extracting climate variables from natural climate-proxy records

The already reasonably coherent picture of Holocene climate change in Europe and Africa presented above bears testimony to the considerable body of new data which have become available in the past decade, stimulated in part by the IGBP-PAGES programme. At the same time, some of the toughest methodological problems faced by palaeoclimatologists today are still the same as those at the start of PAGES in 1990. One fundamental problem pertains to how we extract the fundamental climate variables temperature and rainfall from records of climate-proxy indicators in natural archives. Few proxy indicators are uniquely controlled by or can be directly related to only temperature or precipitation. Instead, most biological and geochemical proxy indicators respond to both, directly or indirectly, and in varying proportions depending on the time frame, the region, or the archive under study. Traditionally, palaeoclimatologists try to build a convincing argument that in the particular circumstances of their study most reconstructed variability in the selected climate-proxy indicator can be attributed only to temperature or only to hydrological (rainfall and/or evaporation) change. However, the validity of this assumption often remains untested. Ideally, climate inference from individual proxy-indicator records should be supported by: (i) process-oriented studies designed to better understand the quantitative relationship between proxy-indicator response and the fundamental climate variables; (ii) historical validation of the climate-proxy record through direct comparison with instrumental climate data; and (iii) forward modelling to test whether indicator response to climate change is adequately understood at the time scale of the intended reconstruction. Development and application of novel climate-proxy indicators over the past decade has produced considerable progress in testing the relative importance of temperature and hydrological change (Barker et al., this volume), and new methods will no doubt continue to do so in the future. However, what is urgently needed are more process-oriented validation studies that can improve understanding of the many climate-proxy indicators already in use, to place climatic inferences based on those indicators on a firmer footing.

High-latitude Europe has been a fertile testing ground for quantitative palaeoecology, in which calibrations of species assemblages of aquatic biota along modern-day environmental gradients are used to infer past changes in climate-driven habitat conditions (Snowball et al., this volume). Even when the direct causal link between species composition and the selected climate variable remains uncertain, these inference models possess great heuristic value to condense complex biological variation into simplified climate trajectories. Similar multivariate statistical methods are now also being applied to calibrate physical climate-proxy indicators (stable-isotope composition of geochemical species, sediment texture, etc.) against regional climate gradients. Temperature-inference models based on diatoms, cladocerans, pollen, and midges now exist for various parts of boreal and mid-latitude Europe. The remaining prediction error (0.6–1.2 °C for pollen, 0.8–1.1 °C for aquatic biota) will be difficult to improve further, however, so that except for broad climate trends such as the early Holocene Optimum, decadal and century-scale Holocene temperature fluctuations will remain difficult to discriminate from background noise.

Given the cold winters, short summer season, and positive water balance in most areas of high-latitude Europe, biological and geochemical climate-proxy indicators there are typically assumed to primarily reflect temperature change. At the other extreme is inter-tropical Africa, where temperature changes are generally assumed to be negligible compared to shifts in the hydrological balance, at least on Holocene time scales. But given that many important African climate records are based on the lake-level history of amplifier lakes, the effects of temperature-driven evaporation changes on reconstructed hydrological variability may well be under-estimated. This is not a trivial problem, because on sub-millennial time scales the usually inferred association of warmth with wetness and cold with drought may not always have applied (Verschuren, this volume).

Palaeoclimatologists working in temperate Europe, the Mediterranean region and subtropical regions of North and South Africa have long understood that moisture effects can seriously confound temperature inferences, and vice versa. For example, biosphere response to the 8.2 kyr event is clearly recorded in tree-ring records throughout Europe, but it is often unclear whether poor local growth was due to cold or to drought. The water balance of oceanic bogs in Ireland and northern England is an integration of temperature and precipitation effects, but evidence so far favours temperature as the main forcing factor (Barber et al., this volume). Periods of wetter bog surfaces most probably reflect summer temperature declines affecting evapotranspiration, rather than irregular changes characterising the precipitation record. Hence, oceanic bogs react sensitively to temperature-driven changes in moisture balance while bogs in continental regions show infrequent sudden changes as if forced over a climatic threshold. In the dry eastern Mediterranean, increasing tree-ring width is justly assumed to reflect primarily improved moisture balance, but this itself may result from reduced evaporation due to lower temperatures (Roberts et al., this volume). Although the climatic record from the Mediterranean sector can be complicated by the longevity of human impact here, this region is nonetheless key to better calibration and validation of climate-proxy indicators, and to integrate the data from different archives. This is because, unusually within the PEP III transect, the major continental climate archives of high-latitude regions (tree rings, peat, hydrologically stable lakes) are there found together with the major archives of low-latitude regions (speleothems, climate-sensitive lakes).

Annual versus non-annually resolved climate archives

Tree-ring records have unique power to provide evidence of very high frequency (seasonal to inter-annual) climate variability over centuries to millennia, with exact calendrical age control more reliable than that of most ice cores, varved lake sediments, or speleothems. Only corals provide age control comparable to that of tree rings, but their temporal range is rather limited. Disadvantages of tree-ring records are that resolution of decade- to century-scale variability is ambiguous, as is assessment of the absolute magnitude of the climate anomalies which resulted in the observed ring-width or wood-density changes (Snowball et al., this volume). One reason for the prominent PAGES focus on annual-resolution records is that they permit rates of change to be established, but these inferred rates are valid only if the relationship between the proxy indicator and the climatic parameter is well understood. Often the nature of a climate archive imposes a non-linear, threshold relationship between the proxy indicator and climate, so that the rate of indicator change does not necessarily reflect the true rate of climate change. There is a real risk to trust intuitively climatic inferences from annual records more than those inferred from non-annually resolved records, even though the information value of climatic inferences from any natural archive depend first and foremost on how well the processes controlling the selected climate-proxy indicator are understood, and is largely independent of dating accuracy. Published examples on both extremes of this gradient exist: rather speculative inferences based on poorly understood climate proxies in annual records with calendrical precision; and highly accurate characterisations of past climatic conditions which, despite poor age control, can be directly used to constrain climate-modelling exercises. Consequently, a major challenge remains how best to distil the strong elements from climate-proxy records with non-annual resolution and dating control, even including non-continuous and truncated records from regions where these are the only natural climate records available.

Dating errors inherent to ^{14}C-based geochronologies complicate the use of regional networks of palaeoclimate records to establish regional synchrony of short, high-frequency events (<100 years). In some regions this problem may be resolved by using identified tephra layers as absolute time markers to link ice, marine, peat and lake records. If this is not an option, then comparison of records must often be based on looking for common temporal patterns, i.e., the number, frequency, and relative magnitude of climatic anomalies in a particular time window. In an exploratory phase, some tuning or wiggle-matching can be allowed to evaluate the relative merits of possible physical mechanisms. However, deliberately over-riding chronological mismatches between records on account of dating uncertainty precludes exploration of real (and possibly significant) lags in the response of different climate-system components, and of individual climate-proxy indicators reflecting the change in each of those components, to the assumed forcing.

Of all continental regions, high-latitude Europe has been most intensely studied in terms of Holocene climate variability, because it is unrivalled in providing both long (>200 years) instrumental data and a great quantity of natural archives with high temporal resolution: ice cores, tree rings, and varved lake sediments. It is also immediately adjacent to the North Atlantic Ocean where much indirect climate forcing is played out; it includes sensitive ecotones along various steep climatic gradients; and it has mountain glaciers with mass balances sensitive to temperature or to precipitation. It thus constitutes an ideal region to test our understanding of the links between climate-proxy indicators and climate, and to guide the combination and integration of climate-inference data from various proxy indicators and natural archives into a coherent picture of past climate change (Snowball et al., this volume).

Optimal use of paleodata in climate modelling

Because future climate change will be the result of interactions between natural climate variability and the effects of human activities on global and regional climates, there is an urgent need to document and understand the characteristics of natural climate variability at the time scales directly relevant to the immediate future of society; to understand how it may interact with anthropogenic climate change; and to improve the performance of climate models so that they provide improved, probabilistic projections of both the modes and rates of future climate change. Virtually all climate-change projections, and the current generation of climate-change detection studies, assume or use a model-based characterisation of natural climate variation. This invariably underestimates the possible range of prediction uncertainties. In order to arrive at realistic scenarios of patterns and rates of future climate change (scenarios that make explicit the uncertainties associated with natural and anthropogenic forcings and their possible interactions), a very detailed picture of natural climate variability and of the forcing mechanisms both external and internal to the Earth system must be incorporated along with the anthropogenically-forced changes. Such a complete, detailed picture can only be assembled by integrating the best empirical data from instrumental, historical, and natural climate archives. The primary challenge for palaeoclimatologists is to move towards unprecedented collaboration across sub-disciplines to combine and harmonise data from the various natural climate archives, in order to develop the multi-proxy, multi-site, multi-archive database that enables data to be manipulated and translated into formats facilitating use in and comparison with climate-model simulations. Syntheses of meteorological and documentary data must be incorporated in an early stage, and feed into the calibration and validation of palaeodata. Interpreted palaeodata from the different natural archives, e.g., tree rings, corals, ice cores, lake and marine sediments, peat, and speleothems, each with their own specific regional concentrations, climate sensitivities and resolvable time-scales, are then integrated to form a high-quality geographical palaeoclimate data set.

Availability of this continuously improving comprehensive data set will encourage comparison between model simulations and palaeoclimate data in order to improve model performance on the one hand and, through model experiments on the other hand, better understand the causes of past climate variability (Braconnot et al., this volume). It must be stressed, however, that data-model comparisons are not straightforward, as the characteristics of model output and palaeodata are very different, especially with respect to scale. Proxy-indicator records integrate climatic information locally over space and time, as if climate history were passed through a low band-pass filter. In contrast, numerical models simulate climatic variables at discrete grid points, tens to hundreds of kilometers apart. Matching the two approaches requires the use of scaling techniques. Upscaling is an approach in which climate history reconstructed from the proxy-indicator data is compared to the climate-model simulations. Downscaling is a forward-modelling approach in which climate-model output drives process-based or empirical models to simulate a proxy-indicator value or time series, which is then compared with the actual proxy-indicator data. It is the upscaling that will benefit most from the availability of a harmonised geographical palaeoclimate data set. The main potential of downscaling is its ability to explore the processes that lie behind the climate-proxy indicator relationship. Results of climate-modelling experiments must then be compared with control (unforced) simulations to identify model-based uncertainties and to identify what climate variables, at what times, in which regions, and on what time-scales, are most sensitive to internal and to external forcings, and to estimate the possible influence of natural forcings on climate projections. Data-model comparisons must be undertaken at different levels of spatial and temporal integration, to optimise the 'signal' characteristics of the data sets and to provide assessments of the veracity of the climate models.

Simulations of global and regional climate changes are produced using a range of distinct simple and complex computer models (Braconnot et al., this volume), including the coupled atmosphere-ocean general circulation models (AOGCMs), Earth system models of intermediate complexity (EMICs), and the more simple energy balance models (EBMs). AOGCMs provide the desired high spatial resolution and include representations of all key processes in the global climate system, but long-term dynamic simulations are still expensive and slow to run. However, the fact that data-model comparisons for the entire Holocene may not be possible for a number of years obviously need not slow down the momentum of proxy-data compilation and synthesis.

Summary and concluding remarks

Since the start of the PAGES programme a decade ago, a reasonably coherent picture of Holocene climate change in Europe and Africa has emerged, showing a long-term trend reflecting precession-driven orbital (Milankovitch) forcing of solar insolation in both Hemispheres, and a sequence of century-scale climate anomalies tentatively linked to variation in solar radiation output. The long-term climate trend is expressed as early-Holocene warming in northern boreal and temperate regions, enhanced monsoonal rainfall in the northern tropics, and relative drought in the Mediterranean (as warming increased Summertime evaporation) and southern Africa (due to anti-phase precessional forcing in the Southern Hemisphere). The superimposed century-scale climatic variability shows minima in solar activity coinciding with cooling of the North Atlantic Ocean and boreal and north-temperate continental regions, increased moisture in temperate Europe and the Mediterranean, and weakened monsoon circulation causing drought in the northern tropics. The Medieval Warm Period and Little Ice Age can be considered as prominent recent expressions of this century-scale climate variability, although in much of inter-tropical Africa the MWP - LIA transition brought at least a few centuries of increased wetness, not drought.

Further progress in the understanding of global and regional climate system dynamics at inter-annual to century time scales will depend as much on finding better ways to exploit the multitude of climate-proxy data already available than on producing new data from ever-more exotic (and sometimes ever-more mundane) localities. Fundamental challenges pertain to how we read climate change in the records of proxy climate indicators, how we determine the relative merits of various natural climate archives and exploit each of them to our maximum advantage, and how the multitude of existing and forthcoming palaeo-data can best be fed into the climate modelling effort.

Yet one other challenge is more philosophical, as it pertains to essential differences in the paradigms guiding Time Stream 1 and Time Stream 2 climate reconstruction. As the focus of international climate-research programmes is increasingly geared to issues requiring climate reconstruction with both high time resolution and dating accuracy, specialists of some traditional natural climate archives must come to grips with the fact that the traditional paradigms guiding their interpretation of these archives, ingrained by decades of climate reconstruction at long time scales, no longer apply. Challenges in Time Stream 2 climate reconstruction mainly revolve around the limited time range of radio-isotope dating methods; the validity of the uniformitarian principle, i.e., to what extent modern relationships between indicators and climate can be extrapolated into deep time; and non-analogue situations due to drastically different boundary conditions or, sometimes, biological evolution. By contrast, high-resolution Time Stream 1 reconstructions involve an altogether different set of challenges: limits on the precision rather than time range of radio-isotope dating methods; limits on the *de facto* time resolution of important natural climate archives due to post-depositional mixing, diffusion, and biological 'memory' effects; the difficulty to distinguish relatively modest climate-change signatures from archive- and indicator-specific noise; and the prevalence of transient conditions related to the lagged and/or non-linear response of proxy indicators to climate change. Even in the best of circumstances, climate reconstruction from natural archives has its limits. We need to constantly probe and test those limits to find the right balance between justified inferences and open-ended speculation.

References

Barber K., Zolitschka B., Tarasov P. and Lotter A.F., this volume. Atlantic to Urals — the Holocene climatic record of Mid-latitude Europe. In: Battarbee R.W., Gasse F. and Stickley C.E. (eds), Past Climate Variability through Europe and Africa. Kluwer Academic Publishers, Dordrecht, the Netherlands, pp. 417–442.

Barker P., Talbot M.R., Street-Perrott F.A., Marret J., Scourse J. and Odada E., this volume. Late Quaternary climatic variability in intertropical Africa. In: Battarbee R.W., Gasse F. and Stickley C.E. (eds), Past Climate Variability through Europe and Africa. Kluwer Academic Publishers, Dordrecht, the Netherlands, pp. 117–138.

Braconnot P., Harrison S.P., Joussaume S., Hewitt C.D., Kitoh A., Kutzbach J.E., Liu Z., Otto-Bliesner B., Syktus J. and Weber S.L., this volume. Evaluation of PMIP coupled ocean-atmosphere simulations of the mid-Holocene. In: Battarbee R.W., Gasse F. and Stickley C.E. (eds), Past Climate Variability through Europe and Africa. Kluwer Academic Publishers, Dordrecht, the Netherlands, pp. 515–533.

deMenocal P.B. 2001. Cultural responses to climate change during the late Holocene. Science 292: 667–673.

Hoelzmann P., Gasse F., Dupont L.M., Salzmann U., Staubwasser M., Leuschner D.C. and Sirocko F., this volume. Palaeoenvironmental changes in the arid and subarid belt (Sahara-Sahel-Arabian Peninsula) from 150 kyr to present. In: Battarbee R.W., Gasse F. and Stickley C.E. (eds), Past Climate Variability through Europe and Africa. Kluwer Academic Publishers, Dordrecht, the Netherlands, pp. 219–256.

Kallel N., Duplessy J., Labeyrie L., Fontugne M. and Paterne M., this volume. Mediterranean Sea palaeohydrology and pluvial periods during the Late Quaternary. In: Battarbee R.W., Gasse F. and Stickley C.E. (eds), Past Climate Variability through Europe and Africa. Kluwer Academic Publishers, Dordrecht, the Netherlands, pp. 307–324.

Roberts N., Stevenson A.C., Davis B., Cheddadi R., Brewer S. and Rosen A., this volume. Holocene climate, environment and cultural change in the circum-Mediterranean region. In: Battarbee R.W., Gasse F. and Stickley C.E. (eds), Past Climate Variability through Europe and Africa. Kluwer Academic Publishers, Dordrecht, the Netherlands, pp. 343–362.

Scott L. and Lee-Thorp J.A., this volume. Holocene climatic trends and rhythms in Southern Africa. In: Battarbee R.W., Gasse F. and Stickley C.E. (eds), Past Climate Variability through Europe and Africa. Kluwer Academic Publishers, Dordrecht, the Netherlands, pp. 69–91.

Snowball I., Korhola A., Briffa K.R. and Koç N., this volume. Holocene climate dynamics in Fennoscandia and the North Atlantic. In: Battarbee R.W., Gasse F. and Stickley C.E. (eds), Past Climate Variability through Europe and Africa. Kluwer Academic Publishers, Dordrecht, the Netherlands, pp. 465–494.

Umer M., Legesse D., Gasse F., Bonnefille R., Lamb H., Leng M.J. and Lamb A., this volume. Late Quaternary climate changes in the Horn of Africa. In: Battarbee R.W., Gasse F. and Stickley C.E. (eds), Past Climate Variability through Europe and Africa. Kluwer Academic Publishers, Dordrecht, the Netherlands, pp. 159–180.

Verschuren D., this volume. Decadal and century-scale climate variability in tropical Africa during the past 2000 years. In: Battarbee R.W., Gasse F. and Stickley C.E. (eds), Past Climate Variability through Europe and Africa. Kluwer Academic Publishers, Dordrecht, the Netherlands, pp. 139–158.

Weiss H. 2000. Beyond the Younger Dryas: Collapse as adaptation to abrupt climate change in ancient West Asia and the Eastern Mediterranean. In: Bawden G. and Reycraft R. (eds), Confronting Natural Disaster: Engaging the Past to Understand the Future. University of New Mexico Press, Albuquerque, pp. 75–98.

28. CLIMATE VARIABILITY IN EUROPE AND AFRICA: A PAGES-PEP III TIME STREAM II SYNTHESIS

TIM C. PARTRIDGE (tcp@iafrica.com)
Climatology Research Group
University of the Witwatersrand
Private Bag 3
WITS 2050, Johannesburg
South Africa

JOHN J. LOWE (j.lowe@rhul.ac.uk)
Department of Geography, Royal Holloway
University of London
Egham, Surrey, TW20 0EX
UK

PHILIP BARKER (p.barker@lancaster.ac.uk)
Department of Geography
Lancaster University
Lancaster, LA1 4YB
UK

PHILIPP HOELZMANN
(phillip.hoelzmann@bgc-jena.mpg.de)
Max-Planck-Institut für Biogeochemie
Postfach 100164, 07701 Jena
Germany

DONATELLA MAGRI (magri@mail.uniroma1.it)
Dipartimento di Biologia Vegetale
Università "La Sapienza"
P.le Aldo Moro, 5, 00185 Roma
Italy

MATTI SAARNISTO (matti.saarnisto.gsf.fi)
Geological Survey of Finland
P.O. Box 96
FIN-02150 Espoo
Finland

R. W. Battarbee et al. (eds) 2004. *Past Climate Variability through Europe and Africa.*
Springer, Dordrecht, The Netherlands.

JEF VANDENBERGHE (vanj@geo.vu.nl)
Faculty of Earth and Life Sciences
Vrije Universiteit
Amsterdam
The Netherlands

F. ALAYNE STREET-PERROTT
(f.a.street-perrott@swansea.ac.uk)
Department of Geography
University of Wales Swansea
Swansea, SA2 8PP
UK

FRANÇOISE GASSE (gasse@cerege.fr)
Centre Européen de Recherche et d'Enseignement
de Géosciences de l'Environnement (CEREGE)
Europole de l'Arbois
B.P. 80, F-13454 Aix-en-Provence cedex 4
France

Keywords: Precessional forcing, Abrupt climatic changes, Last interglacial-glacial cycle, Europe, Africa, Tropical monsoon influences, Antarctic circulation influences, N. Atlantic circulation influences, Climate 'teleconnections'.

Introduction

The PEP III Europe-Africa transect extends from the arctic fringes of NW Eurasia to South Africa. It encompasses the presently temperate sector of mid-latitude Europe, the Mediterranean region, the arid and semi-arid lands of the Sahara, Sahel and the Arabian Peninsula, and the inter-tropical belt of Africa. The palaeoenvironmental evidence available from these regions, which has been summarised in earlier chapters of this volume and which collectively spans the last 250,000 years, clearly bears the stamp of long-term global climate forcing induced by variations in solar insolation. External forcing is ultimately the reason why the Eurasian continental ice sheets waxed and waned repeatedly during the late Quaternary, and why the southerly limit of permafrost migrated southwards across mid-latitude Europe, periodically becoming degraded during warmer episodes. At the same time, pronounced fluctuations in atmospheric and soil moisture have affected the Mediterranean, desert and Sahel regions, while there is abundant evidence from every sector of the PEP III transect for marked migrations of the principal vegetation belts, as well as for other major environmental changes, that are also considered to reflect long-term climate forcing. It is only in the last decade or so, however, that the full complexity of the history of climate changes during the last interglacial-glacial cycle, and their environmental impacts in continental Europe and Africa, have begun to be recognised. The discovery of evidence for the abrupt Dansgaard-Oeschger (D-O) and Heinrich (H) climatic oscillations in Greenland ice-core (Johnsen et al. 1992) and North Atlantic (Bond et al. 1993) records, have prompted a re-examination of the continental record. This, together with a number of technical improvements in field and laboratory equipment, greater access to sites in remote and difficult terrain, diversification in the range of available palaeoecological and geochronological tools, and closer inter-disciplinary collaboration, have led to a more penetrating examination of the field evidence, which has progressed the science considerably. We can now see that the stratigraphical record is much more complex than appreciated hitherto, and more detailed and refined models of past climatic and environmental models are beginning to emerge. There is, for example, a growing body of evidence which suggests that D-O and H events had significant impacts on the environment of Europe and Africa, as well as on the Mediterranean Sea.

We are, however, still a long way from being able to synthesise these data at a continental scale because the information is patchy at present, while much of the available palaeoenvironmental data is unquantified. Continental records are also fragmentary, especially in the glaciated, periglaciated and arid zones, and the records are difficult to date and correlate with adequate precision, particularly those that are older than the limit of radiocarbon calibration. There are, inevitably, major gaps in many of the key stratigraphical records. Exceptions to this rule, however, are continuous lake sediment sequences, especially in lake basins in Southern Europe, though not all of these extend back over the full 250,000 years. Even where long, continuous sedimentary records exist, however, the evidence is not always easy to interpret, and the precise chronology of the sequences is frequently equivocal (see Magri et al. (this volume)).

A second problem is that each sector within the PEP III transect is characterised today by marked climatic variability, which reflects a variety of local influences; presumably the same must have been the case in the past. A regional or global climatic forcing effect could well have been significantly dampened or enhanced by these local influences, but distinguishing between local, regional and global influences on climatic records is difficult. This problem may be compounded by the possibility that regional climatic changes, or, at least, environmental responses to abrupt climatic changes, were measurably time-transgressive. There is evidence to suggest that this may have been the case, for example, in Europe during the last glacial-interglacial transition, an interval for which the evidence can generally be dated more precisely than is the case for earlier intervals (see Vandenberghe et al. (this volume)).

A third issue concerns the manner by which climate mechanisms operate at a global and continental scale. There are presently diverging views as to whether climate changes during the last glacial cycle were synchronous or asynchronous between the two hemispheres (Charles et al. 1996; Blunier et al. 1998; Broecker 1998; Kanfoush et al. 2000). The arguments are made fuzzy, somewhat, by geochronological uncertainties, though a recent study which was based on very precise dating appears to confirm that climatic changes in Japan at the close of the last glacial stage were not synchronous with those in the North Atlantic region (Nakagawa et al. 2003). At issue here is how the major elements of the global climate machine, such as, for example, the Indian Monsoon, North Atlantic circulation, and the 'heat engine' of Tropical Africa, interact, and whether external forcing of the system (caused, for example, by precession) brings about instantaneous changes around the globe. Some assume this to be the case, when, for example, the marine oxygen isotope stratigraphy is used as a basis for inter-regional correlation. Yet evidence presented by Partridge et al. (this volume) indicates that climatic changes in southern Africa, because they were heavily influenced by atmospheric circulation over Antarctica and the Southern Ocean, preceded those in the North Atlantic by some 3 to 4 kyr. Different sectors of the PEP

III transect will have been variously affected by changes in North Atlantic circulation, the Indian Monsoon, Mediterranean circulation, circulation changes in the Southern Ocean, and other components of the global climate system. One should not assume that these different components responded synchronously to external forcing factors; indeed, the evidence seems increasingly to suggest otherwise.

These difficulties notwithstanding, significant progress has been made in understanding the strengths and limitations of the stratigraphical and proxy data at our disposal, while new perspectives are emerging that will surely lead to more effective palaeoclimatic research in the future. Here we summarise the key conclusions to emerge from the reviews of the palaeoenvironmental reconstructions for the PEP III transect during time stream 2, presented in earlier chapters in this volume. We focus especially on those conclusions that may be of most relevance to palaeoclimate modelling.

The NW Eurasian ice-sheets

Ice-sheets play a crucial role in the global climate system, influencing, *inter alia*, albedo, global sea-level, supply of freshwater into the oceans and land drainage. It is extremely important, therefore, to be able to generate reasonable approximations of the dimensions of the ice-sheets that existed at different times in the past, as well as the rate at which these altered. Recent field investigations in NW Eurasia have forced a major re-think on the size and shape of the ice-sheets during the last glacial cycle, and also on how ice-sheets respond to climate (Saarnisto and Lunkka, this volume). One of the crucial conclusions to emerge from recent research is that ice-sheet cover in NW Eurasia may have been greatly over-estimated in earlier models (e.g., Grosswald (1980)), while there is also clear evidence that the dominant centres of ice accumulation shifted across the region during the last interglacial-glacial cycle, presumably in response to local variations in moisture supply.

The ice built up as early as marine isotope stage (MIS) 5d in NW Siberia, while at the same time there was a restricted ice cover in Scandinavia. The centre of ice accumulation then shifted to the Kara-Barents Sea regions during the early part of MIS-4, though this ice mass thinned considerably by the Last Glacial Maximum (MIS-2). In contrast, the Scandinavian ice sheet achieved its maximum extent during MIS-2, at a time when the mainland of Northern Russia was ice-free. It seems, therefore, that the location of the main centres of ice accumulation shifted from the east to the west during the last glacial cycle, possibly as a result of progressive cooling of Siberian and, subsequently, Atlantic coastal waters. As the waters froze, and sea-ice built up, the continental interior became increasingly starved of moisture.

Behind that brief overview lies a much more complex story (Thiede et al. 2001). Until recently our understanding of the behaviour of the Scandinavian ice sheet relied heavily on the analysis of the glacial sedimentary record. This evidence is, of course, very fragmentary; in Scandinavia, for example, because the maximum advance of ice occurred during MIS-2, much of the evidence for earlier episodes has been destroyed, though nearly continuous sequences of glacial diamicts and fossiliferous sediments do appear to have survived in isolated localities, such as Finnish Lapland (Helmens et al. 2000). More continuous records have survived close to, or beyond the limits of the MIS-2 ice sheet, which help to fill in the gaps. These indicate that the Scandinavian ice mass oscillated markedly throughout the last glacial cycle. For example, sedimentary variations in the Norwegian Sea (Baumann et al. 1995) reflect whether the Scandinavian ice mass was distal or proximal to the coast, and this evidence indicates that the Scandinavian Ice Sheet was oscillating in size almost continuously during the last glacial cycle, though the actual dimensions of the ice mass at each stage in the process are difficult to determine. Mangerud et al. (2003) report evidence obtained from laminated clay sequences preserved in caves in Western Norway, which have been dated using palaeomagnetic excursions and cosmogenic nuclide peaks, that suggests that some of the oscillations correlate with Greenland D-O events, though they also conclude that not all of the ice-sheet fluctuations were in phase with Greenland climate oscillations. What is clear, however is that the ice sheets in NW Eurasia were able to respond quickly to changes in climate. For example, the ice advanced from the west coast of Finland to east of lake Onega, a distance of some 900 km, during MIS-2 within ca. 7 kyr, a rate of advance that equals the rate of retreat which took place across that region between 17 and 10 kyr BP.

The NW Eurasian ice masses may well have been responding initially to ocean temperature changes, especially in the North Atlantic. The optimal conditions of the Eemian interglacial (MIS-5e) appear to have been 2–3 °C greater than that of the Holocene, and conditions became more oceanic towards the end of the Eemian. Temperatures, although subject to a number of fluctuations, grew gradually colder during MIS-4 and MIS-3, and the most severe conditions were experienced in MIS-2, when winter temperatures were at least 20 °C, and summer temperatures between 5 and 11 °C, lower than present. An important objective for future research is to determine how much of this temperature decrease was caused by feed-back influences resulting from the growth and configuration of the ice-sheets themselves. Given the history of events summarised above, it seems that with an open ocean and relatively mild conditions (MIS-5e/5d), glacier ice would initially have been restricted to the coldest parts of the arctic, where low winter temperatures were the critical factor. A reduction in regional temperature led to expansion of the arctic ice masses, but as that ice expanded and the adjacent seas became colder, the arctic became progressively starved of moisture. This should have led to a steepening of west-east moisture and temperature gradients, with enhanced continental conditions being experienced in the east.

The waxing and waning of the ice sheets also contributed to significant regional palaeo-hydrological changes that must have had wide climatic effects. For example, the blocking of the northern outflows of the Ob and Yenisei rivers by the growth of the Kara Ice Sheet caused an enormous ice-dammed lake to build up, which eventually occupied the whole of the western Siberian plain (Arkhipov et al. 1995). This lake drained southwards via the Aral Sea to the Caspian, overflowing from there into the Black Sea and the Mediterranean. This influx of freshwater may have altered the balance of exchange between Mediterranean and Atlantic waters, which could be thought of as closing a complex feed-back loop, since it is likely that water mass changes in the North Atlantic initiated the growth of ice in NW Eurasia in the first place. Another potentially pronounced climatic consequence of the creation of such huge ice-dammed lakes would be the sudden influx of cold freshwater into Arctic seas and, via the Baltic, into the North Atlantic, a process that might have had dramatic effects on sea surface temperatures and water mass circulation in the NE Atlantic and Arctic seas. River discharge to the Arctic Ocean is linked to fluctuations in the North Atlantic Oscillation (Peterson et al. 2002), while freshwater forcing of N. Atlantic thermohaline circulation during episodes of catastrophic flooding from ice-dammed lakes in N. America is widely believed to have triggered abrupt climatic changes during the latter stages of the last glacial cycle (Clark et al. 2001; Teller et al. 2002; Broecker 2003).

In view of the potentially important palaeoclimatic significance of such environmental changes, future research should be directed towards generating more reliable, quantified estimates of: (i) variations in ice sheet volume and configuration; (ii) the volumes of stored ice-dammed water; (iii) fluxes of reversed/diverted stream flows; (iv) fluxes of freshwater into the Arctic seas and NE Atlantic during the catastrophic drainage of ice-dammed lakes; and (v) the temperature and moisture gradients that existed over NW Eurasia during the last interglacial-glacial cycle.

The periglaciated zone of Mid-Latitude Europe

Quantified temperature estimates based on palaeobotanical data suggest that the Eemian in temperate Europe was characterised by summer and winter temperatures not much different to those that prevail in the region today (Kühl et al. 2001; Litt et al. 2001). There is, however, some debate over whether the Eemian was a period of uninterrupted warmth, or whether significant cooling episodes occurred (see Vandenberghe et al. (this volume)). This is an important issue, which has significance for understanding the nature of climate deterioration from predominantly 'interglacial' to wholly 'glacial' conditions. There is limited information from this region for conditions during this important transition, and indeed scant information for the early Weichselian in general. Exceptions are the lake sediment sequences at sites like Les Echets and La Grande Pile, but these are extremely rare in mid-latitude Europe, and the palaeoclimatic interpretations available for these records are only crudely quantified at present. A priority for future research in this region, therefore, is to improve the database of information for the Eemian and the Eemian-early Weichselian transition, in order to derive more robust reconstructions of the climate conditions that prevailed during these periods.

By MIS-3, when the Kara-Barents Ice Sheet was dominating in NW Eurasia, much of mid-latitude Europe was frozen. The palaeohydrological consequences of this were severe and widespread, for river régimes are drastically altered by the long seasonal freezing of the ground surface. However, it is difficult to quantify the climatic conditions at this time, as frozen ground phenomena provide indications that temperatures were below certain thresholds, but not by how much the temperatures were below those thresholds. Periglacial sediments also tend to be devoid of biological remains, which limits the range of proxy methods that can be used. Intensely cold conditions did not persist throughout MIS-3, however. There is evidence for short episodes of warming, possibly of the order of only 2 to 3 °C, but these were of sufficient warmth and duration to enable some thermophilous plants and temperate insects to migrate northwards. The evidence for these short warm episodes is extremely fragmentary, however, and their precise chronology remains obscure. It is possible that some of them correlate with Greenland D-O events, but the extent to which this is the case remains to be established.

More detailed information is available for conditions during MIS-2. During this period, when the Scandinavian Ice Sheet was building up to its maximum for the last cold stage, mid-latitude Europe was extremely cold. Widespread periglacial evidence suggests a mean annual air temperature for the coldest parts of the region of about −8 °C, with mean temperatures for the coldest month in some areas as low as −25 °C. There is evidence, however, of a north-to-south temperature gradient, with the mean temperature of the warmest month being ca. 4 °C in the north but as high as 8 °C in the south. The climatic conditions were extremely continental at this time, with an annual temperature amplitude of around 28 to 33 °C (Vandenberghe and Pissart 1993; Huijzer and Vandenberghe 1998). Conditions were also very dry in many districts, especially towards the end of MIS-2, as extensive belts of loess and sand dominated the landscape. Analysis of the regional occurrence of these deposits suggests that they were deposited by winds with a strong northerly component. Most of the rivers incised into their flood-plains at the beginning of MIS-2, while later on they deposited coarse sediments in braided channels as they flooded during the short thaw seasons.

The picture that emerges, therefore, is of a progressive shift towards colder and more arid conditions in mid-latitude Europe as the Scandinavian ice sheet built up during MIS-2. There is clearly a complex set of feed-back mechanisms operating between: (i) the surface temperature conditions and extent of sea-ice cover in the oceans; (ii) the size and location of the NW Eurasian ice sheets; (iii) the local climatic effects (katabatic winds) created by large ice masses; and (iv) the prevailing insolation régime. It should be possible to obtain more precise palaeoenvironmental data to clarify how closely these environmental elements were interconnected during this period.

Warming in mid-latitude Europe at the end of the last cold stage appears to have commenced soon after the LGM, when the European permafrost became degraded, presumably in response to some initial thermal warming. Some further warming may have occurred at ca. 15.0 ^{14}C kyr BP, as suggested by records in southern Europe and NE Atlantic cores (Walker 1995), though this is not clearly reflected in continental records from mid-latitude Europe. The most marked increase in temperature occurred at ca. 13.0 ^{14}C kyr BP, and is most clearly reflected in records from the British Isles and The Netherlands (Coope et al. 1998; Witte et al. 1998). Warming was delayed in Scandinavia, however, probably because of the cooling effects of the residual ice mass (Coope et al. 1998; Witte et al. 1998). Thermal gradients steeper than those of today are therefore suggested by the palaeo-data for some intervals between ca. 15 and 11 ^{14}C kyr BP. Within this interval a series of short-lived episodes of climate cooling occurred, of which the most intense and widely recorded is the event referred to as the 'Younger Dryas'. There is still some dispute about the number of cooling events during this interval, their precise timing, and the geographical areas over which they have left their imprint. It is likely that they reflect the climate oscillations recorded in the Greenland ice-core and North Atlantic records, but the extent to which these were all in phase is not yet clear (Lowe et al. 1995).

The Mediterranean region

Since the Mediterranean region was not glaciated (except in high mountain locations) nor extensively periglaciated during the last glacial cycle, continuous sediment sequences extending from the present time back through the full glacial cycle to the Eemian, and in some cases to much earlier periods, can be found in a number of deep lake basins. Investigations of these sequences have provided a fuller picture of the history of climate events than is generally the case for other sectors of the PEP III transect. There is much to be done to improve the dating of these records, while some of the sequences may be subject to hiatuses and other stratigraphical complications. Furthermore, palaeoenvironmental interpretations based on these records are frequently contradictory (see Magri et al. (this volume)). Nevertheless, they provide probably the best available archives within the PEP

III transect for assessing, at a high resolution, the nuances of climatic change and variability during time stream 2.

Studies of long lake sediment sequences are being complemented by other records that extend through the last interglacial-glacial cycle, from deep marine sediments in the Mediterranean and the eastern fringe of the North Atlantic (e.g., Sánchez-Goñi et al. (1999), Pailler and Bard (2002)) and from cave calcite deposits (e.g., Frumkin et al. (1999)). Collectively, these palaeoenvironmental archives indicate: (i) the imprint of long-term insolation forcing on Mediterranean climate (Tzedakis et al. 1997); (ii) the signature of D-O and H events in marine and continental sequences (e.g., Allen et al. (1999), Cayre et al. (1999), Cacho et al. (2000)); (iii) possible close coupling between marine and continental climate variations (Roucoux et al. 2001; Sánchez-Goñi et al. 2002); and (iv) the importance of Saharan dust transport throughout the Mediterranean and eastern Atlantic during times of enhanced atmospheric circulation in high northern latitudes (e.g., Moreno et al. (2001, 2002), Magri and Parra (2002)).

One of the major palaeoclimatic issues that remains to be resolved, however, is the extent to which climatic changes in the Mediterranean region were in-phase with those in the North Atlantic region, with respect to oscillations of both long (glacial-interglacial cycles) and short (D-O events) amplitude. Tzedakis (2003) argues that long-held assumptions of synchroneity of major climate shifts between, on the one hand, the oceans and the continents, and, on the other, northern and southern Europe, may be erroneous. He concludes that available evidence indicates that the onset of warming which culminated in the Eemian interglacial on land commenced some time after the ocean had warmed at the start of MIS-5e, while cooling at the end of the Eemian appears to have been in-phase with the end of MIS-5e in northern Europe, but considerably delayed in the south. The interpretations are based primarily on palaeobotanical evidence; there is a need to establish the degree to which the inferred vegetational changes represent instantaneous or delayed responses to climatic change.

Establishing the degree to which millennial-scale variations in the Mediterranean were synchronous with the D-O and H events of the North Atlantic is a challenging problem. Magri et al. (this volume) have drawn attention to several difficulties that need to be overcome, before this can be achieved satisfactorily. First is the complex issue of dating the records with the temporal precision required. D-O and H events were short-lived, and were terminated by very abrupt warmings — some transitions (according to the ice-core evidence) taking only a few decades. Ideally, therefore, correlations need to be effected with a decadal precision. The potential exists to achieve this, for some of the lake sequences are annually laminated, and several additional geochronological tools can be applied as independent tests of age-depth models (e.g., Allen et al. (1999)). Nevertheless, considerable uncertainties beset the published age estimates for events within the last interglacial-glacial cycle, and confident correlation at a high (decadal) precision has still to be achieved. Even the boundaries of the youngest abrupt climatic events of the last glacial cycle, such as the 'Younger Dryas' cold event, which lie within the range of radiocarbon dating, cannot yet be defined with a decadal precision because of the problems of calibrating radiocarbon dates that are older than ca. 11,500 cal yr BP (Asioli et al. 1999; Lowe and Walker 2000).

A second issue is that of distinguishing those palaeoenvironmental signals that reflect changes in temperature from those that reflect changes in atmospheric humidity; of course, these two parameters are sure to have varied in concert. The available evidence seems to point to variations in climatic seasonality as having been the key orchestrator of environmental changes in the Mediterranean zone throughout the last interglacial-glacial cycle. These may have been linked in turn (as appears to be increasingly assumed) to changes in North Atlantic circulation, the controlling mechanism being water mass exchange between the Mediterranean and Atlantic through the Strait of Gibraltar and its resultant effect on salinity. However, as Magri et al. point out, there are influences on climate in the Mediterranean area, other than marine ones, such as incursions of Saharan air masses from the south, which may be driven by changes in atmospheric pressure over North Africa, and perturbations of the climatic regime of the Middle East and eastern Mediterranean basin, which reflect variations in the strength of the Indian monsoon pressure cell. Extreme changes in any one of these climate forcing agents or, more likely, enhanced or modulated signals brought about by the interplay between them, probably account for periodic expansions and contractions of forest cover and in the composition of woodland throughout the Mediterranean, for extremely arid conditions during which the forest cover declined abruptly and dust transport from the Sahara became more prevalent, and for marked changes in water density and circulation within the Mediterranean, the most extreme effects of which may have led to the deposition of sapropel layers on the sea floor.

Sahara-Sahel-Arabian Peninsula

Records from this sector of the PEP-III transect for time stream 2 are few in number and fragmentary in nature. Most of the evidence consists of lake sediments laid down intermittently, generally during wetter climatic periods, and from which migrations in vegetation types can be inferred from the pollen records that they contain (e.g., Gasse et al. (1990)). These data can be compared with pollen records obtained from the adjacent oceans, which provide more continuous and longer records (e.g., Prell and van Campo (1986), deMenocal et al. (2000)), and with speleothem records from, for example, Northern Oman (Burns et al. 1998). Evidence of groundwater recharge and increased lake levels, which presumably reflect times of increased atmospheric moisture, can also be inferred from studies of sediment chemistry or gas content (Edmunds et al. 1999). Overall, however, the number of records from which such data have been obtained is very low compared with the size and complexity (topographic and climatic) of the region they are taken to represent, and also compared with the higher density of site records that is available to support reconstructions for the other sectors of the PEP III transect. Furthermore, interpretations of site records from the arid zone are constrained by the limited number and statistical uncertainty of radiometric dates currently available. Radiocarbon dates obtained from deposits laid down in the later phases of the glacial cycle are particularly prone to error where the groundwater is calcareous.

In view of these limitations, considerable care is required when generalising about the history and nature of climatic changes in this sector during time stream 2; more confidence can be attached to Holocene reconstructions, a period for which many more site records are available, a greater diversity of proxy records has been investigated, and more reliable site chronologies have been developed (Hoelzmann et al., this volume; Verschuren et al., this volume).

Some tentative conclusions can be drawn, however. Considerably wetter conditions are inferred for the Arabian Peninsula during MIS-5e, both on the basis of pollen records obtained from marine cores from adjacent seas, and from the speleothem record in northern Oman. A strong south-west monsoon influence is considered to be responsible for this. Climate during the last glacial stage appears to have oscillated between episodes drier and wetter than the present, with some strong indications that the region experienced millennial-scale climatic oscillations which may equate with the sequence of D-O and H events in the North Atlantic region (Schulz et al. 1998; Leuschner and Sirocko 2000). Conditions appear to have been predominantly dry for much of MIS-5d and MIS-4, but significantly wetter than today between about 30 and 19 ^{14}C kyr BP, in both the Arabian Peninsula and the northern Sahara, where lake levels were significantly higher than those of today.

Conditions during the last glacial-interglacial transition (ca. 15–11 ^{14}C kyr BP) appear to have oscillated abruptly in North Africa, though problems of chronology make precise correlations between site records difficult. The majority of records, however, suggest that conditions in North Africa were generally wetter between ca. 14 and 5.5 cal yr BP, the 'African Humid Period' (deMenocal et al. 2000), though not all records accord with this interpretation: either some areas remained dry during this period, or the humid phase was interrupted by brief periods of drier conditions (Hoelzmann et al., this volume).

It is important to improve the overall palaeoenvironmental archive for the Sahara-Sahel-Arabia sector, in several ways: (i) by filling in the gaps in the record for the last glacial stage; (ii) by developing better quantified reconstructions of past climatic conditions, using a more diverse array of proxy indicators; and (iii) by increasing the density of sites that provide data for each important interval, thereby enabling more detailed spatial patterns to be constructed. The strength of the African monsoon is an important climatic parameter, not only controlling the degree of humidity in the arid parts of North Africa and the Arabian Peninsula, but impacting on the climate of the Mediterranean, and playing a key role in the global climate system. DeMenocal et al. (2000) give a pointer as to why this is potentially very significant: they conclude from their study of the marine core off the coast of Mauritania that: (i) variations in the strength of the African monsoon were governed by gradual orbital increases in summer season; (ii) the onset and termination of the 'African Humid Period' were very abrupt; and (iii) that the transitions occurred when summer season insolation crossed a critical threshold, i.e., when it was 4.2% greater than the present value.

Inter-tropical Africa

By contrast with the more arid zones to the north, Tropical Africa provides reasonably abundant opportunities for reconstructing past environmental conditions. The greater humidity and lush vegetation that is characteristic of this sector lead to the generation and preservation of organic sediments in lake basins, while some of the organic detritus is carried by the major rivers to estuaries and the open sea: hence migrations in the forest-savanna boundary and changes in forest composition, as well as changes in river discharge, can be inferred from variations in sediment type and in fossil content of core records obtained from, for example, the Congo Fan (Marret et al. 2001). Past climatic conditions can also be inferred from studies of fluctuations in lake levels; where a number of lake records provide accordant evidence for either a general increase or decrease in lake volume, then this probably reflects a change in regional climatic wetness (Street-Perrott and Perrott 1993). Barker et al. (this volume) draw attention to the growing number of lake and marine records from the inter-tropical sector of the PEP III transect that contain records that extend back to MIS-5e. An

impressive range of proxy indicators are now being studied for many of the key records, and these are enabling the sequence and nature of past climate variations to be reconstructed not only in more detail, but also with a greater degree of confidence, since assessments can often be based on several independent proxy indicators.

What has emerged from these recent studies is the over-riding importance of the 19–23 kyr precession cycle as a driver of climate change in tropical Africa. This cycle is reflected in marine records obtained from localities that lie close to the African coast (e.g., Schneider et al. (1996)), in reconstructions of lake-level variations (e.g., Trauth et al. (2001, 2003)) and in pollen-stratigraphic changes and other stratigraphical records obtained from lake sediments (e.g., Gasse and van Campo (2001)). Tropical Africa forms an important component of the 'heat engine' that drives meridional circulation of the atmosphere, which governs the strength of the SW monsoon circulation. This in turn determines atmospheric moisture patterns over Africa. But the influence of low-latitude monsoon circulation may extend much further. Trauth et al. (2003) report new evidence obtained from Lake Naivasha in Kenya, that leads them to the conclusion that changes in the strength of the African monsoon do not correspond to peaks in summer insolation, but may in fact lead them, and the changes that took place in extra-tropical regions. They also conclude that the data provide evidence for low-latitude forcing of deglaciation in the northern hemisphere at around 135 kyr BP.

Understanding the role played by low-latitude atmospheric components of the global system has emerged as a key objective for future palaeoclimatic research. In addition to clarifying the manner in which the global climate system responds to the precession cycle, Barker et al. (this volume) draw attention to other features of the palaeo-environmental archive from inter-tropical Africa that deserve further attention. First, environmental responses in this region were most often abrupt and irregular, and not gradual, as would be expected if insolation factors alone were driving the regional climatic changes. Clearly, as with other sectors in the PEP III transect, feed-back factors have operated to modulate the influence of insolation changes. Second: although for most of the record for the time stream 2 interval, palaeoenvironmental variations appear to accord with expected precession effects, this is not the case for the Last Glacial Maximum (MIS-2). The implication drawn from this is that direct insolation effects may be over-ridden by the impacts of processes operating in the higher latitudes during times of maximal glaciation (deMenocal et al. 1993). Third, the data for the LGM from inter-tropical Africa provide interesting comparisons with the results of GCM simulations. Barker et al. (this volume) and Barker and Gasse (2003) point out that the general synopsis derived from the palaeo-data records, which indicates that most of inter-tropical Africa experienced drought conditions during the LGM, is best simulated by GCMs that use computed SSTs rather than empirical SST values, as employed in the CLIMAP project (Kutzbach and Guetter 1986). Barker et al. conclude that the data indicate that climate in inter-tropical Africa was closely linked to the temperature of the adjacent oceans during the LGM, which in turn reflected the growth of the polar ice sheets.

Southern Africa

Data assembled on southern African palaeoenvironments during the last two glacial cycles, based on lake-level records, cave sediment sequences and some marine records from adjacent oceans, reinforces much of what has been learned from regions to the north (Partridge et al., this volume). Additional complexity is, however, introduced by the presence of a persistent zone of atmospheric subsidence over the west of the subcontinent, which is itself sufficiently narrow to be influenced strongly by the contrasting current régimes along both coasts. The winter rainfall area of the extreme south-western tip of the continent differed from that receiving precipitation during the summer in displaying inverse responses to major forcings: e.g., the occurrence of a cool, wet Last Glacial Maximum in the zone of influence of the Atlantic westerlies. Most information for the earlier part of the record comes from the Tswaing impact crater (previously Pretoria Saltpan) within the summer rainfall region. Here precessionally-driven changes in moisture receipts dominated the sedimentary record during MIS-6 in conformity with the strength of precessional variance at that time. The sensitivity coefficient between changes in insolation and those inferred for rainfall (estimated by applying a sedimentological transfer function) is 4.5; this compares with a range of 3.5–5.0 estimated for the northern subtropics by Prell and Kutzbach (1987) in model experiments. Evidence for elevated temperatures during the last interglacial, and a $\sim3°$ southward shift of the Miombo/Savannah boundary, comes from Border Cave in the eastern hinterland, which is affected strongly by the warm Agulhas Current. Elsewhere increases in both temperature and precipitation were evidently more modest during the Eemian; however, rainfall at Tswaing appears to have increased substantially thereafter during MIS 5d.

A decline in dominance of orbital precession, as the amplitude of the eccentricity signal lessened, after about 60 kyr was matched by an increase in the evidence for other forcings in the Tswaing rainfall record. A number of arid spikes on either side of the Last Glacial Maximum, which were paralleled by discrete episodes of dune activity in the Kalahari, coincided broadly with Heinrich Events in the North Atlantic; however, the onset of each local event led that of the corresponding Heinrich Event on the basis of high-precision, calibrated ^{14}C dates. The average lead time of 3.1 kyr corresponds closely to the ~3 kyr by which South Atlantic temperature responses led their North Atlantic counterparts (Little et al. 1997) and by which phases of warming in the Antarctic led those in the western Indian Ocean (Sonzogni et al. 1998). Since, in almost every case, the onset of southern African aridity coincided with the end of a period of declining temperatures in the Antarctic ice-core records, an atmospheric link between these events (rather than responses via the thermohaline circulation) must be postulated (Partridge 2002; Partridge et al., this volume). The driving mechanism evidently involved changes in the extent and intensity of the circum-Antarctic atmospheric vortex, which were not only sufficient to be felt over widely separated areas of the southern hemisphere, but appear to have driven moist air across the equator, thereby contributing to the rapid growth and collapse of northern hemisphere ice-sheets.

Several findings reported in this volume add credence to this proposition. Barker et al. remark in their chapter that their record of Holocene fluctuations (especially those closely related to the 8.2 kyr event) suggests that the tropics seem to lead Greenland. Export of heat from the tropics is given as a possible cause. The other seemingly significant observation is that records from the Makapansgat speleothem (South Africa), Sacred Lake (Kenya) and Huascaran (equatorial South America) all suggest a pronounced cooling episode centred on 13.5 kyr; this event coincides closely in time with the Antarctic Cold Reversal that is well represented in the southern ice-core records.

A further point seems worthy of comment: diachronism between atmospheric influences from Antarctica and those conveyed over longer timescales via the thermohaline conveyor, and involving heat transfer via the Agulhas and Benguela current systems, seems to have been manifested in differing ways in the southern African palaeoclimatic record. In some cases the duration of events (e.g., spells of dune building) was extended; in others multiple responses are evident while, in yet others, closely spaced fluctuations were accentuated or suppressed. Considerable modification of signals attributable to primary orbital forcings occurred in the process. It is for this reason that the path of deglaciation in Southern Africa was not intimately connected to that of the northern hemisphere (see also Gasse (2000)).

Conclusions

While the history of climate changes that affected the PEP III transect during the last two glacial-interglacial cycles remains sketchy, especially for the period prior to 40 kyr BP, some key temporal and spatial patterns are nevertheless emerging from recent palaeoenvironmental research. Attention has been drawn in this summary to the growing potential that the palaeo-data provide for establishing, at a millennial to centennial timescale, the leads and lags between the Eurasian ice sheets, North Atlantic circulation, the tropical-monsoon system and the circum-Antarctic atmospheric vortex. The potential, therefore, also exists to resolve the relative importance of the three orbital insolation parameters (precession, obliquity, eccentricity) in the long-term climatic record, and thereby to assess the manner by which the orbital signals become modulated by internal feed-back processes. This last aim is clearly fundamental to global climate theory and modelling. There is presently some uncertainty over whether orbital insolation forces northern ice sheets directly, through variations in summer ablation, or whether a more complex set of events is set in train, with initial summer insolation changes in the northern hemisphere being transmitted to the southern hemisphere through deep flow in the Atlantic, which leads to further changes in the northern hemisphere ice sheets that are driven by atmospheric CO_2 changes and other feed-backs (Shackleton 2000; Ruddiman 2003). Overarching questions such as these are capable of being addressed by examination and synthesis of palaeoenvironmental records from the different sectors of the PEP III transect, but only if we are able to improve the chronology and correlation of the different archives, and to develop more reliable, quantified palaeoclimatic indices.

Until recently, reliance had been placed on the marine isotope scheme or on ice-core stratigraphy as a basis for correlating continental and marine sequences. This review has shown that several problems confound this approach. Firstly, some terrestrial responses have been shown to be asynchronous with those in the oceans: a case in point are the findings of Sánchez Goñi et al. (1999, 2000), which show that terrestrial manifestations of the Eemian, as reflected in pollen records of the south-western Iberian margin, are not coeval with MIS-5e in the same marine sequence. Secondly, responses of terrestrial plant communities to global changes in ice-volume and orbitally modulated variations in receipts of solar variation are not uniform, even within relatively small regions. Thus Tzedakis et al. (1997) and Magri et al. (this volume) have drawn attention to the occurrence of large changes in Mediterranean vegetation at times when fluxes in global ice volume were relatively small. Nor do plant communities necessarily respond similarly during marine stages with comparable isotopic signatures, probably because of differences in the orbital configuration that characterised each separate stage. In terrestrial environments, ecological responses to such differences remain poorly understood. Under these circumstances the identification of leads and lags that may throw light on global feedback mechanisms is, at best, ambivalent prior to about 40 kyr.

The record of changing lake levels, which has the potential to resolve questions about hemispheric and inter-hemispheric teleconnections, has been expanded greatly over the past decade. However, the resolution of regional trends during the deglacial period has proved particularly difficult. Thus, while widespread aridity is evident in the tropics and mid-latitudes of Africa at the time of the Last Glacial Maximum, during the lead-up to the Holocene responses to changing insolation receipts were far from consistent. In tropical Africa an increase in insolation from 22 to 12 kyr should have been associated with an increase in rainfall of 35–45% based on orbital parameters (Barker et al., this volume), but, instead, the record of increasing humidity is stepped and interrupted by several dry spikes. Particularly impressive is evidence of substantial lake regressions over a period of about 1000 yr coinciding closely with the Younger Dryas interval. In southern Africa stable isotopes in speleothems indicate that deglaciation was associated with increasing wetness only after 17.5 kyr; drier intervals are evident around 13.5 kyr and again after 12.7 kyr (Holmgren et al. 2003). The fact that, in tropical Africa, the principal increase in precipitation came somewhat later (after 15 kyr), while the first recharge of aquifers in the western Sahara and Nile valley was more recent still (15–12 kyr), suggests that the onset of wetter conditions tracked the slow northward shift of the Intertropical Convergence Zone with the passage of the precessional cycle.

Although deglacial hydrological records across Africa are broadly comparable in their responses to orbital forcing, important differences are evident which can be attributed, in part, to regional patterns of atmospheric circulation. In western North Africa the hydrology of sites north of 23° varied in response to changes in receipts of precipitation from Atlantic air masses, whereas areas to the south were directly subject to fluctuations in the strength of the West African monsoon. Further to the east the highlands of Ethiopia and the Arabian peninsula were dominated by variations in the Indian Ocean monsoon (Hoelzmann et al., this volume). Monsoonal circulations associated with both Atlantic and Indian Ocean air masses influenced the tropics; their varying seasonal strength is imparted by the twice-annual passage of the Intertropical Convergence Zone, while long-term changes occurred in response to alterations in the zone of influence of the ITCZ induced by orbital precession. In southern Africa similar interactions between Atlantic and Indian Ocean air masses are evident in the palaeo-records, but in contrast to tropical Africa, where the Atlantic circulation is dominant geographically, much of the summer rainfall region of the southern mid-latitudes receives its moisture from the Indian Ocean, and apparently did so during much of the past. What is clear, however, is that the past climatic changes in southern Africa have responded rapidly to changes in the strength and intensity of the circum-Antarctic atmospheric vortex and much more slowly to fluxes in the strength of the N. Atlantic thermohaline circulation, which were transmitted via current systems along both coasts. Expansion of the vortex during cold stadials in Antarctica (which were not always synchronous with those in northern polar regions) caused extension northward of the mid-latitude westerlies and associated equatorward displacement of the subtropical highs with their attendant weather-suppressing subsidence (Partridge 2002). Shifts in the position of the semi-permanent high pressure cell over the western regions of southern Africa are reflected both in lake-levels and in the distribution and alignment of linear dunes.

The Mediterranean Sea appears to have responded to large-scale climatic changes in much the same way as the open oceans, with some amplification of glacial-interglacial $\delta^{18}O$ change as revealed by planktonic foraminifera. Important additional hydrological evidence is forthcoming from the sapropel horizons preserved in Quaternary marine sequences of the eastern Mediterranean. These are indicators of enhanced runoff, which led to stratification within the water column and a reduction in deep water oxygenation (Kallel et al., this volume). Six main periods of increased river discharge are indicated; these centre on 195, 170, 122, 96, 80 and 8 kyr. That at 122 kyr corresponds with the lowest $\delta^{18}O$ and high $\delta^{13}C$ values in Israeli speleothems, confirming that the last interglacial was associated with a marked increase in wetness (Bar-Matthews and Ayalon, this volume). There is evidence to suggest that Heinrich Events were paralleled by cooling and decreased salinity within the Mediterranean, although poor chronological control renders any firm conclusions on the timing of local responses to millennial-scale events premature. There is also a suggestion that climatic seasonality was reduced during cold stadials. Extra-regional climatic influences have, however, resulted in a complex mosaic of local responses, the correlation of which is difficult to resolve.

The interplay between long-range atmospheric circulations is equally apparent across North Africa. Cyclonic disturbances associated with the Atlantic westerlies have been important in bringing moisture to the area north of 23 °N at times in the past. Their influence is apparent in the progressive depletion in ^{18}O in groundwaters eastwards across the Sahara. Here an important period of aquifer recharge occurred from about 40 kyr up to the Last Glacial Maximum; after a prolonged dry interval recharge did not resume until after 15 kyr. The influence of precessional hemi-cycles is evident in the timing of these events. The Ethiopian Highlands were, in contrast, most directly affected by variations in the strength of south-west Indian Ocean monsoon whose intensity in this locality was modulated over much longer timescales. Wet conditions are indicated by speleothem growth in Oman between 125 and 117 kyr, but the Arabian Peninsula remained dry thereafter until about 30 kyr, after which palaeolakes rose until about 19 kyr. After another arid interval during the deglacial period the south-west monsoon strengthened rapidly in several steps from about 12 kyr, reaching its modern intensity at 9.4 kyr. In this predominantly dry area the influence of precession on this circulation system was thus largely subordinate to those associated with the much longer glacial-interglacial cycles. The third major circulation system affecting North Africa is that of the West African monsoon, which modulates the seasonal advection of moisture from the Atlantic Ocean in the area between the tropics and 22 °N. During MIS 5d the Saharan-Sahelian boundary shifted from around 23 °N to 15 °N, with lesser latitudinal fluctuations evident thereafter. High lake levels around 40 kyr gave way to widespread aridity during the Last Glacial Maximum, which was alleviated by rapid stepwise changes towards wetter conditions at 15 kyr and 10.5 kyr (the latter following a well defined arid spike coinciding with the Younger Dryas). According to Hoelzmann et al. (this volume) these changes (and that in the reverse direction which occurred in the mid-Holocene) imply a very strong amplification of weak orbital signals by atmosphere-surface boundary feedbacks.

Tropical Africa has also yielded persuasive evidence of the important influence of biological changes (in the form of atmospheric CO_2 and CH_4 reservoirs) in reinforcing orbital effects. Particularly important among these has been the influence of precessional changes on the intensity of precipitation during the dominant (March) rainy season. However, over longer time-scales the overriding effect of the insolation forcing associated with full glacials is apparent, particularly at times when the precessional signal weakened (e.g., after about 50 kyr). An important sequence at Sacred Lake on Mt. Kenya displays changes in bulk δ^{13}C and concentrations of grass cuticles that indicate a shift towards a dominance of grasses and sedges possessing CO_2 concentrating mechanisms during the Last Glacial Maximum. These changes are consistent with lower CO_2 concentrations, higher aridity and an increased frequency of fires. But, overall, the Sacred Lake sequence is dominated by the precessional cycle and its harmonics.

Future work: recommendations of relevance to the PAGES scientific agenda

This review has drawn attention to the sheer complexity of environmental responses to both long-term climate changes induced by orbital forcing, and to more abrupt events, such as D-O and Heinrich fluctuations. A number of feed-back loops have been suggested to come into play when controlling thresholds are crossed, some of which may act extremely abruptly, such as the catastrophic melting of the Eurasian ice sheets and associated sea-ice cover, or the sudden release of ice-dammed lakes. Others are more gradual in nature, reflecting, for example, migrations of the ITCZ, or changes in continental biomass. There is clearly a need for increased quantification of the reconstructions based on palaeo-data sets, and for clarification as to which of the purported feed-back mechanisms were the most important, and why. Clearly, this is an area which would profit immensely from increased dialogue between, and joint-research agendas involving, the palaeo-data and climate modelling communities. Some of the links which appear to have been important in the PEP III transect during time stream 2, and which need to be examined in greater detail, include:

- The rate of growth and decay, and configuration of, the Eurasian ice sheets

- The size and development of ice-dammed lakes in Northern Eurasia

- The magnitude and timing of catastrophic drainage of the ice-dammed lakes

- Temperature and moisture gradients over different sectors of the PEP III transect during critical transitional periods

- Impact of the African monsoon on the Mediterranean area

- Impact of sudden influx of freshwater into the Mediterranean derived from the northern Eurasian ice-dammed lakes

- The balance of salinity exchange between the Mediterranean and the Atlantic

- The influx of Saharan dust into the Mediterranean and its likely climatic effects

- Regional differences in lake-level variations in Africa, and continental responses to precessional forcing

- The strength of the Antarctic signal on the climate of Southern Africa and on circulation in adjacent oceans

- The continental biomass in Europe and Africa and its implications for global atmospheric CO_2 levels

- The possible lead of the northern hemisphere by the low latitudes during deglaciation

Of course, the most valuable perspective to be gained would be an over-arching synthesis of how *all* of these elements interact. Here there are major questions to be faced, over such matters as: (i) prioritising the research effort into what are believed to be the *key* links and processes; (ii) the feasibility of setting up appropriate databases that can integrate multi-proxy data at the PEP III or even global scale; (iii) ensuring that future investigations lead to palaeo-reconstructions with the spatial coverage and temporal resolution required to meet the needs of the climate modelling community; and (iv) resolving disparities between palaeo-data reconstructions and climate-model simulations for key climatic episodes. These are pressing questions for PAGES, if the momentum and coherency of global palaeoclimate research are to be maintained.

References

Allen J.R., Brandt U., Brauer A., Hubberten H.W., Huntley B., Keller J., Kraml M., Mackensen A., Mingram J., Negendank J.F.W., Nowaczyk N.R., Oberhänsli H., Watts W.A., Wulf S. and Zolitschka B. 1999. Rapid environmental changes in Southern Europe during the last glacial period. Nature 400: 740–743.

Arkhipov S.A., Ehlers J., Johnson R.G. and Wright H.E. Jr. 1995. Glacial drainage towards the Mediterranean during the Middle and Late Pleistocene. Boreas 24: 196–206.

Asioli A., Lowe J.J., Trincardi F. and Oldfield F. 1999. Short-term climate changes during the Last Glacial-Holocene transition: comparison between Mediterranean and North Atlantic records. J. Quat. Sci. 14: 373–381.

Barker P. and Gasse F. 2003. New evidence for a reduced water balance in East Africa during the Last Glacial Maximum: implication for model-data comparison. Quat. Sci. Rev. 22: 823–837.

Barker P., Talbot M.R., Street-Perrott F.A., Marret J., Scourse J. and Odada E., this volume. Late Quaternary climatic variability in intertropical Africa. In: Battarbee R.W., Gasse F. and Stickley C.E. (eds), Past Climate Variability through Europe and Africa. Kluwer Academic Publishers, Dordrecht, the Netherlands, pp. 117–138.

Bar-Matthews M. and Ayalon A., this volume. Speleothems as palaeoclimate indicators, a case study from Soreq cave located in the Eastern Mediterranean Region, Israel. In: Battarbee R.W., Gasse F. and Stickley C.E. (eds), Past Climate Variability through Europe and Africa. Kluwer Academic Publishers, Dordrecht, the Netherlands, pp. 363–391.

Baumann K.-H., Lackschewitz K., Mangerud J., Spielhagen R., Wolf-Welling T., Rüdiger H. and Kassens H. 1995. Reflection of Scandinavian ice sheet fluctuations in Norwegian Sea sediments during the past 150,000 years. Quat. Res. 43: 185–187.

Blunier T., Chappellaz J., Schwander J., Dällenbach A., Stauffer B., Stocker T.F., Raynaud D., Jouzel J., Clausen H.B., Hammer C.U. and Johnsen S.J. 1998. Asynchrony of Antarctic and Greenland climate change during the last glacial period. Nature 394: 739–743.

Bond G., Broecker W., Johnsen S., McManus J., Labeyrie L., Jouzel J. and Bonani G. 1993. Correlations between climate records from North Atlantic sediments and Greenland ice. Nature 365: 143–147.

Broecker W.S. 1998. Paleocean circulation during the last deglaciation: a bipolar seesaw. Paleoceanography 13: 119–121.

Broecker W.S. 2003. Does the trigger for abrupt climate change reside in the ocean or the atmosphere? Science 300: 1519–1522.

Burns S.J., Matter A., Frank N. and Mangini A. 1998. Speleothem-based palaeoclimatic record from Northern Oman. Geology 26: 499–502.

Cacho I., Grimalt J.O., Sierro F.J., Shackleton N.J. and Canals M. 2000. Evidence for enhanced Mediterranean thermohaline circulation during rapid climatic coolings. Ear. Planet. Sci. Lett. 183: 417–429.

Cayre O., Lancelot Y., Vincent E. and Hall M.A. 1999. Paleoceanographic reconstructions from planktonic foraminifera of the Iberian margin: temperature, salinity and Heinrich Events. Paleoceanography 14: 384–396.

Charles C.D., Lynch-Stieglitz J., Ninnemann U.S. and Fairbanks R.G. 1996. Climate connections between the hemispheres revealed by deep-sea sediment core/ice core correlations. Ear. Planet. Sci. Lett. 142: 19–27.

Clark P.U., Marshall S.J., Clarke G.K.C., Hostetler S.W., Licciardi J.M. and Teller J.T. 2001. Freshwater forcing of abrupt climate change during the last glaciation. Science 293: 283–287.

Coope G.R., Lemdahl G., Lowe J.J. and Walkling A. 1998. Temperature gradients in Northern Europe during the last glacial-Holocene transition (14–9 ^{14}C ka BP) interpreted from coleopteran assemblages. J. Quat. Sci. 13: 419–434.

deMenocal P., Ortiz J., Guilderson T., Adkins J., Sarnthein M., Baker L. and Yarusinsky M. 2000. Abrupt onset and termination of the African Humid Period: rapid climate responses to gradual insolation forcing. Quat. Sci. Rev. 19: 347–361.

deMenocal P., Ruddiman W.F. and Pokras E.M. 1993. Influences of high-latitude and low-latitude processes on African terrestrial climate — Pleistocene eolian records from equatorial Atlantic-Ocean Drilling Program Site-663. Paleoceanography 8: 209–242.

Edmunds W.M., Fellman E. and Baha Goni I. 1999. Environmental change, lakes and groundwater in the Sahel of Northern Nigeria. J. Geol. Soc. Lond. 156: 345–356.

Frumkin A., Ford D.C. and Schwarcz H.P. 1999. Continental oxygen isotopic record of the last 170,000 years in Jerusalem. Quat. Res. 51: 317–327.

Gasse F. 2000. Hydrological changes in the African tropics since the Last Glacial Maximum. Quat. Sci. Rev. 19: 189–211.

Gasse F., Téhet R., Durand A., Gilbert E. and Fontes J.-C. 1990. The arid-humid transition in the Saraha and the Sahel during the last deglaciation. Nature 346: 141–146.

Gasse F. and van Campo E. 2001. Late Quaternary environmental changes from a pollen and diatom record in the southern tropics (Lake Tritrivakely, Madagascar). Palaeogeogr. Palaeoclim. Palaeoecol. 167: 287–308.

Grosswald M.G. 1980. Late Weichselian ice sheets of Northern Eurasia. Quat. Res. 13: 1–32.

Johnsen S.J., Clausen H.B., Dansgaard W., Fuhrer K., Stauffer B. and Steffensen J.P. 1992. Irregular glacial interstadials recorded in a new Greenlanc ice core. Nature 359: 311–313.

Helmens K.F., Räsänen M.E., Johansson P.W., Jungner H. and Korjonen K. 2000. The Last Interglacial-Glacial cycle in NE Fennoscandia: a nearly continuous record from Sokli (Finnish Lapland). Quat. Sci. Rev. 19: 1605–1623.

Hoelzmann P., Gasse F., Dupont L.M., Salzmann U., Staubwasser M., Leuschner D.C. and Sirocko F., this volume. Palaeoenvironmental changes in the arid and subarid belt (Sahara-Sahel-Arabian Peninsula) from 150 kyr to present. In: Battarbee R.W., Gasse F. and Stickley C.E. (eds), Past Climate Variability through Europe and Africa. Kluwer Academic Publishers, Dordrecht, the Netherlands, pp. 219–256.

Holmgren K., Lee-Thorp J.A., Cooper G., Lundblad K., Partridge T.C., Scott L., Sithaldeen R., Talma A.S. and Tyson P.D. 2003. Persistent millennial-scale variability over the past 25 thousand years in Southern Africa. Quat. Sci. Rev. 22: 2311–2326.

Huijzer A.S. and Vandenberghe J. 1998. Climatic reconstruction of the Weichselian Pleniglacial in North-Western and Central Europe. J. Quat. Sci. 13: 391–417.

Kallel N., Duplessy J., Labeyrie L., Fontugne M. and Paterne M., this volume. Mediterranean Sea palaeohydrology and pluvial periods during the Late Quaternary. In: Battarbee R.W., Gasse F. and Stickley C.E. (eds), Past Climate Variability through Europe and Africa. Kluwer Academic Publishers, Dordrecht, the Netherlands, pp. 307–324.

Kanfoush S.L., Hodell D.A., Kanfoush S.L., Hodell D.A., Charles C.D., Guilderson T.P., Mortyn P.G. and Ninnemann U.S. 2000. Millennial-scale instability of the Antarctic ice sheet during the Last Deglaciation. Science 288: 1815–1818.

Kühl N., Gebhardt C., Litt T. and Hense A. 2002. Probability density functions as botanical-climatological transfer functions for climatic reconstruction. Quat. Res. 58: 381–392.

Kutzbach J.E. and Guetter P.J. 1986. The influence of changing orbital parameters and surface boundary conditions on climatic simulations for the past 18,000 years. Journal of Atmospheric Sciences 43: 1726–1759.

Leuschner D.C. and Sirocko F. 2000. The low-latitude monsoon climate during Dansgaard-Oescheger cycles and Heinrich Events. Quat. Sci. Rev. 19: 243–254.

Litt T., Brauer A., Goslar T., Merkt J., Balaga K., Müller H., Ralska-Jasiewiczowa M., Stebich M. and Negendank J.F.W. 2001. Correlation and synchronisation of Lateglacial continental sequences in Northern Central Europe based on annually-laminated lacustrine sediments. Quat. Sci. Rev. 20: 1233–1249.

Little M.G., Schneider R.R., Kroon D., Price N.B., Summerhays C. and Segl M. 1997.Trade wind forcing of upwelling seasonality and Heinrich events as a response to sub-Milankovitch climate variability. Palaeoceanography 12: 568–576.

Lowe J.J., Coope G.R., Harkness D.D., Sheldrick C. and Walker M.J.C. 1995. Direct comparison of UK temperatures and Greenland snow accumulation rates, 15–12,000 years ago. J. Quat. Sci. 10: 175–180.

Lowe J.J. and Walker M.J.C. 2000. Radiocarbon dating the last glacial-interglacial transition (ca. 14–9 ^{14}C ka BP) in terrestrial and marine records: the need for new quality assurance protocols. Radiocarbon 42: 53–68.

Magri D. and Parra I. 2002. Late Quaternary W-Mediterranean pollen records and African winds. Ear. Planet. Sci. Lett. 200: 401–408.

Magri D., Kallel N. and Narcisi B., this volume. Palaeoenvironmental changes in the Mediterranean region 250–10 kyr BP. In: Battarbee R.W., Gasse F. and Stickley C.E. (eds), Past Climate Variability through Europe and Africa. Kluwer Academic Publishers, Dordrecht, the Netherlands, pp. 325–341.

Mangerud J., Lovlie R., Gulliksen S., Hufthammer A.-K., Larsen E. and Valen V. 2003. Paleomagnetic correlations between Scandinavian ice-sheet fluctuations and Greenland Dansgaard-Oeschger events, 45,000–25,000 yr BP. Quat. Res. 59: 213–222.

Marret F., Scourse J., Versteegh G., Jansen J.H.F. and Schneider R. 2001. Integrated marine and terrestrial evidence for abrupt Congo River palaeodischarge fluctuations during the last deglaciation. J. Quat. Sci. 16: 761–766.

Moreno A., Targarona J., Henderiks J., Canals M., Freudenthal T. and Meggers H. 2001. Orbital forcing of dust supply to the North Canary basin over the last 250 kyr. Quat. Sci. Rev. 20: 1327–1340.

Moreno A., Cacho I., Canals M., Prins M.A., Sánchez-Goñi M.-F., Grimalt J.O. and Weltje G.J. 2002. Saharan dust transport and high-latitude glacial climatic variability: the Alboran Sea record. Quat. Res. 58: 318–328.

Nakagawa T., Kitagawa H., Yasuda Y., Tarasov P., Nishida P., Gotanda K., Sawai Y. and YRCP members. 2003. Asynchronous climate changes in the North Atlantic and Japan during the Last Termination. Science 299: 688–691.

Huijzer A.S. and Vandenberghe J. 1998. Climatic reconstruction of the Weichselian Pleniglacial in North-Western and Central Europe. J. Quat. Sci. 13: 391–417.

Kallel N., Duplessy J., Labeyrie L., Fontugne M. and Paterne M., this volume. Mediterranean Sea palaeohydrology and pluvial periods during the Late Quaternary. In: Battarbee R.W., Gasse F. and Stickley C.E. (eds), Past Climate Variability through Europe and Africa. Kluwer Academic Publishers, Dordrecht, the Netherlands, pp. 307–324.

Kanfoush S.L., Hodell D.A., Kanfoush S.L., Hodell D.A., Charles C.D., Guilderson T.P., Mortyn P.G. and Ninnemann U.S. 2000. Millennial-scale instability of the Antarctic ice sheet during the Last Deglaciation. Science 288: 1815–1818.

Kühl N., Gebhardt C., Litt T. and Hense A. 2002. Probability density functions as botanical-climatological transfer functions for climatic reconstruction. Quat. Res. 58: 381–392.

Kutzbach J.E. and Guetter P.J. 1986. The influence of changing orbital parameters and surface boundary conditions on climatic simulations for the past 18,000 years. Journal of Atmospheric Sciences 43: 1726–1759.

Leuschner D.C. and Sirocko F. 2000. The low-latitude monsoon climate during Dansgaard-Oescheger cycles and Heinrich Events. Quat. Sci. Rev. 19: 243–254.

Litt T., Brauer A., Goslar T., Merkt J., Balaga K., Müller H., Ralska-Jasiewiczowa M., Stebich M. and Negendank J.F.W. 2001. Correlation and synchronisation of Lateglacial continental sequences in Northern Central Europe based on annually-laminated lacustrine sediments. Quat. Sci. Rev. 20: 1233–1249.

Little M.G., Schneider R.R., Kroon D., Price N.B., Summerhays C. and Segl M. 1997.Trade wind forcing of upwelling seasonality and Heinrich events as a response to sub-Milankovitch climate variability. Palaeoceanography 12: 568–576.

Lowe J.J., Coope G.R., Harkness D.D., Sheldrick C. and Walker M.J.C. 1995. Direct comparison of UK temperatures and Greenland snow accumulation rates, 15–12,000 years ago. J. Quat. Sci. 10: 175–180.

Lowe J.J. and Walker M.J.C. 2000. Radiocarbon dating the last glacial-interglacial transition (ca. 14–9 ^{14}C ka BP) in terrestrial and marine records: the need for new quality assurance protocols. Radiocarbon 42: 53–68.

Magri D. and Parra I. 2002. Late Quaternary W-Mediterranean pollen records and African winds. Ear. Planet. Sci. Lett. 200: 401–408.

Magri D., Kallel N. and Narcisi B., this volume. Palaeoenvironmental changes in the Mediterranean region 250–10 kyr BP. In: Battarbee R.W., Gasse F. and Stickley C.E. (eds), Past Climate Variability through Europe and Africa. Kluwer Academic Publishers, Dordrecht, the Netherlands, pp. 325–341.

Mangerud J., Lovlie R., Gulliksen S., Hufthammer A.-K., Larsen E. and Valen V. 2003. Paleomagnetic correlations between Scandinavian ice-sheet fluctuations and Greenland Dansgaard-Oeschger events, 45,000–25,000 yr BP. Quat. Res. 59: 213–222.

Marret F., Scourse J., Versteegh G., Jansen J.H.F. and Schneider R. 2001. Integrated marine and terrestrial evidence for abrupt Congo River palaeodischarge fluctuations during the last deglaciation. J. Quat. Sci. 16: 761–766.

Moreno A., Targarona J., Henderiks J., Canals M., Freudenthal T. and Meggers H. 2001. Orbital forcing of dust supply to the North Canary basin over the last 250 kyr. Quat. Sci. Rev. 20: 1327–1340.

Moreno A., Cacho I., Canals M., Prins M.A., Sánchez-Goñi M.-F., Grimalt J.O. and Weltje G.J. 2002. Saharan dust transport and high-latitude glacial climatic variability: the Alboran Sea record. Quat. Res. 58: 318–328.

Nakagawa T., Kitagawa H., Yasuda Y., Tarasov P., Nishida P., Gotanda K., Sawai Y. and YRCP members. 2003. Asynchronous climate changes in the North Atlantic and Japan during the Last Termination. Science 299: 688–691.

Pailler D. and Bard E. 2002. High frequency palaeoceanographic changes during the past 140000 yr recorded by the organic matter in sediments of the Iberian margin. Palaeogeogr. Palaeoclim. Palaeoec. 181: 431–452.

Partridge T.C. 2002. Were Heinrich Events forced from the southern hemisphere? South African Journal of Science 98: 43–46.

Partridge T.C., Scott L. and Schneider R.R., this volume. Between Agulhas and Benguela: responses of Southern African climates of the Late Pleistocene to current fluxes, orbital precession and extent of the Circum-Antarctic vortex. In: Battarbee R.W., Gasse F. and Stickley C.E. (eds), Past Climate Variability through Europe and Africa. Kluwer Academic Publishers, Dordrecht, the Netherlands, pp. 45–68.

Peterson B.J., Holmes R.M., McClelland J.W., Vörösmarty C.J., Lammers R.B., Shiklomanov A.I., Shiklomanov I.A. and Rahmstorf S. 2002. Increasing river discharge to the Arctic Ocean. Science 298: 2171–2173.

Prell W.L. and Kutzbach J.E. 1987. Monsoon variability over the past 150 000 years. J. Geophys. Res. 92: 8411–8425.

Prell W.L. and van Campo E. 1986. Coherent response of Arabian Sea upwelling and pollen transport to late Quaternatary monsoonal winds. Nature 323: 526–528.

Roucoux K.H., Shackleton N.J., de Abreu L., Schönfeld J. and Tzedakis P.C. 2001. Combined marine proxy and pollen analyses reveal rapid Iberian vegetation response to North Atlantic millennial-scale climatic oscillations. Quat. Res. 56: 128–132.

Ruddiman W.F. 2003. Orbital insolation, ice volume, and greenhouse gases. Quat. Sci. Rev. 22: 1597–1629.

Saarnisto M. and Lunkka J.P., this volume. Climate variability during the last interglacial-glacial cycle in NW Eurasia. In: Battarbee R.W., Gasse F. and Stickley C.E. (eds), Past Climate Variability through Europe and Africa. Kluwer Academic Publishers, Dordrecht, the Netherlands, pp. 443–464.

Sánchez-Goñi M.F., Eynaud F., Turon J.-L. and Shackleton N.J. 1999. High resolution palynological record off the Iberian margin: direct land-sea correlation for the Last Interglacial complex. Ear. Planet. Sci. Lett. 171: 123–137.

Sánchez-Goñi M.F., Turon J.-L., Eynaud F., Shackleton N.J. and Cayre O. 2002. Direct land/sea correlation of the Eemian and its comparison with the Holocene: a high-resolution palynological record off the Iberian margin. Geologie en Mijnbouw 79: 345–354.

Schneider R.R., Müller P.J., Rulhand G., Meinecke G., Schmidt H. and Wefer G. 1996. Late Quaternary surface temperatures and productivity in the east-equatorial South Atlantic: response to changes in trade/monsoon wind forcing and surface water advection. In: Wefer G., Berger W.H., Siedler G. and Webb D.J. (eds), The South Atlantic: Present and Past Circulation. Springer-Verlag, Berlin, pp. 527–551.

Schulz H., von Rad U. and Erlenkauser H. 1998. Correlation between Arabian Sea and Greenland climate oscillations of the past 110,000 years. Nature 393: 54–57.

Shackleton N.J. 2000. The 100,000-year ice-age cycle identified and found to lag temperature, carbon dioxide and orbital eccentricity. Science 289: 1897–1902.

Sonzogni C., Bard E. and Rostek E. 1998. Tropical sea-surface temperatures during the Last Glacial period: a view based on alkenones in Indian Ocean sediments. Quat. Sci. Rev. 17: 1185–1201.

Street-Perrott F.A. and Perrott R.A. 1993. Holocene vegetation, lake levels and climate of Africa. In: Wright H.E., Kutzbach J.E., Webb III T., Ruddiman W.F., Street-Perrott F.A. and Bartlein P.J. (eds), Global Climates Since the Last Glacial Maximum. University of Minnesota Press, Minneapolis, pp. 318–356.

Teller J.T., Leverington D.W. and Mann J.D. 2002. Freshwater outbursts to the oceans from glacial Lake Agassiz and their role in climate change during the last deglaciation. Quat. Sci. Rev. 21: 879–888.

Thiede J., Bauch H.A., Hjort C. and Mangerud J. (eds) 2001. The Late Quaternary Stratigraphy and Environments of Northern Eurasia and the Adjacent Arctic Seas — New Contributions from QUEEN. Glob. Plan. Chan., Special Issue 31 (1-4): 1–474.

Trauth M.H., Deino A.L. and Strecker M.R. 2001. Response of the East African climate to orbital forcing during the last interglacial (13–117 ka) and the early last glacial (117–60 ka). Geology 29: 499–502.

Trauth M.H., Deino A.L., Bergner A.G.N. and Strecker M.R. 2003. East African climate change and orbital forcing during the last 175 kyr BP. Ear. Planet. Sci. Lett. 206: 297–313.

Tzedakis P.C. 2003. Timing and duration of Last Interglacial conditions in Europe: a chronicle of a changing chronology. Quat. Sci. Rev. 22: 763–768.

Tzedakis P.C., Andrieu V., de Beaulieu J.-L., Crowhurst S., Follieri M., Hooghiemstra H., Magri D., Reille M., Sadori L., Shackleton N.J. and Wijmstra T.A. 1997. Comparison of terrestrial and marine records of changing climate of the last 500,000 years. Ear. Planet. Sci. Lett. 150: 171–176.

Vandenberghe J., Lowe J.J., Coope R., Litt T. and Zoller L., this volume. Climatic and environmental variability in the Mid-Latitude Europe sector during the last interglacial-glacial cycle. In: Battarbee R.W., Gasse F. and Stickley C.E. (eds), Past Climate Variability through Europe and Africa. Kluwer Academic Publishers, Dordrecht, the Netherlands, pp. 393–416.

Vandenberghe J. and Pissart A. 1993. Permafrost changes in Europe during the last glacial. Permafrost and Periglacial Processes 4: 121–135.

Verschuren D., Briffa K.R., Hoelzmann P., Barber K., Barker P., Scott L., Snowball I., Roberts N. and Battarbee R.W., this volume. Holocene climate variability in Europe and Africa: a PAGES-PEP III timestream 1 synthesis. In: Battarbee R.W., Gasse F. and Stickley C.E. (eds), Past Climate Variability through Europe and Africa. Kluwer Academic Publishers, Dordrecht, the Netherlands, pp. 567–582.

Walker M.J.C. 1995. Climatic changes in Europe during the last glacial-interglacial transition. Quaternary International 28: 63–76.

Witte H.J.L., Coope G.R., Lemdahl G. and Lowe J.J. 1998. Regression coefficients of thermal gradients in Northwestern Europe during the last glacial-Holocene transition using beetle MCR data. J. Quat. Sci. 13: 435–446.

List of all abbreviations

ACR: Antarctic Cold Reversal.

AGCM: Atmospheric General Circulation Model.

AMJJ: April, May, June, July.

AMS: Accelerator Mass Spectrometry.

AOGCM: Atmosphere-Ocean General Circulation Model.

AOVGCM: Atmosphere-Ocean-Vegetation General Circulation Model.

BIOME1: Biome Model No 1.

BP: Before Present.

BSW: Bog Surface Wetness.

CALIB: A computer program used to calibrate radiocarbon dates.

COHMAP: Cooperative Holocene Mapping Project.

CI: Continental Intercalaire.

CLIMAP: Climate: Long Range Investigation, Mapping and Prediction.

CLIVAR: Climate Variability and Predictability (World Climate Research Programme).

CSIRO: Commonwealth Scientific and Industrial Research Organisation.

CSM: National Centre for Atmospheric Research (NCAR), Boulder, USA climate model.

CT: Complexe Terminal.

DCA: Detrended Correspondence Analysis.

DJF: December, January, February.

D/O: Dansgaard-Oeschger Event.

EBM: Energy Balance Model.

ECBilt: Dutch intermediate complexity climate model.

ECHAM/LSG: Climate model developped at the Max Planck Institute in Hamburg (Germany).

ECHO-G: European Centre HAMburg/ Hamburg Ocean Primitive Equation model — Global version.

ELA: Equilibrium Line Altitude.

ELDP: European Lake Drilling Programme (European Science Foundation).

EM: Eastern Mediterranean.

EMICs: Earth systems Models of Intermediate Complexity.

ENSO: El Niño Southern Oscillation.

EPECC: European Palaeo-Environmental Climate and Circulation project.

EPD: European Pollen Database.

EPILOG: Environmental Processes of the Ice age: Land, Ocean, Glaciers.

EU: European Union.

FAO: Food and Agriculture Organsiation of the United Nations.

FISSILVA: Dynamics of forest tree biodiversity: linking genetic, paleogenetic and plant historical approaches.

GCM: General Circulation Model.

GDD: Growing Degree Days.

GDP: Gross Domestic Product.

GHG: Greenhouse Gas.

GISP: Greenland Ice-Sheet Project.

GLCC: Global Land Cover Classification.

GLSDB: Global Lake Status Data Base.

GMWL: Global Meteoric Water Line.

GPD: Global Pollen Database.

GPS: Global Positioning System.

GRIP: Greenland Ice-core Project.

GS-1: Greenland Stadial 1.

GSC: Geological Survey of Canada.

H: Heinrich.

HADCM3: United Kingdom Hadley Centre General Circulation Model.

HE: Heinrich Events.

HO: Heinrich Event 0 (=Younger Dryas).

HOLIVAR: Holocene Climate Variability (European Science Foundation).

HTM: Holocene Thermal Maximum.

HYDRA: Hydrological model developed at the University of Wisconsin.

IAEA: International Atomic Energy Agency.

IAEA GNIP: International Atomic Energy Agency Global Network for Isotopes in Precipitation.

ICP-MS: Inductivity Coupled Plasma — Mass Spectrometry.

IDEAL: International Decade for the East African Lakes.

IGBP-DIS AVHRR: International Geosphere Biosphere Programme-advanced Very High Resolution Radiometer.

IGBP: International Geosphere-Biosphere Programme.

IGCP: International Geological Correlation Programme.

IMAGES: International Marine Past Global Changes Study.

INQUA: International Union for Quaternary Studies.

INTIMATE: INTegration of Ice-Core MArine and TErrestrial records of the Last Termination.

IPCC: Intergovernmental Panel on Climate Change.

IRD: Ice Rafted Debris or Institut de Recherche pour le Développement (France).

ISOW: Iceland-Scotland Overflow Water.

ITCZ: Intertropical Convergence Zone.

JJA: June, July, August.

kyr BP: 1000 years before present.

LIA: Little Ice Age.

LGM: Last Glacial Maximum.

LMC: Low Magnesium Calcite.

LRC: Limnological Research Center.

LSR: Linear Sedimentation Rate.

MAAT: Mean Annual Air Temperature.

MAM: March, April, May.

MAR: Mass Accumulation Rate.

MCA: Medieval Climate Anomaly.

MCO: Medieval Climatic Optimum.

MCR: Mutual Climatic Range.

MHO: Mid-Holocene Climatic Optimum.

MIJJ: May, June, July.

MIS: Marine Isotope Stage.

MWL: Meteoric Water Line.

MMWL: Mediterranean Meteoric Water Line.

MSA: Middle Stone Age.

MTCO: Mean Temperature of the Coldest Month.

MWL: Meteoric Water Line.

MWP: Medieval Warm Period.

MWP-1b: Melt-water Pulse 1b.

NADW: North Atlantic Deep Water.

NAP: Non-arboreal Pollen.

NATO: North Atlantic Treaty Organisation.

NAO: North Atlantic Oscillation.

ND: Naivasha Drought.

NDJF: November, December, January, February.

NGRT: Noble Gas Recharge Temperatures.

NEAP: North East Atlantic Palaeoceanography and Climate Change project.

NERC: Natural Environment Research Council.

NIM: Not in Model.

NIST: National Institute of Standards and Technology.

OAGCM: Ocean-Atmosphere General Circulation Model.

OAVGCM: Ocean-Atmosphere-Vegetation General Circulation Model.

OGCM: Oceanic General Circulation Model.

OSL: Optically Stimulated Luminescence.

PAGES: Past Global Changes (IGBP core project).

PANGAEA: PalaeoNetwork for Geological And Environmental datA (information system, AWI, Bremerhaven, Germany).

PASH: Palaeoclimates of the Southern Hemisphere.

PBO: Preboreal Oscillation.

PCA: Principal Components Analysis.

P-E: Precipitation — Evaporation.

PDB: Pee-Dee Belemnite (reference for oxygen and carbon isotopic measurements in carbonates).

PDR: People's Democratic Republic (of Congo).

PEP: Pole-Equator-Pole Transect (PAGES).

PFT: Plant Functional Type.

PMIP: Paleoclimate Modelling Intercomparison Project.

pmc: Percent Modern Carbon.

PMM: Post-Middle Miocene.

PROBE: PRoto-Ocean Basin Evolution.

QM: Mixed Oak Forest (*Quercus, Tilia, Ulmus, Carpinus*).

QUEEN: Quaternary Environment of the Eurasian North.

SEPRO: SEquel to PRObe.

SMOW: Standard Marine Ocean Water (reference for oxygen and carbon isotopic measurements in water).

SPECMAP: Spectral Mapping Project Timescale.

SPEP: Speleothem Pole-Equator-Pole Transect.

SSS: Sea-Surface Salinity.

SST: Sea-Surface Temperatures.

TA: Tropical Africa.

THC: Thermohaline Circulation.

TILIA: A computer program user for plotting pollen diagrams.

TIMS: Thermal Ionisation Mass Spectrometry.

TOC: Total Organic Carbon.

TP: Tropical Pacific.

TP: Total Phosphorus.

TS1: Time Stream 1.

UKMO: United Kingdom Meteorological Office.

UNESCO: United Nations Educational, Scientific and Cultural Organisation.

WHO: World Health Organisation.

WLM: Walton Moss.

WMO: World Meteorological Organisation.

XRF: X-ray Fluorescence.

YD: Younger Dryas.

YD-PB: Younger Dryas — Preboreal Transition.

GENERAL INDEX

ablation 454, 595
Acacia albida 150
Acacia karroo 80, 82
Acrostichum aureu 202, 210
Adansonia digitata 149
Adelomycetes 211
aeolian particles 49, 57, 211, 224, 228,
231, 238, 242, 244, 281, 403, 410, 454,
572
aerosols 469, 503, 506
afforestation 329, 353, 427
African humid period 125, 228, 242, 244,
314, 500, 572, 592
Afrormosia laxiflora 208
agriculture 85, 86, 162, 237, 353, 381, 536,
540-2, 551, 554, 556, 561
Agulhas current 5, 62, 69, 77, 79, 594, 595
Akkadian civilisation 240, 381, 385, 574
albedo 127, 130, 353, 479, 497, 503, 552,
557, 570, 586
Alchornea 188, 201-214, 234, 236
Alchornea cordifolia 202, 203, 211
alkenones 120, 125, 240, 333, 382, 479
Alnus 232, 450, 453
Alstonia 205, 214
amino-acid racemization 51
amplifier lakes 143, 145, 577
Annona senegalensis 203, 211
annually laminated sediments see varves
anoxic conditions 11, 94, 97, 109, 170,
308, 315, 320
Antarctic circumpolar vortex 46, 59, 82,
83, 87, 594-6
Antarctic Cold Reversal 56, 63, 108, 127,
594
Antarctic convergence 56, 62
Antarctic ice-core 38, 55, 83, 86, 103, 594
Antarctic sea ice 46, 51, 56, 62
aquifers 221, 285, 288-9, 291-2, 556
argon 281, 282, 292, 297, 328, 331,409
aridification 56, 77, 83, 169, 174, 230-1,
242-3, 560, 574
Artocarpus communis 203
Ascomycetes 211
Asteraceae 49, 74, 77-80, 82, 87, 207
Atlantic jet stream 288

atmospheric circulation 3, 4, 9, 31, 38, 51,
56, 63, 69, 83, 266, 268, 272, 335, 427,
469, 486, 552, 585, 590, 596, 597
atmospheric composition 34, 38, 39, 118,
119, 124, 129, 190, 282, 299, 311, 353,
364, 469, 484, 557, 570, 574, 593
atmospheric methane 39, 119, 570
Aulacoseira nyassensis 105
Avicennia 209
Avicennia marina 210

Banyoro people 536, 538, 542
Barbados se-level curve 314
barium/calcium quotient 15
Benguela current 5, 34, 45-63, 69, 84, 595
Betula 425, 453, 456, 459, 460, 480
Betula pubescens ssp *tortuosa* 480, 570
biome 59, 69, 70, 81, 184, 243, 348, 350,
497, 498, 506, 524, 527, 588
BIOME 6000 517, 524, 527
biomisation 347, 348-351, 524
biosphere 23, 243, 331, 497, 527, 577
biostratigraphy 380, 450
blue-green algae 105
Bolling-Allerød 103, 405
Bond Cycles 52, 57, 62, 394, 573
Borassus aethiopum 203, 211
boreal taiga 454
Bosquiea angolensis 205
Brachystegia 207
Bridelia ferruginea 203, 205, 211
Bronze Age 344, 351, 354, 355, 357
Buxus 397
Byzovaya dwelling, Russia 450

C3 vegetation 60-61, 69, 74, 75, 78, 79,
86, 123, 281, 364, 372, 383-5
C4 vegetation 60-61, 69, 72, 75-80, 82, 85-
7, 123, 281, 364, 381, 383
Caesalpiniaceae 188
Calamus deëratus 205, 214
calcite: 14, 169, 188, 224, 364, 367, 384,
428, 590
  endogenic 109
  low magnesium 366, 367, 372, 384
calcite-water fractionation 372, 381
calcium carbonate 366, 372

611

GEOGRAPHICAL INDEX

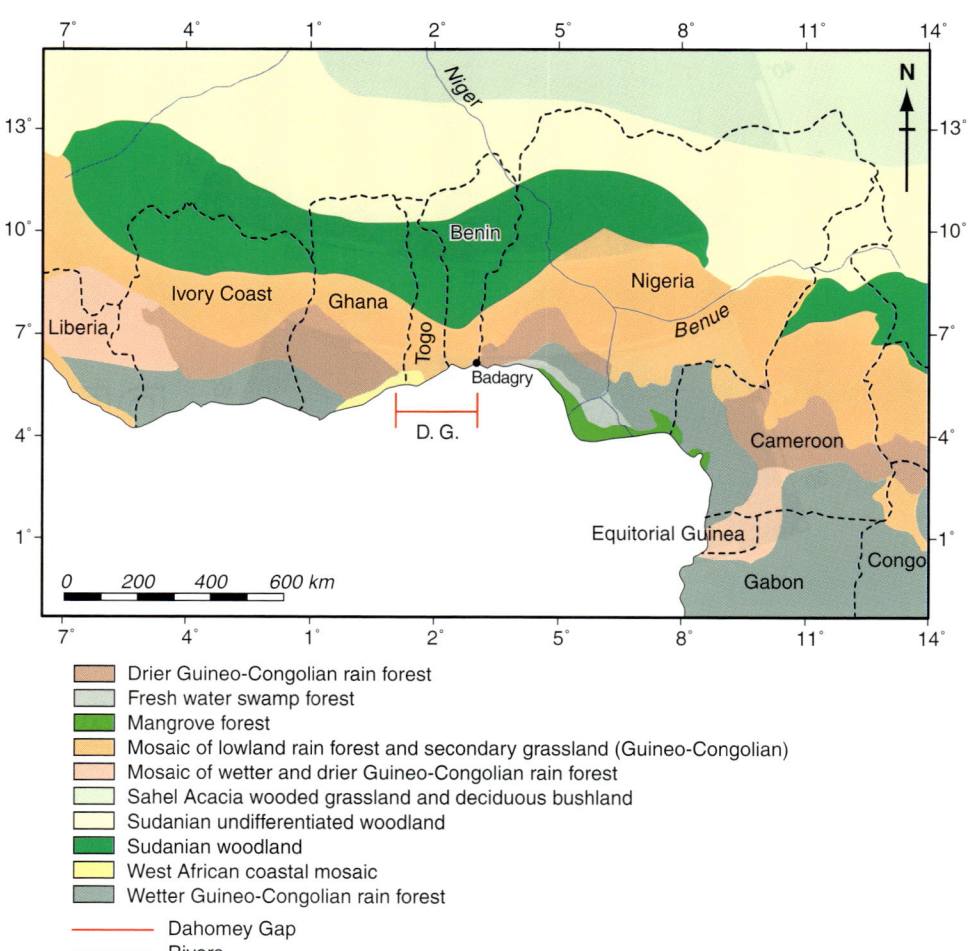

Plate 2. Vegetation map of West Africa (modified after Lawson (1986)).

GEOGRAPHICAL INDEX

APPENDIX: Color Version of Plates 1-11

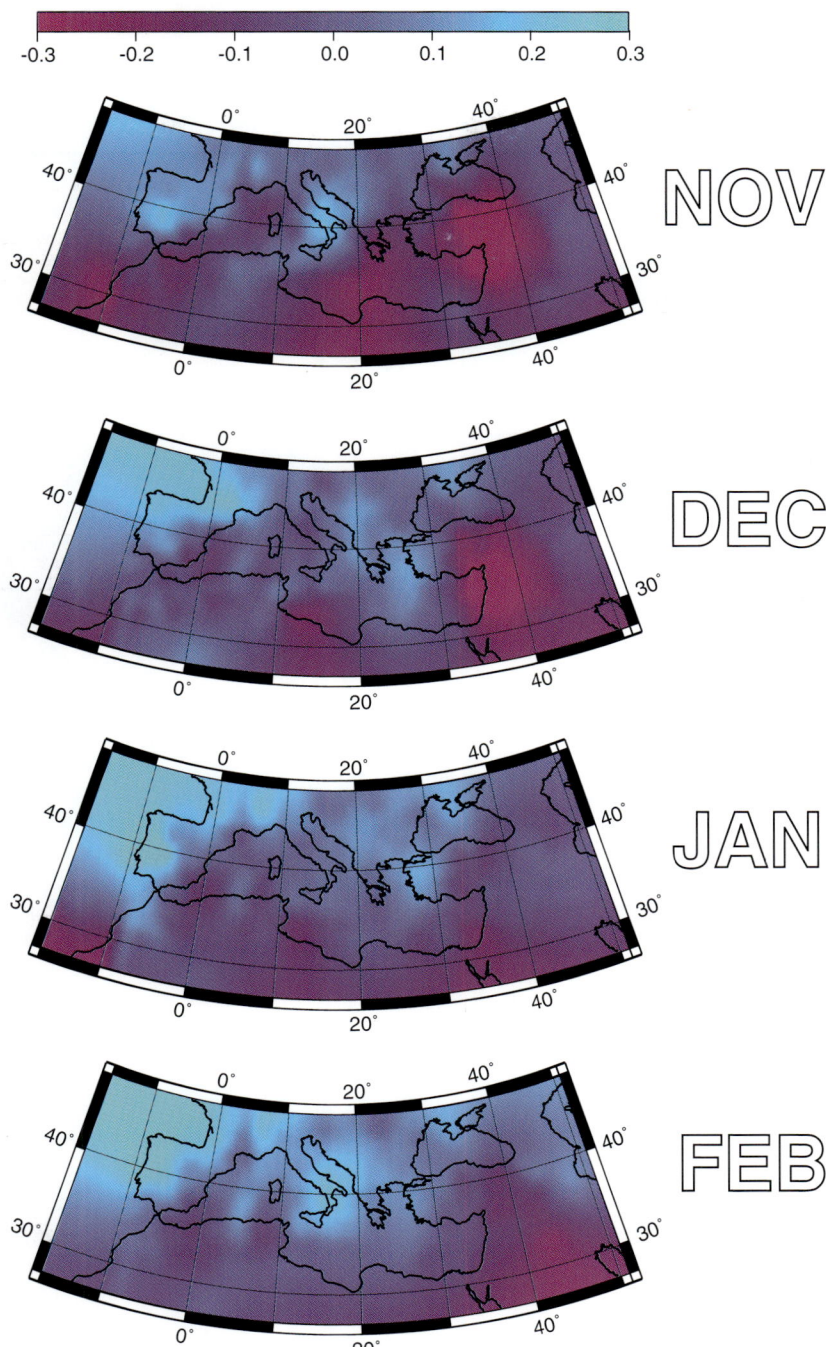

Plate 1. Long-term linear trend in Mediterranean precipitation (mm/month/yr) during the 136 year period (1855 to 1990). Note the drying trend (purple) for all four-winter months in the Near East compared to the increase (blue) for Western Iberia. The null change line runs approximately through Gibraltar-Sicily-Greece-Black Sea. Note how the same east-west teleconnection seesaw is found for precipitation as in the interannual variation of Table 1. (Albers equal area projection).

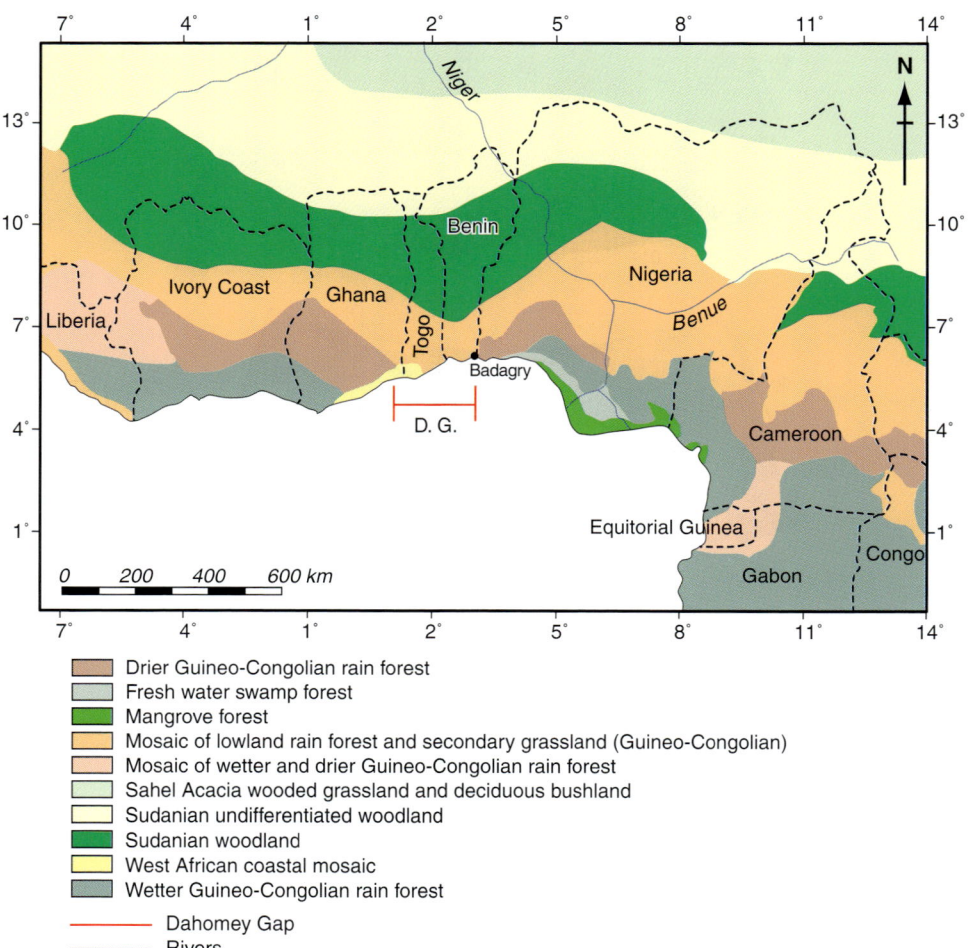

Plate 2. Vegetation map of West Africa (modified after Lawson (1986)).

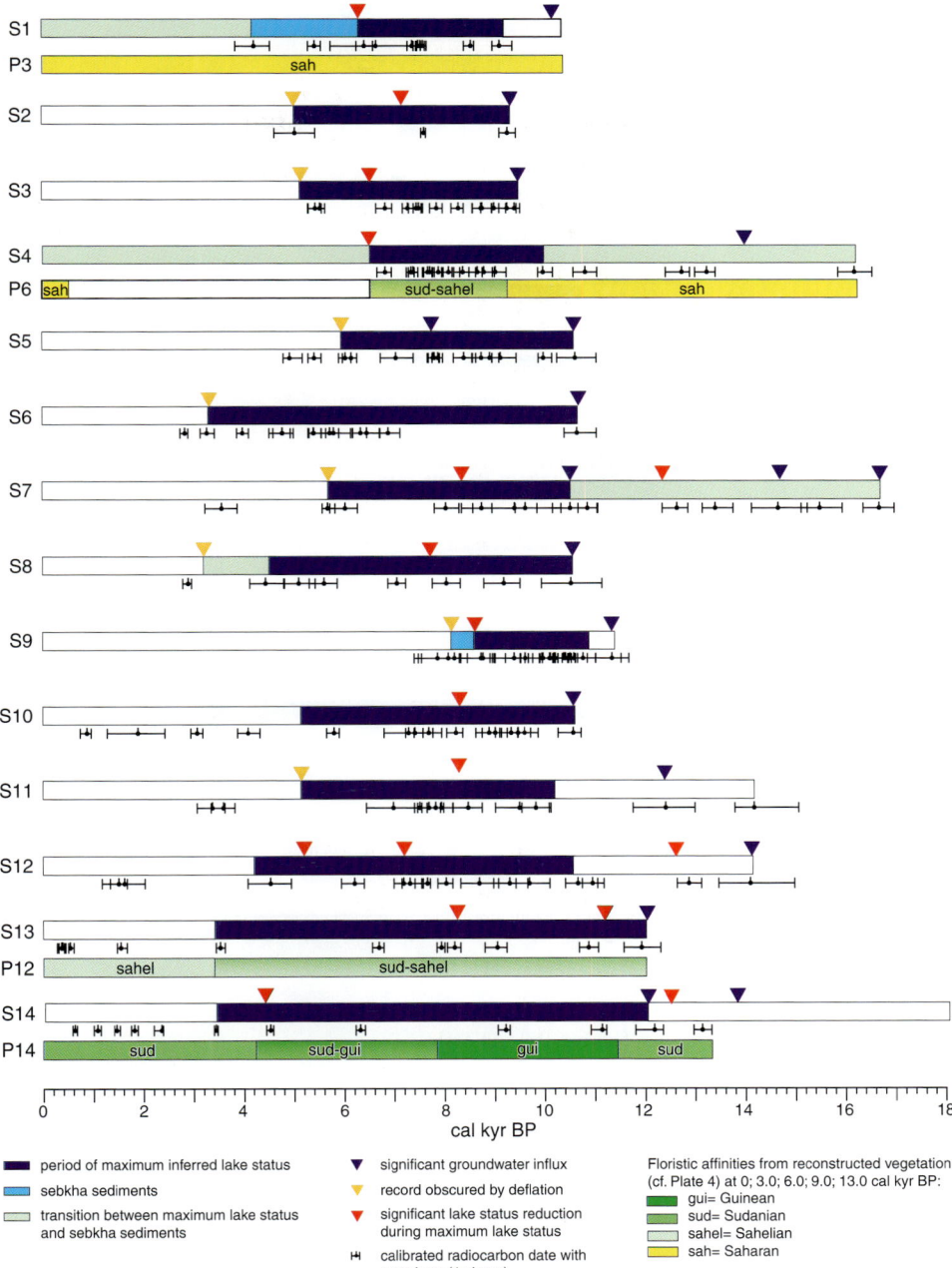

Plate 3a. Observed surface-near palaeohydrological changes over the last 18.0 cal. kyr for selected sites of the Western and Central Sahara/Sahel. Shaded areas represent periods of maximum inferred moisture (dark blue), transition between maximum lake status and sebkha environments (-pale grey,) and sebkha environments (mid blue). Triangles show centre-points of major events: significant groundwater influx (solid dark blue), lake-level reductions (red), and the age of truncation of the record by deflation (solid yellow). Radiocarbon dates from palaeolake records were calibrated according to Stuiver et al. (1998) using CALIB 4.3 (Stuiver and Reimer 1993). The midpoint of the 1 sigma range taken from the probability method is shown and error bars define the age range. Floristic affinities (cf. Plate 4) are also shown for selected palaeolake sites.

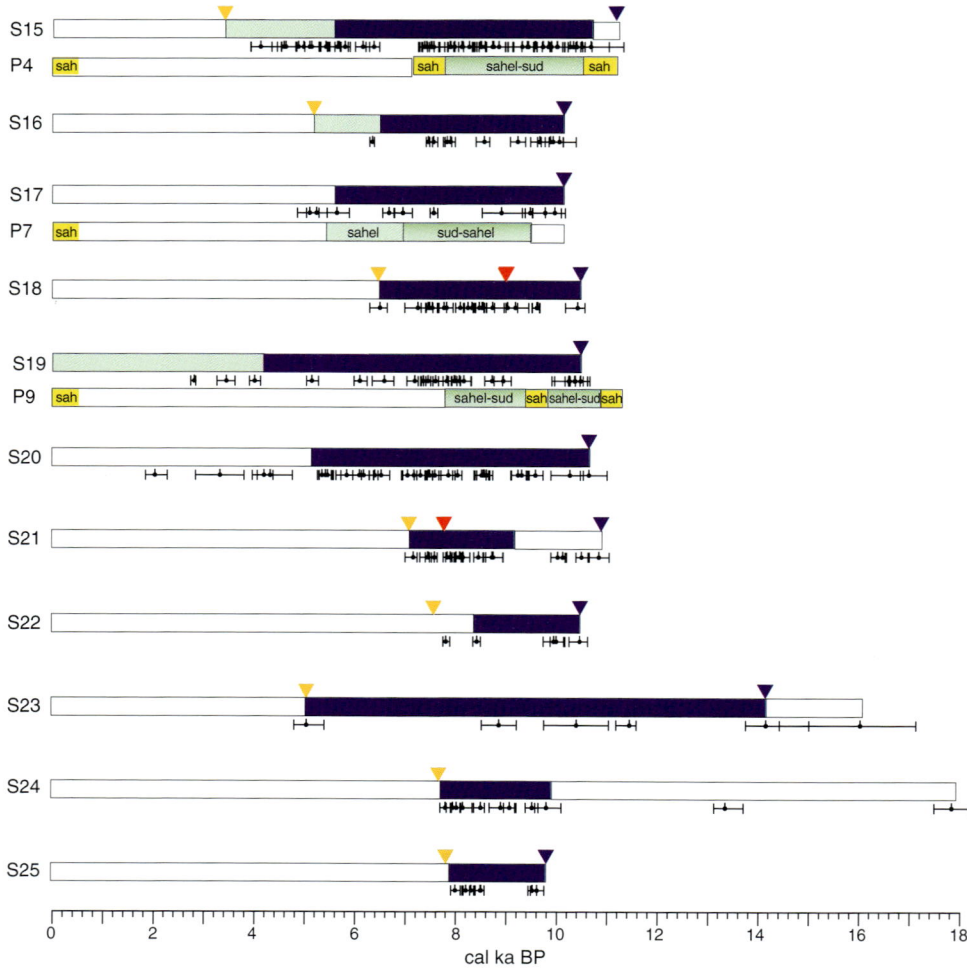

Plate 3b. Same as a) but for Eastern Sahara/Sahel and Arabian Peninsula.

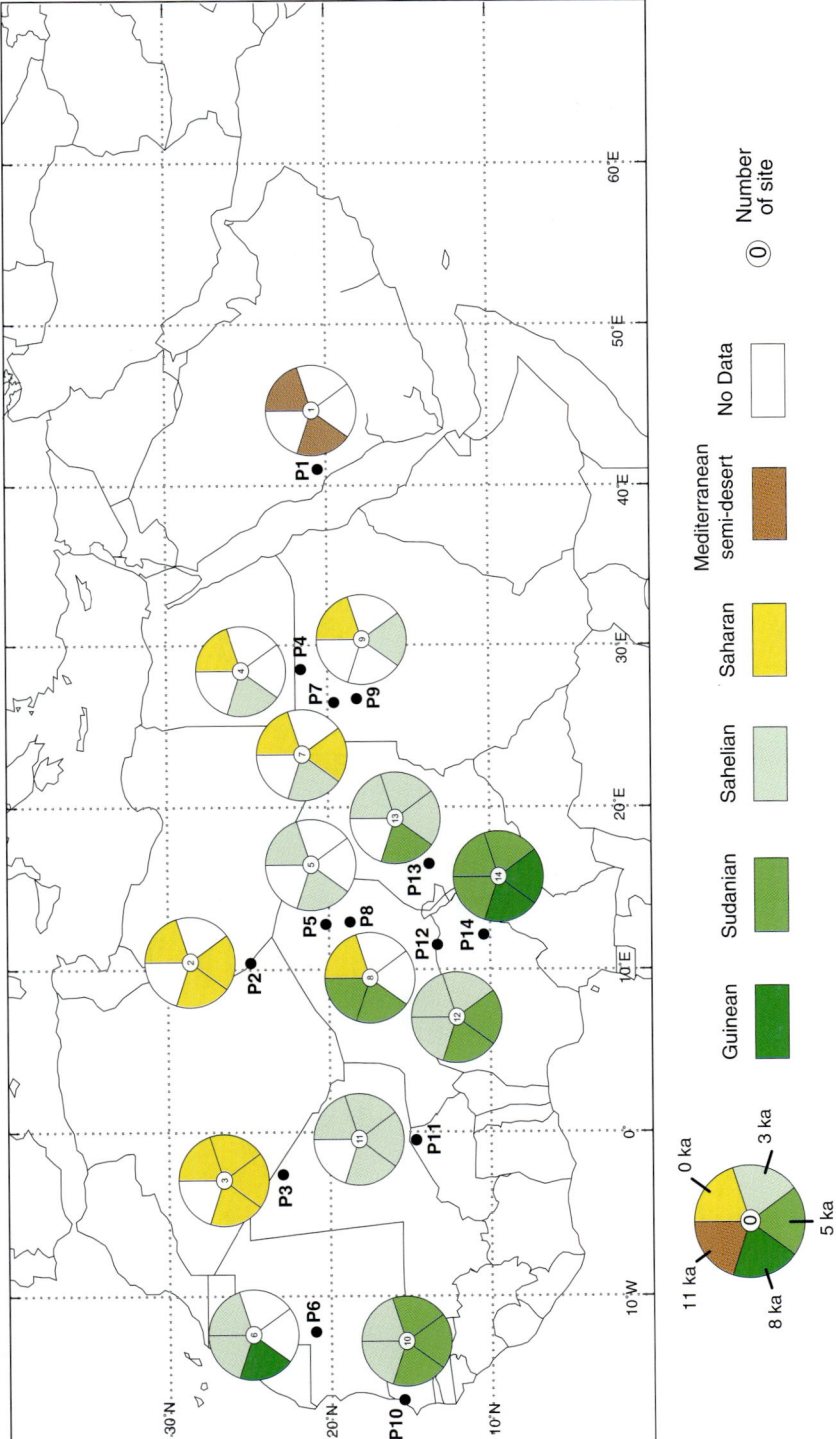

Plate 4. Floristic affinities of reconstructed vegetation from selected terrestrial pollen diagrams for the time slices 0 kyr, 3.0 ^{14}C kyr (ca. 3.2 cal. kyr); 5.0 ^{14}C kyr (ca. 5.7 cal. kyr), 8.0 ^{14}C kyr (ca. 9.0 cal. kyr), and 11.0 ^{14}C kyr (ca. 13.0 cal. kyr).

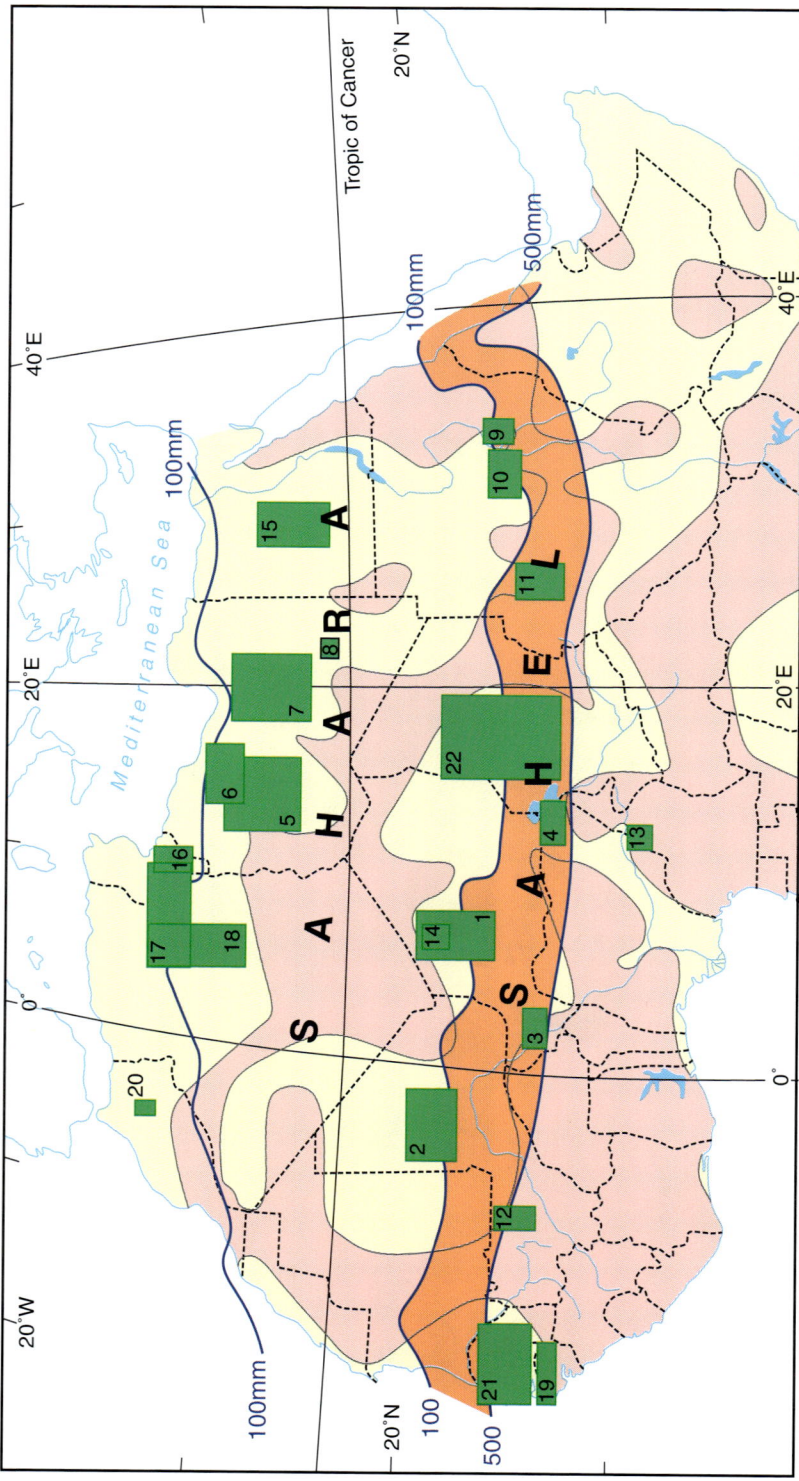

Plate 5. The distribution of sedimentary basins in Africa (pale yellow shading) and the location of studies used as data sources for the present paper: 1) and 2) Azaouad (Fontes et al. 1991); 3) Illumeden (Le Gal La Salle 1992); 4) Chad Basin; 5) Western Libya (Salem et al. 1980); 6) Northern Libya (Srdoč et al. 1980); 7) Sirt Basin (Edmunds and Wright 1979); 8) Kufra Basin (Edmunds and Wright 1979); 9) Butana region (Darling et al. 1991); 10) Kordofan (Groening et al. 1993); 11) Darfur (Groening et al. 1993); 12) Kolokoni-Nara (Dincer et al. 1983); 13) Garoua (Njitchoua et al. 1993); 14) Irhazer (Andrews et al. 1993); 15) Western Desert (Thorweihe 1982); 16) and 17) Continental Intercalaire (Guendouz et al. 1998); 18) Complexe Terminal (Edmunds et al., in press); 19) Casamance (Faye et al. 1993); 20) Saïs (Kabbaj et al. 1978); 21) N Senegal (Faye 1994); 22) Chad basin (UNESCO 1972).

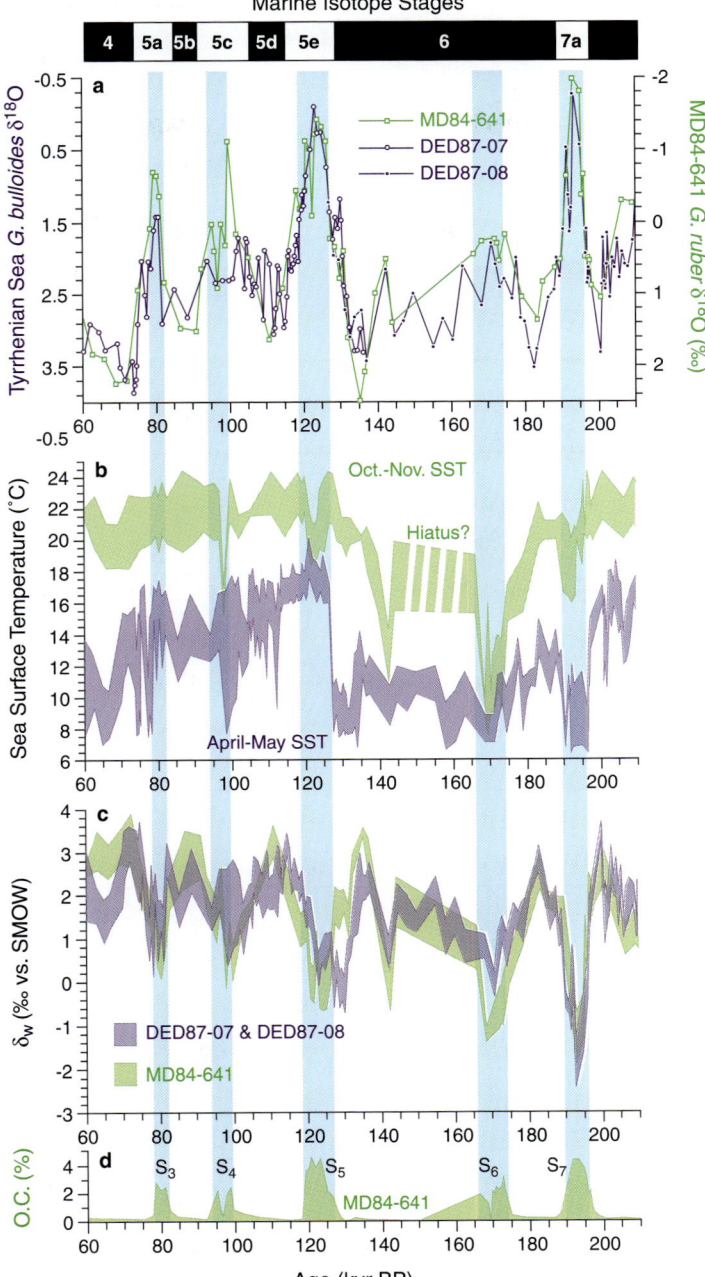

Marine Isotope Stages

Plate 6. Tyrrhenian and Levantine Seas climatic records plotted against time: (a) oxygen isotope records of *G. bulloides* (Tyrrhenian Sea cores DED87-07 & DED87-08) and *G. ruber* (Levantine basin core MD84-641); (b) upper and lower confidence margins of April-May (Tyrrhenian Sea cores DED87-07 and DED-87-08) and October-November (core MD84-641) sea surface temperature (SST) estimates; (c) upper and lower confidence margins (depending on the error on SST estimates and on the foraminiferal isotope analytical error; Kallel et al. (1997b)) of the sea-surface water oxygen isotopic composition (δw) estimates; (d) organic carbon record in the Levantine Sea core MD84-641. Light blue intervals delimit the periods of significant increase in the percentage of the organic carbon (sapropels) in the Eastern Mediterranean Sea.

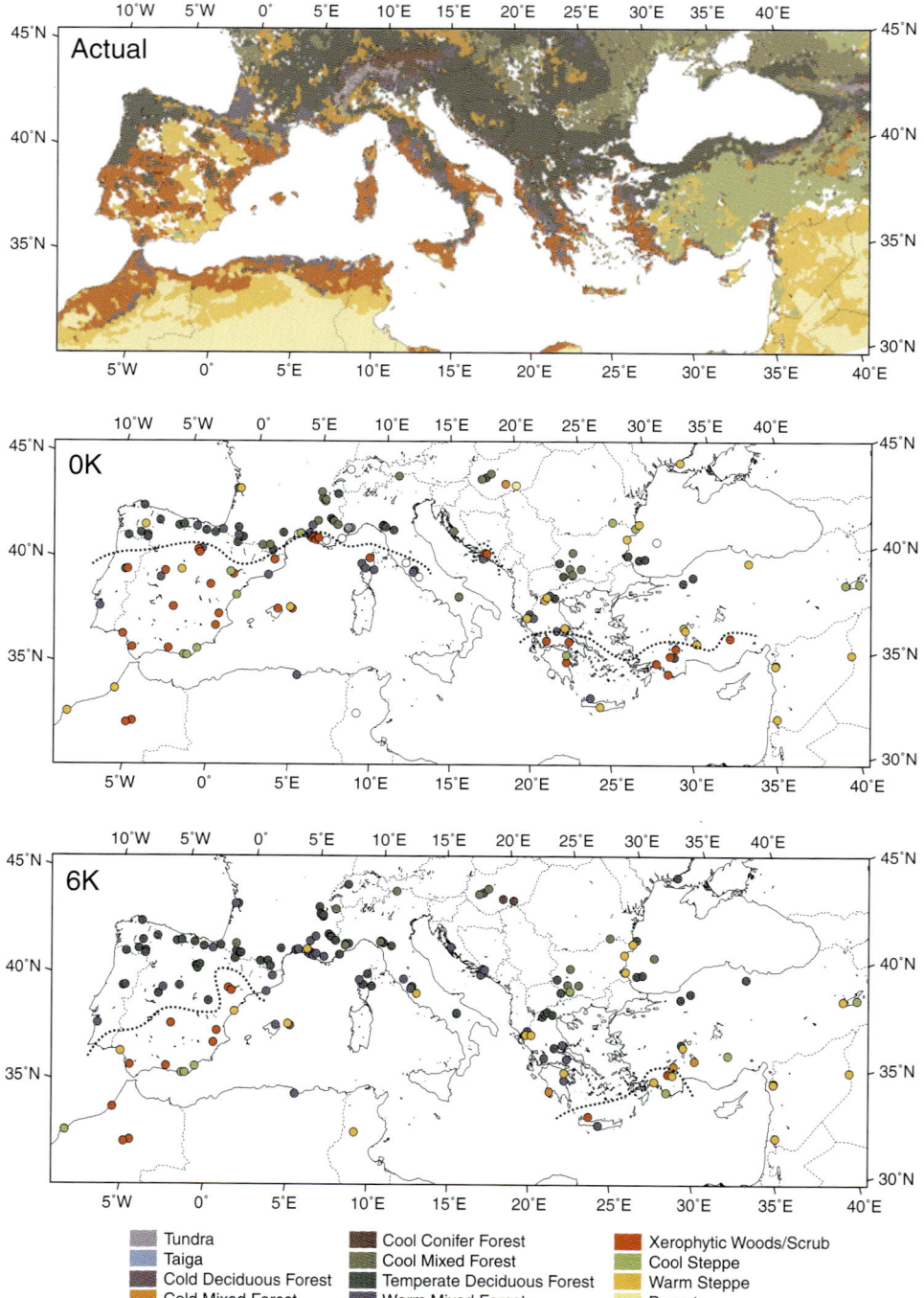

Plate 7. (a) Actual biome distribution from a high-resolution satellite-derived dataset (Olson 1994) global ecosystems legend; (b) Mediterranean biomes for 0–500 BP (0 kyr) based on pollen data biomised according to Peyron et al. (1998); (c) pollen-derived biomes for 6000 ± 500 yr BP (6 kyr). White circles at 0 kyr indicate sites where no pollen data are available for this time period. The dashed line marks the inferred northern limit of xerophytic woodland/scrub.

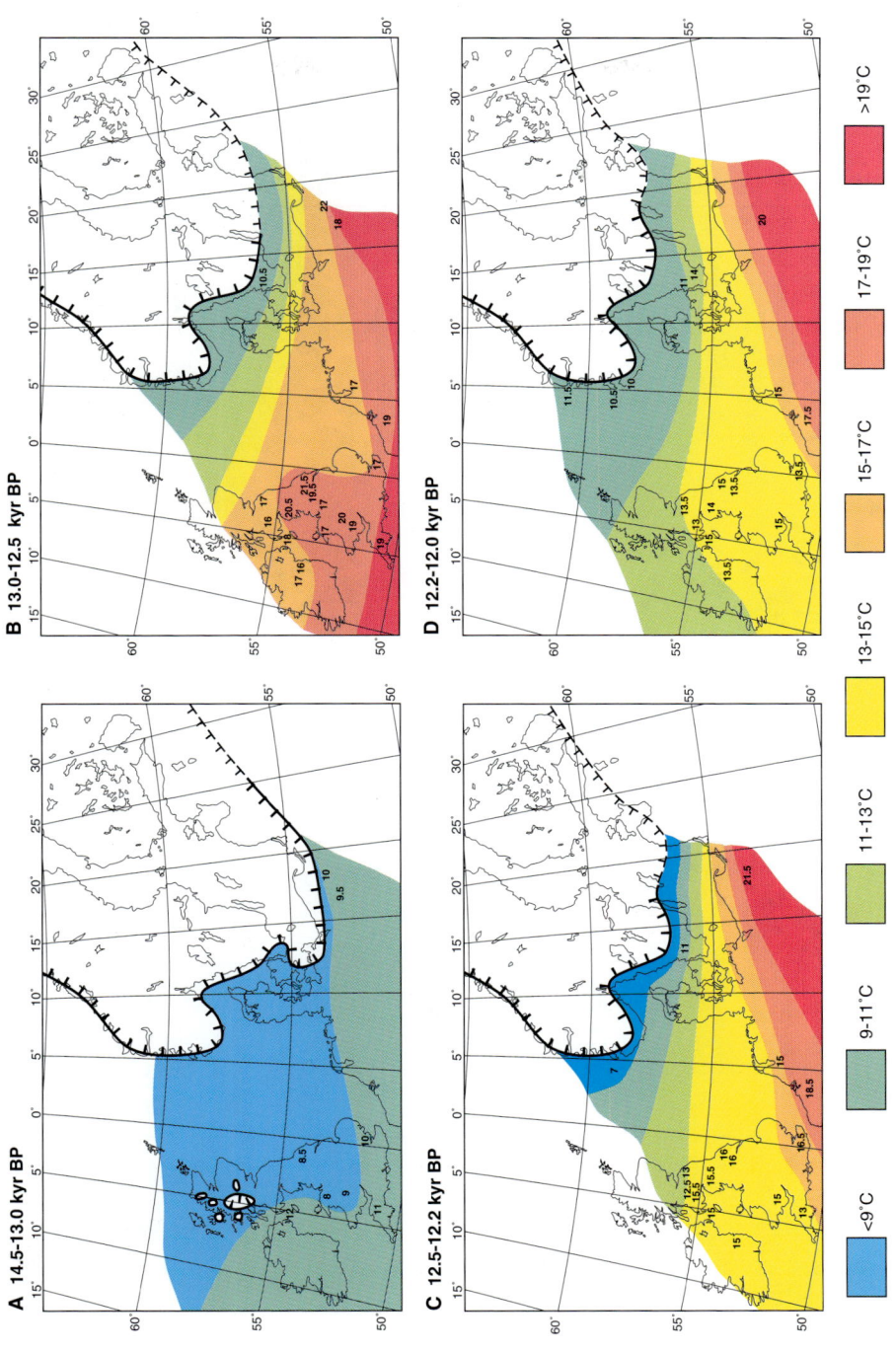

Plate 8. Generalised isotherms for the late-glacial period in Europe, based on beetle MCR interpretations. The diagram shows four of the eight time-slices for which such reconstructions were provided in Coope et al. 1998.

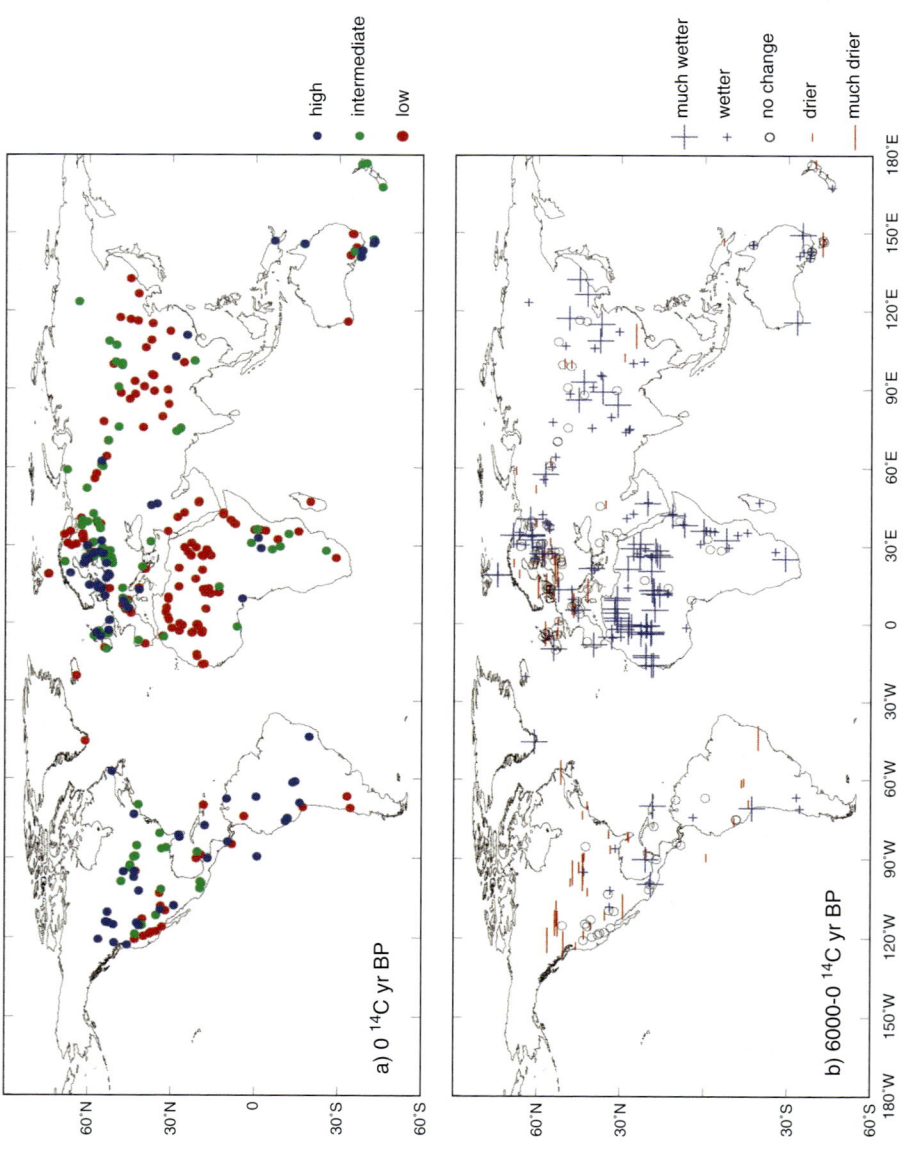

Plate 9. Lake status (a) today and (b) the change in lake status at 6000 ^{14}C yr BP compared to present. The data are derived from the Global Lake Status Data Base (Kohfeld and Harrison 2000; Yu et al. 2001; Harrison et al. 2003). Only sites with dating control ≤4 are included.

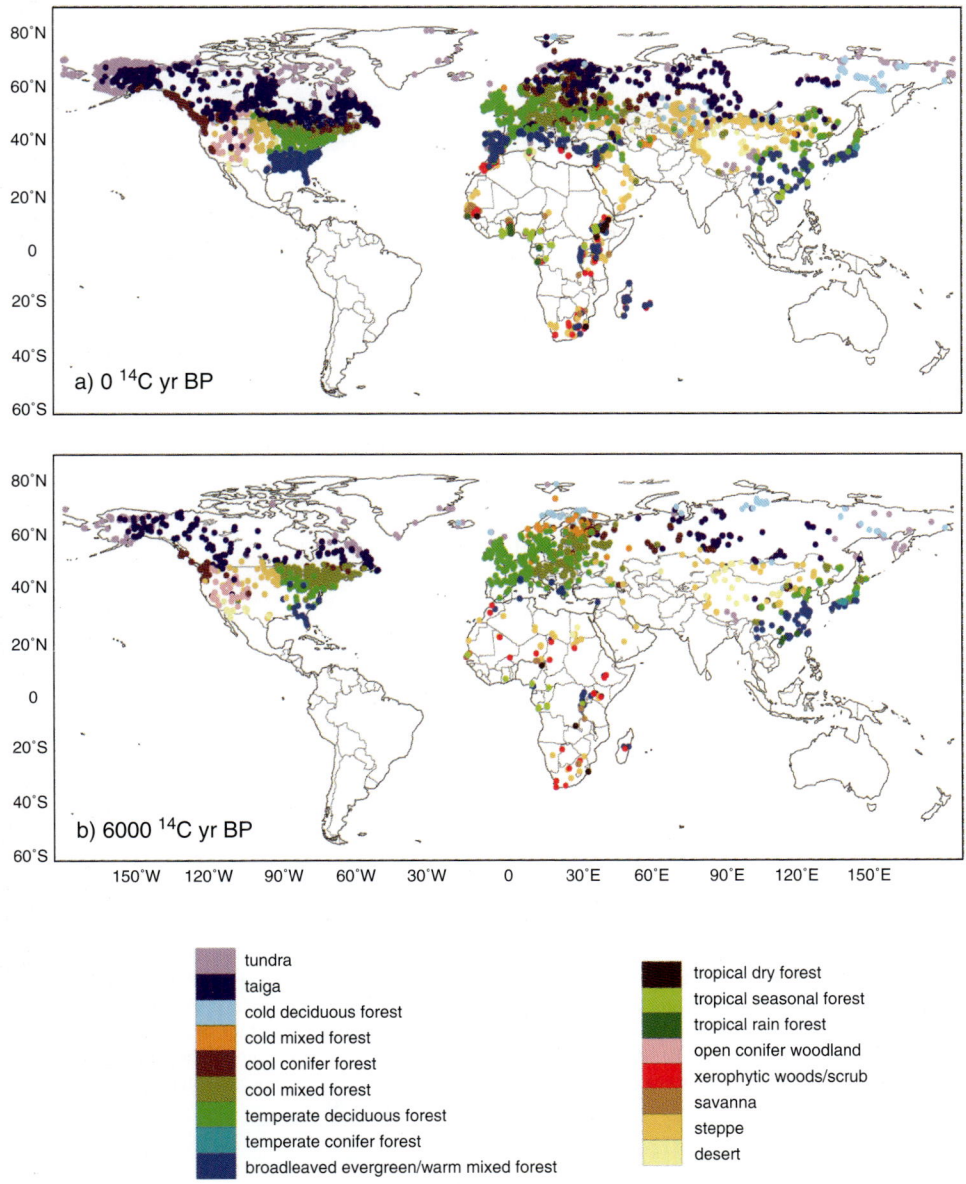

tundra
taiga
cold deciduous forest
cold mixed forest
cool conifer forest
cool mixed forest
temperate deciduous forest
temperate conifer forest
broadleaved evergreen/warm mixed forest

tropical dry forest
tropical seasonal forest
tropical rain forest
open conifer woodland
xerophytic woods/scrub
savanna
steppe
desert

Plate 10. Biome reconstructions for (a) today and (b) 6000 ^{14}C yr BP (from Prentice et al. (2000)).

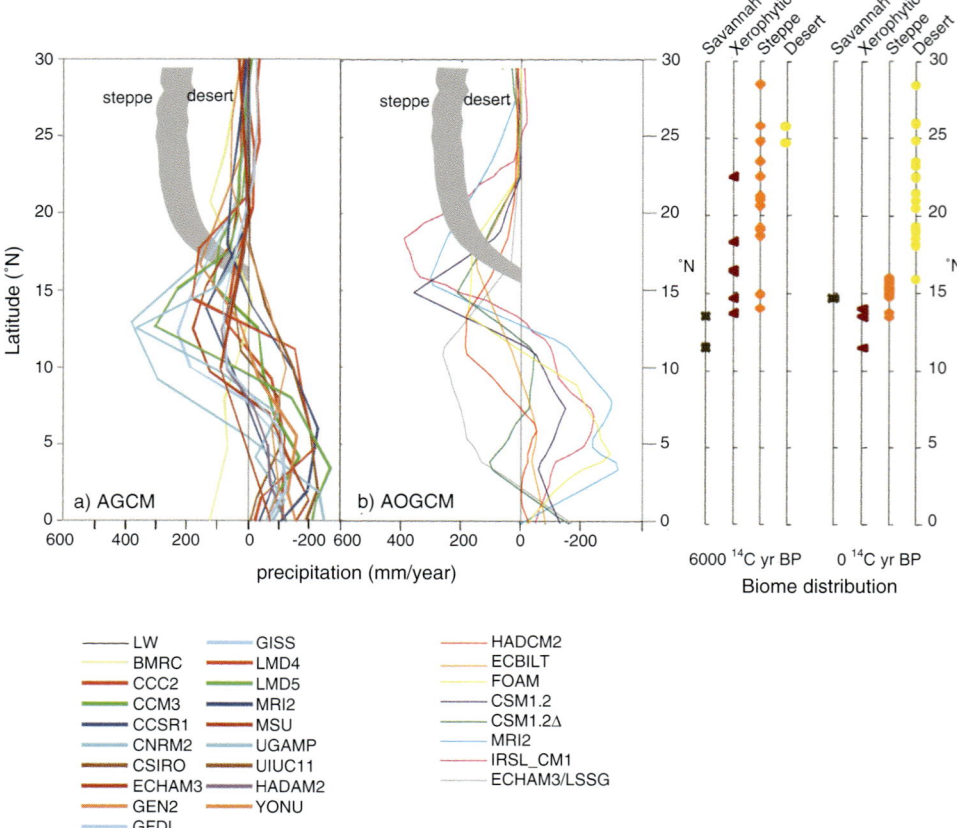

Plate 11. Zonally-averaged simulated annual precipitation anomalies (6000 yr BP minus control) over northern African land grid cells (20 °W–30 °E) vs. latitude (from 0° to 30 °N) as simulated (a) by the 18 AGCMs participating in the PMIP basic experiment, and (b) in a suite of 9 OAGCM experiments. The grey shaded area in both diagrams represents upper and low estimates of the precipitation needed in excess of the modern precipitation to support steppe vegetation at each latitude (Joussaume et al. 1999). The plots at the right of the figure (c) show the latitudinal distribution of vegetation at 6000 ^{14}C yr BP compared to today, and show that steppe and xerophytic vegetation were extended much further north than today at 6000 ^{14}C yr BP.